Roxy Wilson

SOLUTIONS TO EXERCISES

CHEMISTRY

THE CENTRAL SCIENCE

Ninth Edition

Brown LeMay Bursten

Upper Saddle River, NJ 07458

Project Manager: Kristen Kaiser
Senior Editor: Nicole Folchetti
Editor in Chief: John Challice
Executive Managing Editor: Kathleen Schiaparelli
Assistant Managing Editor: Dinah Thong
Production Editor: Natasha Wolfe
Supplement Cover Management/Design: Paul Gourhan
Manufacturing Buyer: Ilene Kahn
Cover Image Credit: Ken Eward/Biografx

Pearson Education, Inc.
Upper Saddle River, NJ 07458

Printed in the United States of America

10 9 8 7 6 5 4 3 2 1

ISBN 0-13-009798-5

Pearson Education Ltd., *London*
Pearson Education Australia Pty. Ltd., *Sydney*
Pearson Education Singapore, Pte. Ltd.
Pearson Education North Asia Ltd., *Hong Kong*
Pearson Education Canada, Inc., *Toronto*
Pearson Educacíon de Mexico, S.A. de C.V.
Pearson Education—Japan, *Tokyo*
Pearson Education Malaysia, Pte. Ltd.
Pearson Education, *Upper Saddle River, New Jersey*

Contents

Introduction

Chemistry: The Central Science, 9th edition, contains nearly 2400 end-of-chapter exercises. Considerable attention has been given to these exercises because one of the best ways for students to master chemistry is by solving problems. Grouping the exercises according to subject matter is intended to aid the student in selecting and recognizing particular types of problems. Within each subject matter group, similar problems are arranged in pairs. This provides the student with an opportunity to reinforce a particular kind of problem. There are also a substantial number of general exercises in each chapter to supplement those grouped by topic. Integrative exercises, which require students to integrate concepts from several chapters, are a continuing feature of the 9th edition. Answers to the odd numbered topical exercises plus selected general and integrative exercises, about 1100 in all, are provided in the text. These appendix answers help to make the text a useful self-contained vehicle for learning.

This manual, **Solutions to Exercises in Chemistry: The Central Science, 9th edition,** was written to enhance the end-of-chapter exercises by provided documented solutions. The manual assists the instructor by saving time spent generating solutions for assigned problem sets and aids the student by offering a convenient independent source to check their understanding of the material. Most solutions have been worked in the same detail as the in-chapter sample exercises to help guide students in their studies.

To reinforce the '*Analyze, Plan, Solve, Check*' problem-solving method used extensively in the text, this strategy has also been incorporated into the Solution Manual. Solutions to most red paired exercises and selected Additional and Integrative exercises feature this four-step approach. We strongly encourage students to master this powerful and totally general method.

When using this manual, keep in mind that the numerical result of any calculation is influenced by the precision of the numbers used in the calculation. In this manual, for example, atomic masses and physical constants are typically expressed to four significant figures, or at least as precisely as the data given in the problem. If students use slightly different values to solve problems, their answers will differ slightly from those listed in the appendix of the text or this manual. This is a normal and a common occurrence when comparing results from different calculations or experiments.

Rounding methods are another source of differences between calculated values. In this manual, when a solution is given in steps, intermediate results will be rounded to the correct number of significant figures; however, unrounded numbers will be used in subsequent calculations. By following this scheme, calculators need not be cleared to re-enter rounded intermediate results in the middle of a calculation sequence. The final answer will appear with the correct number of significant figures. This may result in a small discrepancy in the last significant digit between student-calculated answers and those given in this manual. Variations due to rounding can occur in any analysis of numerical data.

The first step in checking your solution and resolving differences between your answer and the listed value is to look for similarities and differences in problem-solving methods. Ultimately, resolving the small numerical differences described above is less important than understanding the general method for solving a problem. The goal of this manual is to provide a reference for sound and consistent problem-solving methods in addition to accurate answers to text exercises.

Extraordinary efforts have been made to keep this manual as error-free as possible. All exercises were worked and proof-read by at least three chemists to ensure clarity in methods and accuracy in mathematics. The work and advice of Dr. Mary Ellen Biggin, Augustana College and Dr. Angela Manders Cannon, University of Illinois have been invaluable to this project. However, in a written work as technically challenging as this manual, typos and errors inevitably creep in. Please help us find and eliminate them. We hope that both instructors and students will find this manual accurate, helpful and instructive.

Roxy B. Wilson
University of Illinois
School of Chemical Sciences
601 S. Mathews Ave., Box A-2
Urbana, IL 61801
rbwilson@uiuc.edu

1 Introduction: Matter and Measurement

Classification and Properties of Matter

1.1 (a) heterogeneous mixture (b) homogeneous mixture (If there are undissolved particles, such as sand or decaying plants, the mixture is heterogeneous.) (c) pure substance (d) homogeneous mixture

1.2 (a) homogeneous mixture (b) heterogeneous mixture (particles in liquid) (c) pure substance (d) heterogeneous mixture

1.3 (a) Al (b) Na (c) Br (d) Cu (e) Si (f) N (g) Mg (h) He

1.4 (a) C (b) K (c) Cl (d) Zn (e) P (f) Ar (g) Ca (h) Ag

1.5 (a) hydrogen (b) magnesium (c) lead (d) silicon (e) fluorine (f) tin (g) manganese (h) arsenic

1.6 (a) chromium (b) iodine (c) lithium (d) selenium (e) lead (f) vanadium (g) mercury (h) gallium

1.7 $A(s) \xrightarrow{\text{heat}} B(s) + C(g)$

When carbon(s) is burned in excess oxygen the two elements combine to form a gaseous compound, carbon dioxide. Clearly substance C is this compound.

Since C is produced when A is heated in the absence of oxygen (from air), both the carbon and oxygen in C must have been present in A originally. A is, therefore, a compound composed of two or more elements chemically combined. Without more information on the chemical or physical properties of B, we cannot determine absolutely whether it is an element or a compound. However, few if any elements exist as white solids, so B is probably also a compound.

1.8 Before modern instrumentation, the classification of a pure substance as an element was determined by whether it could be broken down into component elements. Scientists subjected the substance to all known chemical means of decomposition, and if the results were negative, the substance was an element. Classification by negative results was somewhat ambiguous, since an effective decomposition technique might exist, but not yet have been discovered.

1.9 1.10

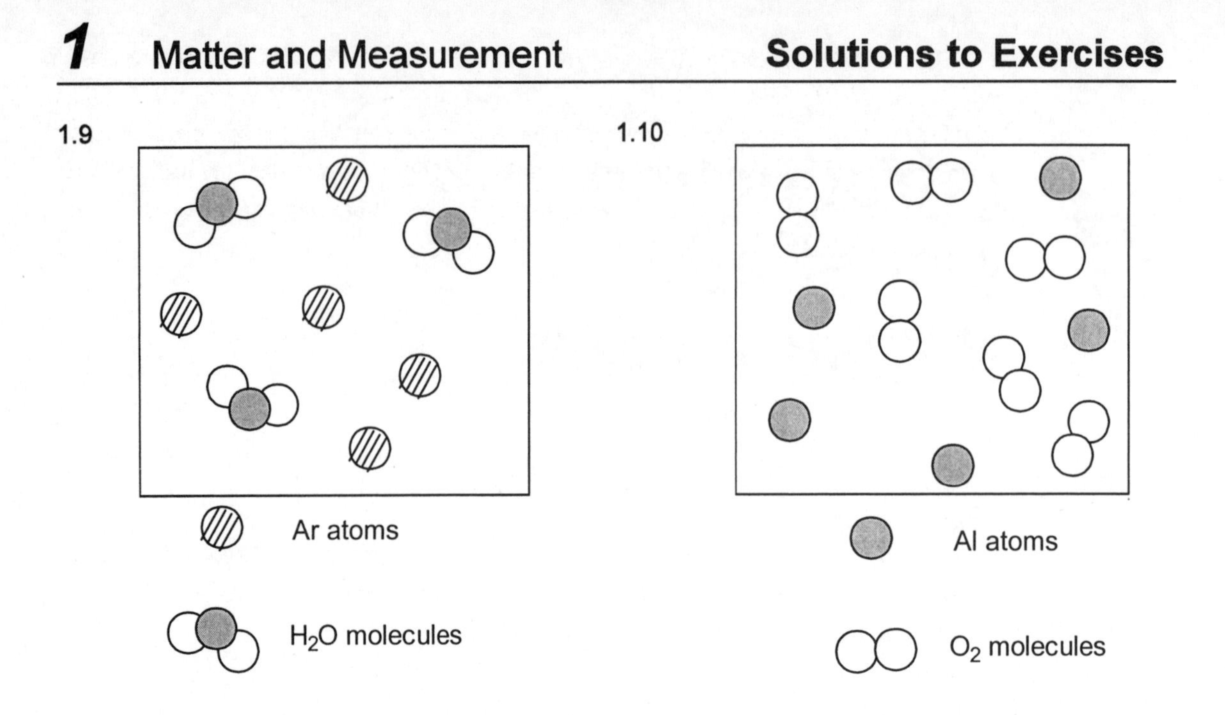

1.11 Physical properties: silvery white (color); lustrous; melting point = 649°C; boiling point = 1105°C; density at 20°C = 1.738 g/cm^3; pounded into sheets (malleable); drawn into wires (ductile); good conductor. Chemical properties: burns in air to give intense white light; reacts with Cl_2 to produce brittle white solid.

1.12 Physical properties: silver-grey (color); melting point = 420°C; hardness = 2.5 Mohs; density = 7.13 g/cm^3 at 25°C. Chemical properties: metal; reacts with sulfuric acid to produce hydrogen gas; reacts slowly with oxygen at elevated temperatures to produce ZnO.

1.13 (a) chemical (b) physical (c) physical (d) chemical (e) chemical

1.14 (a) chemical (b) physical (c) physical (The production of H_2O is a chemical change, but its **condensation** is a physical change.) (d) physical (The production of soot is a chemical change, but its **deposition** is a physical change.)

1.15 Take advantage of differences in physical properties to separate the components of a mixture. First heat the liquid to 100°C to evaporate the water. This is conveniently done in a distillation apparatus (Figure 1.13) so that the water can be collected. After the water is completely evaporated and **if** there is a residue, measure the physical properties of the residue such as color, density and melting point. Compare the observed properties of the residue to those of table salt, NaCl. If the properties match, the colorless liquid contained table salt. If the properties don't match, the liquid contained a different dissolved solid. If there is no residue, no dissolved solid is present.

1.16 (a) Take advantage of the different water solubilities of the two solids. Add water to dissolve the sugar; filter this mixture, collecting the sand on the filter paper and the sugar water in the flask. Evaporate the water from the flask to reproduce solid sugar.

(b) Either the melting-point difference or magnetism difference between iron and sulfur can be used to separate these two elements. Heat the mixture until the sulfur melts, then decant (pour off) the liquid sulfur. Or, use a magnet to attract the iron particles, leaving the solid sulfur behind.

Units and Measurement

1.17 (a) 1×10^{-1} (b) 1×10^{-2} (c) 1×10^{-5} (d) 1×10^{-6} (e) 1×10^{6} (f) 1×10^{3} (g) 1×10^{-9} (h) 1×10^{-3} (i) 1×10^{-12}

1.18 (a) $6.5 \times 10^{-6}\ \text{m} \times \dfrac{1\ \mu\text{m}}{1 \times 10^{-6}\ \text{m}} = 6.5\ \mu\text{m}$

(b) $6.35 \times 10^{-4}\ \text{L} \times \dfrac{1\ \text{mL}}{1 \times 10^{-3}\ \text{L}} = 0.635\ \text{mL}\ (635\ \mu\text{L})$

(c) $2.5 \times 10^{-3}\ \text{L} \times \dfrac{1\ \text{mL}}{1 \times 10^{-3}\ \text{L}} = 2.5\ \text{mL}$

(d) $4.23 \times 10^{-9}\ \text{m}^3 \times \dfrac{1^3\ \text{mm}^3}{(1 \times 10^{-3})^3\ \text{m}^3} = 4.23\ \text{mm}^3\ (4.23\ \mu\text{L};\ 1\ \text{mm}^3 = 1\ \text{uL})$

$4.23\ \text{mm}^3 \times \dfrac{(10^{-1})^3\ \text{cm}^3}{1^3\ \text{mm}^3} \times \dfrac{1\ \text{mL}}{1\ \text{cm}^3} \times \dfrac{1 \times 10^{-3}\ \text{L}}{1\ \text{mL}} \times \dfrac{1\ \mu\text{L}}{1 \times 10^{-6}\ \text{L}} = 4.23\ \mu\text{L}$

(e) $12.5 \times 10^{-8}\ \text{kg} \times \dfrac{1 \times 10^{3}\ \text{g}}{1\ \text{kg}} \times \dfrac{1\ \text{mg}}{1 \times 10^{-3}\ \text{g}} = 0.125\ \text{mg}\ (125\ \mu\text{g})$

(f) $3.5 \times 10^{-11}\ \text{s} \times \dfrac{1\ \text{ps}}{1 \times 10^{-12}\ \text{s}} = 35\ \text{ps}\ (0.035\ \text{ns})$

(g) $6.54 \times 10^{9}\ \text{fs} \times \dfrac{1 \times 10^{-15}\ \text{s}}{1\ \text{fs}} \times \dfrac{1\ \mu\text{s}}{1 \times 10^{-6}\ \text{s}} = 6.54\ \mu\text{s}$

1.19 (a) $25.5\ \text{mg} \times \dfrac{1 \times 10^{-3}\ \text{g}}{1\ \text{mg}} = 0.0255\ \text{g}\ (2.55 \times 10^{-2}\ \text{g})$

(b) $4.0 \times 10^{-10}\ \text{m} \times \dfrac{1\ \text{nm}}{1 \times 10^{-9}\ \text{m}} = 0.40\ \text{nm}$

(c) $0.575\ \text{mm} \times \dfrac{1 \times 10^{-3}\ \text{m}}{1\ \text{mm}} \times \dfrac{1\ \mu\text{m}}{1 \times 10^{-6}\ \text{m}} = 575\ \mu\text{m}$

1.20 (a) $1.48 \times 10^{2}\ \text{kg} \times \dfrac{1 \times 10^{3}\ \text{g}}{1\ \text{kg}} \times = 1.48 \times 10^{5}\ \text{g}$

(b) $0.0023\ \mu\text{m} \times \dfrac{1 \times 10^{-6}\ \text{m}}{1\ \mu\text{m}} \times \dfrac{1\ \text{nm}}{1 \times 10^{-9}\ \text{m}} = 2.3\ \text{nm}$

(c) $7.25 \times 10^{-4}\ \text{s} \times \dfrac{1\ \text{ms}}{1 \times 10^{-3}\ \text{s}} = 0.725\ \text{ms}$

1.21 (a) time (b) density (c) length (d) area (e) temperature
(f) volume (g) temperature

1.22 (a) volume (b) area (c) volume (d) density (e) time
(f) length (g) temperature

1.23 (a) $\text{density} = \dfrac{\text{mass}}{\text{volume}} = \dfrac{39.73\text{ g}}{25.0\text{ mL}} = 1.59\text{ g/mL or } 1.59\text{ g/cm}^3$

(The units cm^3 and mL will be used interchangeably in this manual.)

Carbon tetrachloride, 1.59 g/mL, is more dense than water, 1.00 g/mL; carbon tetrachloride will sink rather than float on water.

(b) $75.00\text{ cm}^3 \times 21.45\dfrac{\text{g}}{\text{cm}^3} = 1.609 \times 10^3\text{ g}\ (1.609\text{ kg})$

(c) $87.50\text{ g} \times \dfrac{1\text{ cm}^3}{1.738\text{ g}} = 50.3452 = 50.35\text{ cm}^3 = 50.35\text{ mL}$

1.24 (a) volume = length^3 (cm^3); density = mass/volume (g/cm^3)

volume = $(1.500)^3\text{ cm}^3 = 3.375\text{ cm}^3$

$\text{density} = \dfrac{76.31\text{ g}}{3.375\text{ cm}^3} = 22.61\text{ g/cm}^3$ osmium

(b) $65.8\text{ mL} \times \dfrac{1\text{ cm}^3}{1\text{ mL}} \times \dfrac{4.51\text{ g}}{1\text{ cm}^3} = 296.758 = 297\text{ g}$ titanium

(c) $0.1500\text{ L} \times \dfrac{1\text{ mL}}{1 \times 10^{-3}\text{ L}} \times \dfrac{0.8787\text{ g}}{1\text{ mL}} = 131.8\text{ g}$ benzene

1.25 (a) $\text{density} = \dfrac{38.5\text{ g}}{45\text{ mL}} = 0.86\text{ g/mL}$

The substance is probably toluene, density = 0.866 g/mL.

(b) $45.0\text{ g} \times \dfrac{1\text{ mL}}{1.114\text{ g}} = 40.4\text{ mL}$ ethylene glycol

(c) $(5.00)^3\text{ cm}^3 \times \dfrac{8.90\text{ g}}{1\text{ cm}^3} = 1.11 \times 10^3\text{ g}\ (1.11\text{ kg})$ nickel

1.26 (a) $\dfrac{21.95\text{ g}}{25.0\text{ mL}} = 0.878\text{ g/mL}$

The tabulated value has 4 significant figures, while the experimental value has 3. The tabulated value rounded to 3 figures is 0.879. The values agree within 1 in the last significant figure of the experimental value; the two results agree. The liquid could be benzene.

(b) $15.0\text{ g} \times \dfrac{1\text{ mL}}{0.7781\text{ g}} = 19.3\text{ mL}$ cyclohexane

(c) $r = d/2 = 5.0 \text{ cm}/2 = 2.5 \text{ cm}$

$V = 4/3\ \pi r^3 = 4/3 \times \pi \times (2.5)^3 \text{ cm}^3 = 65 \text{ cm}^3$

$65.4498 \text{ cm}^3 \times \frac{11.34 \text{ g}}{\text{cm}^3} = 7.4 \times 10^2 \text{ g}$

(The answer has 2 significant figures because the diameter had only 2 figures.)

Note: This is the first exercise where "intermediate rounding" occurs. In this manual, when a solution is given in steps, the intermediate result will be rounded to the correct number of significant figures. However, the **unrounded** number will be used in subsequent calculations. The final answer will appear with the correct number of significant figures. That is, calculators need not be cleared and new numbers entered in the middle of a calculation sequence. This may result in a small discrepancy in the last significant digit between student-calculated answers and those given in the solution manual. These variations occur in any analysis of numerical data.

For example, in this exercise the volume of the sphere, 65.4498 cm^3 is rounded to 65 cm^3, but 65.4498 is retained in the subsequent calculation of mass, 7.4×10^2 g. In this case, 65 cm^3 × 11.34 g/cm^3 also yields 7.4×10^2 g. In other exercises, the correctly rounded results of the two methods may not be identical.

1.27 thickness = volume/area

$\text{volume} = 200 \text{ mg} \times \frac{1 \times 10^{-3} \text{ g}}{1 \text{ mg}} \times \frac{1 \text{ cm}^3}{19.32 \text{ g}} = 0.01035 = 0.0104 \text{ cm}^3$

$\text{area} = 2.4 \text{ ft} \times 1.0 \text{ ft} \times \frac{12^2 \text{ in}^2}{1 \text{ ft}^2} \times \frac{2.54^2 \text{ cm}^2}{\text{in}^2} = 2.23 \times 10^3 = 2.2 \times 10^3 \text{ cm}^2$

$\text{thickness} = \frac{0.01035 \text{ cm}^3}{2{,}230 \text{ cm}^2} \times \frac{1 \times 10^{-2} \text{ m}}{1 \text{ cm}} = 4.6 \times 10^{-8} \text{ m}$

$4.6 \times 10^{-8} \text{ m} \times \frac{1 \text{ nm}}{1 \times 10^{-9} \text{ m}} = 46 \text{ nm thick}$

1.28 Calculate the volume of the rod:

$2.17 \text{ kg} \times \frac{1000 \text{ g}}{1 \text{ kg}} \times \frac{1 \text{ cm}^3}{2.33 \text{ g}} = 931.3 = 931 \text{ cm}^3$

$V = \pi r^2 h;\ d = 2r,\ r = d/2;\ \ V = \pi \left(\frac{d}{2}\right)^2 h;\ d^2 = \frac{4V}{\pi h};\ d = \left(\frac{4V}{\pi h}\right)^{1/2}$

$d = \left(\frac{4\,(931.3) \text{ cm}^3}{\pi\,(16.8) \text{ cm}}\right)^{1/2} = 8.401 = 8.40 \text{ cm}$

1.29 (a) °C = 5/9 (°F - 32°); 5/9 (62 - 32) = 17°C

(b) °F = 9/5 (°C) + 32°; 9/5 (216.7) + 32 = 422.1°F

(c) K = °C + 273.15; 233°C + 273.15 = 506 K

(d) 315 K - 273 = 42°C; 9/5 (42°C) + 32 = 108°F

(e) °C = 5/9 (°F - 32°); 5/9 (2500 - 32) = 1371°C; 1371°C + 273 = 1644 K

(assuming 2500 C has 4 sig figs)

1.30 (a) °C = 5/9 (87°F - 32°) = 31°C

(b) °F = 9/5 (755°C) + 32 = 1391°F

(It could be argued that the result of 9/5 (755) has 3 sig figs, so the final answer should have 3 sig figs, 1390°F.)

(c) °F = 9/5 (- 95°C) + 32 = -139°F (with 2 sig figs, -140°F)

(d) K = 25°C + 273 = 298 K; °F = 9/5 (25°C) + 32 = 77°F

(e) melting point = -248.6°C + 273.15 = 24.6 K
boiling point = -246.1°C + 273.15 = 27.1 K

Uncertainty In Measurement

1.31 Exact: (c), (d), and (f) (All others depend on measurements and standards that have margins of error, e.g., the length of a week as defined by the earth's rotation.)

1.32 Exact: (b), (e) (The number of students is exact on any given day.)

1.33 7.5 cm. There are two significant figures in this measurement; the number of cm can be read precisely but there is some estimating (uncertainty) required to read tenths of a centimeter. Two significant figures is consistent with the convention that measured quantities are reported so that there is uncertainty in only the last digit.

1.34 140°C. The temperature can be read to the nearest 50°C and estimated to the nearest 5-10°C. Since there is uncertainty in the tens digit, the measurement has two significant figures.

1.35 (a) 4 (b) 3 (c) 4 (d) 3 (e) 5

1.36 (a) 4 (b) 3 (c) 4 (d) 5 (e) 6

1.37 (a) 3.002×10^2 (b) 4.565×10^5 (c) 6.543×10^{-3}
(d) 9.578×10^{-4} (e) 5.078×10^4 (f) -3.500×10^{-2}

1.38 (a) 1.44×10^5 (b) 9.75×10^{-2} (c) 8.90×10^5
(d) 6.76×10^4 (e) 3.40×10^4 (f) -6.56

1.39 (a) 27.04 (b) -8.0 (c) 1.84×10^{-3} (d) 7.66×10^{-4}

1.40 (a) -2.3×10^3 (The intermediate result has 2 significant figures, so only the thousand and hundred places in the answer are significant.)

(b) $[285.3 \times 10^5 - 0.01200 \times 10^5] \times 2.8954 = 8.260 \times 10^7$ (Since subtraction depends on decimal places, both numbers must have the same exponent to determine decimal places/sig figs. The intermediate result has 1 decimal place and 4 sig figs, so the answer has 4 sig figs)

(c) $(0.0045 \times 20{,}000.0)$ + (2813×12) $= 3.4 \times 10^4$
2 sig figs /0 dec pl 2 sig figs /first 2 digits

(d) 863 × [1255 - (3.45×108)] $= 7.62 \times 10^5$
3 sig figs /0 dec pl
3 sig figs × 0 dec pl/3 sig figs = 3 sig figs

Dimensional Analysis

1.41 In order to cancel units, the conversion factor must have the unit being canceled opposite the starting position. For example, if the unit cm starts in the numerator, then the conversion factor must have cm in its denominator. However, if the unit cm starts in the denominator, the conversion factor must have cm in the numerator. Ideally, this will lead to the desired units in the appropriate location, numerator or denominator. However, the inverse of the answer can be taken when necessary.

1.42 (a) $\frac{1.6093 \text{ km}}{1 \text{ mi}}$; when converting miles to kilometers, miles goes in the denominator so that it cancels the original unit, leaving km in the numerator.

(b) $\frac{453.59 \text{ g}}{16 \text{ oz}} = \frac{28.349 \text{ g}}{1 \text{ oz}}$ (c) $\frac{1 \text{ L}}{1.0567 \text{ qt}}$

1.43 (a) $0.076 \text{ L} \times \frac{1000 \text{ mL}}{1 \text{ L}} = 76 \text{ mL}$

(b) $5.0 \times 10^{-8} \text{ m} \times \frac{1 \text{ nm}}{1 \times 10^{-9} \text{ m}} = 50. \text{ nm}$

(c) $6.88 \times 10^5 \text{ ns} \times \frac{1 \times 10^{-9} \text{ s}}{1 \text{ ns}} = 6.88 \times 10^{-4} \text{ s}$

(d) $\frac{1.55 \text{ kg}}{\text{m}^3} \times \frac{1000 \text{ g}}{1 \text{ kg}} \times \frac{1 \text{ m}^3}{(10)^3 \text{ dm}^3} \times \frac{1 \text{ dm}^3}{1 \text{ L}} = 1.55 \text{ g/L}$

(e) $\frac{5.850 \text{ gal}}{\text{hr}} \times \frac{3.7854 \text{ L}}{1 \text{ gal}} \times \frac{1 \text{ hr}}{60 \text{ min}} \times \frac{1 \text{ min}}{60 \text{ s}} = 6.151 \times 10^{-3}$ L/s

Estimated answer: 6 × 4 = 24; 24/60 = 0.4; 0.4/60 = 0.0066 = 7 × 10^{-3}. This agrees with the calculated answer of 6.151 × 10^{-3} L/s.

1.44 (a) $\frac{2.998 \times 10^8 \text{m}}{\text{s}} \times \frac{1 \text{ km}}{1000 \text{ m}} \times \frac{60 \text{ s}}{1 \text{ min}} \times \frac{60 \text{ min}}{1 \text{ hr}} = 1.079 \times 10^9$ km/hr

(b) $1.35 \times 10^9 \text{ km}^3 \times \frac{(1 \times 10^3)^3 \text{ m}^3}{1^3 \text{ km}^3} \times \frac{1^3 \text{ dm}^3}{(1 \times 10^{-1})^3 \text{ m}^3} \times \frac{1 \text{ L}}{1 \text{ dm}^3} = 1.35 \times 10^{21}$ L

(c) $\frac{232 \text{ mg cholesterol}}{100 \text{ mL blood}} \times \frac{1 \text{ mL}}{1 \times 10^{-3} \text{ L}} \times 5.2 \text{ L} \times \frac{1 \times 10^{-3} \text{ g}}{1 \text{ mg}} = 12$ g cholesterol

1.45 (a) $5.00 \text{ days} \times \frac{24 \text{ hr}}{1 \text{ day}} \times \frac{60 \text{ min}}{1 \text{ hr}} \times \frac{60 \text{ s}}{1 \text{ min}} = 4.32 \times 10^5$ s

(b) $0.0550 \text{ mi} \times \frac{1.6093 \text{ km}}{\text{mi}} \times \frac{1000 \text{ m}}{1 \text{ km}} = 88.5$ m

(c) $\frac{\$1.89}{\text{gal}} \times \frac{1 \text{ gal}}{3.7854 \text{ L}} = \frac{\$0.499}{\text{L}}$

(d) $\frac{0.510 \text{ in}}{\text{ms}} \times \frac{2.54 \text{ cm}}{1 \text{ in}} \times \frac{1 \times 10^{-2} \text{ m}}{1 \text{ cm}} \times \frac{1 \text{ km}}{1000 \text{ m}} \times \frac{1 \text{ ms}}{1 \times 10^{-3} \text{ s}} \times \frac{60 \text{ s}}{1 \text{ min}} \times \frac{60 \text{ min}}{1 \text{ hr}}$

$= 46.6 \frac{\text{km}}{\text{hr}}$

Estimate: 0.5 × 2.5 = 1.25; 1.25 × 0.01 ≈ 0.01; 0.01 × 60 × 60 ≈ 36 km/hr

(e) $\frac{22.50 \text{ gal}}{\text{min}} \times \frac{3.7854 \text{ L}}{\text{gal}} \times \frac{1 \text{ in}}{60 \text{ s}} = 1.41953 = 1.420$ L/s

Estimate: 20 × 4 = 80; 80/60 ≈ 1.3 L/s

(f) $0.02500 \text{ ft}^3 \times \frac{12^3 \text{ in}^3}{1 \text{ ft}^3} \times \frac{2.54^3 \text{ cm}^3}{1 \text{ in}^3} = 707.9 \text{ cm}^3$

Estimate: 10^3 = 1000; 3^3 = 27; 1000 × 27 = 27,000; 27,000/0.04 ≈ 700 cm^3

1.46 (a) $145.7 \text{ ft} \times \frac{1 \text{ yd}}{3 \text{ ft}} \times \frac{1 \text{ m}}{1.094 \text{ yd}} = 44.39$ m

(b) $0.570 \text{ qt} \times \frac{1 \text{ L}}{1.057 \text{ qt}} \times \frac{1 \text{ mL}}{1 \times 10^{-3} \text{ L}} = 539$ mL

(c) $\frac{3.75 \text{ µm}}{\text{s}} \times \frac{1 \times 10^{-6} \text{ m}}{1 \text{ µm}} \times \frac{1 \text{ km}}{1 \times 10^3 \text{ m}} \times \frac{60 \text{ s}}{1 \text{ min}} \times \frac{60 \text{ min}}{1 \text{ hr}} = 1.35 \times 10^{-5}$ km/hr

(d) $3.977 \text{ yd}^3 \times \frac{1 \text{ m}^3}{(1.0936)^3 \text{ yd}^3} = 3.0407 = 3.041 \text{ m}^3$

(e) $\frac{\$2.99}{\text{lb}} \times \frac{2.205\ \text{lb}}{1\ \text{kg}} = \$6.59/\text{kg}$

(f) $\frac{9.75\ \text{lb}}{\text{ft}^3} \times \frac{453.59\ \text{g}}{1\ \text{lb}} \times \frac{1\ \text{ft}^3}{12^3\ \text{in}^3} \times \frac{1\ \text{in}^3}{2.54^3\ \text{cm}^3} \times \frac{1\ \text{cm}^3}{1\ \text{mL}} = 0.156\ \text{g/mL}$

1.47 (a) $31\ \text{gal} \times \frac{4\ \text{qt}}{1\ \text{gal}} \times \frac{1\ \text{L}}{1.057\ \text{qt}} = 1.2 \times 10^2\ \text{L}$

Estimate: (30 × 4)/1 ≈ 120 L

(b) $\frac{6\ \text{mg}}{\text{kg (body)}} \times \frac{1\ \text{kg}}{2.205\ \text{lb}} \times 150\ \text{lb} = 4 \times 10^2\ \text{mg}$

Estimate: 6/2 = 3; 3 × 150 = 450 mg

(c) $\frac{254\ \text{mi}}{11.2\ \text{gal}} \times \frac{1.609\ \text{km}}{1\ \text{mi}} \times \frac{1\ \text{gal}}{4\ \text{qt}} \times \frac{1.057\ \text{qt}}{1\ \text{L}} = \frac{9.64\ \text{km}}{\text{L}}$

Estimate: 250/10 = 25; 1.6/4 = 0.4; 25 × 0.4 × 1 ≈ 10 km/L

(d) $\frac{50\ \text{cups}}{1\ \text{lb}} \times \frac{1\ \text{qt}}{4\ \text{cups}} \times \frac{1\ \text{L}}{1.057\ \text{qt}} \times \frac{1000\ \text{mL}}{1\ \text{L}} \times \frac{1\ \text{lb}}{453.6\ \text{g}} = \frac{26\ \text{mL}}{\text{g}}$

Estimate: 50/4 = 12; 1000/500 = 2; (12 × 2)/1 ≈ 24 mL/g

1.48 (a) $1486\ \text{mi} \times \frac{1\ \text{km}}{0.62137\ \text{mi}} \times \frac{\text{charge}}{225\ \text{km}} = 10.6\ \text{charges}$

Since charges are integral events, 11 charges are required.

(b) $\frac{14\ \text{m}}{\text{s}} \times \frac{1\ \text{km}}{1 \times 10^3\ \text{m}} \times \frac{1\ \text{mi}}{1.6093\ \text{km}} \times \frac{60\ \text{s}}{1\ \text{min}} \times \frac{60\ \text{min}}{1\ \text{hr}} = 31\ \text{mi/hr}$

(c) $450\ \text{in}^3 \times \frac{(2.54)^3\ \text{cm}^3}{1\ \text{in}^3} \times \frac{1\ \text{mL}}{1\ \text{cm}^3} \times \frac{1 \times 10^{-3}\ \text{L}}{1\ \text{mL}} = 7.37\ \text{L}$

(d) $2.4 \times 10^5\ \text{barrels} \times \frac{42\ \text{gal}}{1\ \text{barrel}} \times \frac{4\ \text{qt}}{1\ \text{gal}} \times \frac{1\ \text{L}}{1.057\ \text{qt}} = 3.8 \times 10^7\ \text{L}$

1.49 $12.5\ \text{ft} \times 15.5\ \text{ft} \times 8.0\ \text{ft} = 1580 = 1.6 \times 10^3\ \text{ft}^3$ (2 sig figs)

$1550\ \text{ft}^3 \times \frac{(1\ \text{yd})^3}{(3\ \text{ft})^3} \times \frac{(1\ \text{m})^3}{(1.0936)^3\ \text{yd}^3} \times \frac{10^3\ \text{dm}^3}{1\ \text{m}^3} \times \frac{1\ \text{L}}{1\ \text{dm}^3} \times \frac{1.19\ \text{g}}{\text{L}} \times \frac{1\ \text{kg}}{1000\ \text{g}} = 52\ \text{kg air}$

Estimate: 1550/30 = 50; (50 × 1)/1 ≈ 50 kg

1.50 $9.0 \text{ ft} \times 14.5 \text{ ft} \times 18.8 \text{ ft} = 2453.4 = 2.5 \times 10^3 \text{ ft}^3$

$$2453.4 \text{ ft}^3 \times \frac{(1 \text{ yd})^3}{(3 \text{ ft})^3} \times \frac{(1 \text{ m})^3}{(1.094 \text{ yd})^3} \times \frac{48 \text{ μg CO}}{1 \text{ m}^3} \times \frac{1 \times 10^{-6} \text{ g}}{1 \text{ μg}} = 3.3 \times 10^{-3} \text{ g CO}$$

1.51 A wire is a very long, thin cylinder of volume, $V = \pi r^2 h$, where h is the length of the wire and πr^2 is the cross-sectional area of the wire.

Strategy: 1) Calculate total volume of copper in cm^3 from mass and density

2) h (length in cm) = $\frac{V}{\pi r^2}$

3) Change cm → ft

$$150 \text{ lb Cu} \times \frac{453.6 \text{ g}}{1 \text{ lb Cu}} \times \frac{1 \text{ cm}^3}{8.94 \text{ g}} = 7.61 \times 10^3 \text{ cm}^3$$

$$r = d/2 = 8.25 \text{ mm} \times \frac{1 \text{ cm}}{10 \text{ mm}} \times \frac{1}{2} = 0.4125 = 0.413 \text{ cm}$$

$$h = \frac{V}{\pi r^2} = \frac{7610.7 \text{ cm}^3}{\pi (0.4125)^2 \text{ cm}^2} = 1.4237 \times 10^4 = 1.42 \times 10^4 \text{ cm}$$

$$1.4237 \times 10^4 \text{ cm} \times \frac{1 \text{ in}}{2.54 \text{ cm}} \times \frac{1 \text{ ft}}{12 \text{ in}} = 467 \text{ ft}$$

(too difficult to estimate)

1.52 (a) $$26.73 \text{ g total} \times \frac{0.90 \text{ g Ag}}{1 \text{ g total}} \times \frac{1 \text{ tr oz}}{31.1 \text{ g Ag}} \times \frac{\$1.18}{1 \text{ tr oz}} = \$0.91$$

(b) $$\$25.00 \times \frac{1 \text{ tr oz}}{\$5.30} \times \frac{31.1 \text{ g}}{1 \text{ tr oz}} \times \frac{1 \text{ g total}}{0.90 \text{ g Ag}} \times \frac{1 \text{ coin}}{26.73 \text{ g}} = 6.1 \text{ coins}$$

Since coins come in integer numbers, 7 coins are required.

1.53 Select a common unit for comparison, in this case the kg.

1 kg > 2 lb, 1 L ≈ 1 qt

5 lb potatoes < 2.5 kg

5 kg sugar = 5 kg

1 gal = 4 qt ≈ 4 L. 1 mL H_2O = 1 g H_2O. 1 L = 1000 g, 4 L = 4000 g = 4 kg

The order of mass from lightest to heaviest is 5 lb potatoes < 1 gal water < 5 kg sugar.

1.54 Select a common unit for comparison, in this case the cm.

1 in ≈ 2.5 cm, 1 m = 100 cm

57 cm = 57 cm

14 in ≈ 35 cm

1.1 m = 110 cm

The order of length from shortest to longest is 14-in shoe < 57-cm string < 1.1-m pipe.

Additional Exercises

1.55 Composition is the contents of a substance, the kinds of elements that are present and their relative amounts. Structure is the arrangement of these contents.

1.56 A gold coin is probably a **solid solution**. Pure gold (element 79) is too soft and too valuable to be used for coinage, so other metals are added. However, the simple term "gold coin" does not give a specific indication of the other metals in the mixture.

A cup of coffee is a **solution** if there are no suspended solids (coffee grounds). It is a heterogeneous mixture if there are grounds. If cream or sugar are added, the homogeneity of the mixture depends on how thoroughly the components are mixed.

A wood plank is a **heterogeneous mixture** of various cellulose components. The different domains in the mixture are visible as wood grain or knots.

The ambiguity in each of these examples is that the name of the substance does not provide a complete description of the material. We must rely on mental images and these vary from person to person.

1.57 (a) A **hypothesis** is a possible explanation for certain phenomena based on preliminary experimental data. A **theory** may be more general, and has a significant body of experimental evidence to support it; a theory has withstood the test of experimentation.

(b) A scientific **law** is a summary or statement of natural behavior; it tells how matter behaves. A **theory** is an explanation of natural behavior; it attempts to explain why matter behaves the way it does.

1.58 Any sample of vitamin C has the same relative amount of carbon and oxygen; the ratio of oxygen to carbon in the isolated sample is the same as the ratio in synthesized vitamin C.

$$\frac{2.00 \text{ g O}}{1.50 \text{ g C}} = \frac{x \text{ g O}}{6.35 \text{ g C}}; \quad x = \frac{(2.00 \text{ g O})(6.35 \text{ g C})}{1.50 \text{ g C}} = 8.47 \text{ g O}$$

This calculation assumes the *law of constant composition.*

1.59 (a) I. $(22.52 + 22.48 + 22.54)/3 = 22.51$

II. $(22.64 + 22.58 + 22.62)/3 = 22.61$

Based on the average, set I is more accurate. That is, it is closer to the true value of 22.52%.

(b) Average deviation = $\sum |\text{ value} - \text{average }|/3$

I. $|22.52 - 22.51| + |22.48 - 22.51| + |22.54 - 22.51|/3 = 0.02$

II. $|22.64 - 22.61| + |22.58 - 22.61| + |22.62 - 22.61|/3 = 0.02$

The two sets display the same precision, even though set I is more accurate.

1.60 (a) Inappropriate - The circulation of a widely read publication would vary over a year's time, and could simply not be counted to the nearest single subscriber. Probably about four significant figures would be appropriate. (b) Appropriate - It might be possible to do better than two significant figures, but an estimate to two significant figures should easily be possible. (c) Rainfall can be measured to within 0.02 in., but it is probably not possible to record an entire year's rainfall to the nearest 0.01 in. Further, the variation from year to year is sufficiently large that it does not make much sense to report the annual average to this number of significant figures. Probably two significant figures would be appropriate. (d) Assuming that the statement is based on survey data from a reasonably large sample, two (6.8%) or three (12.0%) significant figures are appropriate. If the sample contained less than 1000 participants, three significant figures is not appropriate.

1.61 K = °C + 273.15; K = -246.1°C + 273.15 = 27.05 = 27.1 K

°F = 9/5 (°C) + 32; °F = 9/5 (-246.1) + 32 = -411.0°F

(We consider 32 to be exact, so the result has 4 significant figures, as does the datum.)

1.62 (a) $\frac{m}{s^2}$ (b) $\frac{kg \bullet m}{s^2}$ (c) $\frac{kg \bullet m}{s^2} \times m = \frac{kg \bullet m^2}{s^2}$

(d) $\frac{kg \bullet m}{s^2} \times \frac{1}{m^2} = \frac{kg}{m \bullet s^2}$ (e) $\frac{kg \bullet m^2}{s^2} \times \frac{1}{s} = \frac{kg \bullet m^2}{s^3}$

1.63 $\frac{40 \text{ lb peat}}{14 \times 20 \times 30 \text{ in}^3} \times \frac{1 \text{ in}^3}{(2.54)^3 \text{ cm}^3} \times \frac{453.6 \text{ g}}{1 \text{ lb}} = 0.13 \text{ g/cm}^3$ peat

$\frac{40 \text{ lb soil}}{1.9 \text{ gal}} \times \frac{1 \text{ gal}}{4 \text{ qt}} \times \frac{1.057 \text{ qt}}{1 \text{ L}} \times \frac{1 \times 10^{-3} \text{ L}}{1 \text{ mL}} \times \frac{1 \text{ mL}}{1 \text{ cm}^3} \times \frac{453.6 \text{ g}}{1 \text{ lb}} = 2.5 \text{ g/cm}^3$ soil

No. The densities tell us that a certain volume of peat moss is "lighter" (weighs less) than the same volume of top soil. Volume must be specified in order to compare mass.

1.64 Density is the ratio of mass and volume. For substances with different densities, the greater the density the smaller the volume of substance that will contain a certain mass. Since volume is directly related to diameter ($V = 4/3 \pi r^3 = 1/6 \pi d^3$), the more dense the substance, the smaller the diameter of a ball that contains a certain mass. The order of the sphere diameters is the reverse order of densities: Pb < Ag < Al. Mathematically, assume 10.0 g of material.

Pb: $10.0 \text{ g} \times \frac{1 \text{ cm}^3}{11.3 \text{ g}} = 0.88496 = 0.885 \text{ cm}^3$; $d = (6 V/\pi)^{1/3} = 1.19$ cm

Ag: $10.0 \text{ g} \times \frac{1 \text{ cm}^3}{10.5 \text{ g}} = 0.95238 = 0.952 \text{ cm}^3$; d = 1.22 cm

Al: $10.0 \text{ g} \times \frac{1 \text{ cm}^3}{2.70 \text{ g}} = 3.7037 = 3.70 \text{ cm}^3$; d = 1.92 cm

Note that Pb and Ag, with similar densities have similar diameters; Al, with a much smaller density, has a much larger diameter.

1.65 The most dense liquid, Hg, will sink; the least dense, cyclohexane, will float; H_2O will be in the middle.

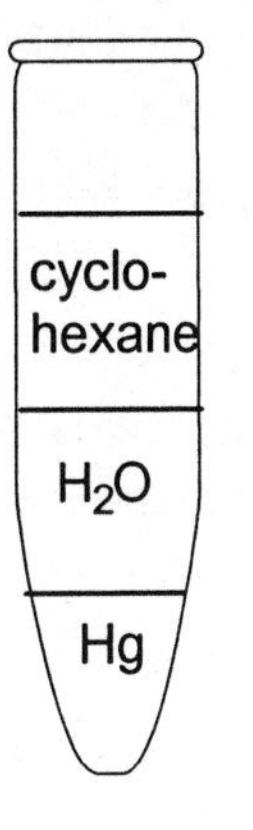

1.66 (a) $23.2 \times 10^9 \text{ lb} \times \frac{453.6 \text{ g}}{1 \text{ lb}} = 1.05235 \times 10^{13} = 1.05 \times 10^{13}$ g NaOH

(b) $1.05235 \times 10^{13} \text{ g} \times \frac{1 \text{ cm}^3}{2.130 \text{ g}} \times \frac{1 \text{ m}^3}{(100)^3 \text{ cm}^3} \times \frac{1 \text{ km}^3}{(1000)^3 \text{ m}^3} = 4.94 \times 10^{-3} \text{ km}^3$

1.67 (a) density = $(35.66 \text{ g} - 14.23 \text{ g})/4.59 \text{ cm}^3 = 4.67 \text{ g/cm}^3$

(b) $34.5 \text{ kg} \times \frac{1000 \text{ g}}{1 \text{ kg}} \times \frac{1 \text{ mL}}{13.6 \text{ g}} \times \frac{1 \text{ L}}{1000 \text{ mL}} = 2.54 \text{ L}$

(c) $V = 4/3 \pi r^3 = 4/3 \pi (28.9 \text{ cm})^3 = 1.0111 \times 10^5 = 1.01 \times 10^5 = 1.01 \times 10^5 \text{ cm}^3$

$1.011 \times 10^5 \text{ cm}^3 \times \frac{3.52 \text{ g}}{\text{cm}^3} = 3.56 \times 10^5 \text{ g}$

The sphere weighs 356 kg or 785 pounds. The student is unlikely to be able to carry the sphere.

1.68 mass of toluene = 58.58 g – 32.65 g = 25.93 g

volume of toluene = $25.93 \text{ g} \times \frac{1 \text{ mL}}{0.864 \text{ g}} = 30.0116 = 30.0 \text{ mL}$

volume of solid = 5.00 mL – 30.0116 mL = 19.9884 = 20.0 mL

density of solid = $\frac{32.65 \text{ g}}{19.9884 \text{ mL}} = 1.63 \text{ g/mL}$

1.69 There are 209.1 degrees between the freezing and boiling points on the Celsius (C) scale and 100 degrees on the glycol (G) scale. Also, -11.5°C = 0°G. By analogy with °F and °C,

$°G = \frac{100}{209.1}(°C + 11.5)$ or $°C = \frac{209.1}{100}(°G) - 11.5$

These equations correctly relate the freezing point and boiling point of ethylene glycol on the two scales.

f.p. of H_2O: $°G = \frac{100}{209.1}(0°C + 11.5) = 5.50°G$

b.p. of H_2O: $°G = \frac{100}{209.1}(100°C + 11.5) = 53.3°G$

1.70 (a) *Plan.* Calculate the total distance in miles, the total time in hours.

$$192 \text{ yd} \times \frac{3 \text{ ft}}{1 \text{ yd}} \times \frac{1 \text{ mi}}{5280 \text{ ft}} = 0.10909 = 0.109 \text{ mi}$$

If the distance can be measured to the nearest yard, or 0.0006 mi, there is uncertainty in the fourth and probably third decimal place, so the total distance is reported to 3 decimal places and 5 sig figs.

[total distance = 13 + 0.109 = 13.109 mi]

$$44 \text{ min} \times \frac{1 \text{ hr}}{60 \text{ min}} = 0.7333 = 0.73 \text{ hr}$$

$$18 \text{ sec} \times \frac{1 \text{ min}}{60 \text{ sec}} \times \frac{1 \text{ hr}}{60 \text{ min}} = 0.0050 \text{ hr}$$

Time can be measured to the nearest second, or 0.0003 hr, so the total time is reported with 4 decimal places and 5 sig figs.

total time = 1 + 0.7333 + 0.0050 = 1.7383 hr

$$\text{avg. speed} = \frac{13.109 \text{ mi}}{1.7383 \text{ hr}} = 7.5413 \text{ mi/hr}$$

(b) $$\frac{1 \text{ hr}}{7.5413 \text{ mi}} \times \frac{60 \text{ min}}{1 \text{ hr}} = 7.9562 \text{ min/mi}$$

$$\frac{1 \text{ hr}}{7.5413 \text{ mi}} \times \frac{60 \text{ min}}{1 \text{ hr}} \times \frac{60 \text{ s}}{1 \text{ min}} = 477.37 \text{ s/mi}$$

1.71 (a) $$2.4 \times 10^5 \text{ mi} \times \frac{1.609 \text{ km}}{1 \text{ mi}} \times \frac{1000 \text{ m}}{1 \text{ km}} = 3.9 \times 10^8 \text{ m}$$

(b) $$2.4 \times 10^5 \text{ mi} \times \frac{1.609 \text{ km}}{1 \text{ mi}} \times \frac{1 \text{ hr}}{2.4 \times 10^3 \text{ km}} \times \frac{60 \text{ min}}{1 \text{ hr}} \times \frac{60 \text{ s}}{1 \text{ min}} = 5.8 \times 10^5 \text{ s}$$

1.72 (a) $$575 \text{ ft} \times \frac{12 \text{ in}}{1 \text{ ft}} \times \frac{2.54 \text{ cm}}{1 \text{ in}} \times \frac{10 \text{ mm}}{1 \text{ cm}} \times \frac{1 \text{ quarter}}{1.55 \text{ mm}} = 1.1307 \times 10^5$$

$$= 1.13 \times 10^5 \text{ quarters}$$

(b) $$1.1307 \times 10^5 \text{ quarters} \times \frac{5.67 \text{ g}}{1 \text{ quarter}} = 6.41 \times 10^5 \text{ g} \quad (641 \text{ kg})$$

(c) $$1.1307 \times 10^5 \text{ quarters} \times \frac{1 \text{ dollar}}{4 \text{ quarters}} = \$28{,}268 = \$2.83 \times 10^4$$

(d) $$\$4.9 \times 10^{12} \times \frac{1 \text{ stack}}{\$28{,}268} = 1.7 \times 10^8 \text{ stacks}$$ (approximately 170 million stacks)

1.73 (a) $$\frac{\$2480}{\text{acre} \bullet \text{ft}} \times \frac{1 \text{ acre}}{4840 \text{ yd}^2} \times \frac{3 \text{ ft}}{1 \text{ yd}} \times \frac{(1.094 \text{ yd})^3}{(1 \text{ m})^3} \times \frac{(1 \text{ m})^3}{(10 \text{ dm})^3} \times \frac{(1 \text{ dm})^3}{1 \text{ L}} =$$

$\$2.013 \times 10^{-3}$/L or 0.2013 ¢ / L (0.201 ¢/L to 3 sig figs)

(b) $\frac{\$2480}{\text{acre}\cdot\text{ft}} \times \frac{1\ \text{acre}\cdot\text{ft}}{2\ \text{households}\cdot\text{year}} \times \frac{1\ \text{year}}{365\ \text{days}} \times 1\ \text{household} = \frac{\$3.397}{\text{day}} = \frac{\$3.40}{\text{day}}$

1.74 (a) volume = $\pi r^2 h = \pi \times (3.55\ \text{cm})^2 \times 75.3\ \text{cm} = 2.98 \times 10^3\ \text{cm}^3$

(b) $r = d/2 = 12.9\ \text{in}/2 = 6.45\ \text{in}$

$$V = \pi\ (6.45\ \text{in})^2 \times \frac{(2.54\ \text{cm})^2}{1\ \text{in}^2} \times \frac{(1\ \text{m})^2}{(100\ \text{cm})^2} \times 22.5\ \text{in} \times \frac{2.54\ \text{cm}}{1\ \text{in}} \times \frac{1\ \text{m}}{100\ \text{cm}}$$

$$= 0.04819 = 0.0482\ \text{m}^3$$

(c) $0.04819\ \text{m}^3 \times \frac{(100\ \text{cm})^3}{(1\ \text{m})^3} \times \frac{13.6\ \text{g Hg}}{1\ \text{cm}^3} \times \frac{1\ \text{kg}}{1000\ \text{g}} = 655\ \text{kg Hg}$

1.75 $11.86\ \text{g ethanol} \times \frac{1\ \text{cm}^3}{0.789\ \text{g ethanol}} = 15.0317 = 15.03\ \text{cm}^3$, volume of cylinder

$V = \pi r^2 h;\ r = (V/\pi h)^{1/2} = \left[\frac{15.0317\ \text{cm}^3}{\pi \times 15.0\ \text{cm}}\right]^{1/2} = 0.5648 = 0.565\ \text{cm}$

$d = 2r = 1.13\ \text{cm}$

1.76 (a) Let x = mass of Au in jewelry

9.85 - x = mass of Ag in jewelry

The total volume of jewelry = volume of Au + volume of Ag

$$0.675\ \text{cm}^3 = x\ \text{g} \times \frac{1\ \text{cm}^3}{19.3\ \text{g}} + (9.85-x)\ \text{g} \times \frac{1\ \text{cm}^3}{10.5\ \text{g}}$$

$0.675 = \frac{x}{19.3} + \frac{9.85-x}{10.5}$ (To solve, multiply both sides by (19.3)(10.5))

$$0.675\ (19.3)(10.5) = 10.5\ x + (9.85 - x)(19.3)$$

$$136.79 = 10.5\ x + 190.105 - 19.3\ x$$

$$-53.315 = -8.8\ x$$

x = 6.06 g Au; 9.85 g total - 6.06 g Au = 3.79 g Ag

$$\text{mass \% Au} = \frac{6.06\ \text{g Au}}{9.85\ \text{g jewelry}} \times 100 = 61.5\%\ \text{Au}$$

(b) 24 karats × 0.615 = 15 karat gold

1.77 A solution can be separated into components by physical means, so separation would be attempted. If the liquid is a solution, the solute could be a solid or a liquid; these two kinds of solutions would be separated differently. Therefore, divide the liquid into several samples and do different tests on each. Try evaporating the solvent from one sample. If a solid remains, the liquid is a solution and the solute is a solid. If the result is negative, try distilling a sample to see if two or more liquids with different boiling points are present. If this result is negative, the liquid is probably a pure substance, but negative results are never entirely conclusive. We might not have tried the appropriate separation technique.

1.78 The separation is successful if two distinct spots are seen on the paper. To quantify the characteristics of the separation, calculate a reference value for each spot that is

$$\frac{\text{distance travelled by spot}}{\text{distance travelled by solvent}}$$

If the values for the two spots are fairly different, the separation is successful. (One could measure the distance between the spots, but this would depend on the length of paper used and be different for each experiment. The values suggested above are independent of the length of paper.)

1.79 The densities are:

carbon tetrachloride (methane, tetrachloro) - 1.5940 g/cm^3

hexane - 0.6603 g/cm^3

benzene - 0.87654 g/cm^3

methylene iodide (methane, diiodo) - 3.3254 g/cm^3

Only methylene iodide will separate the two granular solids. The undesirable solid (2.04 g/cm^3) is less dense than methylene iodide and will float; the desired material is more dense than methylene iodide and will sink. The other three liquids are less dense than both solids and will not produce separation.

1.80 Study (a) is likely to be both precise and accurate, because the errors are carefully controlled. The secondary weight standard will be resistant to chemical and physical changes, the balance is carefully calibrated, and weighings are likely to be made by the same person. The relatively large number of measurements is likely to minimize the effect of random errors on the average value. The accuracy and precision of study (b) depend on the veracity of the participants' responses which cannot be carefully controlled. It also depends on the definition of "comparable lifestyle". The percentages are not precise, because the broad definition of lifestyle leads to a range of results (scatter). The relatively large number of participants improves the precision and accuracy. In general, controlling errors and maximizing the number of data points in a study improves precision and accuracy.

2 Atoms, Molecules, and Ions

Atomic Theory and Atomic Structure

2.1 Postulate 4 of the atomic theory is the *law of constant composition.* It states that the relative number and kinds of atoms in a compound are constant, regardless of the source. Therefore, 1.0 g of pure water should always contain the same relative amounts of hydrogen and oxygen, no matter where or how the sample is obtained.

2.2 (a) 6.500 g compound - 0.384 g hydrogen = 6.116 g sulfur

(b) *Conservation of mass*

(c) According to postulate 3 of the atomic theory, atoms are neither created nor destroyed during a chemical reaction. If 0.384 g of H are recovered from a compound that contains only H and S, the remaining mass must be sulfur.

2.3 (a) $\frac{17.60 \text{ g oxygen}}{30.82 \text{ g nitrogen}} = \frac{0.5711 \text{ g O}}{1 \text{ g N}}$; 0.5711/0.5711 = 1.0

$\frac{35.20 \text{ g oxygen}}{30.82 \text{ g nitrogen}} = \frac{1.142 \text{ g O}}{1 \text{ g N}}$; 1.142/0.5711 = 2.0

$\frac{70.40 \text{ g oxygen}}{30.82 \text{ g nitrogen}} = \frac{2.284 \text{ g O}}{1 \text{ g N}}$; 2.284/0.5711 = 4.0

$\frac{88.00 \text{ g oxygen}}{30.82 \text{ g nitrogen}} = \frac{2.855 \text{ g O}}{1 \text{ g N}}$; 2.855/0.5711 = 5.0

(b) These masses of oxygen per one gram nitrogen are in the ratio of 1:2:4:5 and thus obey the *law of multiple proportions.* Multiple proportions arise because atoms are the indivisible entities combining, so they must combine in ratios of small whole numbers.

2.4 (a) 1: $\frac{3.56 \text{ g fluorine}}{4.75 \text{ g iodine}}$ = 0.749 g fluorine/1 g iodine

2: $\frac{3.43 \text{ g fluorine}}{7.64 \text{ g iodine}}$ = 0.449 g fluorine/1 g iodine

3: $\frac{9.86 \text{ g fluorine}}{9.41 \text{ g iodine}}$ = 1.05 g fluorine/1 g iodine

(b) To look for integer relationships among these values, divide each one by the smallest. If the quotients aren't all integers, multiply by a common factor to obtain all integers.

1: 0.749/0.449 = 1.67; 1.67 × 3 = 5

2: 0.449/0.449 = 1.00; 1.00 × 3 = 3

3: 1.05/0.449 = 2.34; 2.34 × 3 = 7

The ratio of g fluorine to g iodine in the three compounds is 5:3:7. These are in the ratio of small whole numbers and, therefore, obey the *law of multiple proportions*. This integer ratio indicates that the combining fluorine "units" (atoms) are indivisible entities.

2.5 Evidence that cathode rays were negatively charged particles was (1) that electric and magnetic fields deflected the rays in the same way they would deflect negatively charged particles and (2) that a metal plate exposed to cathode rays acquired a negative charge.

2.6 (a) The electron itself has a negative charge so it is repelled by the negatively charged plate and attracted to the positively charged plate.

(b) As the charge on the plates is increased, the respective repulsion and attraction of the electron increases (Coloumb's Law) and the amount of bend should increase.

(c) As the mass of the particle increases, a greater force is required to deflect it. If the strength of the magnetic field is constant, the bend should decrease as the mass increases.

(d) Since the unknown particle is deflected in the opposite direction, it is attracted to the (-) plate and repelled by the (+) plate. The unknown particle is positively charged. The magnitude of the deflection is less than that of the electron, so the unknown particle has greater mass than the electron. The unknown is a positively charged particle of greater mass than the electron.

2.7 (a) In Millikan's oil-drop experiment the X-rays serve as "ionizing radiation". That is, X-rays interact with gaseous atoms or molecules in the chamber in such a way that the particles are ionized. The energy of the X-rays is sufficient to eject electrons from the gaseous particles, forming positive ions and free electrons. The free electrons are then able to recombine with the ions or cling to the oil drops.

(b) If the positive plate were lower than the negative plate, the oil drops "coated" with negatively charged electrons would be attracted to the positively charged plate and would descend much more quickly.

(c) The more times a measurement is repeated, the better the chance of detecting and compensating for experimental errors. That is, if a quantity is measured five times and four measurements agree but one does not, the disagreeable measurement is probably the result of an error. Also, the four agreeable measurements can be averaged to compensate for small random fluctuations. Millikan wanted to demonstrate the validity of his result via its reproducibility.

2.8 (a) The droplets contain different charges because there may be 1, 2, 3 or more excess electrons on the droplet.

(b) The electronic charge is likely to be the lowest common factor in all the observed charges.

(c) Assuming this is so, we calculate the apparent electronic charge from each drop as follows:

A: $1.60 \times 10^{-19} / 1 = 1.60 \times 10^{-19}$ C

B: $3.15 \times 10^{-19} / 2 = 1.58 \times 10^{-19}$ C

C: $4.81 \times 10^{-19} / 3 = 1.60 \times 10^{-19}$ C

D: $6.31 \times 10^{-19} / 4 = 1.58 \times 10^{-19}$ C

The reported value is the average of these four values. Since each calculated charge has three significant figures, the average will also have three significant figures.

$(1.60 \times 10^{-19}\ \text{C} + 1.58 \times 10^{-19}\ \text{C} + 1.60 \times 10^{-19}\ \text{C} + 1.58 \times 10^{-19}\ \text{C}) / 4 = 1.59 \times 10^{-19}\ \text{C}$

2.9 (a) Because γ-rays are not deflected by the electric field, they carry no charge. [No conclusion can be made about their mass, or whether they are, in fact, particles or waves.]

(b) If α and β rays are deflected in opposite directions in an electric field, then they must have opposite electrical charges.

2.10 In the scattering experiment, most alpha particles, massive positively charged helium nuclei, passed directly through the foil, but a few were deflected at large angles. The diffuse positive charge in the 'plum pudding' model would not have produced a repulsion strong enough to deflect the alpha particles at large angles. In Rutherford's model, the positive charge in the atom is concentrated in the small nucleus. If an alpha particle strikes a gold nucleus directly, it is deflected at a large angle.

Modern View of Atomic Structure; Atomic Weights

2.11 (a) $$1.9\ \text{Å} \times \frac{1 \times 10^{-10}\ \text{m}}{1\ \text{Å}} \times \frac{1\ \text{nm}}{1 \times 10^{-9}\ \text{m}} = 0.19\ \text{nm}$$

$$1.9\ \text{Å} \times \frac{1 \times 10^{-10}\ \text{m}}{1\ \text{Å}} \times \frac{1\ \text{pm}}{1 \times 10^{-12}\ \text{m}} = 1.9 \times 10^2 \text{ or } 190\ \text{pm}\ (1\ \text{Å} = 100\ \text{pm})$$

(b) Aligned Kr atoms have **diameters** touching. d = 2r = 2(1.9 Å) = 3.8 Å

$$1.0\ \text{mm} \times \frac{1\ \text{m}}{1000\ \text{mm}} \times \frac{1\ \text{Å}}{1 \times 10^{-10}\ \text{m}} \times \frac{1\ \text{Kr atom}}{3.8\ \text{Å}} = 2.6 \times 10^6\ \text{Kr atoms}$$

(c) $V = 4/3\ \pi r^3$. $r = 1.9\ \text{Å} \times \dfrac{1 \times 10^{-10}\ \text{m}}{1\ \text{Å}} \times \dfrac{100\ \text{cm}}{\text{m}} = 1.9 \times 10^{-8}\ \text{cm}$

$V = (4/3)(\pi)(1.9 \times 10^{-8})^3\ \text{cm}^3 = 2.9 \times 10^{-23}\ \text{cm}^3$

2.12 (a) $r = d/2$; $r = \dfrac{2.5 \times 10^{-8}\ \text{cm}}{2} \times \dfrac{1\ \text{Å}}{1\ 10^{-8}\ \text{cm}} = 1.25 = 1.3\ \text{Å}$

$r = \dfrac{2.5 \times 10^{-8}\ \text{cm}}{2} \times \dfrac{1\ \text{m}}{100\ \text{cm}} = 1.25 \times 10^{-10} = 1.3 \times 10^{-10}\ \text{m}$

(b) Aligned Rh atoms have **diameters** touching. $d = 2.5 \times 10^{-8}\ \text{cm} = 2.5 \times 10^{-10}\ \text{m}$

$6.0\ \mu\text{m} \times \dfrac{1 \times 10^{-6}\ \text{m}}{1\ \mu\text{m}} \times \dfrac{1\ \text{Rh atom}}{2.5 \times 10^{-10}\ \text{m}} = 2.4 \times 10^{-4}$ Rh atoms

(c) $V\ 4/3\ \pi r^3 \quad r = 1.25 \times 10^{-10} = 1.3 \times 10^{-30}\ \text{m}$

$V = (4/3)(\pi(1.25 \times 10^{-10})^3\ \text{m}^3 = 8.2 \times 10^{-8}\ \text{m}^3$

2.13 (a) proton, neutron, electron

(b) proton = +1, neutron = 0, electron = -1

(c) The neutron is most massive, the electron least massive. (The neutron and proton have very similar masses).

2.14 (a) The nucleus has most of the mass **but occupies very little** of the volume of an atom.

(b) True

(c) The number of electrons in an atom is equal to the number of **protons** in the atom.

(d) True

2.15 p = protons, n = neutrons, e = electrons

(a) ^{28}Si has 14 p, 14 n, 14 e (b) ^{60}Ni has 28 p, 32 n, 28 e

(c) ^{85}Rb has 37 p, 48 n, 37 e (d) ^{128}Xe has 54 p, 74 n, 54 e

(e) ^{195}Pt has 78 p, 117 n, 78 e (f) ^{238}U has 92 p, 146 n, 92 e

2.16 (a) ^{32}P has 15 p, 17 n (b) ^{51}Cr has 24 p, 27 n (c) ^{60}Co has 27 p, 33 n

(d) ^{99}Tc has 43 p, 56 n (e) ^{131}I has 53 p, 78 n (f) ^{201}Tl has 81 p, 120 n

2.17

Symbol	^{52}Cr	^{75}As	^{40}Ca	^{222}Rn	^{193}Ir
Protons	24	33	20	86	77
Neutrons	28	42	20	136	116
Electrons	24	33	20	86	77
Mass no.	52	75	40	222	193

2.18

Symbol	^{121}Sb	^{88}Sr	^{182}W	^{139}La	^{239}Pu
Protons	51	38	74	57	94
Neutrons	70	50	108	82	145
Electrons	51	38	74	57	94
Mass No.	121	88	182	139	239

2.19 (a) $^{179}_{72}Hf$ (b) $^{40}_{18}Ar$ (c) $^{4}_{2}He$ (d) $^{115}_{49}In$ (e) $^{28}_{14}Si$

2.20 Since the two nuclides are atoms of the same element, by definition they have the same number of protons, 54. They differ in mass number (and mass) because they have different numbers of neutrons. ^{129}Xe has 75 neutrons and ^{130}Xe has 76 neutrons.

2.21 (a) $^{12}_{6}C$

(b) Atomic weights are really average atomic masses, the sum of the mass of each naturally-occurring isotope of an element times its fractional abundance. Each Cl atom will have the mass of one of the naturally-occurring isotopes, while the "atomic weight" is an average value. The naturally-occurring isotopes of Cl, their atomic masses and relative abundances are: ^{35}Cl, 34.968852, 75.77%; ^{37}Cl, 36.965903, 24.23%.

2.22 (a) 12 amu

(b) The atomic weight of carbon reported on the front-inside cover of the text is the abundance-weighted average of the atomic masses of the two naturally-occurring isotopes of carbon, ^{12}C and ^{13}C. The mass of a ^{12}C atom is exactly 12 amu, but the atomic weight of 12.011 takes into account the presence of some ^{13}C atoms in every natural sample of the element.

2.23 Average atomic mass (atomic weight) = $\sum$ fractional abundance × mass of isotope

Average atomic mass = 0.014 (203.97302) + 0.241 (205.9744) + 0.221 (206.97587) + 0.524 (207.97663) = 207.22 = 207 amu

(The result has 0 decimal places and 3 sig figs because the fourth term in the sum has 3 sig figs and 0 decimal places.)

2.24 Average atomic mass (atomic weight) = $\sum$ fractional abundance × mass of isotope
Average atomic mass = 0.6917(62.9296) + 0.3083(64.9278) = 63.5456 = 63.55 amu

2.25 (a) Compare Figures 2.4 and 2.13, referring to Solution 2.6(c). In Thomson's cathode ray experiments and in mass spectrometry a stream of charged particles is passed through the poles of a magnet. The charged particles are deflected by the magnetic field according to their mass and charge. For a constant magnetic field strength and speed of the particles, the lighter particles experience a greater deflection.

(b) The x-axis label (independent variable) is atomic weight and the y-axis label (dependent variable) is signal intensity.

(c) Uncharged particles are not deflected in a magnetic field. The effect of the magnetic field on moving, *charged* particles is the basis of their separation by mass.

2.26 (a) The purpose of the magnet in the mass spectrometer is to change the path of the moving ions. The magnitude of the deflection is inversely related to mass, which is the basis of the discrimination by mass.

(b) The atomic weight of Cl, 35.5, is an average atomic mass. It is the average of the masses of two naturally occurring isotopes, weighted by their abundances. See Solution 2.21(b).

(c) The single peak at mass 31 in the mass spectrum of phosphorus indicates that the sample contains a single isotope of P, and the mass of this isotope is 31 amu.

2.27 (a) Average atomic mass = 0.7899(23.98504) + 0.1000(24.98584) + 0.1101(25.98259)
= 24.31 amu

(b)

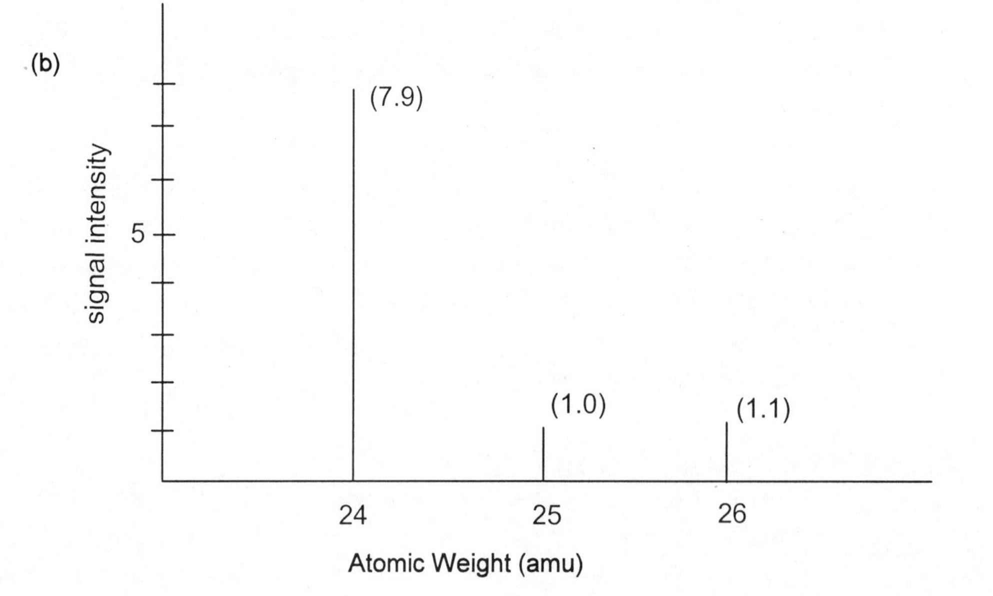

The relative intensities of the peaks in the mass spectrum are the same as the relative abundances of the isotopes. The abundances and peak heights are in the ratio ^{24}Mg: ^{25}Mg: ^{26}Mg as 7.8 : 1.0 : 1.1.

2.28 (a) Three peaks: $^{1}H - {}^{1}H$, $^{1}H - {}^{2}H$, $^{2}H - {}^{2}H$

(b) $^{1}H - {}^{1}H = 2(1.00783) = 2.01566$ amu

$^{1}H - {}^{2}H = 1.00783 + 2.01410 = 3.02193$ amu

$^{2}H - {}^{2}H = 2(2.01410) = 4.02820$ amu

[The mass ratios are 1 : 1.49923 : 1.99845 or 1 : 1.5 : 2]

(c) $^{1}H - {}^{1}H$ is largest, because there is the greatest chance that two atoms of the more abundant isotope will combine.

$^{2}H - {}^{2}H$ is the smallest, because there is the least chance that two atoms of the less abundant isotope will combine.

The Periodic Table; Molecules and Ions

2.29 (a) Ag (metal) (b) He (nonmetal) (c) P (nonmetal) (d) Cd (metal)
(e) Ca (metal) (f) Br (nonmetal) (g) As (metalloid)

2.30 (a) lithium (metal) (b) scandium (metal) (c) germanium (metalloid)
(d) ytterbium (metal) (e) manganese (metal) (f) gold (metal) (g) tellurium (metalloid)

2.31 (a) K, alkali metals (metal) (b) I, halogens (nonmetal) (c) Mg, alkaline earth metals (metal)
(d) Ar, noble gases (nonmetal) (e) S, chalcogens (nonmetal)

2.32 C, carbon, nonmetal; Si, silicon, metalloid; Ge, germanium, metalloid; Sn, tin, metal; Pb, lead, metal

2.33 An empirical formula shows the simplest ratio of the different atoms in a molecule. A molecular formula shows the exact number and kinds of atoms in a molecule. A structural formula shows how these atoms are arranged.

2.34 Compounds with the same empirical but different molecular formulas differ by the integer number of empirical formula units in the respective molecules. Thus, they can have very different molecular structure, size and mass, resulting in very different physical properties.

2.35 A molecular formula contains all atoms in a molecule. An empirical formula shows the simplest ratio of atoms in a molecule or elements in a compound.

(a) molecular formula: C_6H_6; empirical formula: CH

(b) molecular formula: $SiCl_4$; empirical formula: $SiCl_4$ (1:4 is the simplest ratio)

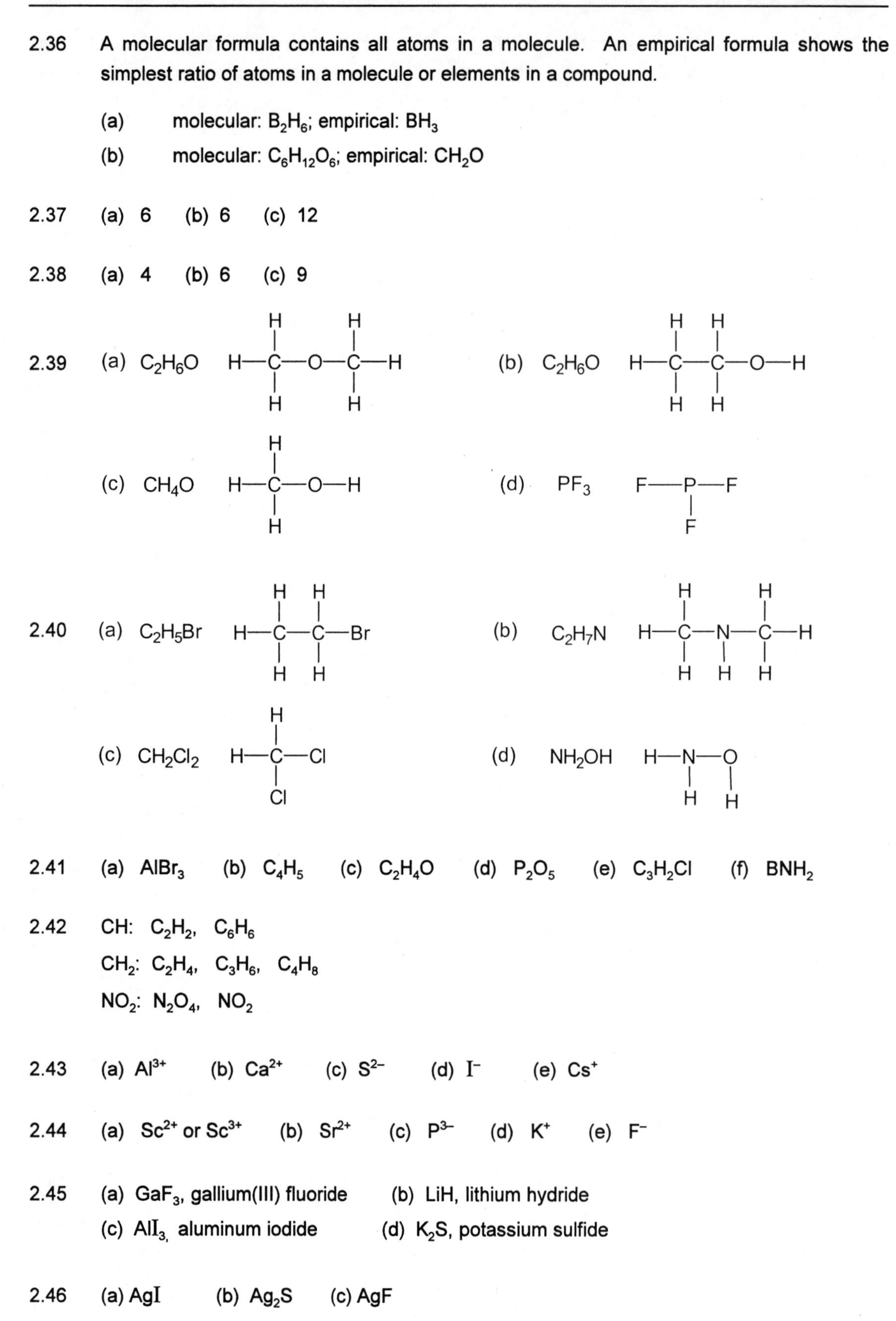

2.36 A molecular formula contains all atoms in a molecule. An empirical formula shows the simplest ratio of atoms in a molecule or elements in a compound.

(a) molecular: B_2H_6; empirical: BH_3

(b) molecular: $C_6H_{12}O_6$; empirical: CH_2O

2.37 (a) 6 (b) 6 (c) 12

2.38 (a) 4 (b) 6 (c) 9

2.39 (a) C_2H_6O (b) C_2H_6O

(c) CH_4O (d) PF_3

2.40 (a) C_2H_5Br (b) C_2H_7N

(c) CH_2Cl_2 (d) NH_2OH

2.41 (a) $AlBr_3$ (b) C_4H_5 (c) C_2H_4O (d) P_2O_5 (e) C_3H_2Cl (f) BNH_2

2.42 CH: C_2H_2, C_6H_6

CH_2: C_2H_4, C_3H_6, C_4H_8

NO_2: N_2O_4, NO_2

2.43 (a) Al^{3+} (b) Ca^{2+} (c) S^{2-} (d) I^- (e) Cs^+

2.44 (a) Sc^{2+} or Sc^{3+} (b) Sr^{2+} (c) P^{3-} (d) K^+ (e) F^-

2.45 (a) GaF_3, gallium(III) fluoride (b) LiH, lithium hydride

(c) AlI_3, aluminum iodide (d) K_2S, potassium sulfide

2.46 (a) AgI (b) Ag_2S (c) AgF

2.47 (a) $CaBr_2$ (b) NH_4Cl (c) $Al(C_2H_3O_2)_3$ (d) K_2SO_4 (e) $Mg_3(PO_4)_2$

2.48 (a) $(NH_4)_2SO_4$ (b) Cu_2S (c) LaF_3 (d) $Ca_3(PO_4)_2$ (e) Hg_2CO_3

2.49 Molecular (all elements are nonmetals): (a) B_2H_6 (b) CH_3OH (f) NOCl (g) NF_3
Ionic (formed by a cation and an anion, usually contains a metal cation): (c) $LiNO_3$, (d) Sc_2O_3, (e) CsBr, (h) Ag_2SO_4

2.50 Molecular (all elements are nonmetals): (a) PF_5 (c) SCl_2
Ionic (formed from ions, usually contains a metal cation): (b) NaI
(d) $Ca(NO_3)_2$ (e) $FeCl_3$ (f) LaP (g) $CoCO_3$ (h) $(NH_4)_2SO_4$

Naming Inorganic Compounds; Organic Molecules

2.51 (a) ClO_2^- (b) Cl^- (c) ClO_3^- (d) ClO_4^- (e) ClO^-

2.52 (a) selenate (b) selenide (c) hydrogen selenide (biselenide)
(d) hydrogen selenite (biselenite)

2.53 (a) aluminum fluoride (b) iron(II) hydroxide (ferrous hydroxide)
(c) copper(II) nitrate (cupric nitrate) (d) barium perchlorate (e) lithium phosphate
(f) mercury(I) sulfide (mercurous sulfide) (g) calcium acetate (h) chromium(III) carbonate (chromic carbonate) (i) potassium chromate (j) ammonium sulfate

2.54 (a) lithium oxide (b) iron(III) carbonate (ferric carbonate)
(c) sodium hypochlorite (d) ammonium sulfite (e) strontium cyanide
(f) chromium(III) hydroxide (chromic hydroxide) (g) cobalt(II) nitrate (cobaltous nitrate)
(h) sodium dihydrogen phosphate (i) potassium permanganate (j) silver dichromate

2.55 (a) Cu_2O (b) K_2O_2 (c) $Al(OH)_3$ (d) $Zn(NO_3)_2$ (e) Hg_2Br_2 (f) $Fe_2(CO_3)_3$ (g) NaBrO

2.56 (a) $K_2Cr_2O_7$ (b) $Co(NO_3)_2$ (c) $Cr(C_2H_3O_2)_3$ (d) NaH (e) $Ca(HCO_3)_2$
(f) $Ba(BrO_3)_2$ (g) $Cu(ClO_4)_2$

2.57 (a) bromic acid (b) hydrobromic acid (c) phosphoric acid (d) HClO (e) HIO_3
(f) H_2SO_3

2.58 (a) HBr (b) H_2S (c) HNO_2 (d) carbonic acid (e) chloric acid (f) acetic acid

2.59 (a) sulfur hexafluoride (b) iodine pentafluoride (c) xenon trioxide (d) N_2O_4 (e) HCN
(f) P_4S_6

2.60 (a) dinitrogen monoxide (b) nitrogen monoxide (c) nitrogen dioxide
(d) dinitrogen pentoxide (e) dinitrogen tetroxide

2.61 (a) $ZnCO_3$, ZnO, CO_2 (b) HF, SiO_2, SiF_4, H_2O (c) SO_2, H_2O, H_2SO_3
(d) H_3P (or PH_3) (e) $HClO_4$, Cd, $Cd(ClO_4)_2$ (f) VBr_3

2.62 (a) $NaHCO_3$ (b) $Ca(OCl)_2$ (c) HCN (d) $Mg(OH)_2$ (e) SnF_2
(f) CdS, H_2SO_4, H_2S

2.63 (a) A hydrocarbon is a compound composed of the elements hydrogen and carbon only.

(b) All alkanes are hydrocarbons, but compounds other than alkanes can also be hydrocarbons.

(c)

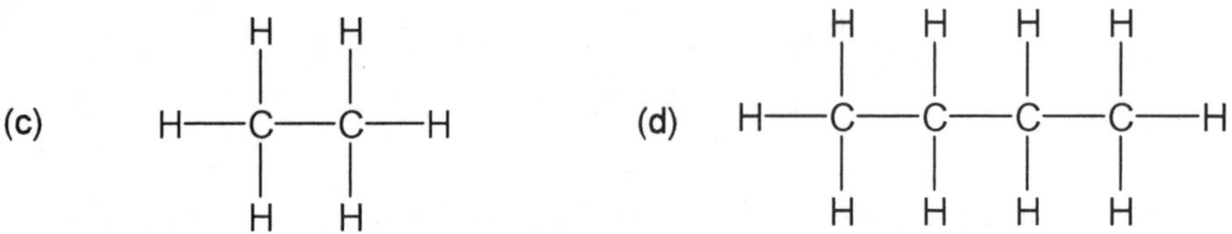

(d)

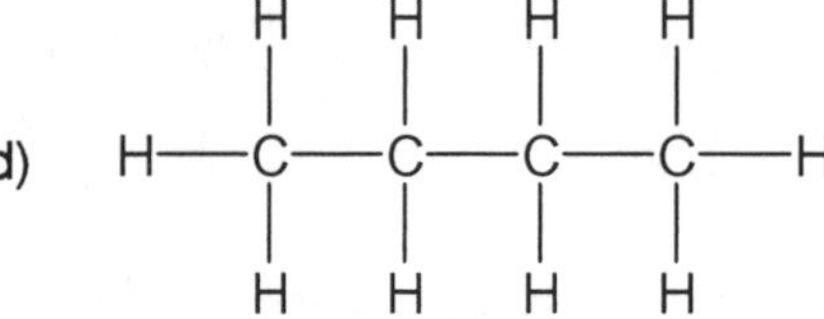

molecular: C_4H_{10}

empirical: C_2H_5

2.64 (a) *-ane*

(b) All alkanes are hydrocarbons; they are composed of hydrogen and carbon only.

(c)

(d) **Hex**ane has 6 carbons in its chain.

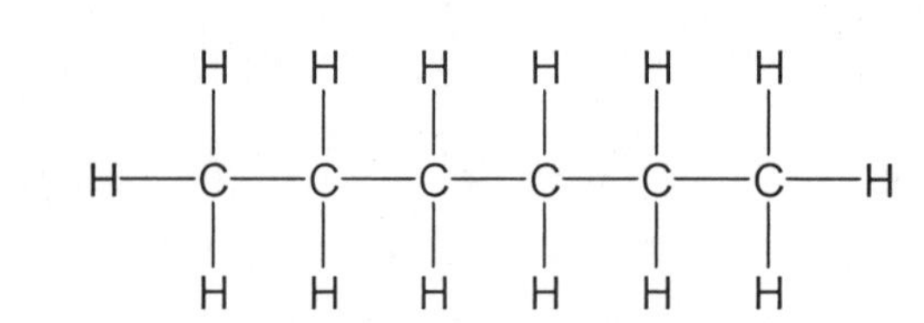

molecular: C_6H_{14}

empirical: C_3H_7

2.65 (a) *Functional groups* are groups of specific atoms that are constant from one molecule to the next. For example, the alcohol functional group is an –OH. Whenever a molecule is called an alcohol, it contains the –OH group.

(b) –OH

(c)

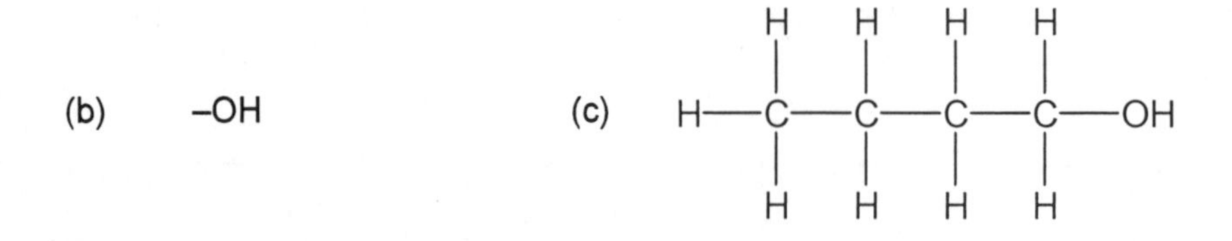

2.66 (a) They all have two carbon atoms in their molecular backbone, or chain.

(b) In 1-propanol one of the H atoms on an outer (terminal) C atom has been replaced by an –OH group.

(c) Ethanoic acid has two carbons in its chain. One of these C atoms is bound to two O atoms in the way that is characteristic of a *carboxylic acid* functional group. Propanoic acid has three C atoms in its chain and one of the outer C atoms is part of the carboxylic acid functional group.

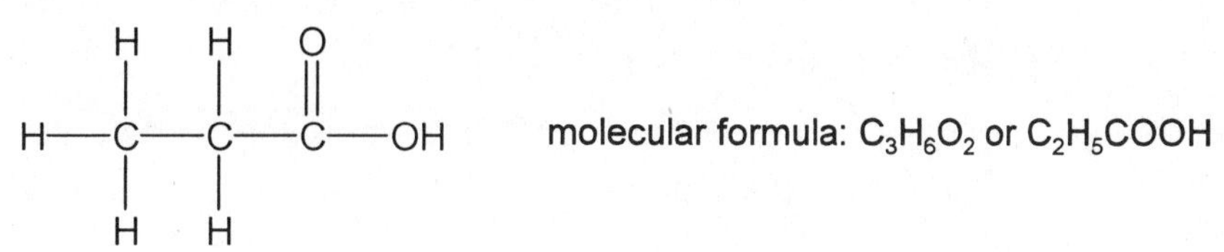

Additional Exercises

2.67 (a) Based on data accumulated in the late eighteenth century on how substances react with one another, **Dalton** postulated the atomic theory. Dalton's theory is based on the indivisible atom as the smallest unit of an element that can combine with other elements.

(b) By determining the effects of electric and magnetic fields on cathode rays, **Thomson** measured the mass-to-charge ratio of the electron. He also proposed the "plum pudding" model of the atom in which most of the space in an atom is occupied by a diffuse positive charge in which the tiny negatively charged electrons are imbedded.

(c) By observing the rate of fall of oil drops in and out of an electric field, **Millikan** measured the charge of an electron.

(d) After observing the scattering of alpha particles at large angles when the particles struck gold foil, **Rutherford** postulated the nuclear atom. In Rutherford's atom, most of the mass of the atom is concentrated in a small dense region called the nucleus and the tiny negatively charged electrons are moving through empty space around the nucleus.

2.68 (a) Droplet D would fall most slowly. It carries the most negative charge, so it would be most strongly attracted to the upper (+) plate and most strongly repelled by the lower (-) plate. These electrostatic forces would provide the greatest opposition to gravity.

(b) Calculate the lowest common factor.

A: $3.84 \times 10^{-8} / 2.88 \times 10^{-8} = 1.33$; $1.33 \times 3 = 4$

B. $4.88 \times 10^{-8} / 2.88 \times 10^{-8} = 1.66$; $1.66 \times 3 = 5$

C. $2.88 \times 10^{-8} / 2.88 \times 10^{-8} = 1.00$; $1.00 \times 3 = 3$

D. $8.64 \times 10^{-8} / 2.88 \times 10^{-8} = 3.00$; $3.00 \times 3 = 9$

The total charge on the drops is in the ratio of 4:5:3:9. Divide the total charge on each drop by the appropriate integer and average the four values to get the charge of an electron in warmombs.

A: $3.84 \times 10^{-8} / 4 = 9.60 \times 10^{-9}$ wa

B: $4.80 \times 10^{-8} / 5 = 9.60 \times 10^{-9}$ wa

C: $2.88 \times 10^{-8} / 3 = 9.60 \times 10^{-9}$ wa

D: $8.64 \times 10^{-8} / 9 = 9.60 \times 10^{-9}$ wa

The charge on an electron is 9.60×10^{-9} wa

(c) The number of electrons on each drop are the integers calculated in part (b). A has 4 e^-, B has 5 e^-, C has 3 e^- and D has 9 e^-.

(d) $$\frac{9.60 \times 10^{-9}\ \text{wa}}{1\ e^-} \times \frac{1\ e^-}{1.60 \times 10^{-16}\ \text{C}} = 6.00 \times 10^{7}\ \text{wa/C}$$

2.69 Radioactivity is the spontaneous emission of radiation from a substance. Becquerel's discovery showed that atoms could decay, or degrade, *implying* that they are not indivisible. However, it wasn't until Rutherford and others characterized the nature of radioactive emissions, especially the particle nature of α and β rays, that the full significance of the discovery was apparent.

2.70 (a) Most of the volume of an atom is empty space in which electrons move. Most alpha particles passed through this space. The path of the massive alpha particle would not be significantly altered by interaction with a "puny" electron.

(b) Most of the mass of an atom is contained in a very small, dense area called the nucleus. The few alpha particles that hit the massive, positively charged gold nuclei were strongly repelled and essentially deflected back in the direction they came from.

(c) The Be nuclei have a much smaller volume and positive charge than the Au nuclei; the charge repulsion between the alpha particles and the Be nuclei will be less, and there will be fewer direct hits because the Be nuclei have an even smaller volume than the Au nuclei. Fewer alpha particles will be scattered in general and fewer will be strongly back scattered.

2.71 (a) 2 protons and 2 neutrons (b) the nuclear strong force

(c) The charge of an α particle is twice the magnitude of the charge of an electron, with the opposite sign. That is, $2(+1.6022 \times 10^{-19})\ \text{C} = +3.2044 \times 10^{-19}\ \text{C}$.

(d) $$\frac{3.2044 \times 10^{-19}\ \text{C}}{4.8224 \times 10^{4}\ \text{g/C}} = 6.6448 \times 10^{-24}\ \text{g}$$

$$6.6448 \times 10^{-24}\ \text{g} \times \frac{1\ \text{amu}}{1.66054 \times 10^{-24}\ \text{g}} = 4.0016\ \text{amu}$$

(e) The sum of the particle masses in an α particle is 2(1.0073 amu) and 2(1.0087) amu = 4.0320 amu. The actual particle mass, 4.0016 amu, is less than the sum of the masses of the components. The difference is the nuclear binding energy, the energy released when protons and neutrons combine to form a nucleus. Mass and energy are interchangeable according to the Einstein relationship $E = mc^2$.

2.72 (a) ^3He has 2 protons, 1 neutron and 2 electrons.

(b) ^3H has 1 proton, 2 neutrons and 1 electron.

^3He: $2(1.6726231 \times 10^{-24}\ g) + 1.6749286 \times 10^{-24}\ g + 2(9.1093897 \times 10^{-28}\ g)$

$= 5.021996 \times 10^{-24}\ g$

^3H: $1.6726231 \times 10^{-24}\ g + 2(1.6749286 \times 10^{-24}\ g) + 9.1093897 \times 10^{-28}\ g$

$= 5.023391 \times 10^{-24}\ g$

Tritium, ^3H, is more massive.

(c) The masses of the two particles differ by 0.0014×10^{-24} g. Each particle loses 1 electron to form the +1 ion, so the difference in the masses of the ions is still 1.4×10^{-27}. A mass spectrometer would need precision to 1×10^{-27} g to differentiate $^3He^+$ and ^3H.

2.73 (a) Calculate the mass of a single gold atom, then divide the mass of the cube by the mass of the gold atom.

$$\frac{197.0\ \text{amu}}{\text{gold atom}} \times \frac{1\ \text{g}}{6.022 \times 10^{23}\ \text{amu}} = 3.2713 \times 10^{-22} = 3.271 \times 10^{-22}\ \text{g/gold atom}$$

$$\frac{19.3\ \text{g}}{\text{cube}} \times \frac{1\ \text{gold atom}}{3.271 \times 10^{-22}\ \text{g}} = 5.90 \times 10^{22}\ \text{Au atoms in the cube}$$

(b) The shape of atoms is spherical; spheres cannot be arranged into a cube so that there is no empty space. The question is, how much empty space is there? We can calculate the two limiting cases; no empty space and maximum empty space. The true diameter will be somewhere in this range.

No empty space: volume cube/number of atoms = volume of one atom

$V = 4/3\pi r^3$; $r = (3\pi V/4)^{1/3}$; $d = 2r$

$$\text{vol. of cube} = (1.0 \times 1.0 \times 1.0) = \frac{1.0\ \text{cm}^3}{5.90 \times 10^{22}\ \text{Au atoms}} = 1.695 \times 10^{-23}$$

$$= 1.7 \times 10^{-23}\ \text{cm}^3$$

$r = [\pi\ (1.695 \times 10^{-23}\ cm^3)/4]^{1/3} = 3.4 \times 10^{-8}$ cm; $d = 2r = 6.8 \times 10^{-8}$ cm

Maximum empty space: assume atoms are arranged in rows in all three directions so they are touching across their diameters. That is, each atom occupies the volume of a cube, with the atomic diameter as the length of the side of the cube. The number of atoms along one edge of the gold cube is then
$(5.90 \times 10^{22})^{1/3} = 3.893 \times 10^7 = 3.89 \times 10^7$ atoms/1.0 cm.

The diameter of a single atom is 1.0 cm/3.89×10^7 atoms $= 2.569 \times 10^{-8} = 2.6 \times 10^{-8}$ cm.

The diameter of a gold atom is between 2.6×10^{-8} cm and 6.8×10^{-8} cm (2.6 - 6.8 Å).

(c) Some atomic arrangement must be assumed, since none is specified. The solid state is characterized by an orderly arrangement of particles, so it isn't surprising that atomic arrangement is required to calculate the density of a solid. A more detailed discussion of solid-state structure and density appears in Chapter 11.

2.74 (a) In arrangement A the number of atoms in 1 cm^2 is just the square of the number that fit linearly in 1 cm.

$$1.0 \text{ cm} \times \frac{1 \text{ atom}}{4.95 \text{ Å}} \times \frac{1 \times 10^{10} \text{ Å}}{1 \text{ m}} \times \frac{1 \text{ m}}{100 \text{ cm}} = 2.02 \times 10^7 = 2.0 \times 10^7 \text{ atoms/cm}$$

$1.0 \text{ cm}^2 = (2.02 \times 10^7)^2 = 4.081 \times 10^{14} = 4.1 \times 10^{14} \text{ atoms/cm}^2$

(b) In arrangement B, the atoms in the horizontal rows are touching along their diameters, as in arrangement A. The number of Rb atoms in a 1.0 cm row is then 2.0×10^7 Rb atoms. Relative to arrangement A, the vertical rows are offset by 1/2 of an atom. Atoms in a 'column' are no longer touching along their vertical diameter. We must calculate the vertical distance occupied by a row of atoms, which is now less than the diameter of one Rb atom.

Consider the triangle shown below. This is an isosceles triangle (equal side lengths, equal interior angles) with a side-length of 2d and an angle of 60°. Drop a bisector to the uppermost angle so that it bisects the opposite side.

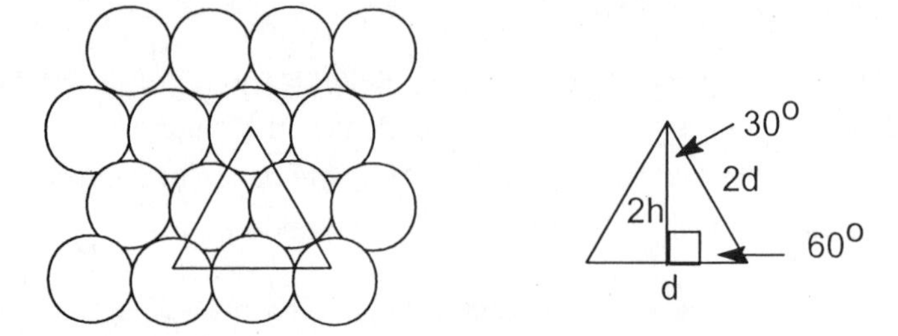

The result is a right triangle with two known side lengths. The length of the unknown side (the angle bisector) is 2h, two times the vertical distance occupied by a row of atoms. Solve for h, the "height" of one row of atoms.

$(2h)^2 + d^2 = (2d)^2; \; 4h^2 = 4d^2 - d^2 = 3d^2; \; h^2 = 3d^2/4$

$h = (3d^2/4)^{1/2} = (3(4.95 \text{ Å})^2/4)^{1/2} = 4.2868 = 4.29 \text{ Å}$

The number of rows of atoms in 1 cm is then

$$1.0 \text{ cm} \times \frac{1 \text{ row}}{4.2868 \text{ Å}} \times \frac{1 \times 10^{10} \text{ Å}}{1 \text{ m}} \times \frac{1 \text{ m}}{100 \text{ cm}} = 2.333 \times 10^7 = 2.3 \times 10^7$$

The number of atoms in a 1.0 cm^2 square area is then

$$\frac{2.020 \times 10^7 \text{ atoms}}{1 \text{ row}} \times 2.333 \times 10^7 \text{ rows} = 4.713 \times 10^{14} = 4.7 \times 10^{14}$$

Note that we have ignored the loss of "1/2" atom at the end of each horizontal row. Out of 2.0×10^7 atoms per row, one atom is not significant.

(c) The ratio of atoms in arrangement B to arrangement A is then 4.713×10^{14} atoms/4.081×10^{14} = 1.555 = 1.2:1. Clearly, arrangement B results in less empty space per unit area or volume. If extended to three dimensions, arrangement B would lead to a greater density for Rb metal.

2.75 (a) diameter of nucleus = 1×10^{-4} Å; diameter of atom = 1 Å

$V = 4/3\ \pi r^3$; $r = d/2$; $r_n = 0.5 \times 10^{-4}$ Å; $r_a = 0.5$ Å

volume of nucleus = $4/3\ \pi\ (0.5 \times 10^{-4})^3$ Å^3

volume of atom = $4/3\ \pi\ (0.5)^3$ Å^3

$$\text{volume fraction of nucleus} = \frac{\text{volume of nucleus}}{\text{volume of atom}} = \frac{4/3\ \pi\ (0.5 \times 10^{-4})^3\ \text{Å}^3}{4/3\ \pi\ (0.5)^3\ \text{Å}^3} = 1 \times 10^{-12}$$

diameter of atom = 5 Å, r_a = 2.5 Å

$$\text{volume fraction of nucleus} = \frac{4/3\ \pi\ (0.5 \times 10^{-4})^3\ \text{Å}^3}{4/3\ \pi\ (2.5)^3\ \text{Å}^3} = 8 \times 10^{-15}$$

Depending on the radius of the atom, the volume fraction of the nucleus is between 1×10^{-12} and 8×10^{-15}; that is, between 1 part in 10^{12} and 8 parts in 10^{15}.

(b) mass of proton = 1.0073 amu

1.0073 amu $\times$ 1.66054×10^{-24} g/amu = 1.6727×10^{-24} g

$$\text{diameter} = 1.0 \times 10^{-15}\ \text{m, radius} = 0.50 \times 10^{-15}\ \text{m} \times \frac{100\ \text{cm}}{1\ \text{m}} = 5.0 \times 10^{-14}\ \text{cm}$$

Assuming a proton is a sphere, $V = 4/3\ \pi r^3$.

$$\text{density} = \frac{\text{g}}{\text{cm}^3} = \frac{1.6727 \times 10^{-24}\ \text{g}}{4/3\ \pi\ (5.0 \times 10^{-14})^3\ \text{cm}^3} = 3.2 \times 10^{15}\ \text{g/cm}^3$$

2.76 (a) ${}^{16}_{8}\text{O}$, ${}^{17}_{8}\text{O}$, ${}^{18}_{8}\text{O}$

(b) All isotopes are atoms of the same element, oxygen, with the same atomic number (Z = 8), 8 protons in the nucleus and 8 electrons. Elements with similar electron arrangements have similar chemical properties (Section 2.5). Since the 3 isotopes all have 8 electrons, we expect their electron arrangements to be the same and their chemical properties to be very similar, perhaps identical. Each has a different number of neutrons (8, 9 or 10), a different mass number (A = 16, 17 or 18) and thus a different atomic mass.

2.77 Weight is the force exerted by the mass of an object under the acceleration due to gravity. The SI unit of weight is the Newton, or kg • m/s^2. Mass is a fundamental quantity with SI units of kg. The weight of an object varies depending on the position of the object relative to Earth, but mass is invariant with position. Atomic 'weights', with units of grams, are the same on the moon as on Earth. Clearly, they are masses rather than weights.

2.78 (a) The 68.926 amu isotope has a mass number of 69, with 31 protons, 38 neutrons and the symbol ${}^{69}_{31}Ga$. The 70.926 amu isotope has a mass number of 71, 31 protons, 40 neutrons, and symbol ${}^{71}_{31}Ga$. (All Ga atoms have 31 protons.)

(b) The average mass of a Ga atom (given on the inside cover of the text) is 69.72 amu. Let x = abundance of the lighter isotope, 1-x = abundance of the heavier isotope. Then x(68.926) + (1-x)(70.925) = 69.723; x = 0.6013, ${}^{69}Ga$ = 60.13%, ${}^{71}Ga$ = 39.87%

2.79 (a) There are 24 known isotopes of Ni, from ${}^{51}Ni$ to ${}^{74}Ni$.

(b) The five most abundant isotopes are

${}^{58}Ni$, 57.935346 amu, 68.077%

${}^{60}Ni$, 59.930788 amu, 26.223%

${}^{62}Ni$, 61.928346 amu, 3.634%

${}^{61}Ni$, 60.931058 amu, 1.140%

${}^{64}Ni$, 63.927968 amu, 0.926%

Data from *Handbook of Chemistry and Physics*, 74th Ed. [Data may differ slightly in other editions.]

2.80 (a) A Br_2 molecule could consist of two atoms of the same isotope or one atom of each of the two different isotopes. This second possibility is twice as likely as the first. Therefore, the second peak (twice as large as peaks 1 and 3) represents a Br_2 molecule containing different isotopes. The mass numbers of the two isotopes are determined from the masses of the two smaller peaks. Since 157.836 ≈ 158, the first peak represents a ${}^{79}Br$–${}^{79}Br$ molecule. Peak 3, 161.832 ≈ 162, represents a ${}^{81}Br$–${}^{81}Br$ molecule. Peak 2 then contains one atom of each isotope, ${}^{79}Br$–${}^{81}Br$, with an approximate mass of 160 amu.

(b) The mass of the lighter isotope is 157.836 amu/2 atoms, or 78.918 amu/atom. For the heavier one, 161.832 amu/2 atoms = 80.916 amu/atom.

(c) The relative size of the three peaks in the mass spectrum of Br_2 indicates their relative abundance. The average mass of a Br_2 molecule is
0.2569(157.836) + 0.4999(159.834) + 0.2431(161.832) = 159.79 amu
(Each product has four significant figures and two decimal places, so the answer has two decimal places.)

(d) $$\frac{159.79 \text{ amu}}{\text{avg. } Br_2 \text{ molecule}} \times \frac{1\ Br_2 \text{ molecule}}{2 \text{ Br atoms}} = 79.895 \text{ amu}$$

(e) Let x = the abundance of ${}^{79}Br$, 1-x = abundance of ${}^{81}Br$. From (b), the masses of the two isotopes are 78.918 amu and 80.916 amu, respectively. From (d), the mass of an average Br atom is 79.895 amu.

x(78.918) + (1 - x)(80.916) = 79.895, x = 0.5110

${}^{79}Br$ = 51.10%, ${}^{81}Br$ = 48.90%

2.81 (a) 5 significant figures. $^{1}H^{+}$ is a bare proton with mass 1.0073 amu. ^{1}H is a hydrogen atom, with 1 proton and 1 electron. The mass of the electron is 5.486×10^{-4} or 0.0005486 amu. Thus the mass of the electron is significant in the fourth decimal place or fifth significant figure in the mass of ^{1}H.

(b) Mass of ^{1}H = 1.0073 amu (proton)
0.0005486 amu (electron)
1.0078 amu (We have not rounded up to 1.0079 since 49 < 50 in the final sum.)

$$\text{Mass \% of electron} = \frac{\text{mass of e}^-}{\text{mass of }^1\text{H}} \times 100 = \frac{5.486 \times 10^{-4}\text{ amu}}{1.0078\text{ amu}} \times 100 = 0.05444\%$$

2.82 copper: Cu, 1B (coinage metals, transition metal)
tin: Sn, 4A
zinc: Zn, 2B
phosphorus: P, 5A
lead: Pb, 4A

2.83 (a) an alkali metal: K (b) an alkaline earth metal: Ca (c) a noble gas: Ar
(d) a halogen: Br (e) a metalloid: Ge (f) a nonmetal in 1A: H
(g) a metal that forms a 3+ ion: Al (h) a nonmetal that forms a 2− ion: O
(i) an element that resembles Al: Ga

2.84 (a) $^{266}_{106}Sg$ has 106 protons, 160 neutrons and 106 electrons
(b) Sg is in Group 6B (or 6) and immediately below tungsten, W. We expect the chemical properties of Sg to most closely resemble those of W.

2.85 (a) chlorine gas: ii (b) propane: v (c) nitrate ion: i
(d) sulfur trioxide: iii (e) methylchloride: iv

2.86

Symbol	$^{102}Ru^{3+}$	$^{80}Se^{2-}$	$^{192}Os^{2+}$	$^{127}I^{-}$	$^{140}Ce^{3+}$
Protons	44	34	76	53	58
Neutrons	58	46	116	74	82
Electrons	41	36	74	54	55
Net Charge	3+	2-	2+	1-	3+

2.87 (a) nickel(II) oxide, 2+ (b) manganese(IV) oxide, 4+ (c) chromium(III) oxide, 3+
(d) molybdenium(VI) oxide, 6+

2.88 (a) IO_3^- (b) IO_4^- (c) IO^- (d) HIO (e) HIO_4 or (H_5IO_6)

2.89 (a) perbromate ion (b) selenite ion (c) AsO_4^{3-} (d) $HTeO_4^-$

2.90 (a) sodium chloride (b) sodium bicarbonate (or sodium hydrogen carbonate)
(c) sodium hypochlorite (d) sodium hydroxide (e) ammonium carbonate
(f) calcium sulfate

2.91 (a) potassium nitrate (b) sodium carbonate (c) calcium oxide
(d) hydrochloric acid (e) magnesium sulfate (f) magnesium hydroxide

2.92 (a) CaS, $Ca(HS)_2$ (b) HBr, $HBrO_3$ (c) AlN, $Al(NO_2)_3$ (d) FeO, Fe_2O_32
(e) NH_3, NH_4^+ (f) K_2SO_3, $KHSO_3$ (g) Hg_2Cl_2 , $HgCl_2$ (h) $HClO_3$, $HClO_4$

2.93

	Formula	Name	Density, g/mL	Melting Point, °C	Boiling Point, °C
(a)	PF_3	phosphorus trifluoride	3.907	-151.5	-101.5
(b)	$SiCl_4$	silicon tetrachloride	1.483	-70	57.57
(c)	C_2H_6O (C_2H_5OH)	ethanol	0.7893	-117.3	78.5

2.94 (a) CH

(b) No. Benzene is not an alkane because alkanes are hydrocarbons with ALL single bonds.

(c) In an alcohol, the –OH group replaces an H atom of the hydrocarbon.

The molecular formula is C_6H_6O or C_6H_5OH. (The OH could go on any one of the six carbon atoms.)

2.95 (a) $\frac{0.0774 \text{ g H}}{0.9226 \text{ g C}} = \frac{\text{x g H}}{1.000 \text{ g C}}$; 0.0839 g H/1 g C

(b) xylene: $\frac{0.0949 \text{ g H}}{0.9051 \text{ g C}} = \frac{\text{x g H}}{1.000 \text{ g C}}$; 0.105 g H/1 g C

biphenyl: $\frac{0.0654 \text{ g H}}{0.9346 \text{ g C}} = \frac{\text{x g H}}{1.000 \text{ g C}}$; 0.0700 g H/1 g C mesitylene: $\frac{0.1006 \text{ g H}}{0.8994 \text{ g C}} = \frac{\text{x g H}}{1.000 \text{ g C}}$; 0.112 g H/1 g C

toluene: $\frac{0.0875 \text{ g H}}{0.9125 \text{ g C}} = \frac{\text{x g H}}{1.000 \text{ g C}}$; 0.0959 g H/1 g C

(c) If 0.0839 g H/1 g C is a 1:1 combining ratio, dividing the g H/1 g C obtained in (b) by 0.083 should indicate the ratio of H:1C in the other compounds.

(d) Empirical formulas follow the combining ratios.

xylene: $\frac{0.105 \text{ g H/1 g C}}{0.0839 \text{ g H/1 g C}}$ = 1.25H : 1C = 5H : 4C; C_4H_5

biphenyl: $\frac{0.0700 \text{ g H/1 g C}}{0.0839 \text{ g H/1 g C}}$ = 0.834H : 1C = 5H : 6C; C_6H_5

mesitylene: $\frac{0.112 \text{ g H/1 g C}}{0.0839 \text{ g H/1 g C}}$ = 1.33H : 1C = 4 H : 3C; C_3H_4

toluene: $\frac{0.0959 \text{g H/1 g C}}{0.0839 \text{ g H/1 g C}}$ = 1.14H: 1C = 8H : 7C = C_7H_8

2.96 (a) In an alkane, all C atoms have 4 single bonds, so each C in the partial structure needs 2 more bonds. All alkanes are hydrocarbons, so 2 H atoms will bind to each C atom in the ring.

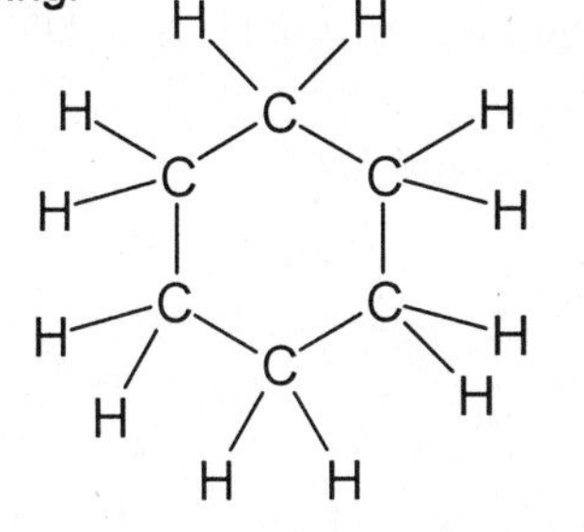

(b) The molecular formula of cyclohexane is C_6H_{12}; the molecular formula of *n*-hexane is C_6H_{14} (see Solution 2.64(d)). Cyclohexane can be thought of as *n*-hexane in which the two outer (terminal) C atoms are joined to each other. In order to form this C-C bond, each outer C atom must lose 1 H atom. The number of C atoms is unchanged and each C atom still has 4 single bonds. The resulting molecular formula is $C_6H_{14-2} = C_6H_{12}$.

(c) On the structure in part (a), replace 1 H atom with an OH group.

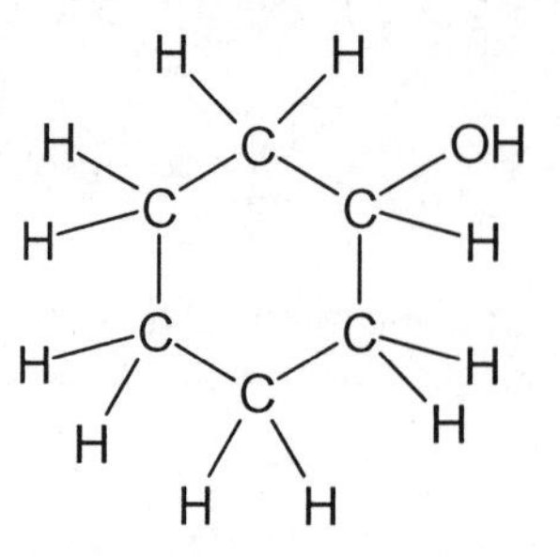

(d) Note that ethene, which contains a C-C double bond, has one fewer H atom per C atom than ethane. The C-C double bond in cyclohexene will have the same effect.

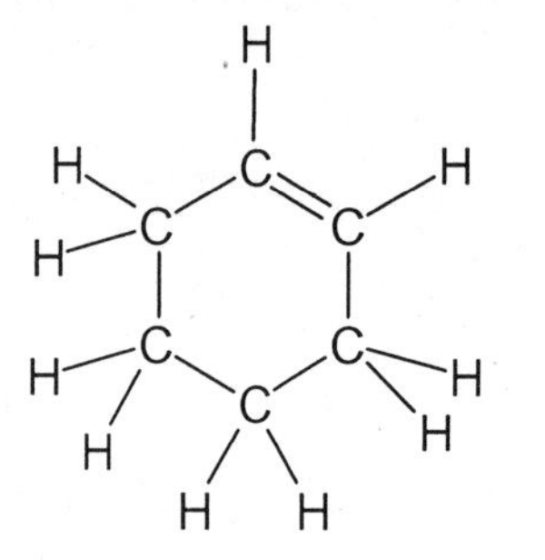

molecular formula: C_6H_{10}

The molecular formula of cyclohex**ene**, C_6H_{10}, is different than the molecular formula of cyclohex**ane**, C_6H_{12}.

2.97 Elements are arranged in the periodic table by increasing atomic number and so that elements with similar chemical and physical properties form a vertical column or group. By its position in the periodic chart, we know whether an element is a metal, nonmetal or metalloid, and the common charge of its ion. Members of a group have the same common ionic charge and combine in similar ways with other elements.

3 Stoichiometry: Calculation with Chemical Formulas and Equations

Balancing Chemical Equations

3.1 (a) In balancing chemical equations, the *law of conservation of mass*, that atoms are neither created nor destroyed during the course of a reaction, is observed. This means that the **number** and **kinds** of atoms on both sides of the chemical equation must be the same.

(b) Subscripts in chemical formulas should not be changed when balancing equations because changing the subscript changes the identity of the compound (*law of constant composition*).

(c) gases - (g); liquids - (l); solids - (s); aqueous solutions - (aq)

3.2 (a) H_2O_2 indicates that there are 2 hydrogen atoms and 2 oxygen atoms bound by chemical bonds into a single molecule (of hydrogen peroxide). $2H_2O$ indicates 2 molecules (of water), each of which contains 2 hydrogen atoms and 1 oxygen atom. The composition of the different molecules, H_2O_2 and H_2O, is different and the physical and chemical properties of the two compounds they constitute are very different. The subscript 2 changes molecular composition and thus properties of the compound. The prefix 2 indicates how many molecules (or moles) of the original compound are under consideration.

(b) No. There are more H and O atoms on the reactant side than the product side. The coefficient 6 before $H_2O(l)$ in the products would balance the equation.

3.3 Equation (a) best fits the diagram.

Overall, 4 A_2 molecules + 4 B atoms → 4 A_2B molecules

Since 4 is a common factor, this equation reduces to equation (a).

3.4

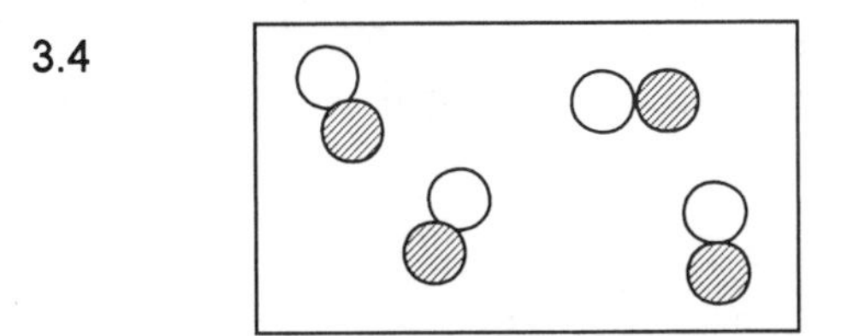

Write the balanced equation for the reaction.

$2H_2 + CO \rightarrow CH_3OH$

The combining ratio of H_2: CO is 2:1. If we have 8 H_2 molecules, 4 CO molecules are required for complete reaction. Alternatively, you could examine the atom ratios in the formula of CH_3OH, but the balanced equation is most direct.

3.5
(a) $2SO_2(g) + O_2(g) \rightarrow 2SO_3(g)$
(b) $P_2O_5(s) + 3H_2O(l) \rightarrow 2H_3PO_4(aq)$
(c) $CH_4(g) + 4Cl_2(g) \rightarrow CCl_4(l) + 4HCl(g)$
(d) $Al_4C_3(s) + 12H_2O(l) \rightarrow 4Al(OH)_3(s) + 3CH_4(g)$
(e) $C_4H_{10}O(l) + 6O_2(g) \rightarrow 4CO_2(g) + 5H_2O(l)$
(f) $2Fe(OH)_3(s) + 3H_2SO_4(aq) \rightarrow Fe_2(SO_4)_3(aq) + 6H_2O(l)$
(g) $Mg_3N_2(s) + 4H_2SO_4(aq) \rightarrow 3MgSO_4(aq) + (NH_4)_2SO_4(aq)$

3.6
(a) $6Li(s) + N_2(g) \rightarrow 2Li_3N(s)$
(b) $TiCl_4(l) + 2H_2O(l) \rightarrow TiO_2(s) + 4HCl(aq)$
(c) $2NH_4NO_3(s) \rightarrow 2N_2(g) + O_2(g) + 4H_2O(g)$
(d) $Ca_3P_2(s) + 6H_2O(l) \rightarrow 3Ca(OH)_2(s) + 2PH_3(g)$
(e) $Al(OH)_3(s) + 3HClO_4(aq) \rightarrow Al(ClO_4)_3(aq) + 3H_2O(l)$
(f) $2AgNO_3(aq) + Na_2SO_4(aq) \rightarrow Ag_2SO_4(s) + 2NaNO_3(aq)$
(g) $2N_2H_4(g) + N_2O_4(g) \rightarrow 4H_2O(g) + 3N_2(g)$

3.7
(a) $CaC_2(s) + 2H_2O(l) \rightarrow Ca(OH)_2(aq) + C_2H_2(g)$
(b) $2KClO_3(s) \xrightarrow{\Delta} 2KCl(s) + 3O_2(g)$
(c) $Zn(s) + H_2SO_4(aq) \rightarrow H_2(g) + ZnSO_4(aq)$
(d) $PCl_3(l) + 3H_2O(l) \rightarrow H_3PO_3(aq) + 3HCl(aq)$
(e) $3H_2S(g) + 2Fe(OH)_3(s) \rightarrow Fe_2S_3(s) + 6H_2O(g)$

3.8
(a) $SO_3(g) + H_2O(l) \rightarrow H_2SO_4(aq)$
(b) $B_2S_3(s) + 6H_2O(l) \rightarrow 2H_3BO_3(aq) + 3H_2S(g)$
(c) $4PH_3(g) + 8O_2(g) \rightarrow 6H_2O(g) + P_4O_{10}(s)$
(d) $2Hg(NO_3)_2(s) \xrightarrow{\Delta} 2HgO(s) + 4NO_2(g) + O_2(g)$
(e) $Cu(s) + 2H_2SO_4(aq) \rightarrow CuSO_4(aq) + SO_2(g) + 2H_2O(l)$

Patterns of Chemical Reactivity

3.9 (a) When a metal reacts with a nonmetal, an ionic compound forms. The combining ratio of the atoms is such that the total positive charge on the metal cation(s) is equal to the total negative charge on the nonmetal anion(s). All ionic compounds are solids.
$2\ Na(s) + Br_2(l) \rightarrow 2NaBr(s)$

(b) The second reactant is oxygen gas from the air, $O_2(g)$. The products are $CO_2(g)$ and $H_2O(l)$. $2C_6H_6(l) + 15O_2(g) \rightarrow 12CO_2(g) + 6H_2O(l)$

3.10 (a) Neutral Ca atom loses $2e^-$ to form Ca^{2+}. Neutral O_2 molecule gains $4e^-$ to form $2O^{2-}$. The formula of the product will be CaO, because the cationic and anionic charges are opposite and equal. $2Ca(s) + O_2(g) \rightarrow 2CaO$

(b) The products are $CO_2(g)$ and $H_2O(l)$. $C_3H_6O(l) + 4O_2(g) \rightarrow 3CO_2(g) + 3H_2O(l)$

3.11 (a) $Mg(s) + Cl_2(g) \rightarrow MgCl_2(s)$

(b) $Ni(OH)_2(s) \xrightarrow{\Delta} NiO(s) + H_2O(g)$

(c) $C_8H_8(l) + \mathbf{10}O_2(g) \rightarrow \mathbf{8}CO_2(g) + \mathbf{4}H_2O(l)$

(d) $\mathbf{2}C_5H_{12}O(l) + \mathbf{15}O_2(g) \rightarrow \mathbf{10}CO_2(g) + \mathbf{12}H_2O(l)$

3.12 (a) $\mathbf{2}Al(s) + \mathbf{3}Br_2(l) \rightarrow \mathbf{2}AlBr_3(s)$

(b) $SrCO_3(s) \xrightarrow{\Delta} SrO(s) + CO_2(g)$

(c) $C_7H_{16}(l) + \mathbf{11}O_2(g) \rightarrow \mathbf{7}CO_2(g) + \mathbf{8}H_2O(l)$

(d) CH_3OCH_3 is C_2H_6O. $C_2H_6O(l) + \mathbf{3}O_2(g) \rightarrow \mathbf{2}CO_2(g) + \mathbf{3}H_2O(l)$

3.13 (a) $\mathbf{2}Al(s) + \mathbf{3}Cl_2(g) \rightarrow \mathbf{2}AlCl_3(s)$ combination

(b) $C_2H_4(g) + \mathbf{3}O_2(g) \rightarrow \mathbf{2}CO_2(g) + \mathbf{2}H_2O(l)$ combustion

(c) $\mathbf{6}Li(s) + N_2(g) \rightarrow \mathbf{2}Li_3N(s)$ combination

(d) $PbCO_3(s) \rightarrow PbO(s) + CO_2(g)$ decomposition

(e) $C_7H_8O_2(l) + \mathbf{8}O_2(g) \rightarrow \mathbf{7}CO_2(g) + \mathbf{4}H_2O(l)$ combustion

3.14 (a) $\mathbf{2}C_3H_6(g) + \mathbf{9}O_2(g) \rightarrow \mathbf{6}CO_2(g) + \mathbf{6}H_2O(l)$ combustion

(b) $NH_4NO_3(s) \rightarrow N_2O(g) + \mathbf{2}H_2O(l)$ decomposition

(c) $C_5H_6O(l) + \mathbf{6}O_2(g) \rightarrow \mathbf{5}CO_2(g) + \mathbf{3}H_2O(l)$ combustion

(d) $N_2(g) + \mathbf{3}H_2(g) \rightarrow \mathbf{2}NH_3(g)$ combination

(e) $K_2O(s) + H_2O(l) \rightarrow \mathbf{2}KOH(aq)$ combination

Formula Weights

3.15 Formula weight (FW) in amu to 1 decimal place (see Sample Exercise 3.5)

(a) H_2S: 2(1.0) + 1(32.1) = 34.1 amu

(b) $NiCO_3$: 1(58.7) + 1(12.0) + 3(16.0) = 118.7 amu

(c) $Mg(C_2H_3O_2)_2$: 1(24.3) + 4(12.0) + 6(1.0) + 4(16.0) = 142.3 amu

(d) $(NH_4)_3SO_4$: 3(14.0) + 12(1.0) + 1(32.1) + 4(16.0) = 150.1 amu

(e) K_3PO_4: 3(39.1) + 1(31.0) + 4(16.0) = 212.3 amu

(f) Fe_2O_3: 2(55.8) + 3(16.0) = 159.6 amu

(g) P_2S_5: 2(31.0) + 5(32.1) = 222.5 amu

3.16 Formula weight in amu to 1 decimal place.

(a) N_2O: FW = 2(14.0) + 1(16.0) = 44.0 amu

(b) $HC_7H_5O_2$: 7(12.0) + 6(1.0) + 2(16.0) = 122.0 amu

(c) $Mg(OH)_2$: 1(24.3) + 2(16.0) + 2(1.0) = 58.3 amu

(d) $(NH_2)_2CO$: 2(14.0) + 4(1.0) + 1(12.0) + 1(16.0) = 60.0 amu

(e) $CH_3CO_2C_5H_{11}$: 7(12.0) + 14(1.0) + 2(16.0) = 130.0 amu

3.17 *Plan.* Calculate the formula weight (FW), then the mass % oxygen in the compound. *Solve*:

(a) SO_2: FW = 1(32.1) + 2(16.0) = 64.1 amu

$$\% \text{ O} = \frac{2(16.0) \text{ amu}}{64.1 \text{ amu}} \times 100 = 49.9\%$$

(b) Na_2SO_4: FW = 2(23.0) + 1(32.1) + 4(16.0) = 142.1 amu

$$\% \text{ O} = \frac{4(16.0) \text{ amu}}{142.1 \text{ amu}} \times 100 = 45.0\%$$

(c) C_2H_5COOH: FW = 3(12.0) + 6(1.0) + 2(16.0) = 74.0 amu

$$\% \text{ O} = \frac{2(16.0) \text{ amu}}{74.0 \text{ amu}} \times 100 = 43.2\%$$

(d) $Al(NO_3)_3$: FW = 1(27.0) + 3(14.0) + 9(16.0) = 213.0 amu

$$\% \text{ O} = \frac{9(16.0) \text{ amu}}{213.0 \text{ amu}} \times 100 = 67.6\%$$

(e) NH_4NO_3: FW = 2(14.0) + 4(1.0) + 3(16.0) = 80.0 amu

$$\% \text{ O} = \frac{3(16.0) \text{ amu}}{80.0 \text{ amu}} \times 100 = 60.0\%$$

3.18 (a) C_2H_2: FW = 2(12.0) + 2(1.0) = 26.0 amu

$$\% \text{ C} = \frac{2(12.0) \text{ amu}}{26.0 \text{ amu}} \times 100 = 92.3\ \%$$

(b) $(NH_4)_2SO_4$: FW = 2(14.0) + 8(1.0) + 1(32.1) + 4(16.0) = 132.1 amu

$$\% \text{ H} = \frac{8(1.0) \text{ amu}}{132.1 \text{ amu}} \times 100 = 6.1\%$$

(c) $HC_6H_7O_6$: FW = 6(12.0) + 8(1.0) + 6(16.0) = 176.0 amu

$$\% \text{ O} = \frac{6(16.0) \text{ amu}}{176.0 \text{ amu}} \times 100 = 54.5\%$$

(d) $PtCl_2(NH_3)_2$: FW = 1(195.1) + 2(35.5) + 2(14.0) + 6(1.0) = 300.1 amu

$$\% \text{ Pt} = \frac{1(195.1) \text{ amu}}{300.1 \text{ amu}} \times 100 = 65.01\%$$

(e) $C_{18}H_{24}O_2$: FW = 18(12.0) + 24(1.0) + 2(16.0) = 272.0 amu

$$\% \text{ C} = \frac{18(12.0) \text{ amu}}{272.0 \text{ amu}} \times 100 = 79.4\%$$

(f) $C_{18}H_{27}NO_3$: FW = 18(12.0) + 27(1.0) + 1(14.0) + 3(16.0) = 305.0 amu

$$\% C = \frac{18(12.0) \text{ amu}}{305.0 \text{ amu}} \times 100 = 70.8\%$$

3.19 *Plan*. Follow the logic for calculating mass % C given in Sample Exercise 3.6. *Solve*:

(a) C_7H_6O: FW = 7(12.0) + 6(1.0) + 1(16.0) = 106.0 amu

$$\%C = \frac{7(12.0) \text{ amu}}{106.0 \text{ amu}} \times 100 = 79.2\%$$

(b) $C_8H_8O_3$: FW = 8(12.0) + 8(1.0) + 3(16.0) = 152.0 amu

$$\% C = \frac{8(12.0) \text{ amu}}{152.0 \text{ amu}} \times 100 = 63.2\%$$

(c) $C_7H_{14}O_2$: FW = 7(12.0) + 14(1.0) + 2(16.0) = 130.0 amu

$$\% C = \frac{7(12.0) \text{ amu}}{130.0 \text{ amu}} \times 100 = 64.6\%$$

3.20 (a) CO_2: FW = 1(12.0) + 2(16.0) = 44.0 amu

$$\% C = \frac{12.0 \text{ amu}}{44.0 \text{ amu}} \times 100 = 27.3\%$$

(b) CH_3OH: FW = 1(12.0) + 4(1.0) + 1(16.0) = 32.0 amu

$$\% C = \frac{12.0 \text{ amu}}{32.0 \text{ amu}} \times 100 = 37.5\%$$

(c) C_2H_6: FW = 2(12.0) + 6(1.0) = 30.0 amu

$$\% C = \frac{2(12.0) \text{ amu}}{30.0 \text{ amu}} \times 100 = 80.0\%$$

(d) $CS(NH_2)_2$: FW = 1(12.0) + 1(32.1) + 2(14.0) + 4(1.0) = 76.1 amu

$$\% C = \frac{12.0 \text{ amu}}{76.1 \text{ amu}} \times 100 = 15.8\%$$

The Mole

3.21 (a) 6.022×10^{23}. This is the number of objects in a mole of anything.

(b) The formula weight of a substance in amu has the same numerical value as the molar mass expressed in grams.

3.22 (a) exactly 12 g (b) 6.0221421×10^{23}, Avogadro's number

3.23 *Plan*. Since the mole is a counting unit, use it as a basis of comparison; determine the total moles of atoms in each given quantity. *Solve*:

23 g Na contains 1 mol of atoms

0.5 mol H_2O contains (3 atoms × 0.5 mol) = 1.5 mol atoms

6.0×10^{23} N_2 molecules contains (2 atoms × 1 mol) = 2 mol atoms

3.24 3.0×10^{23} H_2O_2 molecules contains (4 atoms × 0.5 mol) = 2 mol atoms

32 g O_2 contains (2 atoms × 1 mol) = 2 mol atoms

2.0 mol CH_4 contains (5 atoms × 2 mol) = 10 mol atoms

3.25 *Analyze.* Given: 16 lb/ball; Avogadro's number of balls, 6.022×10^{23} balls. Find: mass in kg of Avogadro's number of balls; compare with mass of Earth.

Plan. balls → mass in lb → mass in kg; mass of balls/mass of Earth

Solve. $6.022 \times 10^{23} \text{ balls} \times \frac{16 \text{ lb}}{\text{ball}} \times \frac{1 \text{ kg}}{2.2046 \text{ lb}} = 4.370 \times 10^{24} = 4.4 \times 10^{24} \text{ kg}$

$\frac{4.370 \times 10^{24} \text{ kg of balls}}{5.98 \times 10^{24} \text{ kg Earth}} =$ 0.73; One mole of shotput balls weighs 0.73 times as much as Earth.

Check. This mass of balls is reasonable since Avogadro's number is large.
Estimate: 16 lb ≈ 7 kg; $6 \times 10^{23} \times 7 = 4.2 \times 10^{24}$ kg

3.26 250 million = $250 \times 10^6 = 2.50 \times 10^8$ people

$$\frac{6.022 \times 10^{23} \text{ ¢}}{2.50 \times 10^8 \text{ people}} \times \frac{\$1}{100 \text{ ¢}} = \frac{\$6.022 \times 10^{21}}{2.50 \times 10^8 \text{ people}} = \$2.41 \times 10^{13}/\text{person}$$

\$5.5 trillion = $\$5.5 \times 10^{12}$ $\qquad \frac{\$2.41 \times 10^{13}}{\$5.5 \times 10^{12}} = 4.4$

Each person would receive an amount that is 4.4 times the dollar amount of the national debt.

3.27 (a) *Analyze.* Given: 1.73 mol CaH_2. Find: mass in g.

Plan. Use molar mass (g/mol) of CaH_2 to find g CaH_2

Solve. molar mass = 1(40.08) + 2(1.008) = 42.096 = 42.10 g/mol CaH_2

$$1.73 \text{ mol } CaH_2 \times \frac{42.096 \text{ g}}{1 \text{ mol}} = 72.8 \text{ g } CaH_2$$

Check. ~2 × 42 = 84 g. The calculated result is reasonable.

(b) *Analyze.* Given: mass. Find: moles. *Plan.* Use molar mass of $Mg(NO_3)_2$.

Solve. molar mass = 1(24.31) + 2(14.01) + 6(16.00) = 148.33 = 148.3

$$3.25 \text{ g } Mg(NO_3)_2 \times \frac{1 \text{ mol}}{148.33 \text{ g}} = 0.0219 \text{ mol } Mg(NO_3)_2$$

Check. 3/150 ≈ 1/50 = 0.02 mol

(c) *Analyze.* Given: moles. Find: molecules. *Plan.* Use Avogadro's number.

Solve. $0.245 \text{ mol } CH_3OH \times \frac{6.022 \times 10^{23} \text{ molecules}}{1 \text{ mol}} = 1.47539 \times 10^{23}$
$= 1.48 \times 10^{23}$ CH_3OH molecules

Check. $(0.25 \times 6 \times 10^{23}) = 1.5 \times 10^{23}$

(d) *Analyze.* Given: mol C_4H_{10}. Find: H atoms.

Plan. mol C_4H_{10} → mol H atoms → H atoms

Solve. $0.585 \text{ mol } C_4H_{10} \times \frac{10 \text{ mol H atoms}}{1 \text{ mol } C_4H_{10}} \times \frac{6.022 \times 10^{23} \text{ atoms}}{1 \text{ mol}}$

$= 3.52 \times 10^{24}$ H atoms

Check. $(0.6 \times 10 \times 10^{23}) = 36 \times 10^{23} = 3.6 \times 10^{24}$.

3.28 (a) molar mass = 1(24.31) + 2(35.45) = 95.21 g

$2.50 \times 10^{-2} \text{ mol } MgCl_2 \times \frac{95.21 \text{ g}}{1 \text{ mol}} = 2.38 \text{ g } MgCl_2$

(b) molar mass = 1(14.01) + 4(1.008) + 1(35.45) = 53.49 g/mol

$76.5 \text{ g } NH_4Cl \times \frac{1 \text{ mol}}{53.49 \text{ g}} = 1.43 \text{ mol } NH_4Cl$

(c) $0.0772 \text{ mol } HCHO_3 \times \frac{6.022 \times 10^{23} \text{ molecules}}{1 \text{ mol}} = 4.65 \times 10^{22}\ HCHO_2$ molecules

(d) $4.88 \times 10^{-3} \text{ mol } Al(NO_3)_3 \times \frac{3 \text{ mol } NO_3^-}{1 \text{ mol } Al(NO_3)_3} \times \frac{6.022 \times 10^{23}\ NO_3^- \text{ ions}}{1 \text{ mol}}$

$= 8.82 \times 10^{23}\ NO_3^-$

3.29 *Analyze/Plan.* See Solution 3.27 for stepwise problem-solving approach. *Solve*:

(a) molar mass = 2(26.98) + 3(32.07) + 12(16.00) = 342.17 = 342.2 g

$2.50 \times 10^{-3} \text{ mol } Al_2(SO_4)_3 \times \frac{342.2 \text{ g } Al_2(SO_4)_3}{1 \text{ mol}} = 0.856 \text{ g } Al_2(SO_4)_3$

(b) molar mass = 26.982 + 3(35.453) = 133.341 = 133.34 g

$0.0750 \text{ g } AlCl_3 \times \frac{1 \text{ mol}}{133.34 \text{ g } AlCl_3} \times \frac{3 \text{ mol } Cl^-}{1 \text{ mol } AlCl_3} = 1.69 \times 10^{-3} \text{ mol } Cl^-$

(c) molar mass = 8(12.01) + 10(1.008) + 4(14.01) + 2(16.00) = 194.20 = 194.2 g

$7.70 \times 10^{20} \text{ molecules} \times \frac{1 \text{ mol}}{6.022 \times 10^{23} \text{ molecules}} \times \frac{194.2 \text{ g } C_8H_{10}N_4O_2}{1 \text{ mol caffeine}}$

$= 0.248 \text{ g } C_8H_{10}N_4O_2$

(d) $\frac{0.406 \text{ g cholesterol}}{0.00105 \text{ mol}} = 387$ g cholesterol/mol

3.30 (a) molar mass = 55.847 + 30.974 + 4(16.00) = 150.82 = 150.8 g

$0.0714 \text{ mol } FePO_4 \times \frac{150.8 \text{ g } FePO_4}{1 \text{ mol}} = 10.77 = 10.8 \text{ g } FePO_4$

(b) molar mass = 2(14.01) + 8(1.008) + 12.01 + 3(16.00) = 96.09 g

$$4.97 \text{ g } (NH_4)_2CO_3 \times \frac{1 \text{ mol}}{96.09 \text{ g } (NH_4)_2CO_3} \times \frac{2 \text{ mol } NH_4^+}{1 \text{ mol } (NH_4)_2CO_3} = 0.0103 \text{ mol } NH_4^+$$

(c) molar mass = 9(12.01) + 8(1.008) + 4(16.00) = 180.154 = 180.2 g

$$6.52 \times 10^{21} \text{ molecules} \times \frac{1 \text{ mol}}{6.022 \times 10^{23} \text{ molecules}} \times \frac{180.2 \text{ g } C_9H_8O_4}{1 \text{ mol aspirin}} = 1.95 \text{ g } C_9H_8O_4$$

(d) $$\frac{15.86 \text{ g Valium}}{0.05570 \text{ mol}} = 284.7 \text{ g Valium/mol}$$

3.31 (a) molar mass = 6(12.01) + 10(1.008) + 1(16.00) + 2(32.07) = 162.28 = 162.3 g

(b) *Plan.* mg → g → mol *Solve:*

$$5.00 \text{ mg allicin} \times \frac{1 \times 10^{-3} \text{ g}}{1 \text{ mg}} \times \frac{1 \text{ mol}}{162.3 \text{ g}} = 3.081 \times 10^{-5} = 3.08 \times 10^{-5} \text{ mol allicin}$$

Check. 5.00 mg is a small mass, so the small answer is reasonable.

$(5 \times 10^{-3})/200 = 2.5 \times 10^{-5}$

(c) *Plan.* Use mol from part (b) and Avogadro's number to calculate molecules.

Solve. $$3.081 \times 10^{-5} \text{ mol allicin} \times \frac{6.022 \times 10^{23} \text{ molecules}}{\text{mol}} = 1.855 \times 10^{19}$$

$= 1.86 \times 10^{19}$ allicin molecules

Check. $(3 \times 10^{-5})(6 \times 10^{23}) = 18 \times 10^{18} = 1.8 \times 10^{19}$

(d) *Plan.* Use molecules from part (c) and molecular formula to calculate S atoms.

Solve. $$1.855 \times 10^{19} \text{ allicin molecules} \times \frac{2 \text{ S atoms}}{1 \text{ allicin molecule}} = 3.71 \times 10^{19} \text{ S atoms}$$

Check. Obvious.

3.32 (a) molar mass = 14(12.01) + 18(1.008) + 2(14.01) + 5(16.00) = 294.30 g

(b) $$1.00 \text{ mg aspartame} \times \frac{1 \times 10^{-3} \text{ g}}{1 \text{ mg}} \times \frac{1 \text{ mol}}{294.3 \text{ g}} = 3.398 \times 10^{-6}$$

$= 3.40 \times 10^{-6}$ mol aspartame

(c) $$3.398 \times 10^{-6} \text{ mol aspartame} \times \frac{6.022 \times 10^{23} \text{ molecules}}{1 \text{ mol}} = 2.046 \times 10^{18}$$

$= 2.05 \times 10^{18}$ aspartame molecules

(d) $$2.046 \times 10^{18} \text{ aspartame molecules} \times \frac{18 \text{ H atoms}}{1 \text{ aspartame molecule}} = 3.68 \times 10^{19} \text{ H atoms}$$

3.33 (a) *Analyze.* Given: $C_6H_{12}O_6$, 5.77×10^{20} C atoms. Find: H atoms.

Plan. Use molecular formula to determine number of H atoms that are present with 5.77×10^{20} C atoms. *Solve:*

$$\frac{12 \text{ H atoms}}{6 \text{ C atoms}} = \frac{2 \text{ H}}{1 \text{ C}} \times 5.77 \times 10^{20} \text{ C atoms} = 1.15 \times 10^{21} \text{ H atoms}$$

Check. $(2 \times 6 \times 10^{20}) = 12 \times 10^{20} = 1.2 \times 10^{21}$

(b) *Plan.* Use molecular formula to find the number of glucose molecules that contain 5.77×10^{20} C atoms. *Solve:*

$$\frac{1\ C_6H_{12}O_6 \text{ molecule}}{6 \text{ C atoms}} \times 5.77 \times 10^{20} \text{ C atoms} = 9.617 \times 10^{19}$$

$$= 9.62 \times 10^{19}\ C_6H_{12}O_6 \text{ molecules}$$

Check. $(6 \times 10^{20}/6) = 1 \times 10^{20} = 1 \times 10^{20} = 10 \times 10^{19}$

(c) *Plan.* Use Avogadro's number to change molecules → mol. *Solve:*

$$9.617 \times 10^{19}\ C_6H_{12}O_6 \text{ molecules} \times \frac{1 \text{ mol}}{6.022 \times 10^{23} \text{ molecules}}$$

$$= 1.597 \times 10^{-4} = 1.60 \times 10^{-4} \text{ mol } C_6H_{12}O_6$$

Check. $(9 \times 10^{19})/(6 \times 10^{23}) = 1.5 \times 10^{-4}$

(d) *Plan.* Use molar mass to change mol → g. *Solve:*

1 mole of $C_6H_{12}O_6$ weighs 180.0 g (Sample Exercise 3.9)

$$1.597 \times 10^{-4} \text{ mol } C_6H_{12}O_6 \times \frac{180.0 \text{ g } C_6H_{12}O_6}{1 \text{ mol}} = 0.0287 \text{ g } C_6H_{12}O_6$$

Check. $1.5 \times 180 = 270;\ 270 \times 10^{-4} = 0.027$

3.34 (a) $$3.08 \times 10^{21} \text{ H atoms} \times \frac{19 \text{ C atoms}}{28 \text{ H atoms}} = 2.09 \times 10^{21} \text{ C atoms}$$

(b) $$3.08 \times 10^{21} \text{ H atoms} \times \frac{1\ C_{19}H_{28}O_2 \text{ molecule}}{28 \text{ H atoms}} = 1.100 \times 10^{20}$$

$$= 1.10 \times 10^{20}\ C_{19}H_{28}O_2 \text{ molecules}$$

(c) $$1.100 \times 10^{20}\ C_{19}H_{28}O_2 \text{ molecules} \times \frac{1 \text{ mol}}{6.022 \times 10^{23} \text{ molecules}} = 1.827 \times 10^{-4}$$

$$= 1.83 \times 10^{-4} \text{ mol } C_{19}H_{28}O_2$$

(d) molar mass = 19(12.01) + 28(1.008) + 2(16.00) = 288.41 = 288.4 g

$$1.827 \times 10^{-4} \text{ mol } C_{19}H_{28}O_2 \times \frac{288.4 \text{ g } C_{19}H_{28}O_2}{1 \text{ mol}} = 0.0527 \text{ g } C_{19}H_{28}O_2$$

3.35 *Analyze.* Given: g C_2H_3Cl/L. Find: mol/L, molecules/L.
Plan. The /L is constant throughout the problem, so we can ignore it. Use molar mass for g → mol, Avogadro's number for mol → molecules. *Solve:*

$$\frac{2.05 \times 10^{-6} \text{ g } C_2H_3Cl}{1 \text{ L}} \times \frac{1 \text{ mol } C_2H_3Cl}{62.50 \text{ g } C_2H_3Cl} = 3.280 \times 10^{-8} = 3.28 \times 10^{-8} \text{ mol } C_2H_3Cl/L$$

$$\frac{3.280 \times 10^{-8} \text{ mol } C_2H_3Cl}{1 \text{ L}} \times \frac{6.022 \times 10^{23} \text{ molecules}}{1 \text{ mol}} = 1.97 \times 10^{16} \text{ molecules/L}$$

Check. $(200 \times 10^{-8})/60 = 2.5 \times 10^{-8}$ mol

$(2.5 \times 10^{-8}) \times (6 \times 10^{23}) = 15 \times 10^{15} = 1.5 \times 10^{16}$

3.36 $$25 \times 10^{-6} \text{ g } C_{21}H_{30}O_2 \times \frac{1 \text{ mol } C_{21}H_{30}O_2}{314.5 \text{ g } C_{21}H_{30}O_2} = 7.95 \times 10^{-8} = 8.0 \times 10^{-8} \text{ mol } C_{21}H_{30}O_2$$

$$7.95 \times 10^{-8} \text{ mol } C_{21}H_{30}O_2 \times \frac{6.022 \times 10^{23} \text{ molecules}}{1 \text{ mol}} = 4.8 \times 10^{16} \; C_{21}H_{30}O_2 \text{ molecules}$$

Empirical Formulas

3.37 (a) There are twice as many O atoms as N atoms, so the empirical formula of the original compound is NO_2.

(b) No, because we have no way of knowing whether the empirical and molecular formulas are the same. NO_2 represents the simplest ratio of atoms in a molecule, but not the only possible molecular formula.

3.38 (a) The box contains 4 C atoms and 16 H atoms, so the empirical formula of the hydrocarbon is CH_4.

(b) If the molecular formula of the hydrocarbon is the same as the empirical formula, the reactants' box would contain 4 CH_4 molecules and 8 O_2 molecules. At this stage, you have no basis for knowing whether or not there might be a hydrocarbon molecule other than CH_4 that has this same empirical formula. It turns out, however, that there are none. In general, the molecular formula of a substance does not have to be identical with its empirical formula.

3.39 (a) *Analyze.* Given: moles. Find: empirical formula.
Plan. Find the **simplest ratio of moles** by dividing by the smallest number of moles present.

Solve. 0.0130 mol C / 0.0065 = 2
0.039 mol H / 0.0065 = 6
0.0065 mol O / 0.0065 = 1

The empirical formula is C_2H_6O.

Check. The subscripts are simple integers.

(b) *Analyze.* Given: grams. Find: empirical formula.

Plan. Calculate the moles of each element present, then the simplest ratio of moles.

Solve. $11.66 \text{ g Fe} \times \frac{1 \text{ mol Fe}}{55.85 \text{ g Fe}} = 0.2088 \text{ mol Fe}; \; 0.2088 / 0.2088 = 1$

$5.01 \text{ g O} \times \frac{1 \text{ mol O}}{16.00 \text{ g O}} = 0.3131 \text{ mol O}; \; 0.3131 / 0.2088 \approx 1.5$

Multiplying by two, the integer ratio is 2 Fe : 3 O; the empirical formula is Fe_2O_3.

Check. The subscripts are simple integers.

(c) *Analyze.* Given: mass %. Find: empirical formulas.

Plan. Assume 100 g sample, calculate moles of each element, find the simplest ratio of moles.

Solve. $40.0 \text{ g C} \times \frac{1 \text{ mol C}}{12.01 \text{ g C}} = 3.33 \text{ mol C}; \; 3.33 / 3.33 = 1$

$6.7 \text{ g H} \times \frac{1 \text{ mol H}}{1.008 \text{ mol H}} = 6.65 \text{ mol H}; \; 6.65 / 3.33 \approx 2$

$53.3 \text{ g O} \times \frac{1 \text{ mol O}}{16.00 \text{ mol O}} = 3.33 \text{ mol O}; \; 3.33 / 3.33 = 1$

The empirical formula is CH_2O.

Check. The subscripts are simple integers.

3.40 (a) Calculate the simplest ratio of moles.

0.104 mol K / 0.052 = 2
0.052 mol C / 0.052 = 1
0.156 mol O / 0.052 = 3

The empirical formula is K_2CO_3.

(b) Calculate moles of each element present, then the simplest ratio of moles.

$5.28 \text{ g Sn} \times \frac{1 \text{ mol Sn}}{118.7 \text{ g Sn}} = 0.04448 \text{ mol Sn}; \; 0.04448 / 0.04448 = 1$

$3.37 \text{ g F} \times \frac{1 \text{ mol F}}{19.00 \text{ g FSn}} = 0.1774 \text{ mol F}; \; 0.1774 / 0.04448 \approx 4$

The integer ratio is 1 Sn : 4 F; the empirical formula is SnF_4.

(c) Assume 100 g sample, calculate moles of each element, find the simplest ratio of moles.

$87.5\% \text{ N} = 87.5 \text{ g N} \times \frac{1 \text{ mol N}}{14.01 \text{ g}} = 6.25 \text{ mol N}; \; 6.25 / 6.25 = 1$

$12.5\% \text{ H} = 12.5 \text{ g H} \times \frac{1 \text{ mol}}{1.008 \text{ g}} = 12.4 \text{ mol H}; \; 12.4 / 6.25 \approx 2$

The empirical formula is NH_2.

3.41 *Analyze/Plan*. The procedure in all these cases is to assume 100 g of sample, calculate the number of moles of each element present in that 100 g, then obtain the ratio of moles as smallest whole numbers. *Solve*:

(a) $10.4 \text{ g C} \times \frac{1 \text{ mol C}}{12.01 \text{ g C}} = 0.866 \text{ mol C}; \ 0.866 / 0.866 = 1$

$27.8 \text{ g S} \times \frac{1 \text{ mol S}}{32.07 \text{ g S}} = 0.867 \text{ mol S}; \ 0.867 / 0.866 \approx 1$

$61.7 \text{ g Cl} \times \frac{1 \text{ mol Cl}}{35.45 \text{ g Cl}} = 1.74 \text{ mol Cl}; \ 1.74 / 0.866 \approx 2$

The empirical formula is $CSCl_2$.

(b) $21.7 \text{ g C} \times \frac{1 \text{ mol C}}{12.01 \text{ g C}} = 1.81 \text{ mol C}; \ 1.81 / 0.600 \approx 3$

$9.6 \text{ g O} \times \frac{1 \text{ mol O}}{16.00 \text{ g O}} = 0.600 \text{ mol O}; \ 0.600 / 0.600 = 1$

$68.7 \text{ g F} \times \frac{1 \text{ mol F}}{19.00 \text{ g F}} = 3.62 \text{ mol F}; \ 3.62 / 0.600 \approx 6$

The empirical formula is C_3OF_6.

(c) $32.79 \text{ g Na} \times \frac{1 \text{ mol Na}}{22.99 \text{ g Na}} = 1.426 \text{ mol Na}; \ 1.426 / 0.4826 \approx 3$

$13.02 \text{ g Al} \times \frac{1 \text{ mol Al}}{26.98 \text{ g Al}} = 0.4826 \text{ mol Al}; \ 0.4826 / 0.4826 = 1$

$54.19 \text{ g F} \times \frac{1 \text{ mol F}}{19.00 \text{ g F}} = 2.852 \text{ mol F}; \ 2.852 / 0.4826 \approx 6$

The empirical formula is Na_3AlF_6.

3.42 See Solution 3.41 for stepwise problem-solving approach.

(a) $55.3 \text{ g K} \times \frac{1 \text{ mol K}}{39.10 \text{ g K}} = 1.414 \text{ mol K}; \ 1.411/0.4714 \approx 3$

$14.6 \text{ g P} \times \frac{1 \text{ mol P}}{30.97 \text{ g P}} = 0.4714 \text{ mol P}; \ 0.4714/0.4714 = 1$

$30.1 \text{ g O} \times \frac{1 \text{ mol O}}{16.00 \text{ g O}} = 1.881 \text{ mol O}; \ 1.881/0.4714 \approx 4$

The empirical formula is K_3PO_4.

(b) $24.5 \text{ g Na} \times \frac{1 \text{ mol Na}}{22.99 \text{ g Na}} = 1.066 \text{ mol Na}$; $1.066/0.5304 \approx 2$

$14.9 \text{ g Si} \times \frac{1 \text{ mol Si}}{28.09 \text{ Si}} = 0.5304 \text{ mol si}$; $0.5304/0.5304 = 1$

$60.6 \text{ g F} \times \frac{1 \text{ mol F}}{19.00 \text{ g F}} = 3.189 \text{ mol F}$; $3.189/0.5304 \approx 6$

The empirical formula is Na_2SiF_6.

c) $62.1 \text{ g C} \times \frac{1 \text{ mol C}}{12.01 \text{ g C}} = 5.17 \text{ mol C}$; $5.17 / 0.864 \approx 6$

$5.21 \text{ g H} \times \frac{1 \text{ mol H}}{1.008 \text{ g H}} = 5.17 \text{ mol O}$; $5.17 / 0.864 \approx 6$

$12.1 \text{ g N} \times \frac{1 \text{ mol N}}{14.01 \text{ g N}} = 0.864 \text{ mol N}$; $0.864 / 0.864 = 1$

$20.7 \text{ g O} \times \frac{1 \text{ mol O}}{16.00 \text{ g O}} = 1.29 \text{ mol O}$; $1.29 / 0.864 \approx 1.5$

Multiplying by two, the empirical formula is $C_{12}H_{12}N_2O_3$.

3.43 *Analyze.* Given: empirical formula, molar mass. Find: molecular formula.

Plan. Calculate the empirical formula weight (FW); divide FW by molar mass ($\mathcal{M}$) to calculate the integer that relates the empirical and molecular formulas. Check. If FW/$\mathcal{M}$ is an integer, the result is reasonable. *Solve:*

(a) FW CH_2 = 12 + 2(1) = 14. $\frac{\mathcal{M}}{FW} = \frac{84}{14} = 6$

The subscripts in the empirical formula are multiplied by 6. The molecular formula is C_6H_{12}.

(b) FW NH_2Cl = 14.01 + 2(1.008) + 35.45 = 51.48. $\frac{\mathcal{M}}{FW} = \frac{51.5}{51.5} = 1$

The empirical and molecular formulas are NH_2Cl.

3.44 (a) FW 12.01 + 1.008 + 2(16.00) = 45.0. $\frac{\mathcal{M}}{FW} = \frac{90.0}{45.0} = 2$

The molecular formula is $H_2C_2O_4$.

(b) FW = 2(12) + 4(1) + 16 = 44. $\frac{\mathcal{M}}{FW} = \frac{88}{44} = 2.$

The molecular formula is $C_4H_8O_2$.

3.45. *Analyze*. Given: mass %, molar mass. Find: molecular formula. *Plan*. Use the plan detailed in Solution 3.41 to find an empirical formula from mass % data. Then use the plan detailed in 3.43 to find the molecular formula. Note that some indication of molar mass must be given, or the molecular formula cannot be determined. *Check*. If there is an integer ratio of moles and $\mathcal{M}$ / FW is an integer, the result is reasonable. *Solve:*

(a)

$$49.5 \text{ g C} \times \frac{1 \text{ mol C}}{12.01 \text{ g C}} = 4.12 \text{ mol C}; \quad 4.12 / 1.03 \approx 4$$

$$5.15 \text{ g H} \times \frac{1 \text{ mol H}}{1.008 \text{ g H}} = 5.11 \text{ mol H}; \quad 5.11 / 1.03 \approx 5$$

$$28.9 \text{ g N} \times \frac{1 \text{ mol N}}{14.01 \text{ g N}} = 2.06 \text{ mol N}; \quad 2.06 / 1.03 \approx 2$$

$$16.5 \text{ g O} \times \frac{1 \text{ mol O}}{16.00 \text{ g O}} = 1.03 \text{ mol O}; \quad 1.03 / 1.03 = 1$$

Thus, $C_4H_5N_2O$, FW = 97. If the molar mass is about 195, a factor of 2 gives the molecular formula $C_8H_{10}N_4O_2$.

(b)

$$35.51 \text{ g C} \times \frac{1 \text{ mol C}}{12.01 \text{ g C}} = 2.96 \text{ mol C}; 2.96/0.592 = 5$$

$$4.77 \text{ g H} \times \frac{1 \text{ mol H}}{1.008 \text{ g H}} = 4.73 \text{ mol H}; 4.73/0.592 = 7.99 \approx 8$$

$$37.85 \text{ g O} \times \frac{1 \text{ mol O}}{16.00 \text{ g O}} = 2.37 \text{ mol O}; 2.37/0.592 = 4$$

$$8.29 \text{ g N} \times \frac{1 \text{ mol N}}{14.01 \text{ g N}} = 0.592 \text{ mol N}; 0.592/0.592 = 1$$

$$13.60 \text{ g Na} \times \frac{1 \text{ mol Na}}{22.99 \text{ g Na}} = 0.592 \text{ mol Na}; 0.592/0.592 = 1$$

The empirical formula is $C_5H_8O_4NNa$, FW = 169 g. Since the empirical formula weight and molar mass are approximately equal, the empirical and molecular formulas are both $NaC_5H_8O_4N$.

3.46 Assume 100 g in the following problems.

(a)

$$75.69 \text{ g C} \times \frac{1 \text{ mol C}}{12.01 \text{ g C}} = 6.30 \text{ mol C}; 6.30/0.969 = 6.5$$

$$8.80 \text{ g H} \times \frac{1 \text{ mol H}}{1.008 \text{ g H}} = 8.73 \text{ mol H}; 8.73/0.969 = 9.0$$

$$15.51 \text{ g O} \times \frac{1 \text{ mol O}}{16.00 \text{ g O}} = 0.969 \text{ mol O}; 0.969/0.969 = 1$$

Multiply by 2 to obtain the integer ratio 13:18:2. The empirical formula is $C_{13}H_{18}O_2$, FW = 206 g. Since the empirical formula weight and the molar mass are equal (206 g), the empirical and molecular formulas are $C_{13}H_{18}O_2$.

(b) $59.0 \text{ g C} \times \frac{1 \text{ mol C}}{12.01 \text{ g C}} = 4.91 \text{ mol C}$; $4.91 / 0.550 \approx 9$

$7.1 \text{ g H} \times \frac{1 \text{ mol H}}{1.008 \text{ g H}} = 7.04 \text{ mol H}$; $7.04 / 0.550 \approx 13$

$26.2 \text{ g O} \times \frac{1 \text{ mol O}}{16.00 \text{ g O}} = 1.64 \text{ mol O}$; $6.64 / 0.550 \approx 3$

$7.7 \text{ g N} \times \frac{1 \text{ mol N}}{14.01 \text{ g N}} = 0.550 \text{ mol N}$; $0.550 / 0.550 = 1$

The empirical formula is $C_9H_{13}O_3N$, FW = 183 amu (or g). Since the molecular weight is approximately 180 amu, the empirical formula and molecular formula are the same, $C_9H_{13}O_3N$.

3.47 (a) *Analyze.* Given: mg CO_2, mg H_2O Find: empirical formula of hydrocarbon, C_xH_y

Plan. Upon combustion, all C → CO_2, all H → H_2O.

mg CO_2 → g CO_2 → mol C; mg H_2O → g H_2O, mol H

Find simplest ratio of moles and empirical formula. *Solve:*

$$5.86 \times 10^{-3} \text{ g } CO_2 \times \frac{1 \text{ mol } CO_2}{44.01 \text{ g } CO_2} \times \frac{1 \text{ mol C}}{1 \text{ mol } CO_2} = 1.33 \times 10^{-4} \text{ mol C.}$$

$$1.37 \times 10^{-3} \text{ g } H_2O \times \frac{1 \text{ mol } H_2O}{18.02 \text{ g } H_2O} \times \frac{2 \text{ mol H}}{1 \text{ mol } H_2O} = 1.52 \times 10^{-4} \text{ mol H.}$$

Dividing both values by 1.33×10^{-4} gives C:H of 1:1.14. This is not "close enough" to be considered 1:1. No obvious multipliers (2, 3, 4) produce an integer ratio. Testing other multipliers (trial and error!), the correct factor seems to be 7. The empirical formula is C_7H_8.

Check. See discussion of C:H ratio above.

(b) *Analyze.* Given: g of menthol, g CO_2, g H_2O, molar mass. Find: molecular formula.

Plan/Solve. Calculate mol C and mol H in the sample.

$$0.2829 \text{ g } CO_2 \times \frac{1 \text{ mol } CO_2}{44.01 \text{ g } CO_2} \times \frac{1 \text{ mol C}}{1 \text{ mol } CO_2} = 0.0064281 = 0.006428 \text{ mol C}$$

$$0.1159 \text{ g } H_2O \times \frac{1 \text{ mol } H_2O}{18.02 \text{ g } H_2O} \times \frac{2 \text{ mol H}}{1 \text{ mol } H_2O} = 0.012863 = 0.01286 \text{ mol H}$$

Calculate g C, g H and get g O by subtraction.

$$0.064281 \text{ mol C} \times \frac{12.01 \text{ g C}}{1 \text{ mol C}} = 0.07720 \text{ g C}$$

$$0.012863 \text{ mol H} \times \frac{1.008 \text{ g H}}{1 \text{ mol H}} = 0.01297 \text{ g H}$$

mass O = 0.1005 g sample - (0.07720 g C + 0.01297 g H) = 0.01033 g O

Calculate mol O and find integer ratio of mol C: mol H: mol O.

$$0.01033 \text{ g O} \times \frac{1 \text{ mol O}}{16.00 \text{ g O}} = 6.456 \times 10^{-4} \text{ mol O}$$

Divide moles by 6.456×10^{-4}.

$$\text{C: } \frac{0.006428}{6.456 \times 10^{-4}} \approx 10; \quad \text{H: } \frac{0.01286}{6.456 \times 10^{-4}} \approx 20; \quad \text{O: } \frac{6.456 \times 10^{-4}}{6.456 \times 10^{-4}} = 1$$

The empirical formula is $C_{10}H_{20}O$.

$$\text{FW} = 10(12) + 20(1) + 16 = 156; \quad \frac{\mathcal{M}}{\text{FW}} = \frac{156}{156} = 1$$

The molecular formula is the same as the empirical formula, $C_{10}H_{20}O$.

Check. The mass of O wasn't negative or greater than the sample mass; empirical and molecular formulas are reasonable.

3.48 (a) *Plan.* Calculate mol C and mol H, then g C and g H; get g O by subtraction. *Solve:*

$$6.32 \times 10^{-3} \text{ g } CO_2 \times \frac{1 \text{ mol } CO_2}{44.01 \text{ g } CO_2} \times \frac{1 \text{ mol C}}{1 \text{ mol } CO_2} = 1.436 \times 10^{-4} = 1.44 \times 10^{-4} \text{ mol C}$$

$$2.58 \times 10^{-3} \text{ g } H_2O \times \frac{1 \text{ mol } H_2O}{18.02 \text{ g } H_2O} \times \frac{2 \text{ mol H}}{1 \text{ mol } H_2O} = 2.863 \times 10^{-4} = 2.86 \times 10^{-4} \text{ mol H}$$

$$1.436 \times 10^{-4} \text{ mol C} \times \frac{12.01 \text{ g C}}{1 \text{ mol C}} = 1.725 \times 10^{-3} \text{ g C} = 1.73 \text{ mg C}$$

$$2.863 \times 10^{-4} \text{ mol H} \times \frac{1.008 \text{ g H}}{1 \text{ mol H}} = 2.886 \times 10^{-4} \text{ g H} = 0.289 \text{ mg H}$$

mass of O = 2.78 mg sample - (1.725 mg C + 0.289 mg H) = 0.77 mg O

$$0.77 \times 10^{-3} \text{ g O} \times \frac{1 \text{ mol O}}{16.00 \text{ g O}} = 4.81 \times 10^{-5} \text{ mol O.}$$ Divide moles by 4.81×10^{-5}.

$$\text{C: } \frac{1.44 \times 10^{-4}}{4.81 \times 10^{-5}} \approx 3; \quad \text{H: } \frac{2.86 \times 10^{-4}}{4.81 \times 10^{-5}} \approx 6; \quad \text{O: } \frac{4.81 \times 10^{-5}}{4.81 \times 10^{-5}} = 1$$

The empirical formula is C_3H_6O.

(b) *Plan.* Calculate mol C and mol H, then g C and g H. In this case, get N by subtraction. *Solve:*

$$14.242 \times 10^{-3} \text{ g } CO_2 \times \frac{1 \text{ mol } CO_2}{44.01 \text{ g } CO_2} \times \frac{1 \text{ mol C}}{1 \text{ mol } CO_2} = 3.2361 \times 10^{-4} = \text{mol C}$$

$$4.083 \times 10^{-3} \text{ g } H_2O \times \frac{1 \text{ mol } H_2O}{18.02 \text{ g } H_2O} \times \frac{2 \text{ mol H}}{1 \text{ mol } H_2O} = 4.5316\ 10^{-4} = 4.532 \times 10^{-4} \text{ mol H}$$

$$3.2361\ 10^{-4} \text{ g mol C} \times \frac{12.01 \text{ g C}}{1 \text{ mol C}} = 3.8866 \times 10^{-3} \text{ g C} = 3.8866 \text{ mg C}$$

$$4.532 \times 10^{-4} \text{ mol H} \times \frac{1.008 \text{ g H}}{1 \text{ mol H}} = 0.45683 \times 10^{-3} \text{ g H} = 0.4568 \text{ mg H}$$

mass of N = 5.250 mg sample – (3.8866 mg C + 0.4568 mg H) = 0.9066 = 0.907 mg N

$0.9066 \times 10^{-3} \text{ g N} \times \frac{1 \text{ mol N}}{14.01 \text{ g N}} = 6.47 \times 10^{-5}$ mol N. Divide moles by 6.47×10^{-5}.

$$\text{C: } \frac{3.24 \times 10^{-4}}{6.47 \times 10^{-5}} \approx 5; \quad \text{H: } \frac{4.53 \times 10^{-4}}{6.47 \times 10^{-5}} \approx 7; \quad \text{N: } \frac{6.47 \times 10^{-5}}{6.47 \times 10^{-5}} = 1$$

The empirical formula is C_5H_7N, FW = 81. A molar mass of 160 ± 5 indicates a factor of 2 and a molecular formula of $C_{10}H_{14}N_2$.

3.49 *Analyze.* Given 2.558 g $Na_2CO_3 \cdot xH_2O$, 0.948 g Na_2CO_3. Find: x.

Plan. The reaction involved is $Na_2CO_3 \cdot xH_2O(s) \rightarrow Na_2CO_3(s) + xH_2O(g)$.
Calculate the mass of H_2O lost and then the mole ratio of Na_2CO_3 and H_2O. *Solve*:

g H_2O lost = 2.558 g sample - 0.948 g Na_2CO_3 = 1.610 g H_2O

$$0.948 \text{ g } Na_2CO_3 \times \frac{1 \text{ mol } Na_2CO_3}{106.0 \text{ g } Na_2CO_3} = 0.00894 \text{ mol } Na_2CO_3$$

$$1.610 \text{ g } H_2O \times \frac{1 \text{ mol } H_2O}{18.02 \text{ g } H_2O} = 0.08935 \text{ mol } H_2O$$

$$\frac{\text{mol } H_2O}{\text{mol } Na_2CO_3} = \frac{0.08935}{0.00894} = 9.99; \quad x = 10.$$

The formula is $Na_2CO_3 \cdot$ **<u>10</u>** H_2O.

Check. x is an integer.

3.50 The reaction involved is $MgSO_4 \cdot xH_2O(s) \rightarrow MgSO_4(s) + xH_2O(g)$. First, calculate the number of moles of product $MgSO_4$; this is the same as the number of moles of starting hydrate.

$$2.472 \text{ g } MgSO_4 \times \frac{1 \text{ mol } MgSO_4}{120.4 \text{ g } MgSO_4} \times \frac{1 \text{ mol } MgSO_4 \cdot x\,H_2O}{1 \text{ mol } MgSO_4} = 0.02053 \text{ mol } MgSO_4 \cdot x\,H_2O$$

Thus, $\frac{5.061 \text{ g } MgSO_4 \cdot x\,H_2O}{0.02053} = 246.5$ g/mol = FW of $MgSO_4 \cdot x\,H_2O$.

FW of $MgSO_4 \cdot x\,H_2O$ = FW of $MgSO_4$ + x(FW of H_2O).

246.5 = 120.4 + x(18.02). x = 6.998. The hydrate formula is $MgSO_4 \cdot$ <u>7</u>H_2O.

Alternatively, we could calculate the number of moles of water represented by weight loss: (5.061 - 2.472) = 2.589 g H_2O lost.

$$2.589 \text{ g } H_2O \times \frac{1 \text{ mol } H_2O}{18.02 \text{ g } H_2O} = 0.1437 \text{ mol } H_2O; \quad \frac{\text{mol } H_2O}{\text{mol } MgSO_4} = \frac{0.1437}{0.02053} = 7.000$$

Again the correct formula is $MgSO_4 \cdot$ <u>7</u>H_2O.

Calculations Based on Chemical Equations

3.51 The mole ratios implicit in the coefficients of a balanced chemical equation express the fundamental relationship between amounts of reactants and products. If the equation is not balanced, the mole ratios will be incorrect and lead to erroneous calculated amounts of products.

3.52 The **integer coefficients** immediately preceding each molecular formula in a chemical equation give information about relative numbers of moles of reactants and products involved in a reaction.

3.53 *Analyze. Given: 4.0 mol CH_4. Find: mol CO and mol H_2.*
Plan. Examine the boxes to determine the CH_4:CO mol ratio and CH_4:H_2O mole ratio.
Solve. There are $2CH_4$ molecules in the reactant box and 2CO molecules in the product box. The mole ratio is 2:2 or 1:1. Therefore, 4.0 mol CH_4 can produce 4.0 mol CO. There are $2CH_4$ molecules in the reactant box and $6H_2$ molecules in the product box. The mole ratio is 2:6 or 1:3. So, 4.0 mol CH_4 can produce 12:0 mol H_2.

Check. Use proportions. 2 mol CH_4/2 mol CO = 4 mol CH_4/4 mol CO;
2 mol CH_4/6 mol H_2 = 4 mol CH_4/12 mol H_2.

3.54 $C_2H_5OH(l) + 3O_2(g) \rightarrow 2CO_2(g) + 3H_2O(g)$
$C_3H_8(g) + 5O_2(g) \rightarrow 3CO_2(g) + 4H_2O(g)$
$CH_3CH_2COCH_3(l) + 11/2\ O_2(g) \rightarrow 4CO_2(g) + 4H_2O(l)$
In a combustion reaction, all H in the fuel is transformed to H_2O in the products. The reactant with most mol H/mol fuel will produce the most H_2O. C_3H_8 and $CH_3CH_2COCH_3$ (C_4H_8O) both have 8 mol H/mol fuel, so 1.5 mol of either fuel will produce the same amount of H_2O.
1.5 mol C_2H_5OH will produce less H_2O.

3.55 $Na_2SiO_3(s) + 8HF(aq) \rightarrow H_2SiF_6(aq) + 2NaF(aq) + 3H_2O(l)$

(a) *Analyze.* Given: mol Na_2SiO_3. Find: mol HF. *Plan.* Use the mole ratio 8HF:1Na_2SiO_3 from the balanced equation to relate moles of the two reactants.
Solve:

$$0.300 \text{ mol } Na_2SiO_3 \times \frac{8 \text{ mol HF}}{1 \text{ mol } Na_2SiO_3} = 2.4 \text{ mol HF}$$

Check. Mol HF should be greater than mol Na_2SiO_3.

(b) *Analyze.* Given: mol HF. Find: g NaF. *Plan.* Use the mole ratio 2NaF:8HF to change mol HF to mol NaF, then molar mass to get NaF. *Solve:*

$$0.500 \text{ mol HF} \times \frac{2 \text{ mol NaF}}{8 \text{ mol HF}} \times \frac{41.99 \text{ g NaF}}{1 \text{ mol NaF}} = 5.25 \text{ g NaF}$$

Check. $(0.5/4) = 0.125$; $0.13 \times 42 > 4$ g NaF

(c) *Analyze.* Given: g HF Find: g Na_2SiO_3.

Plan. g HF → mol HF $\left(\frac{\text{mol}}{\text{ratio}}\right)$ → mol Na_2SiO_3 → g Na_2SiO_3

The mole ratio is at the heart of every stoichiometry problem. Molar mass is used to change to and from grams. *Solve*:

$$0.800 \text{ g HF} \times \frac{1 \text{ mol HF}}{20.01 \text{g HF}} \times \frac{1 \text{ mol } Na_2SiO_3}{8 \text{ mol HF}} \times \frac{122.1 \text{ g } Na_2SiO_3}{1 \text{ mol } Na_2SiO_3} = 0.610 \text{ g } Na_2SiO_3$$

Check. 0.8 (120/160) < 0.75 mol

3.56 $C_6H_{12}O_6(aq) \rightarrow 2C_2H_5OH(aq) + 2CO_2(g)$

(a) $$0.400 \text{ mol } C_6H_{12}O_6 \times \frac{2 \text{ mol } CO_2}{1 \text{ mol } C_6H_{12}O_6} = 0.800 \text{ mol } CO_2$$

(b) $$7.50 \text{ g } C_2H_5OH \times \frac{1 \text{ mol } C_2H_5OH}{46.07 \text{ g } C_2H_5OH} \times \frac{1 \text{ mol } C_6H_{12}O_6}{2 \text{ mol } C_2H_5OH} \times \frac{180.2 \text{ g } C_6H_{12}O_6}{1 \text{ mol } C_6H_{12}O_6}$$

$$= 14.7 \text{ g } C_6H_{12}O_6$$

(c) $$7.50 \text{ g } C_2H_5OH \times \frac{1 \text{ mol } C_2H_5OH}{46.07 \text{ g } C_2H_5OH} \times \frac{2 \text{ mol } CO_2}{2 \text{ mol } C_2H_5OH} \times \frac{44.01 \text{ g } CO_2}{1 \text{ mol } CO_2} = 7.16 \text{ g } CO_2$$

3.57 (a) $Al_2S_3(s) + 6H_2O(l) \rightarrow 2Al(OH)_3(s) + 3H_2S(g)$

(b) *Plan.* g A → mol A → mol B → g B. See Solution 3.55 (c). *Solve*:

$$10.5 \text{ g } Al_2S_3 \times \frac{1 \text{ mol } Al_2S_3}{150.2 \text{ g } Al_2S_3} \times \frac{2 \text{ mol } Al(OH)_3}{1 \text{ mol } Al_2S_3} \times \frac{78.00 \text{ g } Al(OH)_3}{1 \text{ mol } Al(OH)_3}$$

$$= 10.9 \text{ g } Al(OH)_3$$

Check. $10\left(\frac{2 \times 78}{150}\right) \approx 10(1) \approx 10 \text{ g } Al(OH)_3$

3.58 (a) $CaH_2(s) + 2H_2O(l) \rightarrow Ca(OH)_2(aq) + 2H_2(g)$

(b) $$5.00 \text{ g } H_2 \times \frac{1 \text{ mol } H_2}{2.016 \text{ g } H_2} \times \frac{1 \text{ mol } CaH_2}{2 \text{ mol } H_2} \times \frac{42.10 \text{ g } CaH_2}{1 \text{ mol } CaH_2} = 52.2 \text{ g } CaH_2$$

3.59 (a) *Analyze.* Given: mol NaN_3. Find: mol N_2. *Plan.* Use mole ratio from balanced equation. *Solve*:

$$2.50 \text{ mol } NaN_3 \times \frac{3 \text{ mol } N_2}{2 \text{ mol } NaN_3} = 3.75 \text{ mol } N_2$$

Check. The resulting mol N_2 should be greater than mol NaN_3, (the N_2:NaN_3 ratio is > 1), and it is.

(b) *Analyze.* Given: g N_2 Find: g NaN_3. *Plan.* Use molar masses to get from and to grams, mol ratio to relate moles of the two substances. *Solve*:

$$6.00 \text{ g } N_2 \times \frac{1 \text{ mol } N_2}{28.01 \text{ g } N_2} \times \frac{2 \text{ mol } NaN_3}{3 \text{ mol } N_2} \times \frac{65.01 \text{ g } NaN_3}{1 \text{ mol } NaN_3} = 9.28 \text{ g } NaN_3$$

Check. Mass relations are less intuitive than mole relations. Estimating the ratio of molar masses is sometimes useful. In this case, 65 g NaN_3/28 g $N_2 \approx 2.25$ Then, $(6 \times 2/3 \times 2.25) \approx 9$ g NaN_3. The calculated result looks reasonable.

(c) *Analyze.* Given: vol N_2 in ft^3, density N_2 in g/L. Find: g NaN_3. *Plan.* First determine how many g N_2 are in 10.0 ft^3, using the density of N_2. *Solve*:

$$\frac{1.25 \text{ g}}{1 \text{ L}} \times \frac{1 \text{ L}}{1000 \text{ cm}^3} \times \frac{(2.54)^3 \text{ cm}^3}{1 \text{ in}^3} \times \frac{(12)^3 \text{ in}^3}{1 \text{ ft}^3} \times 10.0 \text{ ft}^3 = 354.0 = 354 \text{ g } N_2$$

$$354.0 \text{ g } N_2 \times \frac{1 \text{ mol } N_2}{28.01 \text{ g } N_2} \times \frac{2 \text{ mol } NaN_3}{3 \text{ mol } N_2} \times \frac{65.01 \text{ g } NaN_3}{1 \text{ mol } NaN_3} = 548 \text{ g } NaN_3$$

Check. 1 ft^3 ~ 28 L; 10 ft^3 ~ 280 L; 280 L × 1.25 ~ 350 g N_2

Using the ratio of molar masses from part (b), $(350 \times 2/3 \times 2.25) \approx 525$ g NaN_3

3.60 $2C_8H_{18}(l) + 25O_2(g) \rightarrow 16CO_2(g) + 18H_2O(l)$

(a) $$0.750 \text{ mol } C_8H_{18} \times \frac{25 \text{ mol } O_2}{2 \text{ mol } C_8H_{18}} = 9.375 = 9.38 \text{ mol } O_2$$

(b) $$5.00 \text{ g } C_8H_{18} \times \frac{1 \text{ mol } C_8H_{18}}{114.2 \text{ g } C_8H_{18}} \times \frac{25 \text{ mol } O_2}{2 \text{ mol } C_8H_{18}} \times \frac{32.00 \text{ g } O_2}{1 \text{ mol } O_2} = 17.5 \text{ g } O_2$$

(c) $$1.00 \text{ gal } C_8H_{18} \times \frac{3.7854 \text{ L}}{1 \text{ gal}} \times \frac{1000 \text{ mL}}{1 \text{ L}} \times \frac{0.692 \text{ g}}{1 \text{ mL}} = 2619.5 = 2.62 \times 10^3 \text{ g } C_8H_{18}$$

$$2619.5 \text{ g } C_8H_{18} \times \frac{1 \text{ mol } C_8H_{18}}{114.2 \text{ g } C_8H_{18}} \times \frac{25 \text{ mol } O_2}{2 \text{ mol } C_8H_{18}} \times \frac{32.00 \text{ g } O_2}{1 \text{ mol } O_2} = 9{,}175.1 \text{ g} = 9.18 \times 10^3 \text{ g } O_2$$

3.61 (a) *Analyze.* Given: dimensions of Al foil. Find: mol Al.

Plan. Dimensions $\longrightarrow$ vol $\xrightarrow{\text{density}}$ mass $\xrightarrow[\text{mass}]{\text{molar}}$ mol Al

Solve. $1.00 \text{ cm} \times 1.00 \text{ cm} \times 0.550 \text{ mm} \times \frac{1 \text{ cm}}{10 \text{ mm}} = 0.0550 \text{ cm}^3$ Al

$$0.0550 \text{ cm}^3 \text{ Al} \times \frac{2.699 \text{ g Al}}{1 \text{ cm}^3} \times \frac{1 \text{ mol Al}}{26.98 \text{ g Al}} = 5.502 \times 10^{-3} = 5.50 \times 10^{-3} \text{ mol Al}$$

Check. $2.699/26.98 \approx 0.1$; $(0.055 \text{ cm}^3 \times 0.1) = 5.5 \times 10^{-3}$ mol Al

(b) *Plan.* Write the balanced equation to get a mole ratio; change mol Al → mol $AlBr_3$ → g $AlBr_3$.

Solve. $2Al(s) + 3Br_2(l) \rightarrow 2AlBr_3(s)$

$$5.502 \times 10^{-3} \text{ mol Al} \times \frac{2 \text{ mol AlBr}_3}{2 \text{ mol Al}} \times \frac{266.69 \text{ g AlBr}_3}{1 \text{ mol AlBr}_3} = 1.467 = 1.47 \text{ g AlBr}_3$$

Check. $(0.006 \times 1 \times 270) \approx 1.6$ g $AlBr_3$

3.62 (a) *Plan.* Calculate a "mole ratio" between nitroglycerine and total moles of gas produced. (12 + 6 + 1 + 10) = 29 mol gas; 4 mol nitro: 29 total mol gas. *Solve*:

$$3.00 \text{ mL nitro} \times \frac{1.592 \text{ g}}{\text{mL}} \times \frac{1 \text{ mol nitro}}{227.1 \text{ g nitro}} \times \frac{29 \text{ mol gas}}{4 \text{ mol nitro}} = 0.15247 = 0.152 \text{ mol gas}$$

(b) $$0.15247 \text{ mol gas} \times \frac{55 \text{ L}}{\text{mol}} = 8.3859 = 8.4 \text{ L}$$

(c) $$3.00 \text{ mL nitro} \times \frac{1.592 \text{ g}}{\text{mL}} \times \frac{1 \text{ mol nitro}}{227.1 \text{ g nitro}} \times \frac{6 \text{ mol N}_2}{4 \text{ mol nitro}} \times \frac{28.01 \text{ g N}_2}{1 \text{ mol N}_2} = 0.884 \text{ g N}_2$$

Limiting Reactants, Theoretical Yields

3.63 (a) The *limiting reactant* determines the maximum number of product moles resulting from a chemical reaction; any other reactant is an *excess reactant.*

(b) The limiting reactant regulates the amount of products because it is completely used up during the reaction; no more product can be made when one of the reactants is unavailable.

3.64 (a) *Theoretical yield* is the maximum amount of product possible, as predicted by stoichiometry, assuming that the limiting reactant is converted entirely to product.

Actual yield is the amount of product actually obtained, less than or equal to the theoretical yield. *Percent yield* is the ratio of (actual yield to theoretical yield) × 100.

(b) No reaction is perfect. Not all reactant molecules come together effectively to form products; alternative reaction pathways may produce secondary products and reduce the amount of desired product actually obtained, or it might not be possible to completely isolate the desired product from the reaction mixture. In any case, these factors reduce the actual yield of a reaction.

3.65 $N_2 + 3H_2 \rightarrow 2NH_3$. N_2 = , NH_3 =

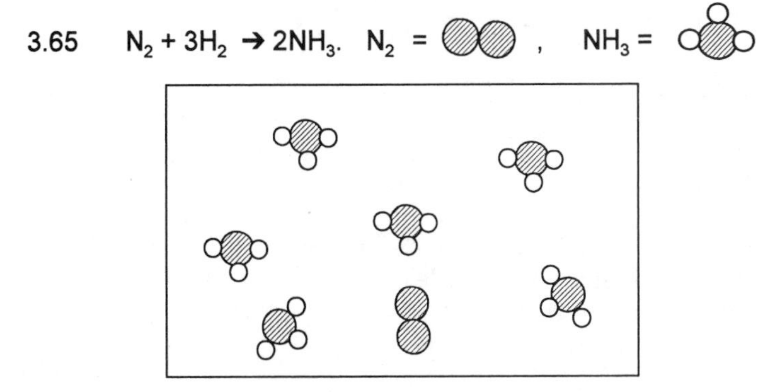

Each N atom (1/2 of an N_2 molecule), reacts with 3 H atoms (1.5 H_2 molecules) to form an NH_3 molecule. Eight N atoms (4 N_2 molecules) require 24 H atoms (12 H_2 molecules) for complete reaction. Only 9 H_2 molecules are available, so H_2 is the limiting reactant. Nine H_2 molecules (18 H atoms) determine that 6 NH_3 molecules are produced. One N_2 molecule is in excess.

3.66 $2NO + O_2 \rightarrow 2NO_2$. O_2 = OO, NO_2 =

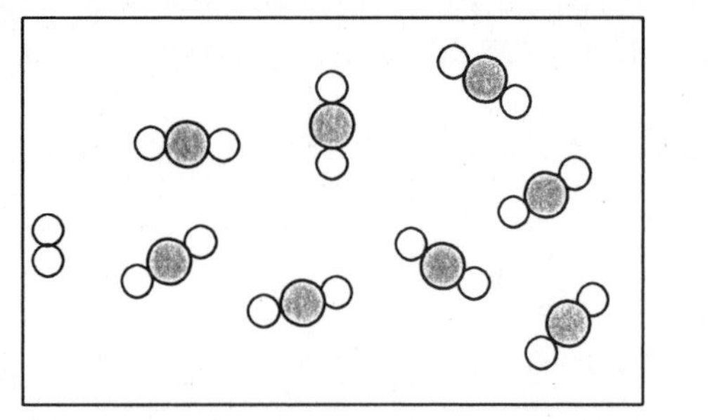

Each NO molecule reacts with 1 O atom (1/2 of an O_2 molecule) to produce 1 NO_2 molecule. Eight NO molecules react with 8 O atoms (4 O_2 molecules) to produce 8 NO_2 molecules. One O_2 molecule doesn't react (is in excess). NO is the limiting reactant.

3.67 (a) Each bicycle needs 2 wheels, 1 frame and 1 set of handlebars. A total of 4250 wheels corresponds to 2125 pairs of wheels. This is fewer than the number of frames or handlebars. The 4250 wheels determine that 2125 bicycles can be produced.

(b) 2755 frames - 2125 bicycles = 630 frames left over

(2255 handlebars - 2125 bicycles) = 130 handlebars left over

(c) The wheels are the "limiting reactant" in that they determine the number of bicycles that can be produced.

3.68 (a) $39{,}375 \text{ L beverage} \times \frac{1 \text{ bottle}}{0.355 \text{ L}} = 110{,}915.49 = 1.11 \times 10^5$ portions of beverage

(The uncertainty in 355 mL limits the precision of the number of portions we can reasonably expect to deliver to 3 significant figures.)

115,350 bottles; 122,500 caps; 111,000 bottles can be filled and capped.

(b) 122,500 caps - 111,000 portions = 11,500 caps remain
115,350 empty bottles - 111,000 portions = 4350 bottles remain

(c) The volume of beverage limits production.

3.69 *Analyze.* Given: 1.70 mol NaOH, 1.00 mol CO_2. Find: mol Na_2CO_3.

Plan. Amounts of more than one reactant are given, so we must determine which reactant regulates (limits) product. Then apply the appropriate mole ratio from the balanced equation.

Solve. The mole ratio is 2NaOH:1CO_2, so 1.00 mol CO_2 requires 2.00 mol NaOH for complete reaction. Less than 2.00 mol NaOH are present, so NaOH is the limiting reactant.

$$1.70 \text{ mol NaOH} \times \frac{1 \text{ mol } Na_2CO_3}{2 \text{ mol NaOH}} = 0.850 \text{ mol } Na_2CO_3 \text{ can be produced}$$

The Na_2CO_3:CO_2 ratio is 1:1, so 0.850 mol Na_2CO_3 produced requires 0.850 mol CO_2 consumed. (Alternately, 1.70 mol NaOH × 1 mol CO_2/2 mol NaOH = 0.850 mol CO_2 reacted). 1.00 mol CO_2 initial - 0.850 mol CO_2 reacted = 0.15 mol CO_2 remain.

Check.

	2NaOH(s) +	CO_2(g) →	Na_2CO_3(s) + H_2O(l)
initial	1.70 mol	1.00 mol	0 mol
change (reaction)	-1.70 mol	-0.85 mol	+0.850 mol
final	0 mol	0.15 mol	0.850 mol

Note that the "change" line (but not necessarily the "final" line) reflects the mole ratios from the balanced equation.

3.70 $$0.450 \text{ mol Al(OH)}_3 \times \frac{3 \text{ mol } H_2SO_4}{2 \text{ mol Al(OH)}_3} = 0.675 \text{ mol } H_2SO_4 \text{ needed for complete reaction}$$

Only 0.550 mol H_2SO_4 available, so H_2SO_4 limits.

$$0.550 \text{ mol } H_2SO_4 \times \frac{1 \text{ mol } Al_2(SO_4)_3}{3 \text{ mol } H_2SO_4} = 0.1833 = 0.183 \text{ mol } Al_2(SO_4)_3 \text{ can form}$$

$$0.550 \text{ mol } H_2SO_4 \times \frac{2 \text{ mol Al(OH)}_3}{3 \text{ mol } H_2SO_4} = 0.3666 = 0.367 \text{ mol Al(OH)}_3 \text{ react}$$

0.450 mol $Al(OH)_3$ initial - 0.367 mol react = 0.083 mol $Al(OH)_3$ remain

(This result has 3 **decimal places** because it was obtained by subtraction, and thus only 2 sig. figs.)

3.71 $3NaHCO_3(aq) + H_3C_6H_5O_7(aq) \rightarrow 3CO_2(g) + 3H_2O(l) + Na_3C_6H_5O_7(aq)$

(a) *Analyze/Plan.* Abbreviate citric acid as H_3Cit. Follow the approach in Sample Exercise 3.19. *Solve:*

$$1.00 \text{ g } NaHCO_3 \times \frac{1 \text{ mol } NaHCO_3}{84.01 \text{ g } NaHCO_3} = 1.190 \times 10^{-2} = 1.19 \times 10^{-2} \text{ mol } NaHCO_3$$

$$1.00\text{g } H_3C_6H_5O_7 \times \frac{1 \text{ mol } H_3Cit}{192.1 \text{ g } H_3Cit} = 5.206 \times 10^{-3} = 5.21 \times 10^{-3} \text{ mol } H_3Cit$$

But $NaHCO_3$ and H_3Cit react in a 3:1 ratio, so 5.21×10^{-3} mol H_3Cit require $3(5.21 \times 10^{-3}) = 1.56 \times 10^{-2}$ mol $NaHCO_3$. We have only 1.19×10^{-2} mol $NaHCO_3$, so $NaHCO_3$ is the limiting reactant.

(b) 1.190×10^{-2} mol $NaHCO_3 \times \frac{3 \text{ mol } CO_2}{3 \text{ mol } NaHCO_3} \times \frac{44.01 \text{ g } CO_2}{1 \text{ mol } CO_2} = 0.524$ g CO_2

(c) 1.190×10^{-2} mol $NaHCO_3 \times \frac{1 \text{ mol } H_3Cit}{3 \text{ mol } NaHCO_3} = 3.968 \times 10^{-3}$

$= 3.97 \times 10^{-3}$ mol H_3Cit react

5.206×10^{-3} mol H_3Cit - 3.968×10^{-3} mol react $= 1.238 \times 10^{-3}$

$= 1.24 \times 10^{-3}$ mol H_3Cit remain

1.238×10^{-3} mol $H_3Cit \times \frac{192.1 \text{ g } H_3Cit}{\text{mol } H_3Cit} = 0.238$ g H_3Cit remain

3.72 $4NH_3(g) + 5O_2(g) \rightarrow 4NO(g) + 6H_2O(g)$

(a) Follow the approach in Sample Exercise 3.19.

$2.25 \text{ g } NH_3 \times \frac{1 \text{ mol } NH_3}{17.03 \text{ g } NH_3} = 0.1321 = 0.132$ mol NH_3

$3.75 \text{ g } O_2 \times \frac{1 \text{ mol } O_2}{32.00 \text{ g } O_2} = 0.1172 = 0.117$ mol O_2

$0.1172 \text{ mol } O_2 \times \frac{4 \text{ mol } NH_3}{5 \text{ mol } O_2} = 0.09375 = 0.0938$ mol NH_3 required

More than 0.0938 mol NH_3 is available, so O_2 is the limiting reactant.

(b) $0.1172 \text{ mol } O_2 \times \frac{4 \text{ mol NO}}{5 \text{ mol } O_2} \times \frac{30.01 \text{ g NO}}{1 \text{ mol NO}} = 2.81$ g NO produced

(c) 0.1321mol NH_3 - 0.0938 mol NH_3 reacted = 0.0383 = 0.038 mol NH_3 remain

$0.0383 \text{ mol } NH_3 \times \frac{17.03 \text{ g } NH_3}{1 \text{ mol } NH_3} = 0.652$ g NH_3 remain

3.73 *Analyze.* Given: initial g Na_2CO_3, g $AgNO_3$. Find: final g Na_2CO_3, $AgNO_3$, Ag_2CO_3, $NaNO_3$

Plan. Write balanced equation; determine limiting reactant; calculate amounts of excess reactant remaining and products, based on limiting reactant.

Solve. $2AgNO_3(aq) + Na_2CO_3(aq) \rightarrow Ag_2CO_3(s) + 2NaNO_3(aq)$

$6.50 \text{ g } Na_2CO_3 \times \frac{1 \text{ mol } Na_2CO_3}{106.0 \text{ g } Na_2CO_3} = 0.06132$ g $= 0.0613$ mol Na_2CO_3

$7.00 \text{ g } AgNO_3 \times \frac{1 \text{ mol } AgNO_3}{169.9 \text{ g } AgNO_3} = 0.04120 = 0.0412$ mol $AgNO_3$

$0.04120 \text{ mol } AgNO_3 \times \frac{1 \text{ mol } Na_2CO_3}{2 \text{ mol } AgNO_3} = 0.02060 = 0.0206$ mol Na_2CO_3 required

$AgNO_3$ is the limiting reactant and Na_2CO_3 is present in excess.

	$2AgNO_3(aq)$ +	$Na_2CO_3(aq)$	→	$Ag_2CO_3(s)$ +	$2NaNO_3(aq)$
initial	0.0412 mol	0.0613 mol		0 mol	0 mol
reaction	-0.0412 mol	-0.0206 mol		+0.0206 mol	+0.0412 mol
final	0 mol	0.0407 mol		0.0206 mol	0.0412 mol

0.04072 mol Na_2CO_3 × 106.0 g/mol = 4.316 = 4.32 g Na_2CO_3

0.02060 mol Ag_2CO_3 × 275.8 g/mol = 5.681 = 5.68 g Ag_2CO_3

0.04120 mol $NaNO_3$ × 85.00 g/mol = 3.502 = 3.50 g $NaNO_3$

Check. The initial mass of reactants was 13.50 g, and the final mass of excess reactant and products is 13.50 g; mass is conserved.

3.74 *Plan.* Write balanced equation; determine limiting reactant; calculate amounts of excess reactant remaining and products, based on limiting reactant.

Solve. $H_2SO_4(aq) + Pb(C_2H_3O_2)_2(aq) \rightarrow PbSO_4(s) + 2HC_2H_3O_2(aq)$

$$7.50 \text{ g } H_2SO_4 \times \frac{1 \text{ mol } H_2SO_4}{98.09 \text{ g } H_2SO_4} = 0.07646 = 0.0765 \text{ mol } H_2SO_4$$

$$7.50 \text{ g } Pb(C_2H_3O_2)_2 \times \frac{1 \text{ mol } Pb(C_2H_3O_2)_2}{325.3 \text{ g } Pb(C_2H_3O_2)_2} = 0.023056 = 0.0231 \text{ mol } Pb(C_2H_3O_2)_2$$

1 mol H_2SO_4:1 mol $Pb(C_2H_3O_2)_2$, so $Pb(C_2H_3O_2)_2$ is the limiting reactant.

0 mol $Pb(C_2H_3O_2)_2$, (0.07646 - 0.023056) = 0.0530 mol H_2SO_4, 0.0231 mol $PbSO_4$, (0.023056 × 2) = 0.0461 mol $HC_2H_3O_2$ are present after reaction

0.053405 mol H_2SO_4 × 98.09 g/mol = 5.2385 = 5.24 g H_2SO_4

0.023056 mol $PbSO_4$ × 303.3 g/mol = 6.9928 = 6.99 g $PbSO_4$

0.046111 mol $HC_2H_3O_2$ × 60.05 g/mol = 2.7690 = 2.77 g $HC_2H_3O_2$

Check. The initial mass of reactants was 15.00 g; and the final mass of excess reactant and products is 15.00 g; mass is conserved.

3.75 *Analyze.* Given: amounts of two reactants. Find: theoretical yield.

Plan. Determine the limiting reactant and the maximum amount of product it could produce. Then calculate % yield. *Solve*:

(a)

$$30.0 \text{ g } C_6H_6 \times \frac{1 \text{ mol } C_6H_6}{78.11 \text{ g } C_6H_6} = 0.3841 = 0.384 \text{ mol } C_6H_6$$

$$65.0 \text{ g } Br_2 \times \frac{1 \text{ mol } Br_2}{159.8 \text{ g } Br_2} = 0.4068 = 0.407 \text{ mol } Br_2$$

Since C_6H_6 and Br_2 react in a 1:1 mole ratio, C_6H_6 is the limiting reactant and determines the theoretical yield.

$$0.3841 \text{ mol } C_6H_6 \times \frac{1 \text{ mol } C_6H_5Br}{1 \text{ mol } C_6H_6} \times 157.0 \text{ g } C_6H_5Br = 60.30 = 60.3 \text{ g } C_6H_5Br$$

Check. 30/78 ~ 3/8 mol C_6H_6 . 65/160 ~ 3/8 mol Br_2. Since moles of the two reactants are similar, a precise calculation is needed to determine the limiting reactant. 3/8 × 160 ≈ 60 g product

(b) $$\% \text{ yield} = \frac{56.7 \text{ g } C_6H_5Br \text{ actual}}{60.3 \text{ g } C_6H_5Br \text{ theoretical}} \times 100 = 94.0\%$$

3.76 (a) $C_2H_6 + Cl_2 \rightarrow C_2H_5Cl + HCl$

$$125 \text{ g } C_2H_6 \times \frac{1 \text{ mol } C_2H_6}{30.07 \text{ g } C_2H_6} = 4.157 = 4.16 \text{ mol } C_2H_6$$

$$255 \text{ g } Cl_2 \times \frac{1 \text{ mol } Cl_2}{70.91 \text{ g } Cl_2} = 3.596 = 3.60 \text{ mol } Cl_2$$

Since the reactants combine in a 1:1 mole ratio, Cl_2 is the limiting reactant. The theoretical yield is:

$$3.596 \text{ mol } Cl_2 \times \frac{1 \text{ mol } C_2H_5Cl}{1 \text{ mol } Cl_2} \times \frac{64.51 \text{ g } C_2H_5Cl}{1 \text{ mol } C_2H_5Cl} = 231.98 = 232 \text{ g } C_2H_5Cl$$

(b) $$\% \text{ yield} = \frac{206 \text{ g } C_2H_5Cl \text{ actual}}{232 \text{ g } C_2H_5Cl \text{ theoretical}} \times 100 = 88.8\%$$

3.77 *Analyze.* Given: g of two reactants, % yield. Find: g Li_3N.

Plan. Determine limiting reactant and theoretical yield. Use definition of % yield to calculate actual yield. *Solve:*

(a) $$5.00 \text{ g Li} \times \frac{1 \text{ mol Li}}{6.941 \text{ g Li}} = 0.7204 = 0.720 \text{ mol Li}$$

$$5.00 \text{ g } N_2 \times \frac{1 \text{ mol } N_2}{28.01 \text{ g } N_2} = 0.1785 = 0.179 \text{ mol } N_2$$

$$0.1785 \text{ mol } N_2 \times \frac{6 \text{ mol Li}}{1 \text{ mol } N_2} = 1.071 = 1.07 \text{ mol Li required}$$

Since there is less than enough Li to react exactly with 0.179 mol N_2, Li is the limiting reactant.

$$0.7204 \text{ mol Li} \times \frac{2 \text{ mol } Li_3N}{6 \text{ mol Li}} \times \frac{34.83 \text{ g } Li_3N}{1 \text{ mol } Li_3N} = 8.363 = 8.36 \text{ g } Li_3N \text{ theoretical yield}$$

Check. 5/7 ≈ mol Li; 5/(4 × 7) ≈ mol N_2. There are 1/4 as many mol N_2 as moles Li, but only 1/6 as many moles N_2 are required for exact reaction. N_2 is in excess and Li limits. 0.7 × (36/3) ≈ 8.4 g Li_3N theoretical

(b) $$\% \text{ yield} = \frac{\text{actual}}{\text{theoretical}} \times 100; \quad \frac{\% \text{ yield} \times \text{theoretical}}{100} = \text{actual yield}$$

$$\frac{80.5\ \%}{100} \times 8.363 \text{ g } Li_3N = 6.7325 = 6.73 \text{ g } Li_3N \text{ actual}$$

3.78 $H_2S(g) + 2NaOH(aq) \rightarrow Na_2S(aq) + 2H_2O(l)$

$$2.00 \text{ g } H_2S \times \frac{1 \text{ mol } H_2S}{34.08 \text{ g } H_2S} = 0.05868 = 0.0587 \text{ mol } H_2S$$

$$2.00 \text{ g NaOH} \times \frac{1 \text{ mol NaOH}}{40.00 \text{ g NaOH}} = 0.0500 \text{ mol NaOH}$$

By inspection, twice as many mol NaOH as H_2S are needed for exact reaction, but mol NaOH given is less than mol H_2S, so NaOH limits.

$$0.0500 \text{ mol NaOH} \times \frac{1 \text{ mol } Na_2S}{2 \text{ mol NaOH}} \times \frac{78.05 \text{ g } Na_2S}{1 \text{ mol } Na_2S} = 1.95125 = 1.95 \text{ g } Na_2S \text{ theoretical}$$

$$\frac{92.0\ \%}{100} \times 1.95125 \text{ g } Na_2S \text{ theoretical} = 1.7951 = 1.80 \text{ g } Na_2S \text{ actual}$$

Additional Exercises

3.79 (a) $C_4H_8O_2(l) + 5O_2(g) \rightarrow 4CO_2(g) + 4H_2O(l)$

(b) $Cu(OH)_2(s) \rightarrow CuO(s) + H_2O(g)$

(c) $Zn(s) + Cl_2(g) \rightarrow ZnCl_2(s)$

3.80 The formulas of the fertilizers are NH_3, NH_4NO_3, $(NH_4)_2SO_4$ and $(NH_2)_2CO$. Qualitatively, the more heavy, non-nitrogen atoms in a molecule, the smaller the mass % of N. By inspection, the mass of NH_3 is dominated by N, so it will have the greatest % N, $(NH_4)_2SO_4$ will have the least. In order of increasing % N: $(NH_4)_2SO_4 < NH_4NO_3 < (NH_2)_2CO < NH_3$.

Check by calculation:

$(NH_4)_2SO_4$: FW = 2(14.0) + 8(1.0) + 1(32.1) + 4(16.0) = 132.1 amu

% N = [2(14.0)/132.1] × 100 = 21.2%

NH_4NO_3: FW = 2(14.0) + 4(1.0) + 3(16.0) = 80.0 amu

% N = [2(14.0)/80.0] × 100 = 35.0 %

$(NH_2)_2CO$: FW = 2(14.0) + 4(1.0) = 1(12.0) + 1(16.0) = 60.0 amu

% N = [2(14.0)/60.0] × 100 = 46.7% N

NH_3: FW = 1(14.0) + 3(1.0) = 17.0

% N = [14.0/17.0] × 100 = 82.4 % N

3.81 (a) $$1.25 \text{ carat} \times \frac{0.200 \text{ g}}{1 \text{ carat}} \times \frac{1 \text{ mol C}}{12.01 \text{ g C}} = 0.020816 = 0.0208 \text{ mol C}$$

$$0.020816 \text{ mol C} \times \frac{6.022 \times 10^{23} \text{ C atoms}}{1 \text{ mol C}} = 1.25 \times 10^{22} \text{ C atoms}$$

(b) $$0.500 \text{ g } C_9H_8O_4 \times \frac{1 \text{ mol } C_9H_8O_4}{180.2 \text{ g } C_9H_8O_4} = 2.7747 \times 10^{-3} = 2.77 \times 10^{-3} \text{ mol } C_9H_8O_4$$

$$0.0027747 \text{ mol } C_9H_8O_4 \times \frac{6.022 \times 10^{23} \text{ molecules}}{1 \text{ mol}} = 1.67 \times 10^{21} \; C_9H_8O_4 \text{ molecules}$$

3.82 (a) $\dfrac{5.342 \times 10^{-21}\text{ g}}{1\text{ molecule penicillin G}} \times \dfrac{6.0221 \times 10^{23}\text{ molecules}}{1\text{ mol}} = 3217\text{ g/mol penicillin G}$

(b) 1.00 g hemoglobin (hem) contains 3.40×10^{-3} g Fe.

$$\frac{1.00\text{ g hem}}{3.40 \times 10^{-3}\text{ g Fe}} \times \frac{55.85\text{ g Fe}}{1\text{ mol Fe}} \times \frac{4\text{ mol Fe}}{1\text{ mol hem}} = 6.57 \times 10^{4}\text{ g/mol hemoglobin}$$

3.83 (a) $1.0000 \times 10^{4}\text{ Si atoms} \times \dfrac{1\text{ mol}}{6.022 \times 10^{23}\text{ atoms}} \times \dfrac{28.0855\text{ g SI}}{1\text{ mol Si}} = 4.6638 \times 10^{-19}\text{ g Si}$

(b) $4.6638 \times 10^{-19}\text{ g Si} \times \dfrac{1\text{ cm}^3\text{ Si}}{2.3\text{ g Si}} = 2.03 \times 10^{-19} = 2.0 \times 10^{-19}\text{ cm}^3$

(c) $V = l^3;\ l = (V)^{1/3} = (2.03 \times 10^{-19}\text{ cm}^3)^{1/3} = 5.9 \times 10^{-7}\text{ cm } (= 5.9\text{ nm})$

3.84 *Plan.* Assume 100 g, calculate mole ratios, empirical formula, then molecular formula from molar mass. *Solve*:

$$68.2\text{ g C} \times \frac{1\text{ mol C}}{12.01\text{ g C}} = 5.68\text{ mol C};\ 5.68/0.568 \approx 10$$

$$6.86\text{ g H} \times \frac{1\text{ mol H}}{1.008\text{ g H}} = 6.81\text{ mol H};\ 6.81/0.568 \approx 12$$

$$15.9\text{ g N} \times \frac{1\text{ mol N}}{14.01\text{ g N}} = 1.13\text{ mol N};\ 1.13/0.568 \approx 2$$

$$9.08\text{ g O} \times \frac{1\text{ mol O}}{16.00\text{ g O}} = 0.568\text{ mol O};\ 0.568/0.0568 = 1$$

The empirical formula is $C_{10}H_{12}N_2O$, FW = 176 amu (or g). Since the molar mass is 176, the empirical and molecular formula are the same, $C_{10}H_{12}N_2O$.

3.85 *Plan.* Assume 1.000 g and get mass O by subtraction. *Solve*:

(a) $0.7787\text{ g C} \times \dfrac{1\text{ mol C}}{12.01\text{ g C}} = 0.06484\text{ mol C}$

$0.1176\text{ g H} \times \dfrac{1\text{ mol H}}{1.008\text{ g H}} = 0.1167\text{ mol H}$

$0.1037\text{ g O} \times \dfrac{1\text{ mol C}}{16.00\text{ g O}} = 0.006481\text{ mol O}$

Dividing through by the smallest of these values we obtain $C_{10}H_{18}O$.

(b) The formula weight of $C_{10}H_{18}O$ is 154. Thus, the empirical formula is also the molecular formula.

3.86 Since all the C in the vanillin must be present in the CO_2 produced, get g C from g CO_2.

$$2.43\text{ g }CO_2 \times \frac{1\text{ mol }CO_2}{44.01\text{ g }CO_2} \times \frac{12.01\text{ g C}}{1\text{ mol C}} = 0.6631 = 0.663\text{ g C}$$

Since all the H in vanillin must be present in the H_2O produced, get g H from g H_2O.

$$0.50 \text{ g } H_2O \times \frac{1 \text{ mol } H_2O}{18.02 \text{ g } H_2O} \times \frac{2 \text{ mol H}}{1 \text{ mol } H_2O} \times \frac{1.008 \text{ g H}}{1 \text{ mol H}} = 0.0559 = 0.056 \text{ g H}$$

Get g O by subtraction. (Since the analysis was performed by combustion, an unspecified amount of O_2 was a reactant, and thus not all the O in the CO_2 and H_2O produced came from vanillin.) 1.05 g vanillin - 0.663 g C - 0.056 g H = 0.331 g O

$$0.6631 \text{ g C} \times \frac{1 \text{ mol C}}{12.01 \text{ g C}} = 0.0552 \text{ mol C}; \ 0.0552 / 0.0207 = 2.67$$

$$0.0559 \text{ g H} \times \frac{1 \text{ mol H}}{1.008 \text{ g H}} = 0.0555 \text{ mol C}; \ 0.0555 / 0.0207 = 2.68$$

$$0.331 \text{ g O} \times \frac{1 \text{ mol O}}{16.00 \text{ g O}} = 0.0207 \text{ mol O}; \ 0.0207 / 0.0207 = 1.00$$

Multiplying the numbers above by **3** to obtain an integer ratio of moles, the empirical formula of vanillin is $C_8H_8O_3$.

3.87 *Plan.* Because different sample sizes were used to analyze the different elements, calculate mass % of each element in the sample.

i. Calculate mass % C from g CO_2.
ii. Calculate mass % Cl from AgCl.
iii. Get mass % H by subtraction.
iv. Calculate mole ratios and the empirical formulas.

Solve:

i. $$3.52 \text{ g } CO_2 \times \frac{1 \text{ mol } CO_2}{44.01 \text{ g } CO_2} \times \frac{1 \text{ mol C}}{1 \text{ mol } CO_2} \times \frac{12.01 \text{ g C}}{1 \text{ mol C}} = 0.9606 = 0.961 \text{ g C}$$

$$\frac{0.9606 \text{ g C}}{1.50 \text{ g sample}} \times 100 = 64.04 = 64.0\% \text{ C}$$

ii. $$1.27 \text{ g AgCl} \times \frac{1 \text{ mol AgCl}}{143.3 \text{ g AgCl}} \times \frac{1 \text{ mol Cl}}{1 \text{ mol AgCl}} \times \frac{35.45 \text{ g Cl}}{1 \text{ mol Cl}} = 0.3142 = 0.314 \text{ g Cl}$$

$$\frac{0.3142 \text{ g Cl}}{1.00 \text{ g sample}} \times 100 = 31.42 = 31.4\% \text{ Cl}$$

iii. % H = 100.0 - (64.04% C + 31.42% Cl) = 4.54 = 4.5% H

iv. Assume 100 g sample.

$$64.04 \text{ g C} \times \frac{1 \text{ mol C}}{12.01 \text{ g C}} = 5.33 \text{ mol C}; \ 5.33 / 0.886 = 6.02$$

$$31.42 \text{ g Cl} \times \frac{1 \text{ mol Cl}}{35.45 \text{ g Cl}} = 0.886 \text{ mol Cl}; \ 0.886 / 0.886 = 1.00$$

$$4.54 \text{ g H} \times \frac{1 \text{ mol H}}{1.008 \text{ g H}} = 4.50 \text{ mol H}; \ 4.50 / 0.886 = 5.08$$

The empirical formula is probably C_6H_5Cl.

The subscript for H, 5.08, is relatively far from 5.00, but C_6H_5Cl makes chemical sense. More significant figures in the mass data are required for a more accurate mole ratio.

3.88 The mass percentage is determined by the relative number of atoms of the element times the atomic weight, divided by the total formula mass. Thus, the mass percent of bromine in $KBrO_x$ is given by $0.5292 = \frac{79.91}{39.10 + 79.91 + x(16.00)}$. Solving for x, we obtain x = 2.00. Thus, the formula is $KBrO_2$.

3.89 (a) Let AW = the atomic weight of X.

According to the chemical reaction, moles XI_3 reacted = moles XCl_3 produced

$$0.5000 \text{ g } XI_3 \times 1 \text{ mol } XI_3 / (AW + 380.71) \text{ g } XI_3 = 0.2360 \text{ g } XCl_3 \times \frac{1 \text{ mol } XCl_3}{(AW + 106.36) \text{ g } XCl_3}$$

0.5000 (AW + 106.36) = 0.2360 (AW + 380.71)

0.5000 AW + 53.180 = 0.2360 AW + 89.848

0.2640 AW = 36.67; AW = 138.9 g

(b) X is lanthanum, La, atomic number 57.

3.90 $O_3(g) + 2NaI(aq) + H_2O(l) \rightarrow O_2(g) + I_2(s) + 2NaOH(aq)$

(a) $3.8 \times 10^{-5} \text{ mol } O_3 \times \frac{2 \text{ mol NaI}}{1 \text{ mol } O_3} = 7.6 \times 10^{-5} \text{ mol NaI}$

(b) $$0.550 \text{ mg } O_3 \times \frac{1 \times 10^{-3} \text{g}}{1 \text{ mg}} \times \frac{1 \text{ mol } O_3}{48.00 \text{ g } O_3} \times \frac{2 \text{ mol NaI}}{1 \text{ mol } O_3} \times \frac{149.9 \text{ g NaI}}{1 \text{ mol NaI}}$$

$$= 3.4352 \times 10^{-3} = 3.44 \times 10^{-3} \text{ g NaI} = 3.44 \text{ mg NaI}$$

3.91 $2NaCl(aq) + 2H_2O(l) \rightarrow 2NaOH(aq) + H_2(g) + Cl_2(g)$

Calculate mol Cl_2 and relate to mol H_2, mol NaOH.

$$1.5 \times 10^6 \text{ kg} \times \frac{1000 \text{ g}}{1 \text{ kg}} \times \frac{1 \text{ mol } Cl_2}{70.91 \text{ g } Cl_2} = 2.115 \times 10^7 = 2.1 \times 10^7 \text{ mol } Cl_2$$

$$2.115 \times 10^7 \text{ mol } Cl_2 \times \frac{1 \text{ mol } H_2}{1 \text{ mol } Cl_2} \times \frac{2.016 \text{ g } H_2}{1 \text{ mol } H_2} = 4.26 \times 10^7 \text{ g } H_2 = 4.3 \times 10^4 \text{ kg } H_2$$

$$4.3 \times 10^7 \text{ g} \times \frac{1 \text{ metric ton}}{1 \times 10^6 \text{ g (1 Mg)}} = 43 \text{ metric tons } H_2$$

$$2.115 \times 10^7 \text{ mol } Cl_2 \times \frac{2 \text{ mol NaOH}}{1 \text{ mol } Cl_2} \times \frac{40.0 \text{ g NaOH}}{1 \text{ mol NaOH}} = 1.69 \times 10^9 = 1.7 \times 10^9 \text{ g NaOH}$$

1.7×10^9 g NaOH = 1.7×10^6 kg NaOH = 1.7×10^3 metric tons NaOH

3.92 $2C_{57}H_{110}O_6 + 163O_2 \rightarrow 114CO_2 + 110H_2O$

molar mass of fat = 57(12.01) + 110(1.008) + 6(16.00) = 891.5

$$1.0 \text{ kg fat} \times \frac{1000 \text{ g}}{1 \text{ kg}} \times \frac{1 \text{ mol fat}}{891.5 \text{ g fat}} \times \frac{110 \text{ mol } H_2O}{2 \text{ mol fat}} \times \frac{18.02 \text{ g } H_2O}{1 \text{ mol } H_2O} \times \frac{1 \text{ kg}}{1000 \text{ g}} = 1.1 \text{ kg } H_2O$$

3.93 (a) *Plan*. Calculate the total mass of C from g CO and g CO_2. Calculate the mass of H from g H_2O. Calculate mole ratios and the empirical formula. *Solve*:

$$0.467 \text{ g CO} \times \frac{1 \text{ mol CO}}{28.01 \text{ g CO}} \times \frac{1 \text{ mol C}}{1 \text{ mol CO}} \times 12.01 \text{ g C} = 0.200 \text{ g C}$$

$$0.733 \text{ g } CO_2 \times \frac{1 \text{ mol } CO_2}{44.01 \text{ g } CO_2} \times \frac{1 \text{ mol C}}{1 \text{ mol } CO_2} \times 12.01 \text{ g C} = 0.200 \text{ g C}$$

Total mass C is 0.200 g + 0.200 g = 0.400 g C.

$$0.450 \text{ g } H_2O \times \frac{1 \text{ mol } H_2O}{18.02 \text{ g } H_2O} \times \frac{2 \text{ mol H}}{1 \text{ mol } H_2O} \times \frac{1.008 \text{ g H}}{1 \text{ mol H}} = 0.0503 \text{ g H}$$

(Since hydrocarbons contain only the elements C and H, g H can also be obtained by subtraction: 0.450 g sample - 0.400 g C = 0.050 g H.)

$$0.400 \text{ g C} \times \frac{1 \text{ mol C}}{12.01 \text{ g C}} = 0.0333 \text{ mol C}; \; 0.0333 / 0.0333 = 1.0$$

$$0.0503 \text{ g H} \times \frac{1 \text{ mol H}}{1.008 \text{ g H}} = 0.0499 \text{ mol H}; \; 0.0499 / 0.0333 = 1.5$$

Multiplying by a factor of 2, the emprical formula is C_2H_3.

(b) Mass is conserved. Total mass products - mass sample = mass O_2 consumed.
0.467 g CO + 0.733 g CO_2 + 0.450 g H_2O - 0.450 g sample = 1.200 g O_2 consumed

(c) For complete combustion, 0.467 g CO must be converted to CO_2.

$2CO(g) + O_2(g) \rightarrow 2CO_2(g)$

$$0.467 \text{ g CO} \times \frac{1 \text{ mol CO}}{28.01 \text{ g C}} \times \frac{1 \text{ mol } O_2}{2 \text{ mol CO}} \times \frac{32.00 \text{ g } O_2}{1 \text{ mol } O_2} = 0.267 \text{ g } O_2$$

The total mass of O_2 required for complete combustion is
1.200 g + 0.267 g = 1.467 g O_2.

3.94 $N_2(g) + 3H_2(g) \rightarrow 2NH_3(g)$

Determine the moles of N_2 and H_2 required to form the 2.0 moles of NH_3 present after the reaction has stopped.

$$2.0 \text{ mol } NH_3 \times \frac{3 \text{ mol } H_2}{2 \text{ mol } NH_3} = 3.0 \text{ mol } H_2 \text{ reacted}$$

$$2.0 \text{ mol } NH_3 \times \frac{1 \text{ mol } N_2}{2 \text{ mol } NH_3} = 1 \text{ mol } N_2 \text{ reacted}$$

mol H_2 initial = 2.0 mol H_2 remain + 3.0 mol H_2 reacted = 5.0 mol H_2
mol N_2 initial = 2.0 mol N_2 remain + 1.0 mol N_2 reacted = 3.0 mol N_2

In tabular form:

	$N_2(g)$	+	$3H_2(g)$	→	$2NH_3(g)$
initial	3.0 mol		5.0 mol		0 mol
reaction	-1.0 mol		-3.0 mol		+2.0 mol
final	2.0 mol		2.0 mol		2.0 mol

(Tables like this will be extremely useful for solving chemical equilibrium problems in Chapter 15.)

3.95 All of the O_2 is produced from $KClO_3$; get g $KClO_3$ from g O_2. All of the H_2O is produced from $KHCO_3$; get g $KHCO_3$ from g H_2O. The g H_2O produced also reveals the g CO_2 from the decomposition of $NaHCO_3$. The remaining CO_2 (13.2 g CO_2 - g CO_2 from $NaHCO_3$) is due to K_2CO_3 and g K_2CO_3 can be derived from it.

$$4.00 \text{ g } O_2 \times \frac{1 \text{ mol } O_2}{32.00 \text{ g } O_2} \times \frac{2 \text{ mol } KClO_3}{3 \text{ mol } O_2} \times \frac{122.6 \text{ g } KClO_3}{1 \text{ mol } KClO_3} = 10.22 = 10.2 \text{ g } KClO_3$$

$$1.80 \; H_2O \times \frac{1 \text{ mol } H_2O}{18.02 \text{ g } H_2O} \times \frac{2 \text{ mol } KHCO_3}{1 \text{ mol } H_2O} \times \frac{100.1 \text{ g } KHCO_3}{1 \text{ mol } KHCO_3} = 20.00 = 20.0 \text{ g } KHCO_3$$

$$1.80 \text{ g } H_2O \times \frac{1 \text{ mol } H_2O}{18.02 \text{ g } H_2O} \times \frac{2 \text{ mol } CO_2}{1 \text{ mol } H_2O} \times \frac{44.01 \text{ g } CO_2}{1 \text{ mol } CO_2} = 8.792 = 8.79 \text{ g } CO_2 \text{ from } KHCO_3$$

13.20 g CO_2 total - 8.792 CO_2 from $KHCO_3$ = 4.408 = 4.41 g CO_2 from K_2CO_3

$$4.408 \text{ g } CO_2 \times \frac{1 \text{ mol } CO_2}{44.01 \text{ g } CO_2} \times \frac{1 \text{ mol } K_2CO_3}{1 \text{ mol } CO_2} \times \frac{138.2 \text{ g } K_2CO_3}{1 \text{ mol } K_2CO_3} = 13.84 = 13.8 \text{ g } K_2CO_3$$

100.0 g mixture - 10.22 g $KClO_3$ - 20.00 g $KHCO_3$ - 13.84 g K_2CO_3 = 56.0 g KCl

3.96 (a) $2C_2H_2(g) + 5O_2(g) \rightarrow 4CO_2(g) + 2H_2O(g)$

(b) Following the approach in Sample Exercise 3.19,

$$10.0 \text{ g } C_2H_2 \times \frac{1 \text{ mol } C_2H_2}{26.04 \text{ g } C_2H_2} \times \frac{5 \text{ mol } O_2}{2 \text{ mol } C_2H_2} \times \frac{32.00 \text{ g } O_2}{1 \text{ mol } O_2} = 30.7 \text{ g } O_2 \text{ required}$$

Only 10.0 g O_2 are available, so O_2 limits.

(c) Since O_2 limits, 0.0 g O_2 remain.

Next, calculate the g C_2H_2 consumed and the amounts of CO_2 and H_2O produced by reaction of 10.0 g O_2.

$$10.0 \text{ g } O_2 \times \frac{1 \text{ mol } O_2}{32.00 \text{ g } O_2} \times \frac{2 \text{ mol } C_2H_2}{5 \text{ mol } O_2} \times \frac{26.04 \text{ g } C_2H_2}{1 \text{ mol } C_2H_2} = 3.26 \text{ g } C_2H_2 \text{ consumed}$$

10.0 g C_2H_2 initial - 3.26 g consumed = 6.74 = 6.7 g C_2H_2 remain

$$10.0 \text{ g } O_2 \times \frac{1 \text{ mol } O_2}{32.00 \text{ g } O_2} \times \frac{4 \text{ mol } CO_2}{5 \text{ mol } O_2} \times \frac{44.01 \text{ g } CO_2}{1 \text{ mol } CO_2} = 11.0 \text{ g } CO_2 \text{ produced}$$

$$10.0 \text{ g } O_2 \times \frac{1 \text{ mol } O_2}{32.00 \text{ g } O_2} \times \frac{2 \text{ mol } H_2O}{5 \text{ mol } O_2} \times \frac{18.02 \text{ g } H_2O}{1 \text{ mol } H_2O} = 2.25 \text{ g } H_2O \text{ produced}$$

3.97 (a) $$1.5 \times 10^2 \text{ g } C_9H_8O_4 \times \frac{1 \text{ mol } C_9H_8O_4}{180.2 \text{ g } C_9H_8O_4} \times \frac{1 \text{ mol } C_7H_6O_3}{1 \text{ mol } C_9H_8O_4} \times \frac{138.1 \text{ g } C_7H_6O_3}{1 \text{ mol } C_7H_6O_3}$$

$$= 1.1496 \times 10^2 = 1.1 \times 10^2 \text{ kg } C_7H_6O_3$$

(b) If only 80 percent of the acid reacts, then we need 1/0.80 = 1.25 times as much to obtain the same mass of product: $1.25 \times 1.15 \times 10^2$ kg = 1.4×10^2 kg.

(c) Calculate the number of moles of each reactant:

$$1.85 \times 10^5 \text{ g } C_7H_6O_3 \times \frac{1 \text{ mol } C_7H_6O_3}{138.1 \text{ g } C_7H_6O_3} = 1.340 \times 10^3 = 1.34 \times 10^3 \text{ mol } C_7H_6O_3$$

$$1.25 \times 10^5 \text{ g } C_4H_6O_3 \times \frac{1 \text{ mol } C_4H_6O_3}{102.1 \text{ g } C_4H_6O_3} = 1.224 \times 10^3 = 1.22 \times 10^3 \text{ mol } C_4H_6O_3$$

We see that $C_4H_6O_3$ limits, because equal numbers of moles of the two reactants are consumed in the reaction.

$$1.224 \times 10^3 \text{ mol } C_4H_6O_3 \times \frac{1 \text{ mol } C_9H_8O_4}{1 \text{ mol } C_7H_6O_3} \times \frac{180.2 \text{ g } C_9H_8O_4}{1 \text{ mol } C_9H_8O_4} = 2.206 \times 10^5$$

$$= 2.21 \times 10^5 \text{ g } C_9H_8O_4$$

(d) $$\text{percent yield} = \frac{1.82 \times 10^5 \text{ g}}{2.206 \times 10^5 \text{ g}} \times 100 = 82.5\%$$

Integrative Exercises

3.98 *Plan.* Volume cube $\xrightarrow{\text{density}}$ mass $CaCO_3$ → moles $CaCO_3$ → moles O → O atoms

Solve. $$(1.25)^3 \text{ in}^3 \times \frac{(2.54)^3 \text{ cm}^3}{1 \text{ in}^3} \times \frac{2.71 \text{ g } CaCO_3}{1 \text{ cm}^3} \times \frac{1 \text{ mol } CaCO_3}{100.1 \text{ g } CaCO_3} \times \frac{3 \text{ mol O}}{1 \text{ mol } CaCO_3}$$

$$\times \frac{6.022 \times 10^{23} \text{ O atoms}}{1 \text{ mol O}} = 1.57 \times 10^{24} \text{ O atoms}$$

3.99 (a) *Plan*: volume of Ag cube $\xrightarrow{\text{density}}$ mass of Ag → mol Ag → Ag atoms

Solve: $$(1.000)^3 \text{ cm}^3 \text{ Ag} \times \frac{10.49 \text{ g Ag}}{1 \text{ cm}^3 \text{ Ag}} \times \frac{1 \text{ mol Ag}}{107.87 \text{ g Ag}} \times \frac{6.022 \times 10^{23} \text{ atoms}}{1 \text{ mol}}$$

$$= 5.8562 \times 10^{22} = 5.856 \times 10^{22} \text{ Ag atoms}$$

(b) $1.000\ cm^3$ cube volume, 74% is occupied by Ag atoms

$0.7400\ cm^3$ = volume of 5.856×10^{22} Ag atoms

$$\frac{0.7400\ cm^3}{5.8562 \times 10^{22}\ \text{Ag atoms}} = 1.2636 \times 10^{-23} = 1.264 \times 10^{-23}\ cm^3 / \text{Ag atom}$$

Since atomic dimensions are usually given in Å, we will show this conversion.

$$1.264 \times 10^{-23}\ cm^3 \times \frac{(1 \times 10^{-2})^3\ m^3}{1\ cm^3} \times \frac{1\ Å^3}{(1 \times 10^{-10})^3\ m^3} = 12.64\ Å^3 / \text{Ag atom}$$

(c) $V = 4/3\ \pi r^3$; $r^3 = 3V/4\pi$; $r = (3V/4\pi)^{1/3}$

$r_{Å} = (3 \times 12.636\ Å^3 / 4\pi)^{1/3} = 1.4449 = 1.445\ Å$

3.100 *Analyze.* Given: gasoline = C_8H_{18}, density = 0.69 g/mL, 19.5 mi/gal, 125 mi. Find: kg CO_2. *Plan.* Write and balance the equation for the combustion of octane. Change mi → gal octane → mL → g octane. Use stoichiometry to calculate g and kg CO_2 from g octane.

Solve. $2C_8H_{18}(l) + 25O_2(g) \rightarrow 16CO_2(g) + 18H_2O(l)$

$$125\ \text{mi} \times \frac{1\ \text{gal}}{19.5\ \text{mi}} \times \frac{3.7854\ L}{1\ \text{gal}} \times \frac{1\ mL}{1 \times 10^{-3}\ L} \times \frac{0.69\ \text{g octane}}{1\ mL} = 1.6743 \times 10^4\ g$$

$$= 17\ \text{kg octane}$$

$$1.6743 \times 10^4\ g\ C_8H_{18} \times \frac{1\ mol\ C_8H_{18}}{114.2\ g\ C_8H_{18}} \times \frac{16\ mol\ CO_2}{2\ mol\ C_8H_{18}} \times \frac{44.01\ g\ CO_2}{1\ mol\ CO_2} = 5.1619 \times 10^4\ g$$

$$= 52\ kg\ CO_2$$

Check. $\left(\frac{125 \times 4 \times 0.7}{20}\right) \times 10^3 = (25 \times 0.7) \times 10^3 = 17.5 \times 10^3\ g = 17.5$ kg octane

$$\frac{44}{114} \approx \frac{1}{3};\ \frac{17\ kg \times 8}{3} \approx 48\ kg\ CO_2$$

3.101 *Plan.* We can proceed by writing the ratio of masses of Ag to $AgNO_3$, where y is the atomic mass of nitrogen. *Solve*:

$$\frac{Ag}{AgNO_3} = 0.634985 = \frac{107.8682}{107.8682 + 3(15.9994) + y}$$

Solve for y to obtain y = 14.0088. This is to be compared with the currently accepted value of 14.0067.

3.102 (a) $S(s) + O_2(g) \rightarrow SO_2(g)$; $SO_2(g) + CaO(s) \rightarrow CaSO_3(s)$

(b)

$$\frac{2000\ \text{tons coal}}{\text{day}} \times \frac{2000\ lb}{1\ ton} \times \frac{1\ kg}{2.20\ lb} \times \frac{1000\ g}{1\ kg} \times \frac{0.025\ g\ S}{1\ g\ coal} \times \frac{1\ mol\ S}{32.1\ g\ S}$$

$$\times \frac{1\ mol\ SO_2}{1\ mol\ S} \times \frac{1\ mol\ CaSO_3}{1\ mol\ SO_2} \times \frac{120\ g\ CaSO_3}{1\ mol\ CaSO_3} \times \frac{1\ kg\ CaSO_3}{1000\ g\ CaSO_3}$$

$$= 1.7 \times 10^5\ kg\ CaSO_3/day$$

This corresponds to about 190 tons of $CaSO_3$ per day as a waste product.

3.103 (a) *Plan.* Calculate the kg of air in the room and then the mass of HCN required to produce a dose of 300 mg HCN/kg air. *Solve*:

12 ft × 15 ft × 8.0 ft = 1440 = 1.4×10^3 ft^3 of air in the room

$$1440\ ft^3\ air \times \frac{(12\ in)^3}{1\ ft^3} \times \frac{(2.54\ cm)^3}{1\ in^3} \times \frac{0.00118\ g\ air}{1\ cm^3\ air} \times \frac{1\ kg}{1000\ g} = 48.12$$

$$= 48\ \text{kg air}$$

$$48.12\ kg\ air \times \frac{300\ mg\ HCN}{1\ kg\ air} \times \frac{1\ g}{1000\ mg} = 14.43 = 14\ g\ HCN$$

(b) $2NaCN(s) + H_2SO_4(aq) \rightarrow Na_2SO_4(aq) + 2HCN(g)$

The question can be restated as: What mass of NaCN is required to produce 14 g of HCN according to the above reaction?

$$14.43\ g\ HCN \times \frac{1\ mol\ HCN}{27.03\ g\ HCN} \times \frac{2\ mol\ NaCN}{2\ mol\ HCN} \times \frac{49.01\ g\ NaCN}{1\ mol\ NaCN} = 26.2 = 26\ g\ NaCN$$

(c)

$$12\ ft \times 15\ ft \times \frac{1\ yd^2}{9\ ft^2} \times \frac{30\ oz}{1\ yd^2} \times \frac{1\ lb}{16\ oz} \times \frac{454\ g}{1\ lb} = 17{,}025$$

$$= 1.7 \times 10^4\ \text{g acrilan in the room}$$

50% of the carpet burns, so the starting amount of CH_2CHCN is 0.50(17,025) = 8,513 = 8.5×10^3 g

$$8{,}513\ g\ CH_2CHCN \times \frac{50.9\ g\ HCN}{100\ g\ CH_2CHCH} = 4333 = 4.3 \times 10^3\ g\ HCN\ possible$$

If the actual yield of combustion is 20%, actual g HCN = 4,333(0.20) = 866.6 = 8.7×10^2 g HCN produced. From part (a), 14 g of HCN is a lethal dose. The fire produces much more than a lethal dose of HCN.

4 Aqueous Reactions and Solution Stoichiometry

Electrolytes

4.1 Tap water contains enough dissolved electrolytes to conduct a significant amount of electricity. Thus, water can complete a circuit between an electrical appliance and our body, producing a shock.

4.2 No. Electrolyte solutions conduct electricity because the dissolved ions carry charge through the solution (from one electrode to the other).

4.3 When CH_3OH dissolves, neutral CH_3OH molecules are dispersed throughout the solution. These electrically neutral particles do not carry charge and the solution is nonconducting. When $HC_2H_3O_2$ dissolves, mostly neutral molecules are dispersed throughout the solution. A few of the dissolved molecules ionize to form $H^+(aq)$ and $C_2H_3O_2^-(aq)$. These few ions carry some charge and the solution is weakly conducting.

4.4 Although H_2O molecules are electrically neutral, there is an unequal distribution of electrons throughout the molecule. There are more electrons near O and fewer near H, giving the O end of the molecule a partial negative charge and the H end of the molecule a partial positive charge. Ionic compounds are composed of positively and negatively charged ions. The partially positive ends of H_2O molecules are attracted to the negative ions (anions) in the solid, while the partially negative ends are attracted to the positive ions (cations). Thus, both cations and anions in an ionic solid are surrounded and separated (dissolved) by H_2O molecules.

4.5 (a) $ZnCl_2(aq) \rightarrow Zn^{2+}(aq) + 2Cl^-(aq)$ (b) $HNO_3(aq) \rightarrow H^+(aq) + NO_3^-(aq)$
(c) $K_2SO_4(aq) \rightarrow 2K^+(aq) + SO_4^{2-}(aq)$ (d) $Ca(OH)_2(aq) \rightarrow Ca^{2+}(aq) + 2OH^-(aq)$

4.6 (a) $MgI_2(aq) \rightarrow Mg^{2+}(aq) + 2I^-(aq)$ (b) $Al(NO_3)_3(aq) \rightarrow Al^{3+}(aq) + 3NO_3^-(aq)$
(c) $HClO_4(aq) \rightarrow H^+(aq) + ClO_4^-(aq)$ (d) $(NH_4)_2SO_4(aq) \rightarrow 2NH_4^+(aq) + SO_4^{2-}(aq)$

4.7 (a) AX is a nonelectrolyte, because no ions form when the molecules dissolve.
(b) AY is a weak electrolyte because a few molecules ionize when they dissolve, but most do not.
(c) AZ is a strong electrolyte because all molecules break-up into ions when they dissolve.

4.8 Both AX and BY are weak electrolytes, because some but not all molecules ionize when they dissolve. BY will be the better conductor of electricity because it produces more ions per mole of dissolved solute.

4.9 When $HCHO_2$ dissolves in water, neutral $HCHO_2$ molecules, H^+ ions and CHO_2^- ions are all present in the solution. $HCHO_2(aq) \rightleftharpoons H^+(aq) + CHO_2^-(aq)$

4.10 (a) acetone (nonelectrolyte): $CH_3COCH_3(aq)$ molecules only; hypochlorous acid (weak electrolyte): $HClO(aq)$ molecules, $H^+(aq)$, ClO^- (aq); ammonium chloride (strong electrolyte): $NH_4^+(aq)$, $Cl^-(aq)$

(b) NH_4Cl, 0.2 mol solute particles; HClO, between 0.1 and 0.2 mol particles; CH_3OCH_3, 0.1 mol of solute particles

Precipitation Reactions and Net Ionic Equations

4.11 *Analyze.* Given: formula of compound. Find: solubility.

Plan. Follow the guidelines in Table 4.1, in light of the anion present in the compound and notable exceptions to the "rules". *Solve*:

(a) $Ni\mathbf{Cl_2}$: soluble (b) $Ag_2\mathbf{S}$: insoluble

(c) $Cs_3\mathbf{PO_4}$: soluble (Cs^+ is an alkali metal cation)

(d) $Sr\mathbf{CO_3}$: insoluble (e) $(NH_4)_2\mathbf{SO_4}$: soluble

4.12 According to Table 4.1:

(a) $Ni(\mathbf{OH})_2$: insoluble

(b) $Pb\mathbf{SO_4}$: insoluble, Pb^{2+} is an exception to soluble sulfates

(c) $Ba(\mathbf{NO_3})_2$: soluble

(d) $Al\mathbf{PO_4}$: insoluble

(e) $Ag\mathbf{C_2H_3O_2}$: soluble

4.13 *Analyze.* Given: formulas of reactants. Find: balanced equation including precipitates.

Plan. Follow the logic in Sample Exercise 4.3.

Solve. In each reaction, the precipitate is in bold type.

(a) $Na_2CO_3(aq) + 2AgNO_3(aq) \rightarrow \mathbf{Ag_2CO_3(s)} + 2NaNO_3(aq)$

(b) No precipitate (all nitrates and most sulfates are soluble).

(c) $FeSO_4(aq) + Pb(NO_3)_2(aq) \rightarrow \mathbf{PbSO_4(s)} + Fe(NO_3)_2(aq)$

4.14 In each reaction, the precipitate is in bold type.

(a) $Sn(NO_3)_2(aq) + 2NaOH(aq) \rightarrow \mathbf{Sn(OH)_2(s)} + 2NaNO_3(aq)$

(b) No precipitate, and therefore, no reaction. There is no chemical change to any of the reactant ions.

(c) $Na_2S(aq) + Cu(C_2H_3O_2)_2(aq) \rightarrow \mathbf{CuS(s)} + 2NaC_2H_3O_2(aq)$

4.15 *Analyze/Plan.* Follow the logic in Sample Exercise 4.4. *Solve:*

(a) $2Na^+(aq) + CO_3^{2-}(aq) + Mg^{2+}(aq) + SO_4^{2-}(aq) \rightarrow MgCO_3(s) + 2Na^+(aq) + SO_4^{2-}(aq)$

$Mg^{2+}(aq) + CO_3^{2-}(aq) \rightarrow MgCO_3(s)$

(b) $Pb^{2+}(aq) + 2NO_3^-(aq) + 2Na^+(aq) + S^{2-}(aq) \rightarrow PbS(s) + 2Na^+(aq) + 2NO_3^-(aq)$

$Pb^{2+}(aq) + S^{2-}(aq) \rightarrow PbS(s)$

(c) $6NH_4^+(aq) + 2PO_4^{3-}(aq) + 3Ca^{2+}(aq) + 6Cl^-(aq) \rightarrow Ca_3(PO_4)_2(s) + 6NH_4^+(aq) + 6Cl^-(aq)$

$3Ca^{2+}(aq) + 2PO_4^{3-}(aq) \rightarrow Ca_3(PO_4)_2(s)$

4.16 Spectator ions are those that do not change during reaction.

(a) $2Cr^{3+}(aq) + 3CO_3^{2-}(aq) \rightarrow Cr_2(CO_3)_3(s)$; spectators: NH_4^+, SO_4^{2-}

(b) $2Ag^+(aq) + SO_4^{2-}(aq) \rightarrow Ag_2SO_4(s)$; spectators: K^+, NO_3^-

(c) $Pb^{2+}(aq) + 2OH^-(aq) \rightarrow Pb(OH)_2(s)$; spectators: K^+, NO_3^-

4.17 *Analyze.* Given: reactions of unknown with HBr, H_2SO_4, NaOH. Find: The unknown contains a single salt. Is K^+ or Pb^{2+} or Ba^{2+} present?

Plan. Analyze solubility guidelines for Br^-, SO_4^{2-} and OH^- and select the cation that produces the observed solubility pattern.

Solve. Pb^{2+} is not present or an insoluble hydroxide would have formed. $BaSO_4$ is insoluble and $Ba(OH)_2$ is soluble, so the solution must contain Ba^{2+}. It could also contain K^+, but since we are dealing with a single salt, we will assume that only Ba^{2+} is present.

4.18 Br^- and NO_3^- can be ruled out because the Ba^{2+} salts are soluble. (Actually all NO_3^- salts are soluble.) CO_3^{2-} forms insoluble salts with the three cations given; it must be the anion in question.

4.19 *Analyze.* Given: $Mg(NO_3)_2(aq)$, $Pb(NO_3)_2(aq)$, $H_2SO_4(aq)$. Find: identify $Mg^{2+}(aq)$ and $Pb^{2+}(aq)$ solutions. *Plan.* Use difference in reactivities with SO_4^{2-} to identify $Pb^{2+}(aq)$ and $Mg^{2+}(aq)$.

Solve. Test a portion of each solution with $H_2SO_4(aq)$. $Pb^{2+}(aq)$ is an exception to the soluble sulfates rule, so $Pb(NO_3)_2(aq)$ will form a precipitate, while $Mg(NO_3)_2(aq)$ will not.

4.20

Compound	**$Ba(NO_3)_2$ result**	**NaCl result**
$AgNO_3(aq)$	no ppt	AgCl ppt
$CaCl_2(aq)$	no ppt	no ppt
$Al_2(SO_4)_3$	$BaSO_4$ ppt	no ppt

This sequence of tests would definitively identify the contents of the bottle.

Acid-Base Reactions

4.21 (a) A *monoprotic acid* has one ionizable (acidic) H and a *diprotic acid* has two.

(b) A *strong acid* is completely ionized in aqueous solution whereas only a fraction of *weak acid* molecules are ionized.

(c) An *acid* is an H^+ donor, a substance that increases the concentration of H^+ in aqueous solution. A *base* is an H^+ acceptor and thus increases the concentration of OH^- in aqueous solution.

4.22 (a) NH_3 produces OH^- in aqueous solution by reacting with H_2O (hydrolysis): $NH_3(aq) + H_2O(l) \rightleftharpoons NH_4^+(aq) + OH^-(aq)$. The OH^- causes the solution to be basic.

(b) The term "weak" refers to the tendency of HF to dissociate into H^+ and F^- in aqueous solution, not its reactivity toward other compounds.

(c) H_2SO_4 is a **diprotic** acid; it has two ionizable hydrogens. The first hydrogen completely ionizes to form H^+ and HSO_4^-, but HSO_4^- only **partially** ionizes into H^+ and SO_4^{2-} (HSO_4^- is a weak electrolyte). Thus, an aqueous solution of H_2SO_4 contains a mixture of H^+, HSO_4^- and SO_4^{2-}, with the concentration of HSO_4^- greater than the concentration of SO_4^{2-}.

4.23 *Analyze.* Given: chemical formulas. Find: acid-base properties.

Plan. Use Table 4.2 to identify common strong acids and bases. If a compound doesn't appear in the table, it is either a weak acid or base, or a nonelectrolyte.

Solve. (a) strong acid (b) weak acid (c) weak base (d) strong base

4.24 Use Table 4.2. (a) strong base (b) weak acid (c) weak acid (d) strong acid

4.25 *Analyze.* Given: chemical formulas. Find: classify as acid, base, salt; strong, weak or nonelectrolyte. *Plan.* Examine formula for: H-first, acid; OH^- anion, base; NH_3, weak base; ionic compound, salt and strong electrolyte; strong acid or base, Table 4.2. *Solve*:

(a) HF: acid, mixture of ions and molecules (weak electrolyte)

(b) CH_3CN: none of the above, entirely molecules (nonelectrolyte)

(c) $NaClO_4$: salt, entirely ions (strong electrolyte)

(d) $Ba(OH)_2$: base, entirely ions (strong electrolyte)

4.26 Since the solution does conduct some electricity, but less than an equimolar NaCl solution (a strong electrolyte) the unknown solute must be a weak electrolyte. The weak electrolytes in the list of choices are NH_3 and H_3PO_3; since the solution is acidic, the unknown must be $\mathbf{H_3PO_3}$.

4.27 *Analyze.* Given: chemical formulas. Find: electrolyte properties.

Plan. In order to classify as electrolytes, formulas must be identified as acids, bases or salts as in Solution 4.25. *Solve:*

(a) H_2SO_3: H first, so acid; not in Table 4.2, so weak acid; therefore, weak electrolyte

(b) C_2H_5OH: not acid, not ionic (no metal cation), contains OH group, but not as anion so not a base; therefore, nonelectrolyte

(c) NH_3: common weak base; therefore, weak electrolyte

(d) $KClO_3$: ionic compound, so strong electrolyte

(e) $Cu(NO_3)_2$: ionic compound, so strong electrolyte

4.28 (a) HBrO: weak (b) HNO_3: strong (c) KOH: strong

(d) CH_3OCH_3: non (e) $CoSO_4$: strong (f) $C_{12}H_{22}O_{11}$: non

4.29 *Plan.* Follow Sample Exercise 4.7. *Solve:*

(a) $2HBr(aq) + Ca(OH)_2(aq) \rightarrow CaBr_2(aq) + 2H_2O(l)$

$H^+(aq) + OH^-(aq) \rightarrow H_2O(l)$

(b) $Cu(OH)_2(s) + 2HClO_4(aq) \rightarrow Cu(ClO_4)_2(aq) + 2H_2O(l)$

$Cu(OH)_2(s) + 2H^+(aq) \rightarrow 2H_2O(l) + Cu^{2+}(aq)$

(c) $Al(OH)_3(s) + 3HNO_3(aq) \rightarrow Al(NO_3)_3(aq) + 3H_2O(l)$

$Al(OH)_3(s) + 3H^+(aq) \rightarrow 3H_2O(l) + Al^{3+}(aq)$

4.30 (a) $HC_2H_3O_2(aq) + KOH(aq) \rightarrow KC_2H_3O_2(aq) + H_2O(l)$

$H^+(aq) + OH^-(aq) \rightarrow H_2O(l)$

(b) $Cr(OH)_3(s) + 3HNO_3(aq) \rightarrow Cr(NO_3)_3(aq) + 3H_2O(l)$

$Cr(OH)_3(s) + 3H^+(aq) \rightarrow 3H_2O(l) + Cr^{3+}(aq)$

(c) $Ca(OH)_2(aq) + 2HClO(aq) \rightarrow Ca(ClO)_2(aq) + 2H_2O(l)$

$OH^-(aq) + H^+(aq) \rightarrow H_2O(l)$

4.31 *Analyze.* Given: names of reactants. Find: gaseous products.

Plan. Write correct chemical formulas for the reactants, complete and balance the metathesis reaction, and identify either H_2S or CO_2 products as gases. *Solve:*

(a) $CdS(s) + H_2SO_4(aq) \rightarrow CdSO_4(aq) + H_2S(g)$

$CdS(s) + 2H^+(aq) \rightarrow H_2S(g) + Cd^{2+}(aq)$

(b) $MgCO_3(s) + 2HClO_4(aq) \rightarrow Mg(ClO_4)_2(aq) + H_2O(l) + CO_2(g)$

$MgCO_3(s) + 2H^+(aq) \rightarrow H_2O(l) + CO_2(g) + Mg^{2+}(aq)$

4.32 (a) $CaCO_3(s) + 2HNO_3(aq) \rightarrow Ca(NO_3)_2(aq) + H_2O(l) + CO_2(g)$

$2H^+(aq) + CaCO_3(s) \rightarrow H_2O(l) + CO_2(g) + Ca^{2+}(aq)$

(b) $FeS(s) + 2HBr(aq) \rightarrow FeBr_2(aq) + H_2S(g)$

$2H^+(aq) + FeS(s) \rightarrow H_2S(g) + Fe^{2+}(aq)$

4.33 *Analyze/Plan.* Given the balanced complete molecular equation, determine the spectator ion(s) and write the net ionic equation. In each case, $HClO_4$ and the metal perchlorate are strong electrolytes, so ClO_4^- (aq) is the only spectator. All other species change form upon reaction. *Solve*:

(a) $FeO(s) + 2H^+(aq) \rightarrow H_2O(l) + Fe^{2+}(aq)$

(b) $NiO(s) + 2H^+(aq) \rightarrow H_2O(l) + Ni^{2+}(aq)$

4.34 $K_2O(aq) + H_2O(l) \rightarrow 2KOH(aq)$, molecular; $O^{2-}(aq) + H_2O(l) \rightarrow 2OH^-(aq)$, net ionic

base: (H^+ ion acceptor) $O^{2-}(aq)$; acid: (H^+ ion donor) $H_2O(aq)$; spectator: K^+

Oxidation-Reduction Reactions

4.35 (a) In terms of electron transfer, *oxidation* is the loss of electrons by a substance, and *reduction* is the gain of electrons (LEO says GER).

(b) Relative to oxidation numbers, when a substance is oxidized, its oxidation number increases. When a substance is reduced, its oxidation number decreases.

4.36 Oxidation and reduction can only occur together, not separately. When a metal reacts with oxygen, the metal atoms lose electrons and the oxygen atoms gain electrons. Free electrons do not exist under normal conditions. If electrons are lost by one substance they must be gained by another, and vice versa.

4.37 The most easily oxidized metals are near the bottom of groups on the left side of the chart, especially groups 1A and 2A. The least easily oxidized metals are on the lower right of the transition metals, particularly those near the bottom of groups 8B and 1B.

4.38 Platinum and gold are called the noble metals because they are especially unreactive and difficult to oxidize. The alkali and alkaline earth metals are called active because they are very easily oxidized and chemically reactive.

4.39 (a) +6 (b) +4 (c) +7 (d) +1 (e) 0 (f) -1 (O_2^{2-} is peroxide ion)

4.40 (a) +4 (b) +4 (c) +3 (d) -3 (e) +3 (f) +6

4.41 *Analyze.* Given: chemical reaction. Find: element oxidized or reduced. *Plan.* Assign oxidation numbers to all species. The element whose oxidation number becomes more positive is oxidized; the one whose oxidation number decreases is reduced. *Solve*:

(a) $Ni \rightarrow Ni^{2+}$, Ni is oxidized; $Cl_2 \rightarrow 2Cl^-$, Cl is reduced

(b) $Fe^{2+} \rightarrow Fe$, Fe is reduced; $Al \rightarrow Al^{3+}$, Al is oxidized

(c) $Cl_2 \rightarrow 2Cl$, Cl is reduced; $2I^- \rightarrow I_2$, I is oxidized

(d) $S^{2-} \rightarrow SO_4^{2-}$ (S, +6), S is oxidized; H_2O_2 (O, -1) $\rightarrow H_2O$ (O, -2); O is reduced

4.42 (a) acid-base reaction

(b) oxidation-reduction reaction; Fe is reduced, C is oxidized

(c) precipitation reaction

(d) oxidation-reduction reaction; Zn is oxidized, N is reduced

4.43 *Analyze.* Given: reactants. Find: balanced molecular and net ionic equations.

Plan. Metals oxidized by H^+ form cations. Predict products by exchanging cations and balance. The anions are the spectator ions and do not appear in the net ionic equations.

Solve:

(a) $Mn(s) + H_2SO_4(aq) \rightarrow MnSO_4(aq) + H_2(g)$; $Mn(s) + 2H^+(aq) \rightarrow Mn^{2+}(aq) + H_2(g)$

(b) $2Cr(s) + 6HBr(aq) \rightarrow 2CrBr_3(aq) + 3H_2(g)$; $2Cr(s) + 6H^+(aq) \rightarrow 2Cr^{3+}(aq) + 3H_2(g)$

(c) $Sn(s) + 2HCl(aq) \rightarrow SnCl_2(aq) + H_2(g)$; $Sn(s) + 2H^+(aq) \rightarrow Sn^{2+}(aq) + H_2(g)$

(d) $2Al(s) + 6HCHO_2(aq) \rightarrow 2Al(CHO_2)_3(aq) + 3H_2(g)$;

$2Al(s) + 6HCHO_2(aq) \rightarrow 2Al^{3+}(aq) + 6CHO_2^-(aq) + 3H_2(g)$

4.44 (a) $2HCl(aq) + Ni(s) \rightarrow NiCl_2(aq) + H_2(g)$; $Ni(s) + 2H^+(aq) \rightarrow Ni^{2+}(aq) + H_2(g)$

(b) $H_2SO_4(aq) + Fe(s) \rightarrow FeSO_4(aq) + H_2(g)$; $Fe(s) + 2H^+(aq) \rightarrow Fe^{2+}(aq) + H_2(g)$

(c) $2HBr(aq) + Mg(s) \rightarrow MgBr_2(aq) + H_2(g)$; $Mg(s) + 2H^+(aq) \rightarrow Mg^{2+}(aq) + H_2(g)$

(d) $2HC_2H_3O_2(aq) + Zn(s) \rightarrow Zn(C_2H_3O_2)_2(aq) + H_2(g)$;

$Zn(s) + 2HC_2H_3O_2(aq) \rightarrow Zn^{2+}(aq) + 2C_2H_3O_2^-(aq) + H_2(g)$

4.45 *Analyze.* Given: a metal and an aqueous solution. Find: balanced equation.

Plan. Use Table 4.5. If the metal is above the aqueous solution, reaction will occur; if the aqueous solution is higher, NR. If reaction occurs, predict products by exchanging cations (a metal ion or H^+), then balance the equation. *Solve:*

(a) $2Al(s) + 3NiCl_2(aq) \rightarrow 2AlCl_3(aq) + 3Ni(s)$

(b) $Ag(s) + Pb(NO_3)_2(aq) \rightarrow$ NR

(c) $2Cr(s) + 3NiSO_4(aq) \rightarrow Cr_2(SO_4)_3(aq) + 3Ni(s)$

(d) $Mn(s) + 2HBr(aq) \rightarrow MnBr_2(aq) + H_2(g)$

(e) $H_2(g) + CuCl_2(aq) \rightarrow Cu(s) + 2HCl(aq)$

4.46 (a) $Fe(s) + Cu(NO_3)_2(aq) \rightarrow Fe(NO_3)_2(aq) + Cu(s)$

(b) $Zn(s) + MgSO_4(aq) \rightarrow$ NR

(c) $Sn(s) + 2HBr(aq) \rightarrow SnBr_2(aq) + H_2(g)$

(d) $H_2(g) + NiCl_2(aq) \rightarrow$ NR

(e) $2Al(s) + 3CoSO_4(aq) \rightarrow Al_2(SO_4)_3(aq) + 3Co(s)$

4.47 (a) i. $Zn(s) + Cd^{2+}(aq) \rightarrow Cd(s) + Zn^{2+}(aq)$

ii. $Cd(s) + Ni^{2+}(aq) \rightarrow Ni(s) + Cd^{2+}(aq)$

(b) According to Table 4.5, the most active metals are most easily oxidized, and Zn is more active than Ni. Observation (i) indicates that Cd is less active than Zn; observation (ii) indicates that Cd is more active than Ni. Cd is between Zn and Ni on the activity series.

(c) Place an iron strip in $CdCl_2(aq)$. If Cd(s) is deposited, Cd is less active than Fe; if there is no reaction, Cd is more active than Fe. Do the same test with Co if Cd is less active than Fe or with Cr if Cd is more active than Fe.

4.48 (a) $Br_2 + 2NaI \rightarrow 2NaBr + I_2$ indicates that Br_2 is more easily reduced than I_2.

$Cl_2 + 2NaBr \rightarrow 2NaCl + Br_2$ shows that Cl_2 is more easily reduced than Br_2.

The order for ease of reduction is $Cl_2 > Br_2 > I_2$. Conversely, the order for ease of oxidation is $I^- > Br^- > Cl^-$.

(b) Since the halogens are nonmetals, they tend to form anions when they react chemically. Nonmetallic character decreases going down a family and so does the tendency to gain electrons during a chemical reaction. Thus, the ease of reduction of the halogen, X_2, decreases going down the family and the ease of oxidation of the halide, X^-, increases going down the family.

(c) $Cl_2 + 2KI \rightarrow 2KCl + I_2$; $Br_2 + LiCl \rightarrow$ no reaction

Solution Composition; Molarity

4.49 (a) Concentration is an intensive property; it is **ratio** of the amount of solute present in a certain quantity of solvent or solution. This ratio remains constant regardless of how much solution is present.

(b) The term *0.50 mol HCl* defines an amount (~18 g) of the pure substance HCl. The term 0.50 *M* HCl is a ratio; it indicates that there are 0.50 mol of HCl solute in 1.0 liter of solution. This same ratio of moles solute to solution volume is present regardless of the volume of solution under consideration.

4.50 (a) The concentration of the remaining solution is unchanged, assuming the original solution was thoroughly mixed. Molar concentration is a **ratio** of moles solute to liters solution. Although there are fewer moles solute remaining in the flask, there is also less solution volume, so the ratio of moles solute/solution volume remains the same.

(b) The second solution is 5 times as concentrated as the first. An equal volume of the more concentrated solution will contain 5 times as much solute (5 times the number of moles and also 5 times the mass) as the 0.50 *M* solution. Thus, the mass of solute in the 2.50 *M* solution is 5 × 4.5 g = 22.5 g.

Mathematically:

$$\frac{\dfrac{2.50 \text{ mol solute}}{1 \text{ L solution}}}{\dfrac{0.50 \text{ mol solute}}{1 \text{ L solution}}} = \frac{\text{x grams solute}}{4.5 \text{ g solute}}$$

$$\frac{2.50 \text{ mol solute}}{0.50 \text{ mol solute}} = \frac{\text{x g solute}}{4.5 \text{ g solute}}; \quad 5.0(4.5 \text{ g solute}) = 23 \text{ g solute}$$

The result has 2 sig figs; 22.5 rounds to 23 g solute

4.51 *Analyze/Plan.* Follow the logic in Sample Exercise 4.11. *Solve*:

(a) $M = \frac{\text{mol solute}}{\text{L solution}}; \quad \frac{0.0345 \text{ mol } NH_4Cl}{400 \text{ mL}} \times \frac{1000 \text{ mL}}{1 \text{ L}} = 0.0863\ M\ NH_4Cl$

Check. Check. $(0.035 \times 0.4) \approx 0.09\ M$

(b) $\text{mol} = M \times \text{L}; \quad \frac{2.20 \text{ mol } HNO_3}{1 \text{ L}} \times 0.0350 \text{ L} = 0.0770 \text{ mol } HNO_3$

Check. $(2 \times 0.035) \approx 0.07\ M$

(c) $\text{L} = \frac{\text{mol}}{M}; \quad \frac{0.125 \text{ mol KOH}}{1.50 \text{ mol KOH/L}} = 0.0833 \text{ L or } 83.3 \text{ mL of } 1.50\ M \text{ KOH}$

Check. (0.125/1.5) is greater than 0.06 and less than 0.12, $\approx 0.08\ M$.

4.52 (a) $M = \frac{\text{mol solute}}{\text{L solution}}; \quad \frac{0.145 \text{ mol } Na_2SO_4}{0.750 \text{ L}} = 0.193\ M\ Na_2SO_4$

(b) $\text{mol} = M \times \text{L}; \quad \frac{0.0850 \text{ mol } KMnO_4}{1 \text{ L}} \times 0.125 \text{ L} = 1.06 \times 10^{-2} \text{ mol } KMnO_4$

(c) $\text{L} = \frac{\text{mol}}{M}; \quad \frac{0.255 \text{ mol HCl}}{11.6 \text{ mol HCl/L}} = 2.20 \times 10^{-2} \text{ L or } 22.0 \text{ mL}$

4.53 Plan. Proceed as in Sample Exercise 4.11.

$M = \frac{\text{mol}}{\text{L}}; \quad \text{mol} = \frac{\text{g}}{\mathcal{M}}$ ($\mathcal{M}$ is the symbol for molar mass in this manual.)

(a) $\frac{0.150\ M\ \text{KBr}}{1\ \text{L}} \times 0.250\ \text{L} \times \frac{119.0\ \text{g KBr}}{1\ \text{mol KBr}} = 4.46\ \text{g KBr}$

Check. (0.15 × 120) ≈ 18; 18 × 0.25 = 18/4 ≈ 4.5 g KBr

(b) $4.75\ \text{g}\ Ca(NO_3)_2 \times \frac{1\ \text{mol}\ Ca(NO_3)_2}{164.1\ \text{g}\ Ca(NO_3)_2} \times \frac{1}{0.200\ \text{L}} = 0.145\ M\ Ca(NO_3)_2$

Check. (4.8/0.2) ≈ 24; 24/160 = 3/20 ≈ 0.15 M $Ca(NO_3)_2$

(c) $5.00\ \text{g}\ Na_3PO_4 \times \frac{1\ \text{mol}\ Na_3PO_4}{163.9\ \text{g}\ Na_3PO_4} \times \frac{1\ \text{L}}{1.50\ \text{mol}\ Na_3PO_4} \times \frac{1000\ \text{mL}}{1\ \text{L}}$

$= 20.3\ \text{mL solution}$

Check. [5/(160 × 1.5)] ≈ 5/240 ≈ 1/50 ≈ 0.02 L = 20 mL

4.54 $M = \frac{\text{mol}}{\text{L}}$; $\text{mol} = \frac{\text{g}}{\mathcal{M}}$ ($\mathcal{M}$ is the symbol for molar mass in this manual.)

(a) $\frac{0.850\ \text{mol}\ K_2Cr_2O_7}{1\ \text{L}} \times 50.0\ \text{mL} \times \frac{1\ \text{L}}{1000\ \text{mL}} \times \frac{294.2\ \text{g}\ K_2Cr_2O_7}{1\ \text{mol}\ K_2Cr_2O_7} = 12.5\ \text{g}\ K_2Cr_2O_7$

(b) $2.50\ \text{g}\ (NH_4)_2SO_4 \times \frac{1\ \text{mol}\ (NH_4)_2SO_4}{132.2\ \text{g}\ (NH_4)_2SO_4} \times \frac{1}{250.\ \text{mL}} \times \frac{1000\ \text{mL}}{1\ \text{L}} = 0.0756\ M\ (NH_4)_2SO_4$

(c) $1.00\ \text{g}\ CuSO_4 \times \frac{1\ \text{mol}\ CuSO_4}{159.6\ \text{g}\ CuSO_4} \times \frac{1\ \text{L}}{0.387\ \text{mol}\ CuSO_4} \times \frac{1000\ \text{mL}}{1\ \text{L}} = 16.2\ \text{mL solution}$

4.55 *Analyze.* Given: formula and concentration of each solute. Find: concentration of K^+ in each solution. *Plan.* Note mol K^+/mol solute and compare concentrations or total moles. *Solve*:

(a) $KCl \rightarrow K^+ + Cl^-$; 0.20 M KCl = 0.20 M K^+

$K_2CrO_4 \rightarrow \mathbf{2}\ K^+ + CrO_4^{2-}$; 0.15 M K_2CrO_4 = 0.30 M K^+

$K_3PO_4 \rightarrow \mathbf{3}\ K^+ + PO_4^{3-}$; 0.080 M K_3PO_4 = 0.24 M K^+

0.15 M K_2CrO_4 has the highest K^+ concentration.

(b) K_2CrO_4: 0.30 M K^+ × 0.0300 L = 0.0090 mol K^+

K_3PO_4: 0.24 M K^+ × 0.0250 L = 0.0060 mol K^+

30.0 mL of 0.15 M K_2CrO_4 has more K^+ ions.

4.56 (a) 0.10 mol NaCl/250 mL = 0.40 mol NaCl/1000 mL = 0.40 M NaCl

0.15 M KCl < 0.10 M $CaCl_2$ < 0.10 mol NaCl in 250 mL

(b) $NaCl \rightarrow Na^+ + Cl^-$; $0.35\ M \times 0.040\ L = 0.014$ mol Cl^-

$CaCl_2 \rightarrow Ca^{2+} + 2Cl^-$; $0.25\ M \times 0.025\ L = 0.00625$ mol $CaCl_2 \times 2 = 0.0125$

$= 0.013$ mol Cl^-

The NaCl solution has slightly more moles Cl^-.

4.57 *Analyze.* Given: formula and concentration of each solute. Find: concentration of each species in solution. *Plan.* Decide whether the solute is a strong, weak or nonelectrolyte, which species are in solution, and concentrations. *Solve*:

(a) $0.14\ M\ Na^+$, $0.14\ M\ OH^-$

(b) $0.25\ M\ Ca^{2+}$, $0.50\ M\ Br^-$

(c) $0.25\ M$ (CH_3OH is a molecular solute)

(d) Mixing two solutions is, in effect, a dilution, Equation 4.35.

$M_2 = M_1V_1/V_2$, where V_2 is the total solution volume.

$$K^+: \frac{0.10\ M \times 0.050\ L}{0.075\ L} = 0.00667 = 0.067\ M$$

ClO_3^-: concentration ClO_3^- = concentration $K^+ = 0.067\ M$

$$SO_4^{2-}: \frac{0.20\ M \times 0.0250\ L}{0.075\ L} = 0.0667 = 0.067\ M\ SO_4^{2-}$$

Na^+: concentration $Na^+ = 2 \times$ concentration $SO_4^{2-} = 0.13\ M$

4.58 (a)

$$H^+: \frac{0.100\ M \times 20.0\ mL + 0.500\ M \times 10.0\ mL}{30.0\ mL} = 0.233\ M\ H^+$$

Cl^-: concentration Cl^- = concentration $H^+ = 0.233\ M\ Cl^-$

(b)

$$Na^+: \frac{2(0.300\ M \times 15.0\ mL)}{25.0\ mL} = 0.360\ M;\ K^+: = \frac{0.200\ M \times 10.0\ mL}{25.0\ mL} = 0.0800\ M$$

$$SO_4^{2-}: \frac{0.300\ M \times 15.0\ mL}{25.0\ mL} = 0.180\ M;\ Cl^-: \frac{0.200\ M \times 10.0\ mL}{25.0\ mL} = 0.0800\ M$$

(c)

$$Na^+: \frac{3.50\ g\ NaCl}{0.050\ L} \times \frac{1\ mol}{58.44\ g} = 1.198\ M;\ Ca^{2+}: 0.500\ M$$

Cl^-: $1.198M$ (from NaCl(s)) + $1.000\ M$ (from $CaCl_2(aq)$) = $2.198\ M$

4.59 *Analyze/Plan.* Follow the logic of Sample Exercise 4.14. *Solve*:

(a)

$$V_1 = M_2V_2/M_1;\ \frac{0.250\ M\ NH_3 \times 100.0\ mL}{14.8\ M\ NH_3} = 1.689 = 1.69\ mL\ 14.8\ M\ NH_3$$

Check. $250/15 \approx 1.5$ mL

(b) $M_2 = M_1V_1/V_2$; $\frac{14.8\ M\ NH_3 \times 10.0\ mL}{250\ mL} = 0.592\ M\ NH_3$

Check. 150/250 ≈ 0.60 *M*

4.60 (a) $V_1 = M_2V_2/M_1$; $\frac{0.500\ M\ HNO_3 \times 500\ mL}{12.0\ M\ HNO_3} = 20.833 = 20.8\ mL$ conc. HNO_3

(b) $M_2 = M_1V_1/V_2$; $\frac{12.0\ M\ HNO_3 \times 25.0\ mL}{500\ mL} = 0.600\ M\ HNO_3$

4.61 (a) *Plan/Solve.* Follow the logic in Sample Exercise 4.13. The number of moles of sucrose needed is $\frac{0.150\ mol}{1\ L} \times 0.125\ L = 0.01875 = 0.0188\ mol$

Weigh out 0.01875 mol $C_{12}H_{22}O_{11} \times \frac{342.3\ g\ C_{12}H_{22}O_{11}}{1\ mol\ C_{12}H_{22}O_{11}} = 6.42\ g\ C_{12}H_{22}O_{11}$

Add this amount of solid to a 125 mL volumetric flask, dissolve in a small volume of water, and add water to the mark on the neck of the flask. Agitate thoroughly to ensure total mixing.

(b) *Plan/Solve.* Follow the logic in Sample Exercise 4.14. Calculate the moles of solute present in the final 400.0 mL of 0.100 *M* $C_{12}H_{22}O_{11}$ solution:

moles $C_{12}H_{22}O_{11} = M \times L = \frac{0.100\ mol\ C_{12}H_{22}O_{11}}{1\ L} \times 0.4000\ L = 0.0400\ mol\ C_{12}H_{22}O_{11}$

Calculate the volume of 1.50 M glucose solution that would contain 0.04000 mol $C_{12}H_{22}O_{11}$:

L = moles/*M*; 0.04000 mol $C_{12}H_{22}O_{11} \times \frac{1\ L}{1.50\ mol\ C_{12}H_{22}O_{11}} = 0.02667 = 0.0267\ L$

$0.02667\ L \times \frac{1000\ mL}{1\ L} = 26.7\ mL$

Thoroughly rinse, clean and fill a 50 mL buret with the 1.50 *M* $C_{12}H_{22}O_{11}$. Dispense 26.7 mL of this solution into a 400 mL volumetric container, add water to the mark and mix thoroughly. (26.7 mL is a difficult volume to measure with a pipette.)

4.62 (a) The amount of $AgNO_3$ needed is: $0.200\ M \times 0.1000\ L = 0.0200\ mol\ AgNO_3$

$0.0200\ mol\ AgNO_3 \times \frac{169.88\ g\ AgNO_3}{1\ mol\ AgNO_3} = 3.3976 = 3.40\ g\ AgNO_3$

Add this amount of solid to a 100 mL volumetric flask, dissolve in a small amount of water, bring the total volume to exactly 100 mL and agitate well.

(b) Dilute the 6.0 *M* HNO_3 to prepare 250 mL of 1.0 *M* HNO_3. To determine the volume of 6.0 *M* HNO_3 needed, calculate the moles HNO_3 present in 250 mL of 1.0 *M* HNO_3 and then the volume of 6.0 *M* solution that contains this number of moles.

0.250 L × 1.0 *M* = 0.250 mol HNO_3 needed;

$$L = \frac{mol}{M};\ L\ 6.0\ M\ HNO_3 = \frac{0.250\ mol\ needed}{6.0\ M} = 0.0417\ L = 42\ mL$$

Thoroughly clean, rinse and fill a 50 mL buret with the 6.0 *M* HNO_3, taking precautions appropriate for working with a relatively concentrated acid. Dispense 42 mL of the 6.0 *M* acid into a 250 mL volumetric flask, add water to the mark and mix thoroughly.

4.63 *Analyze.* Given: density of pure acetic acid, volume pure acetic acid, volume new solution. Find: molarity of new solution. *Plan.* Calculate the mass of acetic acid, $HC_2H_3O_2$, present in 20.0 mL of the pure liquid. *Solve*:

$$20.00\ mL\ acetic\ acid \times \frac{1.049\ g\ acetic\ acid}{1\ mL\ acetic\ acid} = 20.98\ g\ acetic\ acid$$

$$20.98\ g\ HC_2H_3O_2 \times \frac{1\ mol\ HC_2H_3O_2}{60.05\ g\ HC_2H_3O_2} = 0.349375 = 0.3494\ mol\ HC_2H_3O_2$$

$$M = mol/L = \frac{0.349375\ mol\ HC_2H_3O_2}{0.2500\ L\ solution} = 1.39750 = 1.398\ M\ HC_2H_3O_2$$

Check. (20 × 1) ≈ 20 g acid; (20/60) ≈ 0.33 mol acid; (0.33/0.25 = 0.33 × 4) ≈ 1.33 *M*

4.64 $$50.000\ mL\ glycerol \times \frac{1.2656\ g\ glycerol}{1\ mL\ glycerol} = 63.280\ g\ glycerol$$

$$63.280\ g\ C_3H_8O_3 \times \frac{1\ mol\ C_3H_8O_3}{92.094\ g\ C_3H_8O_3} = 0.687124 = 0.68712\ mol\ C_3H_8O_3$$

$$M = \frac{0.687124\ mol\ C_3H_8O_3}{0.25000\ L\ solution} = 2.7485\ M\ C_3H_8O_3$$

Solution Stoichiometry; Titrations

4.65 *Analyze.* Given: volume and molarity $AgNO_3$. Find: mass NaCl.
Plan. *M* × L = mol $AgNO_3$ = mol Ag^+; balanced equation gives ratio mol NaCl/mol $AgNO_3$; mol NaCl → g NaCl. *Solve*:

$$\frac{0.100\ mol\ AgNO_3}{1\ L} \times 0.0200\ L = 2.00 \times 10^{-3}\ mol\ AgNO_3(aq)$$

$AgNO_3(aq) + NaCl(aq) \rightarrow AgCl(s) + NaNO_3(aq)$

mol NaCl = mol $AgNO_3$ = 2.00×10^{-3} mol NaCl

$$2.00 \times 10^{-3}\ mol\ NaCl \times \frac{58.44\ g\ NaCl}{1\ mol\ NaCl} = 0.117\ g\ NaCl$$

Check. (0.1 × 0.02) = 0.002 mol; (0.002 × 60) ≈ 0.12 g NaCl

4.66 *Plan.* $M \times L$ = mol $Fe(NO_3)_2$; balanced equation → mol ratio → mol NaOH → g NaOH

Solve. $\frac{0.500 \text{ mol } Fe(NO_3)_2}{1 \text{ L}} \times 0.02500 \text{ L} = 0.0125 \text{ mol } Fe(NO_3)_2$

$Fe(NO_3)_2(aq) + 2NaOH(aq) \rightarrow Fe(OH)_2(s) + 2NaNO_3(aq)$

$0.0125 \text{ mol } Fe(NO_3)_2 \times \frac{2 \text{ mol NaOH}}{1 \text{ mol } Fe(NO_3)_2} \times \frac{40.00 \text{ g NaOH}}{1 \text{ mol NaOH}} = 1.0 \text{ g NaOH}$

4.67 (a) *Analyze.* Given: *M* and vol base, *M* acid. Find: vol acid

Plan/Solve. Write the balanced equation for the reaction in question:
$HClO_4(aq) + NaOH(aq) \rightarrow NaClO_4(aq) + H_2O(l)$

Calculate the moles of the known substance, in this case NaOH.

moles NaOH = $M \times L = \frac{0.0875 \text{ mol NaOH}}{1 \text{ L}} \times 0.0500 \text{ L} = 0.004375$
$= 0.00438$ mol NaOH

Apply the mole ratio (mol unknown/mol known) from the chemical equation.

$0.004375 \text{ mol NaOH} \times \frac{1 \text{ mol } HClO_4}{1 \text{ mol NaOH}} = 0.004375 \text{ mol } HClO_4$

Calculate the desired quantity of unknown, in this case the volume of 0.115 *M* $HClO_4$ solution.

L = mol/*M*; $L = 0.004375 \text{ mol } HClO_4 \times \frac{1 \text{ L}}{0.115 \text{ mol } HClO_4} = 0.0380 \text{ L} = 38.0 \text{ mL}$

Check. $(0.09 \times 0.045) = 0.0045$ mol; $(0.0045/0.11) \approx 0.040 \text{ L} \approx 40$ mL

(b) Following the logic outlined in part (a):

$2HCl(aq) + Mg(OH)_2(s) \rightarrow MgCl_2(aq) + 2H_2O(l)$

$2.87 \text{ g } Mg(OH)_2 \times \frac{1 \text{ mol } Mg(OH)_2}{58.32 \text{ g } Mg(OH)_2} = 0.049211 = 0.0492 \text{ mol } Mg(OH)_2$

$0.0492 \text{ mol } Mg(OH)_2 \times \frac{2 \text{ mol HCl}}{1 \text{ mol } Mg(OH)_2} = 0.0984 \text{ mol HCl}$

$L = \text{mol}/M = 0.09840 \text{ mol HCl} \times \frac{1 \text{ L HCl}}{0.128 \text{ mol HCl}} = 0.769 \text{ L} = 769 \text{ mL}$

(c) $AgNO_3(aq) + KCl(aq) \rightarrow AgCl(s) + KNO_3(aq)$

$785 \text{ mg KCl} \times \frac{1 \times 10^{-3} \text{ g}}{1 \text{ mg}} \times \frac{1 \text{ mol KCl}}{74.55 \text{ g KCl}} \times \frac{1 \text{ mol } AgNO_3}{1 \text{ mol KCl}} = 0.01053$
$= 0.0105 \text{ mol } AgNO_3$

$M = \text{mol/L} = \frac{0.01053 \text{ mol } AgNO_3}{0.0258 \text{ L}} = 0.408 \, M \, AgNO_3$

(d) $HCl(aq) + KOH(aq) \rightarrow KCl(aq) + H_2O(l)$

$$\frac{0.108 \text{ mol HCl}}{1 \text{ L}} \times 0.0453 \text{ L} \times \frac{1 \text{ mol KOH}}{1 \text{ mol HCl}} \times \frac{56.11 \text{ g KOH}}{1 \text{ mol KOH}} = 0.275 \text{ g KOH}$$

4.68 (a) $2HCl(aq) + Ba(OH)_2(aq) \rightarrow BaCl_2(aq) + 2H_2O(l)$

$$\frac{0.101 \text{ mol Ba(OH)}_2}{1 \text{ L Ba(OH)}_2} \times 0.0500 \text{ L Ba(OH)}_2 \times \frac{2 \text{ mol HCl}}{1 \text{ mol Ba(OH)}_2} \times \frac{1 \text{ L HCl}}{0.120 \text{ mol HCl}} = 0.0842 \text{ L or } 84.2 \text{ mL HCl soln}$$

(b) $H_2SO_4(aq) + 2NaOH(aq) \rightarrow Na_2SO_4(aq) + 2H_2O(l)$

$$0.200 \text{ g NaOH} \times \frac{1 \text{ mol NaOH}}{40.00 \text{ g NaOH}} \times \frac{1 \text{ mol H}_2\text{SO}_4}{2 \text{ mol NaOH}} \times \frac{1 \text{ L H}_2\text{SO}_4}{0.125 \text{ mol H}_2\text{SO}_4} = 0.0200 \text{ L or } 20.0 \text{ mL H}_2\text{SO}_4 \text{ soln}$$

(c) $BaCl_2(aq) + Na_2SO_4(aq) \rightarrow BaSO_4(s) + 2NaCl(aq)$

$$752 \text{ mg} = 0.752 \text{ g Na}_2\text{SO}_4 \times \frac{1 \text{ mol Na}_2\text{SO}_4}{142.1 \text{ g Na}_2\text{SO}_4} \times \frac{1 \text{ mol BaCl}_2}{1 \text{ mol Na}_2\text{SO}_4} \times \frac{1}{0.0558 \text{ L}} = 0.0948 \ M \text{ BaCl}_2$$

(d) $2HCl(aq) + Ca(OH)_2(aq) \rightarrow CaCl_2(aq) + 2H_2O(l)$

$$0.0427 \text{ L HCl} \times \frac{0.208 \text{ mol HCl}}{1 \text{ L HCl}} \times \frac{1 \text{ mol Ca(OH)}_2}{2 \text{ mol HCl}} \times \frac{74.10 \text{ g Ca(OH)}_2}{1 \text{ mol Ca(OH)}_2} = 0.329 \text{ g Ca(OH)}_2$$

4.69 *Analyze/Plan.* See Exercise 4.67(a) for a more detailed approach. *Solve*:

$$\frac{6.0 \text{ mol H}_2\text{SO}_4}{1 \text{ L}} \times 0.027 \text{ L} \times \frac{2 \text{ mol NaHCO}_3}{1 \text{ mol H}_2\text{SO}_4} \times \frac{84.01 \text{ g NaHCO}_3}{1 \text{ mol NaHCO}_3} = 27 \text{ g NaHCO}_3$$

4.70 See Exercise 4.67 (a) for a more detailed approach.

$$\frac{0.102 \text{ mol NaOH}}{1 \text{ L}} \times 0.0355 \text{ L} \times \frac{1 \text{ mol HC}_2\text{H}_3\text{O}_2}{1 \text{ mol NaOH}} \times \frac{60.05 \text{ g HC}_2\text{H}_3\text{O}_2}{1 \text{ mol HC}_2\text{H}_3\text{O}_2} = 0.21744 = 0.217 \text{ g HC}_2\text{H}_3\text{O}_2 \text{ in } 2.50 \text{ mL}$$

$$1.00 \text{ qt vinegar} \times \frac{1 \text{ L}}{1.057 \text{ qt}} \times \frac{1000 \text{ mL}}{1 \text{ L}} \times \frac{0.21744 \text{ g HC}_2\text{H}_3\text{O}_2}{2.50 \text{ mL vinegar}} = 82.3 \text{ g HC}_2\text{H}_3\text{O}_2\text{/qt}$$

4.71 *Analyze.* Given: *M* and vol HBr, vol $Ca(OH)_2$. Find: *M* $Ca(OH)_2$, g $Ca(OH)_2$/100 mL soln

Plan. Write balanced equation;

mol HBr $\xrightarrow[\text{ratio}]{\text{mol}}$ mol $Ca(OH)_2$ $\rightarrow$ *M* $Ca(OH)_2$; $\rightarrow$ g $Ca(OH)_2$/100 mL

Solve. The neutralization reaction here is:

$2HBr(aq) + Ca(OH)_2(aq) \rightarrow CaBr_2(aq) + 2H_2O(l)$

$$0.0488 \text{ L HBr soln} \times \frac{5.00 \times 10^{-2} \text{ mol HBr}}{1 \text{ L soln}} \times \frac{1 \text{ mol Ca(OH)}_2}{2 \text{ mol HBr}} \times \frac{1}{0.100 \text{ L of Ca(OH)}_2}$$

$$= 1.220 \times 10^{-2} = 1.22 \times 10^{-2}\ M \text{ Ca(OH)}_2$$

From the molarity of the saturated solution, we can calculate the gram solubility of $Ca(OH)_2$ in 100 mL of H_2O.

$$0.100 \text{ L soln} \times \frac{1.220 \times 10^{-2} \text{ mol Ca(OH)}_2}{1 \text{ L soln}} \times \frac{74.10 \text{ g Ca(OH)}_2}{1 \text{ mol Ca(OH)}_2}$$

$$= 0.0904 \text{ g Ca(OH)}_2 \text{ in 100 mL soln}$$

Check. (0.05 × 0.05/0.2) = 0.0125 M; (0.1 × 0.0125 × 64) ≈ 0.085 g/100 mL

4.72 The balanced equation for the titration is:

$Sr(NO_3)_2(aq) + Na_2CrO_4(aq) \rightarrow SrCrO_4(s) + 2NaNO_3(aq)$

Beginning with a 0.100 L sample, we can do the following conversions:

volume soln → g $Sr(NO_3)_2$ → mol $Sr(NO_3)_2$ → mol Na_2CrO_4 → vol Na_2CrO_4 soln

$$0.100 \text{ L soln} \times \frac{7.52 \text{ g Sr(NO}_3)_2}{0.750 \text{ L soln}} \times \frac{1 \text{ mol Sr(NO}_3)_2}{211.6 \text{ g Sr(NO}_3)_2} \times \frac{1 \text{ mol Na}_2\text{CrO}_4}{1 \text{ mol Sr(NO}_3)_2}$$

$$\times \frac{1 \text{ L soln}}{0.0425 \text{ mol Na}_2\text{CrO}_4} = 0.111 \text{ L Na}_2\text{CrO}_4 \text{ soln}$$

4.73 (a) $NiSO_4(aq) + 2KOH(aq) \rightarrow Ni(OH)_2(s) + K_2SO_4(aq)$

(b) The precipitate is $Ni(OH)_2$.

(c) *Plan.* Compare mol of each reactant; mol = M × L

Solve. 0.200 M KOH × 0.1000 L KOH = 0.0200 mol KOH
0.150 M $NiSO_4$ × 0.2000 L KOH = 0.0300 mol $NiSO_4$

1 mol $NiSO_4$ requires 2 mol KOH, so 0.0300 mol $NiSO_4$ requires 0.0600 mol KOH. Since only 0.0200 mol KOH is available, KOH is the limiting reactant.

(d) *Plan.* The amount of the limiting reactant (KOH) determines amount of product, in this case $Ni(OH)_2$.

$$\textit{Solve.}\ 0.0200 \text{ mol KOH} \times \frac{1 \text{ mol Ni(OH)}_2}{2 \text{ mol KOH}} \times \frac{92.71 \text{ g Ni(OH)}_2}{1 \text{ mol Ni(OH)}_2} = 0.927 \text{ g Ni(OH)}_2$$

(e) *Plan/Solve.* Limiting reactant: OH^-: no excess OH^- remains in solution.

Excess reactant: Ni^{2+}: M Ni^{2+} remaining = mol Ni^{2+} remaining/L solution

0.0300 mol Ni^{2+} initial - 0.0100 mol Ni^{2+} reacted = 0.0200 mol Ni^{2+} remaining

0.0200 mol Ni^{2+}/0.3000 L = 0.0667 M Ni^{2+}(aq)

Spectators: SO_4^{2-}, K^+. These ions do not react, so the only change in their concentration is dilution. The final volume of the solution is 0.3000 L.

$M_2 = M_1V_1/V_2$: 0.200 M K^+ × 0.1000 L / 0.3000 L = 0.0667 M K^+(aq)

0.150 M SO_4^{2-} × 0.2000 L/0.3000 L = 0.100 M SO_4^{2-} (aq)

4.74 (a) HNO_3(aq) + NaOH(s) → $NaNO_3$(aq) + H_2O(l)

(b) Determine the limiting reactant, then the identity and concentration of ions remaining in solution. Assume that the H_2O(l) produced by the reaction does **not** increase the total solution volume.

$$12.0 \text{ g NaOH} \times \frac{1 \text{ mol NaOH}}{40.00 \text{ g NaOH}} = 0.300 \text{ mol NaOH}$$

0.200 M HNO_3 × 0.0750 L HNO_3 = 0.0150 mol HNO_3.

The mol ratio is 1:1, so HNO_3 is the limiting reactant. No excess H^+ remains in solution. The remaining ions are: OH^- (excess reactant), Na^+ and NO_3^- (spectators).

OH^-: 0.300 mol OH^- initial - 0.0150 mol OH^- react = 0.285 mol OH^- remain

0.285 mol OH^- / 0.0750 L soln = 3.80 M OH^-(aq)

Na^+: 0.300 mol Na^+ / 0.0750 L soln = 4.00 M Na^+(aq)

NO_3^-: 0.0150 mol NO_3^- / 0.0750 L = 0.200 M NO_3^- aq)

(c) The resulting solution is **basic** because of the large excess of OH^-(aq).

4.75 *Analyze*. Given: mass impure $Mg(OH)_2$; M and vol **excess** HCl; M and vol NaOH. Find: mass % $Mg(OH)_2$ in sample. *Plan/Solve*. Write balanced equations.

$Mg(OH)_2$(s) + 2HCl(aq) → $MgCl_2$(aq) + $2H_2O$(l)

HCl(aq) + NaOH(aq) → NaCl(aq) + $2H_2O$(l)

Calculate total moles HCl = M HCl × L HCl

$$\frac{0.2050 \text{ mol HCl}}{1 \text{ L soln}} \times 0.1000 \text{ L} = 0.02050 \text{ mol HCl total}$$

mol excess HCl = mol NaOH used = M NaOH × L NaOH

$$\frac{0.1020 \text{ mol NaOH}}{1 \text{ L soln}} \times 0.01985 \text{ L} = 0.0020247 = 0.002025 \text{ mol NaOH}$$

mol HCl reacted with $Mg(OH)_2$ = total mol HCl - excess mol HCl

0.02050 mol total - 0.0020247 mol excess = 0.0184753 = 0.01848 mol HCl reacted

(The result has 5 decimal places and 4 sig. figs.)

Use mol ratio to get mol $Mg(OH)_2$ in sample, then molar mass of $Mg(OH)_2$ to get g pure $Mg(OH)_2$.

$$0.0184753 \text{ mol HCl} \times \frac{1 \text{ mol Mg(OH)}_2}{2 \text{ mol HCl}} \times \frac{58.32 \text{ g Mg(OH)}_2}{1 \text{ mol Mg(OH)}_2} = 0.5387 \text{ Mg(OH)}_2$$

$$\text{mass \% Mg(OH)}_2 = \frac{\text{g Mg(OH)}_2}{\text{g sample}} \times 100 = \frac{0.5388 \text{ g Mg(OH)}_2}{0.5895 \text{ g sample}} \times 100 = 91.40\% \text{ Mg(OH)}_2$$

4.76 *Plan.* $CaCO_3(s) + 2HCl(aq) \rightarrow CaCl_2(aq) + H_2O(l) + CO_2(g)$

$HCl(aq) + NaOH(aq) \rightarrow NaCl(aq) + H_2O(l)$

total mol HCl - excess mol HCl = mol HCl reacted; mol $CaCO_3$ = (mol HCl)/2;

g $CaCO_3$ = mol $CaCO_3$ × molar mass; mass % = (g $CaCO_3$/g sample) × 100

Solve:

$$\frac{1.035 \text{ mol HCl}}{1 \text{ L soln}} \times 0.02500 \text{ L} = 0.025875 = 0.02588 \text{ mol HCl total}$$

$$\frac{0.1010 \text{ mol NaOH}}{1 \text{ L soln}} \times 0.01525 \text{ L} = 0.00154025 = 0.001540 \text{ mol HCl excess}$$

0.025875 total - 0.001540 excess = 0.02433475 = 0.02433 mol HCl reacted

$$0.243745 \text{ mol HCl} \times \frac{1 \text{ mol CaCO}_3}{2 \text{ mol HCl}} \times \frac{100.09 \text{ g CaCO}_3}{1 \text{ mol CaCO}_3} = 1.21783 = 1.218 \text{ g CaCO}_3$$

$$\text{mass \% CaCO}_3 = \frac{\text{g CaCO}_3}{\text{g rock}} \times 100 = \frac{1.21783}{1.452} \times 100 = 83.87\%$$

Additional Exercises

4.77 The precipitate is CdS(s). $Na^+(aq)$ and $NO_3^-(aq)$ are spectator ions and remain in solution. Any excess reactant ions also remain in solution. The net ionic equation is:

$Cd^{2+}(aq) + S^{2-}(aq) \rightarrow CdS(s)$.

4.78 The two precipitates formed are due to AgCl(s) and $SrSO_4(s)$. Since no precipitate forms on addition of hydroxide ion to the remaining solution, the other two possibilities, Ni^{2+} and Mn^{2+}, are absent.

4.79 (a,b)
Expt. 1 No reaction
Expt. 2 $2Ag^+(aq) + CrO_4^{2-}(aq) \rightarrow Ag_2CrO_4(s)$ red precipitate
Expt. 3 No reaction
Expt. 4 $2Ag^+(aq) + C_2O_4^{2-}(aq) \rightarrow Ag_2C_2O_4(s)$ white precipitate
Expt. 5 $Ca^{2+}(aq) + C_2O_4^{2-}(aq) \rightarrow CaC_2O_4(s)$ white precipitate
Expt. 6 $Ag^+(aq) + Cl^-(aq) \rightarrow AgCl(s)$ white precipitate

(c) The silver salts of both ions are insoluble, but many silver salts are insoluble (Expt. 6). The calcium salt of CrO_4^{2-} is soluble (Expt. 3), while the calcium salt of $C_2O_4^{2-}(aq)$ is insoluble (Expt. 5). Thus, chromate salts appear more soluble than oxalate salts.

4.80 (a) $Al(OH)_3(s) + 3H^+(aq) \rightarrow Al^{3+}(aq) + 3H_2O(l)$

(b) $Mg(OH)_2(s) + 2H^+(aq) \rightarrow Mg^{2+}(aq) + 2H_2O(l)$

(c) $MgCO_3(s) + 2H^+(aq) \rightarrow Mg^{2+}(aq) + H_2O(l) + CO_2(g)$

(d) $NaAl(CO_3)(OH)_2(s) + 4H^+(aq) \rightarrow Na^+(aq) + Al^{3+}(aq) + 3H_2O(l) + CO_2(g)$

(e) $CaCO_3(s) + 2H^+(aq) \rightarrow Ca^{2+}(aq) + H_2O(l) + CO_2(g)$

[In (c), (d) and (e), one could also write the equation for formation of bicarbonate, e.g., $MgCO_3(s) + H^+(aq) \rightarrow Mg^{2+} + HCO_3^-(aq)$.]

4.81 (a) $2H^+(aq) + SO_3^{2-}(aq) \rightarrow H_2\mathbf{SO_3}(aq)$; sulfurous acid

(b) $H_2SO_3(aq) \rightarrow H_2O(l) + \mathbf{SO_2}(g)$; sulfur dioxide

(c) The boiling point of $SO_2(g)$ is -10°C. It is a gas at room temperature (23°C) and pressure (1 atm).

(d) (i) $Na_2SO_3(aq) + 2HCl(aq) \rightarrow 2NaCl(aq) + H_2O(l) + SO_2(g)$

$SO_3^{2-}(aq) + 2H^+(aq) \rightarrow H_2O(l) + SO_2(g)$

(ii) $Ag_2SO_3(s) + 2HCl(aq) \rightarrow 2AgCl(s) + H_2O(l) + SO_2(g)$

$Ag_2SO_3(s) + 2H^+(aq) + 2Cl^-(aq) \rightarrow 2AgCl(s) + H_2O(l) + SO_2(g)$

(iii) $KHSO_3(s) + HCl(aq) \rightarrow KCl(aq) + H_2O(l) + SO_2(g)$

$KHSO_3(s) + H^+(aq) \rightarrow K^+(aq) + H_2O(l) + SO_2(g)$

(iv) $ZnSO_3(aq) + 2HCl(aq) \rightarrow ZnCl_2(aq) + H_2O(l) + SO_2(g)$

$SO_3^{2-}(aq) + 2H^+(aq) \rightarrow H_2O(l) + SO_2(g)$

4.82

$$\overset{-3}{4NH_3(g)} + \overset{0}{5O_2(g)} \longrightarrow \overset{+2}{4NO(g)} + \overset{-2}{6H_2O(g)}.$$

(a) redox reaction
(b) N is oxidized, O is reduced

$$\overset{+2}{2NO(g)} + \overset{0}{O_2(g)} \longrightarrow \overset{+4\ -2}{2NO_2(g)}.$$

(a) redox reaction
(b) N is oxidized, O is reduced

$$\overset{+4}{3NO_2(g)} + H_2O(l) \longrightarrow \overset{+5}{HNO_3(aq)} + \overset{+2}{NO(g)}.$$

(a) redox reaction

(b) N is oxidized ($NO_2 \rightarrow HNO_3$), N is reduced ($NO_2 \rightarrow NO$). A reaction where the same element is both oxidized and reduced is called disproportionation.

4.83 A metal on Table 4.5 is able to displace the metal cations below it from their compounds. That is, zinc will reduce the cations below it to their metals.

(a) $Zn(s) + Na^+(aq) \rightarrow$ no reaction

(b) $Zn(s) + Pb^{2+}(aq) \rightarrow Zn^{2+}(aq) + Pb(s)$

(c) $Zn(s) + Mg^{2+}(aq) \rightarrow$ no reaction

(d) $Zn(s) + Fe^{2+}(aq) \rightarrow Zn^{2+}(aq) + Fe(s)$

(e) $Zn(s) + Cu^{2+}(aq) \rightarrow Zn^{2+}(aq) + Cu(s)$

(f) $Zn(s) + Al^{3+}(aq) \rightarrow$ no reaction

4.84 (a) $2Ti^{4+}(aq) + Zn(s) \rightarrow 2Ti^{3+}(aq) + Zn^{2+}(aq)$

The coefficients are required because the number of electrons lost by Zn ($2e^-$) must equal the number of electrons gained by Ti^{+4} ($2 \times 1e^-$).

(b) This reaction does not indicate the position of Ti on the activity series in Table 4.5. All reactions in Table 4.5 involve oxidation of an elemental metal to a metal ion or compound. This reaction does not produce Ti metal and thus cannot be compared to the other reactions.

4.85 (a)

A : La_2O_3 Metals often react with the oxygen in air to produce metal oxides.

B : $La(OH)_3$ When metals react with water (HOH) to form H_2, OH^- remains.

C : $LaCl_3$ Most chlorides are soluble.

D : $La_2(SO_4)_3$ Sulfuric acid provides SO_4^{2-} ions.

(b) $4La(s) + 3O_2(g) \rightarrow 2La_2O_3(s)$

$2La(s) + 6HOH(l) \rightarrow 2La(OH)_3(s) + 3H_2(g)$

(There are no spectator ions in either of these reactions.)

molecular: $La_2O_3(s) + 6HCl(aq) \rightarrow 2LaCl_3(aq) + 3H_2O(l)$

net ionic: $La_2O_3(s) + 6H^+(aq) \rightarrow 2La^{3+}(aq) + 3H_2O(l)$

molecular: $La(OH)_3(s) + 3HCl(aq) \rightarrow LaCl_3(aq) + 3H_2O(l)$

net ionic: $La(OH)_3(s) + 3H^+(aq) \rightarrow La^{3+}(aq) + 3H_2O(l)$

molecular: $2LaCl_3(aq) + 3H_2SO_4(aq) \rightarrow La_2(SO_4)_3(s) + 6HCl(aq)$

net ionic: $2La^{3+}(aq) + 3SO_4^{2-}(aq) \rightarrow La_2(SO_4)_3(s)$

(c) La metal is oxidized by water to produce $H_2(g)$, so La is definitely above H on the activity series. In fact, since an acid is not required to oxidize La, it is probably one of the more active metals.

4.86 *Plan.* Calculate moles KBr from the two quantities of solution (mol = *M* × L), then new molarity (*M* = mol/L). KBr is nonvolatile, so no solute is lost when the solution is evaporated to reduce the total volume. *Solve*:

1.00 *M* KBr × 0.0250 L = 0.0250 mol KBr; 0.800 *M* KBr × 0.0750 L = 0.0600 mol KBr

0.0250 mol KBr + 0.0600 mol KBr = 0850 mol KBr total

$$\frac{0.0850 \text{ mol KBr}}{0.0500 \text{ L soln}} = 1.70\ M \text{ KBr}$$

4.87 (a)

$$0.0500 \text{ L soln} \times \frac{0.200 \text{ mol NaCl}}{1 \text{ L soln}} = 1.00 \times 10^{-2} \text{ mol NaCl}$$

$$0.0750 \text{ L soln} \times \frac{0.100 \text{ mol NaCl}}{1 \text{ L soln}} = 7.50 \times 10^{-3} \text{ mol NaCl}$$

Total moles NaCl = 1.75×10^{-2}, total volume = 0.0500 L + 0.0750 L = 0.1250 L

$$\text{Molarity} = \frac{1.75 \times 10^{-2} \text{ mol}}{0.1250 \text{ L}} = 0.140\ M$$

(b) $0.025 \text{ L soln} \times \dfrac{1.50 \text{ mol NaOH}}{1 \text{ L soln}} = 0.03675 = 0.0368 \text{ mol NaOH}$

$0.0255 \text{ L soln} \times \dfrac{0.750 \text{ mol NaOH}}{1 \text{ L soln}} = 0.019125 = 0.0191 \text{ mol NaOH}$

Total moles NaOH = 0.055875 = 0.0559, total volume = 0.0500 L

$\text{Molarity} = \dfrac{0.055875 \text{ mol NaOH}}{0.0500 \text{ L}} = 1.12\ M$

4.88 (a) $\dfrac{50 \text{ pg}}{1 \text{ mL}} \times \dfrac{1 \times 10^{-12} \text{ g}}{1 \text{ pg}} \times \dfrac{1 \times 10^{3} \text{ mL}}{\text{L}} \times \dfrac{1 \text{ mol Na}}{23.0 \text{ g Na}} = 2.17 \times 10^{-9} = 2.2 \times 10^{-9}\ M$

(b) $\dfrac{2.17 \times 10^{-9} \text{ mol Na}}{1 \text{ L soln}} \times \dfrac{1 \text{ L}}{1 \times 10^{3} \text{ cm}^3} \times \dfrac{6.02 \times 10^{23} \text{ Na atom}}{1 \text{ mol Na}}$

$= 1.3 \times 10^{12}$ atoms or Na^+ ions/cm^3

4.89 Na^+ must replace the total positive (+) charge due to Ca^{2+} and Mg^{2+}. Think of this as moles of charge rather than moles of particles.

$\dfrac{0.010 \text{ mol Ca}^{2+}}{1 \text{ L water}} \times 1.0 \times 10^{3} \text{ L} \times \dfrac{2 \text{ mol + charge}}{1 \text{ mol Ca}^{2+}} = 20 \text{ mol of + charge}$

$\dfrac{0.0050 \text{ mol Mg}^{2+}}{1 \text{ L water}} \times 1.0 \times 10^{3} \text{ L} \times \dfrac{2 \text{ mol + charge}}{1 \text{ mol Mg}^{2+}} = 10 \text{ mol of + charge}$

30 moles of + charge must be replaced; 30 mol Na^+ are needed.

4.90 $H_2C_4H_4O_6 + 2OH^-(aq) \rightarrow C_4H_4O_6^{2-}(aq) + 2H_2O(l)$

$0.02262 \text{ L NaOH soln} \times \dfrac{0.2000 \text{ mol NaOH}}{1 \text{ L}} \times \dfrac{1 \text{ mol } H_2C_4H_4O_6}{2 \text{ mol NaOH}} \times \dfrac{1}{0.04000 \text{ L } H_2C_4H_4O_6}$

$= 0.05655\ M\ H_2C_4H_4O_6$ soln

4.91 *Plan.* mol MnO_4^- = M × L → mol ratio → mol H_2O_2 → M H_2O_2. *Solve*:

$2MnO_4^-(aq) + 5H_2O_2(aq) + 6H^+ \rightarrow 2Mn^{2+}(aq) + 5O_2(aq) + 8H_2O(l)$

$\dfrac{0.109 \text{ mol } MnO_4^-}{\text{L}} \times 0.0135 \text{ L } MnO_4^- \times \dfrac{5 \text{ mol } H_2O_2}{2 \text{ mol } MnO_4^-} \times \dfrac{1}{0.0100 \text{ L } H_2O_2}$

$= 0.3679 \text{ mol } H_2O_2 / \text{L} = 0.368\ M\ H_2O_2$

4.92 mol OH^- from NaOH(aq) + mol OH^- from $Zn(OH)_2$(s) = mol H^+ from HBr

mol H^+ = M HBr × L HBr = 0.500 M HBr × 0.400 L HBr = 0.200 mol H^+

mol OH^- from NaOH = M NaOH × L NaOH = 0.500 M NaOH × 0.0985 L NaOH

= 0.04925 = 0.0493 mol OH^-

mol OH^- from $Zn(OH)_2$(s) = 0.200 mol H^+ - 0.04925 mol OH^- from NaOH = 0.15075

= 0.151 mol OH^- from $Zn(OH)_2$

$0.15075 \text{ mol } OH^- \times \dfrac{1 \text{ mol } Zn(OH)_2}{2 \text{ mol } OH^-} \times \dfrac{99.41 \text{ g } Zn(OH)_2}{1 \text{ mol } Zn(OH)_2} = 7.49 \text{ g } Zn(OH)_2$

Integrative Exercises

4.93 *Plan.* $M \times L$ = mol Na_3PO_4 → mol Na^+ → Na^+ ions. *Solve*:

$$\frac{0.0100 \text{ mol } Na_3PO_4}{1 \text{ L solution}} \times 1.00 \text{ mL} \times \frac{1 \text{ L}}{1000 \text{ mL}} \times \frac{3 \text{ mol } Na^+}{1 \text{ mol } Na_3PO_4} \times \frac{6.022 \times 10^{23} \, Na^+ \text{ ions}}{1 \text{ mol } Na^+} = 1.81 \times 10^{19} \, Na^+ \text{ ions}$$

4.94 (a) At the equivalence point of a titration, mol NaOH added = mol H^+ present

$$M_{NaOH} \times L_{NaOH} = \frac{\text{g acid}}{\mathcal{M} \text{ acid}} \text{ (for an acid with 1 acidic hydrogen)}$$

$$\mathcal{M} \text{ acid} = \frac{\text{g acid}}{M_{NaOH} \times L_{NaOH}} = \frac{0.2053 \text{ g}}{0.1008 \, M \times 0.0150 \text{ L}} = 136 \text{ g/mol}$$

(b) Assume 100 g of acid.

$$70.6 \text{ g C} \times \frac{1 \text{ mol C}}{12.01 \text{ g C}} = 5.88 \text{ mol C}; \; 5.88 / 1.47 \approx 4$$

$$5.89 \text{ g H} \times \frac{1 \text{ mol H}}{1.008 \text{ g H}} = 5.84 \text{ mol H}; \; 5.84 / 1.47 \approx 4$$

$$23.5 \text{ g O} \times \frac{1 \text{ mol O}}{16.00 \text{ g O}} = 1.47 \text{ mol O}; \; 1.47 / 1.47 = 1$$

The empirical formula is C_4H_4O.

$$\frac{\mathcal{M}}{\text{FW}} = \frac{136}{68.1} = 2; \text{ the molecular formula is } 2 \times \text{ the empirical formula.}$$

The molecular formula is $C_8H_8O_2$.

4.95 $Ba^{2+}(aq) + SO_4^{2-}(aq) \rightarrow BaSO_4(s)$

$$0.4123 \text{ g } BaSO_4 \times \frac{137.3 \text{ g Ba}}{233.4 \text{ g } BaSO_4} = 0.2425 \text{ g Ba}$$

$$\text{mass \%} = \frac{\text{g Ba}}{\text{g sample}} \times 100 = \frac{0.24254 \text{ g Ba}}{6.977 \text{ g sample}} \times 100 = 3.476\% \text{ Ba}$$

4.96 *Plan.* Write balanced equation.

$$\text{mass } H_2SO_4 \text{ soln} \xrightarrow{\text{mass \%}} \text{mass } H_2SO_4 \rightarrow \text{mol } H_2SO_4 \rightarrow \text{mol } Na_2CO_3 \rightarrow \text{mass } Na_2CO_3$$

Solve. $H_2SO_4(aq) + Na_2CO_3(s) \rightarrow Na_2SO_4(aq) + H_2O(l) + CO_2(g)$

$$5.0 \times 10^3 \text{ kg conc. } H_2SO_4 \times \frac{0.950 \text{ kg } H_2SO_4}{1.00 \text{ kg conc. } H_2SO_4} = 4.75 \times 10^3 = 4.8 \times 10^3 \text{ kg } H_2SO_4$$

$$4.75 \times 10^3 \text{ kg } H_2SO_4 \times \frac{1 \times 10^3 \text{ g}}{1 \text{ kg}} \times \frac{1 \text{ mol } H_2SO_4}{98.08 \text{ g } H_2SO_4} \times \frac{1 \text{ mol } Na_2CO_3}{1 \text{ mol } H_2SO_4}$$

$$\times \frac{105.99 \text{ g } NaHCO_3}{1 \text{ mol } NaHCO_3} \times \frac{1 \text{ kg}}{1 \times 10^3 \text{ g}} = 5.133 \times 10^3 = 5.1 \times 10^3 \text{ kg } Na_2CO_3$$

4.97 (a) $Mg(OH)_2(s) + 2HNO_3(aq) \rightarrow Mg(NO_3)_2(aq) + 2H_2O(l)$

(b) $$5.53 \text{ g } Mg(OH)_2 \times \frac{1 \text{ mol } Mg(OH)_2}{58.32 \text{ g } Mg(OH)_2} = 0.09482 = 0.0948 \text{ mol } Mg(OH)_2$$

$0.200\ M\ HNO_3 \times 0.0250 \text{ L} = 0.00500 \text{ mol } HNO_3$

The 0.00500 mol HNO_3 would neutralize 0.00250 mol $Mg(OH)_2$ and much more $Mg(OH)_2$ is present, so HNO_3 is the limiting reactant.

(c) Since HNO_3 limits, 0 mol HNO_3 is present after reaction.

0.00250 mol $Mg(NO_3)_2$ is produced.

0.09482 mol $Mg(OH)_2$ initial - 0.00250 mol $Mg(OH)_2$ react = 0.0923 mol $Mg(OH)_2$ remain

4.98 (a) $Na_2SO_4(aq) + Pb(NO_3)_2(s) \rightarrow PbSO_4(s) + 2NaNO_3(aq)$

(b) Calculate mol of each reactant and compare.

$$1.50 \text{ g } Pb(NO_3)_2 \times \frac{1 \text{ mol } Pb(NO_3)_2}{331.2 \text{ g } Pb(NO_3)_2} = 0.004529 = 4.53 \times 10^{-3} \text{ mol } Pb(NO_3)_2$$

$0.100\ M\ Na_2SO_4 \times 0.125 \text{ L} = 0.0125 \text{ mol } Na_2SO_4$

Since the reactants combine in a 1:1 mol ratio, $Pb(NO_3)_2$ is the limiting reactant.

(c) $Pb(NO_3)_2$ is the limiting reactant, so no Pb^{2+} remains in solution. The remaining ions are: SO_4^{2-} (excess reactant), Na^+ and NO_3^- (spectators).

SO_4^{2-}: 0.0125 mol SO_4^{2-} initial - 0.00453 mol SO_4^{2-} reacted

= 0.00797 = 0.0080 mol SO_4^{2-} remain

0.00797 mol SO_4^{2-} / 0.125 L soln = 0.064 M SO_4^{2-}

Na^+: Since the total volume of solution is the volume of $Na_2SO_4(aq)$ added, the concentration of Na^+ is unchanged.

$0.100\ M\ Na_2SO_4 \times (2 \text{ mol } Na^+ / 1 \text{ mol } Na_2SO_4) = 0.200\ M\ Na^+$

NO_3^-: 4.53×10^{-3} mol $Pb(NO_3)_2 \times 2$ mol NO_3^- / 1 mol $Pb(NO_3)_2$

= 9.06×10^{-3} mol NO_3^-

9.06×10^{-3} mol NO_3^- / 0.125 L = 0.0725 M NO_3^-

4.99 *Plan.* Cl^- is present in NaCl and $MgCl_2$; using mass %, calculate mass NaCl and $MgCl_2$ in mixture, mol Cl^- in each, then molarity of Cl^- in 0.500 L solution. *Solve*:

$$7.50 \text{ mixture} \times \frac{0.890 \text{ g NaCl}}{1.00 \text{ g mixture}} \times \frac{1 \text{ mol NaCl}}{58.44 \text{ g NaCl}} \times \frac{1 \text{ mol Cl}^-}{1 \text{ mol NaCl}} = 0.1142 = 0.114 \text{ mol Cl}^-$$

$$7.50 \text{ mixture} \times \frac{0.015 \text{ g MgCl}_2}{1.00 \text{ g mixture}} \times \frac{1 \text{ mol MgCl}_2}{95.21 \text{ g MgCl}_2} \times \frac{2 \text{ mol Cl}^-}{1 \text{ mol MgCl}_2} = 0.00236 = 0.0024 \text{ mol Cl}^-$$

$$\text{mol Cl}^- = 0.1142 + 0.00236 = 0.11656 = 0.117 \text{ mol Cl}^-;\ M = \frac{0.11656 \text{ mol Cl}^-}{0.5000 \text{ L}} = 0.233\ M \text{ Cl}^-$$

4.100 *Plan.* $M = \frac{\text{mol Br}^-}{\text{L sea water}}$; mg Br^- → g Br^- → mol Br^-;

1 kg sea water → g $\xrightarrow{\text{density}}$ water mL water → L sea water

Solve. $65 \text{ mg Br}^- \times \frac{1 \text{ g Br}^-}{1000 \text{ mg Br}^-} \times \frac{1 \text{ mol Br}^-}{79.90 \text{ g Br}^-} = 8.135 \times 10^{-4} = 8.1 \times 10^{-4} \text{ mol Br}^-$

$$1 \text{ kg sea water} \times \frac{1000 \text{ g}}{1 \text{ kg}} \times \frac{1 \text{ mL water}}{1.025 \text{ g water}} \times \frac{1 \text{ L}}{1000 \text{ mL}} = 0.9756 \text{ L}$$

$$M \text{ Br}^- = \frac{8.135 \times 10^{-4} \text{ mol Br}^-}{0.9756 \text{ L sea water}} = 8.3 \times 10^{-4}\ M \text{ Br}^-$$

4.101 $Ag^+(aq) + Cl^-(aq) \rightarrow AgCl(s)$

$$\frac{0.2997 \text{ mol Ag}^+}{1 \text{ L}} \times 0.04258 \text{ L} \times \frac{1 \text{ mol Cl}^-}{1 \text{ mol Ag}^+} \times \frac{35.453 \text{ g Cl}^-}{1 \text{ mol Cl}^-} = 0.45242 = 0.4524 \text{ g Cl}^-$$

$$25.00 \text{ mL sea water} \times \frac{1.025 \text{ g}}{\text{mL}} = 25.625 = 25.63 \text{ g sea water}$$

$$\text{mass \% Cl}^- = \frac{0.45242 \text{ g Cl}^-}{25.625 \text{ g sea water}} \times 100 = 1.766\% \text{ Cl}^-$$

4.102 (a) AsO_4^{3-}; +5

(b) Ag_3PO_4 is silver phosphate; Ag_3AsO_4 is silver arsenate

(c) $$0.0250 \text{ L soln} \times \frac{0.102 \text{ mol Ag}^+}{1 \text{ L soln}} \times \frac{1 \text{ mol Ag}_3\text{AsO}_4}{3 \text{ mol Ag}^+} \times \frac{1 \text{ mol As}}{1 \text{ mol Ag}_3\text{AsO}_4} \times \frac{74.92 \text{ g As}}{1 \text{ mol As}}$$

$$= 0.06368 = 0.0637 \text{ g As}$$

$$\text{mass percent} = \frac{0.06368 \text{ g As}}{1.22 \text{ g sample}} \times 100 = 5.22\% \text{ As}$$

4.103 (a) mol OH^- in tablet = mol H^+ total - mol H^+ reacted with NaOH

mol H^+ total = 0.500 M HCl × 0.0500 L = 0.0250 mol H^+ total

mol H^+ w/NaOH = 0.255 M NaOH × 0.0309 L = 0.007880 = 0.00788 mol OH^-

= 0.00788 mol H+

mol OH^- in tablet = 0.0250 - 0.00788 = 0.01712 = 0.0171 mol OH^- in tablet

(b) mass $[Mg(OH)_2 + Al(OH)_3]$ = 500 mg × 0.950 = 475 mg = 0.475 g

g $Mg(OH)_2$ = x; g $Al(OH)_3$ = 0.475 - x

$$\frac{x}{58.32\ g\ Mg(OH)_2} \times \frac{2\ mol\ OH^-}{1\ mol\ Mg(OH)_2} = mol\ OH^-\ from\ Mg(OH)_2$$

$$\frac{0.475 - x}{78.01\ g\ Al(OH)_3} \times \frac{3\ mol\ OH^-}{1\ mol\ Al(OH)_3} = mol\ OH^-\ from\ Al(OH)_3$$

$$\frac{2x}{58.32} + \frac{1.425 - 3x}{78.01} = 0.01712\ mol\ OH^-$$

$$\frac{156.02x + 83.106 - 174.96x}{58.32(78.01)} = 0.01712$$

-18.94 x = 77.8882 - 83.106 = -5.2178 = -5.2

x = 0.27549 = 0.275 g $Mg(OH)_2$; 0.475 - 0.275 = 0.200 g $Al(OH)_3$

(Strictly speaking, both masses have 2 sig figs because the difference (77.9 - 83.11) has 1 decimal place and 2 sig figs.)

4.104 (a) mol HCl initial - mol NH_3 from air = mol HCl remaining

= mol NaOH required for titration

mol NaOH = 0.0588 *M* × 0.0131 L = 7.703×10^{-4} = 7.70×10^{-4} mol NaOH

= 7.70×10^{-4} mol HCl remain

mol HCl initial - mol HCl remaining = mol NH_3 from air

(0.0105 M HCl × 0.100 L) - 7.703×10^{-4} mol HCl = mol NH_3

10.5×10^{-4} mol HCl - 7.703×10^{-4} mol HCl = 2.80×10^{-4} = 2.8×10^{-4} mol NH_3

$$2.8 \times 10^{-4}\ mol\ NH_3 \times \frac{17.03\ g\ NH_3}{1\ mol\ NH_3} = 4.77 \times 10^{-3} = 4.8 \times 10^{-3}\ g\ NH_3$$

(b) ppm is defined as molecules of NH_3/1 × 10^6 molecules in air.
Calculate molecules NH_3 from mol NH_3.

$$2.80 \times 10^{-4}\ mol\ NH_3 \times \frac{6.022 \times 20^{23}\ molecules}{1\ mol} = 1.686 \times 10^{20}$$

$= 1.7 \times 10^{20}$ NH_3 molecules

Calculate total volume of air processed, then g air using density, then molecules air using molar mass.

$$\frac{10.0\ L}{1\ min} \times 10.0\ min \times \frac{1.20\ g\ air}{1\ L\ air} \times \frac{1\ mol\ air}{29.0\ g\ air} \times \frac{6.022 \times 10^{23}\ molecules}{1\ mol}$$

$= 2.492 \times 10^{24} = 2.5 \times 10^{24}$ air molecules

$$ppm\ NH_3 = \frac{1.686 \times 10^{20}\ NH_3\ molecules}{2.492 \times 10^{24}\ air\ molecules} \times 1 \times 10^6 = 68\ ppm\ NH_3$$

(c) 68 ppm > 50 ppm. The manufacturer is **not** in compliance.

5 Thermochemistry

Nature of Energy

5.1 An object can possess energy by virtue of its motion or position. Kinetic energy, the energy of motion, depends on the mass of the object and its velocity. Potential energy, stored energy, depends on the position of the object relative to the body with which it interacts.

5.2 (a) The kinetic energy of the ball **decreases** as it moves higher. As the ball moves higher and opposes gravity, kinetic energy is changed into potential energy.

(b) The potential energy of the ball **increases** as it moves higher.

(c) The heavier ball would go **half as high** as the tennis ball. At the apex of the trajectory, all initial kinetic energy has been changed into potential energy. The magnitude of the change in potential energy is $m\ g\ \Delta h$, which is equal to the energy initially imparted to the ball. If the same amount of energy is imparted to a ball with twice the mass, m doubles so Δh is half as large.

5.3 (a) *Analyze.* Given: mass and speed of ball. Find: kinetic energy.
Plan. Since $1\ J = 1\ kg \bullet m^2/s^2$, convert g → kg to obtain E_k in joules.

Solve. $E_k = 1/2\ mv^2 = 1/2 \times 45\ g \times \frac{1\ kg}{1000\ g} \times \left(\frac{61\ m}{1\ s}\right)^2 = \frac{84\ kg \bullet m^2}{1\ s^2} = 84\ J$

Check. $1/2(45 \times 3600/1000) \approx 1/2(40 \times 4) \approx 80\ J$

(b) $83.72\ J \times \frac{1\ cal}{4.184\ J} = 20\ cal$

(c) As the ball hits the sand, its speed (and hence its kinetic energy) drops to zero. Most of the kinetic energy is transferred to the sand, which deforms when the ball lands. Some energy is released as heat through friction between the ball and the sand.

5.4 (a) *Plan.* Convert lb → kg, mi/hr → m/s.

Solve. $950\ lb \times \frac{1\ kg}{2.205\ lb} = 430.84 = 431\ kg$

$$\frac{68\text{ mi}}{1\text{ hr}} \times \frac{1.6093\text{ km}}{1\text{ mi}} \times \frac{1000\text{ m}}{1\text{ km}} \times \frac{1\text{ hr}}{60\text{ min}} \times \frac{1\text{ min}}{60\text{ sec}} = 30.398 = 30\text{ m/s}$$

$E_k = 1/2\ mv^2 = 1/2 \times 430.84\text{ kg} \times (30.398)^2\ \text{m}^2/\text{s}^2 = 2.0 \times 10^5\text{ J}$

(b) E_k is proportional to v^2, so if speed decreases by a factor of 2, kinetic energy decreases by a factor of 4.

(c) Brakes stop a moving vehicle, so the kinetic energy of the motorcycle is primarily transferred to friction between brakes and wheels, and somewhat to deformation of the tire and friction between the tire and road.

5.5 *Analyze.* Given: heat capacity of water = 1 Btu/lb•°F Find: J/Btu

Plan. heat capacity of water = 4.184 J/g•°C; $\frac{\text{J}}{\text{g}\bullet{}^\circ\text{C}} \rightarrow \frac{\text{J}}{\text{lb}\bullet{}^\circ\text{F}} \rightarrow \frac{\text{J}}{\text{Btu}}$

This strategy requires changing °F to °C. Since this involves the magnitude of a degree on each scale, rather than a specific temperature, the 32 in the temperature relationship is not needed.

100 °C = 180 °F; 5 °C = 9 °F

Solve. $\frac{4.184\text{ J}}{\text{g}\bullet{}^\circ\text{C}} \times \frac{453.6\text{ g}}{\text{lb}} \times \frac{5\ {}^\circ\text{C}}{9\ {}^\circ\text{F}} \times \frac{1\text{ lb}\bullet{}^\circ\text{F}}{1\text{ Btu}} = 1054\text{ J/Btu}$

5.6 *Analyze.* Given: 1 kwh; 1 watt = 1 J/s; 1 watt • s = 1 J.

Find: conversion factor for joules and kwh.

Plan. kwh → wh → ws → J

Solve. 1 kwh × $\frac{1000\text{ w}}{1\text{ kw}} \times \frac{60\text{ min}}{\text{h}} \times \frac{60\text{ s}}{\text{min}} \times \frac{1\text{ J}}{1\text{ w}\bullet\text{s}} = 3.6 \times 10^6\text{ J}$

1 kwh = 3.6×10^6 J

5.7 *Analyze.* Given: 100 watt bulb. Find: heat in kcal radiated by bulb or person in 24 hr.

Plan. 1 watt = 1 J/s; 1 kcal = 4.184×10^3 J; watt → J/s → J → kcal. *Solve*:

$100\text{ watt} = \frac{100\text{ J}}{1\text{ s}} \times \frac{60\text{ sec}}{\text{min}} \times \frac{60\text{ min}}{\text{hr}} \times 24\text{ hr} \times \frac{1\text{ kcal}}{4.184 \times 10^3\text{ J}} = 2065 = 2.1 \times 10^3\text{ kcal}$

24 hr has 2 sig figs, but 100 watt is ambiguous. The answer to 1 sig fig would be 2×10^3 kcal.

Check. $(1 \times 10^2 \times 6 \times 10^1 \times 6/10^3) \approx 6^3 \times 10 \approx 2000$ kcal

5.8 The energy source of a 100 watt light bulb is electrical current from household wiring. Current passes through and heats a tungsten filament (thin wire) in the bulb. The energy is radiated in the form of heat and visible light.

The energy source for an adult person is food. When a person eats, the food undergoes a complex series of chemical reactions that release the potential energy stored in chemical bonds. Some of this energy is transferred as electrical impulses that trigger muscle action and become kinetic energy. Some is released as heat.

In both cases, the energy must travel through a network (house wiring and lamp or human body) and undergo several changes in form before it is in the correct location and form to accomplish the desired task. In both cases, the energy given off as heat is wasted; it cannot be applied to the tasks of producing light or motion.

5.9 The air gun imparts a certain amount of kinetic energy to the pellet. As the pellet rises against the force of gravity, kinetic energy is changed to potential energy. When all kinetic energy has been transferred to potential energy (or lost as heat through friction) the pellet stops rising and falls to earth. In principle, if enough kinetic energy could be imparted to the pellet, it could escape the force of gravity and move into space. For an air gun and a pellet, this is 'practically' impossible.

5.10 The change in potential energy equals $m \times g \times \Delta h$. The gravitational force on the moon is much smaller than that on earth, so the change in potential energy is much smaller on the moon, given equal m and Δh.

5.11 (a) In thermodynamics, the *system* is the well-defined part of the universe whose energy changes are being studied.

(b) A closed system can exchange heat but not mass with its surroundings.

5.12 (a) The system is not closed, because it is exchanging mass with the surroundings. That is, solution flows into and out of the flask.

(b) If the system is defined as shown, it can be closed by blocking the flow in and out, but leaving the flask full of solution.

5.13 (a) *Work* is a force applied over a distance.

(b) The amount of work done is the magnitude of the force times the distance over which it is applied. $\boldsymbol{w = F \times d}$.

5.14 (a) If energy is the capacity to do work or transfer heat, then heat is energy because it can do work. For example, heat causes a gas to expand inside a cylinder and move a piston, doing work.

(b) Heat is transferred from one object (system) to another until the two objects (systems) are at the same temperature.

5.15 (a) Gravity; work is done because the force of gravity is opposed and the pencil is lifted.

(b) Mechanical force; work is done because the force of the coiled spring is opposed as the spring is compressed over a distance.

5.16 (a) Electrostatic attraction; no work is done because the particles are held stationary.

(b) Magnetic attraction; work is done because the nail is moved a distance.

The First Law of Thermodynamics

5.17 (a) In any chemical or physical change, energy can be neither created nor destroyed, but it can be changed in form.

(b) The total *internal energy* (E) of a system is the sum of all the kinetic and potential energies of the system components.

(c) The internal energy of a system increases when work is done on the system by the surroundings and/or when heat is transferred to the system from the surroundings (the system is heated).

5.18 (a) $\Delta E_{sys} = -\Delta E_{surr}$; $\Delta E_{sys} = q + w$

(b) In applying the first law, we need only to measure changes in internal energy, not absolute values of E. This is because doing work or transferring heat involves changes in energy, not absolute values. This is convenient, since it is almost impossible to measure the internal energy of a system because it has so many components.

(c) The quantities q and w are negative when the system loses heat to the surroundings (it cools), or does work on the surroundings.

5.19 *Analyze.* Given: heat and work. Find: magnitude and sign of ΔE.

Plan. In each case, evaluate q and w in the expression $\Delta E = q + w$. For an exothermic process, q is negative; for an endothermic process, q is positive. *Solve*:

(a) q is negative because the system loses heat and w is negative because the system does work. $\Delta E = -113 \text{ kJ} - 39 \text{ kJ} = -152 \text{ kJ}$. The process is exothermic.

(b) $\Delta E = +1.62 \text{ kJ} - 847 \text{ J} = +1.62 \text{ kJ} - 0.847 \text{ kJ} = +0.746 = +0.75 \text{ kJ}$. The process is endothermic.

(c) q is positive because the system gains heat and w is negative because the system does work. $\Delta E = +77.5 \text{ kJ} - 63.5 \text{ kJ} = +14.0 \text{ kJ}$. The process is endothermic.

5.20 In each case, evaluate q and w in the expression $\Delta E = q + w$. For an exothermic process, q is negative; for an endothermic process, q is positive.

(a) q is positive and w is negative. $\Delta E = +900 \text{ J} - 422 \text{ J} = +478 \text{ J}$. The process is endothermic.

(b) q is negative and w is essentially zero. $\Delta E = -3140 \text{ J}$. The process is exothermic.

(c) q is negative and w is zero. $\Delta E = -8.65$ kJ. The process is exothermic.

5.21 (a) For an endothermic process, the sign of q is positive; the system gains heat. This is true only for system (iii).

(b) In order for ΔE to be less than 0, there is a net transfer of heat or work from the system to the surroundings. The magnitude of the quantity leaving the system is greater than the magnitude of the quantity entering the system. In system (i), the magnitude of the heat leaving the system is less than the magnitude of the work done on the system. In system (iii), the magnitude of the work done by the system is greater than the magnitude of the heat entering the system. $\Delta E < 0$ for system (iii) only.

(c) In order for ΔE to be greater than 0, there is a net transfer of work or heat to the system from the surroundings. In system (i), the magnitude of the work done on the system is greater than the magnitude of the heat leaving the system. In system (ii), work is done on the system with no change in heat. $\Delta E > 0$ for systems (i) and (ii).

5.22 (a)

$q > 0$

$w < 0$

(b) ΔE will be positive for this process if the magnitude of w is greater than the magnitude of q.

(c) ΔE will be negative for this process if the magnitude of q is greater than the magnitude of w.

5.23 *Analyze.* How do the different physical situations (cases) affect the changes to heat and work of the system upon addition of 100 J of energy? *Plan.* Use the definitions of heat and work and the First Law to answer the questions. *Solve*:

If the piston is allowed to move, case (1), the heated gas will expand and push the piston up, doing work on the surroundings. If the piston is fixed, case (2), most of the electrical energy will be manifested as an increase in heat of the system.

(a) Since little or no work is done by the system in case (2), the gas will absorb most of the energy as heat; the case (2) gas will have the higher temperature.

(b) In case (2), $w \approx 0$ and $q \approx 100$ J. In case (1), a significant amount of energy will be used to do work on the surroundings (-w), but some will be absorbed as heat (+q). [The transfer of electrical energy into work is never completely efficient!]

(c) ΔE is greater for case (2), because the entire 100 J increases the internal energy of the system, rather than a part of the energy doing work on the surroundings.

5.24 $E_{el} = \dfrac{\kappa Q_1 Q_2}{r^2}$ For two oppositely charged particles, the sign of E_{el} is negative; the closer the particles, the greater the magnitude of E_{el}.

(a) The potential energy becomes less negative as the particles are separated (r increases)

(b) ΔE for the process is positive; the internal energy of the system increases as the oppositely charged particles are separated.

(c) Work is done on the system to separate the particles so w is positive. We have no direct knowledge of the change in q, except that it cannot be large and negative, because overall $\Delta E = q + w$ is positive.

5.25 (a) A *state function* is a property of a system that depends only on the physical state (pressure, temperature, etc.) of the system, not on the route used by the system to get to the current state.

(b) Internal energy and enthalpy are state functions; work is not a state function.

(c) Temperature is a state function; regardless of how hot or cold the sample has been, the temperature depends only on its present condition.

5.26 (a) Independent. Potential energy is a state function.

(b) Dependent. Some of the energy released could be employed in performing work, as is done in the body when sugar is metabolized; heat is not a state function.

(c) Dependent. The work accomplished depends on whether the gasoline is used in an engine, burned in an open flame, or in some other manner. Work is not a state function.

Enthalpy

5.27 (a) For the many laboratory and real world processes that occur at constant atmospheric pressure, the enthalpy change is a meaningful measure of the energy change associated with the process. At constant pressure, most of the energy change is transferred as heat ($\Delta H = q_p$), even if gases are involved in the process.

(b) Only under conditions of constant pressure is ΔH for a process equal to the heat transferred during the process.

(c) If ΔH is negative, the enthalpy of the system decreases and the process is exothermic.

5.28 (a) When a process occurs under constant external pressure, the enthalpy change (ΔH) equals the amount of heat transferred. $\Delta H = q_p$.

(b) State functions are particularly useful because they are totally defined by the current conditions (state) of the system, not the history of how the system arrived at its current state. Changes to state functions, like ΔH can be calculated from knowledge of initial and final states, without details of how the change takes place.

(c) $\Delta H = q_p$. If the system absorbs heat, q and ΔH are positive and the enthalpy of the system increases.

5.29 (a) $HC_2H_3O_2(l) + 2O_2(g) \rightarrow 2H_2O(l) + 2CO_2(g)$

$\Delta H = -871.7$ kJ

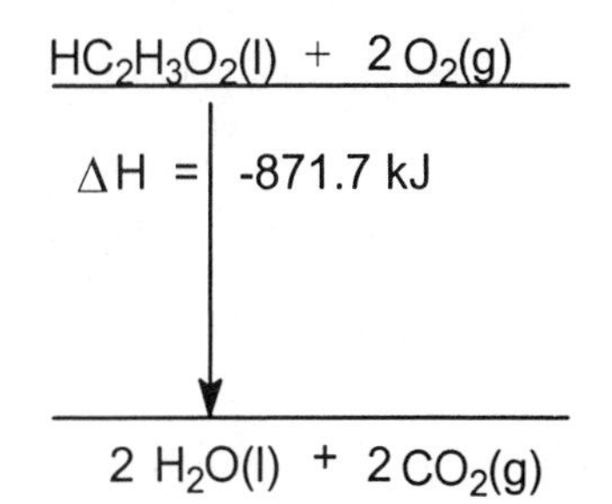

(b) *Analyze.* How are reactants and products arranged on an enthalpy diagram?

Plan. The substances (reactants or products, collectively) with higher enthalpy are shown on the upper level, and those with lower enthalpy are shown on the lower level.

Solve. For this reaction, ΔH is negative, so the products have lower enthalpy and are shown on the lower level; reactants are on the upper level. The arrow points in the direction of reactants to products and is labeled with the value of ΔH.

5.30 (a) $ZnCO_3(s) \rightarrow ZnO(s) + CO_2(g)$

$\Delta H = 71.5$ kJ

(b)

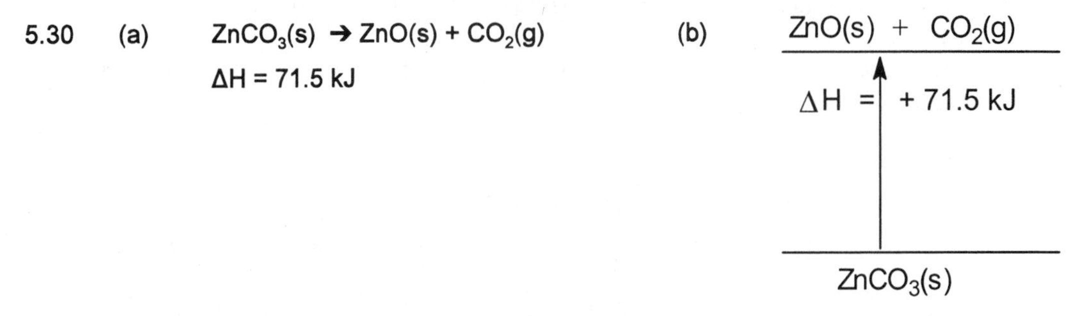

5.31 *Plan.* Consider the sign of ΔH. *Solve*:

Since ΔH is negative, the reactants, 2Cl(g) have the higher enthalpy.

5.32 *Plan.* Consider the sign of an enthalpy change that would convert one of the substances into the other. *Solve*:

(a) $CO_2(s) \rightarrow CO_2(g)$. This change is sublimation, which is endothermic, +ΔH. $CO_2(g)$ has the higher enthalpy.

(b) $H_2 \rightarrow 2H$. Breaking the H–H bond requires energy, so the process is endothermic, +ΔH. Two moles of H atoms have higher enthalpy.

(c) $H_2O(g) \rightarrow H_2(g) + 1/2\ O_2(g)$. Decomposing H_2O into its elements requires energy and is endothermic, +ΔH. One mole of $H_2(g)$ and 0.5 mol $O_2(g)$ at 25°C have the higher enthalpy.

(d) $N_2(g)$ at 100° $\rightarrow$ $N_2(g)$ at 300°. An increase in the temperature of the sample requires that heat is added to the system, +q and +ΔH. $N_2(g)$ at 300° has the higher enthalpy.

5.33 *Analyze/Plan.* Follow the strategy in Sample Exercise 5.5. *Solve*:

(a) Exothermic (ΔH is negative)

(b) $2.4 \text{ g Mg} \times \frac{1 \text{ mol Mg}}{24.305 \text{ g Mg}} \times \frac{-1204 \text{ kJ}}{2 \text{ mol Mg}} = -59 \text{ kJ}$ heat transferred

Check. The units of kJ are correct for heat. The negative sign indicates heat is evolved.

(c) $-96.0 \text{ kJ} \times \frac{2 \text{ mol MgO}}{-1204 \text{ kJ}} \times \frac{40.30 \text{ g MgO}}{1 \text{ mol Mg}} = 6.43 \text{ g MgO}$ produced

Check. Units are correct for mass. $(100 \times 2 \times 40/1200) \approx (8000/1200) \approx 6.5$ g

(d) $2MgO(s) \rightarrow 2Mg(s) + O_2(g) \qquad \Delta H = +1204 \text{ kJ}$

This is the reverse of the reaction given above, so the sign of ΔH is reversed.

$7.50 \text{ g MgO} \times \frac{1 \text{ mol MgO}}{40.30 \text{ g MgO}} \times \frac{1204 \text{ kJ}}{2 \text{ mol MgO}} = +112 \text{ kJ}$ heat absorbed

Check. The units are correct for energy. $(\sim 9000/80) \approx 110$ kJ)

5.34 (a) The reaction is endothermic, so heat is absorbed by the system during the course of reaction.

(b) $1.60 \text{ kg } CH_3OH \times \frac{1000 \text{ g}}{1 \text{ kg}} \frac{1 \text{ mol } CH_3OH}{32.04 \text{ g } CH_3OH} \times \frac{90.7 \text{ kJ}}{1 \text{ mol } CH_3OH}$

$= +4.53 \times 10^3 \text{ kJ}$ heat transferred (absorbed)

(c) $64.7 \text{ kJ} \times \frac{2 \text{ mol } H_2}{90.7 \text{ kJ}} \times \frac{2.016 \text{ g } H_2}{1 \text{ mol } H_2} = 2.88 \text{ g } H_2$ produced

(d) The sign of ΔH is reversed for the reverse reaction: $\Delta H = -90.7$ kJ

$32.0 \text{ g CO} \times \frac{1 \text{ mol CO}}{28.01 \text{ g CO}} \times \frac{-90.7 \text{ kJ}}{1 \text{ mol CO}} = -104 \text{ kJ}$ heat transferred (evolved)

5.35 *Analyze.* Given: balanced thermochemical equation, various quantities of substances and/or enthalpy. *Plan.* Enthalpy is an extensive property; it is "stoichiometric". Use the mole ratios implicit in the balanced thermochemical equation to solve for the desired quantity. Use molar masses to change mass to moles and vice versa where appropriate. *Solve*:

(a) $0.540 \text{ mol AgCl} \times \frac{-65.5 \text{ kJ}}{1 \text{ mol AgCl}} = -35.4 \text{ kJ}$

Check. Units are correct; sign indicates heat evolved.

(b) $1.66 \text{ g AgCl} \times \frac{1 \text{ mol AgCl}}{143.3 \text{ g AgCl}} \times \frac{-65.5 \text{ kJ}}{1 \text{ mol AgCl}} = -0.759 \text{ kJ}$

Check. Units correct; sign indicates heat evolved.

(c) $0.188 \text{ mmol AgCl} \times \frac{1 \times 10^{-3} \text{ mol}}{1 \text{ mmol}} \times \frac{+65.5 \text{ kJ}}{1 \text{ mol AgCl}} = +0.0123 \text{ kJ} = +12.3 \text{ J}$

Check. Units correct; sign of ΔH reversed; sign indicates heat is absorbed during the reverse reaction.

5.36 (a) $4.34 \text{ mol } O_2 \times \frac{-89.4 \text{ kJ}}{3 \text{ mol } O_2} = -129.33 = -129 \text{ kJ}$

(b) $200.8 \text{ g KCl} \times \frac{1 \text{ mol KCl}}{74.55 \text{ g KCl}} \times \frac{-89.4 \text{ kJ}}{2 \text{ mol KCl}} = -120.40 = -120 \text{ kJ}$

(c) Since the sign of ΔH is reversed for the reverse reaction, it seems reasonable that other characteristics would be reversed, as well. If the forward reaction proceeds spontaneously, the reverse reaction is probably not spontaneous. Also, we know from experience that KCl(s) does not spontaneously react with atmospheric $O_2(g)$, even at elevated temperature.

5.37 At constant pressure, $\Delta E = \Delta H - P\Delta V$. In order to calculate ΔE, more information about the conditions of the reaction must be known. For an ideal gas at constant pressure and temperature, $P\Delta V = RT\Delta n$. The values of either P and ΔV or T and Δn must be known to calculate ΔE from ΔH.

5.38 At constant volume ($\Delta V = 0$), $\Delta E = q_v$. According to the definition of enthalpy, $H = E + PV$, so $\Delta H = \Delta E + \Delta(PV)$. For an ideal gas at constant temperature and volume $\Delta PV = V\Delta P = RT\Delta n$. For this reaction, there are 2 mol of gaseous product and 3 mol of gaseous reactants, so

$\Delta n = -1$. Thus $V\Delta P$ or $\Delta(PV)$ is negative. Since $\Delta H = \Delta E + \Delta(PV)$, the negative $\Delta(PV)$ term means that ΔH will be smaller or more negative than ΔE.

5.39 *Analyze/Plan.* q = -89 kJ (heat is given off by the system), w = -36 kJ (work is done by the system). *Solve*:

$\Delta E = q + w = -89 \text{ kJ} - 36 \text{ kJ} = -125 \text{ kJ}$. $\Delta H = q = -89$ kJ (at constant pressure).

Check. The reaction is exothermic.

5.40 The gas is the system. If 518 J of heat is added, q = +518 J. Work done by the system decreases the overall energy of the system, so w = -127 J.

$\Delta E = q + w = +518 \text{ J} - 127 \text{ J} = +391 \text{ J}$. $\Delta H = q = +518$ J (at constant pressure).

5.41 *Analyze.* Given: balanced thermochemical equation. *Plan.* Follow the guidelines given in Section 5.4 for evaluating thermochemical equations. *Solve*:

(a) When a chemical equation is reversed, the sign of ΔH is reversed.

$CO_2(g) + 2H_2O(l) \rightarrow CH_3OH(l) + 3/2\ O_2(g) \qquad \Delta H = +726.5 \text{ kJ}$

(b) Enthalpy is extensive. If the coefficients in the chemical equation are multiplied by 2 to obtain all integer coefficients, the enthalpy change is also multiplied by 2.

$2CH_3OH(l) + 3O_2(g) \rightarrow 2CO_2(g) + 4H_2O(l) \qquad \Delta H = 2(-726.5) \text{ kJ} = -1453 \text{ kJ}$

(c) The exothermic forward reaction is more likely to be thermodynamically favored.

(d) Vaporization (liquid $\rightarrow$ gas) is endothermic. If the product were $H_2O(g)$, the reaction would be more endothermic and would have a smaller negative ΔH. (Depending on temperature, the enthalpy of vaporization for 2 mol H_2O is about +88 kJ, not large enough to cause the overall reaction to be endothermic.)

5.42 (a) $C_2H_2(g) \rightarrow 1/3\ C_6H_6(l)$ $\Delta H = -210$ kJ

(b) $C_6H_6(l) \rightarrow 3C_2H_2(g)$ $\Delta H = 3(+210)$ kJ = +630 kJ

(c) The exothermic reverse reaction is more likely to be thermodynamically favored.

(d)

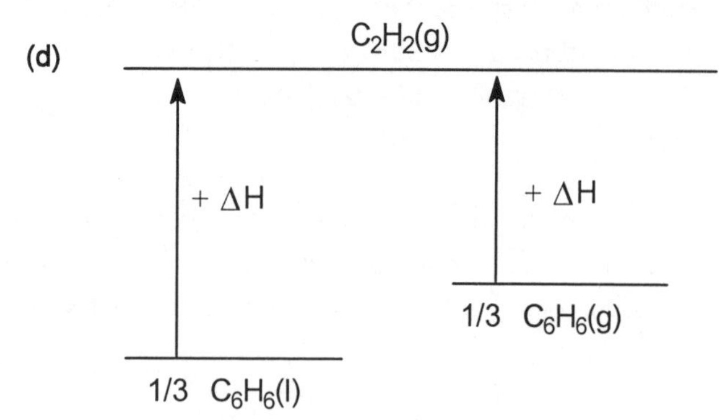

If the reactant is in the higher enthalpy gas phase, the overall ΔH for the reaction has a smaller positive value.

Calorimetry

The specific heat of water to four significant figures, **4.184 J/g • K**, will be used in many of the following exercises; temperature units of K and °C will be used interchangeably.

5.43 (a) J/°C or J/K. Heat capacity is the amount of heat in J required to raise the temperature of an object or a certain amount of a substance 1°C or 1 K. Since the amount is defined, units of amount are not included.

(b) $\frac{J}{g \cdot °C}$ or $\frac{J}{g \cdot °K}$ Specific heat is a particular kind of heat capacity where the amount of substance is 1 g.

5.44 *Analyze.* Both objects are heated to 100°C. The two hot objects are placed in the same amount of cold water at the same temperature. Object A raises the water temperature more than object B. *Plan.* Apply the definition of heat capacity to heating the water and heating the objects to determine which object has the greater heat capacity. *Solve*:

(a) Both beakers of water contain the same mass of water, so they both have the same heat capacity. Object A raises the temperature of its water more than object B, so more heat was transferred from object A than from object B. Since both objects were heated to the same temperature initially, object A must have absorbed more heat to reach the 100° temperature. The greater the heat capacity of an object, the greater the heat required to produce a given rise in temperature. Thus, object A has the greater heat capacity.

(b) Since no information about the masses of the objects is given, we cannot compare or determine the specific heats of the objects.

5.45 *Plan.* Manipulate the definition of specific heat to solve for the desired quantity, paying close attention to units. specific heat = $q/(m \times \Delta t)$. *Solve:*

(a) $\dfrac{4.184\ J}{1\ g \bullet K}$ or $\dfrac{4.184\ J}{1\ g \bullet {}^\circ C}$ (b) $\dfrac{185\ g\ H_2O \times 4.184\ J}{1\ g \bullet {}^\circ C} = 774\ J/{}^\circ C$

(c) $10.00\ kg\ H_2O \times \dfrac{1000\ g}{1\ kg} \times \dfrac{4.184\ J}{1\ g \bullet {}^\circ C} \times \dfrac{1\ kJ}{1000\ J} \times (46.2^\circ C - 24.6^\circ C) = 904\ kJ$

Check. $(10 \times 4 \times 20) \approx 800$ kJ; the units are correct. Note that the conversion factors for kg → g and J → kJ cancel. An equally correct form of specific heat would be kJ/kg • C°.

5.46 (a) $\dfrac{4.184\ J}{1\ g \bullet {}^\circ C} \times \dfrac{18.02\ g\ H_2O}{1\ mol\ H_2O} = \dfrac{75.40\ J}{mol \bullet {}^\circ C}$

(b) $8.42\ mol\ H_2O \times \dfrac{75.40\ J}{mol \bullet {}^\circ C} = 635\ J/{}^\circ C$

(c) $2.56\ kg\ H_2O \times \dfrac{1000\ g}{1\ kg} \times \dfrac{4.84\ J}{g \bullet {}^\circ C} \times \dfrac{1\ kJ}{1000\ J} \times (92.0^\circ C - 44.8^\circ C) = 505.56 = 506\ kJ$

5.47 *Analyze/Plan.* Follow the logic in Sample Exercise 5.6. *Solve:*

$1.42\ kg\ Cu \times \dfrac{1000\ g}{1\ kg} \times \dfrac{0.385\ J}{g \bullet K} \times (88.5^\circ C - 25.0^\circ C) = 3.47 \times 10^4\ J$ (or 34.7 kJ)

5.48 $62.0\ g\ C_7H_8 \times \dfrac{1.13\ J}{g \bullet K} \times (38.8^\circ C - 16.3^\circ C) = 1.58 \times 10^3\ J$

5.49 *Analyze.* Since the temperature of the water increases, the dissolving process is exothermic and the sign of ΔH is negative. The heat lost by the NaOH(s) dissolving equals the heat gained by the solution.

Plan/Solve. Calculate the heat gained by the solution. The temperature change is 47.4 - 23.6 = 23.8°C. The total mass of solution is (100.0 g H_2O + 9.55 g NaOH) = 109.55 = 109.6 g.

$109.55\ g\ solution \times \dfrac{4.184\ J}{1\ g \bullet {}^\circ C} \times 23.8^\circ C \times \dfrac{1\ kJ}{1000\ J} = 10.909 = 10.9\ kJ$

This is the amount of heat lost when 9.55 g of NaOH dissolves.

The heat loss per mole NaOH is

$\dfrac{-10.909\ kJ}{9.55\ g\ NaOH} \times \dfrac{40.00\ g\ NaOH}{1\ mol\ NaOH} = -45.7\ kJ/mol \quad \Delta H = q_p = -45.7\ kJ/mol\ NaOH$

Check. $(-11/9 \times 40) \approx -45$ kJ; the units and sign are correct.

5.50 Following the logic in Solution 5.49, the dissolving process is endothermic, ΔH is positive. The total mass of the solution is (60.0 g H_2O + 3.88 g NH_4NO_3) = 63.88 = 63.9 g. The temperature change of the solution is 23.0 - 18.4 = 4.6°C. The heat lost by the water is

$$63.88 \text{ g } H_2O \times \frac{4.184 \text{ J}}{1 \text{ g} \cdot °C} \times 4.6°C \times \frac{1 \text{ kJ}}{1000 \text{ J}} = 1.229 = 1.2 \text{ kJ}$$

Thus, 1.2 kJ is absorbed when 3.88 g $NH_4NO_3(s)$ dissolves.

$$\frac{+1.229 \text{ kJ}}{3.88 \; NH_4NO_3} \times \frac{80.05 \text{ g } NH_4NO_3}{1 \text{ mol } NH_4NO_3} = +25.36 = +25 \text{ kJ/mol } NH_4NO_3$$

5.51 *Analyze/Plan*. Follow the logic in Sample Exercise 5.8. *Solve*:

$q_{bomb} = -q_{rxn}$; $\Delta T = 30.57°C - 23.44°C = 7.13°C$

$$q_{bomb} = \frac{7.854 \text{ kJ}}{1°C} \times 7.13°C = 56.00 = 56.0 \text{ kJ}$$

At constant volume, $q_V = \Delta E$. ΔE and ΔH are very similar.

$$\Delta H_{rxn} \approx \Delta E_{rxn} = q_{rxn} = -q_{bomb} = \frac{-56.0 \text{ kJ}}{2.20 \text{ g } C_6H_4O_2} = -25.454 = -25.5 \text{ kJ/g } C_6H_4O_2$$

$$\Delta H_{rxn} = \frac{-25.454 \text{ kJ}}{1 \text{ g } C_6H_4O_2} \times \frac{108.1 \text{ g } C_6H_4O_2}{1 \text{ mol } C_6H_4O_2} = -2.75 \times 10^3 \text{ kJ/mol } C_6H_4O_2$$

5.52 (a) $C_6H_5OH(s) + 7O_2(g) \rightarrow 6CO_2(g) + 3H_2O(l)$

(b) $q_{bomb} = -q_{rxn}$; $\Delta T = 26.37°C - 21.36°C = 5.01°C$

$$q_{bomb} = \frac{11.66 \text{ kJ}}{1°C} \times 5.01°C = 58.417 = 58.4 \text{ kJ}$$

At constant volume, $q_v = \Delta E$. ΔE and ΔH are very similar.

$$\Delta H_{rxn} \approx \Delta E_{rxn} = q_{rxn} = -q_{bomb} = \frac{-58.417 \text{ kJ}}{1.800 \text{ g } C_6H_5OH} = -32.454 = -32.5 \text{ kJ/g } C_6H_5OH$$

$$\Delta H_{rxn} = \frac{-32.454 \text{ kJ}}{1 \text{ g } C_6H_5OH} \times \frac{94.11 \text{ g } C_6H_5OH}{1 \text{ mol } C_6H_5OH} = \frac{-3.054 \times 10^3 \text{ kJ}}{\text{mol } C_6H_5OH}$$

$$= -3.05 \times 10^3 \text{ kJ/mol } C_6H_5OH$$

5.53 *Analyze*. Given: specific heat and mass of glucose, ΔT for calorimeter. Find: heat capacity, C, of calorimeter. *Plan*. All heat from the combustion raises the temperature of the calorimeter. Calculate heat from combustion of glucose, divide by ΔT for calorimeter to get kJ/°C. *Solve*:

(a) $$C_{total} = 2.500 \text{ g glucose} \times \frac{15.57 \text{ kJ}}{1 \text{ g glucose}} \times \frac{1}{2.70°C} = 14.42 = 14.4 \text{ kJ/°C}$$

(b) Qualitatively, assuming the same exact initial conditions in the calorimeter, twice as much glucose produces twice as much heat, which raises the calorimeter temperature by twice as many °C. Quantitatively,

$$5.000 \text{ g glucose} \times \frac{15.57 \text{ kJ}}{1 \text{ g glucose}} \times \frac{1\,°\text{C}}{14.42 \text{ kJ}} = 5.40°\text{C}$$

Check. Units are correct. ΔT is twice as large as in part (a). The result has 3 sig figs, because the heat capacity of the calorimeter is known to 3 sig figs.

5.54 (a) $C = 1.640 \text{ g } HC_7H_5O_2 \times \frac{26.38 \text{ kJ}}{1 \text{ g } HC_7H_5O_2} \times \frac{1}{4.95°\text{C}} = 8.740 = 8.74 \text{ kJ/°C}$

(b) $\frac{8.740 \text{ kJ}}{°\text{C}} \times 4.68°\text{C} \times \frac{1}{1.320 \text{ g sample}} = 30.99 = 31.0 \text{ kJ/g sample}$

(c) If water is lost from the calorimeter, there is less water to heat, so the same amount of heat (kJ) from a reaction would cause a larger increase in the calorimeter temperature. The calorimeter constant, kJ/°C would decrease, because °C is in the denominator of the expression.

Hess's Law

5.55 If a reaction can be described as a series of steps, ΔH for the reaction is the sum of the enthalpy changes for each step. As long as we can describe a route where ΔH for each step is known, ΔH for any process can be calculated.

5.56 Hess's Law is a consequence of the fact that enthalpy is a state function. Since ΔH is independent of path, we can describe a process by any series of steps that add up to the overall process and ΔH for the process is the sum of the ΔH values for the steps.

5.57 (a) *Analyze/Plan.* Arrange the reactions so that in the overall sum, B appears in both reactants and products and can be canceled. This is a general technique for using Hess's Law. *Solve:*

A → B	ΔH = +30 kJ
B → C	ΔH = +60 kJ
A → C	ΔH = +90 kJ

(b)

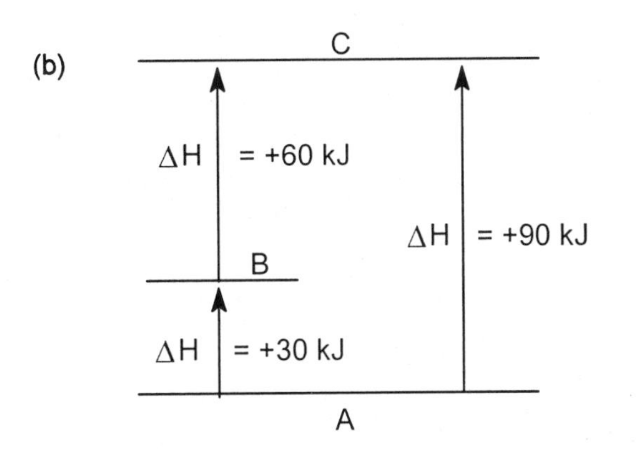

Check. The process of A forming C can be described as A forming B and B forming C.

5.58 (a)

Y → X	ΔH = +35 kJ
X → Z	ΔH = +90 kJ
Y → Z	ΔH = +125 kJ

(b)

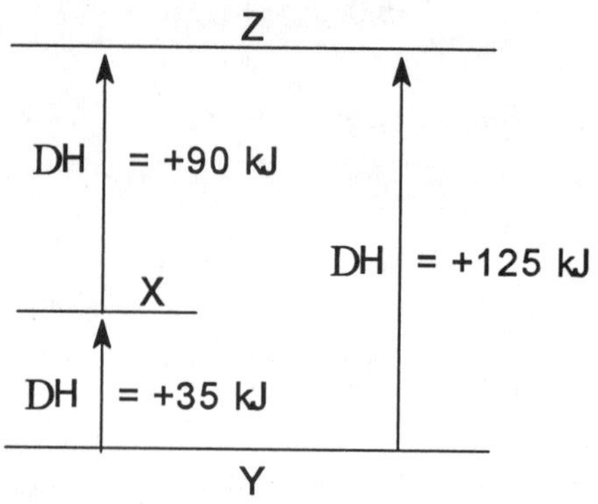

The process of Y forming Z can be described as Y forming X and X forming Z.

(c) No. Any temperature change is accompanied by a change in heat. Comparing reactions and enthalpies of reaction at different temperatures doesn't take into account the change in heat to go from one temperature to the other.

5.59 *Analyze/Plan.* Follow the logic in Sample Exercise 5.9. Manipulate the equations so that "unwanted" substances can be canceled from reactants and products. Adjust the corresponding sign and magnitude of ΔH. *Solve*:

$P_4O_6(s) \rightarrow P_4(s) + 3O_2(g)$	ΔH = 1640.1 kJ
$P_4(s) + 5O_2(g) \rightarrow P_4O_{10}(s)$	ΔH = -2940.1 kJ
$P_4O_6(s) + 2O_2(g) \rightarrow P_4O_{10}(s)$	ΔH = -1300.0 kJ

Check. We have obtained the desired reaction.

5.60

$3H_2(g) + 3/2\ O_2(g) \rightarrow 3H_2O(g)$	ΔH = 3/2(-483.6 kJ)
$O_3(g) \rightarrow 3/2\ O_2(g)$	ΔH = 1/2(-284.6 kJ)
$3H_2(g) + O_3(g) \rightarrow 3H_2O(g)$	ΔH = -867.7 kJ

5.61 *Analyze/Plan.* Follow the logic in Sample Exercise 5.9. Manipulate the equations so that "unwanted" substances can be canceled from reactants and products. Adjust the corresponding sign and magnitude of ΔH. *Solve*:

$C_2H_4(g) \rightarrow 2\ H_2(g) + 2C(s)$	ΔH = -52.3 kJ
$2C(s) + 4F_2(g) \rightarrow 2CF_4(g)$	ΔH = 2(-680 kJ)
$2H_2(g) + 2F_2(g) \rightarrow 4HF(g)$	ΔH = 2(-537 kJ)
$C_2H_4(g) + 6F_2(g) \rightarrow 2CF_4(g) + 4HF(g)$	ΔH = -2.49×10^3 kJ

Check. We have obtained the desired reaction.

5.62

$N_2O(g) \rightarrow N_2(g) + 1/2\ O_2(g)$	$\Delta H = 1/2\ (-163.2\ kJ)$
$NO_2(g) \rightarrow NO(g) + 1/2\ O_2(g)$	$\Delta H = 1/2(113.1\ kJ)$
$N_2(g) + O_2(g) \rightarrow 2NO(g)$	$\Delta H = 180.7\ kJ$
$N_2O(g) + NO_2(g) \rightarrow 3NO(g)$	$\Delta H = 155.7\ kJ$

Enthalpies of Formation

5.63 (a) *Standard conditions* for enthalpy changes are usually P = 1 atm and T = 298 K. For the purpose of comparison, standard enthalpy changes, ΔH°, are tabulated for reactions at these conditions.

(b) *Enthalpy of formation* , ΔH_f, is the enthalpy change that occurs when a compound is formed from its component elements.

(c) Standard enthalpy of formation, ΔH_f° is the enthalpy change that accompanies formation of one mole of a substance from elements in their standard states.

5.64 (a) Tables of ΔH_f° are useful because, according to Hess's law, the standard enthalpy of any reaction can be calculated from the standard enthalpies of formation for the reactants and products.

$$\Delta H_{rxn}^\circ = \sum \Delta H_f^\circ \text{ (products)} - \sum \Delta H_f^\circ \text{ (reactants)}$$

(b) The standard enthalpy of formation for any element in its standard state is zero. Elements in their standard states are the reference point for the enthalpy of formation scale.

5.65 Yes, it would still be possible to have tables of standard enthalpies of formation like Table 5.3. Standard enthalpies of formation are the overall **enthalpy difference** between a compound and its component elements in their standard states. Regardless of the value of the enthalpy of formation of the elements, the magnitude of the difference in enthalpies should be the same (assuming the same reaction stoichiometry).

5.66 $C_{12}H_{22}O_{11}(s) + H_2O(l) \rightarrow 2C_6H_{12}O_6(s)$

$$\Delta H_{rxn}^\circ = 2\Delta H^\circ_f\ C_6H_{12}O_6(s) - \Delta H_f^\circ\ C_{12}H_{22}O_{11}(s) - \Delta H_f^\circ\ H_2O(l)$$
$$= 2(-1273) - (-2221) - (-285.83) = -39\ kJ$$

The reaction is slightly exothermic.

5.67

(a)	$1/2\ N_2(g) + 3/2\ H_2(g) \rightarrow NH_3(g)$	ΔH_f° = ~~-80.29 kJ~~ -46.19
(b)	$1/8\ S_8(s) + O_2(g) \rightarrow SO_2(g)$	$\Delta H_f^\circ = -269.9\ kJ$
(c)	$Rb(s) + 1/2\ Cl_2(g) + 3/2\ O_2(g) \rightarrow RbClO_3(s)$	$\Delta H_f^\circ = -392.4\ kJ$
(d)	$N_2(g) + 2H_2(g) + 3/2\ O_2(g) \rightarrow NH_4NO_3(s)$	$\Delta H_f^\circ = -365.6\ kJ$

5.68 (a) $1/2\ H_2(g) + 1/2\ Br_2(l) \rightarrow HBr(g)$ $\Delta H^\circ_f = -36.23$ kJ

(b) $Ag(s) + 1/2\ N_2(g) + 3/2\ O_2(g) \rightarrow AgNO_3(s)$ $\Delta H^\circ_f = -124.4$ kJ

(c) $2Hg(l) + Cl_2(g) \rightarrow Hg_2Cl_2(s)$ $\Delta H^\circ_f = -264.9$ kJ

(d) $2C(s, gr) + 1/2\ O_2(g) + 3H_2(g) \rightarrow C_2H_5OH(l)$ $\Delta H^\circ_f = -277.7$ kJ

5.69 *Plan.* $\Delta H^\circ_{rxn} = \Sigma n\Delta H^\circ_f$ (products) - $\Sigma n\Delta H^\circ_f$ (reactants). Be careful with coefficients, states and signs. *Solve*:

$\Delta H^\circ_{rxn} = \Delta H^\circ_f\ Al_2O_3(s) + 2\Delta H^\circ_f\ Fe(s) - \Delta H^\circ_f\ Fe_2O_3 - 2\Delta H^\circ_f\ Al(s)$

$\Delta H^\circ_{rxn} = (-1669.8 \text{ kJ}) + 2(0) - (-822.16 \text{ kJ}) - 2(0) = -847.6 \text{ kJ}$

5.70 Use heats of formation to calculate ΔH° for the combustion of butane.

$C_4H_{10}(l) + 13/2\ O_2(g) \rightarrow 4CO_2(g) + 5H_2O(l)$

$\Delta H^\circ_{rxn} = 4\Delta H\ CO_2(g) + 5\Delta H^\circ_f\ H_2O(l) - \Delta H^\circ_f\ C_4H_{10}(l) - 13/2\ \Delta H^\circ_f\ O_2(g)$

$\Delta H^\circ_{rxn} = 4(-393.5 \text{ kJ}) + 5(-285.83 \text{ kJ}) - (-147.6 \text{ kJ}) - 13/2(0) = -2855.6 = -2856 \text{ kJ/mol } C_4H_{10}$

$$1.0 \text{ g } C_4H_{10} \times \frac{1 \text{ mol } C_4H_{10}}{58.123 \text{ g } C_4H_{10}} \times \frac{-2855.6 \text{ kJ}}{1 \text{ mol } C_4H_{10}} = -49 \text{ kJ}$$

5.71 *Plan.* $\Delta H^\circ_{rxn} = \Sigma n\Delta H^\circ_f$ (products) - $\Sigma n\Delta H^\circ_f$ (reactants). Be careful with coefficients, states and signs. *Solve*:

(a) $\Delta H^\circ_{rxn} = 2\Delta H^\circ_f\ SO_3(g) - 2\Delta H^\circ_f\ SO_2(g) - \Delta H^\circ_f\ O_2(g)$

$= 2(-395.2 \text{ kJ}) - 2(-296.9 \text{ kJ}) - 0 = -196.6 \text{ kJ}$

(b) $\Delta H^\circ_{rxn} = \Delta H^\circ_f\ MgO(s) + \Delta H^\circ_f\ H_2O(l) - \Delta H^\circ_f\ Mg(OH)_2(s)$

$= -601.8 \text{ kJ} + (-285.83 \text{ kJ}) - (-924.7 \text{ kJ}) = 37.1 \text{ kJ}$

(c) $\Delta H^\circ_{rxn} = 2\Delta H^\circ_f\ Fe_2O_3(s) - 4\Delta H^\circ_f\ FeO(s) - \Delta H^\circ_f\ O_2(g)$

$= 2(-822.16 \text{ kJ}) - 4(-271.9 \text{ kJ}) - 0 = -556.7 \text{ kJ}$

(d) $\Delta H^\circ_{rxn} = \Delta H^\circ_f\ SiO_2(s) + 4\Delta H^\circ_f\ HCl(g) - \Delta H^\circ_f\ SiCl_4(l) - 2\ \Delta H^\circ_f\ H_2O(l)$

$= -910.9 \text{ kJ} + 4(-92.30 \text{ kJ}) - (-640.1 \text{ kJ}) - 2(-285.83 \text{ kJ}) = -68.3 \text{ kJ}$

5.72 (a) $\Delta H^\circ_{rxn} = 2\Delta H^\circ_f\ H_2O(g) + \Delta H^\circ_f\ N_2(g) - \Delta H^\circ_f\ N_2O_4(g) - \Delta H^\circ_f\ H_2(g)$

$= 2(-241.82) + 0 - (9.66) - 0 = -493.30 \text{ kJ}$

(b) $\Delta H^\circ_{rxn} = \Delta H^\circ_f\ K_2CO_3(s) + \Delta H^\circ_f\ H_2O(g) - 2\Delta H^\circ_f\ KOH(s) - \Delta H^\circ_f\ CO_2(g)$

$= -1150.18 \text{ kJ} - 241.82 \text{ kJ} - 2(-424.7) \text{ kJ} - (-393.5 \text{ kJ}) = -149.1 \text{ kJ}$

(c) $\Delta H^\circ_{rxn} = 3/8\Delta H^\circ_f\ S_8(s) + 2\Delta H^\circ_f\ H_2O(g) - \Delta H^\circ_f\ SO_2(g) - 2\Delta H^\circ_f\ H_2S(g)$

$= 3/8(0) + 2(-241.82) - (-269.9) - 2(-20.17) = -173.4$ kJ

(d) $\Delta H^\circ_{rxn} = 2\Delta H^\circ_f\ FeCl_3(s) + 3\Delta H^\circ_f\ H_2O(g) - \Delta H^\circ_f\ Fe_2O_3(s) - 6\Delta H^\circ_f\ HCl(g)$

$= 2(-400$ kJ$) + 3(-241.82$ kJ$) - (-822.16$ kJ$) - 6(-92.30$ kJ$) = -149.5$ kJ

5.73 *Analyze.* Given: combustion reaction, enthalpy of combustion, enthalpies of formation for most reactants and products. Find: enthalpy of formation for acetone. *Plan.* Rearrange the expression for enthalpy of reaction to calculate the desired enthalpy of formation. *Solve*:

$\Delta H^\circ_{rxn} = 3\Delta H^\circ_f\ CO_2(g) + 3\Delta H^\circ_f\ H_2O(l) - \Delta H^\circ_f\ C_3H_6O(l)$

-1790 kJ $= 3(-393.5$ kJ$) + 3(-285.83$ kJ$) - \Delta H^\circ_f\ C_3H_6O(l)$

$\Delta H^\circ_f\ C_3H_6O(l) = -248$ kJ

5.74 $\Delta H^\circ_{rxn} = \Delta H^\circ_f\ Ca(OH)_2(s) + \Delta H^\circ_f\ C_2H_2(g) - 2\Delta H^\circ_f\ H_2O(l) - \Delta H^\circ_f\ CaC_2(s)$

-127.2 kJ $= -986.2$ kJ $+ 226.7$ kJ $- 2(-285.83$ kJ$) - \Delta H^\circ_f\ CaC_2(s)$

ΔH°_f for $CaC_2(s) = -60.6$ kJ

5.75 *Plan.* Use Hess's Law to arrange the given reactions so the overall sum is the formation reaction for $Mg(OH)_2(s)$. Adjust the corresponding ΔH values and calculate ΔH°_f for $Mg(OH)_2(s)$. *Solve*:

$Mg(s) + 1/2\ O_2(g) \rightarrow MgO(s)$	$\Delta H^\circ = 1/2(-1203.6$ kJ$)$
$MgO(s) + H_2O(l) \rightarrow Mg(OH)_2(s)$	$\Delta H^\circ = -(37.1$ kJ$)$
$H_2(g) + 1/2\ O_2(g) \rightarrow H_2O(l)$	$\Delta H^\circ = 1/2(-571.7$ kJ$)$
$Mg(s) + O_2(g) + H_2(g) \rightarrow Mg(OH)_2(s)$	$\Delta H^\circ_f = -924.8$ kJ

Check. The overall reaction is correct.

5.76 (a)

$2B(s) + 3/2\ O_2(g) \rightarrow B_2O_3(s)$	$\Delta H^\circ = 1/2(-2509.1$ kJ$)$
$3H_2(g) + 3/2\ O_2(g) \rightarrow 3H_2O(l)$	$\Delta H^\circ = 3/2(-571.7$ kJ$)$
$B_2O_3(s) + 3H_2O(l) \rightarrow B_2H_6(g) + 3O_2(g)$	$\Delta H^\circ = -(-2147.5$ kJ$)$
$2B(s) + 3H_2(g) \rightarrow B_2H_6(g)$	$\Delta H^\circ_f = +35.4$ kJ

(b) If, like B_2H_6, the combustion of B_5H_9 produces B_2O_3 as the boron-containing product, the heat of combustion of B_5H_9 in addition to data given in part (a) would enable calculation of the heat of formation of B_5H_9.

The combustion reaction is: $B_5H_9(l) + 6O_2(g) \rightarrow 5/2\ B_2O_3(s) + 9/2\ H_2O(l)$

$5/4\ [4B(s) + 3O_2(g) \rightarrow 2B_2O_3(s)]$	$\Delta H = 5/4(-2509.1\ kJ)$
$9/4\ [2H_2(g) + O_2(g) \rightarrow 2H_2O(l)]$	$\Delta H = 9/4\ (-571.7\ kJ)$
$5/2\ B_2O_3(s) + 9/2\ H_2O(l) \rightarrow B_5H_9(l) + 6O_2(g)$	ΔH = - (heat of combustion)
$5B(s) + 9/2\ H_2(g) \rightarrow B_5H_9(l)$	ΔH_f° of $B_5H_9(l)$

[ΔH_f° B_5H_9 (l) = - [heat of combustion of $B_5H_9(l)$] - 3136.4 kJ - 1286 kJ]

We need to measure the heat of combustion of $B_5H_9(l)$.

5.77 (a) $C_8H_{18}(l) + 25/2\ O_2(g) \rightarrow 8CO_2(g) + 9H_2O(g)$ $\quad \Delta H^\circ = -5069\ kJ$

(b) $8C(s, gr) + 9H_2(g) \rightarrow C_8H_{18}(l)$ $\quad \Delta H_f^\circ = ?$

(c) *Plan.* Follow the logic in Solution 5.73. *Solve:*

$$\Delta H_{rxn}^\circ = 8\Delta H_f^\circ\ CO_2(g) + 9\Delta H_f^\circ\ H_2O(g) - \Delta H_f^\circ\ C_8H_{18}(l) - 25/2\ \Delta H_f^\circ\ O_2(g)$$

$$-5069\ kJ = 8(-393.5\ kJ) + 9(-241.82\ kJ) - \Delta H_f^\circ\ C_8H_{18}(l) - 25/2(0)$$

$$\Delta H_f^\circ\ C_8H_{18}(l) = 8(-393.5\ kJ) + 9(-241.82\ kJ) + 5069\ kJ = -255\ kJ$$

5.78 (a) $10C(s) + 4H_2(g) \rightarrow C_{10}H_8(s)$ $\quad$ formation

$C_{10}H_8(s) + 12O_2(g) \rightarrow 10CO_2(g) + 4H_2O(l)$ $\quad$ combustion, $\Delta H^\circ = -5154\ kJ$

(b) $$\Delta H_{rxn}^\circ = 10\Delta H_f^\circ\ CO_2(g) + 4\Delta H_f^\circ\ H_2O(l) - \Delta H_f^\circ\ C_{10}H_8 - 12\Delta H_f^\circ\ O_2(g)$$

$$-5154 = 10(-393.5\ kJ) + 4(-285.83\ kJ) - \Delta H_f^\circ\ C_{10}H_8(s) - 12(0)$$

$$\Delta H_f^\circ\ C_{10}H_8(s) = 10(-393.5\ kJ) + 4(-285.83\ kJ) + 5154\ kJ = 76\ kJ$$

Check. The result has 0 decimal places because the heat of combustion has 0 decimal places.

Foods and Fuels

5.79 (a) *Fuel value* is the amount of heat produced when 1 gram of a substance (fuel) is combusted.

(b) Glucose, $C_6H_{12}O_6$, is referred to as *blood sugar.* It is important because glucose is the fuel that is carried by blood to cells and combusted to produce energy in the body.

(c) The fuel value of fats is 9 kcal/g and of carbohydrates is 4 kcal/g. Therefore, 5 g of fat produce 45 kcal, while 9 g of carbohydrates produce 36 kcal; 5 g of fat are a greater energy source.

5.80 (a) Fats are appropriate for fuel storage because they are insoluble in water (and body fluids) and have a high fuel value.

(b) Combustion analysis is a poor measure of the fuel value of proteins, since the products of protein combustion in the body are different than the products of combustion in a calorimeter. That is, we are not measuring ΔH for the reaction that occurs in the body.

(c) $25 \text{ g fat} \times \frac{38 \text{ kJ}}{\text{g fat}} = x \text{ g protein} \times \frac{17 \text{ kJ}}{\text{g protein}}$; $x = 56$ g protein

5.81 *Plan.* Calculate the Cal (kcal) due to each nutritional component of the Campbell's® soup, then sum. *Solve.*

$$9 \text{ g carbohydrates} \times \frac{17 \text{ kJ}}{1 \text{ g carbohydrate}} = 153 \text{ or } 2 \times 10^2 \text{ kJ}$$

$$1 \text{ g protein} \times \frac{17 \text{ kJ}}{1 \text{ g protein}} = 17 \text{ or } 0.2 \times 10^2 \text{ kJ}$$

$$7 \text{ g fat} \times \frac{38 \text{ kJ}}{1 \text{ g fat}} = 266 \text{ or } 3 \times 10^2 \text{ kJ}$$

total energy = 153 kJ + 17 kJ + 266 kJ = 436 or 4×10^2 kJ

$$436 \text{ kJ} \times \frac{1 \text{ kcal}}{4.184 \text{ kJ}} \times \frac{1 \text{ Cal}}{1 \text{kcal}} = 104 \text{ or } 1 \times 10^2 \text{ Cal/serving}$$

Check. 100 Cal/serving is a reasonable result; units are correct. The data and the result have 1 sig fig.

5.82 Calculate the fuel value in a pound of M&M® candies.

$$96 \text{ fat} \times \frac{38 \text{ kJ}}{1 \text{ g fat}} = 3648 \text{ kJ} = 3.6 \times 10^3 \text{ kJ}$$

$$320 \text{ g carbohydrate} \times \frac{17 \text{ kJ}}{1 \text{ g carbohydrate}} = 5440 \text{ kJ} = 5.4 \times 10^3 \text{ kJ}$$

$$21 \text{ g protein} \times \frac{17 \text{ kJ}}{1 \text{ g protein}} = 357 \text{ kJ} = 3.6 \times 10^2 \text{ kJ}$$

total fuel value = 3648 kJ + 5440 kJ + 357 kJ = 9445 kJ = 9.4×10^3 kJ/lb

$$\frac{9445 \text{ kJ}}{\text{lb}} \times \frac{1 \text{ lb}}{453.6 \text{ g}} \times \frac{42 \text{ g}}{\text{serving}} = 874.5 \text{ kJ} = 8.7 \times 10^2 \text{ kJ/serving}$$

$$\frac{874.5 \text{ kJ}}{\text{serving}} \times \frac{1 \text{ kcal}}{4.184 \text{ kJ}} \times \frac{1 \text{ Cal}}{1 \text{ kcal}} = 209.0 \text{ Cal} = 2.1 \times 10^2 \text{ Cal/serving}$$

Check. 210 Cal is the approximate food value of a candy bar so the result is reasonable.

5.83 *Plan.* g → mol → kJ → Cal *Solve*:

$$16.0 \text{ g } C_6H_{12}O_6 \times \frac{1 \text{ mol } C_6H_{15}O_6}{180.2 \text{ g } C_6H_{12}O_6} \times \frac{2812 \text{ kJ}}{\text{mol } C_6H_{12}O_6} \times \frac{1 \text{ Cal}}{4.184 \text{ kJ}} = 59.7 \text{ Cal}$$

Check. 60 Cal is a reasonable result for most of the food value in an apple.

5.84 $$177 \text{ mL} \times \frac{1.0 \text{ g wine}}{1 \text{ mL}} \times \frac{0.106 \text{ g ethanol}}{1 \text{ g wine}} \times \frac{1 \text{ mol ethanol}}{46.1 \text{ g ethanol}} \times \frac{1367 \text{ kJ}}{1 \text{ mol ethanol}} \times \frac{1 \text{ Cal}}{4.184 \text{ kJ}}$$

$$= 133 = 1.3 \times 10^2 \text{ Cal}$$

Check. A "typical" 6 oz. glass of wine has 150-250 Cal, so this is a reasonable result. Note that alcohol is responsible for most of the food value of wine.

5.85 *Plan.* Use enthalpies of formation to calculate molar heat (enthalpy) of combustion using Hess's Law. Use molar mass to calculate heat of combustion per kg of hydrocarbon. *Solve*:

Propyne: $C_3H_4(g) + 4O_2(g) \rightarrow 3CO_2(g) + 2H_2O(g)$

(a) $\Delta H = 3(-393.5 \text{ kJ}) + 2(-241.82 \text{ kJ}) - (185.4 \text{ kJ}) - 4(0) = -1849.5 = -1850 \text{ kJ/mol } C_3H_4$

(b) $$\frac{-1849.5 \text{ kJ}}{1 \text{ mol } C_3H_4} \times \frac{1 \text{ mol } C_3H_4}{40.065 \text{ g } C_3H_4} \times \frac{1000 \text{ g } C_3H_4}{1 \text{ kg } C_3H_4} = -4.616 \times 10^4 \text{ kJ/kg } C_3H_4$$

Propylene: $C_3H_6(g) + 9/2\ O_2(g) \rightarrow 3CO_2(g) + 3H_2O(g)$

(a) $\Delta H = 3(-393.5 \text{ kJ}) + 3(-241.82 \text{ kJ}) - (20.4 \text{ kJ}) - 9/2(0) = -1926.4 = -1926 \text{ kJ/mol } C_3H_6$

(b) $$\frac{-1926.4 \text{ kJ}}{1 \text{ mol } C_3H_6} \times \frac{1 \text{ mol } C_3H_6}{42.080 \text{ g } C_3H_6} \times \frac{1000 \text{ g } C_3H_6}{1 \text{ kg } C_3H_6} = -4.578 \times 10^4 \text{ kJ/kg } C_3H_6$$

Propane: $C_3H_8(g) + 5O_2(g) \rightarrow 3CO_2(g) + 4H_2O(g)$

(a) $\Delta H = 3(-393.5 \text{ kJ}) + 4(-241.82 \text{ kJ}) - (-103.8 \text{ kJ}) - 5(0) = -2044.0 = -2044 \text{ kJ/mol } C_3H_8$

(b) $$\frac{-2044.0 \text{ kJ}}{1 \text{ mol } C_3H_8} \times \frac{1 \text{ mol } C_3H_8}{44.096 \text{ g } C_3H_8} \times \frac{1000 \text{ g } C_3H_8}{1 \text{ kg } C_3H_8} = -4.635 \times 10^4 \text{ kJ/kg } C_3H_8$$

(c) These three substances yield nearly identical quantities of heat per unit mass, but propane is marginally higher than the other two.

5.86 $$\Delta H^\circ_{rxn} = \Delta H^\circ_f\ CO_2(g) + 2\Delta H^\circ_f\ H_2O(g) - \Delta H^\circ_f\ CH_4(g) - \Delta H^\circ_f\ O_2(g)$$

$$= -393.5 \text{ kJ} + 2(-241.82 \text{ kJ}) - (-74.8 \text{ kJ}) - 0 \text{ kJ} = -802.3 \text{ kJ}$$

$$\Delta H^\circ_{rxn} = \Delta H^\circ_f\ CF_4(g) + 4\Delta H^\circ_f\ HF(g) - \Delta H^\circ_f\ CH_4(g) - 4\Delta H^\circ_f\ F_2(g)$$

$$= -679.9 \text{ kJ} + 4(-268.61 \text{ kJ}) - (-74.8 \text{ kJ}) - 0 \text{ kJ} = -1679.5 \text{ kJ}$$

The second reaction is twice as exothermic as the first. The "fuel values" of hydrocarbons in a fluorine atmosphere are approximately twice those in an oxygen atmosphere. Note that the difference in ΔH° values for the two reactions is in the ΔH°_f for the products, since the ΔH°_f for the reactants are identical.

Additional Exercises

5.87 (a) mi/hr → m/s

$$1050 \frac{mi}{hr} \times \frac{1.6093\ km}{1\ mi} \times \frac{1000\ m}{1\ km} \times \frac{1\ hr}{3600\ s} = 469.38 = 469.4\ m/s$$

(b) Find the mass of one N_2 molecule in kg.

$$\frac{28.0134\ g\ N_2}{1\ mol} \times \frac{1\ mol}{6.022 \times 10^{23}\ molecules} \times \frac{1\ kg}{1000\ g} = 4.6518 \times 10^{-26}$$

$$= 4.652 \times 10^{-26}\ kg$$

$E_k = 1/2\ mv^2 = 1/2 \times 4.6518 \times 10^{-26}\ kg \times (469.38\ m/s)^2$

$$= 5.1244 \times 10^{-21} \frac{kg \bullet m^2}{s^2} = 5.124 \times 10^{-21}\ J$$

(c) $$\frac{5.1244 \times 10^{21}\ J}{molecule} \times \frac{6.022 \times 10^{23}\ molecules}{1\ mol} = 3086\ J/mol = 3.086\ kJ/mol$$

5.88 (a) $E_p = mgh = 52.0\ kg \times 9.81\ m/s^2 \times 10.8\ m = 5509.3\ J = 5.51\ kJ$

(b) $E_k = 1/2\ mv^2;\ v = (2E_k/m)^{1/2} = \left(\frac{2 \times 5509\ kg \bullet m^2/s^2}{52.0\ kg}\right)^{1/2} = 14.6\ m/s$

(c) Yes, the diver does work on entering (pushing back) the water in the pool.

5.89 In the process described, one mole of solid CO_2 is converted to one mole of gaseous CO_2. The volume of the gas is much greater than the volume of the solid. Thus the system (that is, the mole of CO_2) must work against atmospheric pressure when it expands. To accomplish this work while maintaining a constant temperature requires the absorption of additional heat beyond that required to increase the internal energy of the CO_2. The remaining energy is turned into work.

5.90 Like the combustion of $H_2(g)$ and $O_2(g)$ described in Section 5.4, the reaction that inflates airbags is spontaneous after initiation. Spontaneous reactions are usually exothermic, $-\Delta H$. The airbag reaction occurs at constant atmospheric pressure, $\Delta H = q_p$; both are likely to be large and negative. When the bag inflates, work is done by the system on the surroundings so the sign of w is negative.

5.91 Freezing is an exothermic process (the opposite of melting, which is clearly endothermic). When the system, the soft drink, freezes, it releases energy to the surroundings, the can. Some of this energy does the work of splitting the can.

5.92 Equation 5.5 describes energy changes to the system. Specifically, it says any change in the internal energy of the system is manifested as heat transfer or work. The first law requires that an equal but opposite change occur to the surroundings.

$$\Delta E_{sys} = -\Delta E_{surr}$$

$$q_{sys} + w_{sys} = -(q_{surr} + w_{surr})$$

5.93 (a) $q = 0$, $w > 0$ (work done to system), $\Delta E > 0$

(b) Since the system (the gas) is losing heat, the sign of q is negative. The changes in state described in cases (a) and (b) are identical and ΔE is the same in both cases. The distribution of energy transferred as either work or heat is different in the two scenarios. In case (b), more work is required to compress the gas because some heat is lost to the surroundings. [The moral of this story is that the more energy lost by the system as heat, the greater the work on the system required to accomplish the desired change.]

5.94 $\Delta E = q + w = +38.95 \text{ kJ} - 2.47 \text{ kJ} = +36.48 \text{ kJ}$

$\Delta H = q_p = +38.95 \text{ kJ}$

5.95 If a function sometimes depends on path, then it is simply not a state function. Enthalpy is a state function, so ΔH for the two pathways leading to the same change of state pictured in Figure 5.10 must be the same. However, q is not the same for both. Our conclusion must be that $\Delta H \neq q$ for these pathways. The condition for $\Delta H = q_p$ (other than constant pressure) is that the only possible work on or by the system is pressure-volume work. Clearly, the work being done in this scenario is not pressure-volume work, so $\Delta H \neq q$, even though the two changes occur at constant pressure.

5.96 Find the heat capacity of 1.7×10^3 gal H_2O.

$$C_{H_2O} = 1.7 \times 10^3 \text{ gal } H_2O \times \frac{4 \text{ qt}}{1 \text{ gal}} \times \frac{1 \text{ L}}{1.057 \text{ qt}} \times \frac{1 \times 10^3 \text{ cm}^3}{1 \text{ L}} \times \frac{1 \text{ g}}{1 \text{ cm}^3} \times \frac{4.184 \text{ J}}{1 \text{ g} \cdot {}^\circ\text{C}}$$

$= 2.692 \times 10^7 \text{ J/}^\circ\text{C} = 2.7 \times 10^4 \text{ kJ/}^\circ\text{C}$; then,

$$\frac{2.692 \times 10^7 \text{ J}}{1^\circ\text{C}} \times \frac{1^\circ\text{C} \cdot \text{g}}{0.85 \text{ J}} \times \frac{1 \text{ kg}}{1 \times 10^3 \text{ g}} \times \frac{1 \text{ brick}}{1.8 \text{ kg}} = 1.8 \times 10^4 \text{ or 18,000 bricks}$$

Check. $(1.7 \times {\sim}16 \times 10^6) / ({\sim}1.6 \times 10^3) \approx 17 \times 10^3$ bricks; the units are correct.

5.97 (a) $$q_{Cu} = \frac{0.385 \text{ J}}{\text{g} \cdot \text{K}} \times 121.0 \text{ g Cu} \times (30.1^\circ\text{C} - 100.4^\circ\text{C}) = -3274.9 = -3.27 \times 10^3 \text{ J}$$

The negative sign indicates the 3.27×10^3 J are lost by the Cu block.

(b) $$q_{H_2O} = \frac{4.184 \text{ J}}{\text{g} \cdot \text{K}} \times 150.0 \text{ g } H_2O \times (30.1^\circ\text{C} - 25.1^\circ\text{C}) = 3138 = +3.1 \times 10^3 \text{ J}$$

The positive sign indicates that 3.14×10^3 J are gained by the H_2O.

(c) The difference in the heat lost by the Cu and the heat gained by the water is $3.275 \times 10^3 \text{ J} - 3.138 \times 10^3 \text{ J} = 0.137 \times 10^3 \text{ J} = 1 \times 10^2$ J. The temperature change of the calorimeter is 5.0°C. The heat capacity of the calorimeter in J/K is $0.137 \times 10^3 \text{ J} \times \frac{1}{5.0^\circ\text{C}} = 27.4 = 3 \times 10 \text{ J/K}$.

Since q_{H_2O} is known to 1 dec. place, the difference has 1 dec. place and the result has 1 sig fig.

If the rounded results from (a) and (b) are used, $C_{calorimeter} = \frac{0.2 \times 10^3 \text{ J}}{5.0\ °\text{C}} = 4 \times 10$ J/K.

(d) $q_{H_2O} = 3.275 \times 10^3 \text{ J} = \frac{4.184 \text{ J}}{\text{g}\bullet\text{K}} \times 150.0 \text{ g} \times (\Delta T)$

$\Delta T = 5.22°C$; $T_f = 25.1°C + 5.22°C = 30.3°C$

5.98 (a) From the mass of benzoic acid that produces a certain temperature change, we can calculate the heat capacity of the calorimeter.

$$\frac{0.235 \text{ g benzoic acid}}{1.642°\text{C change observed}} \times \frac{26.38 \text{ kJ}}{1 \text{ g benzoic acid}} = 3.7755 = 3.78 \text{ kJ/°C}$$

Now we can use this experimentally determined heat capacity with the data for caffeine.

$$\frac{1.525°\text{C rise}}{0.265 \text{ g caffeine}} \times \frac{3.7755 \text{ kJ}}{1°\text{C}} \times \frac{194.2 \text{ g caffeine}}{1 \text{ mol caffeine}} = 4.22 \times 10^3 \text{ kJ/mol caffeine}$$

(b) The overall uncertainty is approximately equal to the sum of the uncertainties due to each effect. The uncertainty in the mass measurement is 0.001/0.235 or 0.001/0.265, about 1 part in 235 or 1 part in 265. The uncertainty in the temperature measurements is 0.002/1.642 or 0.002/1.525, about 1 part in 820 or 1 part in 760. Thus the uncertainty in heat of combustion from each measurement is

$$\frac{4220}{235} = 18 \text{ kJ}; \quad \frac{4220}{265} = 16 \text{ kJ}; \quad \frac{4220}{820} = 5 \text{ kJ}; \quad \frac{4220}{760} = 6 \text{ kJ}$$

The sum of these uncertainties is 45 kJ. In fact, the overall uncertainty is less than this because independent errors in measurement do tend to partially cancel.

5.99 $\Delta E_p = m\, g\, \Delta h$. Be careful with units. $1 \text{ J} = 1 \text{ kg} \bullet \text{m}^2/\text{s}^2$

$$200 \text{ lb} \times \frac{2.205 \text{ kg}}{1 \text{ lb}} \times \frac{9.81 \text{ m}}{\text{s}^2} \times \frac{45 \text{ ft}}{\text{time}} \times \frac{1 \text{ yd}}{3 \text{ ft}} \times \frac{1 \text{ m}}{1.0936 \text{ yd}} \times 20 \text{ times}$$

$$= 1.87 \times 10^6 \text{ kg} \bullet \text{m}^2/\text{s}^2 = 1.187 \times 10^6 \text{ J} = 1.2 \times 10^3 \text{ kJ}$$

1 Cal = 1 kcal = 4.184 kJ

$$1.187 \times 10^3 \text{ kJ} \times \frac{1 \text{ Cal}}{4.184 \text{ kJ}} = 283.6 = 2.8 \times 10^2 \text{ Cal}$$

If all work is used to increase the man's potential energy, 20 rounds of stair-climbing should compensate for one extra order of 245 Cal fries. In fact, more than 280 Cal of work will be required to climb the stairs, because some energy will be lost as heat (see Solution 5.93).

5.100 (a) For comparison, balance the equations so that 1 mole of CH_4 is burned in each.

$CH_4(g) + O_2(g) \rightarrow C(s) + 2H_2O(l)$ $\Delta H^\circ = -496.9$ kJ

$CH_4(g) + 3/2\ O_2(g) \rightarrow CO(g) + 2H_2O(l)$ $\Delta H^\circ = -607.4$ kJ

$CH_4(g) + 2O_2(g) \rightarrow CO_2(g) + 2H_2O(l)$ $\Delta H^\circ = -890.4$ kJ

(b) $\Delta H^\circ_{rxn} = \Delta H^\circ_f\ C(s) + 2\Delta H^\circ_f\ H_2O(l) - \Delta H^\circ_f\ CH_4(g) - \Delta H^\circ_f\ O_2(g)$

$= 0 + 2(-285.83 \text{ kJ}) - (-74.8) - 0 = -496.9 \text{ kJ}$

$\Delta H^\circ_{rxn} = \Delta H^\circ_f\ CO(g) + 2\Delta H^\circ_f\ H_2O(l) - \Delta H^\circ_f\ CH_4(g) - 3/2\ \Delta H^\circ_f\ O_2(g)$

$= (-110.5 \text{ kJ}) + 2(-285.83 \text{ kJ}) - (-74.8 \text{ kJ}) - 3/2(0) = -607.4 \text{ kJ}$

$\Delta H^\circ_{rxn} = \Delta H^\circ_f\ CO_2(g) + 2\Delta H^\circ_f\ H_2O(l) - \Delta H^\circ_f\ CH_4(g) - 2\Delta H^\circ_f\ O_2(g)$

$= -393.5 \text{ kJ} + 2(-285.83 \text{ kJ}) - (-74.8 \text{ kJ}) - 2(0) = -890.4 \text{ kJ}$

(c) Assuming that $O_2(g)$ is present in excess, the reaction that produces $CO_2(g)$ represents the most negative ΔH per mole of CH_4 burned. More of the potential energy of the reactants is released as heat during the reaction to give products of lower potential energy. The reaction that produces $CO_2(g)$ is the most "downhill" in enthalpy.

5.101 For nitroethane:

$$\frac{1368 \text{ kJ}}{1 \text{ mol } C_2H_5NO_2} \times \frac{1 \text{ mol } C_2H_5NO_2}{75.072 \text{ g } C_2H_5NO_2} \times \frac{1.052 \text{ g } C_2H_5NO_2}{1 \text{ cm}^3} = 19.17 \text{ kJ/cm}^3$$

For ethanol:

$$\frac{1367 \text{ kJ}}{1 \text{ mol } C_2H_5OH} \times \frac{1 \text{ mol } C_2H_5OH}{46.069 \text{ g } C_2H_5OH} \times \frac{0.789 \text{ g } C_2H_5OH}{1 \text{ cm}^3} = 23.4 \text{ kJ/cm}^3$$

For methylhydrazine:

$$\frac{1305 \text{ kJ}}{1 \text{ mol } CH_6N_2} \times \frac{1 \text{ mol } CH_6N_2}{46.072 \text{ g } CH_6N_2} \times \frac{0.874 \text{ g } CH_6N_2}{1 \text{ cm}^3} = 24.8 \text{ kJ/cm}^3$$

Thus, **methylhydrazine** would provide the most energy per unit volume, with ethanol a close second.

5.102 (a) $3C_2H_2(g) \rightarrow C_6H_6(l)$

$\Delta H^\circ_{rxn} = \Delta H^\circ_f\ C_6H_6(l) - 3\Delta H^\circ_f\ C_2H_2(g) = 49.0 \text{ kJ} - 3(226.7 \text{ kJ}) = -631.1 \text{ kJ}$

(b) Since the reaction is exothermic (ΔH is negative), the product, 1 mole of $C_6H_6(l)$, has less enthalpy than the reactants, 3 moles of $C_2H_2(g)$.

(c) The fuel value of a substance is the amount of heat (kJ) produced when 1 gram of the substance is burned. Calculate the molar heat of combustion (kJ/mol) and use this to find kJ/g of fuel.

$C_2H_2(g) + 5/2\ O_2(g) \rightarrow 2CO_2(g) + H_2O(l)$

$\Delta H^\circ_{rxn} = 2\Delta H^\circ_f\ CO_2(g) + \Delta H^\circ_f\ H_2O(l) - \Delta H^\circ_f\ C_2H_2(g) - 5/2\ \Delta H^\circ_f\ O_2(g)$

$= 2(-393.5\text{ kJ}) + (-285.83\text{ kJ}) - 226.7\text{ kJ} - 5/2\ (0) = -1299.5\text{ kJ/mol } C_2H_2$

$$\frac{-1299.5\text{ kJ}}{1\text{ mol } C_2H_2} \times \frac{1\text{ mol } C_2H_2}{26.036\text{ g } C_2H_2} = 49.912 = 50\text{ kJ/g}$$

$C_6H_6(g) + 15/2\ O_2(g) \rightarrow 6CO_2(g) + 3H_2O(l)$

$\Delta H^\circ_{rxn} = 6\Delta H^\circ_f\ CO_2(g) + 3\Delta H^\circ_f\ H_2O(l) - \Delta H^\circ_f\ C_6H_6(l) - 15/2\ \Delta H^\circ_f\ O_2(g)$

$= 6(-393.5\text{ kJ}) + 3(-285.83\text{ kJ}) - 49.0\text{ kJ} - 15/2\ (0) = -3267.5\text{ kJ/mol } C_6H_6$

$$\frac{-3267.5\text{ kJ}}{1\text{ mol } C_6H_6} \times \frac{1\text{ mol } C_6H_6}{78.114\text{ g } C_6H_6} = 41.830 = 42\text{ kJ/g}$$

5.103 **1,3-butadiene**, C_4H_6, $\mathcal{M} = 54.092$ g/mol

(a) $C_4H_6(g) + 11/2\ O_2(g) \rightarrow 4CO_2(g) + 3H_2O(l)$

$\Delta H^\circ_{rxn} = 4\Delta H^\circ_f\ CO_2(g) + 3\Delta H^\circ_f\ H_2O(l) - \Delta H^\circ_f\ C_4H_6(g) - 11/2\ \Delta H^\circ_f\ O_2(g)$

$= 4(-393.5\text{ kJ}) + 3(-285.83\text{ kJ}) - 111.9\text{ kJ} + 11/2\ (0) = -2543.4\text{ kJ/mol } C_4H_6$

(b) $$\frac{-2543.4\text{ kJ}}{1\text{ mol } C_4H_6} \times \frac{1\text{ mol } C_4H_6}{54.092\text{ g}} = 47.020 \rightarrow 47\text{ kJ/g}$$

(c) $$\%\ H = \frac{6(1.008)}{54.092} \times 100 = 11.18\%\ H$$

1-butene, C_4H_8, $\mathcal{M} = 56.108$ g/mol

(a) $C_4H_8(g) + 6O_2(g) \rightarrow 4CO_2(g) + 4H_2O(l)$

$\Delta H^\circ_{rxn} = 4\Delta H^\circ_f\ CO_2(g) + 4\Delta H^\circ_f\ H_2O(l) - \Delta H^\circ_f\ C_4H_8(g) - 6\Delta H^\circ_f\ O_2(g)$

$= 4(-393.5\text{ kJ}) + 4(-285.83\text{ kJ}) - 1.2\text{ kJ} - 6(0) = -2718.5\text{ kJ/mol } C_4H_8$

(b) $$\frac{-2718.5\text{ kJ}}{1\text{ mol } C_4H_8} \times \frac{1\text{ mol } C_4H_8}{56.108\text{ g } C_4H_8} = 48.451 \rightarrow 48\text{ kJ/g}$$

(c) $$\%\ H = \frac{8(1.008)}{56.108} \times 100 = 14.37\%\ H$$

n-butane, $C_4H_{10}(g)$, $\mathcal{M} = 58.124$ g/mol

(a) $C_4H_{10}(g) + 13/2\ O_2(g) \rightarrow 4CO_2(g) + 5H_2O(l)$

$\Delta H^\circ_{rxn} = 4\Delta H^\circ_f\ CO_2(g) + 5\Delta H^\circ_f\ H_2O(l) - \Delta H^\circ_f\ C_4H_{10}(g) - 13/2\ \Delta H^\circ_f\ O_2(g)$

$= 4(-393.5\text{ kJ}) + 5(-285.83\text{ kJ}) - (-124.7\text{ kJ}) - 13/2(0) = -2878.5\text{ kJ/mol } C_4H_{10}$

(b) $$\frac{-2878.5\text{ kJ}}{1\text{ mol } C_4H_{10}} \times \frac{1\text{ mol } C_4H_{10}}{58.124\text{ g } C_4H_{10}} = 49.523 \rightarrow 50\text{ kJ/g}$$

(c) $\% H = \frac{10(1.008)}{58.124} \times 100 = 17.34\% H$

(d) It is certainly true that as the mass % H increases, the fuel value (kJ/g) of the hydrocarbon increases, given the same number of C atoms. A graph of the data in parts (b) and (c) (see below) suggests that mass % H and fuel value are directly proportional when the number of C atoms is constant.

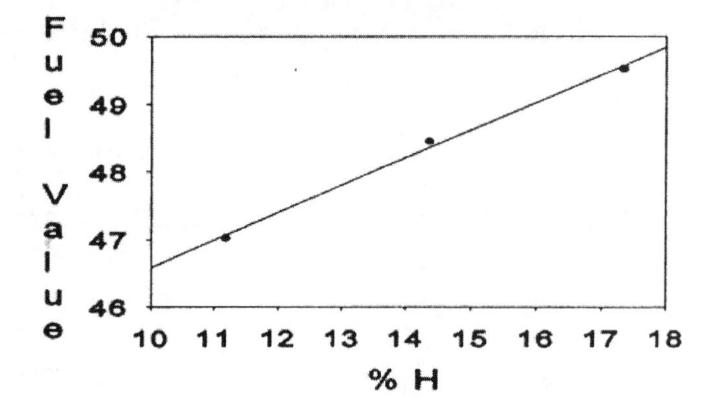

5.104 (a) $C_6H_{12}O_6(s) + 6O_2(g) \rightarrow 6CO_2(g) + 6H_2O(l)$

$\Delta H^{\circ}_{rxn} = 6\Delta H^{\circ}_f\ CO_2(g) + 6\Delta H^{\circ}_f\ H_2O(l) - \Delta H^{\circ}_f\ C_6H_{12}O_6(s) - 6\Delta H^{\circ}_f\ O_2(g)$

$= 6(-393.5\ kJ) + 6(-285.83\ kJ) - (-1273\ kJ) - 6(0)$

$= -2803\ kJ/mol\ C_6H_{12}O_6$

$C_{12}H_{22}O_{11}(s) + 12O_2(g) \rightarrow 12CO_2(g) + 11H_2O(l)$

$\Delta H^{\circ}_{rxn} = 12\Delta H^{\circ}_f\ CO_2(g) + 11\Delta H^{\circ}_f\ H_2O(l) - \Delta H^{\circ}_f\ C_{12}H_{22}O_{11}(s) - 12\Delta H^{\circ}_f\ O_2(g)$

$= 12(-393.5\ kJ) + 11(-285.83\ kJ) - (-2221\ kJ) -12(0)$

$= -5645\ kJ/mol\ C_{12}H_{22}O_{11}$

(b) $\frac{-2803\ kJ}{1\ mol\ C_6H_{12}O_6} \times \frac{1\ mol\ C_6H_{12}O_6}{180.2\ g\ C_6H_{12}O_6} = -\frac{15.55\ kJ}{1\ g\ C_6H_{12}O_6} \rightarrow 16\ kJ/g\ C_6H_{12}O_6$ (fuel value)

$\frac{-5645\ kJ}{1\ mol\ C_{12}H_{22}O_{11}} \times \frac{1\ mol\ C_{12}H_{22}O_{11}}{342.3\ g\ C_{12}H_{22}O_{11}} = -\frac{16.49\ kJ}{1\ g\ C_{12}H_{22}O_{11}} \rightarrow 16\ kJ/g\ C_{12}H_{22}O_{11}$ (fuel value)

(c) The average fuel value of carbohydrates (Section 5.8) is 17 kJ/g. These two carbohydrates have fuel values (16 kJ/g) slightly lower but in line with this average. (More complex carbohydrates supply more energy and raise the average value.)

5.105 (a) $6CO_2(g) + 6H_2O(l) \rightarrow C_6H_{12}O_6(s) + 6O_2(g)$, $\Delta H^{\circ} = 2803\ kJ$

This is the reverse of the combustion of glucose (Section 5.8 and Solution 5.104), so $\Delta H^{\circ} = -(-2803)\ kJ = +2803\ kJ$.

$\frac{5.5 \times 10^{16}\ g\ CO_2}{yr} \times \frac{1\ mol\ CO_2}{44.01\ g\ CO_2} \times \frac{2803\ kJ}{6\ mol\ CO_2} = 5.838 \times 10^{17} = 5.8 \times 10^{17}\ kJ$

(b) $1\ W = 1\ J/s;\ 1\ W \bullet s = 1\ J$

$$\frac{5.838 \times 10^{17}\ kJ}{yr} \times \frac{1000\ J}{kJ} \times \frac{1\ yr}{365\ d} \times \frac{1\ d}{24\ hr} \times \frac{1\ hr}{60\ min} \times \frac{1\ min}{60\ s} \times \frac{1\ W \bullet s}{J} \times \frac{1\ MW}{1 \times 10^5\ W} = 1.851 \times 10^7\ MW = 1.9 \times 10^7\ MW$$

$$1.9 \times 10^7\ MW \times \frac{1\ plant}{10^3\ MW} = 1.9 \times 10^4 = 19{,}000 \text{ nuclear power plants}$$

5.106 The reaction for which we want ΔH is:

$$4NH_3(l) + 3O_2(g) \rightarrow 2N_2(g) + 6H_2O(g)$$

Before we can calculate ΔH for this reaction, we must calculate ΔH_f for $NH_3(l)$.

We know that ΔH_f for $NH_3(g)$ is -46.2 kJ, and that for $NH_3(l) \rightarrow NH_3(g)$, $\Delta H = 23.2$ kJ

Thus, $\Delta H_{vap} = \Delta H_f\ NH_3(g) - \Delta H_f\ NH_3(l)$.

$$23.2\ kJ = -46.2\ kJ - \Delta H_f\ NH_3(l);\ \ \Delta H_f\ NH_3(l) = -69.4\ kJ/mol$$

Then for the overall reaction, the enthalpy change is:

$$\Delta H_{rxn} = 6\Delta H_f\ H_2O(g) + 2\Delta H_f\ N_2(g) - 4\Delta H_f\ NH_3(l) - 3\Delta H_f\ O_2$$

$$= 6(-241.82\ kJ) + 2(0) - 4(-69.4\ kJ) - 3(0) = -1173.3\ kJ$$

$$\frac{-1173.3\ kJ}{4\ mol\ NH_3} \times \frac{1\ mol\ NH_3}{17.0\ g\ NH_3} \times \frac{0.81\ g\ NH_3}{1\ cm^3} \times \frac{1000\ cm^3}{1\ L} = \frac{1.4 \times 10^4\ kJ}{L\ NH_3}$$

(This result has 2 significant figures because the density is expressed to 2 figures.)

$$2CH_3OH(l) + 3O_2(g) \rightarrow 2CO_2(g) + 4H_2O(g)$$

$$\Delta H = 2(-393.5\ kJ) + 4(-241.82\ kJ) - 2(-239\ kJ) - 3(0) = -1276\ kJ$$

$$\frac{-1276\ kJ}{2\ mol\ CH_3OH} \times \frac{1\ mol\ CH_3OH}{32.04\ g\ CH_3OH} \times \frac{0.792\ g\ CH_3OH}{1\ cm^3} \times \frac{1000\ cm^3}{1\ L} = \frac{1.58 \times 10^4\ kJ}{1\ L\ CH_3OH}$$

In terms of heat obtained per unit volume of fuel, methanol is a slightly better fuel than liquid ammonia.

Integrative Exercises

5.107 (a) $CH_4(g) + 2O_2(g) \rightarrow CO_2(g) + 2H_2O(l)$

$$\Delta H^\circ = \Delta H^\circ_f\ CO_2(g) + 2\Delta H^\circ_f\ H_2O(l) - \Delta H^\circ_f\ CH_4(g) - 2\Delta H^\circ_f\ O_2(g)$$

$$= -393.5\ kJ + 2(-285.83\ kJ) - (-74.8\ kJ) - 2(0) = -890.36 = -890.4\ kJ/mol\ CH_4$$

$$\frac{-890.36\ kJ}{mol\ CH_4} \times \frac{1000\ J}{1\ kJ} \times \frac{1\ mol}{6.022 \times 10^{23}\ molecules\ CH_4} = 1.4785 \times 10^{-18}$$

$$= 1.479 \times 10^{-18}\ J/molecule$$

(b) 1eV = 96.485 kJ/mol

$$8\ \text{keV} \times \frac{1000\ \text{eV}}{1\ \text{keV}} \times \frac{96.485\ \text{kJ}}{\text{eV} \bullet \text{mol}} \times \frac{1\ \text{mol}}{6.022 \times 10^{23}} \times \frac{1000\ \text{J}}{\text{kJ}} = 1.282 \times 10^{-15}$$

$$= 1 \times 10^{-15}\ \text{J/X-ray}$$

The X-ray has approximately 1000 times more energy than is produced by the combustion of 1 molecule of $CH_4(g)$.

5.108 The situation in Figure 4.3 is a more complex version of that pictured in Exercise 5.24 and discussed in Solution 5.24. An ionic solid is an orderly arrangement of closely spaced ions, oppositely charged particles. The potential energy of any pair of these ions is described as $E_{el} = \kappa\ Q_1Q_2/r^2$. Separating these oppositely charged particles leads to an increase in the energy of the system, + ΔE. Work is done to the NaCl by the water molecules. Since both NaCl and water are part of the system, the net amount of work is zero. Since ΔE = q + w, ΔE = q and both are positive. The dissolving process typically takes place at constant atmospheric pressure, so ΔH = q and ΔH is also positive.

To verify this conclusion, carry out the dissolution of NaCl in a constant pressure calorimeter. Begin with 1 L of H_2O, record the temperature, add 0.1 mol NaCl and dissolve completely; record the final temperature. If ΔH is positive, the temperature will increase.

5.109 (a),(b) $Ag^+(aq) + Li(s) \rightarrow Ag(s) + Li^+(aq)$

$\Delta H^\circ = \Delta H^\circ_f\ Li^+(aq) - \Delta H^\circ_f\ Ag^+(aq)$

$= -278.5\ \text{kJ} - 105.90\ \text{kJ} = -384.4\ \text{kJ}$

$Fe(s) + 2Na^+(aq) \rightarrow Fe^{2+}(aq) + 2Na(s)$

$\Delta H^\circ = \Delta H^\circ_f\ Fe^{2+}(aq) - 2\Delta H^\circ_f\ Na^+(aq)$

$= -87.86\ \text{kJ} - 2(-240.1\ \text{kJ}) = +392.3\ \text{kJ}$

$2K(s) + 2H_2O(l) \rightarrow 2KOH(aq) + H_2(g)$

$\Delta H^\circ = 2\Delta H^\circ_f\ KOH(aq) - 2\Delta H^\circ_f\ H_2O(l)$

$= 2(-482.4\ \text{kJ}) - 2(-285.83\ \text{kJ}) = -393.1\ \text{kJ}$

(c) Exothermic reactions are more likely to be favorable, so the first and third reactions should be favorable and the second reaction should be unfavorable.

(d) In the activity series of metals, Table 4.5, any metal can be oxidized by the cation of a metal below it on the table.

Ag^+ is below Li, so the first reaction will occur.

Na^+ is above Fe, so the second reaction will not occur.

H^+ (formally in H_2O) is below K, so the third reaction will occur.

These predictions agree with those in part (c).

5.110 (a) $\Delta H^\circ = \Delta H_f^\circ\ NaNO_3(aq) + \Delta H_f^\circ\ H_2O(l) - \Delta H_f^\circ\ HNO_3(aq) - \Delta H_f^\circ\ NaOH(aq)$

$\Delta H^\circ = -446.2\ kJ - 285.83\ kJ - (-206.6\ kJ) - (-469.6\ kJ) = -55.8\ kJ$

$\Delta H^\circ = \Delta H_f^\circ\ NaCl(aq) + \Delta H_f^\circ\ H_2O(l) - \Delta H_f^\circ\ HCl(aq) - \Delta H_f^\circ\ NaOH(aq)$

$\Delta H^\circ = -407.1\ kJ - 285.83\ kJ - (-167.2\ kJ) - (-469.6\ kJ) = -56.1\ kJ$

$\Delta H^\circ = \Delta H_f^\circ\ NH_3(aq) + \Delta H_f^\circ\ Na^+(aq) + \Delta H_f^\circ\ H_2O(l) - \Delta H_f^\circ\ NH_4^+(aq) - \Delta H_f^\circ\ NaOH(aq)$

$= -80.29\ kJ - 240.1\ kJ - 285.83\ kJ - (-132.5\ kJ) - (-469.6\ kJ) = -4.1\ kJ$

(b) $H^+(aq) + OH^-(aq) \rightarrow H_2O(l)$ is the net ionic equation for both reactions.

(c) The ΔH° values for the first two reactions are nearly identical, -55.9 kJ and -56.2 kJ. The spectator ions by definition do not change during the course of a reaction, so ΔH° is the enthalpy change for the net ionic equation. Since the first two reactions have the same net ionic equation, it is not surprising that they have the same ΔH°.

(d) Strong acids are more likely than weak acids to donate H^+. The neutralization of the two strong acids is energetically favorable, while the third reaction is not. $NH_4^+(aq)$ is probably a weak acid.

5.111 (a) mol Cu = $M \times L$ = 1.00 M × 0.0500 L = 0.0500 mol

g = mol × $\mathcal{M}$ = 0.0500 × 63.546 = 3.1773 = 3.18 g Cu

(b) The precipitate is copper(II) hydroxide, $Cu(OH)_2$.

(c) $CuSO_4(aq) + 2KOH(aq) \rightarrow Cu(OH)_2(s) + K_2SO_4(aq)$, complete

$Cu^{2+}(aq) + 2OH^-(aq) \rightarrow Cu(OH)_2(s)$, net ionic

(d) The temperature of the calorimeter rises, so the reaction is exothermic and the sign of q is negative.

$$q = -6.2°C \times 100\ g \times \frac{4.184\ J}{1\ g \cdot °C} = -2.6 \times 10^3\ J = -2.6\ kJ$$

The reaction as carried out involves only 0.050 mol of $CuSO_4$ and the stoichiometrically equivalent amount of KOH. On a molar basis,

$$\Delta H = \frac{-2.6\ kJ}{0.050\ mol} = -52\ kJ$$ for the reaction as written in part (c)

5.112 (a) $AgNO_3(aq) + NaCl(aq) \rightarrow NaNO_3(aq) + AgCl(s)$

net ionic equation: $Ag^+(aq) + Cl^-(aq) \rightarrow AgCl(s)$

$\Delta H^\circ = \Delta H_f^\circ\ AgCl(s) - \Delta H_f^\circ\ Ag^+(aq) - \Delta H_f^\circ\ Cl^-(aq)$

$\Delta H^\circ = -127.0\ kJ - (105.90\ kJ) - (-167.2\ kJ) = -65.7\ kJ$

(b) ΔH° for the complete molecular equation will be the same as ΔH° for the net ionic equation. $Na^+(aq)$ and $NO_3^-(aq)$ are spectator ions; they appear on both sides of the chemical equation. Since the overall enthalpy change is the enthalpy of the products minus the enthalpy of the reactants, the contributions of the spectator ions cancel.

(c) $\Delta H^\circ = \Delta H_f^\circ\ NaNO_3(aq) + \Delta H_f^\circ\ AgCl(s) - \Delta H_f^\circ\ AgNO_3(aq) - \Delta H_f^\circ\ NaCl(aq)$

$\Delta H_f^\circ\ AgNO_3(aq) = \Delta H_f^\circ\ NaNO_3(aq) + \Delta H_f^\circ\ AgCl(s) - \Delta H_f^\circ\ NaCl(aq) - \Delta H^\circ$

$\Delta H_f^\circ\ AgNO_3(aq) = -446.2\ kJ + (-127.0\ kJ) - (-407.1\ kJ) - (-65.7\ kJ)$

$\Delta H_f^\circ\ AgNO_3(aq) = -100.4\ kJ/mol$

5.113 (a) $21.83\ g\ CO_2 \times \frac{1\ mol\ CO_2}{44.01\ g\ CO_2} \times \frac{1\ mol\ C}{1\ mol\ CO_2} \times \frac{12.01\ g\ C}{1\ mol\ C} = 5.9572 = 5.957\ g\ C$

$4.47\ g\ H_2O \times \frac{1\ mol\ H_2O}{18.02\ g\ H_2O} \times \frac{2\ mol\ H}{1\ mol\ H_2O} \times \frac{1.008\ g\ H}{mol\ H} = 0.5001 = 0.500\ g\ H$

The sample mass is (5.9572 + 0.5001) = 6.457 g

(b) $5.957\ g\ C \times \frac{1\ mol\ C}{12.01\ g\ C} = 0.4960\ mol\ C;\ 0.4960/0.496 = 1$

$0.500\ g\ H \times \frac{1\ mol\ H}{1.008\ g\ H} = 0.496\ mol\ H;\ 0.496/0.496 = 1$

The empirical formula of the hydrocarbon is CH.

(c) Calculate the ΔH_f° for 6.457 g of the sample.

$6.457\ g\ sample + O_2(g) \rightarrow 21.83\ g\ CO_2(g) + 4.47\ g\ H_2O(g),\ \Delta H^\circ = -311\ kJ$

$\Delta H^\circ_{comb} = \Delta H_f^\circ\ CO_2(g) + \Delta H_f^\circ\ H_2O(g) - \Delta H_f^\circ\ sample - \Delta H_f^\circ\ O_2(g)$

$\Delta H_f^\circ\ sample = \Delta H_f^\circ\ CO_2(g) + \Delta H_f^\circ\ H_2O(g) - \Delta H^\circ_{comb}$

$\Delta H_f^\circ\ CO_2(g) = 21.83\ g\ CO_2 \times \frac{1\ mol\ CO_2}{44.01\ g\ CO_2} \times \frac{-393.5\ kJ}{mol\ CO_2} = -195.185 = -195.2\ kJ$

$\Delta H_f^\circ\ H_2O(g) = 4.47\ g\ H_2O \times \frac{1\ mol\ H_2O}{18.02\ g\ H_2O} \times \frac{-241.82\ kJ}{mol\ H_2O} = 59.985 = -60.0\ kJ$

$\Delta H_f^\circ\ sample = -195.185\ kJ - 59.985\ kJ - (-311\ kJ) = 55.83 = 56\ kJ$

$\frac{55.83\ kJ}{6.457\ g\ sample} \times \frac{13.02\ g}{CH\ unit} = 112.6 = 1.1 \times 10^2\ kJ/CH\ unit$

(d) The hydrocarbons in Appendix C with empirical formula CH are C_2H_2 and C_6H_6.

substance	ΔH°_f / mol	ΔH°_f / CH unit
$C_2H_2(g)$	226.7 kJ	113.4 kJ
$C_6H_6(g)$	82.9 kJ	13.8 kJ
$C_6H_6(l)$	49.0 kJ	8.17 kJ
sample		1.1×10^2 kJ

The calculated value of ΔH°_f/CH unit for the sample is a good match with acetylene, $C_2H_2(g)$.

5.114 (a) $CH_4(g) \rightarrow C(g) + 4H(g)$ (i) reaction given

$CH_4(g) \rightarrow C(s) + 2H_2(g)$ (ii) reverse of formation

The differences are: the state of C in the products; the chemical form, atoms or diatomic molecules, of H in the products.

(b) i. $\Delta H^\circ = \Delta H^\circ_f\ C(g) + 4\Delta H^\circ_f\ H(g) - \Delta H^\circ_f\ CH_4(g)$

$= 718.4\ kJ + 4(217.94)\ kJ - (-74.8)\ kJ = 1665.0\ kJ$

ii. $\Delta H^\circ = \Delta H^\circ_f\ CH_4 = -(-74.8)\ kJ = 74.8\ kJ$

The rather large difference in ΔH° values is due to: the enthalpy difference between isolated gaseous C atoms and the orderly, bonded array of C atoms in graphite, C(s); and the enthalpy difference between isolated H atoms and H_2 molecules. In other words, the difference in the enthalpy stored in chemical bonds in C(s) and $H_2(g)$ versus the corresponding isolated atoms.

(c) $CH_4(g) + 4F_2(g) \rightarrow CF_4(g) + 4HF(g)$ $\Delta H^\circ = -1679.5\ kJ$

The ΔH° value for this reaction was calculated in Solution 5.86.

$$3.45\ g\ CH_4 \times \frac{1\ mol\ CH_4}{16.04\ g\ CH_4} = 0.21509 = 0.215\ mol\ CH_4$$

$$1.22\ g\ F_2 \times \frac{1\ mol\ F_2}{38.00\ g\ F_2} = 0.03211 = 0.0321\ mol\ F_2$$

There are fewer mol F_2 than CH_4, but 4 mol F_2 are required for every 1 mol of CH_4 reacted, so clearly F_2 is the limiting reactant.

$$0.03211\ mol\ F_2 \times \frac{-1679.5\ kJ}{4\ mol\ F_2} = -13.48 = -13.5\ kJ\ heat\ evolved$$

6 Electronic Structure of Atoms

Radiant Energy

6.1 (a) meters (m) (b) 1/seconds (s^{-1}) (c) meters/second ($m{\bullet}s^{-1}$ or m/s)

6.2 (a) Wavelength (λ) and frequency (ν) are inversely proportional; the proportionality constant is the speed of light (c). $\nu = c/\lambda$.

(b) Light in the 210-230 nm range is in the ultraviolet region of the spectrum. These wavelengths are slightly shorter than the 400 nm short-wavelength boundary of the visible region.

6.3 (a) True.
(b) False. The frequency of radiation decreases as the wavelength increases.
(c) False. Ultraviolet light has shorter wavelengths than visible light. [See Solution 6.2(b)]
(d) False. Electromagnetic radiation and sound waves travel at different speeds.

6.4 (a) False. Electromagnetic radiation passes through water. See Figure 13.27.
(b) True.
(c) False. Infrared light has lower frequencies than visible light.
(d) False. A foghorn blast is a form of sound waves.

6.5 *Analyze/Plan*. Use the electromagnetic spectrum in Figure 6.4 to determine the wavelength of each type of radiation; put them in order from shortest to longest wavelength.

Solve. Wavelength of X-rays < ultraviolet < green light < red light < infrared < radio waves

Check. These types of radiation should read from left to right on Figure 6.4

6.6 Wavelength of (a) gamma rays < (d) yellow (visible) light < (e) red (visible) light < (b) 93.1 MHz FM (radio) waves < (c) 680 kHz or 0.680 MHz AM (radio) waves

6.7 *Analyze/Plan*. These questions involve relationships between wavelength, frequency and the speed of light. Manipulate the equation $\nu = c/\lambda$ to obtain the desired quantities, paying attention to units. *Solve*:

(a) $\nu = c/\lambda; \quad \frac{2.998 \times 10^8 \text{ m}}{\text{s}} \times \frac{1}{0.452 \text{ pm}} \times \frac{1 \text{ pm}}{1 \times 10^{-12} \text{ m}} = 6.63 \times 10^{20} \text{ s}^{-1}$

(b) $\lambda = c/\nu$; $\frac{2.998 \times 10^8 \text{ m}}{\text{s}} \times \frac{1 \text{ s}}{2.55 \times 10^{16}} = 1.18 \times 10^{-8}$ m (11.8 nm)

(c) No. The radiation in (a) is gamma rays and in (b) is ultraviolet. Neither is visible to humans.

(d) $7.50 \text{ ms} \times \frac{1 \text{ s}}{1 \times 10^3 \text{ ms}} \times \frac{2.998 \times 10^8 \text{ m}}{\text{s}} = 2.25 \times 10^6$ m

Check. Confirm that powers of 10 make sense and units are correct.

6.8 (a) $\nu = c/\lambda$; $\frac{2.998 \times 10^8 \text{ m}}{\text{s}} \times \frac{1}{589 \text{ nm}} \times \frac{1 \text{ nm}}{1 \times 10^{-9} \text{ m}} = 5.10 \times 10^{14} \text{ s}^{-1}$

(b) $\lambda = c/\nu$; $\frac{2.998 \times 10^8 \text{ m}}{\text{s}} \times \frac{1 \text{ s}}{1.2 \times 10^{13}} = 2.5 \times 10^{-5}$ m

(c) Yes. The radiation in (b) is in the infrared range.

(d) $10 \ \mu\text{s} \times \frac{1 \times 10^{-6} \text{ s}}{1 \ \mu\text{s}} \times \frac{2.998 \times 10^8 \text{ m}}{\text{s}} = 3.0 \times 10^3$ m (3.0 km)

6.9 *Analyze/Plan.* $\nu = c/\lambda$; change nm → m.

Solve: $\nu = c/\lambda$; $\frac{2.998 \times 10^8 \text{ m}}{1 \text{ s}} \times \frac{1}{436 \text{ nm}} \times \frac{1 \text{ nm}}{1 \times 10^{-9} \text{ m}} = 6.88 \times 10^{14} \text{ s}^{-1}$

The color is blue.

Check. $(3000 \times 10^5 / 500 \times 10^{-9}) = 6 \times 10^{14} \text{ s}^{-1}$; units are correct.

6.10 $\nu = c/\lambda$; $\frac{2.998 \times 10^8 \text{ m}}{1 \text{ s}} \times \frac{1}{489 \text{ nm}} \times \frac{1 \text{ nm}}{1 \times 10^{-9} \text{ m}} = 6.13 \times 10^{14} \text{ s}^{-1}$

The laser emits visible light; the color is green to blue-green.

Quantized Energy and Photons

6.11 (a) *Quantization* means that energy can only be absorbed or emitted in specific amounts or multiples of these amounts. This minimum amount of energy is called a quantum and is equal to a constant times the frequency of the radiation absorbed or emitted. $E = h\nu$.

(b) In everyday activities, we deal with macroscopic objects such as our bodies or our cars, which gain and lose total amounts of energy much larger than a single quantum, $h\nu$. The gain or loss of the relatively minuscule quantum of energy is unnoticed.

6.12 Planck's original hypothesis was that energy could only be gained or lost in discreet amounts (quanta) with a certain minimum size. The size of the minimum energy change is related to the frequency of the radiation absorbed or emitted, $\Delta E = h\nu$, and energy changes occur only in multiples of $h\nu$.

Einstein postulated that light itself is quantized, that the minimum energy of a photon (a quantum of light) is directly proportional to its frequency, E = hν. If a photon that strikes a metal surface has less than the threshold energy, no electron is emitted from the surface. If the photon has energy equal to or greater than the threshold energy, an electron is emitted and any excess energy becomes the kinetic energy of the electron.

6.13 *Analyze/Plan.* These questions deal with the relationships between energy, wavelength and frequency. Use the relationships E = hν = hc/λ to calculate the desired quantities. Pay attention to units. *Solve*:

(a) $E = h\nu = hc/\lambda = 6.626 \times 10^{-34}\,\text{J}\cdot\text{s} \times \frac{2.998 \times 10^{8}\,\text{m}}{1\,\text{s}} \times \frac{1}{812\,\text{nm}} \times \frac{1\,\text{nm}}{1 \times 10^{-9}\,\text{m}}$

$= 2.45 \times 10^{-19}\,\text{J}$

(b) $E = h\nu = 6.626 \times 10^{-34}\,\text{J}\cdot\text{s} \times \frac{2.72 \times 10^{13}}{1\,\text{s}} = 1.80 \times 10^{-20}\,\text{J}$

(c) $\lambda = hc/E = 6.626 \times 10^{-34}\,\text{J}\cdot\text{s} \times \frac{2.998 \times 10^{8}\,\text{m}}{1\,\text{s}} \times \frac{1}{7.84 \times 10^{-18}\,\text{J}} = 2.53 \times 10^{-8}\,\text{m}$

$= 25.3\,\text{nm}$

This radiation is in the ultraviolet region.

Check. Units are correct and powers of 10 are reasonable.

6.14 (a) $E = hc/\lambda = 6.626 \times 10^{-34}\,\text{J}\cdot\text{s} \times \frac{2.998 \times 10^{8}\,\text{m}}{1\,\text{s}} \times \frac{1}{3.80\,\text{mm}} \times \frac{1\,\text{mm}}{1 \times 10^{-3}\,\text{m}}$

$= 5.23 \times 10^{-23}\,\text{J}$

(b) $E = h\nu = 6.626 \times 10^{-34}\,\text{J}\cdot\text{s} \times \frac{80.5 \times 10^{6}}{1\,\text{s}} = 5.33 \times 10^{-26}\,\text{J}$

(c) $\nu = E/h = \frac{1.77 \times 10^{-19}\,\text{J}}{6.626 \times 10^{-34}\,\text{J}\cdot\text{s}} = 2.67 \times 10^{14}\,\text{s}^{-1}$

$\lambda = hc/E = 1.12 \times 10^{-6}$ m; the radiation is infrared but near the visible "edge."

6.15 *Analyze/Plan.* Use E = hc/λ; pay close attention to units. *Solve*:

(a) $E = hc/\lambda = 6.626 \times 10^{-34}\,\text{J}\cdot\text{s} \times \frac{2.998 \times 10^{8}\,\text{m}}{1\,\text{s}} \times \frac{1}{3.3\,\mu\text{m}} \times \frac{1\,\mu\text{m}}{1 \times 10^{-6}\,\text{m}}$

$= 6.0 \times 10^{-20}\,\text{J}$

$E = hc/\lambda = 6.626 \times 10^{-34}\,\text{J}\cdot\text{s} \times \frac{2.998 \times 10^{8}\,\text{m}}{1\,\text{s}} \times \frac{1}{0.154\,\text{nm}} \times \frac{1\,\text{nm}}{1 \times 10^{-9}\,\text{m}}$

$= 1.29 \times 10^{-15}\,\text{J}$

Check. $(6.6 \times 3/3.3) \times (10^{-34} \times 10^{8}/10^{-6}) \approx 6 \times 10^{-20}\,\text{J}$

$(6.6 \times 3/0.15) \times (10^{-34} \times 10^{8}/10^{-9}) \approx 120 \times 10^{-17} \approx 1.2 \times 10^{-15}\,\text{J}$

The results are reasonable. We expect the longer wavelength 3.3 μm radiation to have the lower energy.

(b) The 3.3 μm photon is in the infrared and the 0.154 nm (1.54×10^{-10} m) photon is in the X-ray region; the X-ray photon has the greater energy.

6.16 $E = h\nu$

AM: $6.626 \times 10^{-34}\ \text{J}\cdot\text{s} \times \dfrac{1440 \times 10^{3}}{1\ \text{s}} = 9.54 \times 10^{-28}\ \text{J}$

FM: $6.626 \times 10^{-34}\ \text{J}\cdot\text{s} \times \dfrac{94.5 \times 10^{6}}{1\ \text{s}} = 6.26 \times 10^{-26}\ \text{J}$

The FM photon has about 66 times more energy than the AM photon.

6.17 *Analyze/Plan.* Use $E = hc/\lambda$ to calculate J/photon; Avogadro's number to calculate J/mol; photon/J (the result from part (a)) to calculate photons in 1.00 mJ. Pay attention to units. *Solve*:

(a) $E_{photon} = hc/\lambda = \dfrac{6.626 \times 10^{-34}\ \text{J}\cdot\text{s}}{325 \times 10^{-9}\ \text{m}} \times \dfrac{2.998 \times 10^{8}\ \text{m}}{\text{s}} = 6.1122 \times 10^{-19}$

$= 6.11 \times 10^{-19}$ J/photon

(b) $\dfrac{6.1122 \times 10^{-19}\ \text{J}}{1\ \text{photon}} \times \dfrac{6.022 \times 10^{23}\ \text{photons}}{1\ \text{mol}} = 3.68 \times 10^{5}\ \text{J/mol} = 368\ \text{kJ/mol}$

(c) $\dfrac{1\ \text{photon}}{6.1122 \times 10^{-19}\ \text{J}} \times 1.00\ \text{mJ} \times \dfrac{1 \times 10^{-3}}{1\ \text{mJ}} = 1.64 \times 10^{15}\ \text{photons}$

Check. Powers of 10 (orders of magnitude) and units are correct.

6.18 $\dfrac{495 \times 10^{3}\ \text{J}}{\text{mol } O_2} \times \dfrac{1\ \text{mol}}{6.022 \times 10^{23}\ \text{photons}} = 8.220 \times 10^{-19} = 8.22 \times 10^{-19}$ J/photon

$\lambda = hc/E = \dfrac{6.626 \times 10^{-34}\ \text{J}\cdot\text{s}}{8.220 \times 10^{-19}\ \text{J}} \times \dfrac{2.998 \times 10^{8}\ \text{m}}{1\ \text{s}} = 2.42 \times 10^{-7}\ \text{m} = 242\ \text{nm}$

According to Figure 6.4, this is ultraviolet radiation.

6.19 *Analyze/Plan.* $E = hc/\lambda$ gives J/photon. Use this result with J/s (given) to calculate photons/s. *Solve*:

$E_{photon} = hc/\lambda = \dfrac{6.626 \times 10^{-34}\ \text{J}\cdot\text{s}}{987 \times 10^{-9}\ \text{m}} \times \dfrac{2.998 \times 10^{8}\ \text{m}}{1\ \text{s}} = 2.0126 \times 10^{-19} = 2.01 \times 10^{-19}$ J/photon

$\dfrac{0.52\ \text{J}}{32\ \text{s}} \times \dfrac{1\ \text{photon}}{2.0126 \times 10^{-19}\ \text{J}} = 8.1 \times 10^{16}$ photons/s

Check. $(7 \times 3/1000) \times (10^{-34} \times 10^{8}/10^{-9}) \approx 21 \times 10^{-20} \approx 2.1 \times 10^{-19}$ J/photon

$(0.5/30/2) \times (1/10^{-19}) = 0.008 \times 10^{19} = 8 \times 10^{16}$ photons/s

Units are correct; powers of 10 are reasonable.

6.20 $E_{photon} = hc/\lambda = \frac{6.626 \times 10^{-34}\ J \cdot s}{1350 \times 10^{-9}\ m} \times \frac{2.998 \times 10^{8}\ m}{1\ s} = 1.4715 \times 10^{-19} = 1.47 \times 10^{-19}$ J/photon

$$\frac{1.4715 \times 10^{-19}\ J}{1\ photon} \times \frac{8 \times 10^{7}\ photons}{1\ s} \times \frac{60\ s}{1\ min} \times \frac{60\ min}{1\ hr} = 4.24 \times 10^{-8} = 4 \times 10^{-8}\ J/hr$$

6.21 *Analyze/Plan.* Use $E = h\nu$ and $\nu = c/\lambda$. Calculate the desired characteristics of the photons. Compare E_{min} and E_{120} to calculate maximum kinetic energy of the emitted electron. *Solve*:

(a) $E = h\nu = 6.626 \times 10^{-34}\ J \cdot s \times 1.09 \times 10^{15}\ s^{-1} = 7.22 \times 10^{-19}\ J$

(b) $\lambda = c/\nu = \frac{2.998 \times 10^{8}\ m}{1\ s} \times \frac{1\ s}{1.09 \times 10^{15}} = 2.75 \times 10^{-7}\ m = 275\ nm$

(c) $E_{120} = hc/\lambda = 6.626 \times 10^{-34}\ J \cdot s \times \frac{2.998 \times 10^{8}\ m}{1\ s} \times \frac{1}{120\ nm} \times \frac{1\ nm}{1 \times 10^{-9}\ m}$

$= 1.655 \times 10^{-18} = 1.66 \times 10^{-18}\ J$

The excess energy of the 120 nm photon is converted into the kinetic energy of the emitted electron.

$E_k = E_{120} - E_{min} = 16.55 \times 10^{-19}\ J - 7.22 \times 10^{-19}\ J = 9.3 \times 10^{-19}$ J/electron

Check. E_{120} must be greater than E_{min} in order for the photon to impart kinetic energy to the emitted electron. Our calculations are consistent with this requirement.

6.22 (a) $\nu = E/h = \frac{4.41 \times 10^{-19}\ J}{6.626 \times 10^{-34}\ J \cdot s} = 6.6556 \times 10^{14} = 6.66 \times 10^{14}\ s^{-1}$

(b) $\lambda = hc/E = \frac{6.626 \times 10^{-34}\ J \cdot s}{4.41 \times 10^{-19}\ J} \times \frac{2.998 \times 10^{8}\ m}{s} = 4.50 \times 10^{-7}\ m = 450\ nm$

(c) $E_{439} = hc/\lambda = \frac{6.626 \times 10^{-34}\ J \cdot s}{439 \times 10^{-9}\ m} \times \frac{2.998 \times 10^{8}\ m}{s} = 4.525 \times 10^{-19} = 4.53 \times 10^{-19}\ J$

$E_K = E_{439} - E_{min} = 4.525 \times 10^{-19}\ J - 4.41 \times 10^{-19}\ J = 0.115 \times 10^{-19} = 1.1 \times 10^{-20}\ J$

(d) One electron is emitted per photon. Calculate the number of 439 nm photons in 1.00 μJ. The excess energy in each photon will become the kinetic energy of the electron; it cannot be "pooled" to emit additional electrons.

$$1.00\ \mu J \times \frac{1 \times 10^{-6}\ J}{\mu J} \times \frac{1\ photon}{4.525 \times 10^{-19}\ J} \times \frac{1\ e^-}{1\ photon} = 2.21 \times 10^{12}\ electrons$$

Bohr's Model; Matter Waves

6.23 When applied to atoms, the notion of quantized energies means that only certain energies can be gained or lost, only certain values of ΔE are allowed. The allowed values of ΔE are represented by the lines in the emission spectra of excited atoms.

6.24 (a) According to Bohr theory, when hydrogen emits radiant energy, electrons are moving from a higher allowed energy state to a lower one. Since only certain energy states are allowed, only certain energy changes can occur. These allowed energy changes correspond ($\lambda = hc/\Delta E$) to the wavelengths of the lines in the emission spectrum of hydrogen.

(b) When a hydrogen atom changes from the ground state to an excited state, the single electron moves further away from the nucleus, so the atom "expands".

6.25 *Analyze/Plan.* An isolated electron is assigned an energy of zero; the closer the electron comes to the nucleus, the more negative its energy. Thus, as an electron moves closer to the nucleus, the energy of the electron decreases and the excess energy is emitted. Conversely, as an electron moves further from the nucleus, the energy of the electron increases and energy must be absorbed. *Solve*:

(a) As the principle quantum number decreases, the electron moves toward the nucleus and energy is **emitted**.

(b) An increase in the radius of the orbit means the electron moves away from the nucleus; energy is **absorbed**.

(c) An isolated electron is assigned an energy of zero. As the electron moves to the n = 3 state closer to the H^+ nucleus, its energy becomes more negative (decreases) and energy is **emitted**.

6.26 (a) absorbed (b) emitted (c) absorbed.

6.27 *Analyze/Plan.* Equation 6.5: $E = (-2.18 \times 10^{-18}\ J)(1/n^2)$. *Solve*:

$E_2 = -2.18 \times 10^{-18}\ J/(2)^2 = -5.45 \times 10^{-19}\ J$

$E_6 = -2.18 \times 10^{-18}\ J/(6)^2 = -6.0556 \times 10^{-20} = -0.606 \times 10^{-19}\ J$

$\Delta E = E_6 - E_2 = (-0.606 \times 10^{-19}\ J) - (-5.45 \times 10^{-19}\ J) = 4.844 \times 10^{-19}\ J = 4.84 \times 10^{-19}\ J$

$$\lambda = hc/\Delta E = \frac{6.626 \times 10^{-34}\ J \bullet s}{4.844 \times 10^{-19}\ J} \times \frac{2.998 \times 10^8\ m}{s} = 4.10 \times 10^{-7}\ m = 410\ nm$$

The visible range is 400 - 700 nm, so this line is visible; the observed color is violet.

Check. We expect E_6 to be a more positive (or less negative) than E_2, and it is. ΔE is positive, which indicates emission. The orders of magnitude make sense and units are correct.

6.28 (a)
$$\Delta E = -2.18 \times 10^{-18}\ J \left[\frac{1}{n_f^2} - \frac{1}{n_i^2}\right] = -2.18 \times 10^{-18}\ J\ (1/1 - 1/25) = -2.093 \times 10^{-18}$$
$$= -2.09 \times 10^{-18}\ J$$

$$\nu = E/h = \frac{2.093 \times 10^{-18}\ J}{6.626 \times 10^{-34}\ J \bullet s} = 3.158 \times 10^{15} = 3.16 \times 10^{15}\ s^{-1}$$

$$\lambda = c/\nu = \frac{2.998 \times 10^8 \text{ m}}{1 \text{ s}} \times \frac{1 \text{ s}}{3.158 \times 10^{15}} = 9.49 \times 10^{-8} \text{ m}$$

Since the sign of ΔE is negative, radiation is emitted.

(b) $\Delta E = -2.18 \times 10^{-18}$ J(1/4 - 1/16) = -4.0875×10^{-19} = -4.09×10^{-19} J

$$\nu = \frac{4.0875 \times 10^{-19} \text{ J}}{6.626 \times 10^{-34} \text{ J}\bullet\text{s}} = 6.1689 \times 10^{14} = 6.17 \times 10^{14} \text{ s}^{-1}\ ;\ \lambda = \frac{2.998 \times 10^8 \text{ m/s}}{6.1689 \times 10^{14}/\text{s}}$$

$\lambda = 4.86 \times 10^{-7}$ m. Visible radiation is emitted.

(c) $\Delta E = -2.18 \times 10^{-18}$ J (1/36 - 1/16) = 7.5694×10^{-20} = 7.57×10^{-20} J

$$\nu = \frac{7.5694 \times 10^{-20} \text{ J}}{6.626 \times 10^{-34} \text{ J}\bullet\text{s}} = 1.1424 \times 10^{14} = 1.14 \times 10^{14} \text{ s}^{-1}\ ;\ \lambda = \frac{2.998 \times 10^8 \text{ m/s}}{1.1424 \times 10^{14}/\text{s}}$$

$\lambda = 2.62 \times 10^{-6}$ m. Radiation is absorbed.

6.29 (a) Only lines with $n_f = 2$ represent ΔE values and wavelengths that lie in the visible portion of the spectrum. Lines with $n_f = 1$ have larger ΔE values and shorter wavelengths that lie in the ultraviolet. Lines with $n_f > 2$ have smaller ΔE values and lie in the lower energy longer wavelength regions of the electromagnetic spectrum.

(b) *Analyze/Plan*. Use Equation 6.7 to calculate ΔE, then $\lambda = hc/\Delta E$. *Solve*:

$$n_i = 3,\ n_f = 2;\ \Delta E = -2.18 \times 10^{-18} \text{ J} \left[\frac{1}{n_f^2} - \frac{1}{n_i^2}\right] = -2.18 \times 10^{-18} \text{ J } (1/4 - 1/9)$$

$$\lambda = hc/E = \frac{6.626 \times 10^{-34} \text{ J}\bullet\text{s} \times 2.998 \times 10^8 \text{ m/s}}{-2.18 \times 10^{-18} \text{ J } (1/4 - 1/9)} = 6.56 \times 10^{-7} \text{ m}$$

This is the red line at 656 nm.

$$n_i = 4,\ n_f = 2;\ \lambda = hc/E = \frac{6.626 \times 10^{-34} \text{ J}\bullet\text{s} \times 2.998 \times 10^8 \text{ m/s}}{-2.18 \times 10^{-18} \text{ J } (1/4 - 1/16)} = 4.86 \times 10^{-7} \text{ m}$$

This is the blue line at 486 nm.

$$n_i = 5,\ n_f = 2;\ \lambda = hc/E = \frac{6.626 \times 10^{-34} \text{ J}\bullet\text{s} \times 2.998 \times 10^8 \text{ m/s}}{-2.18 \times 10^{-18} \text{ J } (1/4 - 1/25)} = 4.34 \times 10^{-7} \text{ m}$$

This is the violet line at 434 nm.

Check. The calculated wavelengths correspond well to 3 lines in the H emission spectrum in Figure 6.12, so the results are sensible.

6.30 (a) Transitions with $n_f = 1$ have larger ΔE values and shorter wavelengths than those with $n_f = 2$. These transitions will lie in the ultraviolet region.

(b) $n_i = 2, n_f = 1$; $\lambda = hc/E = \dfrac{6.626 \times 10^{-34}\ J\bullet s \times 2.998 \times 10^{8}\ m/s}{-2.18 \times 10^{-18}\ J\,(1/1 - 1/4)} = 1.22 \times 10^{-7}\ m$

$n_i = 3, n_f = 1$; $\lambda = hc/E = \dfrac{6.626 \times 10^{-34}\ J\bullet s \times 2.998 \times 10^{8}\ m/s}{-2.18 \times 10^{-18}\ J\,(1/1 - 1/9)} = 1.03 \times 10^{-7}\ m$

$n_i = 4, n_f = 1$; $\lambda = hc/E = \dfrac{6.626 \times 10^{-34}\ J\bullet s \times 2.998 \times 10^{8}\ m/s}{-2.18 \times 10^{-18}\ J\,(1/1 - 1/16)} = 0.972 \times 10^{-7}\ m$

6.31 (a) $93.8\ nm \times \dfrac{1 \times 10^{-9}\ m}{1\ nm} = 9.38 \times 10^{-8}\ m$; this line is in the ultraviolet region.

(b) *Analyze/Plan.* Only lines with $n_f = 1$ have a large enough ΔE to lie in the ultraviolet region (see Solution 6.29 and 6.30). Solve Equation 6.7 for n_i, recalling that ΔE is negative for emission. *Solve:*

$$\frac{-hc}{\lambda} = -2.18 \times 10^{-18}\ J \left[\frac{1}{n_f^2} - \frac{1}{n_i^2}\right]; \quad \frac{hc}{\lambda(2.18 \times 10^{-18}\ J)} = \left[1 - \frac{1}{n_i^2}\right]$$

$$-\frac{1}{n_i^2} = \left[\frac{hc}{\lambda(2.18 \times 10^{-18}\ J)} - 1\right]; \quad \frac{1}{n_i^2} = \left[1 - \frac{hc}{\lambda(2.18 \times 10^{-18}\ J)}\right]$$

$$n_i^2 = \left[1 - \frac{hc}{\lambda(2.18 \times 10^{-18}\ J)}\right]^{-1}; \quad n_i = \left[1 - \frac{hc}{\lambda(2.18 \times 10^{-18}\ J)}\right]^{-1/2}$$

$$n_i = \left(1 - \frac{6.626 \times 10^{-34}\ J\bullet s \times 2.998 \times 10^{8}\ m/s}{9.38 \times 10^{-8}\ m \times 2.18 \times 10^{-18}\ J}\right)^{-1/2} = 6 \text{ (}n\text{ values must be integers)}$$

$n_i = 6, n_f = 1$

Check. From Solution 6.30, we know that $n_i > 4$ for $\lambda = 93.8$ nm. The calculated result is close to 6, so the answer is reasonable.

6.32 (a) $4055\ nm \times \dfrac{1 \times 10^{-9}\ m}{1\ nm} = 4.055 \times 10^{-6}\ m$; this line is in the infrared.

(b) Absorption lines with $n_i = 1$ are in the ultraviolet and with $n_i = 2$ are in the visible. Thus, $n_i \geq 3$, but we do not know the exact value of n_i. Calculate the longest wavelength with $n_i = 3$ ($n_f = 4$). If this is less than 4055 nm, $n_i > 3$.

$$\lambda = hc/E = \frac{6.626 \times 10^{-34}\ J\bullet s \times 2.998 \times 10^{8}\ m/s}{-2.18 \times 10^{-18}\ J\,(1/9 - 1/16)} = 1.875 \times 10^{-6}\ m$$

This wavelength is shorter than 4.055×10^{-6} m, so $n_i > 3$; try $n_i = 4$ and solve for n_f as in Solution 6.31.

$$n_f = \left(\frac{1}{n_i^2} - \frac{hc}{\lambda(2.18 \times 10^{-18}\ J)}\right)^{-1/2} = \left(1/16 - \frac{6.626 \times 10^{-34}\ J\bullet s \times 2.998 \times 10^{8}\ m/s}{4.055 \times 10^{-6}\ m \times 2.18 \times 10^{-18}\ J}\right)^{-1/2} = 5$$

$n_f = 5, n_i = 4$

6.33 *Analyze/Plan.* $\lambda = \frac{h}{mv}$; $1\ J = \frac{1\ kg \cdot m^2}{s^2}$; Change mass to kg and velocity to m/s in each case. *Solve*:

(a) $\frac{50\ km}{1\ hr} \times \frac{1000\ m}{1\ km} \times \frac{1\ hr}{60\ min} \times \frac{1\ min}{60\ s} = 13.89 = 14\ m/s$

$$\lambda = \frac{6.626 \times 10^{-34}\ kg \cdot m^2 \cdot s}{1\ s^2} \times \frac{1}{85\ kg} \times \frac{1\ s}{13.89\ m} = 5.6 \times 10^{-37}\ m$$

(b) $10.0\ g \times \frac{1\ kg}{1000\ g} = 0.0100\ kg$

$$\lambda = \frac{6.626 \times 10^{-34}\ kg \cdot m^2 \cdot s}{1\ s^2} \times \frac{1}{0.0100\ kg} \times \frac{1\ s}{250\ m} = 2.65 \times 10^{-34}\ m$$

(c) We need to calculate the mass of a single Li atom in kg.

$$\frac{6.94\ g\ Li}{1\ mol\ Li} \times \frac{1\ kg}{1000\ g} \times \frac{1\ mol}{6.022 \times 10^{23}\ Li\ atoms} = 1.152 \times 10^{-26} = 1.15 \times 10^{-26}\ kg$$

$$\lambda = \frac{6.626 \times 10^{-34}\ kg \cdot m^2 \cdot s}{1\ s^2} \times \frac{1}{1.152 \times 10^{-26}\ kg} \times \frac{1\ s}{2.5 \times 10^5\ m} = 2.3 \times 10^{-13}\ m$$

6.34 $\lambda = h/mv$; change mass to kg and velocity to m/s

mass of muon $= 206.8 \times 9.1094 \times 10^{-28}\ g \times \frac{1\ kg}{1000\ g} = 1.8838 \times 10^{-28} = 1.88 \times 10^{-28}\ kg$

$$\lambda = \frac{6.626 \times 10^{-34}\ kg \cdot m^2 \cdot s}{1\ s^2} \times \frac{1}{1.8838 \times 10^{-28}\ kg} \times \frac{1\ s}{8.85 \times 10^3\ m/s} = 3.97 \times 10^{-10}\ m = 3.97\ \text{Å}$$

6.35 *Analyze/Plan.* Use $v = h/m\lambda$; change wavelength to meters and mass of neutron (back-inside cover) to kg. *Solve*:

$\lambda = 0.955\ \text{Å} \times \frac{1 \times 10^{-10}\ m}{1\ \text{Å}} = 0.955 \times 10^{-10}\ m$; $m = 1.6749 \times 10^{-27}\ kg$

$$v = \frac{6.626 \times 10^{-34}\ kg \cdot m^2 \cdot s}{1\ s^2} \times \frac{1}{1.6749 \times 10^{-27}\ kg} \times \frac{1}{0.955 \times 10^{-10}\ m} = 4.14 \times 10^3\ m/s$$

Check. $(6.6/1.6/1) \times (10^{-34}/10^{-27}/10^{-10}) \approx 4 \times 10^3\ m/s$

6.36 $m_e = 9.1094 \times 10^{-31}\ kg$ (back-inside cover of text)

$$\lambda = \frac{6.626 \times 10^{-34}\ kg \cdot m^2 \cdot s}{1\ s^2} \times \frac{1}{9.1094 \times 10^{-31}\ kg} \times \frac{1\ s}{5.93 \times 10^6\ m} = 1.23 \times 10^{-10}\ m$$

$$1.23 \times 10^{-10}\ m \times \frac{1\ \text{Å}}{1 \times 10^{-10}\ m} = 1.23\ \text{Å}$$

Since atomic radii and interatomic distances are on the order of 1-5 Å (Section 2.3), the wavelength of this electron is comparable to the size of atoms.

6.37 *Analyze/Plan.* Use $\Delta x \geq h/4\pi m \Delta v$, paying attention to appropriate units. Note that the uncertainty in speed of the particle (Δv) is important, rather than the speed itself. *Solve*:

(a) $$m = 1.50 \text{ mg} \times \frac{1 \text{ g}}{1000 \text{ mg}} \times \frac{1 \text{ kg}}{1000 \text{ g}} = 1.50 \times 10^{-6} \text{ kg}; \Delta v = 0.01 \text{ m/s}$$

$$\Delta x \geq = \frac{6.626 \times 10^{-34} \text{ J}\bullet\text{s}}{4\pi(1.50 \times 10^{-6} \text{ kg})(0.01 \times 10^{4} \text{ m/s})} \geq 3.52 \times 10^{-27} = 4 \times 10^{-27} \text{ m}$$

(b) $m = 1.673 \times 10^{-24}$ g $= 1.673 \times 10^{-27}$ kg; $\Delta v = 0.01 \times 10^{4}$ m/s

$$\Delta x \geq = \frac{6.626 \times 10^{-34} \text{ J}\bullet\text{s}}{4\pi(1.673 \times 10^{-27} \text{ kg})(0.01 \times 10^{4} \text{ m/s})} \geq 3 \times 10^{-10} \text{ m}$$

Check. The more massive particle in (a) has a much smaller uncertainty in position.

6.38 $\Delta x \geq = h/4\pi m \Delta v$; use masses in kg, Δv in m/s.

(a) $$\frac{6.626 \times 10^{-34} \text{ J}\bullet\text{s}}{4\pi(9.109 \times 10^{-31} \text{ kg})(0.01 \times 10^{5} \text{ m/s})} = 6 \times 10^{-8} \text{ m}$$

(b) $$\frac{6.626 \times 10^{-34} \text{ J}\bullet\text{s}}{4\pi(1.675 \times 10^{-27} \text{ kg})(0.01 \times 10^{5} \text{ m/s})} = 3 \times 10^{-11} \text{ m}$$

(c) For particles moving with the same uncertainty in velocity, the more massive neutron has a much smaller uncertainty in position than the lighter electron. In our model of the atom, we know where the massive particles in the nucleus are located, but we cannot know the location of the electrons with any certainty, if we know their speed.

Quantum Mechanics and Atomic Orbitals

6.39 The Bohr model states with 100% certainty that the electron in hydrogen can be found 0.53 Å from the nucleus. The quantum mechanical model, taking the wave nature of the electron and the uncertainty principle into account, is a statistical model that states the probability of finding the electron in certain regions around the nucleus. While 0.53 Å might be the radius with highest probability, that probability would always be less than 100%.

6.40 (a) The square of the wave function has the physical significance of an amplitude, or probability. The quantity ψ^2 at a given point in space is the probability of locating the electron within a small volume element around that point at any given instant. The total probability, that is, the sum of ψ^2 over all the space around the nucleus, must equal 1.

(b) Electron density is a way to express the probability of finding an electron in a region of space. The greater the probability (the value of ψ^2), the greater the electron density.

(c) An orbital is one of the wave functions, ψ, that is a solution to Schrödinger's equation for a particular atom. Each orbital has a distinctive energy and shape or electron density distribution.

6.41 (a) The possible values of l are (n-1) to 0. $n = 4$, $l = 3, 2, 1, 0$

(b) The possible values of m_l are $-l$ to $+l$. $l = 2$, $m_l = -2, -1, 0, 1, 2$

6.42 (a) For $n = 3$, there are 3 l values (2, 1, 0) and 9 m_l values ($l = 2$; $m_l = -2, -1, 0, 1, 2$; $l = 1$, $m_l = -1, 0, 1$; $l = 0$, $m_l = 0$).

(b) For $n = 5$, there are 5 l values (4, 3, 2, 1, 0) and 25 m_l values ($l = 4$, $m_l = -4$ to $+4$; $l = 3$, $m_l = -3$ to $+3$; $l = 2$, $m_l = -2$ to $+2$; $l = 1$, $m_l = -1$ to $+1$; $l = 0$, $= 0$).

In general, for each principle quantum number n there are n l-values and n^2 m_l-values. For each shell, there are n kinds of orbitals and n^2 total orbitals.

6.43 (a) 3p: $n = 3$, $l = 1$ (b) 2s: $n = 2$, $l = 0$ (c) 4f: $n = 4$, $l = 3$ (d) 5d: $n = 5$, $l = 2$

6.44 (a) 2, 1, 1; 2, 1, 0; 2, 1 -1

(b) 5, 2, 2; 5, 2, 1; 5, 2, 0; 5, 2, -1; 5, 2, -2

6.45 impossible: (a) 1p, only $l = 0$ is possible for $n = 1$; (d) 2d, for $n = 2$, $l = 1$ or 0, but not 2

6.46 (a) permissible, 2p (b) forbidden, for $l = 0$, m_l can only equal 0
(c) permissible, 4d (d) forbidden, for n = 3, the largest l value is 2

6.47 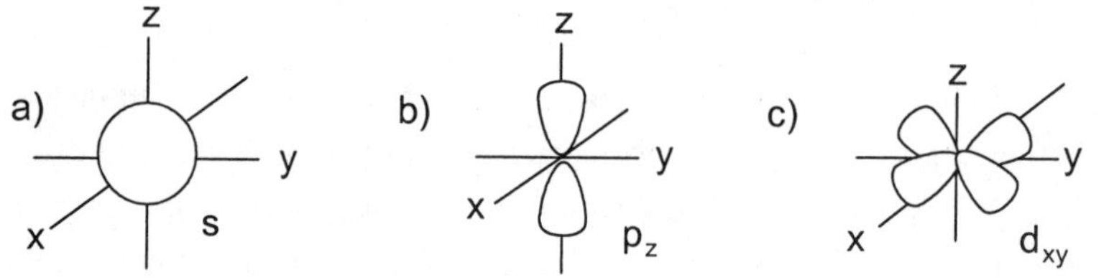 a) s b) p_z c) d_{xy}

6.48 a) p_x b) d_{z^2} c) $d_{x^2-y^2}$

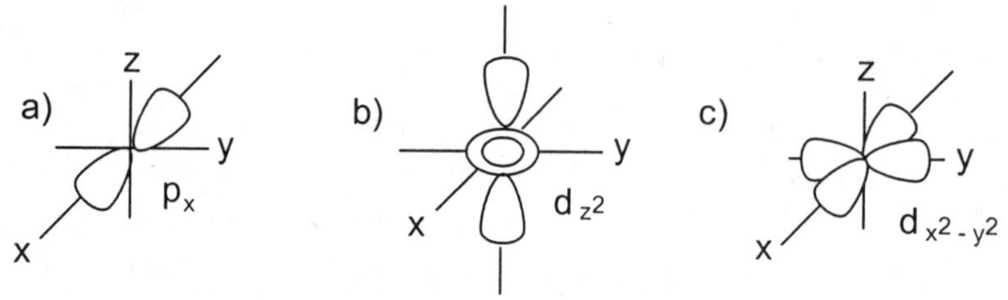

6.49 (a) The 1s and 2s orbitals of a hydrogen atom have the same overall spherical shape. The 2s orbital has a larger radial extension and one node, while the 1s orbital has continuous electron density. Since the 2s orbital is "larger", there is greater probability of finding an electron further from the nucleus in the 2s orbital.

(b) A single 2p orbital is directional in that its electron density is concentrated along one of the three cartesian axes of the atom. The $d_{x^2-y^2}$ orbital has electron density along both the x- and y-axes, while the p_x orbital has density only along the x-axis.

(c) The average distance of an electron from the nucleus in a 3s orbital is greater than for an electron in a 2s orbital. In general, for the same kind of orbital, the larger the n value, the greater the average distance of an electron from the nucleus of the atom.

(d) 1s < 2p < 3d < 4f < 6s. In the hydrogen atom, orbitals with the same *n* value are degenerate and energy increases with increasing *n* value. Thus, the order of increasing energy is given above.

6.50 (a) In an s orbital, there are (*n* - 1) nodes.

(b) The $2p_x$ orbital has one node (the yz plane passing through the nucleus of the atom). The 3s orbital has 2 nodes.

(c) The nodes in p orbitals are planes.

(d) 2s = 2p < 3s < 4d < 5s. In the hydrogen atom, orbitals with the same *n* value are degenerate and energy increases with increasing *n* value.

Many-Electron Atoms and Electron Configurations

6.51 (a) In the hydrogen atom, orbitals with the same principle quantum number, *n*, have the same energy; they are degenerate.

(b) In a many-electron atom, for a given *n*-value, orbital energy increases with increasing *l*-value: s < p < d < f

6.52 (a) The electron with the greater average distance from the nucleus feels a smaller attraction for the nucleus and is higher in energy. Thus the 3p is higher in energy than 3s.

(b) Because it has a larger *n* value, a 3s electron has a greater average distance from the chlorine nucleus than a 2p electron. The 3s electron experiences a smaller attraction for the nucleus and requires less energy to remove from the chlorine atom.

6.53 (a) +1/2, - 1/2

(b) Electrons with opposite spins are affected differently by a strong inhomogeneous magnetic field. An apparatus similar to that in Figure 6.24 can be used to distinguish electrons with opposite spins.

(c) The Pauli exclusion principle states that no two electrons can have the same four quantum numbers. Two electrons in a 1s orbital have the same, *n*, *l* and m_l values. They must have different m_s values.

6.54 (a) The Pauli exclusion principle states that no two electrons can have the same four quantum numbers.

(b) An alternate statement of the Pauli exclusion principle is that a single orbital can hold a maximum of two electrons. Thus, the Pauli principle limits the maximum number of electrons in a main shell and its subshells, which determines when a new row of the periodic table begins.

6.55 Each subshell has an *l*-value associated with it. For a particular *l*-value, permissible m_l-values are $-l$ to $+l$. Each m_l-value represents an orbital, which can hold two electrons.

(a) 10 (b) 2 (c) 6 (d) 14

6.56 (a) 4 (b) 14 (c) 2 (d) 2

6.57 (a) Each box represents an orbital.

(b) Electron spin is represented by the direction of the half-arrows.

(c) No. The electron configuration of Be is $1s^2 2s^2$. There are no electrons in subshells that have degenerate orbitals, so Hund's rule is not used.

6.58 (a) "Outer-shell electrons" are those beyond the previous noble-gas or core electron configuration.

(b) "Unpaired electrons" are electrons that occupy orbitals singly. That is, when there is only one electron in an orbital, this electron is "unpaired."

(c) A Si atom has 4 outer-shell electrons: $3s^2 3p^2$. Two of them (those in the degenerate 3p orbitals) are unpaired.

6.59 (a) Cs: $[Xe]6s^1$ (b) Ni: $[Ar]4s^2 3d^8$ (c) Se: $[Ar]4s^2 3d^{10} 4p^4$

(d) Cd: $[Kr]5s^2 4d^{10}$ (e) Ac: $[Rn]7s^2 6d^1$ (f) Pb: $[Xe]6s^2 4f^{14} 5d^{10} 6p^2$

6.60 (a) Al: $[Ne]3s^2 3p^1$ (b) Sc: $[Ar]4s^2 3d^1$ (c) Co: $[Ar]4s^2 3d^7$

(d) Br: $[Ar]4s^2 3d^{10} 4p^5$ (e) Ba: $[Xe]6s^2$ (f) Re: $[Xe]6s^2 4f^{14} 5d^5$

(g) Lu: $[Xe]6s^2 4f^{14} 5d^1$

6.61 (a) S: [Ne] [↑↓] (3s) [↑↓][↑][↑] (3p) 2 unpaired electrons

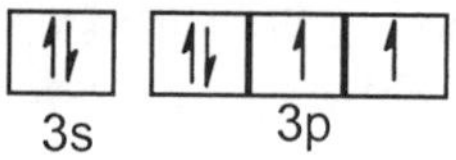

(b) Sr: [Kr] [↑↓] (5s) 0 unpaired electrons

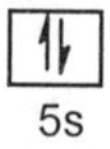

(c) Fe: [Ar] [↑↓] (4s) [↑↓][↑][↑][↑][↑] (3d) 4 unpaired electrons

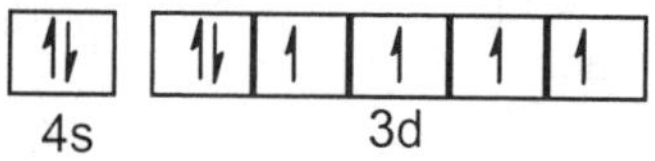

(d) Zr: [Kr] [↑↓] (5s) [↑][↑][ ][ ][ ] (4d) 2 unpaired electrons

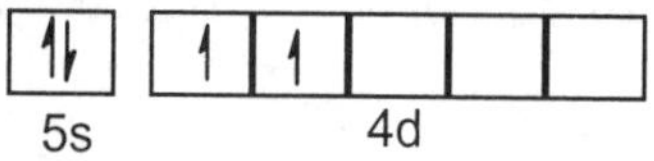

(e) Sb: [Kr] [↑↓] (5s) [↑↓][↑↓][↑↓][↑↓][↑↓] (4d) [↑][↑][↑] (5p) 3 unpaired electrons

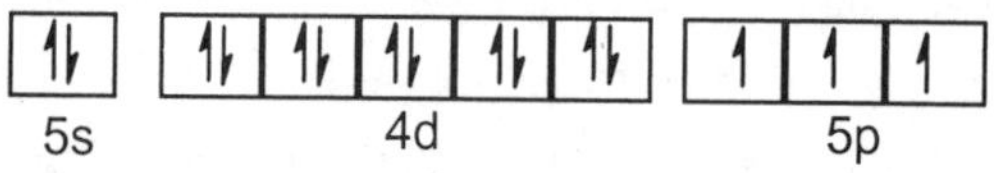

(f) U: [Rn] [↑↓] (7s) [↑][↑][↑][ ][ ][ ][ ] (5f) [↑][ ][ ][ ][ ] (6d)

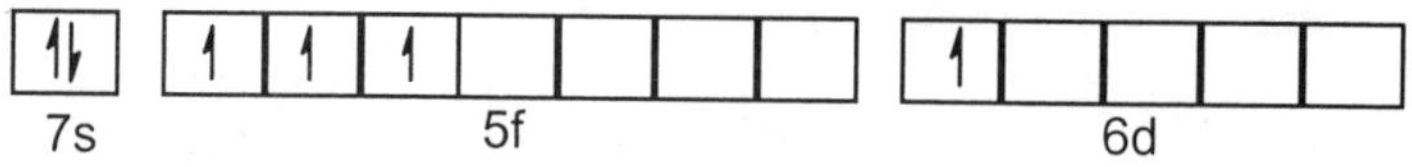

4 unpaired electrons

6.62

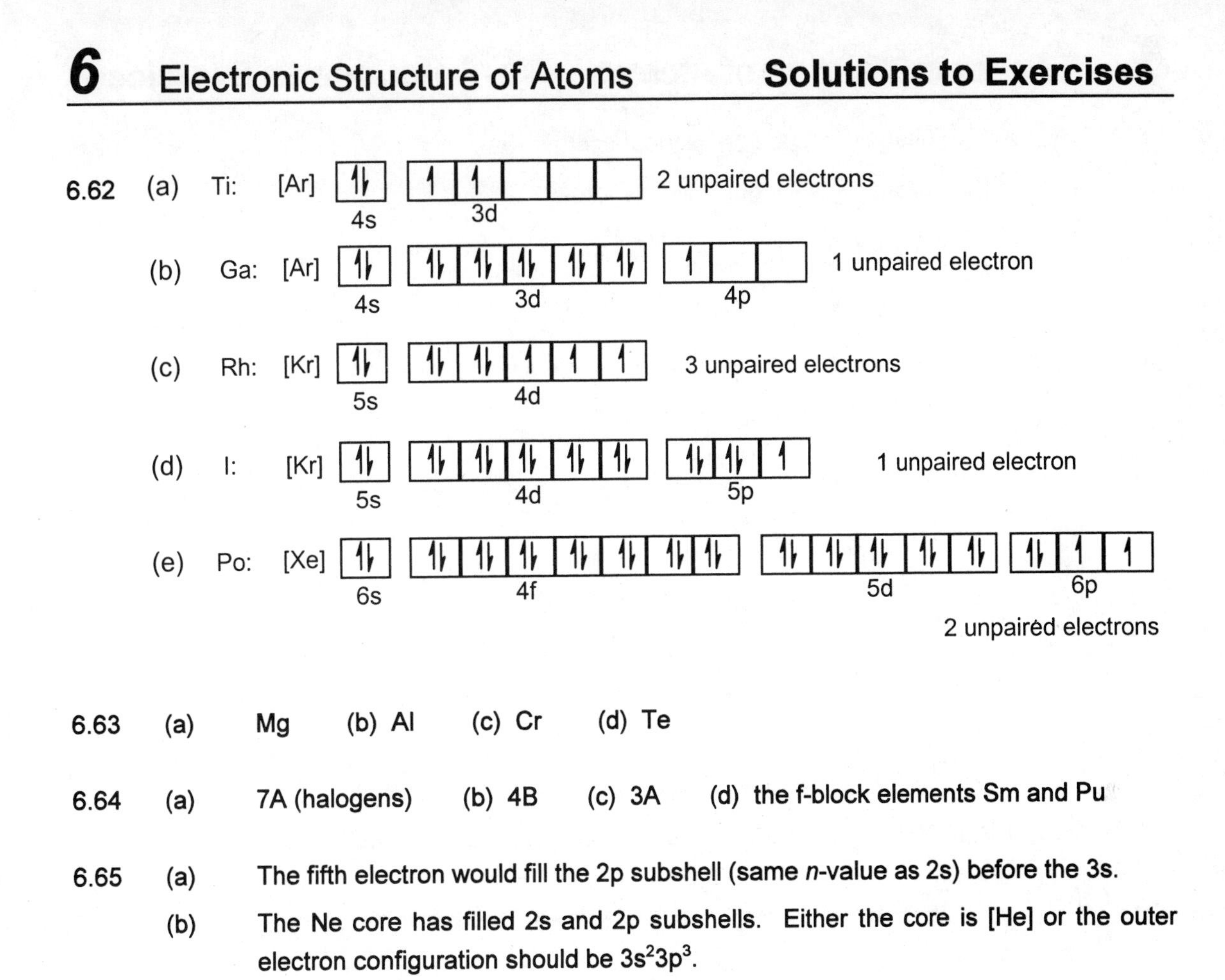

6.63 (a) Mg (b) Al (c) Cr (d) Te

6.64 (a) 7A (halogens) (b) 4B (c) 3A (d) the f-block elements Sm and Pu

6.65 (a) The fifth electron would fill the 2p subshell (same *n*-value as 2s) before the 3s.

(b) The Ne core has filled 2s and 2p subshells. Either the core is [He] or the outer electron configuration should be $3s^2 3p^3$.

(c) The 3p subshell would fill before the 3d because it has the lower *l*-value and the same *n*-value.

6.66 Count the total number of electrons to assign the element.

(a) N: $[He]2s^2 2p^3$ (b) Se: $[Ar]4s^2 3d^{10} 4p^4$ (c) Rh: $[Kr]5s^2 4d^7$

Additional Exercises

6.67 (a) $\lambda_A = 1.6 \times 10^{-7}$ m / 4.5 = $3.56 \times 10^{-8} = 3.6 \times 10^{-8}$ m

$\lambda_B = 1.6 \times 10^{-7}$ m / 2 = 8.0×10^{-8} m

(b) $$\nu = c/\lambda;\ \nu_A = \frac{2.998 \times 10^8 \text{ m}}{1 \text{ s}} \times \frac{1}{3.56 \times 10^{-8} \text{ m}} = 8.4 \times 10^{15} \text{ s}^{-1}$$

$$\nu_B = \frac{2.998 \times 10^8 \text{ m}}{1 \text{ s}} \times \frac{1}{8.0 \times 10^{-8} \text{ m}} = 3.7 \times 10^{15} \text{ s}^{-1}$$

(c) A: ultraviolet, B: ultraviolet

6.68 (a) Elements that emit in the visible: Ba (dark blue), Ca (dark blue), K (dark blue), Na (yellow/orange). (The other wavelengths are in the ultraviolet.)

(b) Au: shortest wavelength, highest energy

Na: longest wavelength, lowest energy

(c) $\lambda = c/\nu = \dfrac{2.998 \times 10^8 \text{ m/s}}{6.59 \times 10^{14}\text{/s}} \times \dfrac{1 \text{ nm}}{1 \times 10^{-9} \text{ m}} = 455 \text{ nm}$, **Ba**

6.69 All electromagnetic radiation travels at the same speed, 2.998×10^8 m/s. Change miles to meters and seconds to some appropriate unit of time.

$$522 \times 10^6 \text{ mi} \times \frac{1.6093 \text{ km}}{1 \text{ mi}} \times \frac{1000 \text{ m}}{1 \text{ km}} \times \frac{1 \text{ s}}{2.998 \times 10^8 \text{ m}} \times \frac{1 \text{ min}}{60 \text{ s}} = 46.7 \text{ min}$$

6.70 (a) $\nu = c/\lambda = \dfrac{2.998 \times 10^8 \text{ m/s}}{320 \text{ nm}} \times \dfrac{1 \text{ nm}}{1 \times 10^{-9} \text{ m}} = 9.37 \times 10^{14} \text{ s}^{-1}$

(b) $E = hc/\lambda = \dfrac{6.626 \times 10^{-34} \text{ J}\bullet\text{s} \times 2.998 \times 10^8 \text{ m/s}}{3.20 \times 10^{-7} \text{ m}} \times \dfrac{1 \text{ kJ}}{1000 \text{ J}} \times \dfrac{6.022 \times 10^{23} \text{ photons}}{\text{mole}}$

$= 374$ kJ/mol

(c) UV-B photons have shorter wavelength and higher energy.

(d) Yes. The higher energy UV-B photons would be more likely to cause sunburn.

6.71 $E = hc/\lambda$ → J/photon; total energy = power × time; photons = total energy / J / photon

$$E = \frac{6.626 \times 10^{-34} \text{ J}\bullet\text{s} \times 2.998 \times 10^8 \text{ m/s}}{780 \times 10^{-9} \text{ m}} = 2.5468 \times 10^{-19} = 2.55 \times 10^{-19} \text{ J/photon}$$

$$0.10 \text{ mW} = \frac{0.10 \times 10^{-3} \text{ J}}{1 \text{ s}} \times 69 \text{ min} \times \frac{60 \text{ s}}{1 \text{ min}} = 0.4140 = 0.41 \text{ J}$$

$$0.4140 \text{ J} \times \frac{1 \text{ photon}}{2.5468 \times 10^{-19} \text{ J}} = 1.626 \times 10^{18} = 1.6 \times 10^{18} \text{ photons}$$

6.72 E/photon = hc/λ

$$E = \frac{6.626 \times 10^{-34} \text{ J}\bullet\text{s} \times 2.998 \times 10^8 \text{ m/s}}{455 \times 10^{-9} \text{ m}} \times \frac{6.022 \times 10^{23} \text{ photons}}{\text{mol}} = 2.63 \times 10^5 \text{ J/mol}$$

$= 263$ kJ/mol

6.73 $\dfrac{5.8 \times 10^{-13} \text{ C}}{1 \text{ s}} \times \dfrac{1e^-}{1.602 \times 10^{-19} \text{ C}} \times \dfrac{1 \text{ photon}}{1 e^-} = 3.620 \times 10^6 = 3.6 \times 10^6$ photons/s

$$\frac{E}{\text{photon}} = hc/\lambda = \frac{6.626 \times 10^{-34} \text{ J}\bullet\text{s}}{550 \text{ nm}} \times \frac{2.998 \times 10^8 \text{ m}}{1 \text{ s}} \times \frac{1 \text{ nm}}{1 \times 10^{-9} \text{ m}} \times \frac{3.620 \times 10^6 \text{ photon}}{\text{s}}$$

$= 1.3 \times 10^{-12}$ J/s

6.74 (a) $\dfrac{2.00 \times 10^5 \text{ J}}{\text{mol}} \times \dfrac{1 \text{ mol photons}}{6.022 \times 10^{23} \text{ photons}} = 3.321 \times 10^{-19} = 3.32 \times 10^{-19}$ J/photon

(b) $\lambda = \frac{hc}{E} = \frac{6.626 \times 10^{-34}\ J\cdot s \times 2.998 \times 10^{8}\ m/s}{3.321 \times 10^{-19}\ J} = 5.98 \times 10^{-7}\ m$

(c) 5.98×10^{-7} m = 598 nm is well within the visible portion of the electromagnetic spectrum and corresponds to yellow or yellow-orange light. Red light, with wavelengths near or greater than 700 nm, does not have sufficient energy to initiate electron transfer and darken the film.

6.75 (a) Gaseous atoms of various elements in the sun's atmosphere typically have ground state electron configurations. When these atoms are exposed to radiation from the sun, the electrons change from the ground state to one of several allowed excited states. Atoms absorb the wavelengths of light which correspond to these allowed energy changes. All other wavelengths of solar radiation pass through the atmosphere unchanged. Thus, the dark lines are the wavelengths that correspond to allowed energy changes in atoms of the solar atmosphere. The continuous background is all other wavelengths of solar radiation.

(b) The scientist should record the absorption spectrum of pure neon or other elements of interest. The black lines should appear at the same wavelengths regardless of the source of neon.

6.76 (a) He^+ is hydrogen-like because it is a one-electron particle. An He atom has two electrons. The Bohr model is based on the interaction of a single electron with the nucleus, but does not accurately account for additional interactions when two or more electrons are present.

(b) Divide each energy by the smallest value to find the integer relationship.

H: $2.18 \times 10^{-18}/2.18 \times 10^{-18} = 1;\ Z = 1$

He^+: $8.72 \times 10^{-18}/2.18 \times 10^{-18} = 4;\ Z = 2$

Li^{2+}: $1.96 \times 10^{-17}/2.18 \times 10^{-18} = 9;\ Z = 3$

The ground-state energies are in the ratio of 1:4:9, which is also the ratio Z^2, the square of the nuclear charge for each particle.

The ground state energy for hydrogen-like particles is:
$E = R_H Z^2$. (By definition, n = 1 for the ground state of a one-electron particle.)

(c) Z = 6 for C. $E = -2.18 \times 10^{-18}\ J\ (6)^2 = -7.85 \times 10^{-17}\ J$

6.77 $\lambda = h/mv$; $v = h/m\lambda$. $\lambda = 0.711\ Å \times \frac{1 \times 10^{-10}\ m}{1\ Å} = 7.11 \times 10^{-11}$ m; $m_e = 9.1094 \times 10^{-31}$ kg

$$v = \frac{6.626 \times 10^{-34}\ J\cdot s}{9.1094 \times 10^{-31}\ kg \times 7.11 \times 10^{-11}\ m} \times \frac{1\ kg\cdot m^2/s^2}{1\ J} = 1.02 \times 10^{7}\ m/s$$

6.78 *Plan.* Change keV to J/electron. Calculate v from kinetic energy. $\lambda = h/mv$. *Solve:*

$$82.4 \text{ keV} \times \frac{1000 \text{ eV}}{\text{keV}} \times \frac{96.485 \text{ kJ}}{1 \text{ eV} \bullet \text{mol}} \times \frac{1000 \text{ J}}{1 \text{ kJ}} \times \frac{1 \text{ mol}}{6.022 \times 10^{23} \text{ electrons}}$$

$$= 1.320 \times 10^{-14} = 1.32 \times 10^{-14} \text{ J/electron}$$

$E_k = mv^2/2$; $v^2 = 2E_k/m$; $v = \sqrt{2E_k/m}$

$$v = \left(\frac{2 \times 1.320 \times 10^{-14} \text{ kg} \bullet \text{m}^2/\text{s}^2}{9.1094 \times 10^{-31} \text{ kg}}\right)^{1/2} = 1.703 \times 10^8 = 1.70 \times 10^8 \text{ m/s}$$

$$\lambda = h/mv = \frac{6.626 \times 10^{-34} \text{ J} \bullet \text{s}}{9.1094 \times 10^{-31} \text{ kg} \times 1.703 \times 10^8 \text{ m/s}} \times \frac{1 \text{ kg} \bullet \text{m}^2/\text{s}^2}{1 \text{ J}} = 4.27 \times 10^{-12} \text{ m} = 4.27 \text{ pm}$$

6.79 (a) l (b) n and l (c) m_s (d) m_l

6.80 (a) 3p (b) 6g (c) 2s (d) 4f

6.81 (a) 1 (b) 3 (c) 5 (d) 9

6.82 What the noble gas elements have in common are completed ns and np subshells. Since the Pauli principle limits the number of electrons per orbital to two, this leads to the first three magic numbers, $2(1s^2)$, $10(1s^22s^22p^6)$ and $18(1s^22s^22p^63s^23p^6)$. In the fourth row, (n-1) d orbitals begin to fill as their energy falls below that of the np orbitals. This leads to the next two magic numbers, $36(1s^22s^22p^63s^23p^64s^23d^{10}4p^6)$ and $54(1s^22s^22p^63s^23p^64s^23d^{10}4p^65s^24d^{10}5p^6)$. In the sixth row, the energy of the 4f orbitals falls below that of the (n-1)d and np subshells, and it fills. This explains the final magic number, $86(1s^22s^22p^63s^23p^64s^23d^{10}4p^65s^24d^{10}5p^66s^24f^{14}5d^{10}6p^6)$.

6.83 (a) The p_z orbital has a nodal plane where z = 0. This is the xy plane.

(b) The d_{xy} orbital has 4 lobes and 2 nodal planes, the two planes where x = 0 and y = 0. These are the yz and xz planes.

(c) The $d_{x^2-y^2}$ has 4 lobes and 2 nodal planes, the planes where $x^2 - y^2 = 0$. These are the planes that bisect the x and y axes and contain the z axis.

6.84 (a) Se: [Ar] $4s^23d^{10}4p^4$ (b) Rh: [Kr] $5s^24d^7$ (c) Si: [Ne] $3s^23p^2$

(d) Hg: [Xe] $6s^24f^{14}5d^{10}$ (e) Hf: [Xe] $6s^24f^{14}5d^2$

6.85 Mt: [Rn] $7s^25f^{14}6d^7$

6.86 The core would be the electron configuration of element 118. If no new subshell begins to fill, the condensed electron configuration of element 126 would be similar to those of elements vertically above it on the periodic chart, Pu and Sm. The condensed configuration would be $[118]8s^2 6f^6$. On the other hand, the 5g subshell could begin to fill after 8s, resulting in the condensed configuration $[118]8s^2 5g^6$. Exceptions are also possible.

Integrative Exercises

6.87 We know the wavelength of microwave radiation, the volume of coffee to be heated and the desired temperature change. Assume the density and heat capacity of coffee are the same as pure water. We need to calculate: (i) the total energy required to heat the coffee and (ii) the energy of a single photon in order to find (iii) the number of photons required.

(i) From Chapter 5, the heat capacity of liquid water is 4.184 J/g°C.

To find the mass of 200 mL of coffee at 23°C, use the density of water given in Appendix B.

$$200 \text{ mL} \times \frac{0.997 \text{ g}}{1 \text{ mL}} = 199.4 = 199 \text{ g coffee}$$

$$\frac{4.184 \text{ J}}{1 \text{ g °C}} \times 199.4 \text{ g} \times (60°\text{C} - 23°\text{C}) = 3.087 \times 10^4 \text{ J} = 31 \text{ kJ}$$

(ii) $$E = hc/\lambda = 6.626 \times 10^{-34} \text{ J} \bullet \text{s} \times \frac{2.998 \times 10^8 \text{ m}}{1 \text{ s}} \times \frac{1}{0.112 \text{ m}} = \frac{1.77 \times 10^{-24} \text{ J}}{1 \text{ photon}}$$

(iii) $$3.087 \times 10^4 \text{ J} \times \frac{1 \text{ photon}}{1.774 \times 10^{-24} \text{ J}} = 1.7 \times 10^{28} \text{ photons}$$

(The answer has 2 sig figs because the temperature change, 43°C, has 2 sig figs)

6.88 $$\Delta H^\circ_{rxn} = \Delta H^\circ_f \ O_2(g) + \Delta H^\circ_f \ O(g) - \Delta H^\circ_f \ O_3(g)$$

$$\Delta H^\circ_{rxn} = 0 + 247.5 \text{ kJ} - 142.3 \text{ kJ} = +105.2 \text{ kJ}$$

$$\frac{105.2 \text{ kJ}}{\text{mol } O_3} \times \frac{1 \text{ mol } O_3}{6.022 \times 10^{23} \text{ molecules}} \times \frac{1000 \text{ J}}{1 \text{ kJ}} = \frac{1.747 \times 10^{-19} \text{ J}}{O_3 \text{ molecule}}$$

$$\Delta E = hc/\lambda \ ; \ \lambda = \frac{hc}{\Delta E} = \frac{6.626 \times 10^{-34} \text{ J} \bullet \text{s} \times 2.998 \times 10^8 \text{ m/s}}{1.747 \times 10^{-19} \text{ J}} = 1.137 \times 10^{-6} \text{ m}$$

Radiation with this wavelength is in the infrared portion of the spectrum. (Clearly, processes other than simple photodissociation cause O_3 to absorb ultraviolet radiation.)

6.89 (a) The electron configuration of Zr is $[Kr]\ 5s^2 4d^2$ and that of Hf is $[Xe]\ 6s^2 4f^{14} 5d^2$. Although Hf has electrons in f orbitals as the rare earth elements do, the 4f subshell in Hf is filled, and the 5d electrons primarily determine the chemical properties of the element. Thus, Hf should be chemically similar to Zr rather than the rare earth elements.

(b) $ZrCl_4(s) + 4Na(l) \rightarrow Zr(s) + 4NaCl(s)$

This is an oxidation-reduction reaction; Na is oxidized and Zr is reduced.

(c) $2ZrO_2(s) + 4Cl_2(g) + 3C(s) \rightarrow 2ZrCl_4(s) + CO_2(g) + 2CO(g)$

$$55.4 \text{ g } ZrO_2 \times \frac{1 \text{ mol } ZrO_2}{123.2 \text{ g } ZrO_2} \times \frac{2 \text{ mol } ZrCl_4}{2 \text{ mol } ZrO_2} \times \frac{233.0 \text{ g } ZrCl_4}{1 \text{ mol } ZrCl_4} = 105 \text{ g } ZrCl_4$$

(d) In ionic-compounds of the type MCl_4 and MO_2, the metal ions have a 4+ charge, indicating that the neutral atoms have lost 4 electrons. Zr, [Kr] $5s^24d^2$, loses the 4 electrons beyond its Kr core configuration. Hf, [Xe]$6s^24f^{14}5d^2$, similarly loses its four 6s and 5d electrons, but not electrons from the "complete" 4f subshell.

6.90 (a) Each oxide ion, O^{2-}, carries a 2- charge. Each metal oxide is a neutral compound, so the metal ion or ions must adopt a total positive charge equal to the total negative charge of the oxide ions in the compound. The table below lists the electron configuration of the neutral metal atom, the positive charge of **each** metal ion in the oxide, and the corresponding electron configuration of the metal ion.

i.	K:	[Ar] $4s^1$	1+	[Ar]
ii.	Ca:	[Ar] $4s^2$	2+	[Ar]
iii.	Sc:	[Ar] $4s^23d^1$	3+	[Ar]
iv.	Ti:	[Ar] $4s^23d^2$	4+	[Ar]
v.	V:	[Ar] $4s^23d^3$	5+	[Ar]
vi.	Cr:	[Ar]$4s^13d^5$	6+	[Ar]

Each metal atom loses all (valence) electrons beyond the Ar core configuration. In K_2O, Sc_2O_3 and V_2O_5, where the metal ions have odd charges, two metal ions are required to produce a neutral oxide.

(b) i. potassium oxide ii. calcium oxide iii. scandium(III) oxide

iv. titanium (IV) oxide v. vanadium (V) oxide vi. chromium(VI) oxide

(Roman numerals are required to specify the charges on the transition metal ions, because more than one stable ion may exist.)

(c) Recall that $\Delta H_f^\circ = 0$ for elements in their standard states. In these reactions, M(s) and $H_2(g)$ are elements in their standard states.

i. $K_2O(s) + H_2(g) \rightarrow 2K(s) + H_2O(g)$

$\Delta H^\circ = \Delta H_f^\circ\ H_2O(g) + 2\Delta H_f^\circ\ K(s) - \Delta H\ K_2O(s) - \Delta H_f^\circ\ H_2(g)$

$\Delta H^\circ = -241.82 \text{ kJ} + 2(0) - (-363.2 \text{ kJ}) - 0 = 121.4 \text{ kJ}$

ii. $CaO(s) + H_2(g) \rightarrow Ca(s) + H_2O(g)$

$\Delta H^\circ = \Delta H_f^\circ\ H_2O(g) + \Delta H_f^\circ\ Ca(s) - \Delta H_f^\circ\ CaO(s) - \Delta H_f^\circ\ H_2(g)$

$\Delta H^\circ = -241.82 \text{ kJ} + 0 - (-635.1 \text{ kJ}) - 0 = 393.3 \text{ kJ}$

iv. $TiO_2(s) + 2H_2(g) \rightarrow Ti(s) + 2H_2O(g)$

$\Delta H^\circ = 2\Delta H_f^\circ\ H_2O(g) + \Delta H_f^\circ\ Ti(s) - \Delta H_f^\circ\ TiO_2(s) - 2\Delta H_f^\circ\ H_2(g)$

$= 2(-241.82) + 0 - (-938.7) - 2(0) = 455.1 \text{ kJ}$

v. $V_2O_5(s) + 5H_2(g) \rightarrow 2V(s) + 5H_2O(g)$

$\Delta H° = 5\Delta H_f^\circ\ H_2O(g) + 2\Delta H_f^\circ\ V(s) - \Delta H_f^\circ\ V_2O_5(s) - 5\Delta H_f^\circ\ H_2(g)$

$= 5(-241.82) + 2(0) - (-1550.6) - 5(0) = 341.5$ kJ

(d) ΔH_f° becomes more negative moving from left to right across this row of the periodic chart. Since Sc lies between Ca and Ti, the median of the two ΔH_f° values is approximately -785 kJ/mol. However, the trend is clearly not linear. Dividing the ΔH_f° values by the positive charge on the pertinent metal ion produces the values -363, -318, -235 and -310. The value between Ca^{2+} (-318) and Ti^{4+} (-235) is Sc^{3+} (-277). Multiplying (-277) by 3, a value of approximately -830 kJ results. A reasonable range of values for ΔH_f° of $Sc_2O_3(s)$ is then -785 to -830 kJ/mol.

6.91 (a) Bohr's theory was based on the Rutherford "nuclear" model of the atom. That is, Bohr theory assumed a dense positive charge at the center of the atom and a diffuse negative charge (electrons) surrounding it. Bohr's theory then specified the nature of the diffuse negative charge. The prevailing theory before the nuclear model was Thomson's plum pudding or watermelon model, with discrete electrons scattered about a diffuse positive charge cloud. Bohr's theory could not have been based on the Thomson model of the atom.

(b) deBroglie's hypothesis is that electrons exhibit both particle and wave properties. Thomson's conclusion that electrons have mass is a particle property, while the nature of cathode rays is a wave property. de Broglie's hypothesis actually rationalizes these two seemingly contradictory observations about the properties of electrons.

7 Periodic Properties of the Elements

Periodic Table; Effective Nuclear Charge

7.1 Mendeleev insisted that elements with similar chemical and physical properties be placed within a family or column of the table. Since many elements were as yet undiscovered, Mendeleev left blanks. He predicted properties for the "blanks" based on properties of other elements in the family.

7.2 (a) The verification of the existence of many new elements by accurately measuring their atomic weights spurred interest in a classification scheme. Mendeleev (and Meyer) noted that certain chemical and physical properties recur periodically when the elements are arranged by increasing atomic weight. The accurate atomic weights provided a common property on which to base a classification scheme of the elements.

(b) Moseley realized that the characteristic X-ray frequencies emitted by each element were related to a unique integer that he assigned to each element. We now know this integer as the atomic number, the number of protons in the nucleus of an atom. In general, atomic weight increases as atomic number increases, but there are a few exceptions. If elements are arranged by increasing atomic number, a few seeming contradictions in the Mendelev table (the positions of Ar and K or Te and I) are eliminated.

7.3 (a) Effective nuclear charge, Z_{eff}, is a representation of the average electrical field experienced by a single electron. It is the average environment created by the nucleus and the other electrons in the molecule, expressed as a **net** positive charge at the nucleus. It is approximately the nuclear charge, Z, minus the number of core electrons.

(b) Going from left to right across a period, nuclear charge increases while the number of electrons in the core is constant. This results in an increase in Z_{eff}.

7.4 (a) Electrostatic attraction for the nucleus lowers the energy of an electron, while electron-electron repulsions increase this energy. The concept of effective nuclear charge allows us to model this increase in the energy of an electron as a smaller net attraction to a nucleus with a smaller positive charge, Z_{eff}.

(b) In Be (or any element), the 1s electrons are not shielded by any core electrons, so they experience a much greater Z_{eff} than the 2s electrons.

7.5 *Analyze/Plan.* The problem states that shielding, S, is exactly equal to the number of electrons in the core. Z_{eff} = Z - # of core electrons. *Solve*:

(a) K: 19 - 18 = +1 (b) Br: 35 - 28 = +7

7.6 (a) Z_{eff} = Z - # of core electrons = 13 - 10 = +3.

(b) An estimate of Z_{eff} based on the assumption in part (a) will always be lower than the value based on detailed calculations. Because outer electrons have some probability of being in the core, the core electrons are never 100% effective at shielding; the number of core electrons represents an upper limit for S.

7.7 Krypton has a larger nuclear charge (Z = 36) than argon (Z = 18). The shielding of electrons in the *n* = 3 shell by the 1s and 2s core electrons in the two atoms is approximately equal, so the *n* = 3 electrons in Kr experience a greater effective nuclear charge and are thus situated closer to the nucleus.

7.8 Mg < P < K < Ti < Rh. The shielding of electrons in the *n* = 3 shell by 1s and 2s core electrons in these elements is approximately equal, so the effective nuclear charge increases as Z increases.

Atomic and Ionic Radii

7.9 Atomic radii are determined by distances between atoms (interatomic distances) in various situations. **Bonding radii** are calculated from the internuclear separation of two atoms joined by a chemical bond. **Nonbonding radii** are calculated from the internuclear separation between two gaseous atoms that collide and move apart, but do not bond.

7.10 (a) Since the quantum mechanical description of the atom does not specify the exact location of electrons, there is no specific distance from the nucleus where the last electron can be found. Rather, the electron density decreases gradually as the distance from the nucleus increases. There is no quantum mechanical "edge" of an atom.

(b) When nonbonded atoms touch, it is their electron clouds that interact. These interactions are primarily repulsive because of the negative charges of electrons. Thus, the size of the electron clouds determines the nuclear approach distance of nonbonded atoms.

7.11 The atomic radius of Au is the interatomic Au-Au distance divided by 2, 2.88 Å/2 = 1.44 Å.

7.12 The distance between Si atoms in solid silicon is 2 times the bonding atomic radius from Figure 7.5. the Si-Si distance is 2 × 1.11 Å = 2.22 Å.

7.13 From atomic radii, As-I = 1.19 Å + 1.33 Å = 2.52 Å. This is very close to the experimental value of 2.55 Å.

7.14 (a) The estimated distances in the table below are the sum of the radii of the group 5A elements and H from Table 7.5.

bonded atoms	estimated distance	measured distance
P – H	1.43	1.419
As – H	1.56	1.519
Sb – H	1.75	1.707

In general, the estimated distances are a bit longer than the measured distances. This probably shows a systematic bias in either the estimated radii or in the method of obtaining the measured values.

(b) The principle quantum number of the outer electrons and thus the average distance of these electrons from the nucleus increases from P (n = 3) to As (n = 4) to Sb (n = 5). This causes the systematic increase in M-H distance.

7.15 (a) Atomic radii **decrease** moving from left to right across a row and (b) **increase** from top to bottom within a group.

(c) F < S < P <As. The order is unambiguous according to the trends of increasing atomic radius moving down a column and to the left in a row of the table.

7.16 (a) The vertical difference in radius is due to a change in principle quantum number of the outer electrons. The horizontal difference in radius is due to the change in electrostatic attraction between the outer electron and a nucleus with one more or one fewer proton. Adding or subtracting a proton has a much smaller radius effect than moving from one principle quantum level to the next.

(b) S < Si < Se < Ge. This order is predicted by the trends in increasing atomic radius moving to the left in a row and down a column of the periodic chart, assuming that changes moving down a column are larger [see part (a)]. That is, the order above assumes that the change from S to Se is larger than the change from S to Si. This order is confirmed by the values in Figure 7.5.

7.17 *Plan*. Locate each element on the periodic charge and use trends in radii to predict their order. *Solve*:

(a) Be < Mg < Ca (b) Br < Ge < Ga (c) Si < Al < Tl

7.18 (a) K < Cs < Rb (b) Te < Sn < In (c) Cl < P < Sr

7.19 (a) Electrostatic repulsions are reduced by removing an electron from a neutral atom, Z_{eff} increases, and the cation is smaller.

(b) The additional electrostatic repulsion produced by adding an electron to a neutral atom causes the electron cloud to expand, so that the radius of the anion is larger than the radius of the neutral atom.

(c) Going down a column, the *n* value of the valence electrons increases and they are further from the nucleus. Thus, the size of particles with like charge increases.

7.20 (a) As Z stays constant and the number of electrons increases, the electron-electron repulsions increase, the electrons spread apart and the ions become larger.
$I^- > I > I^+$

(b) Going down a column, the increasing average distance of the outer electrons from the nucleus causes the size of particles with like charge to increase. $Ca^{2+} > Mg^{2+} > Be^{2+}$

(c) Fe: $[Ar]4s^2 3d^6$; Fe^{2+}: $[Ar]3d^6$; Fe^{3+}: $[Ar]3d^5$. The 4s valence electrons in Fe are on average further from the nucleus than the 3d electrons, so Fe is larger than Fe^{2+}. Since there are five 3d orbitals, in Fe^{2+} at least one orbital must contain a pair of electrons. Removing one electron to form Fe^{3+} significantly reduces repulsion, increasing the nuclear charge experienced by each of the other d electrons and decreasing the size of the ion. $Fe > Fe^{2+} > Fe^{3+}$

7.21 The size of the blue sphere decreases on reaction, so it loses one or more electrons and becomes a cation. Metals lose electrons when reacting with nonmetals, so the blue sphere represents a metal. The size of the red sphere increases on reaction, so it gains one or more electrons and becomes an anion. Nonmetals gain electrons when reacting with metals, so the red sphere represents a nonmetal.

7.22 The order of radii is $Ca > Ca^{2+} > Mg^{2+}$, so the largest sphere is Ca, the intermediate one is Ca^{2+} and the smallest is Mg^{2+}.

7.23 (a) An isoelectronic series is a group of atoms or ions that have the same number of electrons, and thus the same electron configuration.

(b) (i) Cl^- : **Ar** (ii) Se^{2-} : **Kr** (iii) Mg^{2+} : **Ne**

7.24 (a) K^+, Ca^{2+} (b) Ca^{2+} , Sc^{3+} (c) S^{2-} , Ar (d) Co^{3+} , Fe^{2+}

7.25 (a) Since the electron configurations of the ions in an isoelectronic series are the same, shielding effects do not vary for the different particles. As Z increases, Z_{eff} increases, the valence electrons are more strongly attracted to the nucleus and the size of the particle decreases.

(b) Because F^-, Ne and Na^+ have the same electron configuration, the 2p electron in the particle with the largest Z experiences the largest effective nuclear charge. A 2p electron in Na^+ experiences the greatest effective nuclear charge.

7.26 (a) Cl < S < K (b) $K^+ < Cl^- < S^{2-}$

(c) Even though K has the largest Z value, the *n*-value of the outer electron is larger than the *n*-value of valence electrons in S and Cl so K atoms are largest. When the 4s electron is removed, K^+ is isoelectronic with Cl^- and S^{2-}. The larger Z value causes the 3p electrons in K^+ to experience the largest effective nuclear charge and K^+ is the smallest ion.

7.27 *Plan.* Use relative location on periodic chart and trends in ionic radii to establish the order.

Solve: (a) $Se < Se^{2-} < Te^{2-}$ (b) $Co^{3+} < Fe^{3+} < Fe^{2+}$ (c) $Ti^{4+} < Sc^{3+} < Ca$ (d) $Be^{2+} < Na^+ < Ne$

7.28 (a) Cl^- is larger than Cl because the increase in electron repulsions that accompany addition of an electron cause the electron cloud to expand.

(b) S^{2-} is larger than O^{2-}, because for particles with like charges, size increases going down a family.

(c) K^+ is larger than Ca^{2+} because the two ions are isoelectronic and K^+ has the larger Z.

Ionization Energies; Electron Affinities

7.29 $Te(g) \rightarrow Te^+(g) + 1e^-$; $Te^+(g) \rightarrow Te^{2+}(g) + 1e^-$; $Te^{2+}(g) \rightarrow Te^{3+}(g) + 1e^-$

7.30 (a) $Ga(g) \rightarrow Ga^+(g) + 1e^-$; $Ga^+(g) \rightarrow Ga^{2+}(g) + 1e^-$ (b) $Rh^{3+}(g) \rightarrow Rh^{4+}(g) + 1e^-$

7.31 (a) According to Coulomb's law, the energy of an electron in an atom is negative, because of the electrostatic attraction of the electron for the nucleus. In order to overcome this attraction, remove the electron and increase it's energy; energy must be added to the atom. Ionization energy, ΔE for this process, is positive, regardless of the magnitude of Z or the quantum numbers of the electron.

(b) F has a greater first ionization energy than O, because F has a greater Z_{eff} and the outer electrons in both elements are approximately the same distance from the nucleus.

(c) The second ionization energy of an element is greater than the first because Z_{eff} is larger for the +1 cation than the neutral atom; more energy is required to overcome the larger Z_{eff}.

7.32 (a) The effective nuclear charges of Li and Na are similar, but the outer electron in Li has a smaller *n*-value and is closer to the nucleus than the outer electron in Na. More energy is needed to overcome the greater attraction of the Li electron for the nucleus.

(b) Sc: [Ar] $4s^23d^1$; Ti: [Ar] $4s^23d^2$. The fourth ionization of titanium involves removing a 4s outer electron, while the fourth ionization of Sc requires removing a 3p electron from the [Ar] core. The effective nuclear charges experienced by the two 4s electrons in Ti are much more similar than the effective nuclear charges of a 4s outer electron and a 3p core electron in Sc. Thus, the difference between the third and fourth ionization energies of Sc is much larger.

(c) The electron configuration of Li^+ is $1s^2$ or [He] and that of Be^+ is $[He]2s^1$. Be^+ has one more valence electron to lose while Li^+ has the stable noble gas configuration of He. It requires much more energy to remove a 1s core electron close to the nucleus of Li^+ than a 2s valence electron further from the nucleus of Be^+.

7.33 (a) In general, the smaller the atom, the larger its first ionization energy.

(b) According to Figure 7.10, He has the largest and Cs the smallest first ionization energy of the nonradioactive elements.

7.34 (a) Moving from F to I in group 7A, first ionization energies decrease and atomic radii increase. The greater the atomic radius, the smaller the electrostatic attraction of an outer electron for the nucleus and the smaller the ionization energy of the element.

(b) First ionization energies increase slightly going from K to Ar and atomic sizes decrease. As valence electrons are drawn closer to the nucleus (atom size decreases), it requires more energy to completely remove them from the atom (first ionization energy increases). Each trend has a discontinuity at Ga, owing to the increased shielding of the 4p electrons by the filled 3d subshell.

7.35 *Plan*. Use periodic trends in first ionization energy. *Solve*:

(a) Ne (b) Mg (c) Cr (d) Br (e) Ge

7.36 (a) Cd. As the effective nuclear charge increases in moving from left to right in the fifth row, the energy required to remove an electron increases.

(b) C. Valence electrons in C are closer to the nucleus ($n = 2$) and are shielded only by the [He] core, so they experience greater attraction for the nucleus and have a higher ionization energy.

(c) I. In and I compare as two elements in the same row, with the effects of increasing effective nuclear charge dominating their periodic properties.

(d) Xe. Sn and Xe are in the same row, so Xe with the larger Z experiences a greater effective nuclear charge and a larger ionization energy.

7.37 *Plan*. Follow the logic of Sample Exercise 7.7. *Solve:*

(a) Sb^{3+}: $[Kr]5s^24d^{10}$ (b) Ga^+: $[Ar]4s^23d^{10}$ (c) P^{3-}: $[Ne]3s^23p^6$ or [Ar]

(d) Cr^{3+}: $[Ar]3d^3$ (e) Zn^{2+}: $[Ar]3d^{10}$ (f) Ag^+: $[Kr]4d^{10}$

7.38 (a) Mn^{3+} $[Ar]3d^4$ (b) Se^{2-} $[Ar]4s^23d^{10}4p^6$ = [Kr], noble-gas configuration

(c) Sc^{3+}: [Ar], noble-gas configuration (d) Ru^{2+}: $[Kr]4d^6$

(e) Tl^+: $[Xe]6s^24f^{14}5d^{10}$ (f) Au^+: $[Xe]4f^{14}5d^{10}$

7.39 *Plan.* Follow the logic in Sample Exercise 7.7. Construct a mental box diagram for the outer electrons to determine how many are unpaired.

(a) Co^{2+}: $[Ar]3d^7$, 3 unpaired electrons

(b) In^+: $[Kr]5s^24d^{10}$, 0 unpaired electrons

7.40 (a) Cr^{3+}, 3 unpaired electrons

(b) Sn^{2+}, 0 unpaired electrons

7.41 Ionization energy: $Se(g) \rightarrow Se^+(g) + 1e^-$

$[Ar]4s^23e^{10}4p^4 \quad [Ar]4s^23d^{10}4p^3$

Electron affinity: $Se(g) + 1e^- \rightarrow Se^-(g)$

$[Ar]4s^23d^{10}4p^4 \quad [Ar]4s^23d^{10}4p^5$

7.42 Electron affinity $Br(g) + 1e^- \rightarrow Br^-(g)$

$[Ar]4s^23d^{10}4p^5 \quad [Ar]4s^23d^{10}4p^6$

When a Br atom gains an electron, the Br^- ion adopts the stable electron configuration of Kr. Since the electron is added to the same 4p subshell as other outer electrons, it experiences essentially the same attraction for the nucleus. Thus, the energy of the Br^- ion is lower than the total energy of a Br atom and an isolated electron, and electron affinity is negative.

Electron affinity: $Kr(g) + 1e^- \rightarrow Kr^-(g)$

$[Ar]4s^23d^{10}4p^6 \quad [Ar]4s^23d^{10}4p^65s^1$

Energy is required to add an electron to a Kr atom; Kr^- has a higher energy than the isolated Kr atom and free electron. In Kr^- the added electron would have to occupy the higher energy 5s orbital; a 5s electron is farther from the nucleus and effectively shielded by the spherical Kr core and is not stabilized by the nucleus.

7.43 $Li + 1e^- \rightarrow Li^-$; $Be + 1e^- \rightarrow Be^-$

$[He]2s^1 \quad [He]2s^2 \quad [He]2s^2 \quad [He]2s^22p^1$

Adding an electron to Li completes the 2s subshell. The added electron experiences essentially the same effective nuclear charge as the other valence electron, except for the repulsion of pairing electrons in a orbital. There is an overall stabilization; ΔE is negative.

An extra electron in Be would occupy the higher energy 2p subshell. This electron is shielded from the full nuclear charge by the 2s electrons and does not experience a stabilization in energy; ΔE is positive.

7.44 $Mg^+(g) + 1e^- \rightarrow Mg(g)$

$[Ne]\ 3s^1 \quad [Ne]\ 3s^2$

This process is the reverse of the first ionization of Mg. The magnitude of the energy change for this process is the same as the magnitude of the first ionization energy of Mg, 738 kJ/mol.

Properties of Metals and Nonmetals

7.45 The smaller the first ionization energy of an element, the greater the metallic character of that element.

7.46 P < Sb < Ag. P is a nonmetal, Sb is a metalloid and Ag is a metal. We expect that electrical conductivity increases as metallic character increases. Since metallic character increases going down a column and to the left in a row, the order of increasing electrical conductivity is as shown above.

7.47 *Analyze/Plan.* Metallic character increases moving down a family and to the left in a period. Use these trends to select the element with greater metallic character. *Solve*:

(a) Li (b) Na (c) Sn (d) Al

7.48 (a) In general, ionization energy decreases as metallic character increases. According to Figure 7.10 the ionization energies of the group 5A elements decreases as atomic weight increases (going down the column). Therefore, metallic character of the group 5A elements increases with increasing atomic weight.

(b) Ag < Sn < Se < C < F. Nonmetallic character increases going up and to the right on the periodic chart. Vertical trends dominate the relationship between Se and C.

7.49 Analyze/Plan. Ionic compounds are formed by combining a metal and a nonmetal; molecular compounds are formed by two or more nonmetals. *Solve*:

Ionic: MgO, Li_2O, Y_2O_3; molecular: SO_2, P_2O_5, N_2O, XeO_3

7.50 (a) metal oxides: $Li_2O(s) + H_2O(l) \rightarrow 2\ LiOH(aq)$

$BaO(s) + H_2O(l) \rightarrow Ba(OH)_2(aq)$

nonmetal oxides: $P_4O_{10}(s) + 6H_2O(l) \rightarrow 4H_3PO_4(aq)$

$SO_3(g) + H_2O(l) \rightarrow H_2SO_4(aq)$

(b) Metals have lower ionization energies than nonmetals, so they tend to form ionic oxides, while nonmetals form molecular oxides. Ionic compounds, in this case oxides, dissociate into ions when they dissolve in water. The reactive oxide ion ends up as hydroxide, separated from the metal ion. Molecular oxides do not ionize upon dissolution, so the oxygen remains bound to the nonmetal.

7.51 (a) When dissolved in water, an "acidic oxide" produces an acidic (pH < 7) solution. A "basic oxide" dissolved in water produces a basic (pH > 7) solution.

(b) Oxides of nonmetals are acidic. Example: $SO_3(g) + H_2O(l) \rightarrow H_2SO_4(aq)$. Oxides of metals are basic. Example: CaO (quick lime). $CaO(s) + H_2O(l) \rightarrow Ca(OH)_2(aq)$

7.52 The more **nonmetallic** the central atom, the more acidic the oxide. In order of increasing acidity: $CaO < Al_2O_3 < SiO_2 < CO_2 < P_2O_5 < SO_3$

7.53 (a) $BaO(s) + H_2O(l) \rightarrow Ba(OH)_2(aq)$

(b) $FeO(s) + 2HClO_4(aq) \rightarrow Fe(ClO_4)_2(aq) + H_2O(l)$

(c) $SO_3(g) + H_2O(l) \rightarrow H_2SO_4(aq)$

(d) $CO_2(g) + 2NaOH(aq) \rightarrow Na_2CO_3(aq) + H_2O(l)$

7.54 (a) $K_2O(s) + H_2O(l) \rightarrow 2KOH(aq)$

(b) $P_2O_3(s) + 3H_2O(l) \rightarrow 2H_3PO_3(aq)$

(c) $Cr_2O_3(s) + 6HCl(aq) \rightarrow 2CrCl_3(aq) + 3H_2O(l)$

(d) $SeO_2(s) + 2KOH(aq) \rightarrow K_2SeO_3(aq) + H_2O(l)$

Group Trends in Metals and Nonmetals

7.55

	Na	Mg
(a)	[Ne] $3s^1$	[Ne] $3s^2$
(b)	+1	+2
(c)	+496 kJ/mol	+738 kJ/mol
(d)	very reactive	reacts with steam, but not $H_2O(l)$
(e)	1.54 Å	1.30 Å

(b) When forming ions, both adopt the stable configuration of Ne, but Na loses one electron and Mg two electrons to achieve this configuration.

(c),(e) The nuclear charge of Mg (Z = 12) is greater than that of Na, so it requires more energy to remove a valence electron with the same *n* value from Mg than Na. It also means that the 2s electrons of Mg are held closer to the nucleus, so the atomic radius (e) is smaller than that of Na.

(d) Mg is less reactive because it has a filled subshell and it has a higher ionization energy.

7.56 (a) Rb: [Kr] $5s^1$, r = 2.11 Å Ag: [Kr] $5s^1 4d^{10}$, r = 1.53 Å

The electron configurations both have a [Kr] core and a single 5s electron; Ag has a completed 4d subshell as well. The radii are very different because the 5s electron in Ag experiences a much greater effective nuclear charge. Ag has a much larger Z (47 vs. 37), and although the 4d electrons in Ag shield the 5s electron somewhat, the increased shielding does not compensate for the large increase in Z.

(b) Ag is much less reactive (less likely to lose an electron) because its 5s electron experiences a much larger effective nuclear charge and is more difficult to remove.

7.57 (a) Ca and Mg are both metals; they tend to lose electrons and form cations when they react. Ca is more reactive because it has a lower ionization energy than Mg. The Ca valence electrons in the 4s orbital are less tightly held because they are farther from the nucleus than the 3s valence electrons of Mg.

(b) K and Ca are both metals; they tend to lose electrons and form cations when they react. K is more reactive because it has a lower ionization energy. The 4s valence electron in K is less tightly held because it experiences a smaller nuclear charge ($Z = 19$ for K versus $Z = 20$ for Ca) with similar shielding effects than the 4s valence electrons of Ca.

7.58 (a) Cs is much more reactive than Li toward H_2O because its valence electron is less tightly held (greater *n* value), and Cs is more easily oxidized.

(b) The purple flame indicates that the metal is potassium (see Figure 7.22).

(c) $K_2O_2(s) + H_2O(l) \rightarrow H_2O_2(aq) + K_2O(aq)$
potassium peroxide → hydrogen peroxide

7.59 (a) $2K(s) + Cl_2(g) \rightarrow 2KCl(s)$

(b) $SrO(s) + H_2O(l) \rightarrow Sr(OH)_2(aq)$

(c) $4Li(s) + O_2(g) \rightarrow 2Li_2O(s)$

(d) $2Na(s) + S(l) \rightarrow Na_2S(s)$

7.60 (a) $2K(s) + 2H_2O(l) \rightarrow 2KOH(aq) + H_2(g)$

(b) $Ba(s) + 2H_2O(l) \rightarrow Ba(OH)_2(aq) + H_2(g)$

(c) $6Li(s) + N_2(g) \rightarrow 2Li_3N(s)$

(d) $2Mg(s) + O_2(g) \rightarrow 2MgO(s)$

7.61 H: $1s^1$; Li: [He] $2s^1$; F: [He] $2s^22p^5$. Like Li, H has only one valence electron, and its most common oxidation number is +1, which both H and Li adopt after losing the single valence electron. Like F, H needs only one electron to adopt the stable electron configuration of the nearest noble gas. Both H and F can exist in the -1 oxidation state, when they have gained an electron to complete their valence shells.

7.62 (a) The reactions of the alkali metals with hydrogen and with a halogen are redox reactions. In both classes of reaction, the alkali metal loses electrons and is oxidized. Both hydrogen and the halogen gain electrons and are reduced. The product is an ionic solid, where either hydride ion, H^-, or a halide ion, X^-, is the anion and the alkali metal is the cation.

(b) $Ca(s) + F_2(g) \rightarrow CaF_2(s)$
$Ca(s) + H_2(g) \rightarrow CaH_2(s)$

Both products are ionic solids containing Ca^{2+} and the corresponding anion in a 1:2 ratio.

7.63

	F	Cl
(a)	[He] $2s^22p^5$	[Ne] $3s^23p^5$
(b)	-1	-1

	F	Cl
(c)	1681 kJ/mol	1251 kJ/mol
(d)	reacts exothermically to form HF	reacts slowly to form HCl
(e)	-328 kJ/mol	-349 kJ/mol
(f)	0.71 Å	0.99 Å

(b) F and Cl are in the same group, have the same valence electron configuration and common ionic charge.

(c),(f) The $n = 2$ valence electrons in F are closer to the nucleus and more tightly held than the $n = 3$ valence electrons in Cl. Therefore, the ionization energy of F is greater, and the atomic radius is smaller.

(d) In its reaction with H_2O, F is reduced; it gains an electron. Although the electron affinity, a gas phase single atom property, of F is less negative than that of Cl, the tendency of F to hold its own electrons (high ionization energy) coupled with a relatively large exothermic electron affinity makes it extremely susceptible to reduction and chemical bond formation. Cl is unreactive to water because it is less susceptible to reduction.

(e) While F has approximately the same Z_{eff} as Cl, its small atomic radius gives rise to large repulsions when an extra electron is added, so the overall electron affinity of F is smaller (less exothermic) than that of Cl.

(f) The $n = 2$ valence electrons in F are closer to the nucleus so the atomic radius is smaller than that of Cl.

7.64 *Plan*. Predict the physical and chemical properties of At based on the trends in properties in the halogen (7A) family. *Solve*:

(a) F, at the top of the column, is a gas; I, immediately above At, is a solid; the melting points of the halogens increase going down the column. At is likely to be a solid at room temperature.

(b) All halogens form ionic compounds with Na; they have the generic formula NaX. The compound formed by At will have the formula NaAt.

7.65 Under ambient conditions, the Group 8A elements are all gases that are extremely unreactive, owing to their stable core electron configurations. Thus, the name "inert gases" seemed appropriate.

In the 1960s, scientists discovered that Xe, which has the lowest ionization energy of the nonradioactive Noble gases, would react with substances having a strong tendency to remove electrons, such as PtF_6 or F_2. Thus, the term "inert" no longer described all the Group 8A elements. (Kr also reacts with F_2, but reactions of Ar, Ne and He are as yet unknown.)

7.66 (a) When the noble gases react with fluorine, they (at least partially) lose electrons. This is related to ionization energy; the smaller the ionization energy the more reactive the noble-gas element. Moving down the family, the *n* value and the average distance of the outer electrons from the nucleus increase. Outer electrons experience a smaller attraction for the nucleus, ionization energy decreases and reactivity towards fluorine increases. Xe is most reactive toward F_2, followed by Kr.

(b) Although it has a relatively large negative electron affinity, Cl does not have the same ability to remove electrons from other atoms as F has. At least in part, this is due to the small size of F atoms, which can approach the electron clouds of the noble gases more closely than Cl atoms.

7.67 (a) $2O_3(g) \rightarrow 3O_2(g)$

(b) $Xe(g) + F_2(g) \rightarrow XeF_2(g)$

$Xe(g) + 2F_2(g) \rightarrow XeF_4(s)$

$Xe(g) + 3F_2(g) \rightarrow XeF_6(s)$

(c) $S(s) + H_2(g) \rightarrow H_2S(g)$

(d) $2F_2(g) + 2H_2O(l) \rightarrow 4HF(aq) + O_2(g)$

7.68 (a) $Cl_2(g) + H_2O(l) \rightarrow HCl(aq) + HOCl(aq)$

(b) $Ba(s) + H_2(g) \rightarrow BaH_2(s)$

(c) $2Li(s) + S(s) \rightarrow Li_2S(s)$

(d) $Mg(s) + F_2(g) \rightarrow MgF_2(s)$

7.69 (a) Te has more metallic character and is a better electrical conductor.

(b) At room temperature, oxygen molecules are diatomic and exist in the gas phase. Sulfur molecules are 8-membered rings and exist in the solid state.

(c) Chlorine is generally more reactive than bromine because Cl atoms have a greater (more exothermic) electron affinity than Br atoms.

7.70 (a) $S(s) + 2F_2(g) \rightarrow SF_4(g)$

(b) O_2, oxygen; O_3, ozone

(c) Fluorine can remove electrons from silica glass (SiO_2) according to the reaction $SiO_2(s) + 2F_2(g) \rightarrow SiF_4(g) + O_2(g)$. Thus, fluorine gas would "dissolve" the glass container. This process is also known as etching.

Additional Exercises

7.71 Up to Z = 83, there are three instances where atomic weights are reversed relative to atomic numbers: Ar and K; Co and Ni; Te and I.

In each case, the most abundant isotope of the element with the larger atomic number (Z) has one more proton, but fewer neutrons than the element with the smaller atomic number. The smaller number of neutrons causes the element with the larger Z to have a smaller than expected atomic weight.

7.72 (a) *eka - aluminum* is gallium (Ga), Z = 31.

(b) From the **Handbook of Chemistry and Physics**, 74th edition, the physical properties of Ga are:

atomic weight = 69.92 g/mol	m.p. = 29.78°C
density = 5.904 g/mL (s)	b.p. = 2403°C
= 6.095 g/mL (l)	oxide = Ga_2O_3, Ga_2O

The atomic weight, density and formula of oxide all agree with Mendeleev's predictions. Gallium does have a low melting point and a high boiling point.

7.73 (a) Na. In an isoelectronic series, all electronic effects (shielding and repulsion) are the same, so the particle with the smallest Z will have the smallest effective nuclear charge.

(b) Si^{3+}. Si has the largest Z and effective nuclear charge.

(c) The greater the effective nuclear charge experienced by a valence electron, the larger the ionization energy for that electron. According to Table 7.2, I_1 for Na is 496 kJ/mol. I_4 for Si is 4360 kJ/mol.

7.74 (a) P: (+15 - 10) = +5

(b) The 3s electrons have an approximately spherical distribution and provide additional shielding, beyond the Ne core to the 3p electrons. That is, S is greater for 3p electrons, owing to the presence of the 3s electrons, so Z - S(3p) is less than Z - S(3s).

(c) The electron configuration of P is $[Ne]3s^23p^3$. The 3p electrons are the outermost electrons; they experience a smaller Z_{eff} than 3s electrons and thus a smaller attraction for the nucleus, given equal *n*-values. The first electron lost is a 3p electron. Each 3p orbital holds one electron, so there is no preference as to which 3p electron will be lost.

7.75 Moving from left to right across the representative elements, electrons are added to *n*p orbitals. These are valence or outer electrons, so their distance from the nucleus directly determines atomic size.

Moving from left to right across the transition elements, electrons are added to an (*n*-1)d subshell, which is not the outermost shell. These (*n*-1)d electrons modify the shielding experienced by the outer or valence electrons, but this has a much smaller effect on atomic size.

7.76 (a) Mo – F distance = $r_{Mo} + r_F = 1.45 + 0.71 = 2.16$ Å

(b) S – F distance = $r_S + r_F = 1.02 + 0.71 = 1.73$ Å

(c) Cl – F distance = $r_{Cl} + r_F = 0.99 + 0.71 = 1.70$ Å

7.77 Ge – H distance = $r_{Ge} + r_H = 1.22 + 0.37 = 1.59$ Å

Ge – Cl distance = $r_{Ge} + r_{Cl} = 1.22 + 0.99 = 2.21$ Å

7.78 Close approach by two positively charged nuclei is impossible because of the large electrostatic repulsion between like-charged particles at small distances. The additional space between the nuclei in a molecule like F_2 is occupied by bonding electrons, which are electrostatically stabilized by attraction to both nuclei. The electrons also provide a buffer between the two nuclei.

7.79 Y: $[Kr]5s^24d^1$, Z = 39; La: $[Xe]6s^25d^1$, Z = 57; Zr: $[Kr]\ 5s^24d^2$, Z = 40; Hf: $[Xe]\ 6s^24f^{14}5d^2$, Z = 72. The completed 4f subshell in Hf leads to a much larger change in Z going from Zr to Hf (72 - 40 = 32) than in going from Y to La (57 - 39 = 18). The 4f electrons in Hf do not completely shield the 5d valence electrons, so there is also a larger increase in Z_{eff}. This larger increase in Z_{eff} going from Zr to Hf leads to a smaller increase in atomic radius than in going from Y to La.

7.80 C: $1s^22s^22p^2$. I_1 through I_4 represent loss of the 2p and 2s electrons in the outer shell of the atom. The values of $I_1 - I_4$ increase as expected. The nuclear charge is constant, but removing each electron reduces repulsive interactions between the remaining electrons, so effective nuclear charge increases and ionization energy increases. I_5 and I_6 represent loss of the 1s core electrons. These 1s electrons are much closer to the nucleus and experience the **full nuclear charge (they are not shielded)**, so the values of I_5 and I_6 are significantly greater than $I_1 - I_4$. I_6 is larger than I_5 because all repulsive interactions have been eliminated.

7.81 (a) Ca: $[Ar]\ 4s^2$; Ca^{2+}: [Ar]. Zn: $[Ar]\ 4s^23d^{10}$; Zn^{2+}: $[Ar]\ 3d^{10}$. In both cases, forming ions involves losing the only electrons in the *n* = 4 shell. The average distance from the nucleus of *n* = 3 electrons is less, so the radii of the ions are smaller.

(b) The atomic radius of Ca is greater than that of Zn because Zn has a significantly greater Z and Z_{eff}, which draws the 4s valence electrons closer to the nucleus.

(c) In both Ca and Zn atoms, the outermost electrons are in the 4s subshell. The much larger Z_{eff} for Zn causes it to have a much smaller radius. In the +2 ions, both have lost their 4s electrons. The outermost electrons in Zn are in the 3d sublevel and are significantly shielded by the [Ar] core electrons. Thus, the radius of Zn^{2+} is closer to the radius of Ca^{2+} than the radii of the neutral ions.

7.82 Ionization energy of F^-: $F^-(g) \rightarrow F(g) + 1e^-$

Electron affinity of F: $F(g) + 1e^- \rightarrow F^-(g)$

The two processes are the reverse of each other. The energies are equal in magnitude but opposite in sign. $I_1\ (F^-) = -E\ (F)$

7.83 The statement is somewhat true, but more accurate if changed to read: "A negative value for the electron affinity of an atom occurs when the outermost electrons only incompletely shield **the added electron** from the nucleus." This new statement totally explains the negative electron affinity of Br and the positive value for Kr. For Br^-, the electron is added to the 4p subshell and is incompletely shielded by the "other" 4s and 4p electrons. For Kr^-, the electron is added to the 5s subshell, which is effectively shielded by the spherical Kr core.

7.84 O: $[He]2s^22p^4$

O^{2-}: $[He]2s^22p^6$ = [Ne]

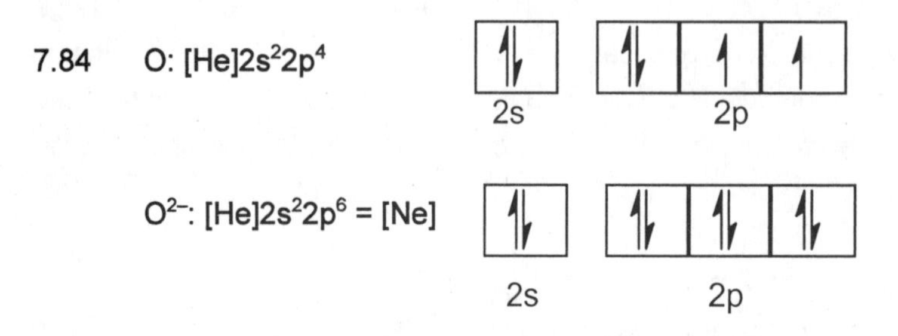

O^{3-}: $[Ne]3s^1$ The third electron would be added to the 3s orbital, which is further from the nucleus and more strongly shielded by the [Ne] core. The overall attraction of this 3s electron for the O nucleus is not large enough for O^{3-} to be a stable particle.

7.85 (a) P: [Ne] $3s^23p^3$; S: [Ne] $3s^23p^4$. In P, each 3p orbital contains a single electron, while in S one 3p orbital contains a pair of electrons. Removing an electron from S eliminates the need for electron pairing and reduces electrostatic repulsion, so the overall energy required to remove the electron is smaller than in P, even though Z is greater.

(b) C: [He] $2s^22p^2$; N: [He] $2s^22p^3$; O: [He] $2s^22p^4$. An electron added to a N atom must be paired in a relatively small 2p orbital, so the additional electron-electron repulsion more than compensates for the increase in Z and the electron affinity is smaller (less exothermic) than that of C. In an O atom, one 2p orbital already contains a pair of electrons, so the additional repulsion from an extra electron is offset by the increase in Z and the electron affinity is greater (more exothermic). Note from Figure 7.11 that the electron affinity of O is only slightly more exothermic than that of C, although the value of Z has increased by 2.

(c) O^+: [Ne] $2s^22p^3$; O^{2+}: [Ne] $2s^22p^2$; F^+: [Ne] $2s^22p^4$; F^{2+}: [Ne] $2s^22p^3$. The decrease in electron-electron repulsion going from F^+ to F^{2+} energetically favors ionization and causes it to be less endothermic than the corresponding process in O, where there is no significant decrease in repulsion.

(d) Mn^{2+}: $[Ar]3d^5$; Mn^{3+}: [Ar] $3d^4$; Cr^{2+}: [Ar] $3d^4$; Cr^{3+}: [Ar] $3d^3$; Fe^{2+}: [Ar] $3d^6$; Fe^{3+}: [Ar] $3d^5$. The third ionization energy of Mn is expected to be larger than that of Cr because of the larger Z value of Mn. The third ionization energy of Fe is **less** than that of Mn because going from $3d^6$ to $3d^5$ reduces electron repulsions, making the process less endothermic than predicted by nuclear charge arguments.

7.86 (a) The group 2B metals have complete (n-1)d subshells. An additional electron would occupy an np subshell and be substantially shielded by both ns and (n-1)d electrons. Overall this is not a lower energy state than the neutral atom and a free electron.

(b) Valence electrons in Group 1B elements experience a relatively large effective nuclear charge due to the build-up in Z with the filling of the (n-1)d subshell. Thus, the electron affinities are large and negative. Group 1B elements are exceptions to the usual electron filling order and have the generic electron configuration $ns^1(n\text{-}1)d^{10}$. The additional electron would complete the ns subshell and experience repulsion with the other ns electron. Going down the group, size of the ns subshell increases and repulsion effects decrease. That is, effective nuclear charge is greater going down the group because it is less diminished by repulsion, and electron affinities become more negative.

7.87 (a) For both H and the alkali metals, the added electron will complete an ns subshell (1s for H and ns for the alkali metals) so shielding and repulsion effects will be similar. For the halogens, the electron is added to an np subshell, so the energy change is likely to be quite different.

(b) True. Only He has a smaller estimated "bonding" atomic radius, and no known compounds of He exist. The electron configuration of H is $1s^1$. The single 1s electron experiences no repulsion from other electrons and feels the full unshielded nuclear charge. It is held very close to the nucleus. The outer electrons of all other elements that form compounds are shielded by a spherical inner core of electrons and are less strongly attracted to the nucleus, resulting in larger atomic radii.

(c) Ionization is the process of removing an electron from an atom. For the alkali metals, the ns electron being removed is effectively shielded by the core electrons, so ionization energies are low. For the halogens, a significant increase in nuclear charge occurs as the np orbitals fill, and this is not offset by an increase in shielding. The relatively large effective nuclear charge experienced by np electrons of the halogens is similar to the unshielded nuclear charge experienced by the H 1s electron. Both H and the halogens have large ionization energies.

7.88 Since Xe reacts with F_2, and O_2 has approximately the same ionization energy as Xe, O_2 will probably react with F_2. Possible products would be O_2F_2, analogous to XeF_2, or OF_2.

$$O_2(g) + F_2(g) \rightarrow O_2F_2(g)$$

$$O_2(g) + 2F_2(g) \rightarrow 2OF_2(g)$$

7.89 $O_2 < Br_2 < K < Mg$. O_2 and Br_2 are (nonpolar) nonmetals. We expect O_2, with the much lower molar mass, to have the lower melting point. This is confirmed by data in Tables 7.6 and 7.7. K and Mg are metallic solids (all metals are solids), with higher melting points than the two nonmetals. Since alkaline earth metals (Mg) are typically harder, more dense and higher melting than alkali metals (K), we expect Mg to have the highest melting point of the group. This is confirmed by data in Tables 7.4 and 7.5.

7.90 (a) Li has a relatively small, positive ionization energy while O and F have relatively large negative electron affinities. When Li reacts with the nonmetals, it loses electrons while O and F gain electrons. Two Li atoms are required to provide the two electrons needed by O to complete its 2p subshell and form Li_2O. One Li is required to complete the 2p subshell of F and form LiF.

(b) The radius of F, 0.71 Å, is only slightly smaller than the radius of O, 0.73 Å. Because F has a greater nuclear charge, we might anticipate a larger difference. However, electron-electron repulsions are so great in the relatively cramped 2p orbitals of F that the larger Z does not lead to a significantly smaller radius.

(c) When nonmetals react, they tend to gain electrons, a process related to electron affinity. Fluorine has a more negative (more exothermic) electron affinity, so it is the more reactive nonmetal.

7.91 Moving one place to the right in a horizontal row of the table, for example, from Li to Be, there is an increase in ionization energy. Moving downward in a given family, for example from Be to Mg, there is usually a decrease in ionization energy. Similarly, atomic size decreases in moving one place to the right and increases in moving downward. Thus, two elements such as Li and Mg that are diagonally related tend to have similar ionization energies and atomic sizes. This in turn gives rise to some similarities in chemical behavior. Note, however, that the valences expected for the elements are not the same. That is, lithium still appears as Li^+, magnesium as Mg^{2+}.

7.92 Fr (1A), Ra (2A), Po (6A), At (7A), Rn (8A)

(a) most metallic character: **Fr** (Metallic character decreases from left to right in a row.)
(b) most nonmetallic character: **Rn**
(c) largest ionization energy: **Rn** (Ionization energy increases from left to right in a row.)
(d) smallest ionization energy: **Fr**
(e) greatest electron affinity: **At** (Electron affinity becomes more exothermic from left to right in a row.)
(f) largest atomic radius: **Fr** (Size decreases from left to right in a row.)
(g) appears least like the element above it: **Fr** (According to the trends in melting point, Fr may be a liquid or a gas at room temperature, while Cs is a solid. The others should be in the same state as the element above them.)
(h) highest melting point: **Ra** (The melting points of the group 2A metals are much higher than those of the other groups, even though the values decrease going down the group.)
(i) react most readily with H_2O: **Fr** (It has the lowest ionization energy.)

7.93 (a) *Plan.* Use qualitative physical (bulk) properties to narrow the range of choices, then match melting point and density to identify the specific element. *Solve*:

Hardness varies widely in metals and nonmetals, so this information is not too useful. The relatively high density, appearance and ductility indicate that the element is probably less metallic than copper. Focus on the block of nine main group elements centered around Sn. Pb is not a possibility because it was used as a comparison standard. The melting point of the five elements closest to Pb are:

Tl, 303.5°C; In, 156.1°C; Sn, 232°C ; Sb, 630.5°C; Bi, 271.3°C

The best match is In. To confirm this identification, the density of In is 7.3 g/cm^3, also a good match to properties of the unknown element.

(b) In order to write the correct balanced equation, determine the formula of the oxide product from the mass data, assuming the unknown is In.

5.08 g oxide - 4.20 g In = 0.88 g O

4.20 g In/114.82 g/mol = 0.0366 mol In; 0.0366/0.0366 = 1

0.88g O/16.00 g/mol = 0.0550 mol O; 0.0550/0.0366 = 1.5

Multiplying by 2 produces an integer ratio of 2 In: 3 O and a formula of In_2O_3. The balanced equation is: $4\ In(s) + 3O_2(g) \rightarrow 2\ In_2O_3(s)$

(c) According to Figure 7.2, the element In was discovered between 1843 –1886. The investigator who first recorded this data in 1822 could have been the first to discover In.

7.94 Ionic "inorganic" halogen compounds are formed when a metal with low ionization energy and small negative electron affinity combines with a halogen with large ionization energy and large negative electron affinity. That is, it is relatively easy to remove an electron from a metal, and there is only a small energy payback if a metal gains an electron. The opposite is true of a halogen; it is hard to remove an electron and there is a large energy advantage if a halogen gains an electron. Thus, the metal "gives up" an electron to the halogen and an ionic compound is formed. Carbon, on the other hand, is much closer in ionization energy and electron affinity to the halogens. Carbon has a much greater tendency than a metal to keep its own electrons and at least some attraction for the electrons of other elements. Thus, compounds of carbon and the halogens are molecular, rather than ionic.

Integrative Exercises

7.95 (a) $\nu = c/\lambda$; 1 Hz = 1s^{-1}

$$\text{Ne: } \nu = \frac{2.998 \times 10^8 \text{ m/s}}{14.610 \text{ Å}} \times \frac{1 \text{ Å}}{1 \times 10^{-10} \text{ m}} = 2.052 \times 10^{17} \text{ s}^{-1} = 2.052 \times 10^{17} \text{ Hz}$$

$$\text{Ca: } \nu = \frac{2.998 \times 10^8 \text{ m/s}}{3.358 \times 10^{-10} \text{ m}} = 8.928 \times 10^{17} \text{ Hz}$$

Zn: $\nu = \dfrac{2.998 \times 10^8 \text{ m/s}}{1.435 \times 10^{-10} \text{ m}} = 20.89 \times 10^{17}$ Hz

Zr: $\nu = \dfrac{2.998 \times 10^8 \text{ m/s}}{0.786 \times 10^{-10} \text{ m}} = 38.14 \times 10^{17} = 38.1 \times 10^{17}$ Hz

Sn: $\nu = \dfrac{2.998 \times 10^8 \text{ m/s}}{0.491 \times 10^{-10} \text{ m}} = 61.06 \times 10^{17} = 61.1 \times 10^{17}$ Hz

(b)

Element	Z	ν	$\nu^{1/2}$
Ne	10	2.052×10^{17}	4.530×10^8
Ca	20	8.928×10^{17}	9.449×10^8
Zn	30	20.89×10^{17}	14.45×10^8
Zr	40	38.14×10^{17}	19.5×10^8
Sn	50	61.06×10^{17}	24.7×10^8

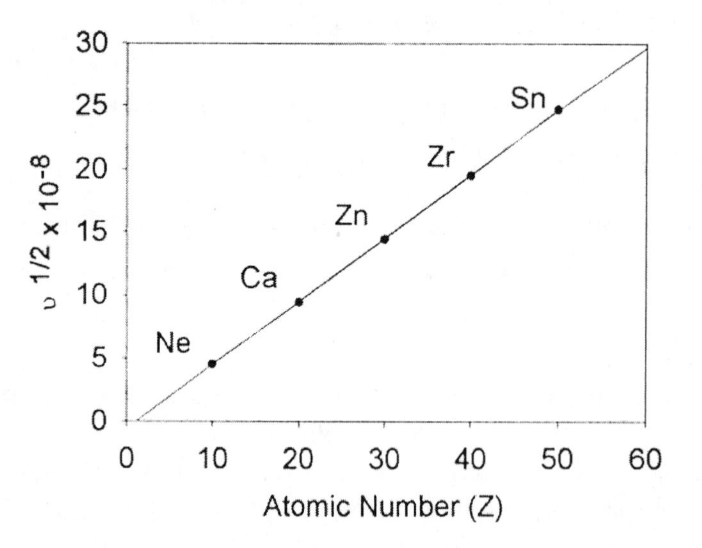

(c) The plot in part (b) indicates that there is a linear relationship between atomic number and the square root of the frequency of the X-rays emitted by an element. Thus, elements with each integer atomic number should exist. This relationship allowed Moseley to predict the existence of elements that filled "holes" or gaps in the periodic table.

(d) For Fe, Z = 26. From the graph, $\nu^{1/2} = 12.5 \times 10^8$, $\nu = 1.56 \times 10^{18}$ Hz.

$$\lambda = c/\nu = \frac{2.998 \times 10^8 \text{ m/s}}{1.56 \times 10^{18} \text{ s}^{-1}} \times \frac{1 \text{ Å}}{1 \times 10^{-10} \text{ m}} = 1.92 \text{ Å}$$

(e) $\lambda = 0.980$ Å $= 0.980 \times 10^{-10}$ m

$$\nu = c/\lambda = \frac{2.998 \times 10^8 \text{ m/s}}{0.980 \times 10^{-10} \text{ m}} = 30.6 \times 10^{17} \text{ Hz}; \ \nu^{1/2} = 17.5 \times 10^8$$

From the graph, $\nu^{1/2} = 17.5 \times 10^8$, Z = 36. The element is krypton, Kr.

7.96 (a) Li: [He]$2s^1$. Assume that the [He] core is 100% effective at shielding the 2s valence electron $Z_{eff} = Z - S \approx 3 - 2 = +1$.

(b) The first ionization energy represents loss of the 2s electron.

ΔE = energy of free electron (n = ∞) - energy of electron in ground state (n = 2)

$\Delta E = I_1 = [-2.18 \times 10^{-18} \text{ J } (Z^2 / \infty 2)] - [-2/18 \times 10^{-18} \text{ J}(Z^2/2^2]$

$\Delta E = I_1 = 0 + 2.18 \times 10^{-18} \text{ J } (Z^2/2^2)$

For Li, which is not a one-electron particle, let $Z = Z_{eff}$.

$\Delta E \approx 2.18 \times 10^{-18} \text{ J } (+1^2 / 4) \approx 5.45 \times 10^{-19}$ J/atom

(c) Change the result from part (b) to kJ/mol so it can be compared to the value in Table 7.4. $5.45 \times 10^{-19} \dfrac{\text{J}}{\text{atom}} \times \dfrac{6.022 \times 10^{23} \text{ atom}}{\text{mol}} \times \dfrac{1 \text{ kJ}}{1000 \text{ J}} = 328$ kJ/mol

The value in Table 7.4 is 520 kJ/mol. This means that our estimate for Z_{eff} was a lower limit; that the [He] core electrons do not perfectly shield the 2s electron from the nuclear charge.

(d) From Table 7.4, I_1 = 520 kJ/mol.

$$\frac{520 \text{ kJ}}{\text{mol}} \times \frac{1000 \text{ J}}{\text{kJ}} \times \frac{1 \text{ mol}}{6.022 \times 10^{23} \text{ atoms}} = 8.6350 \times 10^{-19} \text{ J/atom}$$

Use the relationship for I_1 and Z_{eff} developed in part (b).

$$Z_{eff}^2 = \frac{4(8.6350 \times 10^{-19} \text{ J})}{2.18 \times 10^{-18} \text{ J}} = 1.5844 = 1.58; \; Z_{eff} = 1.26$$

This value, Z_{eff} = 1.26, based on the experimental ionization energy, is greater than our estimate from part (a), which is consistent with the explanation in part (c).

7.97 (a) $E = hc/\lambda$; 1 nm = 1×10^{-9} m; 58.4 nm = 58.4×10^{-9} m;

1 eV = 96.485 kJ/mol, 1 eV • mol = 96.485 kJ

$$E = \frac{6.626 \times 10^{-34} \text{ J}\cdot\text{s} \times 2.998 \times 10^{8} \text{ m/s}}{58.4 \times 10^{-9} \text{ m}} = 3.4015 \times 10^{-18} = 3.40 \times 10^{-18} \text{ J/photon}$$

$$\frac{3.4015 \times 10^{-18} \text{ J}}{\text{photon}} \times \frac{1 \text{ kJ}}{1000 \text{ J}} \times \frac{6.022 \times 10^{23} \text{ photons}}{\text{mol}} \times \frac{1 \text{ eV} \cdot \text{mol}}{96.485 \text{ kJ}} = 21.230$$

$= 21.2$ eV

(b) $Hg(g) \rightarrow Hg^+(g) + 1e^-$

(c) $I_1 = E_{58.4} - E_K$ = 21.23 eV - 10.75 eV = 10.48 = 10.5 eV

$$10.48 \text{ eV} \times \frac{96.485 \text{ kJ}}{1 \text{ eV} \cdot \text{mol}} = 1.01 \times 10^3 \text{ kJ/mol}$$

(d) From Figure 7.10, iodine (I) appears to have the ionization energy closest to that of Hg, approximately 1000 kJ/mol.

7.98 (a)

$Na(g) \rightarrow Na^+(g) + 1e^-$ (ionization energy of Na)

$Cl(g) + 1e^- \rightarrow Cl^-(g)$ (electron affinity of Cl)

$Na(g) + Cl(g) \rightarrow Na^+(g) + Cl^-(g)$

(b) $\Delta H = I_1$ (Na) + E_1(Cl) = +496 kJ - 349 kJ = +147 kJ, endothermic

(c) The reaction $2Na(s) + Cl_2(g) \rightarrow 2NaCl(s)$ involves many more steps than the reaction in part (a). One important difference is the production of NaCl(s) versus NaCl(g). The condensation $NaCl(g) \rightarrow NaCl(s)$ is very exothermic and is the step that causes the reaction of the elements in their standard states to be exothermic, while the gas phase reaction is endothermic.

7.99 (a) Mg_3N_2

(b) $Mg_3N_2(s) + 3H_2O(l) \rightarrow 3MgO(s) + 2NH_3(g)$
The driving force is the production of $NH_3(g)$.

(c) After the second heating, all the Mg is converted to MgO.
Calculate the initial mass Mg.

$$0.486 \text{ g MgO} \times \frac{24.305 \text{ g Mg}}{40.305 \text{ g MgO}} = 0.293 \text{ g Mg}$$

x = g Mg converted to MgO; y = g Mg converted to Mg_3N_2; x = 0.293 - y

$$\text{g MgO} = x\left(\frac{40.305 \text{ g MgO}}{24.305 \text{ g Mg}}\right); \text{ g } Mg_3N_2 = y\left(\frac{100.929 \text{ g } Mg_3N_2}{72.915 \text{ g Mg}}\right)$$

g MgO + g Mg_3N_2 = 0.470

$$(0.293 - y)\left(\frac{40.305}{24.305}\right) + y\left(\frac{100.929}{72.915}\right) = 0.470$$

$$(0.293 - y)(1.6583) + y(1.3842) = 0.470$$

$$-1.6583\, y + 1.3842\, y = 0.470 - 0.48588$$

$$-0.2741\, y = -0.016$$

$$y = 0.05794 = 0.058 \text{ g Mg in } Mg_3N_2$$

$$\text{g } Mg_3N_2 = 0.05794 \text{ g Mg} \times \frac{100.929 \text{ g } Mg_3N_2}{72.915 \text{ g Mg}} = 0.0802 = 0.080 \text{ g } Mg_3N_2$$

$$\text{mass \% } Mg_3N_2 = \frac{0.0802 \text{ g } Mg_3N_2}{0.470 \text{ g } (MgO + Mg_3N_2)} \times 100 = 17\%$$

(The final mass % has 2 sig figs because the mass of Mg obtained from solving simultaneous equations has 2 sig figs.)

(d) $3Mg(s) + 2NH_3(g) \rightarrow Mg_3N_2(s) + 3H_2(g)$

$$6.3 \text{ g Mg} \times \frac{1 \text{ mol Mg}}{24.305 \text{ g Mg}} = 0.2592 = 0.26 \text{ mol Mg}$$

$$2.57 \text{ g } NH_3 \times \frac{1 \text{ mol } NH_3}{17.031 \text{ g } NH_3} = 0.1509 = 0.16 \text{ mol } NH_3$$

$$0.2592 \text{ mol Mg} \times \frac{2 \text{ mol } NH_3}{3 \text{ mol Mg}} = 0.1728 = 0.17 \text{ mol } NH_3$$

0.26 mol Mg requires more than the available NH_3 so NH_3 is the limiting reactant.

$$0.1509 \text{ mol } NH_3 \times \frac{3 \text{ mol } H_2}{2 \text{ mol } NH_3} \times \frac{2.016 \text{ g } H_2}{\text{mol } H_2} = 0.4563 = 0.46 \text{ g } H_2$$

(e) $\Delta H^\circ_{rxn} = \Delta H^\circ_f \; Mg_3N_2(s) + 3\Delta H^\circ_f \; H_2(g) - 3\Delta H^\circ_f \; Mg(s) - 2\Delta H^\circ_f \; NH_3(g)$

$= -461.08 \text{ kJ} + 0 - 3(0) - 2(-46.19) = -368.70 \text{ kJ}$

7.100 (a) $r_{Bi} = r_{BiBr_3} - r_{Br} = 2.63 \text{ Å} - 1.14 \text{ Å} = 1.49 \text{ Å}$

(b) $Bi_2O_3(s) + 6HBr(aq) \rightarrow 2BiBr_3(aq) + 3H_2O(l)$

(c) Bi_2O_3 is soluble in acid solutions because it acts as a base and undergoes acid-base reactions like the one in part (b). It is insoluble in base because it cannot act as an acid. Thus, Bi_2O_3 is a basic oxide, the oxide of a metal. Based on the properties of its oxide, Bi is characterized as a metal.

(d) Bi: $[Xe]6s^2 4f^{14} 5d^{10} 6p^3$. Bi has 5 outer electrons in the 6p and 6s subshells. If all 5 electrons participate in bonding, compounds such as BiF_5 are possible. Also, Bi has a large enough atomic radius (1.49 Å) and low-energy orbitals available to accommodate more than four pairs of bonding electrons.

(e) The high ionization energy and relatively large negative electron affinity of F, coupled with its small atomic radius, make it the most electron withdrawing of the halogens. BiF_5 forms because F has the greatest tendency to attract electrons from Bi. Also, the small atomic radius of F reduces repulsions between neighboring bonded F atoms. The strong electron withdrawing properties of F are also the reason that only F compounds of Xe are known (Solution 7.66).

8 Basic Concepts of Chemical Bonding

Lewis Symbols and Ionic Bonding

8.1 (a) Valence electrons are those that take part in chemical bonding, those in the outermost electron shell of the atom. This usually means the electrons beyond the core noble-gas configuration of the atom, although it is sometimes only the outer shell electrons.

(b) N: [He] $2s^22p^3$ (valence electrons) A nitrogen atom has 5 valence electrons.

(c) $1s^22s^22p^6$ [Ne] $3s^23p^2$ (valence electrons) The atom (Si) has 4 valence electrons.

8.2 (a) Atoms will gain, lose or share electrons to achieve the nearest noble-gas electron configuration. Except for H and He, this corresponds to eight electrons in the valence shell, thus the term **octet** rule.

(b) S: $[Ne]3s^23p^4$ A sulfur atom has 6 valence electrons, so it must gain 2 electrons to achieve an octet.

(c) $1s^22s^22p^3 = [He]2s^22p^3$ The atom (N) has 5 valence electrons and must gain 3 electrons to achieve an octet.

8.3 P: $1s^22s^22p^63s^23p^3$. A 3s electron is a valence electron; a 2s (or 1s) electron is a non-valence electron. The 3s valence electron is involved in chemical bonding, while the 2s or 1s non-valence electron is not.

8.4 Sc: $1s^22s^22p^63s^23p^64s^23d^1 = [Ar]4s^23d^1$. Scandium has three (3) valence electrons. These valence electrons are available for chemical bonding, while the core electrons do not participate in bonding.

8.5 (a) $\dot{\text{Ca}}\cdot$ (b) $\cdot\ddot{\text{P}}\cdot$ (with one dot below) (c) $:\ddot{\text{Ne}}:$ (with two dots below) (d) $\cdot\dot{\text{B}}\cdot$

8.6 (a) $\cdot\text{Mg}\cdot$ (b) $\cdot\ddot{\text{As}}\cdot$ (with one dot below) (c) $[:\ddot{\text{Sc}}:]^{3+}$ or Sc^{3+} (d) $[:\ddot{\text{Se}}:]^{2-}$ or Se^{2-}

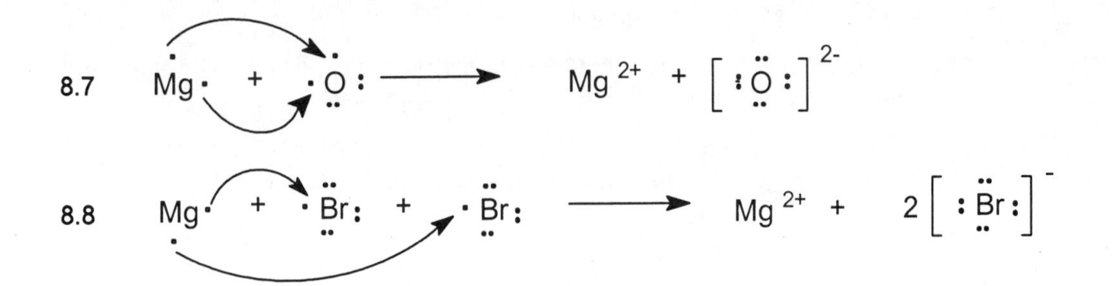

8.9 Potassium has a single valence electron, which it tends to lose to achieve a completed octet. Calcium has two valence electrons, which it tends to lose to achieve a completed octet. Thus, K loses a single electron while Ca loses two electrons when reacting with Cl. Removing one electron from the core of either K^+ or Ca^{2+} would be energetically unfavorable, because the core electrons are stabilized by a strong electrostatic attraction for the nucleus. Even a large lattice energy is not enough to promote removal of a core electron.

8.10 Bromine does not accept more than one electron, because the additional electron(s) would have a higher principle quantum number than the valence electrons in the completed octet of Br^-. Electrons beyond the n = 4 level in a bromine anion would be substantially shielded from the nucleus by the core electrons and are further from the nucleus than the other electrons. In other words, the energy of the 5s orbital in Br^- is sufficiently high that Br^{2-} or other anions with a negative charge greater than -1 do not form.

8.11 (a) AlF_3 (b) K_2S (c) Y_2O_3 (d) Mg_3N_2

8.12 (a) Rb_2O (b) BaI_2 (c) Li_2O (d) $MgCl_2$

8.13 (a) Sr^{2+}: [Kr], noble-gas configuration (b) Ti^{2+}: $[Ar]3d^2$

(b) Se^{2-}: $[Ar]4s^23d^{10}4p^6$ = [Kr], noble-gas configuration (d) Ni^{2+}: $[Ar]3d^8$

(c) Br^-: $[Ar]4s^23d^{10}4p^6$ = [Kr], noble-gas configuration (f) Mn^{3+}: $[Ar]3d^4$

8.14 (a) Zn^{2+}: $[Ar]3d^{10}$ (b) Te^{2-}: $[Kr]5s^24d^{10}5p^6$ = [Xe], noble-gas configuration

(c) Se^{3+}: $[Ar]4s^23d^{10}4p^1$ (This is a very unlikely ion. A more stable and commonly-found ion would be Sc^{3+}.) Sc^{3+}: [Ar], noble-gas configuration

(d) Ru^{2+}: $[Kr]4d^6$ (e) Tl^+: $[Xe]6s^24f^{14}5d^{10}$ (f) $Au^+[Xe]4f^{14}5d^{10}$

8.15 (a) *Lattice energy* is the energy required to totally separate one mole of solid ionic compound into its gaseous ions.

(b) The magnitude of the lattice energy depends on the magnitudes of the charges of the two ions, their radii and the arrangement of ions in the lattice. The main factor is the charges, because the radii of ions do not vary over a wide range.

8.16 (a) NaF, 910 kJ/mol; MgO, 3795 kJ/mol

The two factors that affect lattice energies are charge and ionic radii. The Na–F and Mg–O separations are similar (Na^+ is larger than Mg^{2+}, but F^- is smaller than O^{2-}). The charges on Mg^{2+} and O^{2-} are twice those of Na^+ and F^-, so according to Equation 8.4, the lattice energy of MgO is approximately four times that of NaF.

(b) $MgCl_2$, 2326 kJ/mol; $SrCl_2$, 2127 kJ/mol

The two factors that affect lattice energies are charge and ionic radii. The ionic charges are the same in the two compounds. The ionic radius of Mg^{2+} is smaller than that of Sr^{2+} so the Mg–Cl distance is slightly smaller than the Sr–Cl distance. Since lattice energy is inversely proportional to the ion separation, the lattice energy of $MgCl_2$ is slightly larger than that of $SrCl_2$.

8.17 KF, 808 kJ/mol; CaO, 3414 kJ/mol; ScN, 7547 kJ/mol

The sizes of the ions vary as follows: $Sc^{3+} < Ca^{2+} < K^+$ and $F^- < O^{2-} < N^{3-}$. Therefore, the interionic distances are similar. According to Coulomb's law for compounds with similar ionic separations, the lattice energies should be related as the product of the charges of the ions. The lattice energies above are approximately related as (1)(1): (2)(2): (3)(3) or 1 : 4 : 9. Slight variations are due to the small differences in ionic separations.

8.18 (a) According to Equation 8.4, electrostatic attraction increases with increasing charges of the ions and decreases with increasing radius of the ions. Thus, lattice energy (i) **increases** as the charges of the ions increase and (ii) **decreases** as the sizes of the ions increase.

(b) KBr < KCl < LiCl < CaO. This order is confirmed by the lattice energies given in Table 8.2. CaO has the highest lattice energy because the ions have 2+ and 2- charges. The other compounds have cations with 1+ charges and anions with 1- charges. They are placed in order of decreasing ionic separation. K^+ and Br^- have the largest radii, Cl^- is smaller than Br^-, and Li^+ is smaller than K^+.

8.19 Since the ionic charges are the same in the two compounds, the K–Br and Cs–Cl separations must be approximately equal. Since the radii are related as $Cs^+ > K^+$ and $Br^- > Cl^-$, the difference between Cs^+ and K^+ must be approximately equal to the difference between Br^- and Cl^-. This is somewhat surprising, since K^+ and Cs^+ are two rows apart and Cl^- and Br^- are only one row apart.

8.20 (a) In MgO, the magnitude of the charges on both ions is 2; in $MgCl_2$, the magnitudes of the charges are 2 and 1. Also, the Cl^- ion is larger than the O^{2-} ion, so the charge separation is greater in $MgCl_2$. Thus, the lattice energy of MgO is greater, because the product of the ionic charges is greater and the ion separation is smaller.

(b) The ions have +1 and -1 charger in all three compounds. In NaCl the cationic and anionic radii are smaller than in the other two compounds, so it has the largest lattice energy. In RbBr and CsBr, the anion is the same, but the Cs cation is larger, so CsBr has the smaller lattice energy.

(c) In BaO, the magnitude of the charges of both ions is 2; in KF, the magnitudes are 1. Charge considerations alone predict that BaO will have the higher lattice energy. The distance effect is less clear; O^{2-} and F^- are isoelectronic, so F^-, with the larger Z, has a slightly smaller radius. Ba^{2+} is two rows lower on the periodic chart than K^+, but it has a greater positive charge, so the radii are probably similar. In any case, the ionic separations in the two compounds are not very different, and the charge effect dominates.

8.21. Equation 8.4 predicts that as the oppositely charged ions approach each other, the energy of interaction will be large and negative. This more than compensates for the energy required to form Ca^{2+} and O^{2-} from the neutral atoms (see Figure 8.4 for the formation of NaCl).

8.22 $Ca(s) \rightarrow Ca(g)$; $Br_2(l) \rightarrow 2Br(g)$; $Ca(g) \rightarrow Ca^+(g) + 1e^-$;
$Ca^+(g) \rightarrow Ca^{2+}(g) + 1e^-$; $2Br(g) + 2e^- \rightarrow 2Br^-(g)$, exothermic;
$Ca^{2+}(g) + 2Br^-(g) \rightarrow CaBr_2(s)$, exothermic

8.23 $RbCl(s) \rightarrow Rb^+(g) + Cl^-(g)$ ΔH (lattice energy) = ?

By analogy to NaCl, Figure 8.4, the lattice energy is

$$\Delta H_{latt} = -\Delta H_f^\circ\ RbCl(s) + \Delta H_f^\circ\ Rb(g) + \Delta H_f^\circ\ Cl(g) + I_1\ (Rb) + E\ (Cl)$$

$$= -(-430.5\ kJ) + 85.8\ kJ + 121.7\ kJ + 403\ kJ + (-349\ kJ) = +692\ kJ$$

This value is smaller than that for NaCl (+788 kJ) because Rb^+ has a larger ionic radius than Na^+. This means that the value of d in the denominator of Equation 8.4 is larger for RbCl, and the potential energy of the electrostatic attraction is smaller.

8.24 By analogy to Figure 8.4:

$$\Delta H_{latt} = -\Delta H_f^\circ\ CaCl_2 + \Delta H_f^\circ\ Ca(g) + 2\Delta H_f^\circ\ Cl(g) + I_1\ (Ca) + I_2\ (Ca) + 2E\ (Cl)$$

$$= -(-795.8\ kJ) + 179.3\ kJ + 2(121.7\ kJ) + 590\ kJ + 1145\ kJ + 2(-349\ kJ) = +2256\ kJ$$

From Table 8.2, the lattice energy of NaCl, +788 kJ/mol, is considerably less than that of CaF_2. The 2+ charge of Ca^{2+} leads to much greater electrostatic attractions and a higher lattice energy.

Covalent Bonding, Electronegativity and Bond Polarity

8.25 (a) A *covalent bond* is the bond formed when two atoms share one or more pairs of electrons.

(b) The ionic bonding in NaCl is due to strong electrostatic attraction between oppositely charged Na^+ and Cl^- ions. The covalent bonding in Cl_2 is due to sharing of a pair of electrons by two neutral chlorine atoms.

8.26 K and Ar. K is an active metal with one valence electron. It is most likely to achieve an octet by losing this single electron and to participate in ionic bonding. Ar has a stable octet of valence electrons; it is not likely to form chemical bonds of any type.

8.27. *Analyze/Plan*. Follow the logic in Sample Exercise 8.3. *Solve:*

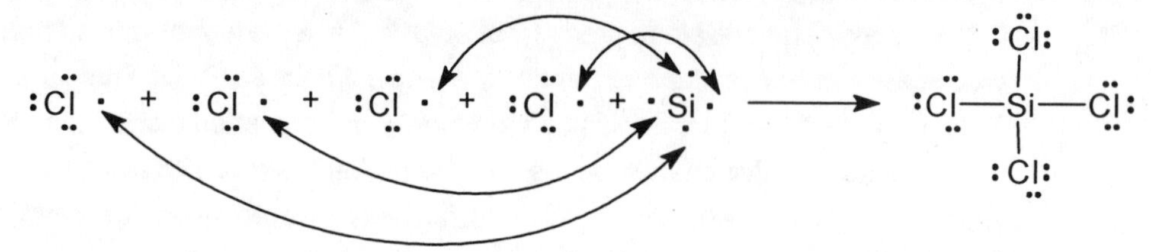

Check. Each pair of shared electrons in $SiCl_4$ is shown as a line; each atom is surrounded by an octet of electrons.

8.28 :Cl· + :Cl· + :Cl· + ·N: ⟶ Cl—N(—Cl)—Cl

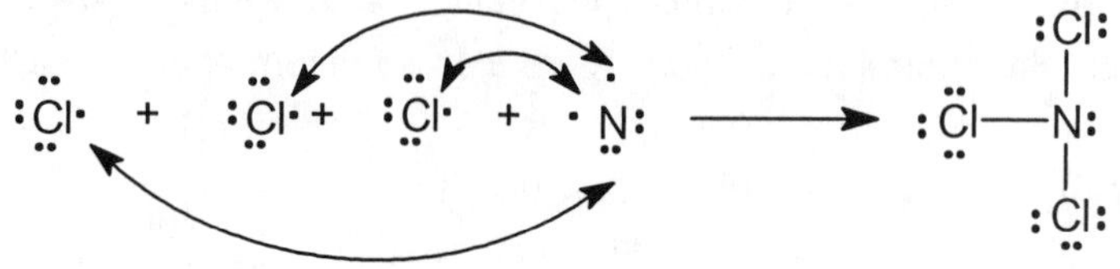

8.29 (a) :O=O:

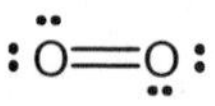

(b) A double bond is required because there are not enough electrons to satisfy the octet rule with single bonds and unshared pairs.

(c) The greater the number of shared electron pairs between two atoms, the shorter the distance between the atoms. If O_2 has a double bond, the O–O distance will be shorter than the O–O single bond distance.

8.30 S=C=S

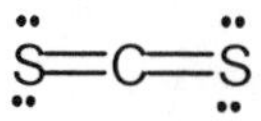

The C–S bonds in CS_2 are double bonds, so the C–S distances will be shorter than a C–S single bond distance.

8.31 (a) *Electronegativity* is the ability of an atom in a molecule (a bonded atom) to attract electrons to itself.

(b) The range of electronegativities on the Pauling scale is 0.7-4.0.

(c) Fluorine, F, is the most electronegative element.

(d) Cesium, Cs, is the least electronegative element that is not radioactive.

8.32 (a) The electronegativity of the elements increases going from left to right across a row of the periodic chart.

(b) Electronegativity decreases going down a family of the periodic chart.

(c) Generally, the trends in electronegativity are the same as those in ionization energy and opposite those in electron affinity. That is, the more positive the ionization energy and the more negative the electron affinity (omitting a few exceptions), the greater the electronegativity of an element.

8.33 *Plan*. Electronegativity increases going up and to the right in the periodic table. *Solve*:

(a) S (b) C (c) As (d) Mg

Check. The electronegativity values in Figure (8.9) confirm these selections.

8.34 Electronegativity increases going up and to the right in the periodic table.

(a) O (b) Al (c) Cl (d) F

8.35 The bonds in (a), (b) and (d) are polar because the atoms involved differ in electronegativity. The more electronegative element in each polar bond is: (a) O (b) F (d) O

8.36 The more different the electronegativity values of the two elements, the more polar the bond.

(a) O–F < C–F < Be–F. This order is clear from the periodic trend.

(b) Br–N < P–Br < Br–O. Refer to the electronegativity values in Figure 8.9 to confirm the order of bond polarity. The electronegativity order of the four elements is P < Br < N < O. Br is significantly to the right and below the other three elements, but these directions have conflicting trends. Even though Br and N are farthest apart, their electronegativity values are nearly equal. The Br-N bond is least polar, with Br the less electronegative element.

(c) C–S < N–O < B–F. You might predict that N–O is least polar since the elements are adjacent on the table. However, the big decrease going from the second row to the third means that the electronegativity of S is not only less than that of O, but essentially the same as that of C. C–S is the least polar.

8.37 (a) A polar molecule has a measurable dipole moment; its centers of positive and negative charge do not coincide. A nonpolar molecule has a zero net dipole moment; its centers of positive and negative charge do coincide.

(b) Yes. If X and Y have different electronegativities, they have different attractions for the electrons in the molecule. The electron density around the more electronegative atom will be greater, producing a charge separation or dipole in the molecule.

(c) $\mu = Qr$. The dipole moment, μ, is the product of the magnitude of the separated charges, Q, and the distance between them, r.

8.38 (a) Non-zero. Cl and F have different electronegativities. In a diatomic molecule, a polar bond means a polar molecule.

(b) Non-zero. C and O have different electronegativities. In a diatomic molecule, a polar bond means a polar molecule.

(c) Zero. Even though the C–O bond is polar in this linear molecule, the two equal C–O bond dipoles lie in opposite directions and cancel each other. This orientation results in a nonpolar molecule.

(d) Non-zero. The two O-H bonds are polar and the molecule is bent (non-linear), so the bond dipoles do not cancel, resulting in a net dipole moment and a polar molecule.

8.39 *Analyze/Plan.* Q is the charge at either end of the dipole. Q = μ/r. From Table 8.3, the values for HF are μ = 1.82 D and r = 0.92 Å. Change Å to m and use the definition of the Debye and the charge of an electron to calculate the charge in units of *e*. *Solve*:

$$Q = \frac{\mu}{r} = \frac{1.82\ \text{D}}{0.92\ \text{Å}} \times \frac{1\ \text{Å}}{1 \times 10^{-10}\ \text{m}} \times \frac{3.34 \times 10^{-30}\ \text{C} \bullet \text{m}}{1\ \text{D}} \times \frac{1\ e}{1.60 \times 10^{-19}\ \text{C}} = 0.41\ e$$

Check. The calculated charge on H and F is 0.41 *e*. This can be thought of as the amount of charge "transferred" from H to F. This value is consistent with our idea that HF is a polar covalent molecule; the bonding electron pair is unequally shared, but not totally transferred from H to F.

8.40 (a) The more electronegative element, Br, will have a stronger attraction for the shared electrons and adopt a partial negative charge.

(b) Q is the charge at either end of the dipole.

$$Q = \frac{\mu}{r} = \frac{1.21\ \text{D}}{2.49\ \text{Å}} \times \frac{1\ \text{Å}}{1 \times 10^{-10}\ \text{m}} \times \frac{3.34 \times 10^{-30}\ \text{C} \bullet \text{m}}{1\ \text{D}} \times \frac{1\ e}{1.60 \times 10^{-19}\ \text{C}}$$

$$= 0.1014 = 0.101\ e$$

The charges on I and Br are 0.101 *e*.

8.41 *Analyze/Plan.* Generally, compounds formed by a metal and a nonmetal are described as ionic, while compounds formed from two or more nonmetals are covalent. *Solve*:

(a) MnO_2, ionic

(b) Ga_2S_3, ionic (Although their electronegativities are similar, Ga is a metal and S is a nonmetal. Use of a roman numeral usually presumes an ionic compound.)

(c) CoO, ionic (d) copper(I) sulfide, ionic (e) chlorine trifluoride, covalent

(f) vanadium(V) fluoride, ionic

8.42 Generally, compounds formed by a metal and a nonmetal are described as ionic, while compounds formed from two or more nonmetals are covalent.

(a) MnF_3, ionic (b) CrO_3, ionic

(c) $AsBr_5$, ionic (Although As is a metalloid, use of a roman numeral usually presumes an ionic compound.)

(d) sulfur tetrafluoride, covalent (e) molybdenum(IV) chloride, ionic

(f) scandium(III) chloride, ionic

Lewis Structures; Resonance Structures

8.43 *Analyze*. Counting the **correct number of valence electrons** is the foundation of every Lewis structure. *Plan/Solve*:

(a) Count valence electrons: $4 + (4 \times 1) = 8\ e^-$, $4\ e^-$ pairs. Follow the procedure in Sample Exercise 8.6.

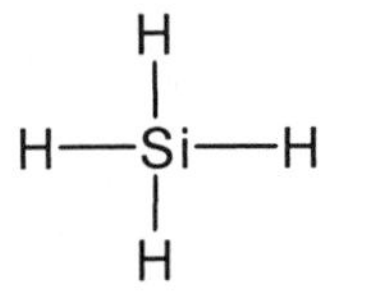

(b) Valence electrons: $4 + 6 = 10\ e^-$, $5\ e^-$ pairs

:C≡O:

(c) Valence electrons: $[6 + (2 \times 7)] = 20\ e^-$, $10\ e^-$ pairs.

:F̤̈—S̤̈—F̤̈:

i. Place the S atom in the middle and connect each F atom with a single bond; this requires $2\ e^-$ pairs.

ii. Complete the octets of the F atoms with nonbonded pairs of electrons; this requires an additional $6\ e^-$ pairs.

iii. The remaining $2\ e^-$ pairs complete the octet of the central S atom.

(d) 32 valence e^-, $16\ e^-$ pairs

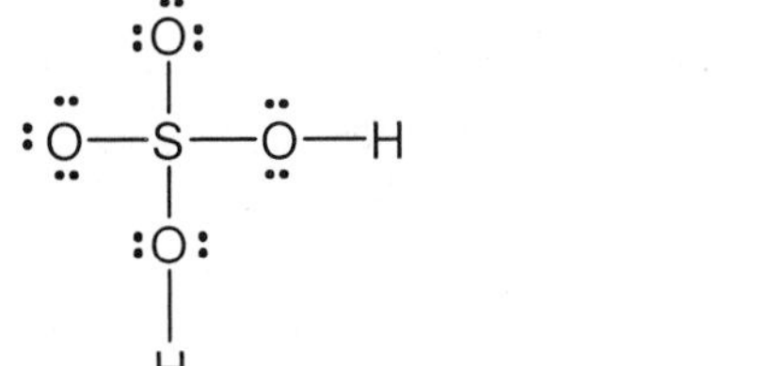

(Choose the Lewis structure that obeys the octet rule, Section 8.7.)

(e) Follow Sample Exercise 8.8. 20 valence e^-, $10\ e^-$ pairs

[:Ö̤—C̤̈l—Ö̤:]⁻

(f) 14 valence e^-, $7\ e^-$ pairs

H—N̈(—H)—Ö̤—H

Check. In each molecule, bonding e^- pairs are shown as lines, and each atom is surrounded by an octet of electrons (duet for H).

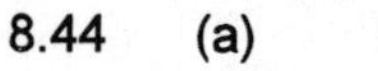

8.44 (a) 12 e^-, 6 e^- pairs

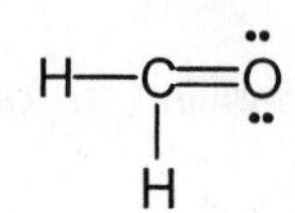

(b) 14 valence e^-, 7 e^- pairs

H—Ö—Ö—H

(c) 50 valence e^-, 25 e^- pairs

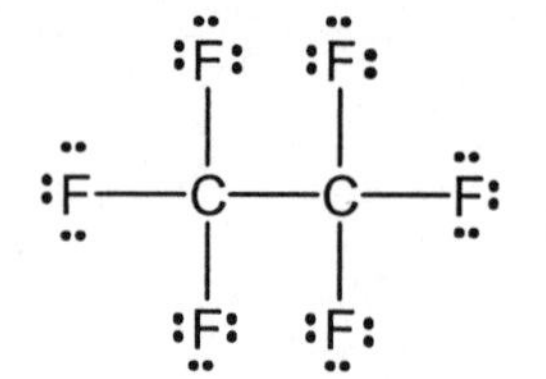

(d) 26 valence e^-, 13 e^- pairs

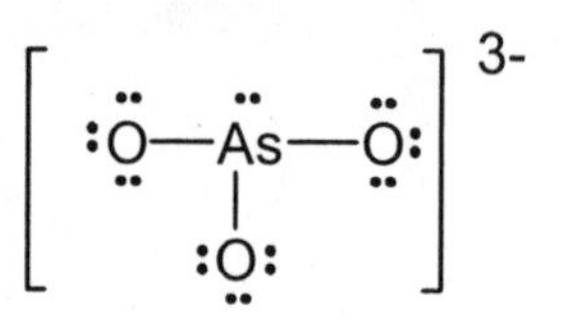

(Choose the Lewis structure that obeys the octet rule, Section 8.7)

(e) 26 valence e^- 13 e^- pairs

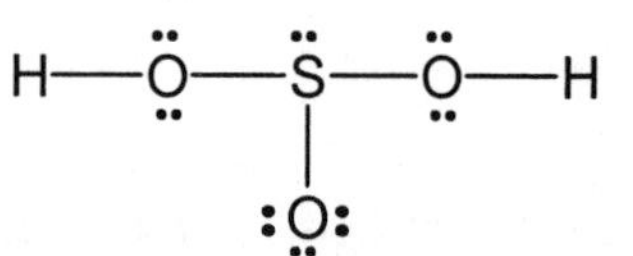

(Choose the Lewis structure that obeys they octet rule, Section 8.7.)

(f) 10 e^-, 5 e^- pairs

H—C≡C—H

8.45 *Analyze/Plan.* Draw the correct Lewis structure: count valence electrons in each atom, sum, determine electron pairs in the molecule; connect bonded atoms with a line, place the remaining e^- pairs as needed, in nonbonded pairs or multiple bonds, so that each atom is surrounded by an octet (or duet for H). Assign formal charges: assign electrons to individual atoms [nonbonding e^- + 1/2 (bonding e^-)]; formal charge = valence electrons - assigned electrons. *Solve:*

(a) 10 e^-, 5 e^- pairs

[:N≡O:]$^+$
0 +1

(b) 32 valence e^-, 16 e^- pairs

-1 :Ö: ; 0 :C̈l—P—C̈l: 0 ; +1 (P) ; :C̈l: 0

(c) 32 valence e^-, 16 e^- pairs

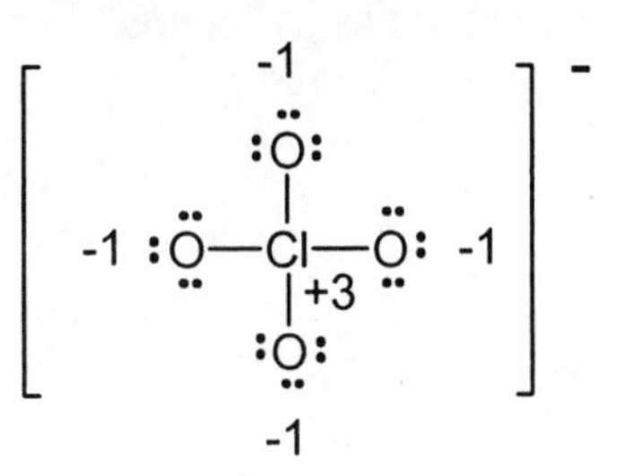

(d) 26 valence e^-, 13 e^- pairs

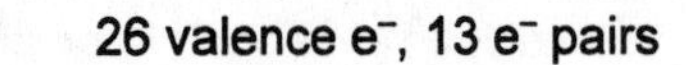

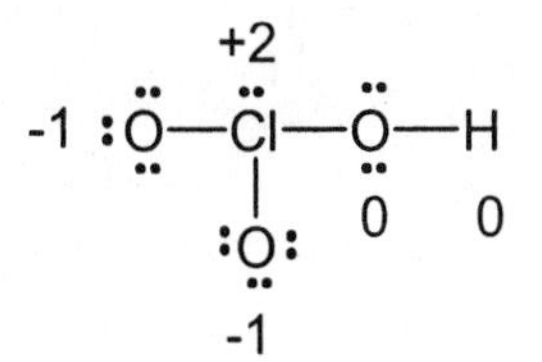

Check. Each atom is surrounded by an octet (or duet) and the sum of the formal charges in particle is the charge on the particle.

8.46 (a) 18 e^-, 9 e^- pairs (b) 24 e^-, 12 e^- pairs

(c) 26 e^-, 13 e^- pairs (d) 32 e^-, 16 e^- pairs

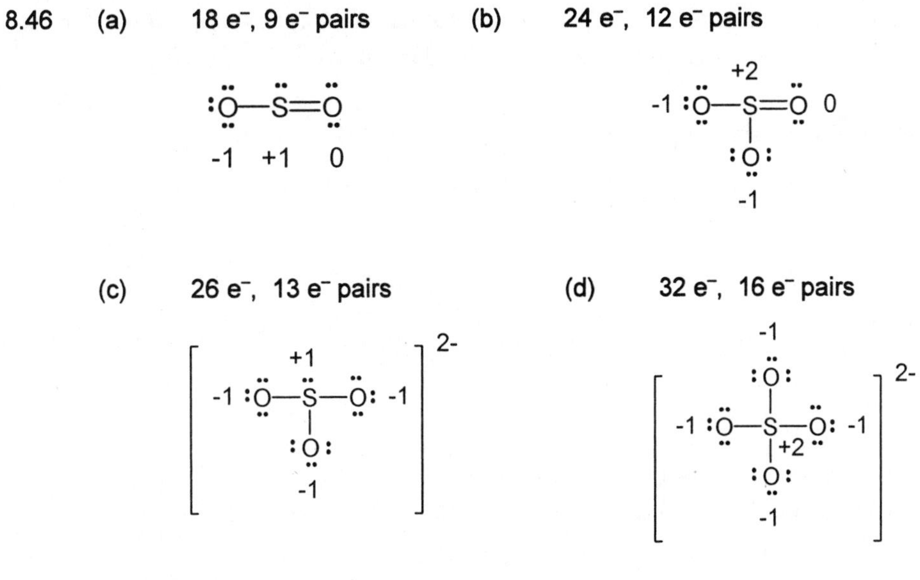

8.47 (a) *Plan.* Count valence electrons, draw all possible correct Lewis structures, taking note of alternate placements for multiple bonds. *Solve:*

18 e^-, 9 e^- pairs

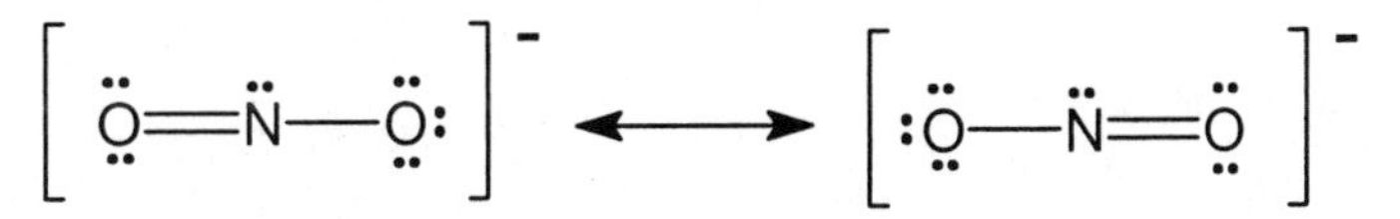

Check. The octet rule is satisfied.

(b) *Plan.* Isoelectronic species have the same number of valence electrons and the same electron configuration. *Solve:*

A single O atom has 6 valence electrons, so the neutral ozone molecule O_3 is isoelectronic with NO_2^-.

O=O—O ⟷ O—O=O

Check. The octet rule is satisfied.

(c) Since each N–O bond has partial double bond character, the N–O bond length in NO_2^- should be shorter than in species with formal N–O single bonds.

8.48 (a) 16 e^-, 8 e^- pairs

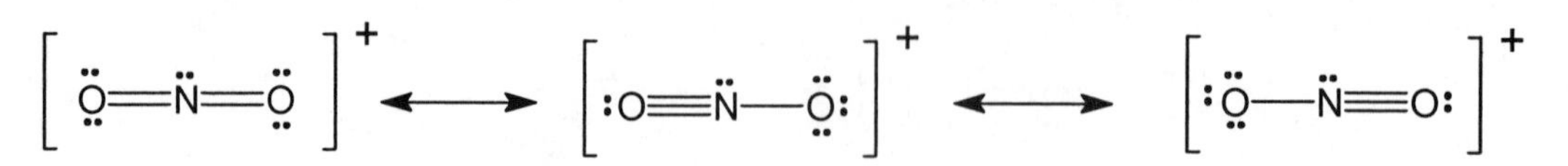

(b) More than one correct Lewis structure can be drawn, so resonance structures are needed to accurately describe the structure.

(c) NO_2^+ has 16 valence electrons. Consider other triatomic molecules involving second-row nonmetallic elements. O_3^{2+} or C_3^{4-} are not 'common' (or stable). CO_2 is common and matches the description (as does N_3^-, azide ion).

8.49 *Plan/Solve*. The Lewis structures are as follows:

5 e^- pairs　　8 e^- pairs

:C≡O:　　O=C=O

12 e^- pairs

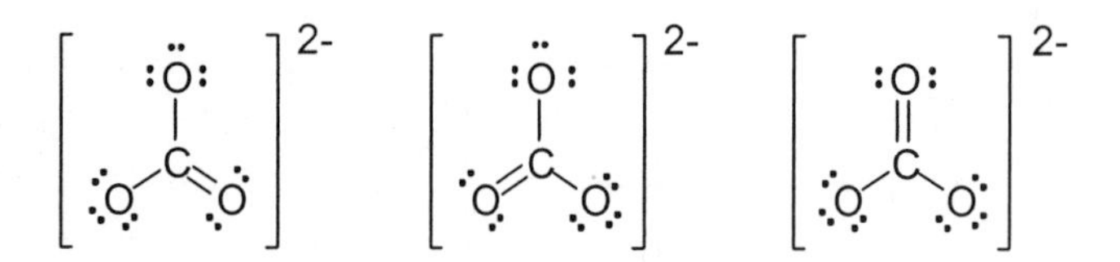

The more pairs of electrons shared by two atoms, the shorter the bond between the atoms. The average number of electron pairs shared by C and O in the three species is 3 for CO, 2 for CO_2 and 1.33 for CO_3^{2-}. This is also the order of increasing bond length: CO < CO_2 < CO_3^{2-}.

8.50 The Lewis structures are as follows:

5 e^- pairs　　9 e^- pairs

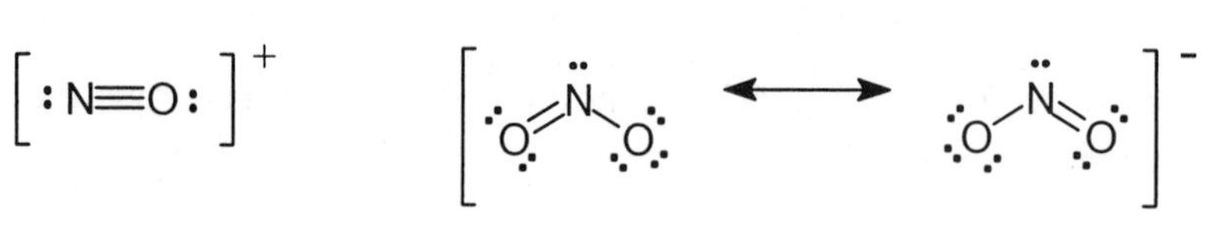

12 e^- pairs

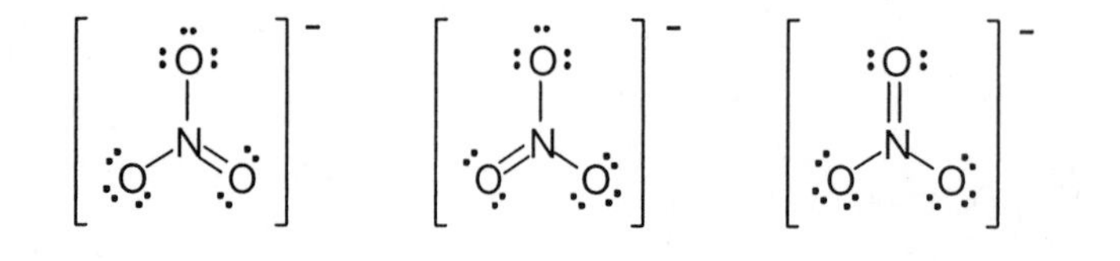

The average number of electron pairs in the N–O bond is 3.0 for NO^+, 1.5 for NO_2^- and 1.33 for NO_3^-. The more electron pairs shared between two atoms, the shorter the bond. Thus the N–O bond lengths vary in the order $NO^+ < NO_2^- < NO_3^-$.

8.51 (a) Two equally valid Lewis structures can be drawn for benzene.

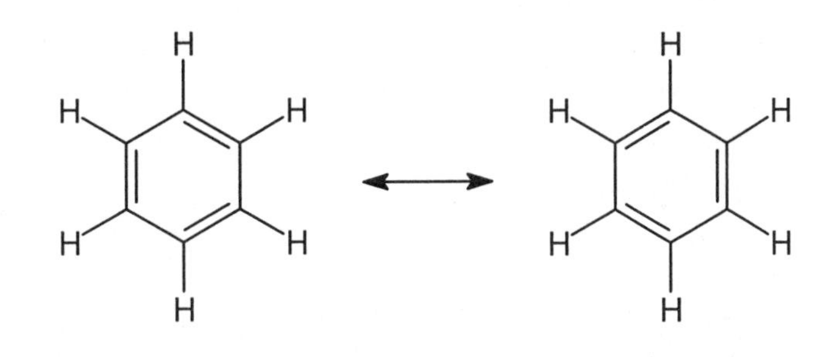

Each structure consists of alternating single and double C–C bonds; a particular bond is single in one structure and double in the other. The concept of resonance dictates that the true description of bonding is some hybrid or blend of the two Lewis structures. The most obvious blend of these two resonance structures is a molecule with six equivalent C–C bonds, each with some but not total double-bond character. If the molecule has six equivalent C–C bonds, the lengths of these bonds should be equal.

(b) The resonance model described in (a) has 6 equivalent C–C bonds, each with some double bond character. That is, more than 1 pair but less than 2 pairs of electrons is involved in each C–C bond. This model predicts a uniform C–C bond length that is shorter than a single bond but longer than a double bond.

8.52 (a)

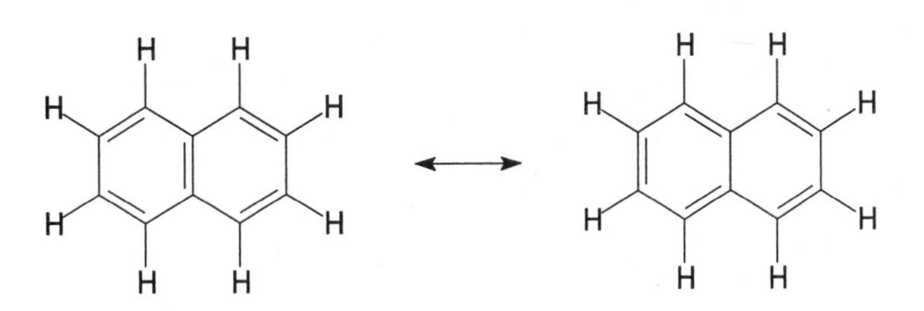

(b) The resonance model of this molecule has bonds that are neither single nor double, but somewhere in between. This results in bond lengths that are intermediate between C–C single and C=C double bond lengths.

(c)

Exceptions to the Octet Rule

8.53 (a) The *octet rule* states that atoms will gain, lose or share electrons until they are surrounded by eight valence electrons.

(b) The octet rule applies to the individual ions in an ionic compound. That is, the cation has lost electrons to achieve an octet and the anion has gained electrons to achieve an octet. For example, in $MgCl_2$, Mg loses 2 e^- to become Mg^{2+} with the electron configuration of Ne. Each Cl atom gains one electron to form Cl^- with the electron configuration of Ar.

8.54 Carbon, in group 14, needs to form four single bonds to achieve an octet, as in CH_4. Nitrogen, in group 15, needs to form three, as in NH_3. If G = group number and n = the number of single bonds, G + n = 18 is a general relationship for the representative non-metals.

Check: O as in H_2O (G = 16) + (n = 2 bonds) = 18

8.55 The most common exceptions to the octet rule are molecules with more than eight electrons around one or more atoms, usually the central atom. Examples: SF_6, PF_5

8.56 In the third period, atoms have the space and available orbitals to accommodate extra electrons. Since atomic radius increases going down a family, elements in the third period and beyond are less subject to destabilization from additional electron-electron repulsions. Also, the third shell contains d orbitals that are relatively close in energy to 3s and 3p orbitals (the ones that accommodate the octet) and provide an allowed energy state for the extra electrons.

8.57 (a) 24 e^-, 12 e^- pairs

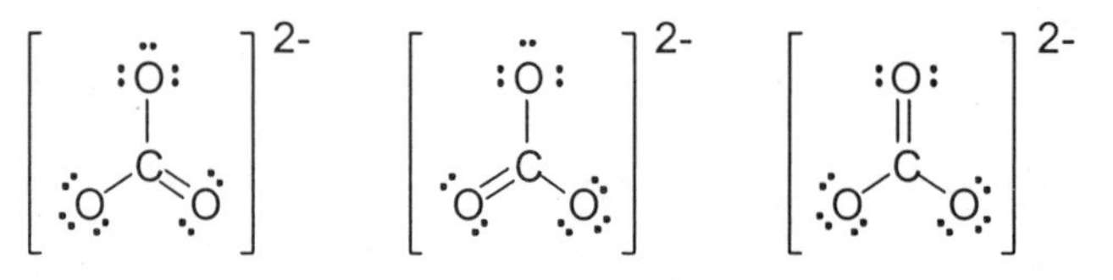

CO_3^{2-} has three resonance structures, but all obey the octet rule.

(b) 6 e^-, 3 e^- pairs, impossible to satisfy octet rule with only 6 valence electrons

H—B—H
|
H

6 electrons around B

(c) 22 e^-, 11 e^- pairs

[:Ï—Ï—Ï:]⁻

10 e^- around central I

(d) 32 e⁻, 16 e⁻ pairs

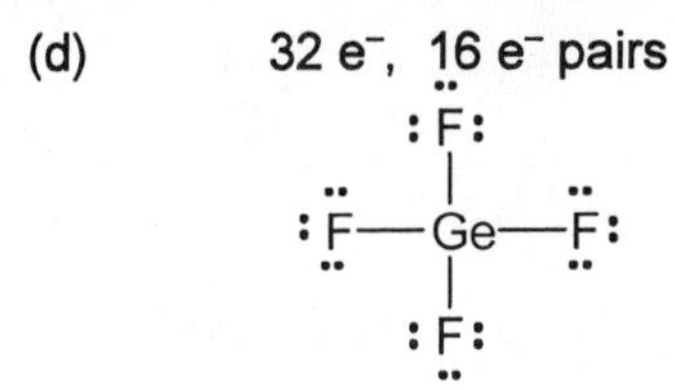

obeys the octet rule

(e) 48 e⁻, 24 e⁻ pairs

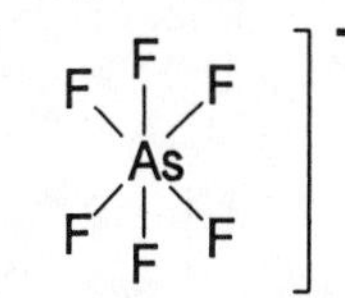

12 e⁻ around As; three nonbonded pairs on each F have been omitted

8.58 (a) 11 e⁻, 5.5 e⁻ pairs; odd electron molecule

(N=O)

The odd electron is probably on N because it is less electronegative than O.

(b) 22 e⁻ 11 e⁻ pairs

[:Cl—I—Cl:]⁻

10 e⁻ around central I

(c) 18 e⁻, 9 e⁻ pairs

O=S—O: ⟷ :O—S=O

Structures with expanded octets around S can be drawn, but are not preferred (Section 8.7).

(d) 24 e⁻, 12 e⁻ pairs

:Cl—B—Cl:
|
:Cl:

6 e⁻ around B

Structures such as the one below can be drawn, but are not likely because of high formal charges on B and Cl.

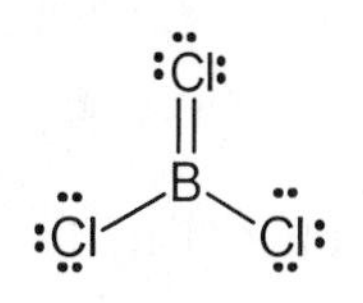

(e) 36 e⁻, 18 e⁻ pairs

:F:
:F—Xe—F:
:F:

12 e⁻ around the central Xe

(F cannot expand beyond an octet.)

8.59 (a) 16 e⁻, 8 e⁻ pairs :Cl—Be—Cl:

This structure violates the octet rule; Be has only 4 e⁻ around it.

(b) Cl=Be=Cl ⟷ :Cl—Be≡Cl: ⟷ :Cl≡Be—Cl:

(c) The formal charges on each of the atoms in the four resonance structures are:

:Cl—Be—Cl:			Cl=Be=Cl			:Cl—Be≡Cl:			:Cl≡Be—Cl:		
0	0	0	+1	-2	+1	0	-2	+2	+2	-2	0

Since formal charges are minimized on the structure that violates the octet rule, this form is probably most important.

8.60 (a) 19 e^-, 9.5 e^- pairs, odd electron molecule

:Ö—Ċl—Ö: ⟷ :Ö—C̈l—Ö· ⟷ ·Ö—C̈l—Ö:

(b) None of the structures satisfies the octet rule. In each structure, one atom has only 7 e^- around it. If a molecule has an odd number of electrons in the valence shell, no Lewis structure can satisfy the octet rule.

(c) :Ö—Ċl—Ö: ⟷ :Ö—C̈l—Ö· ⟷ ·Ö—C̈l—Ö:

-1 +2 -1 -1 +1 0 0 +1 -1

Formal charge arguments predict that the two resonance structures with the odd electron on O are most important. This contradicts electronegativity arguments, which would predict that the less electronegative atom, Cl, would be more likely to have fewer than 8 e^- around it.

Bond Enthalpies

8.61 *Analyze*. Given: structural formulas. Find: enthalpy of reaction.

Plan. Count the number and kinds of bonds that are broken and formed by the reaction. Use bond enthalpies from Table 8.4 and Equation 8.12 to calculate the overall enthalpy of reaction, ΔH. *Solve*:

(a) $\Delta H = 2D(\text{O-H}) + D(\text{O-O}) + 4D(\text{C-H}) + D(\text{C=C})$

$- 2D(\text{O-H}) - 2D(\text{O-C}) - 4D(\text{C-H}) - D(\text{C-C})$

$\Delta H = D(\text{O-O}) + D(\text{C=C}) - 2D(\text{O-C}) - D(\text{C-C})$

$= 146 + 614 - 2(358) - 348 = -304$ kJ

(b) $\Delta H = 5D(\text{C-H}) + D(\text{C}\equiv\text{N}) + D(\text{C=C}) - 5D(\text{C-H}) - D(\text{C}\equiv\text{N}) - 2D(\text{C-C})$

$= D(\text{C=C}) - 2D(\text{C-C}) = 614 - 2(348) = -82$ kJ

(c) $\Delta H = 6D(\text{N-Cl}) - 3D(\text{Cl-Cl}) - D(\text{N}\equiv\text{N})$

$= 6(200) - 3(242) - 941 = -467$ kJ

8.62 (a) $\Delta H = 3D(\text{C-Br}) + D(\text{C-H}) + D(\text{Cl-Cl}) - 3D(\text{C-Br}) - (\text{C-Cl}) - D(\text{H-Cl})$

$= D(\text{C-H}) + D(\text{Cl-Cl}) - D(\text{C-Cl}) - D(\text{H-Cl})$

$\Delta H = 413 + 242 - 328 - 431 = -104$ kJ

(b) $\Delta H = 4D(\text{C-H}) + 2D(\text{C-S}) + 2D(\text{S-H}) + D(\text{C-C}) + D(\text{H-Br})$

$-4D(\text{S-H}) - D(\text{C-C}) - 2D(\text{C-Br}) - 4D(\text{C-H})$

$= 2D(\text{C-S}) + D(\text{H-Br}) - 2D(\text{S-H}) - 2D(\text{C-Br})$

$\Delta H = 2(259) + 366 - 2(339) - 2(276) = -346$ kJ

(c) $\Delta H = 4D(\text{N-H}) + D(\text{N-N}) + D(\text{Cl-Cl}) - 4D(\text{N-H}) - D(\text{N-Cl})$

$= D(\text{N-N}) + D(\text{Cl-Cl}) - D(\text{N-Cl})$

$\Delta H = 163 + 242 - 200 = 205$ kJ

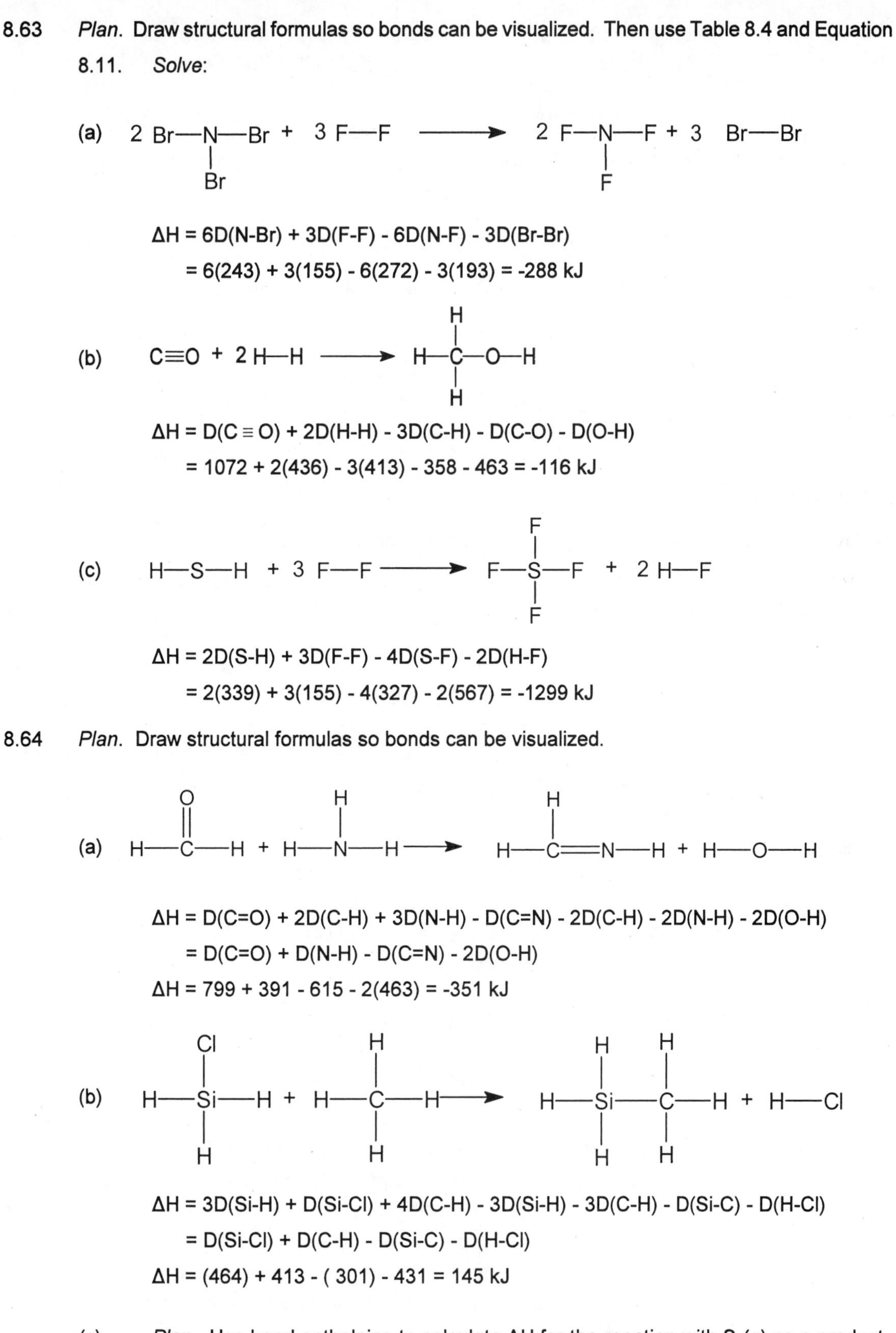

8.63 *Plan*. Draw structural formulas so bonds can be visualized. Then use Table 8.4 and Equation 8.11. *Solve*:

(a)

$\Delta H = 6D(N\text{-}Br) + 3D(F\text{-}F) - 6D(N\text{-}F) - 3D(Br\text{-}Br)$

$= 6(243) + 3(155) - 6(272) - 3(193) = -288$ kJ

(b)

$\Delta H = D(C \equiv O) + 2D(H\text{-}H) - 3D(C\text{-}H) - D(C\text{-}O) - D(O\text{-}H)$

$= 1072 + 2(436) - 3(413) - 358 - 463 = -116$ kJ

(c)

$\Delta H = 2D(S\text{-}H) + 3D(F\text{-}F) - 4D(S\text{-}F) - 2D(H\text{-}F)$

$= 2(339) + 3(155) - 4(327) - 2(567) = -1299$ kJ

8.64 *Plan*. Draw structural formulas so bonds can be visualized.

(a)

$\Delta H = D(C{=}O) + 2D(C\text{-}H) + 3D(N\text{-}H) - D(C{=}N) - 2D(C\text{-}H) - 2D(N\text{-}H) - 2D(O\text{-}H)$

$= D(C{=}O) + D(N\text{-}H) - D(C{=}N) - 2D(O\text{-}H)$

$\Delta H = 799 + 391 - 615 - 2(463) = -351$ kJ

(b)

$\Delta H = 3D(Si\text{-}H) + D(Si\text{-}Cl) + 4D(C\text{-}H) - 3D(Si\text{-}H) - 3D(C\text{-}H) - D(Si\text{-}C) - D(H\text{-}Cl)$

$= D(Si\text{-}Cl) + D(C\text{-}H) - D(Si\text{-}C) - D(H\text{-}Cl)$

$\Delta H = (464) + 413 - (301) - 431 = 145$ kJ

(c) *Plan*. Use bond enthalpies to calculate ΔH for the reaction with $S_8(g)$ as a product. Then,

$8H_2S(g) \rightarrow 8H_2(g) + S_8(g)$ ΔH

$S_8(g) \rightarrow S_8(s)$ $-\Delta H_f^\circ$ for $S_8(g)$

$8H_2S(g) \rightarrow 8H_2(g) + S_8(s)$ $\Delta H_{rxn} = [\Delta H - \Delta H_f^\circ\ S_8(g)]$

$\Delta H = 16D(S\text{-}H) - 8(H\text{-}H) - 8(S\text{-}S)$

$= 16(339) - 8(436) - 8(266) = -192$ kJ

$\Delta H_{rxn} = \Delta H - \Delta H_f^\circ\ S_8(g) = -192\ \text{kJ} - 102.3\ \text{kJ} = -294.3 = -294$ kJ

8.65 *Plan.* Draw structural formulas so bonds can be visualized. Then use Table 8.4 and Equation 8.12. *Solve*:

(a) 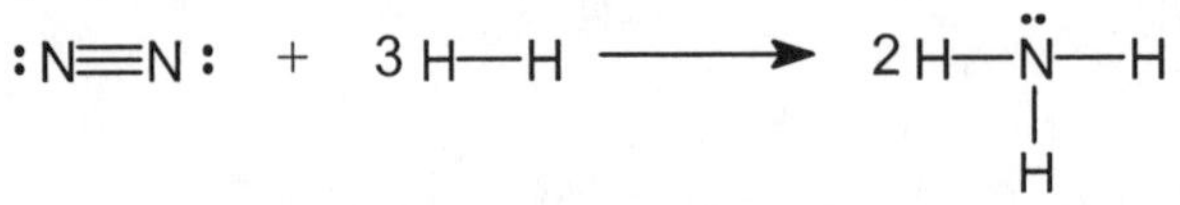

$\Delta H = D(N \equiv N) + 3D(H\text{-}H) - 6(N\text{-}H) = 941\ \text{kJ} + 3(436\ \text{kJ}) - 6(391\ \text{kJ})$

$= -97$ kJ / 2 mol NH_3 ; **exothermic**

(b) Plan. Use Eq. 5.31 to calculate ΔH_{rxn} from ΔH_f° values.

$\Delta H_{rxn}^\circ = \Sigma n\,\Delta H_f^\circ(\text{products}) - \Sigma n\,\Delta H_f^\circ(\text{reactants})$. $\Delta H_f^\circ\ NH_3(g) = -46.19$ kJ. *Solve*:

$\Delta H_{rxn}^\circ = 2\,\Delta H_f^\circ\ NH_3(g) - 3\,\Delta H_f^\circ\ H_2(g) - \Delta H_f^\circ\ N_2(g)$

$\Delta H_{rxn}^\circ = 2(-46.19) - 3(0) - 0 = -92.38$ kJ/2 mol NH_3

The ΔH calculated from bond enthalpies is slightly more exothermic (more negative) than that obtained using ΔH_f° values.

8.66 (a) 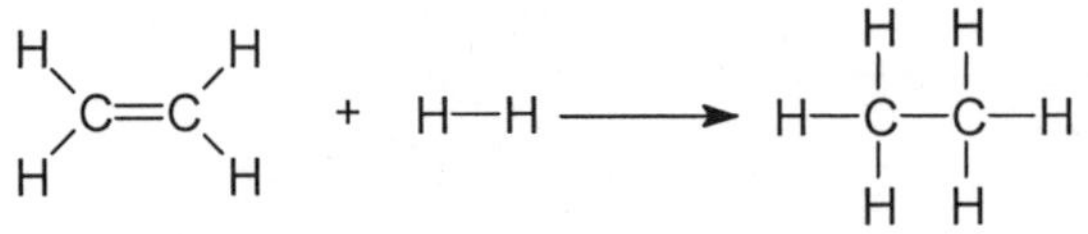

$\Delta H = 4D(C\text{-}H) + D(C{=}C) + D(H\text{-}H) - 6D(C\text{-}H) - D(C\text{-}C)$

$= D(C{=}C) + D(H\text{-}H) - 2D(C\text{-}H) - D(C\text{-}C)$

$\Delta H = 614 + 436 - 2(413) - 348 = -124$ kJ

(b) $\Delta H^\circ = \Delta H_f^\circ\ C_2H_6(g) - \Delta H_f^\circ\ C_2H_4(g) - \Delta H_f^\circ\ H_2(g)$

$= -84.68 - 52.30 - 0 = -136.98$ kJ

The values of ΔH for the reaction differ because the bond enthalpies used in part (a) are average values that can differ from one compound to another. For example, the exact enthalpy of a C–H bond in C_2H_4 is probably not equal to the enthalpy of a C–H bond in C_2H_6. Thus, reaction enthalpies calculated from average bond enthalpies are estimates. On the other hand, standard enthalpies of formation are measured quantities and should lead to accurate reaction enthalpies. The advantage of average bond enthalpies is that they can be used for reactions where no measured enthalpies of formation are available.

8.67 The average Ti–Cl bond enthalpy is just the average of the four values listed, 430 kJ/mol.

8.68 (a) (i) $C + 2\,F{-}F \longrightarrow CF_4$ (F—C—F with F above and below C)

$\Delta H = 2D(F{-}F) - 4D(C{-}F) = 2(155) - 4(485) = -1630$ kJ

(ii) $C{\equiv}O + 3\,F{-}F \longrightarrow CF_4 + F{-}O{-}F$

$\Delta H = D(C{\equiv}O) + 3D(F{-}F) - 4D(C{-}F) - 2D(D{-}F)$
$= 1072 + 3(155) - 4(485) - 2(190) = -783$ kJ

(iii) $O{=}C{=}O + 4\,(F{-}F) \longrightarrow CF_4 + 2\,F{-}O{-}F$

$\Delta H = 2D(C{=}O) + 4D(F{-}F) - 4D(C{-}F) - 4D(O{-}F)$
$= 2(799) + 4(155) - 4(485) - 4(190) = -482$ kJ

(b) The more oxygen atoms bound to carbon, the less exothermic the reaction in this series.

Additional Exercises

8.69 (a) Group 14 or 4A (b) Group 2 or 2A (c) Group 15 or 5A
(These are the appropriate groups in the s and p blocks, where Lewis symbols are most useful.)

8.70 (a) Lattice energy is proportional to Q_1Q_2 / d. For each of these compounds, Q_1Q_2 is the same. The anion H^- is present in each compound, but the ionic radius of the cation increases going from Be to Ba. Thus, the value of d (the cation-anion separation) increases and the ratio Q_1Q_2 / d decreases. This is reflected in the decrease in lattice energy going from BeH_2 to BaH_2.

(b) Again, Q_1Q_2 for ZnH_2 is the same as that for the other compounds in the series and the anion is H^-. The lattice energy of ZnH_2, 2870 kJ, is closest to that of MgH_2, 2791 kJ. The ionic radius of Zn^{2+} is similar to that of Mg^{2+}.

8.71 $E = \frac{-8.99 \times 10^9\ J \bullet m}{C^2} \times \frac{(1.60 \times 10^{-19}\ C)^2}{(1.33 + 1.33) \times 10^{-10}\ m} = -8.652 \times 10^{-19} = -8.65 \times 10^{-19}\ J$

On a molar basis: $(-8.652 \times 10^{-19}\ J)(6.022 \times 10^{23}) = -5.21 \times 10^5\ J = -521\ kJ$

Note that its absolute value is less than the lattice energy, 808 kJ/mol. The difference represents the added energy of putting all the K^+F^- ion pairs together in a three-dimensional array, similar to the one in Figure 8.3.

8.72 $E = Q_1Q_2 / d$; $k = 8.99 \times 10^9\ J \bullet m/coul^2$

(a) $E = \frac{-8.99 \times 10^9\ J \bullet m}{C^2} \times \frac{(1.60 \times 10^{-19}\ C)^2}{(0.97 + 1.96) \times 10^{-10}\ m} = -7.8547 \times 10^{-19} = -7.85 \times 10^{-19}\ J$

The sign of E is negative because one of the interacting ions is an anion; this is an attractive interaction.

On a molar basis: $-7.855 \times 10^{-19} \times 6.022 \times 10^{23} = -4.73 \times 10^5\ J = -473\ kJ$

(b) $E = \frac{-8.99 \times 10^9\ J \bullet m}{C^2} \times \frac{(1.60 \times 10^{-19}\ C)^2}{(1.47 + 1.96) \times 10^{-10}\ m} = -6.71 \times 10^{-19}\ J$

On a molar basis: $-4.04 \times 10^5\ J = -404\ kJ$

(c) $E = \frac{-8.99 \times 10^9\ J \bullet m}{C^2} \times \frac{(2 \times 1.60 \times 10^{-19}\ C)^2}{(1.13 + 1.84) \times 10^{-10}\ m} = -3.10 \times 10^{-18}\ J$

On a molar basis: $-1.87 \times 10^6\ J = -1.87 \times 10^3\ kJ$

8.73 (a)

	Compound	Lattice Energy (kJ)			Compound	Lattice Energy (kJ)	
106 kJ	NaCl	788	56 kJ	104 kJ	LiCl	834	55 kJ
	NaBr	732			**LiBr**	**779**	
	NaI	682			LiI	730	

The difference in lattice energy between LiCl and LiI is 104 kJ. The difference between NaCl and NaI is 106 kJ; the difference between NaCl and NaBr is 56 kJ, or 53% of the difference between NaCl and NaI. Applying this relationship to the Li salts, 0.53(104 kJ) = 55 kJ difference between LiCl and LiBr. The approximate lattice energy of LiBr is (834 - 55) kJ = 779 kJ.

(b)

	Compound	Lattice Energy (kJ)			Compound	Lattice Energy (kJ)	
106 kJ	NaCl	788	56 kJ	57 kJ	CsCl	657	30 kJ
	NaBr	732			**CsBr**	**627**	
	NaI	682			CsI	600	

By analogy to the Na salts, the difference between lattice energies of CsCl and CsBr should be approximately 53% of the difference between CsCl and CsI. The lattice energy of CsBr is approximately 627 kJ.

(c)

	Compound	Lattice Energy (kJ)			Compound	Lattice Energy (kJ)	
578 kJ	MgO	3795	381 kJ	199 kJ	$MgCl_2$	2326	131 kJ
	CaO	3414			**$CaCl_2$**	2195	
	SrO	3217			$SrCl_2$	2127	

By analogy to the oxides, the difference between the lattice energies of $MgCl_2$ and $CaCl_2$ should be approximately 66% of the difference between $MgCl_2$ and $SrCl_2$. That is, 0.66(199 kJ) = 131 kJ. The lattice energy of $CaCl_2$ is approximately (2326 - 131) kJ = 2195 kJ.

8.74 Rh: $[Kr]5s^2 4d^7$. In order to achieve the electron configuration of Kr as a positive ion, Rh would have to lose it's 5s valence electrons and all seven 4d electrons resulting in a Rh^{9+} ion. The 4d electrons are closer to the nucleus than n = 5 electrons and are incompletely shielded from the nucleus by the 4s and 4p electrons. Thus, I_3 through I_9 are likely to be quite large. Even though the lattice energy of an ionic compound involving Rb^{9+} would be large, it could not compensate for the total ionization energy required to form Rh^{9+}. It is unlikely that Rh can adapt the noble-gas configuration of Kr by forming Rh^{9+}.

8.75 (a) 12 + 3 + 15 = 30 valence e^-, 15 e^- pairs.

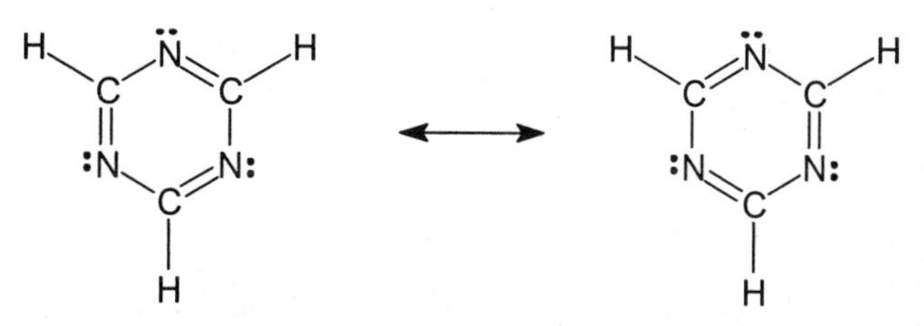

Structures with H bound to N and nonbonded electron pairs on C can be drawn, but the structures above minimize formal charges on the atoms.

(b) The resonance structures indicate that triazine will have six equal C–N bond lengths, intermediate between C–N single and C–N double bond lengths. (See Solutions 8.51 and 8.52). From Table 8.5, an average C–N length is 1.43 Å, a C=N length is 1.38 Å. The average of these two lengths is 1.405 Å. The C–N bond length in triazine should be in the range 1.40 - 1.41 Å.

8.76 Molecule (b) H_2S and ion (c) NO_2^- contain polar bonds. The atoms that form the bonds (H–S) and N–O) have different electronegativity values.

8.77 (a) B–O. the most polar bond will be formed by the two elements with the greatest difference in electronegativity. Since electronegativity increases moving right and up on the periodic chart, the possibilities are B–O and Te–O. These two bonds are likely to have similar electronegativitiy differences (3 columns apart vs. 3 rows apart). Values from Figure 8.6 confirm the similarity, and show that B–O is slightly more polar.

(b) Te–I. Both are in the fifth row of the periodic chart and have the two largest covalent radii among this group of elements.

(c) TeI_2. Te needs to participate in two covalent bonds to satisfy the octet rule, and each I atom needs to participate in one bond, so by forming a TeI_2 molecule, the octet rule can be satisfied for all three atoms.

$$:\ddot{\underset{..}{I}}-\ddot{\underset{..}{Te}}-\ddot{\underset{..}{I}}:$$

(d) B_2O_3. Although this is probably not a purely ionic compound, it can be understood in terms of gaining and losing electrons to achieve a noble-gas configuration. If each B atom were to lose 3 e^- and each O atom were to gain 2 e^-, charge balance and the octet rule would be satisfied.

P_2O_3. Each P atom needs to share 3 e^- and each O atom 2 e^- to achieve an octet. Although the correct number of electrons seem to be available, a correct Lewis structure is difficult to imagine. In fact, phosphorus (III) oxide exists as P_4O_6 rather than P_2O_3 (Chapter 22).

8.78 Use the method detailed in Section 8.5, *A Closer Look,* to estimate partial charges from electronegativity values. From Figure 8.6, the electronegativity of F is 4.0 and of Cl is 3.0.

F has 4.0 / (3.0 + 4.0) = 0.57 of the charge of the bonding e^- pair.
Cl has 3.0 / (3.0 + 4.0) = 0.43 of the charge of the bonding e^- pair.

This amounts to 0.57 × 2*e* = 1.14*e* on F or 0.14*e* more than a neutral F atom. This implies a -0.14 charge on F and +0.14 charge on Cl.

From Figure 7.5, the covalent radius of F is 0.71 Å and of Cl is 0.99 Å. The F–Cl separation is 1.70 Å.

$$\mu = Qr = 0.14e \times \frac{1.60 \times 10^{-19}\ \text{C}}{e} \times 1.70\ \text{Å} \times \frac{1 \times 10^{-10}\ \text{m}}{\text{Å}} \times \frac{1\ \text{D}}{3.34 \times 10^{-30}\ \text{C} \bullet \text{m}} = 1.1\ \text{D}$$

Clearly, this method is approximate. The estimated dipole moment of 1.1 D is within 30% of the measured value of 0.88 D.

8.79 Formal charge (FC) = # valence e^- - (# nonbonding e^- + 1/2 # bonding e^-)

(a) 18 e^-, 9 e^- pairs

$$:\ddot{\underset{..}{O}}-\ddot{O}=\ddot{\underset{..}{O}} \longleftrightarrow \ddot{\underset{..}{O}}=\ddot{O}-\ddot{\underset{..}{O}}:$$

FC for the central O = 6 - [2 + 1/2 (6)] = +1

(b) 48 e^-, 24 e^- pairs

$$\left[PF_6\right]^-$$

FC for P = 5 - [0 + 1/2 (12)] = -1

The three nonbonded pairs on each F have been omitted.

(c) 17 e^-; 8 e^- pairs, 1 odd e^-

O=N—O: ⟷ :O—N=O

The odd electron is probably on N because it is less electronegative than O.
Assuming the odd electron is on N, FC for N = 5 - [1+ 1/2 (6)] = +1.
If the odd electron is on O, FC for N = 5 - [2 + 1/2 (6)] = 0.

(d) 28 e^-, 14 e^- pairs

:Cl—I—Cl:
with :Cl: bonded below I

FC for I = 7 - [4 + 1/2 (6)] = 0

(e) 32 e^-, 16 e^- pairs

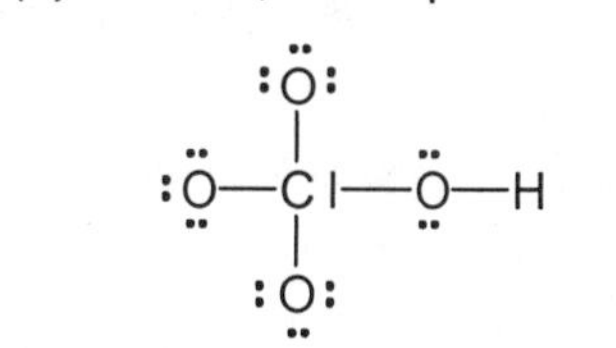

FC for Cl = 7 - [0 + 1/2 (8)] = +3

8.80 (a) 14e^-, 7 e^- pairs

[:Cl—O:]$^-$

FC on Cl = 7 - [6 + 1/2(2)]
= 0

32 e^-, 16 e^- pairs

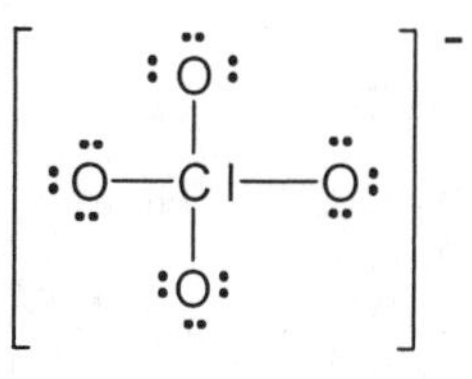

FC on Cl = 7 - [0 + 1/2(8)]
= +3

(b) The oxidation number of Cl is +1 in ClO^- and +7 in ClO_4^-.

(c) The definition of formal charge assumes that all bonding pairs of electrons are equally shared by the two bonded atoms, that all bonds are purely covalent. The definition of oxidation number assumes that the more electronegative element in the bond gets all of the bonding electrons, that the bonds are purely ionic. These two definitions represent the two extremes of how electron density is distributed between bonded atoms.

In ClO^- and ClO_4^-, Cl is the less electronegative element, so the oxidation numbers have a higher positive value than the formal charges. The true description of the electron density distribution is somewhere between the extremes indicated by formal charge and oxidation number.

8.81 (a) :N≡N—O: ⟷ :N—N≡O: ⟷ :N=N=O:

0 +1 -1 -2 +1 +1 -1 +1 0

In the leftmost structure, the more electronegative O atom has the negative formal charge, so this structure is likely to be most important.

(b) In general, the more shared pairs of electrons between two atoms, the shorter the bond, and vice versa. That the N–N bond length in N_2O is slightly longer than the typical N≡N indicates that the middle and right resonance structures where the N atoms share less than 3 electron pairs are contributors to the true structure. That the N-O bond length is slightly shorter than a typical N=O indicates that the middle structure, where N and O share more than 2 electron pairs, does contribute to the true structure. This physical data indicates that while formal charge can be used to predict which resonance form will be more important to the observed structure, the influence of minor contributors on the true structure cannot be ignored.

8.82 I_3^- has a Lewis structure with an expanded octet of electrons around the central I.

$$:\ddot{\underset{..}{I}}-\ddot{\underset{..}{I}}-\ddot{\underset{..}{I}}:$$

F cannot accommodate an expanded octet because it is too small and has no available d orbitals in its valence shell.

8.83 ΔH = 8D(C-H) - D(C-C) - 6D(C-H) - D(H-H)

= 2D(C-H) - D(C-C) - D(H-H)

= 2(413) - 348 - 436 = +42 kJ

ΔH = 8D(C-H) + 1/2 D(O=O) - D(C-C) - 6D(C-H) - 2D(O-H)

= 2D(C-H) + 1/2 D(O=O) - D(C-C) - 2D(O-H)

= 2(413) + 1/2 (495) - 348 - 2(463) = -200 kJ

The fundamental difference in the two reactions is the formation of 1 mol of H–H bonds versus the formation of 2 mol of O–H bonds. The latter is much more exothermic, so the reaction involving oxygen is more exothermic.

8.84 (a) ΔH = 5D(C-H) + D(C-C) + D(C-O) + D(O-H) - 6D(C-H) - 2D(C-O)

= D(C-C) + D(O-H) - D(C-H) - D(C-O)

= 348 kJ + 463 kJ - 413 kJ - 358 kJ

ΔH = +40 kJ; ethanol has the lower enthalpy

(b) ΔH = 4D(C-H) + D(C-C) + 2D(C-O) - 4D(C-H) - D(C-C) - D(C=O)

= 2D(C-O) - D(C=O)

= 2(358 kJ) - 799 kJ

ΔH = -83 kJ; acetaldehyde has the lower enthalpy

(c) ΔH = 8D(C-H) + 4D(C-C) + D(C=C) - 8D(C-H) - 2D(C-C) - 2D(C=C)

= 2D(C-C) - D(C=C)

= 2(348 kJ) - 614 kJ

ΔH = +82 kJ; cyclopentene has the lower enthalpy

(d) $\Delta H = 3D(C\text{-}H) + D(C\text{-}N) + D(C \equiv N) - 3D(C\text{-}H) - D(C\text{-}C) - D(C \equiv N)$

$= D(C\text{-}N) - D(C\text{-}C)$

$= 293\text{ kJ} - 348\text{ kJ}$

$\Delta H = -55\text{ kJ}$; acetonitrile has the lower enthalpy

8.85 (a) 4 [H–C(H)(O–NO₂)–C(H)(O–NO₂)–C(H)(O–NO₂)–H] ⟶ 6 N≡N + 12 O=C=O + 10 H—O—H + O=O

nitroglycerine

$\Delta H = 20D(C\text{-}H) + 8D(C\text{-}C) + 12D(C\text{-}O) + 24D(O\text{-}N) + 12D(N{=}O)$

$- [6D(N{\equiv}N) + 24D(C{=}O) + 20D(H\text{-}O) + D(O{=}O)]$

$\Delta H = 20(413) + 8(348) + 12(358) + 24(201) + 12(607)$

$- [6(941) + 24(799) + 20(463) + 495]$

$= -7129\text{ kJ}$

$$1.00\text{ g } C_3H_5N_3O_9 \times \frac{1\text{ mol } C_3H_5N_3O_9}{227.1\text{ g } C_3H_5N_3O_9} \times \frac{-7129\text{ kJ}}{4\text{ mol } C_3H_5N_3O_9} = 7.85\text{ kJ/g } C_3H_5N_3O_9$$

(b) $4C_7H_5N_3O_6(s) \rightarrow 6N_2(g) + 7CO_2(g) + 10H_2O(g) + 21C(s)$

8.86

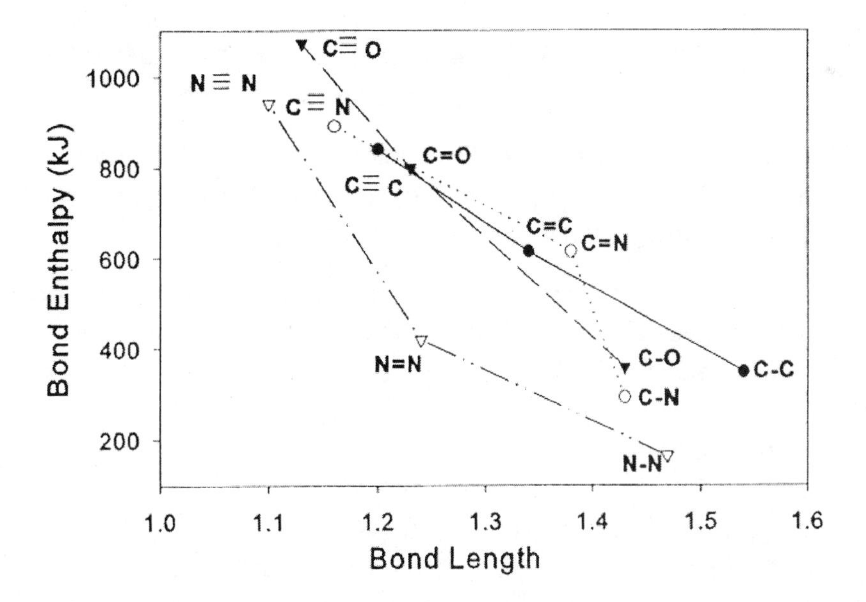

When comparing the same pair of bonded atoms (C–N vs. C=N vs. C≡N), the shorter the bond the greater the bond energy, but the two quantities are not necessarily directly proportional. The plot clearly shows that there are no simple length/strength correlations for single bonds alone, double bonds alone, triple bonds alone, or among different pairs of bonded atoms (all C–C bonds vs. all C–N bonds, etc.).

8.87 (a) S–N ≈ 1.77 Å (sum of the bonding atomic radii from figure 7.5).

(b) S–O ≈ 1.75 Å (the sum of the bonding atomic radii from figure 7.5.) Alternatively, half of the S–S distance in S_8 (1.02) plus half of the O–O distance from Table 8.5 (0.74) is 1.76 Å.

(c) Owing to the resonance structures for SO_2 (see Solution 8.58(c)), we assume that the S–O bond in SO_2 is intermediate between a double and single bond, so the distance of 1.43 Å should be significantly shorter than an S–O single bond distance, 1.75 Å.

(d) 54 e^-, 27 e^- pair

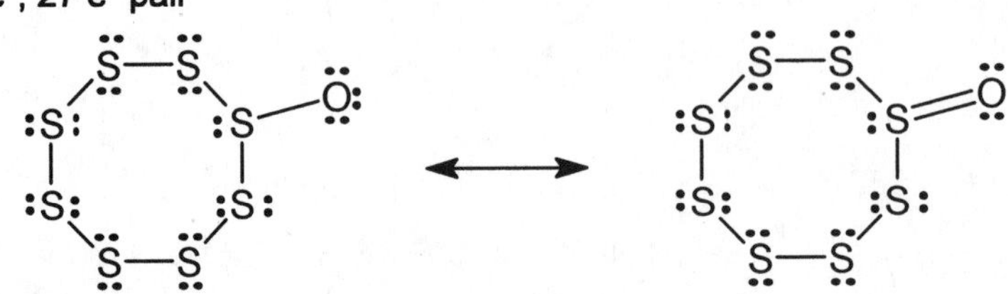

The observed S–O bond distance, 1.48 Å, is similar to that in SO_2, 1.43 Å, which can be described by resonance structures showing both single and double S–O bonds. Thus, S_8O must have resonance structures with both single and double S–O bonds. The structure with the S=O bond has 5 e^- pairs about this S atom. To the extent that this resonance form contributes to the true structure, the S atom bound to O has more than an octet of electrons around it.

Integrative Exercises

8.88 (a) Ti^{2+} : [Ar]$3d^2$; Ca : [Ar]$4s^2$. Yes. The 2 valence electrons in Ti^{2+} and Ca are in different principle quantum levels and different subshells.

(b) According to the Aufbau Principle, valence electrons will occupy the lowest energy empty orbital. Thus, in Ca the 4s is lower in energy than the 3d, while in Ti^{2+}, the 3d is lower in energy than the 4s.

(c) Since there is only one 4s orbital, the 2 valence electrons in Ca are paired. There are 5 degenerate 3d orbitals, so the 2 valence electrons in Ti^{2+} are unpaired. Ca has no unpaired electrons, Ti^{2+} has 2.

8.89 (a)

Sr(s) → Sr(g)	ΔH_f° Sr(g) [ΔH_{sub}° Sr(s)]
Sr(g) → $Sr^+(g) + 1\ e^-$	I_1 Sr
$Sr^+(g) \rightarrow Sr^{2+}(g) + 1\ e^-$	I_2 Sr
$Cl_2(g) \rightarrow 2Cl(g)$	2 ΔH_f° Cl(g) [$D(Cl_2)$]
$2Cl(g) + 2\ e^- \rightarrow 2Cl^-(g)$	$2E_1$ Cl
$SrCl_2(s) \rightarrow Sr(s) + Cl_2(g)$	$-\Delta H_f^\circ$ $SrCl_2$
$SrCl_2(s) \rightarrow Sr^{2+}(g) + 2Cl^-(g)$	ΔH_{latt}

(b) $\Delta H^\circ_f\ SrCl_2(s) = \Delta H^\circ_f\ Sr(g) + I_1(Sr) + I_2(Sr) + 2\Delta H^\circ_f\ Cl(g) + 2E(Cl) - \Delta H_{latt}\ SrCl_2$

$\Delta H^\circ_f\ SrCl_2(s) = 164.4\ kJ + 549\ kJ + 1064\ kJ + 2(121.7)\ kJ + 2(-349)\ kJ - 2127\ kJ$

$= -804\ kJ$

8.90 The pathway to the formation of K_2O can be written:

$2K(s) \rightarrow 2K(g)$	$2\Delta H^\circ_f\ K(g)$
$2K(g) \rightarrow 2K^+(g) + 2\ e^-$	$2\ I_1(K)$
$1/2\ O_2(g) \rightarrow O(g)$	$\Delta H^\circ_f\ O(g)$
$O(g) + 1\ e^- \rightarrow O^-(g)$	$E_1(O)$
$O^-(g) + 1\ e^- \rightarrow O^{2-}(g)$	$E_2(O)$
$2K^+(g) + O^{2-}(g) \rightarrow K_2O(s)$	$-\Delta H_{latt}\ K_2O(s)$
$2K(s) + 1/2\ O_2(g) \rightarrow K_2O(s)$	$\Delta H^\circ_f\ K_2O(s)$

$\Delta H^\circ_f\ K_2O(s) = 2\Delta H^\circ_f\ K(g) + 2\ I_1(K) + \Delta H^\circ_f\ O(g) + E_1(O) + E_2(O) - \Delta H_{latt}\ K_2O(s)$

$E_2(O) = \Delta H^\circ_f\ K_2O(s) + \Delta H_{latt}\ K_2O(s) - 2\Delta H^\circ_f\ K(g) - 2\ I_1(K) - \Delta H^\circ_f\ O(g) - E_1(O)$

$E_2(O) = -363.2\ kJ + 2238\ kJ - 2(89.99)\ kJ - 2(419)\ kJ - 247.5\ kJ - (-141)\ kJ$

$= +750\ kJ$

8.91 (a) Assume 100 g.

$$14.52\ g\ C \times \frac{1\ mol}{12.011\ g\ C} = 1.209\ mol\ C;\ 1.209 / 1.209 = 1$$

$$1.83\ g\ H \times \frac{1\ mol}{1.008\ g\ H} = 1.816\ mol\ H;\ 1.816 / 1.209 = 1.5$$

$$64.30\ g\ Cl \times \frac{1\ mol}{35.453\ g\ Cl} = 1.814\ mol\ Cl;\ 1.814 / 1.209 = 1.5$$

$$19.35\ g\ O \times \frac{1\ mol}{15.9994\ g\ O} = 1.209\ mol\ O;\ 1.209 / 1.209 = 1.0$$

Multiplying by 2 to obtain an integer ratio, the empirical formula is $C_2H_3Cl_3O_2$.

(b) The empirical formula weight is 2(12.0) + 3(1.0) + 3(35.5) + 2(16) = 165.5. The empirical formula is the molecular formula.

(c) 44 e^-, 22 e^- pairs

```
   :Cl:  :O—H
    |     |
:Cl—C—————C—O—H
    |     |
   :Cl:   H
```

8.92 (a) C_2H_2: 10 e^-, 5 e^- pair

H—C≡C—H

N_2: 10 e^-, 5 e^- pair

:N≡N:

(b) N_2 is an extremely stable, unreactive compound. Under appropriate conditions, it can be either oxidized (Section 22.7) or reduced (Sections 14.6 and 15.1). C_2H_2 is a reactive gas, used in combination with O_2 for welding and as starting material for organic synthesis (Section 25.4).

(c) $2N_2(g) + 5O_2(g) \rightarrow 2N_2O_5(g)$

$2C_2H_2(g) + 5O_2(g) \rightarrow 4CO_2(g) + 2H_2O(g)$

(d) $\Delta H^\circ_{rxn}\ (N_2) = 2\Delta H^\circ_f\ N_2O_5(g) - 2\Delta H^\circ_f\ N_2(g) - 5\Delta H^\circ_f\ O_2(g)$

$= 2(11.30) - 2(0) - 5(0) = 22.60$ kJ

$\Delta H^\circ_{ox} = 11.30$ kJ/mol N_2

$\Delta H^\circ_{rxn}\ (C_2H_2) = 4\Delta H^\circ_f\ CO_2(g) + 2\Delta H^\circ_f\ H_2O(g) - 2\Delta H^\circ_f\ C_2H_2(g) - 5\Delta H^\circ_f\ O_2(g)$

$= 4(-393.5 \text{ kJ}) + 2(-241.82 \text{ kJ}) - 2(226.7 \text{ kJ}) - 5(0)$

$= -2511.0$ kJ

$\Delta H^\circ_{ox}\ (C_2H_2) = -1255.5$ kJ/mol C_2H_2

The oxidation of C_2H_2 is highly exothermic, which means that the energy state of the combined products is much lower than that of the reactants. The reaction is "down hill" in an energy sense, and occurs readily. The oxidation of N_2 is mildly endothermic (energy of products higher than reactants) and the reaction does not readily occur. This is in agreement with the general reactivities from part (b).

(e) Referring to bond enthalpies in Table 8.4, when the C–H bonds are taken into account, even more energy is required for bond breaking in the oxidation of C_2H_2 than in the oxidation of N_2. The difference seems to be in the enthalpies of formation of the products. $CO_2(g)$ and $H_2O(g)$ have extremely exothermic ΔH°_f values, which cause the oxidation of C_2H_2 to be energetically favorable. $N_2O_5(g)$ has an endothermic ΔH°_f values, which causes the oxidation of N_2 to be energetically unfavorable.

8.93 (a) Assume 100 g.

$$62.04 \text{ g Ba} \times \frac{1 \text{ mol}}{137.33 \text{ g Ba}} = 0.4518 \text{ mol Ba};\ 0.4518 / 0.4518 = 1.0$$

$$37.96 \text{ g N} \times \frac{1 \text{ mol}}{14.007 \text{ g N}} = 2.710 \text{ mol N};\ 2.710 / 0.4518 = 6.0$$

The empirical formula is BaN_6. Ba has an ionic charge of 2+, so there must be two 1- azide ions to balance the charge. The formula of each azide ion is N_3^-.

(b) 16 e^-, 8 e^- pairs

$$[:\ddot{N}=N=\ddot{N}:]^- \longleftrightarrow [:N\equiv N-\ddot{\underset{..}{N}}:]^- \longleftrightarrow [:\ddot{\underset{..}{N}}-N\equiv N:]^-$$

-1 +1 -1   0 +1 -2   -2 +1 0

(c) The left structure minimizes formal charges and is probably the main contributor.

(d) The two N–N bond lengths will be equal. The two minor contributors would individually cause unequal N–N distances, but collectively they contribute equally to the lengthening and shortening of each bond. The N–N distance will be approximately 1.24 Å, the average N=N distance.

8.94 (a) Assume 100 g of compound

$$69.6 \text{ g S} \times \frac{1 \text{ mol S}}{32.07 \text{ g}} = 2.17 \text{ mol S}$$

$$30.4 \text{ g N} \times \frac{1 \text{ mol N}}{14.01 \text{ g}} = 2.17 \text{ mol N}$$

S and N are present in a 1:1 mol ratio, so the empirical formula is SN. The empirical formula weight is 46. $\mathcal{M}$/FW = 184.3/46 = 4 The molecular formula is S_4N_4.

(b) 44 e^-, 22 e^- pairs. Because of its small radius, N is unlikely to have an expanded octet. Begin with alternating S and N atoms in the ring. Try to satisfy the octet rule with single bonds and lone pairs. At least two double bonds somewhere in the ring are required.

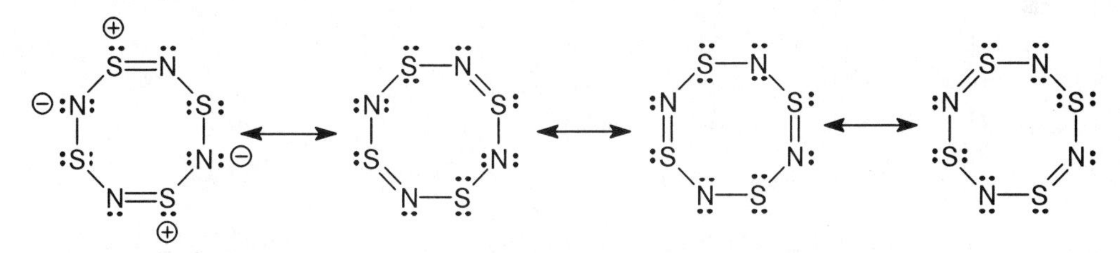

These structures carry formal charges on S and N atoms as shown. Other possibilities include:

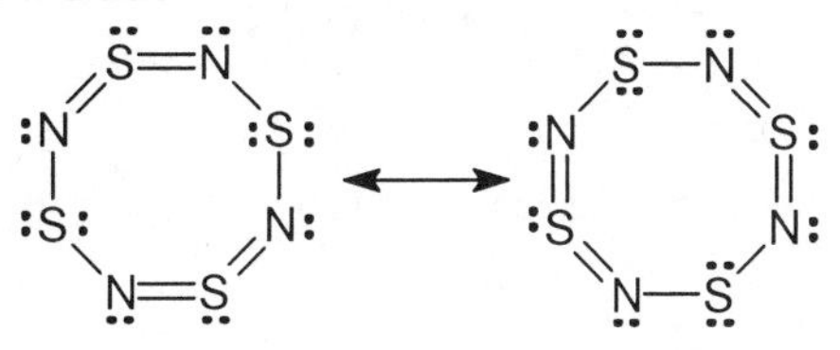

These structures have zero formal charges on all atoms and are likely to contribute to the true structure. Note that the S atoms that are shown with two double bonds are not necessarily linear, because S has an expanded octet. Other resonance structures with four double bonds are.

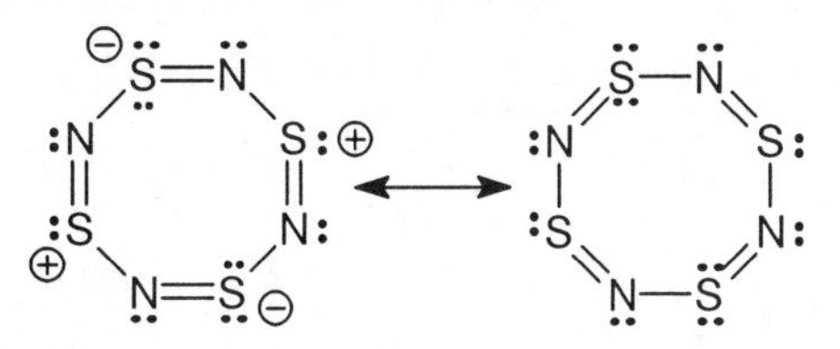

In either resonance structure, the two 'extra' electron pairs can be placed on any pair of S atoms in ring, leading to a total of 10 resonance structures. The sulfur atoms alternately carry formal charges of +1 and -1. Without further structural information, it is not possible to eliminate any of the above structures. Clearly, the S_4N_4 molecule stretches the limits of the Lewis model of chemical bonding.

(c) Each resonance structure has 8 total bonds and more than 8 but less than 16 bonding e^- pairs, so an "average" bond will be intermediate between a S–N single and double bond. We estimate an average S–N single bond length to be 1.77 Å (sum of bonding atomic radii from figure 7.5). We do not have a direct value for a S–N double bond length. Comparing double and single bond lengths for C–C (1.34 Å, 1.54 Å), N–N (1.24 Å, 1.47 Å) and O–O (1.21 Å, 1.48 Å) bonds from Table 8.5, we see that, on average, a double bond is approximately 0.23 Å shorter than a single bond. Applying this difference to the S–N single bond length, we estimate the S–N double bond length as 1.54 Å. Finally, the intermediate S–N bond length in S_4N_4 should be between these two values, approximately 1.60-165 Å. [The measured bond length is 1.62 Å.]

(d) $S_4N_4 \rightarrow 4S(g) + 4N(g)$

$\Delta H = 4\Delta H^\circ_f\ S(g) + 4\Delta H^\circ_f\ N(g) - \Delta H^\circ_f\ S_4N_4$

$\Delta H = 4(222.8\ kJ) + 4(472.7\ kJ) - 480\ kJ = 2302\ kJ$

This energy, 2302 kJ, represents the dissociation of 8 S–N bonds in the molecule; the average dissociation energy of one S-N bond in S_4N_4 is then 2302 kJ/8 bonds = 287.8 kJ.

8.95 (a)

Reaction		
$HF(g) \rightarrow H(g) + F(g)$	D (H-F)	567 kJ
$H(g) \rightarrow H^+(g) + 1\ e^-$	I (H)	1312 kJ
$F(g) + 1\ e^- \rightarrow F^-(g)$	E (F)	-328 kJ
$HF(g) \rightarrow H^+(g) + F^-(g)$	ΔH	1551 kJ

(b) $\Delta H = D(H\text{-}Cl) + I(H) + E(Cl)$

$\Delta H = 431\ kJ + 1312\ kJ + (-349)\ kJ = 1394\ kJ$

(c) $\Delta H = D(H\text{-}Br) + I(H) + E(Br)$

$\Delta H = 366\ kJ + 1312\ kJ + (-325)\ kJ = 1353\ kJ$

8.96 (a) $C_6H_6(g) \rightarrow 6H(g) + 6C(g)$

$\Delta H^\circ = 6\Delta H^\circ_f\ H(g) + 6\Delta H^\circ_f\ C(g) - \Delta H^\circ_f\ C_6H_6(g)$

$\Delta H^\circ = 6(217.94)\ kJ + 6(718.4)\ kJ - 82.9\ kJ = 5535\ kJ$

(b) $C_6H_6(g) \rightarrow 6CH(g)$

(c)

Reaction		
$C_6H_6(g) \rightarrow 6H(g) + 6C(g)$	ΔH°	5535 kJ
$6H(g) + 6C(g) \rightarrow 6CH(g)$	-6D(C-H)	-6(413) kJ
$C_6H_6(g) \rightarrow 6CH(g)$		3057 kJ

3057 kJ is the energy required to break the 6 C–C bonds in $C_6H_6(g)$. The average bond dissociation energy for one carbon-carbon bond in $C_6H_6(g)$ is $\frac{3057\ kJ}{6\ C\text{–}C\ bonds}$ = 509.5 kJ.

(d) The value of 509.5 kJ is between the average value for a C–C single bond (348 kJ) and a C=C double bond (614 kJ). It is somewhat greater than the average of these two values, indicating that the carbon-carbon bond in benzene is a bit stronger than we might expect.

8.97 (a) $Br_2(l) \rightarrow 2Br(g)$ $\Delta H^\circ = 2\Delta H^\circ_f\ Br(g) = 2(111.8)\ kJ = 223.6\ kJ$

(b) $CCl_4(l) \rightarrow C(g) + 4Cl(g)$

$\Delta H^\circ = \Delta H^\circ_f\ C(g) + 4\Delta H^\circ_f\ Cl(g) - \Delta H^\circ_f\ CCl_4(l)$

$= 718.4\ kJ + 4(121.7)\ kJ - (-139.3)\ kJ = 1344.5$

$$\frac{1344.5\ kJ}{4\ C\text{-}Cl\ bonds} = 336.1\ kJ$$

(c) $H_2O_2(l) \rightarrow 2H(g) + 2O(g)$

$2H(g) + 2O(g) \rightarrow 2OH(g)$

$H_2O_2(l) \rightarrow 2OH(g)$

$D(O\text{-}O)(l) = 2\Delta H^\circ_f\ H(g) + 2\Delta H^\circ_f\ O(g) - \Delta H^\circ_f\ H_2O_2(l) - 2D(O\text{-}H)(g)$

$= 2(217.94)\ kJ + 2(247.5)\ kJ - (-187.8)\ kJ - 2(463)\ kJ$

$= 193\ kJ$

(d) The data are listed below.

bond	**D gas kJ/mol**	**D liquid kJ/mol**
Br–Br	193	223.6
C–Cl	328	336.1
O–O	146	192.7

Breaking bonds in the liquid requires more energy than breaking bonds in the gas phase. For simple molecules, bond dissociation from the liquid phase can be thought of in two steps:

molecule (l) → molecule (g)

molecule (g) → atoms (g)

The first step is evaporation or vaporization of the liquid and the second is bond dissociation in the gas phase. Average bond enthalpy in the liquid phase is then the sum of the enthalpy of vaporization for the molecule and the gas phase bond dissociation enthalpies, divided by the number of bonds dissociated. This is greater than the gas phase bond dissociation enthalpy owing to the contribution from the enthalpy of vaporization.

8.98 (a) Assume 100 g.

A: 87.7 g In / 114.82 = 0.764 mol In; 0.764 / 0.384 ≈ 2
12.3 g S / 32.07 = 0.384 mol S; 0.384 / 0.384 = 1

B: 78.2 g In / 114.82 = 0.681 mol In; 0.681 / 0.680 ≈ 1
21.8 g S / 32.07 = 0.680 mol S; 0.680 / 0.680 = 1

C: 70.5 g In / 114.82 = 0.614 mol In; 0.614 / 0.614 = 1
29.5 g S / 32.07 = 0.920 mol S; 0.920 / 0.614 = 1.5

A: In_2S; B: InS; C: In_2S_3

(b) A: In(I); B: In(II); C: In(III)

(c) In(I) : $[Kr]5s^24d^{10}$; In(II) : $[Kr]5s^14d^{10}$; In(III) : $[Kr]4d^{10}$
None of these are noble-gas configurations.

(d) The ionic radius of In^{3+} in compound C will be smallest. Removing successive electrons from an atom reduces electron repulsion, increases the effective nuclear charge experienced by the valence electrons and decreases the ionic radius. The higher the charge on a cation, the smaller the radius.

(e) Lattice energy is directly related to the charge on the ions and inversely related to the interionic distance. Only the charge and size of the In varies in the three compounds. In(I) in compound A has the smallest charge and the largest ionic radius, so compound A has the smallest lattice energy and the lowest melting point. In(III) in compound C has the greatest charge and the smallest ionic radius, so compound C has the largest lattice energy and highest melting point.

9 Molecular Geometry and Bonding Theories

Molecular Shapes; the VSEPR Model

9.1 Yes. This description means that the three terminal atoms point toward the corners of an equilateral triangle and the central atom is in the plane of this triangle. Only 120° bond angles are possible in this arrangement.

9.2 In a symmetrical tetrahedron, the four bond angles are equal to each other, with values of 109.5°. The H–C–H angles in CH_4 and the O–Cl–O angles in ClO_4^- will have values close to 109.5°.

9.3 (a) An *electron domain* is a region in a molecule where electrons are most likely to be found.

(b) Each balloon in Figure 9.5 occupies a volume of space. The best arrangement is one where each balloon has its "own" space, where they are as far apart as possible and repulsions are minimized. Electron domains are negatively charged regions, so they also adopt an arrangement where repulsions are minimized.

9.4 (a) The number of electron domains in a molecule or ion is the number of bonds (double and triple bonds count as one domain) **plus** the number of nonbonding (lone) electron pairs.

(b) A *bonding electron domain* is a region between two bonded atoms that contains one or more pairs of bonding electrons. A *nonbonding electron domain* is localized on a single atom and contains one pair of nonbonding electrons (a lone pair).

9.5 *Analyze/Plan.* See Table 9.1. *Solve*:

(a) trigonal planar (b) tetrahedral (c) trigonal bipyramidal (d) octahedral

9.6 (a) 3 (or 5 if 90° angles are also present)
(b) 2 (or 5 if 120° angles are also present, 6 if more than one 180° angle is present)
(c) 4
(d) 6 (or 5 if 120° angles are also present)

9.7 The electron-domain geometry indicated by VSEPR describes the arrangement of all bonding and nonbonding electron domains. The molecular geometry describes just the atomic positions. NH_3 has the Lewis structure given below; there are four electron domains around nitrogen so the electron-domain geometry is tetrahedral, but the molecular geometry of the four atoms is trigonal pyramidal.

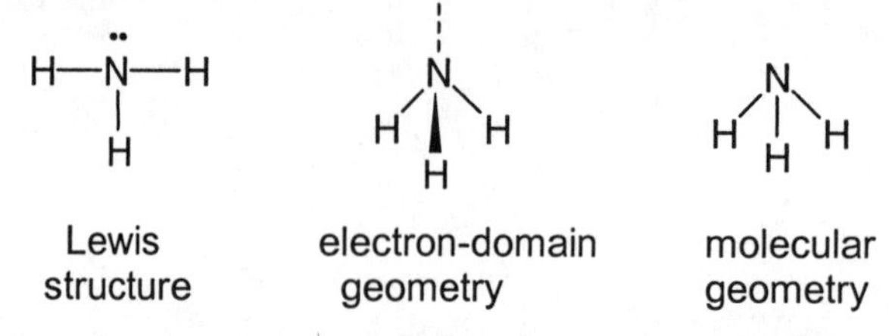

Lewis structure — electron-domain geometry — molecular geometry

9.8 If the electron-domain geometry is trigonal bipyramidal, there are 5 total electron domains around the central atom. An AB_3 molecule has 3 bonding domains, so there must be 2 nonbonding domains on A.

9.9 *Analyze/Plan*. See Table 9.3. *Solve*:

Lewis structure — electron-domain geometry — molecular geometry

(a)

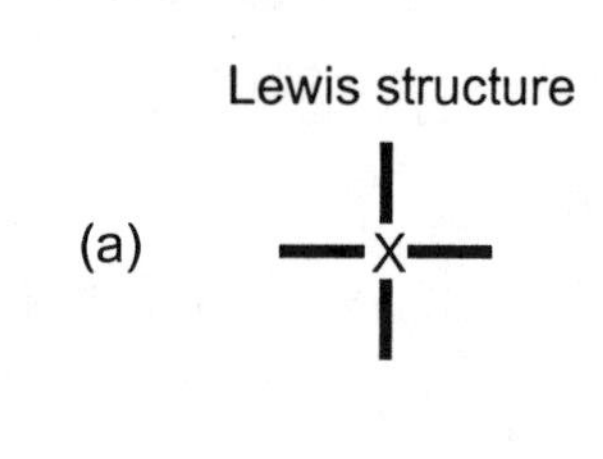

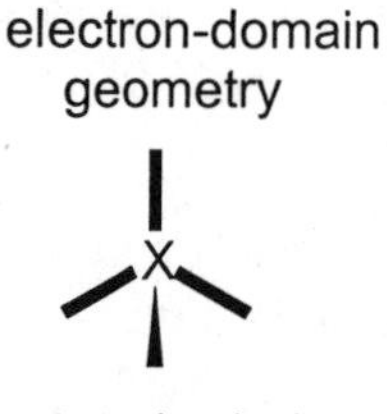

tetrahedral — tetrahedral

(b)

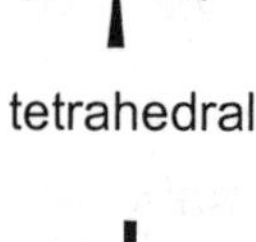

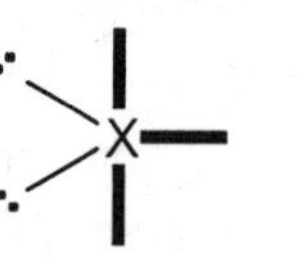

trigonal bipyramidal — T-shaped

(c)

octahedral — square pyramidal

Lewis structure — electron-domain geometry — molecular geometry

9.10 (a)

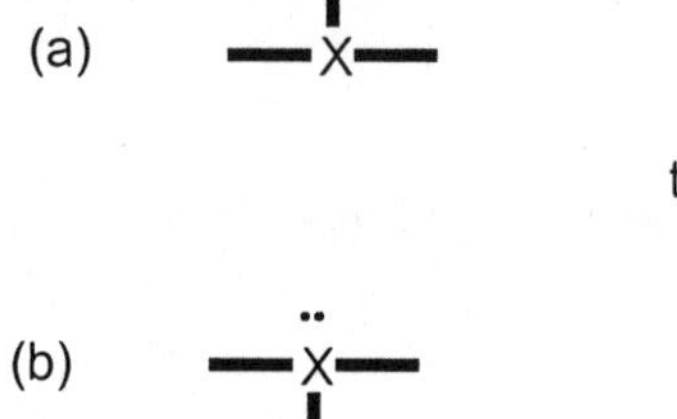

trigonal planar — trigonal planar

(b)

tetrahedral — trigonal pyramidal

9.10 (continued)

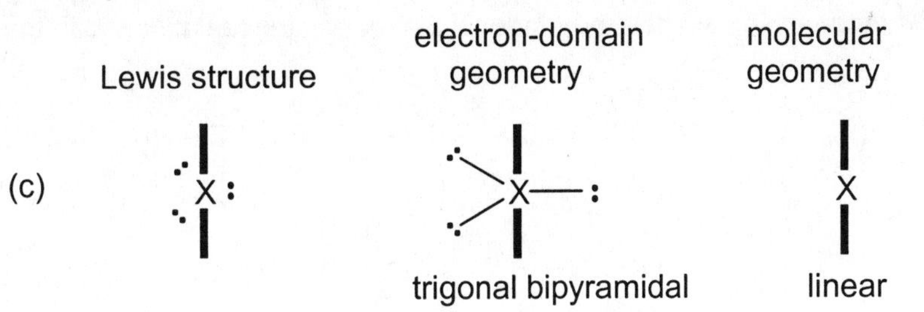

9.11 *Analyze/Plan*. Follow the logic in Sample Exercise 9.1. *Solve:*
bent (b), linear (l), octahedral (oh), seesaw (ss) square pyramidal (sp), square planar (spl), tetrahedral (td), trigonal bipyramidal (tbp), trigonal planar (tr), trigonal pyramidal (tp), T-shaped (T)

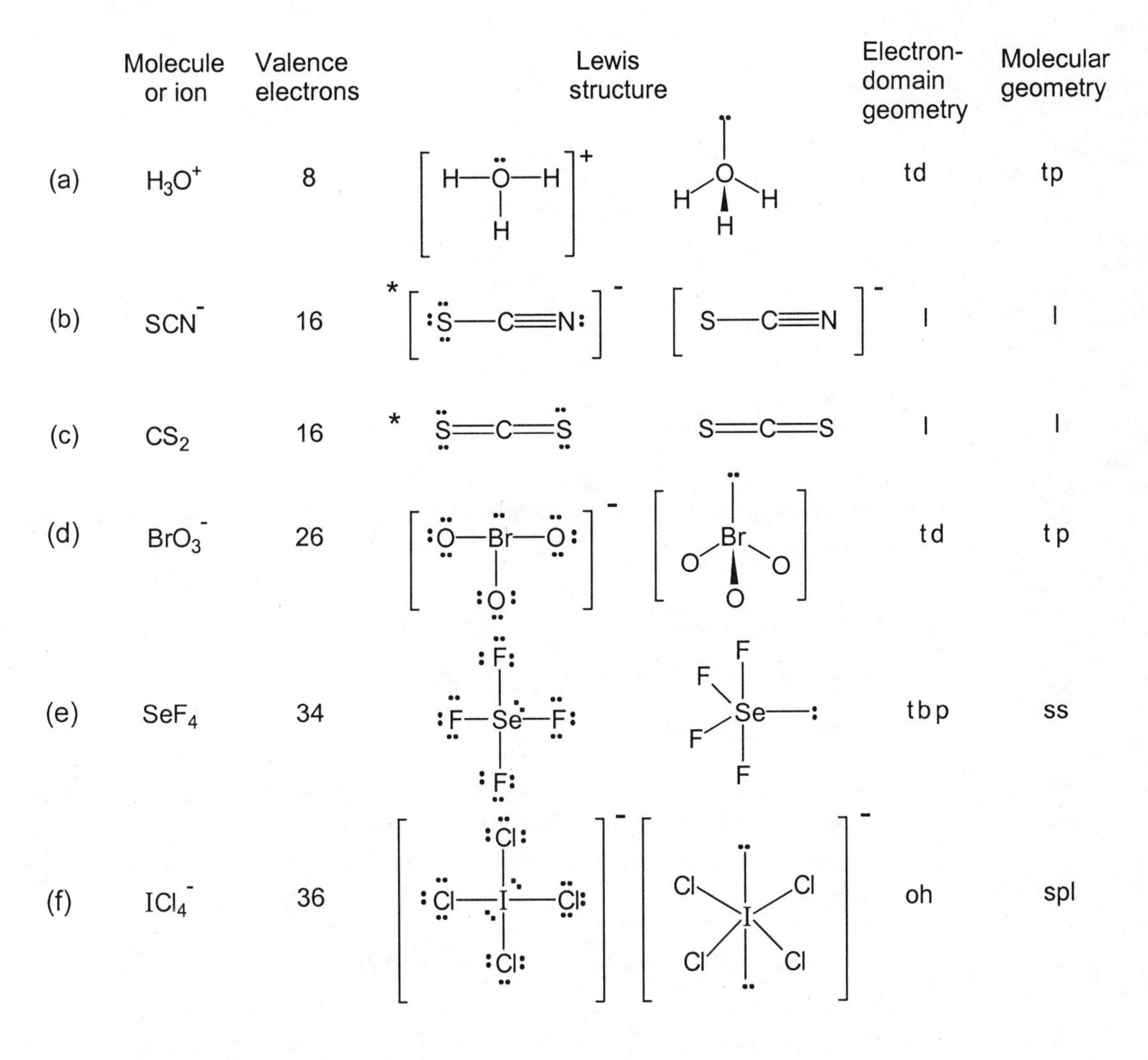

	Molecule or ion	Valence electrons	Lewis structure	Electron-domain geometry	Molecular geometry
(a)	H_3O^+	8		td	tp
(b)	SCN^-	16	*	l	l
(c)	CS_2	16	*	l	l
(d)	BrO_3^-	26		td	tp
(e)	SeF_4	34		tbp	ss
(f)	ICl_4^-	36		oh	spl

* More than 1 resonance structure is possible. All equivalent resonance structures predict the same molecular geometry.

9.12 bent (b), linear (l), octahedral (oh), seesaw (ss) square pyramidal (sp), square planar (spl), tetrahedral (td), trigonal bipyramidal (tbp), trigonal planar (tr), trigonal pyramidal (tp), T-shaped (T)

	Molecule or ion	Valence electrons	Lewis structure		Electron-domain geometry	Molecular geometry
(a)	N_2O	16	*:N≡N—Ö:	N≡N—O	l	l
(b)	SO_3	24	:Ö=S—Ö: / :Ö:	O=S(O)O	tr	tr
(c)	PCl_3	26	:Cl—P—Cl: / :Cl:	P, Cl, Cl, Cl	td	tp
(d)	NH_2Cl	14	H—N—Cl: / H	N, H, Cl, H	td	tp
(e)	BrF_5	42	:F, F:, Br, :F, F:, :F:	F, F, Br, F, F, F	oh	sp
(f)	KrF_2	22	:F—Kr—F:	F, Kr, F	tbp	l

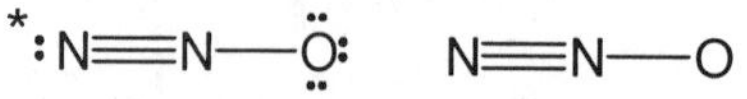

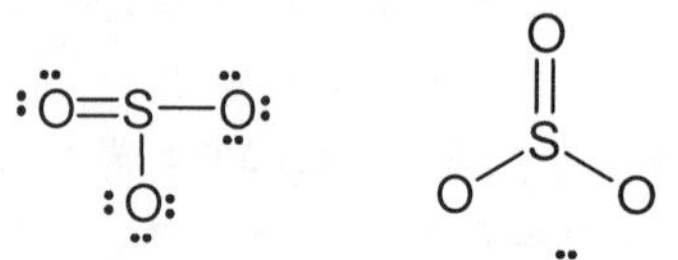

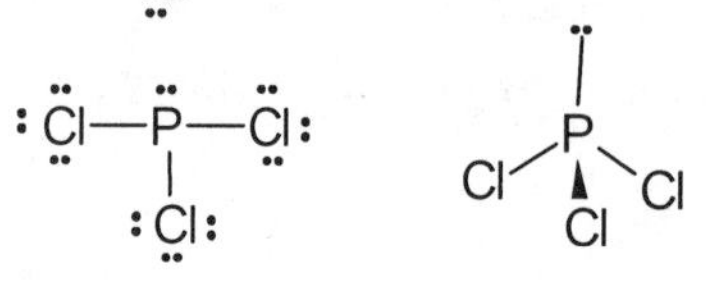

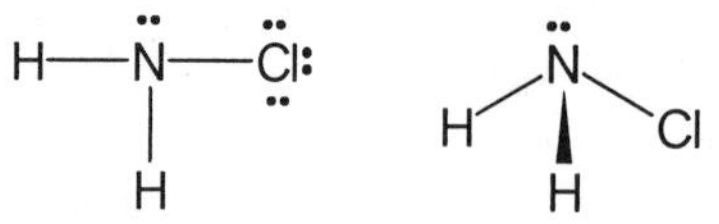

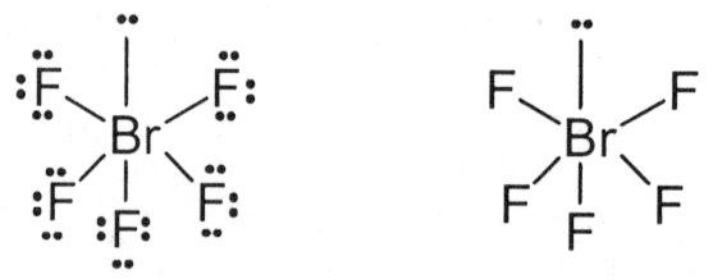

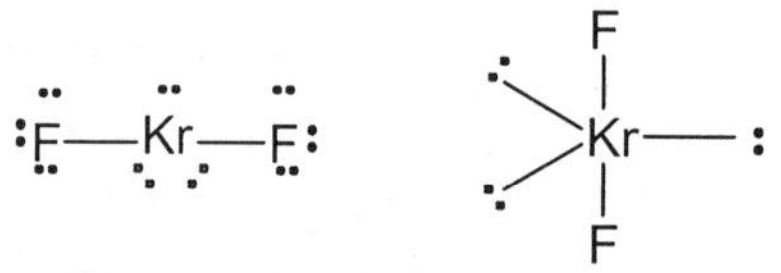

* More than 1 resonance structure is possible. All equivalent resonance structures predict the same molecular geometry.

9.13 *Analyze/Plan*. Work backwards from molecular geometry, using Tables 9.2 and 9.3. *Solve*:

(a) electron-domain geometries: i, trigonal planar; ii, tetrahedral; iii, trigonal bipyramidal

(b) nonbonding electron domains: i, 0; ii, 1; iii, 2

(c) N and P. Shape ii has 3 bonding and 1 nonbonding electron domains. Li and Al would form ionic compounds with F, so there would be no nonbonding electron domains. Assuming that F always has 3 nonbonding domains, BF_3 and ClF_3 would have the wrong number of nonbonding domains to produce shape ii.

(d) Cl (also Br and I, since they have 7 valence electrons). This T-shaped molecular geometry arises from a trigonal bipyramidal electron-domain geometry with 2 nonbonding domains (Table 9.3). Assuming each F atom has 3 nonbonding domains and forms only single bonds with A, A must have 7 valence electrons to produce these electron-domain and molecular geometries. It must be in or below the third row of the periodic table, so that it can accommodate more than 4 electron domains.

9.14 (a) electron-domain geometries: i, octahedral; ii, tedrahedral; iii, trigonal bipyramidal

(b) nonbonding electron domains: i, 2; ii, 0; iii, 1

(c) S or Se. Shape iii has 5 electron domains, so A must be in or below the third row of the periodic table. This eliminates Be and C. Assuming each F atom has 3 nonbonding electron domains and forms only single bonds with A, A must have 6 valence electrons to produce these electron-domain and molecular geometries.

(d) Xe. (See Table 9.3) Assuming F behaves typically, A must be in or below the third row and have 8 valence electrons. Only Xe fits this description. (Noble gas elements above Xe have not been shown to form molecules of the type AF_4. See Section 7.8.)

9.15 *Analyze/Plan.* Follow the logic in Sample Exercise 9.3. *Solve:*

(a) 1 – 109°, 2 – 109° (b) 3 – 109°, 4 – 109°

(c) 5 – 180° (d) 6 – 120°, 7 – 109°, 8 – 109°

9.16 (a) 1 – 109°, 2 – 120° (b) 3 – 109°, 4 – 120°

(c) 5 – 109°, 6 – 109° (d) 7 – 180°, 8 – 109°

9.17 *Analyze.* Given: molecular formulas. Find: explain features of molecular geometries.

Plan. Draw the correct Lewis structures for the molecules and use VSEPR to predict and explain observed molecular geometry. *Solve:*

(a) BrF_4^- 36 e^-, 18 e^- pr

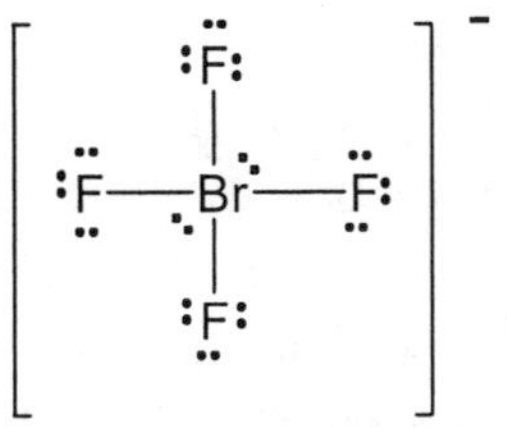

```
      ..         -
    [ :F:         ]
  ..  |  ..  ..
[ :F—Br—F: ]
  ..  |  ..  ..
      :F:
      ..
```

6 e^- pairs around Br
octahedral e^- domain geometry
square planar molecular geometry

BF_4^- 32 e^-, 16 e^- pr

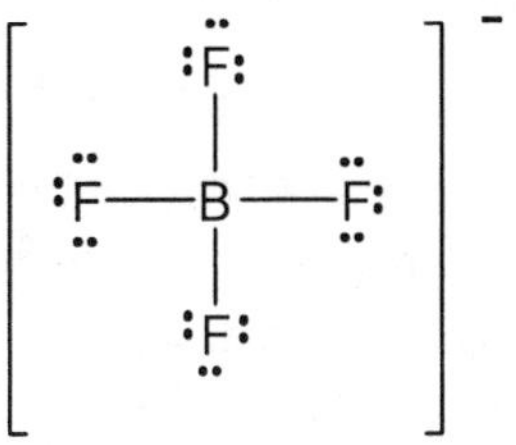

```
      ..         -
    [ :F:         ]
  ..  |     ..
[ :F—B—F: ]
  ..  |     ..
      :F:
      ..
```

4 e^- pairs around B
tetrahedral e^- domain geometry
tetrahedral molecular geometry

The fundamental feature that determines molecular geometry is the number of electron domains around the central atom, and the number of these that are bonding domains. Although BrF_4^- and BF_4^- are both of the form AX_4^-, the central atoms and thus the number of valence electrons in the two ions are different. This leads to different numbers of e^- domains about the two central atoms. Even though both ions have four bonding electron domains, the six total domains around Br require octahedral domain geometry and square planar molecular geometry, while the four total domains about B lead to tetrahedral domain and molecular geometry.

(b) CF_4 32 e^-, 16 e^- pr

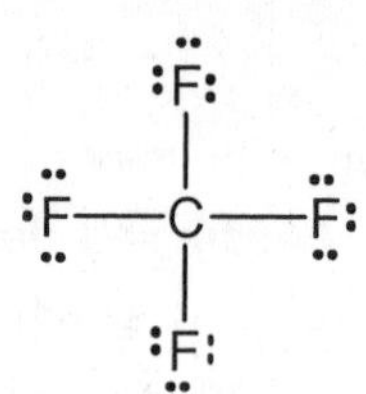

4 e^- domains around C
tetrahedral e^- domain geometry
tetrahedral molecular geometry

SF_4 34 e^-, 17 e^- pr

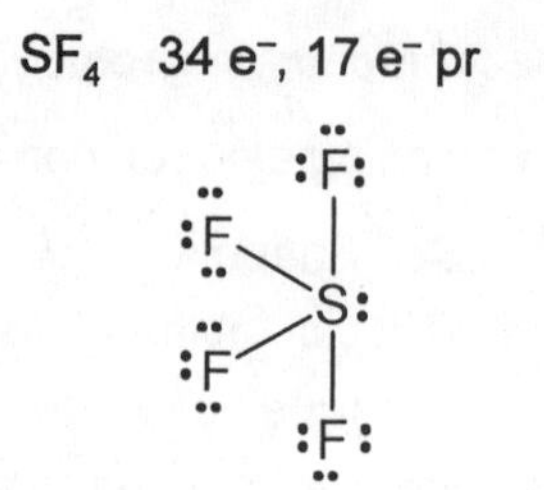

5 e^- domains around S
trigonal bipyramidal e^- domain geometry
see-saw molecular geometry

CF_4 will have bond angles closest to the value predicted by VSEPR, because there are no nonbonding e^- domains around C. The four bonding domains in CF_4 are equivalent and lead to the balance of repulsions implicit in VSEPR theory. In SF_4, one of the e^- domains is nonbonding. A nonbonding domain is surely not equivalent to a bonding domain; we expect it to be more diffuse. That is, nonbonding domains will occupy more space, "push back" the bonding domains, and lead to bond angles that are nonideal.

9.18 (a) ClO_2^- 20 e^-, 10 e^- pr

[:O—Cl—O:]⁻

4 e^- domains around Cl
tetrahedral e^- domain geometry
bent molecular geometry
bond angle $\lesssim$ 109.5°

NO_2^- 18 e^-, 9 e^- pr

[O═N—O:]⁻ ⟷ [:O—N═O]⁻

3 e^- domains about N
(both resonance structures)
trigonal planar e^- domain geometry
bent molecular geometry
bond angle $\lesssim$ 120°

Both molecular geometries are described as 'bent" because both molecules have two nonlinear bonding electron domains. The bond angles (the angle between the two bonding domains) in the two ions are different because the total number of electron domains, and thus the electron domain geometries are different.

(b) XeF_2 22 e^-, 11 e^- pr

:F—Xe—F:

5 e^- domains around Xe
trigonal bipyramidal e^- domain geometry
linear molecular geometry

The question here really is: why do the three nonbonding domains all occupy the equatorial plane of the trigonal bipyramid? In a tbp, there are several different kinds of repulsions, bonding domain-bonding domain (bd-bd), bonding domain-nonbonding domain (bd-nd) and nonbonding domain-nonbonding domain (nd-nd). Each of these can have 90°, 120° or 180° geometry. Since nonbonding domains occupy more space, 90° nd-nd repulsions are most significant and least desirable. The various electron domains arrange themselves to minimize these 90° nd-nd interactions. The arrangement shown above has no 90° nd-nd repulsions. An arrangement with one or two nonbonding domains in axial positions would lead to at least two 90° nd-nd repulsions, a less stable situation. [To convince yourself, tabulate the number and kinds of repulsions for each possible tbp arrangement of 2bd's and 3nd's.]

9.19 *Analyze/Plan.* Given the formula of each molecule or ion, draw the correct Lewis structure and use principles of VSEPR to answer the question. *Solve*:

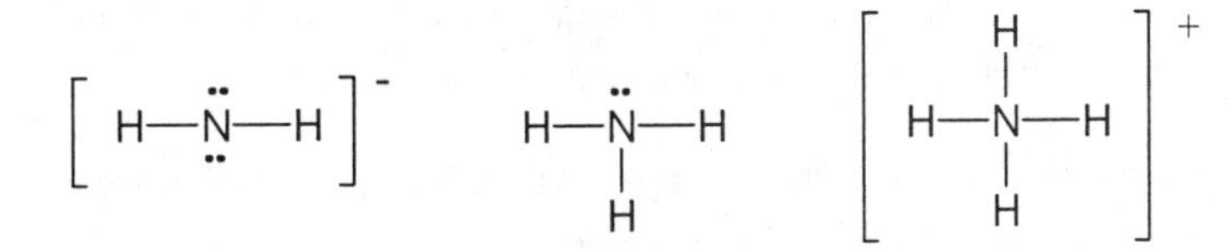

Each species has 4 electron domains around the N atom, but the number of nonbonding domains decreases from 2 to 0, going from NH_2^- to NH_4^+. Since nonbonding domains occupy more space than bonding domains, the bond angles expand as the number of nonbonding domains decreases.

9.20

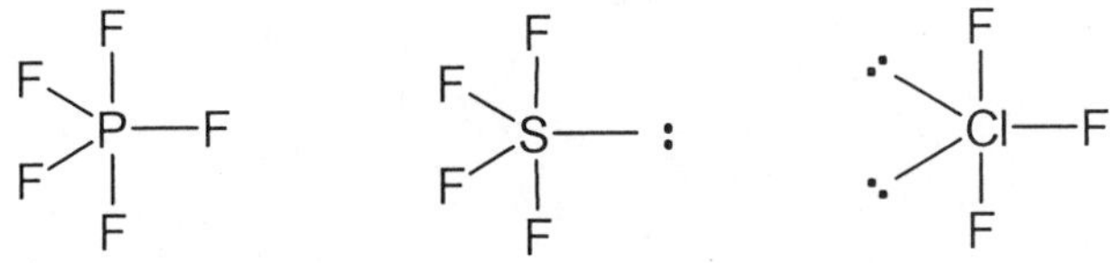

The three nonbonded electron pairs on each F atom have been omitted for clarity.

The F (axial) –A –F (equatorial) angle is largest in PF_5 and smallest in ClF_3. As the number of nonbonding domains in the equatorial plane increases, they push back the axial A–F bonds, decreasing the F (axial) –A –F (equatorial) bond angles.

Polarity of Polyatomic Molecules

9.21 See Sample Exercise 9.4(b) for the correct resonance structures and analysis of S–O bond dipoles. According to the electron density model, the net dipole moment vector points along the O–S–O angle bisector with the negative end pointing away from S. the magnitude of this vector is 1.63 D.

9.22 If H_2O were linear, H–O–H, the two O–H bond dipoles would cancel, and the net dipole moment would be zero. Since H_2O is polar, the O–H and dipoles cannot be directly opposed to each other; the molecule cannot be linear.

9.23 (a) In Exercise 9.13, molecules ii and iii will have nonzero dipole moments. Molecule i has no nonbonding electron pairs on A, and the 3 A-F dipoles are oriented so that the sum of their vectors is zero (the bond dipoles cancel). Molecules ii and iii have nonbonding electron pairs on A and their bond dipoles do not cancel. A nonbonding electron pair (or pairs) on a central atom guarantees at least a small molecular dipole moment, because no bond dipole exactly cancels a nonbonding pair.

(b) AF_4 molecules will have a zero dipole moment if there are no nonbonding electron pairs on the central atom and the 4 A-F bond dipoles are arranged (symmetrically) so that they cancel. Therefore, in Exercise 9.14, molecules i and ii have zero dipole moments and are nonpolar.

9.24 (a) For a molecule with polar bonds to be nonpolar, the polar bonds must be (symmetrically) arranged so that the bond dipoles cancel. In most cases, nonbonding e^- domains must be absent from the central atom. Square planar structures may not meet the second condition.

(b) AB_2: linear e^- domain geometry (edg), linear molecular geometry (mg), trigonal bipyramidal edg, linear mg

AB_3: trigonal planar edg, trigonal planar mg

AB_4: tetrahedral edg, tetrahedral mg; octahedral edg, square planar mg

9.25 *Analyze/Plan.* Given molecular formulas, draw correct Lewis structures, determine molecular structure and polarity. *Solve.*

Nonpolar, in a symmetrical trigonal planar molecule, the bond dipoles cancel.

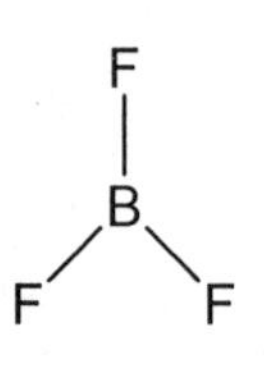

Polar, $\Delta EN > 0$

$C \equiv O$

Nonpolar, in a symmetrical tetrahedral structure (Figure 9.1) the bond dipoles cancel.

Polar, there is an unequal charge distribution due to the nonbonded electron pair on N.

Polar, there is an unequal charge distribution because of the nonbonded electron pairs on S, and the S-F bond dipoles do not cancel.

9.26 (a) **Polar**, ΔEN > 0

I-F

(b) **Nonpolar**, the molecule is linear and the bond dipoles cancel.

S=C=S

(c) **Nonpolar,** in a symmetrical trigonal planar structure [Exercise (9.12 (b)], the bond dipoles cancel.

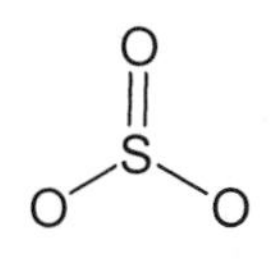

(d) **Polar**, although the bond dipoles are essentially zero, there is an unequal charge distribution due to the nonbonded electron pair on P.

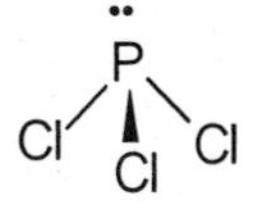

(e) **Nonpolar**, symmetrical octahedron

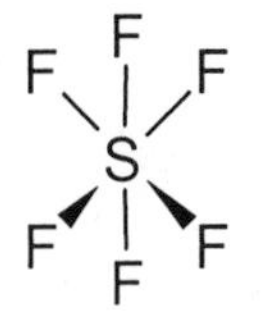

(f) **Polar**, square pyramidal molecular geometry, bond dipoles do not cancel

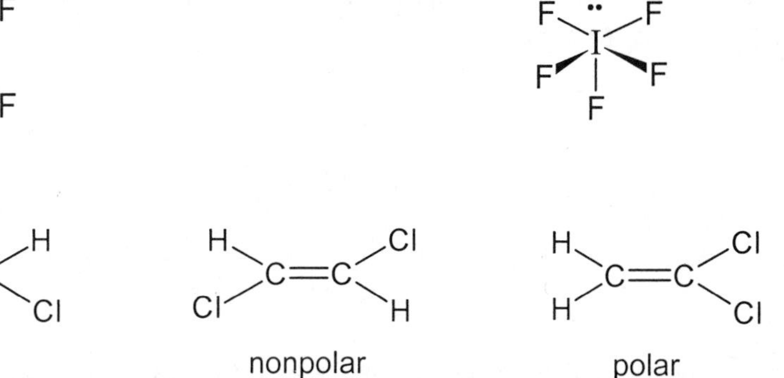

9.27

polar nonpolar polar

All three isomers are planar. The molecules on the left and right are polar because the C–Cl bond dipoles do not point in opposite directions. In the middle isomer, the C–Cl bonds and dipoles are pointing in opposite directions (as are the C–H bonds), the molecule is nonpolar and has a measured dipole moment of zero.

9.28 Each C–Cl bond is polar. The question is whether the vector sum of the C–Cl bond dipoles in each molecule will be nonzero. In the *ortho* and *meta* isomers, the C–Cl vectors are at 60° and 120° angles, respectively, and their resultant dipole moments are nonzero. In the *para* isomer, the C–Cl vectors are opposite, at an angle of 180°, with a resultant dipole moment of zero. The *ortho* and *meta* isomers are polar, the *para* isomer is nonpolar.

Orbital Overlap; Hybrid Orbitals

9.29 (a) *Orbital overlap* occurs when a valence atomic orbital on one atom shares the same region of space with a valence atomic orbital on an adjacent atom.

(b) In valence bond theory, overlap of orbitals allows the two electrons in a chemical bond to mutually occupy the space between the bonded nuclei.

(c) Valence bond theory is a combination of the atomic orbital concept with the Lewis model of electron pair bonding.

9.30 (a) (b) (c)

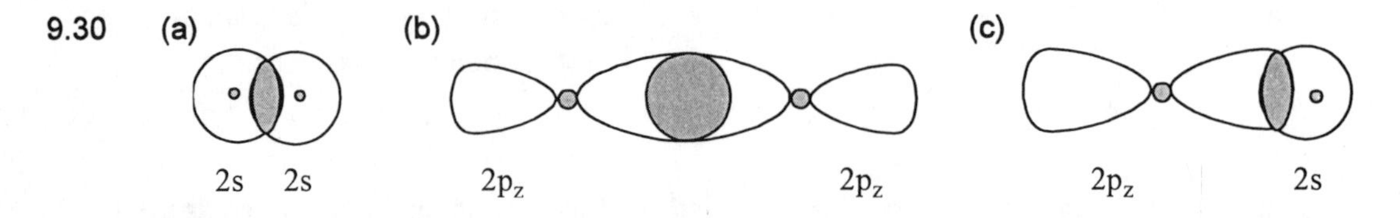

9.31 *Analyze/Plan.* Given electron domain geometry, list the appropriate orbital hybridization and associated bond angles; refer to Table 9.4. *Solve.*

(a) sp -- 180° (b) sp^3 -- 109° (c) sp^2 -- 120°
(d) sp^3d^2 -- 90° and 180° (e) sp^3d -- 90°, 120° and 180°

9.32 (a) sp^2 -- 120° angles in a plane
(b) sp^3d -- 90°, 120° and 180° bond angles (trigonal bipyramid)
(c) sp^3d^2 -- 90° and 180° bond angles (octahedron)

9.33 *Analyze/Plan.* Follow the logic in Sample Exercise 9.5. *Solve:*
SO_3^{2-}, 26 e^-, 13 e^- pr

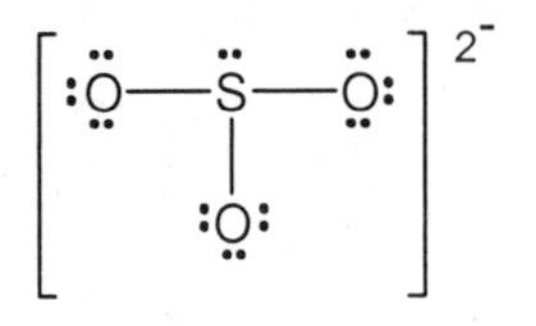

4 e^- domains around S; tetrahedral e^- domain geometry; trigonal pyramidal molecular geometry; sp^3 hybrid orbitals (based on e^- domain geometry) "ideal" O–S–O angle ~107° (The nonbonding e^- domain will close down the tetrahedral angles somewhat.)

9.34 The maximum number of hybrid orbitals on C is 4 (sp^3); the minimum number is 2(sp). Carbon is in the second row of the periodic chart, and has no d-orbitals of sufficiently low energy to allow for an expanded octet or hydrization involving d-orbitals. Thus, the maximum hybridization is when the 2s and all 2p orbitals are involved (sp^3); the minimum is when the 2s and a single 2p are hybridized (sp).

9.35 (a) B: $[He]2s^22p^1$

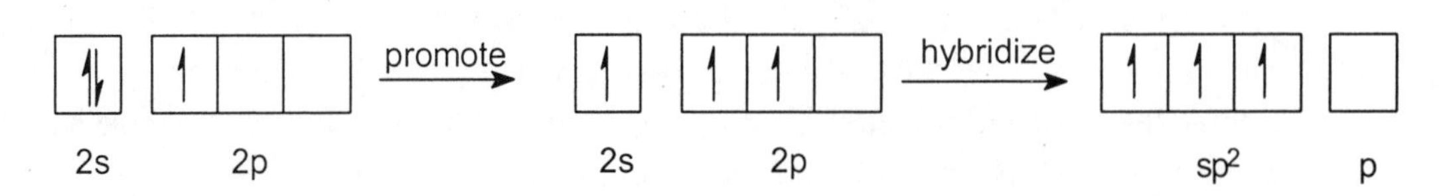

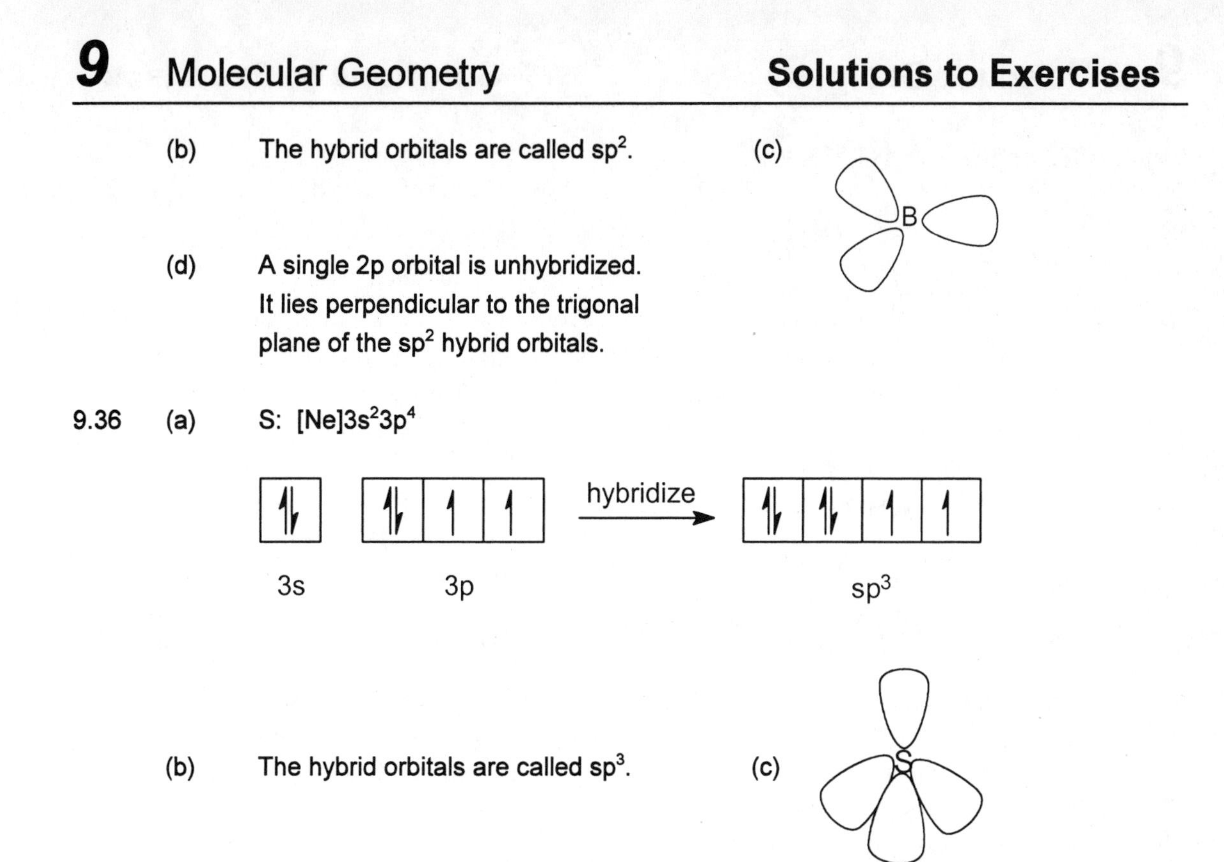

(b) The hybrid orbitals are called sp^2.

(c)

(d) A single 2p orbital is unhybridized. It lies perpendicular to the trigonal plane of the sp^2 hybrid orbitals.

9.36 (a) S: $[Ne]3s^23p^4$

(b) The hybrid orbitals are called sp^3.

(c)

(d) The hybrid orbitals formed in (a) would not be appropriate for SF_4. There are 5 electron domains in SF_4, 4 bonding and one nonbonding, so 5 hybrid orbitals are required. A set of 4 sp^3 hybrid orbitals could not accommodate all the electron pairs around S.

9.37 *Analyze/Plan*. Given the molecular (or ionic) formula, draw the correct Lewis structure and determine the electron domain geometry, which determines hybridization. *Solve*:

(a) 24 e^-, 12 e^- pairs

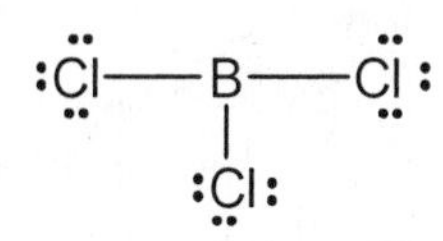

3 e^- pairs around B
trigonal planar e^- domain geometry, sp^2 hybrid orbitals

(b) 32 e^-, 16 e^- pairs

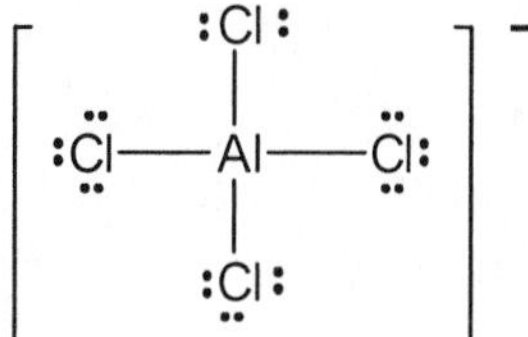

4 e^- domains around Al
tetrahedral e^- domain geometry
sp^3 hybrid orbitals

(c) 16 e^-, 8 e^- pairs

S=C=S

2 e^- domains around C
linear e^- domain geometry
sp hybrid orbitals

(d) 22 e^-, 11 e^- pairs

:F—Kr—F:

5 e^- pairs around Kr
trigonal bipyramidal e^- domain geometry, sp^3d hybrid orbitals

(e) 48 e^-, 24 e^- pairs

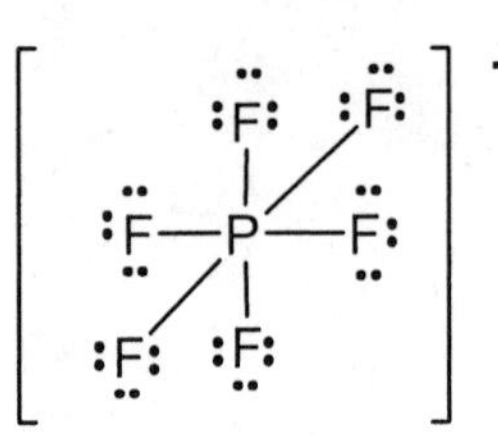

6 e^- pairs around P, octahedral e^- domain geometry, sp^3d^2 orbitals

9.38 (a) 32 e^-, 16 e^- pairs

4 e^- pairs around Si
tetrahedral e^- domain geometry
sp^3 hybrid orbitals

(b) 10 e^-, 5 e^- pairs

H—C≡N:

2 e^- domains around C
linear e^- domain geometry
sp hybrid orbitals

(c) 24 e^-, 12 e^- pairs

(other resonance structures are possible)

3 e^- domains around S
trigonal planar e^- domain geometry, sp^2 hybrid orbitals

(d) 22 e^-, 11 e^- pairs

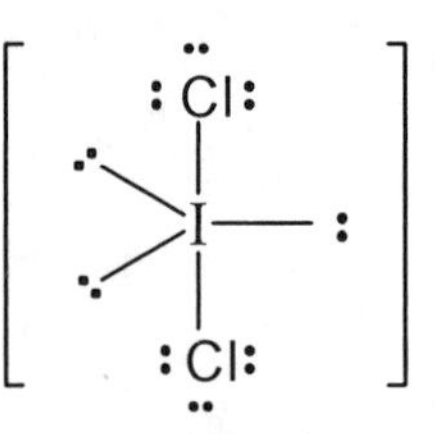

5 e^- domains around I, trigonal bipyramidal e^- domain geometry
sp^3d hybrid orbitals (in a trigonal bipyramid, placing nonbonding e^- pairs in the equatorial position minimizes repulsion

(e) 36 e^-, 18 e^- pairs

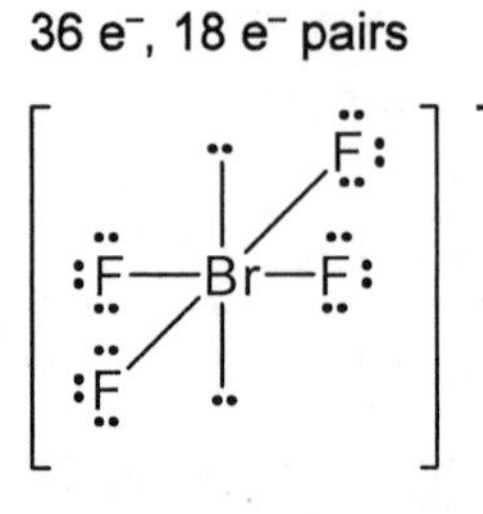

6 e^- domains around Br
octahedral e^- domain geometry, sp^3d^2 hybrid orbitals

Multiple Bonds

9.39 (a) (b)

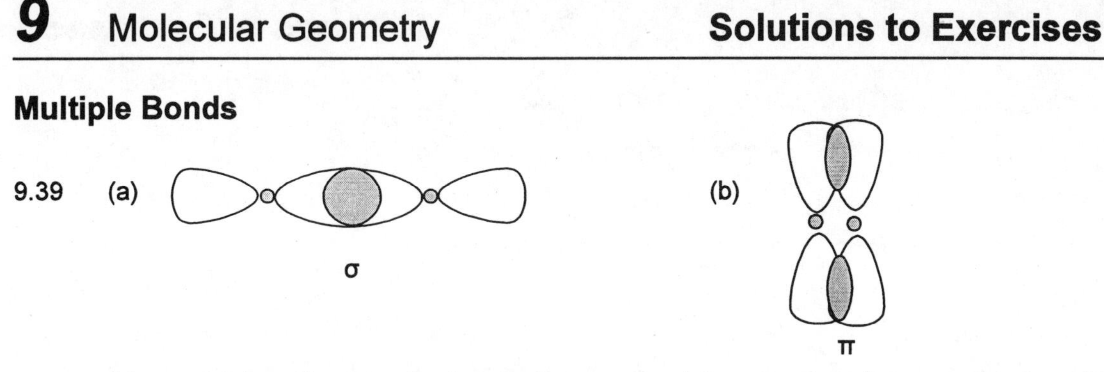

(c) A σ bond is generally stronger than a π bond, because there is more extensive orbital overlap.

9.40 (a) Two unhybridized p orbitals remain, and the atom can form two pi bonds.

(b) A triple bond is composed of 1 σ and 2 π bonds.

(c) There is free rotation of attached groups around a σ bond, but not around a π bond. Rotation about a π bond would require that the bond be broken; the p orbitals would no longer be in the correct orientation for π overlap. The π overlap that is part of all multiple bonds introduces rigidity into molecules.

9.41 (a)

```
    H              O
    |              ||
H—C—H          C
    |            /   \
    H          H      H
```

(b) sp^3 sp^2

(c) The C atom in CH_4 is sp^3 hybridized; there are no unhybridized p orbitals available for the π overlap required by multiple bonds. In CH_2O, the C atom is sp^2 hybridized, with 1 p atomic orbital available to form the π overlap in the C=O double bond.

9.42

```
     ..   ..
H—N—N—H        :N≡N:
     |    |
     H   H
```

The N atoms in N_2H_4 are sp^3 hybridized; there are no unhybridized p orbitals available for π bonding. In N_2, the N atoms are sp hybridized, with 2 unhybridized p orbitals on each N atom available to form the 2 π bonds in the N≡N triple bond.

9.43 *Analyze/Plan*. Single bonds are σ bonds, double bonds consist of 1 σ and 1 π bond. Each bond is formed by a pair of valence electrons. *Solve*:

(a) C_3H_6O has 3(4) + 6(1) + 6 = 24 valence electrons

(b) 9 pairs or 18 total valence electrons form σ bonds

(c) 1 pair or 2 total valence electrons form π bonds

(d) 2 pairs or 4 total valence electrons are nonbonding

(e) The central C atom is sp^2 hybridized

9.44 (a) The C bound to H has 3 electron domains and is sp^2 hybridized.
The C bound to O has 2 electron domains and is sp hybridized.

(b) C_2H_2O has 2(4) + 2(1) + 6 = 16 valence electrons.

(c) 4 pairs or 8 total valence electrons form σ bonds

(d) 2 pairs or 4 total valence electrons form π bonds

(e) 2 pairs or 4 total valence electrons are nonbonding

9.45 *Analyze/Plan.* Given the correct Lewis structure, analyze the electron domain geometry at **each central atom**. This determines the hybridization and bond angles at that atom. *Solve:*

(a) ~109° about the left most C, sp^3; ~120° about the right-hand C, sp^2

(b) The doubly bonded O can be viewed as sp^2, the other as sp^3; the nitrogen is sp^3 with approximately 109° bond angles.

(c) nine σ bonds, one π bond

9.46 (a) 1, 120°; 2, 120°; 3, 109° (b) 1, sp^2; 2, sp^2; 3, sp^3 (c) 21 σ bonds

9.47 (a) In a localized π bond, the electron density is concentrated strictly between the two atoms forming the bond. In a delocalized π bond, parallel p orbitals on more than two adjacent atoms overlap and the electron density is spread over all the atoms that contribute p orbitals to the network. There are still two regions of overlap, above and below the σ framework of the molecule.

(b) The existence of more than one resonance form is a good indication that a molecule will have delocalized π bonding.

(c) [O=N–O]⁻ ⟷ [O–N=O]⁻

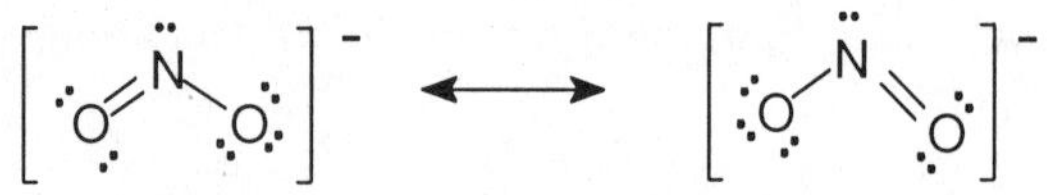

The existence of more than one resonance form for NO_2^- indicates that the π bond is delocalized. From an orbital perspective, the electron-domain geometry around N is trigonal planar, so the hybridization at N is sp^2. This leaves a p orbital on N and one on each O atom perpendicular to the trigonal plane of the molecule, in the correct orientation for delocalized π overlap. Physically, the two N-O bond lengths are equal, indicating that the two N-O bonds are equivalent, rather than one longer single bond and one shorter double bond.

9.48 (a) 24 e^-, 12 e^- pairs (b)

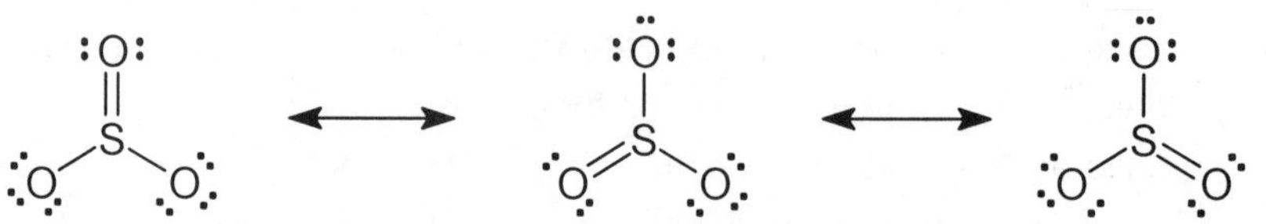

SO_3 resonance structures (S=O double bond to each O in turn) ⟷ ⟷

3 electron domains around S, trigonal planar electron-domain geometry, sp^2 hybrid orbitals

(c) The multiple resonance structures indicate delocalized π bonding. All four atoms lie in the trigonal plane of the sp^2 hybrid orbitals. On each atom there is a p atomic orbital perpendicular to this plane in the correct orientation for π overlap. The resulting delocalized π electron cloud is Y-shaped (the shape of the molecule) and has electron density above and below the plane of the molecule.

Molecular Orbitals

9.49 (a) Both atomic and molecular orbitals have a characteristic energy and shape (region where there is a high probability of finding an electron). Each atomic or molecular orbital can hold a maximum of two electrons. Atomic orbitals are localized on single atoms and their energies are the result of interactions between the subatomic particles in a single atom. MOs can be delocalized over several or even all the atoms in a molecule and their energies are influenced by interactions between electrons on several atoms.

(b) There is a net stabilization (lowering in energy) that accompanies bond formation because the bonding electrons in H_2 are strongly attracted to both H nuclei.

(c) 2

9.50 (a) In the σ anti-bonding MO, electron density is concentrated away from the nuclei; an electron in this orbital experiences less stabilization by the nucleus than an electron in an isolated atom.

(b) Yes. The Pauli principle, that no two electrons can have the same 4 quantum numbers, means that an orbital can hold at most 2 electrons. (Since n, l and m_l are the same for a particular orbital and m_s has only 2 possible values, an orbital can hold at most 2 electrons). This is true for atomic and molecular orbitals.

(c) 4. When AOs combine to form MOs, the total number of orbitals is conserved. Combination of 4 AOs must result in formation of 4 MOs. Both can accommodate 8 electrons.

9.51 (a)

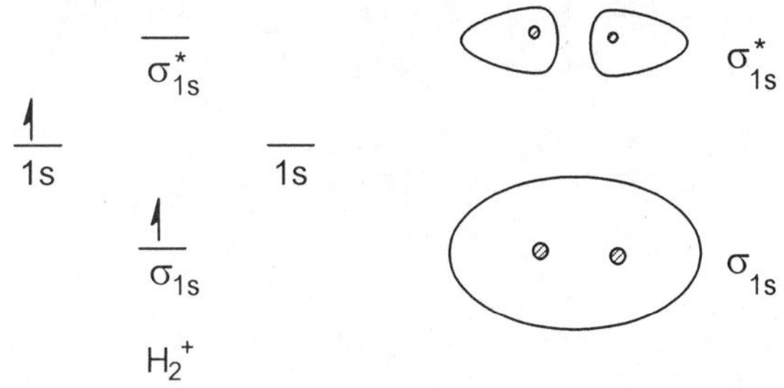

σ^*_{1s}

1s ↿ 1s

↿ σ_{1s}

H_2^+

σ^*_{1s}

σ_{1s}

(b) There is 1 electron in H_2^+.

(c) [ ] σ^*_{1s}

[↿] σ_{1s}

(d) Bond order = 1/2 (1-0) = 1/2

(e) Yes. The stability of H_2^+ is due to the lower energy state of the σ bonding molecular orbital relative to the energy of a H 1s atomic orbital. If the single electron in H_2^+ is excited to the σ^*_{1s} orbital, its energy is higher than the energy of a H 1s atomic orbital and H_2^+ will decompose into a hydrogen atom and a hydrogen ion.

$$H_2^+ \xrightarrow{h\nu} H + H^+.$$

9.52 (a)

1s		1s
	↿ σ^*_{1s}	
↿⇂		↿
	↿⇂ σ_{1s}	
	H_2^-	

σ^*_{1s}

σ_{1s}

(b) [↿] σ^*_{1s}

[↿⇂] σ_{1s}

(c) Bond order = 1/2 (2-1) = 1/2

(d) If 1 electron moves from σ_{1s} to σ^*_{1s}, the bond order becomes -1/2. There is a net increase in energy relative to isolated H atoms, so the molecule will decompose.

$$H_2^- \xrightarrow{h\nu} H + H^-.$$

9.53 *Analyze/Plan.* In a σ molecular orbital, the electron density is spherically symmetric about the internuclear axis and is concentrated along this axis. In a π MO, the electron density is concentrated above and below the internuclear axis and zero along it. *Solve*:

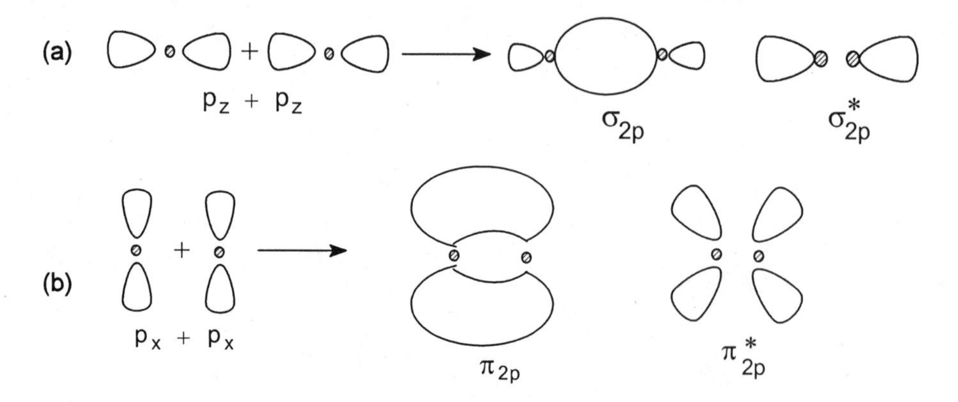

(c) σ_{2p} is lower in energy than π_{2p} due to greater extent of orbital overlap in the σ MO. $\sigma_{2p} < \pi_{2p} < \pi^*_{2p} < \sigma^*_{2p}$

9.54 (a) zero

(b) The 2 π_{2p} molecular orbitals are degenerate; they have the same energy, but they have different spatial orientations 90° apart.

(c) In the bonding MO the electrons are stabilized by both nuclei. In an antibonding MO, the electrons are directed away from the nuclei, so π_{2p} is lower in energy than π^*_{2p}.

9.55 (a) When comparing the same two bonded atoms, the greater the bond order, the shorter the bond length and the greater the bond energy. That is, bond order and bond energy are directly related, while bond order and bond length are inversely related. When comparing different bonded nuclei, there are no simple relationships (see Solution 8.86).

(b) Be_2, 4 e^- Be_2^+, 3 e^-

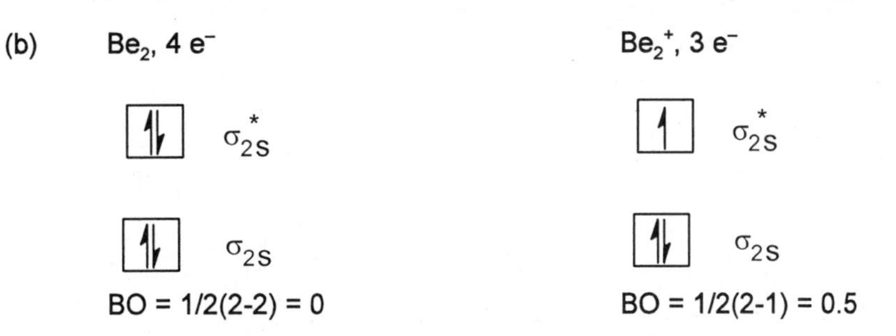

BO = 1/2(2-2) = 0 BO = 1/2(2-1) = 0.5

Be_2 has a bond order of zero and is not energetically favored over isolated Be atoms; it is not expected to exist. Be_2^+ has a bond order of 0.5 and is slightly lower in energy than isolated Be atoms. It will probably exist under special experimental conditions, but be unstable.

9.56 (a) O_2^{2-} has a bond order of 1.0, while O_2^- has a bond order of 1.5. For the same bonded atoms, the greater the bond order the shorter the bond, so O_2^- has the shorter bond.

(b) The two possible orbital energy level diagrams are:

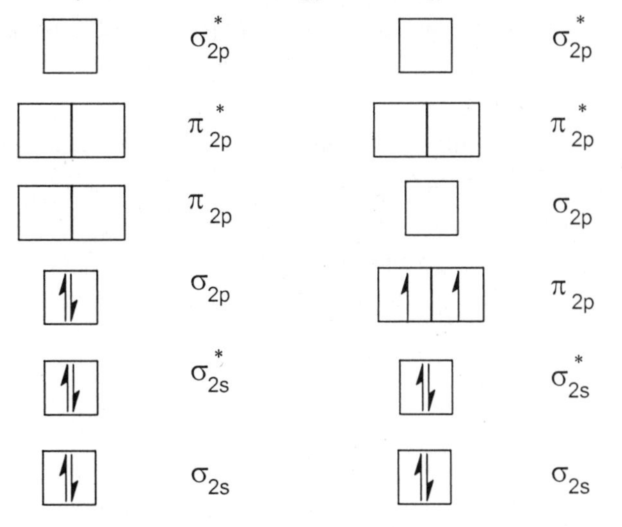

If the σ_{2p} molecular orbital is lower in energy than the π_{2p} orbitals, there are no unpaired electrons, and the molecule is diamagnetic. Switching the order of σ_{2p} and π_{2p} gives 1 unpaired electron in each degenerate π_{2p} orbital and explains the observed paramagnetism of B_2 (see Figure 9.41).

9.57 (a), (b) Substances with no unpaired electrons are weakly repelled by a magnetic field. This property is called *diamagnetism*.

(c) O_2^{2-}, Be_2^{2+} (see Figure 9.41)

9.58 (a) Substances with unpaired electrons are attracted into a magnetic field. This property is called *paramagnetism*.

(b) Weigh the substance normally and in a magnetic field. Paramagnetic substances appear to have a larger mass when weighed in a magnetic field.

(c) See Figures 9.36 and 9.41. O_2^+, 1 unpaired electron; N_2^{2-}, 2 unpaired electrons; Li_2^+, 1 unpaired electron

9.59

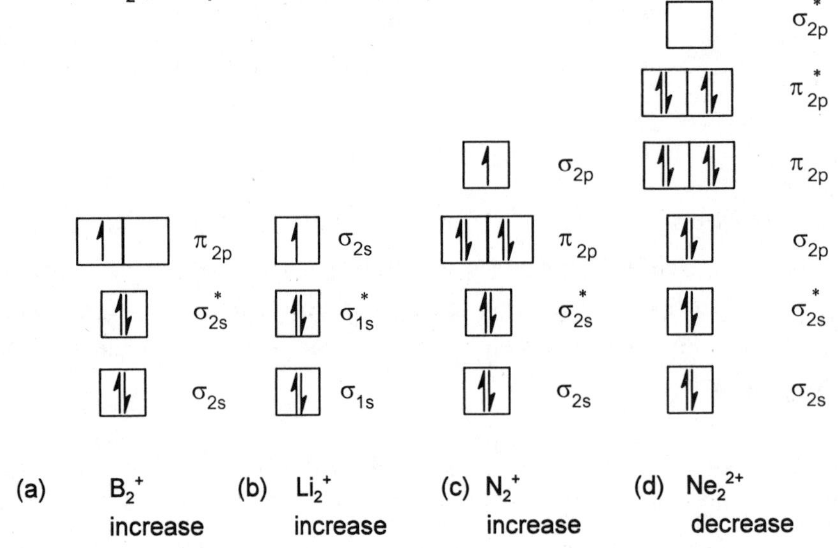

(a) B_2^+ increase (b) Li_2^+ increase (c) N_2^+ increase (d) Ne_2^{2+} decrease

Addition of an electron increases bond order if it occupies a bonding MO and decreases stability if it occupies an antibonding MO.

9.60 Determine the number of 'valence' (non-core) electrons in each molecule or ion. Use the homonuclear diatomic MO diagram from Figure 9.38 (shown below) to calculate bond order and magnetic properties of each species. The electronegativity difference between heteroatomics increases the energy difference between the 2s AO on one atom and the 2p AO on the other, rendering the 'no interaction' MO diagram in Figure 9.38 appropriate.

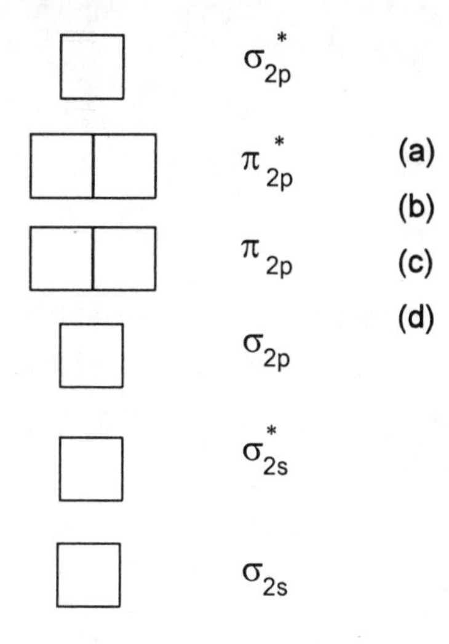

(a) CO: 10 e^-, B.O. = (8-2) / 2 = 3.0, diamagnetic
(b) NO^-: 12 e^-, B.O. = (8-4) / 2 = 2.0, paramagnetic
(c) OF^+ : 12 e^-, B.O. = (8-4) / 2 = 2.0, paramagnetic
(d) NeF^+: 14 e^-, B.O. = (8-6) / 2 = 1.0, diamagnetic

9.61 *Analyze/Plan*. Determine the number of 'valence' (non-core) electrons in each molecule or ion. Use the homonuclear diatomic MO diagram from Figure 9.38 (shown below) to calculate bond order and magnetic properties of each species. The electronegativity difference between heteroatomics increases the energy difference between the 2s AO on one atom and the 2p AO on the other, rendering the 'no interaction' MO diagram in Figure 9.38 appropriate. *Solve*:

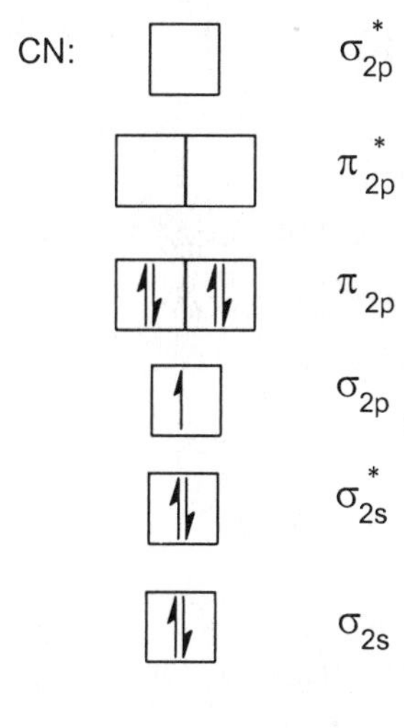

CN: 9 e^-, B.O. = (7-2) / 2 = 2.5, paramagnetic
CN^+ : 8 e^-, B.O. = (6-2) / 2 = 2.0, diamagnetic
CN^- : 10 e^-, B.O. = (8-2) / 2 = 3.0, diamagnetic

9.62 (a) The bond order of NO is [1/2 (8 - 3)] = 2.5. The electron that is lost is in an antibonding molecular orbital, so the bond order in NO^+ is 3.0. The increase in bond order is the driving force for the formation of NO^+.

(b) To form NO^-, an electron is added to an antibonding orbital, and the new bond order is [1/2 (8 - 4)] = 2. The order of increasing bond order and bond strength is: $NO^- < NO < NO^+$.

(c) NO^+ is isoelectronic with N_2, and NO^- is isoelectronic with O_2.

9.63 (a) 3s, $3p_x$, $3p_y$, $3p_z$ (b) π_{3p} (c) 2

(d) If the MO diagram for P_2 is similar to that of N_2, P_2 will have no unpaired electrons and be diamagnetic.

9.64 (a) I: 5s, $5p_x$, $5p_y$, $5p_z$; Br: 4s, $4p_x$, $4p_y$, $4p_z$

(b) By analogy to F_2, the BO of IBr will be 1.

(c) I and Br have valence atomic orbitals with different principle quantum numbers. This means that the radial extensions (sizes) of the valence atomic orbital that contribute to the MO are different. The n = 5 valence AOs on I are larger than the n = 4 valence AOs on Br.

(d) σ^*_{np} (e) none

Additional Exercises

9.65 (a) 8 e⁻, 4 e⁻ pairs

H—Se—H

4 e⁻ domains
tetrahedral domain geometry
bent molecular geometry

(b) 32 e⁻, 16 e⁻ pairs

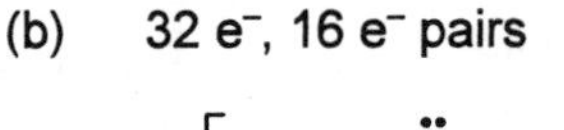

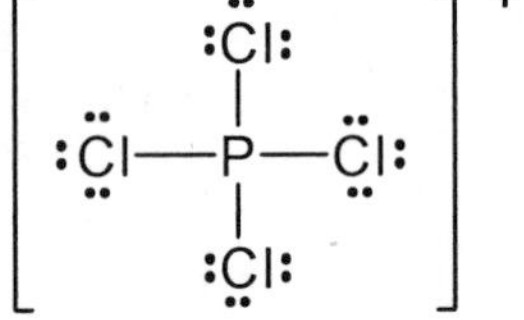

4 e⁻ domains
tetrahedral domain and molecular geometry

(c) 18 e⁻, 9 e⁻ pairs

3 e⁻ domains
trigonal planar domain geometry
bent molecular geometry

(d) 28 e⁻, 14 e⁻ pairs

:F—Br—F:
|
:F:

5 e⁻ domains
trigonal bipyramidal e⁻ domain geometry, T-shaped molecular geometry (in a trigonal bipyramid, nonbonded domains lie in the trigonal plane)

(e) 22 e⁻ 11 e⁻ pairs

[:I—I—I:]⁻

5 e⁻ domains
trigonal bipyramidal e⁻ domain geometry
linear molecular geometry (In a trigonal bipyramid, nonbonded domains lie in the trigonal plane.)

9.66 (a) The physical basis of VSEPR is the electrostatic repulsion of like-charged particles, in this case groups or domains of electrons. That is, owing to electrostatic repulsion, electron domains will arrange themselves to be as far apart as possible.

(b) The σ-bond electrons are localized in the region along the internuclear axes. The positions of the atoms and geometry of the molecule are thus closely tied to the locations of these electron pairs. Because the π-bond electrons are distributed above and below the plane that contains the σ bonds, these electron pairs do not, in effect, influence the geometry of the molecule. Thus, all σ- and π-bond electrons localized between two atoms are counted in the same electron domain.

9.67

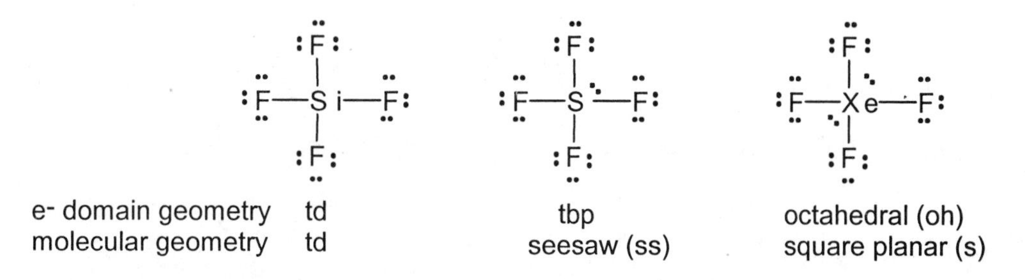

e- domain geometry	td	tbp	octahedral (oh)
molecular geometry	td	seesaw (ss)	square planar (s)

Although there are four bonding electron domains in each molecule, the number of nonbonding domains is different in each case. The bond angles and thus the molecular shape are influenced by the total number of electron domains.

9.68 For any triangle, the law of cosines gives the length of side c as $c^2 = a^2 + b^2 - 2ab \cos\theta$.

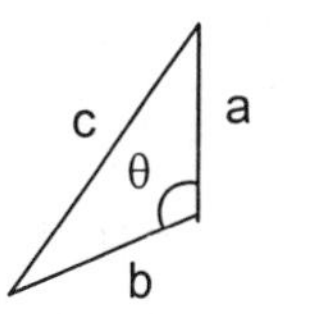

Let the edge length of the cube (uy = vy = vz) = X

The length of the face diagonal (uv) is

$(uv)^2 = (uy)^2 + (vy)^2 - 2(uy)(vy) \cos 90$

$(uv)^2 = X^2 + X^2 - 2(X)(X) \cos 90$

$(uv)^2 = 2X^2$; $uv = \sqrt{2}X$

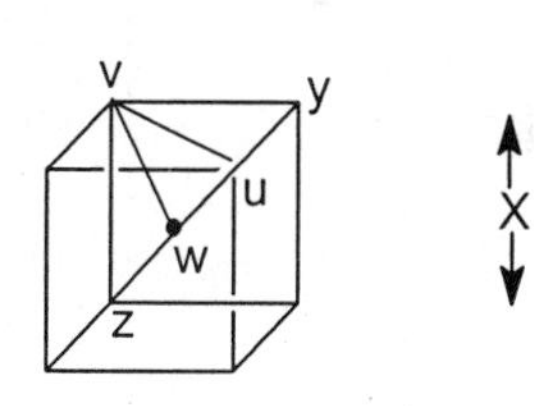

The length of the body diagonal (uz) is

$(uz)^2 = (vz^2) + (uv)^2 - 2(vz)(uv) \cos 90$

$(uz)^2 = X^2 + (\sqrt{2}X)^2 - 2(X)(\sqrt{2}X) \cos 90$

$(uz)^2 = 3X^2$; $uz = \sqrt{3}X$

For calculating the characteristic tetrahedral angle, the appropriate triangle has vertices u, v and w. Theta, θ, is the angle formed by sides wu and wv and the hypotenous is side uv.

$wu = wv = uz/2 = \sqrt{3}/2X$; $uv = \sqrt{2}X$

$(\sqrt{2}X)^2 = (\sqrt{3}/2X)^2 + (\sqrt{3}/2X)^2 - 2(\sqrt{3}/2X)(\sqrt{3}/2) \cos\theta$

$2X^2 = 3/4\ X^2 + 3/4\ X^2 - 3/2\ X^2 \cos\theta$

$2X^2 = 3/2\ X^2 - 3/2\ X^2 \cos\theta$

$1/2\ X^2 = -3/2\ X^2 \cos\theta$

$\cos\theta = -(1/2\ X^2) / (3/2\ X^2) = -1/3 = -0.3333$

$\theta = 109.47°$

9.69 (a) CO_2, 16 valence e^- (b) NCS^-, 16 valence e^-

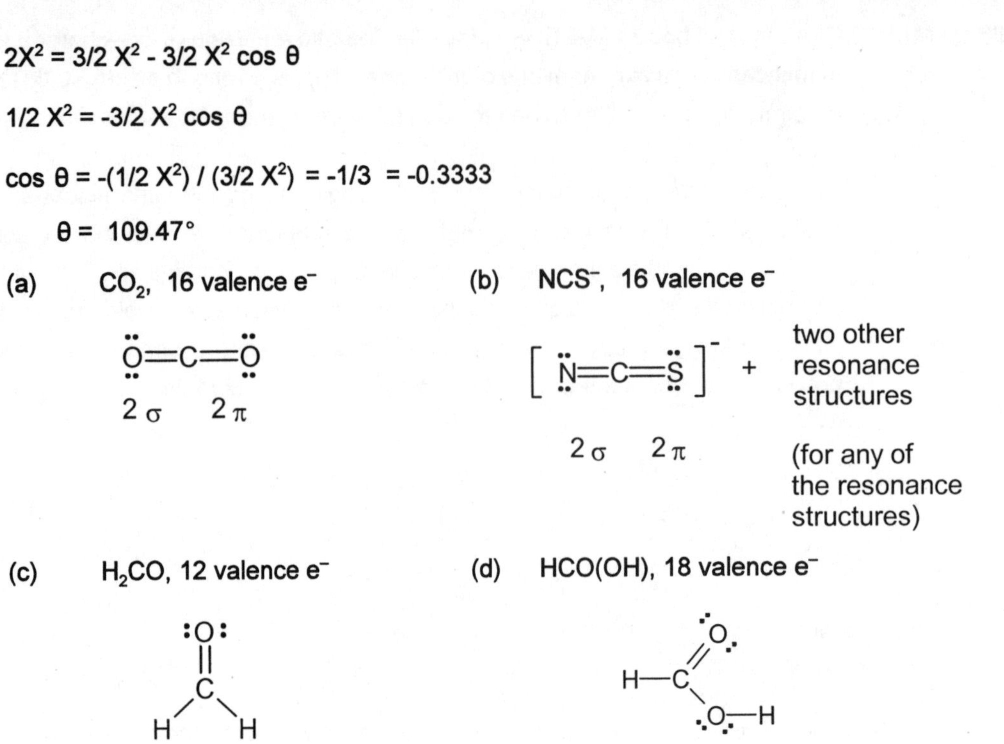

(c) H_2CO, 12 valence e^- (d) HCO(OH), 18 valence e^-

3 σ, 1 π 4 σ, 1 π

9.70

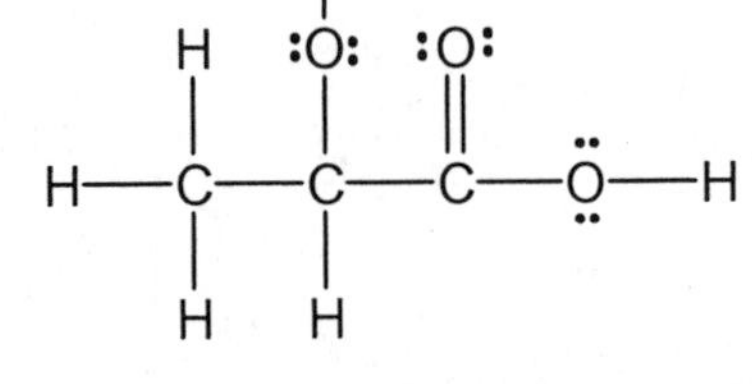

$3(4) + 3(6) + 6(1) = 36\ e^-$, $18\ e^-$ pr
There are 11 σ and 1 π bonds.
The C=O on the right-most C atom is shortest. This C has 3 e^- domains, so the hybridization is sp^2; bond angles about this C atom are approximately 120°. The middle and left-hand C atoms both have 4 e^- domains, are sp^3 hybridized and have bond angles of approximately 109°.

9.71

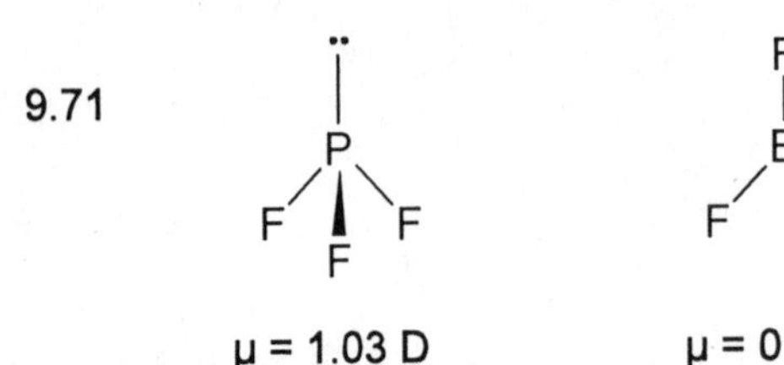

μ = 1.03 D μ = 0

BF_3 is a trigonal planar molecule with the central B atom symmetrically surrounded by the three F atoms (Figure 9.13). The individual B–F bond dipoles cancel, and the molecule has a net dipole moment of zero. PF_3 has tetrahedral electron-domain geometry with one of the positions in the tetrahedron occupied by a nonbonding electron pair. The individual P–F bond dipoles do not cancel and the presence of a nonbonding electron pair ensures an asymmetrical electron distribution; the molecule is polar.

9.72 The compound on the right has a dipole moment. In the square planar *trans* structure on the left, all equivalent bond dipoles can be oriented opposite each other, for a net dipole moment of zero.

9.73 

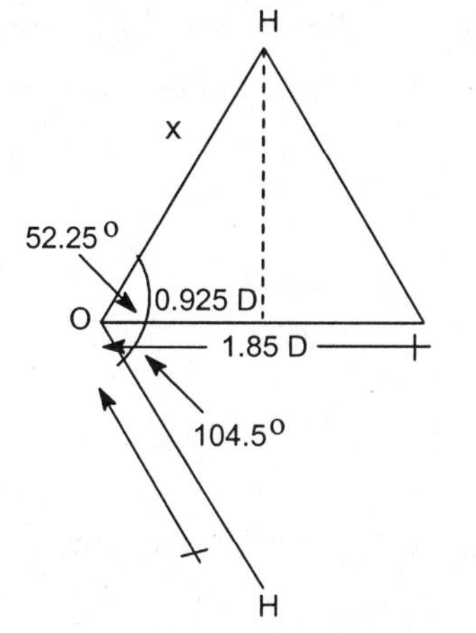

(a) The bond dipoles in H_2O lie along the O–H bonds with the positive end at H and the negative end at O. The dipole moment vector of the H_2O molecule is the resultant (vector sum) of the two bond dipoles. This vector bisects the H–O–H angle and has a magnitude of 1.85 D with the negative end pointing toward O.

(b) Since the dipole moment vector bisects the H–O–H bond angle, the angle between one H-O bond and the dipole moment vector is 1/2 the H–O–H bond angle, 52.25°. Dropping a perpendicular line from H to the dipole moment vector creates the right triangle pictured. If x = the magnitude of the O-H bond dipole, x cos (52.25) = 0.925 D. **x = 1.51 D.**

(c) The X–H bond dipoles (Table 8.3) and the electronegativity values of X (Figure 8.6) are

	Electronegativity	Bond dipole
F	4.0	1.82
O	3.5	1.51
Cl	3.0	1.08

Since the electronegativity of O is midway between the values for F and Cl, the O–H bond dipole should be approximately midway between the bond dipoles of HF and HCl. The value of the O–H bond dipole calculated in part (b) is consistent with this prediction.

9.74

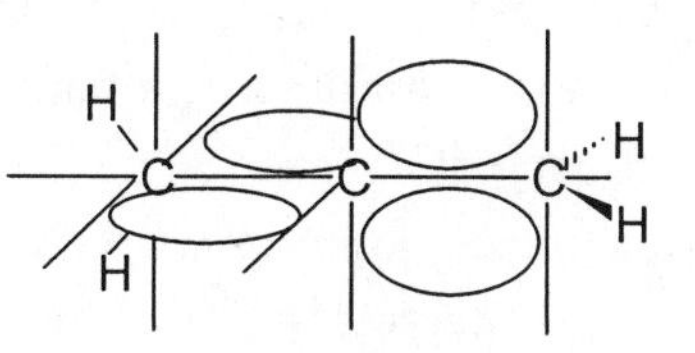

(a) The molecule is nonplanar. The CH_2 planes at each end are twisted 90° from one another.

(b) Allene has no dipole moment.

(c) The bonding in allene would not be described as delocalized. The π electron clouds of the two adjacent C=C are mutually perpendicular. The mechanism for delocalization of π electrons is mutual overlap of parallel p atomic orbitals on adjacent atoms. If adjacent π electron clouds are mutually perpendicular, there is no overlap and no delocalization of π electrons.

9.75 (a) XeF_6 50 e^-, 25 e^- pairs

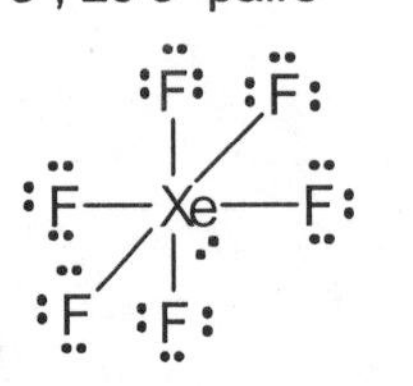

(b) There are 7 VSEPR e^- pairs around Xe, and the maximum number of e^- pairs in Table 9.3 is 6.

(c) Tie 7 balloons together and see what arrangement they adopt (seriously! see Figure 9.5). Alternatively, go to the chemical literature where VSEPR was first proposed and see if there is a preferred orientation for 7 e^- pairs.

(d) Since the hybrid orbitals for 5 VSEPR e^- pairs involve one d orbital and for 6 pairs, two d orbitals, a reasonable suggestion would be sp^3d^3.

(e) One of the 7 VSEPR pairs is a nonbonded pair. The question is whether it occupies an axial or equatorial position. The equatorial plane of a pentagonal bipyramid has F–Xe–F angles of 72°. Placing the nonbonded pair in the equatorial plane would create severe repulsions between it and the adjacent bonded pairs. Thus, the nonbonded pair will reside in the axial position. The molecular structure is a pentagonal pyramid.

9.76 (a) 16 e^-, 8 e^- pairs

$$[\ddot{\underset{..}{N}}=N=\ddot{\underset{..}{N}}]^- \longleftrightarrow [:N\equiv N-\ddot{\underset{..}{N}}:]^- \longleftrightarrow [:\ddot{\underset{..}{N}}-N\equiv N:]^-$$

(b) The observed bond length of 1.16 Å is intermediate between the values for N=N, 1.24 Å, and N≡N, 1.10 Å. This is consistent with the resonance structures, which indicate contribution from formally double and triple bonds to the true bonding picture in N_3^-.

(c) In each resonance structure, the central N has two electron domains, so it must be sp hybridized. It is difficult to predict the hybridization of terminal atoms in molecules where there are resonance structures because there are a different number of electron domains around the terminal atoms in each structure. Since the "true" electronic arrangement is a combination of all resonance structures, we will assume that the terminal N–N bonds have some triple bond character and that the terminal N atoms are sp hybridized. (There is no experimental measure of hybridization at terminal atoms, since there are no bond angles to observe.)

(d) In each resonance structure, N–N σ bonds are formed by sp hybrids and π bonds are formed by unhybridized p orbitals. Nonbonding e^- pairs can reside in sp hybrids or p atomic orbitals.

(e) Recall that electrons in 2s orbitals are on the average closer to the nucleus than electrons in 2p orbitals. Since sp hybrids have greater s orbital character, it is resonable to expect the radial extension of sp orbitals to be smaller than that of sp^2 or sp^3 orbitals and σ bonds formed by sp orbitals to be slightly shorter than those formed by other hybrid orbitals, assuming the same bonded atoms.

There are no solitary σ bonds in N_3^-. That is, the two σ bonds in N_3^- are each accompanied by at least one π bond between the bonding pair of atoms. Sigma bonds that are part of a double or triple bond must be shorter so that the p orbitals can overlap enough for the π bond to form. Thus, the observation is not applicable to this molecule. (Comparison of C–H bond lengths in C_2H_2, C_2H_4, C_2H_6 and related molecules would confirm or deny the observation.)

9.77 (a) $\ddot{\mathrm{O}}=\ddot{\mathrm{O}}-\ddot{\mathrm{O}}: \longleftrightarrow :\ddot{\mathrm{O}}-\ddot{\mathrm{O}}=\ddot{\mathrm{O}}$

To accommodate the π bonding by all 3 O atoms indicated in the resonance structures above, all O atoms are sp^2 hybridized.

(b) For the first resonance structure, both sigma bonds are formed by overlap of sp^2 hybrid orbitals, the π bond is formed by overlap of atomic p orbitals, one of the nonbonded pairs on the right terminal O atom is in a p atomic orbital, and the remaining 5 nonbonded pairs are in sp^2 hybrid orbitals.

(c) Only unhybridized p atomic orbitals can be used to form a delocalized π system.

(d) The unhybridized p orbital on each O atom is used to form the delocalized π system, and in both resonance structures one nonbonded electron pair resides in a p atomic orbital. The delocalized π system then contains 4 electrons, 2 from the π bond and 2 from the nonbonded pair in the p orbital.

9.78 (a) Each C atom is surrounded by 3 electron domains (2 single bonds and 1 double bond), so bond angles at each C atom will be approximately 120°.

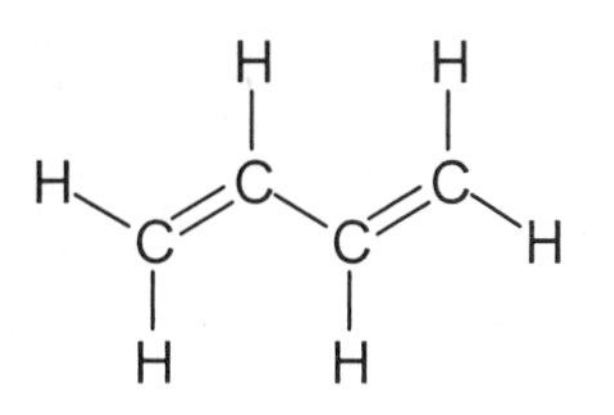

Since there is free rotation around the central C–C single bond, other conformations are possible.

(b) According to Table 8.5, the average C–C length is 1.54 Å, and the average C=C length is 1.34 Å. While the C=C bonds in butadiene appear "normal," the central C–C is significantly shorter than average. Examination of the bonding in butadiene reveals that each C atom is sp^2 hybridized and the π bonds are formed by the remaining unhybridized 2p orbital on each atom. If the central C–C bond is rotated so that all four C atoms are coplanar, the four 2p orbitals are parallel and some delocalization of the π electrons occurs.

9.79

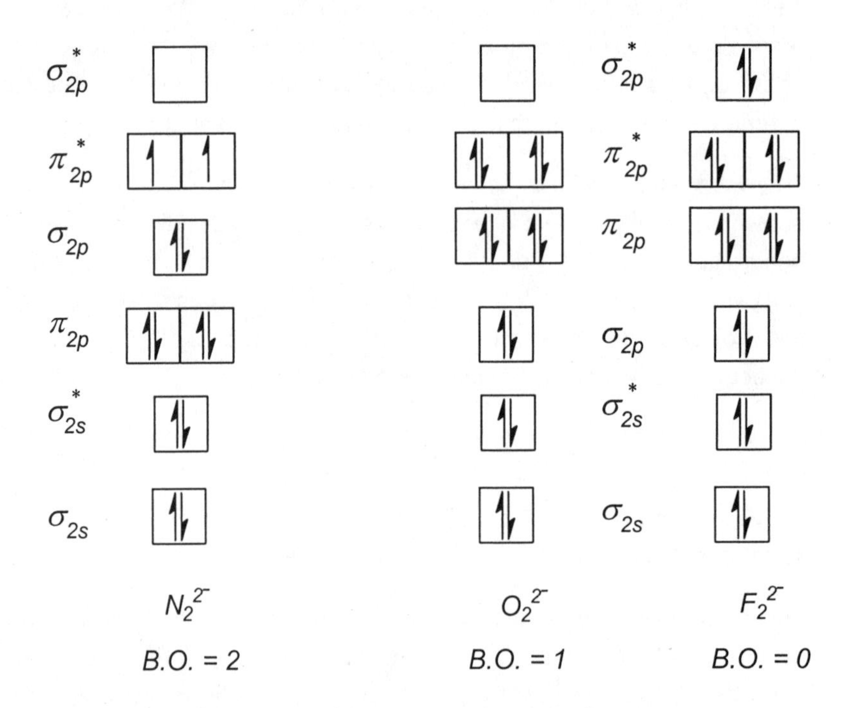

N_2^{2-} and O_2^{2-} are likely to be stable species, F_2^{2-} is not.

9.80

σ^*_{2p} π^*_{2p} σ_{2p} π_{2p} σ^*_{2s} σ_{2s}

N_2 in the ground state has a B.O. of 3; in the first excited state (at left) it has a B.O. of 2. Owing to the reduction in bond order, N_2 in the first excited state would be more reactive, have a smaller bond energy and a longer N–N separation.

9.81 [We will refer to azo benzene (on the left) as A and hydrazobenzene (on the right) as H.]

(a) A: sp^2; H: sp^3

(b) A: Each N and C atom has one unhybridized p orbital. H: Each C atom has one unhybridized p orbital, but the N atoms have no unhybridized p orbitals.

(c) A: 120°; H: 109°

(d) Since all C and N atoms in A have unhybridized p orbitals, all can participate in delocalized π bonding. The delocalized π system extends over the entire molecule, including both benzene rings and the azo "bridge". In H, the N atoms have no unhybridized p orbitals, so they cannot participate in delocalized π bonding. Each of the benzene rings in H is delocalized, but the network cannot span the N atoms in the bridge.

(e) This is consistent with the answer to (d). In order for the unhybridized p orbitals in A to overlap, they must be parallel. This requires a planar σ bond framework where all atoms in the molecule are coplanar.

(f) In order for a substance to appear colored, it must absorb light in the visible region of the electromagnetic spectrum. This requires a HOMO-LUMO energy gap in the visible region. For organic molecules, the size of the gap is related to the number of conjugated π bonds; the more conjugated π bonds, the smaller the gap and the more likely the molecule is to be colored. Azobenzene has 7 conjugated π bonds (π network delocalized over entire molecule) while hydrazobenzene has only 3 (π network on benzene rings only). Thus, the size of the HOMO-LUMO energy gap is smaller in A, and A is more likely than H to appear colored.

Integrative Exercises

9.82 (a) Assume 100 g of compound

$$2.1 \text{ g H} \times \frac{1 \text{ mol H}}{1.008 \text{ g H}} = 2.1 \text{ mol H}; \ 2.1 / 2.1 = 1$$

$$29.8 \text{ g N} \times \frac{1 \text{ mol N}}{14.01 \text{ g N}} = 2.13 \text{ mol N}; \ 2.13 / 2.1 \approx 1$$

$$68.1 \text{ g O} \times \frac{1 \text{ mol O}}{16.00 \text{ g O}} = 4.26 \text{ mol O}; \ 4.26 / 2.1 \approx 2$$

The empirical formula is HNO_2; formula weight = 47. Since the approximate molecular weight is 50, the **molecular formula is HNO_2.**

(b) Assume N is central, since it is unusual for O to be central, and part (d) indicates as much. HNO_2: 18 valence e^-

$$\ddot{\underset{\cdot\cdot}{O}}{=}\ddot{N}{-}\ddot{\underset{\cdot\cdot}{O}}{-}H \longleftrightarrow :\ddot{\underset{\cdot\cdot}{O}}{-}\ddot{N}{=}\ddot{O}{-}H$$

-1 0 +1

The second resonance form is a minor contributor due to unfavorable formal charges.

(c) The electron pair geometry around N is trigonal planar; if the resonance structure on the right makes a significant contribution to the molecular structure, all 4 atoms would lie in a plane. If only the left structure contributes, the H could rotate in and out of the molecular plane. The relative contributions of the two resonance structures could be determined by measuring the O–N–O and N–O–H bond angles.

(d) 3 VSEPR e^- domains around N, sp^2 hybridization

(e) 3 σ, 1 π for both structures (or for H bound to N).

9.83 (a) $2SF_4(g) + O_2(g) \rightarrow 2OSF_4(g)$

(b) 40 e^-, 20 e^- pairs

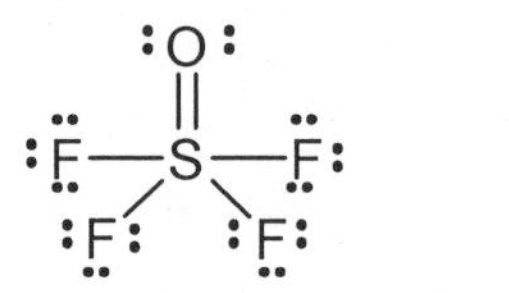

There must be a double bond drawn between O and S in order for their formal charges to be zero.

(c) $\Delta H = 8D(S\text{-}F) + D(O{=}O) - 8D(S\text{-}F) - 2D(S{=}O)$
$\Delta H = D(O{=}O) - 2D(S{=}O) = 495 - 2(523) = -551$ kJ, exothermic

(d) trigonal bipyramidal electron-domain geometry

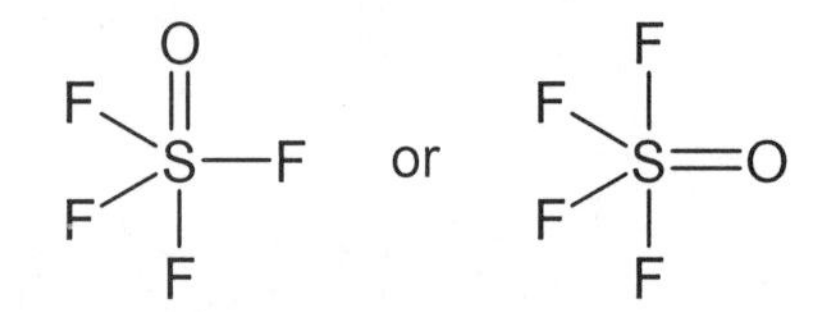

(e) Because F is more electronegative than O, the structure that minimizes S–F repulsions is more likely (see Solution 9.84). That is, the structure with fewer 90° F–S–F angles and more 120°F–S–F angles is favored. The structure on the left, with O in the axial position is more likely. Note that a double bond involving an atom with an expanded octet of electrons, such as the S=O in this molecule, does not have the same geometric implications as a double bond to a first row element.

9.84 (a) PX_3, 26 valence e^-, 13 e^- pairs

:X—P—X:
|
:X:

4 electron domains around P
tetrahedral e^- domain geometry
107° bond angles (less than 109° due to repulsion with nonbonded pair)

(b) As electronegativity increases (I < Br < Cl < F), the X–P–X angles decreases.

(c) For P–I, ΔEN is (2.5 - 2.1) = 0.4 and the bond dipole is small. For P–F, ΔEN is (4.0 - 2.1) = 1.9 and the bond dipole is large. The greater the ΔEN and bond dipole, the larger the magnitude of negative charge centered on X. The more negative charge centered on X, the greater the repulsion between X and the nonbonding electron pair on P and the smaller the bond angle. (Since electrostatic attractions and repulsions vary as 1/r, the shorter P-F bond may also contribute to this effect.)

(d) $PBrCl_4$, 40 valence electrons, 20 e^- pairs. The molecule will have trigonal bipyramidal electron-domain geometry (similar to PCl_5 in Table 9.3.) Based on the argument in part (c), the P–Br bond will have smaller repulsions with P–Cl bonds than P–Cl bonds have with each other. Therefore, the Br will occupy an axial position in the trigonal bipyramid, so that the more unfavorable P–Cl to P–Cl repulsions can be situated at larger angles in the equatorial plane.

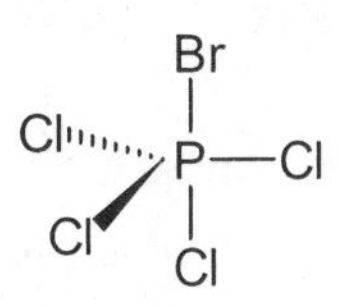

9.85 (a) 3 electron domains around each central C atom, sp^2 hybridization.

(b) A 180° rotation around the C=C is required to convert the *trans* isomer into the *cis* isomer. A 90° rotation around the C=C double bond eliminates all overlap of the p orbitals that form the π bond and the π bond is broken.

(c)

	average bond enthalpy
C=C	614 kJ/mol
C–C	348 kJ/mol

The difference in these values, 266 kJ/mol, is the average bond enthalpy of a C–C π bond. This is the amount of energy required to break 1 mol of C–C π bonds. The energy per molecule is

$$266 \text{ kJ/mol} \times \frac{1000 \text{ J}}{1 \text{ kJ}} \times \frac{1 \text{ mol}}{6.022 \times 10^{23} \text{ molecules}} = 4.417 \times 10^{-19}$$

$$= 4.42 \times 10^{-19} \text{ J/molecule}$$

(d) $$\lambda = hc/E = \frac{6.626 \times 10^{-34} \text{ J} \bullet \text{s} \times 2.998 \times 10^{8} \text{ m/s}}{4.417 \times 10^{-19} \text{ J}} = 4.50 \times 10^{-7} \text{ m} = 450 \text{ nm}$$

(e) Yes, 450 nm light is in the visible portion of the spectrum. A *cis-trans* isomerization in the retinal portion of the large molecule rhodopsin is the first step in a sequence of molecular transformations in the eye that leads to vision. The sequence of events enables the eye to detect visible photons, in other words, to see.

9.86 (a)

C≡C	839 kJ/mol	(1 σ, 2 π)
C=C	614 kJ/mol	(1 σ, 1 π)
C–C	348 kJ/mol	(1 σ)

The contribution from 1 π bond would be (614 - 348) 266 kJ/mol. From a second π bond, (839 - 614), 225 kJ/mol. An average π bond contribution would be (266 + 225)/2 = 246 kJ/mol.

This is $\frac{246 \text{ kJ}/\pi \text{ bond}}{348 \text{ kJ}/\sigma \text{ bond}} \times 100$ = 71% of the average enthalpy of a σ bond.

(b) N≡N 941 kJ/mol

N=N 418 kJ/mol

N–N 163 kJ/mol

first π = (418 - 163) = 255 kJ/mol

second π = (941 - 418) = 523 kJ/mol

average π bond enthalpy = (255 + 523) / 2 = 389 kJ/mol

This is $\frac{389 \text{ kJ}/\pi \text{ bond}}{163 \text{ kJ}/\sigma \text{ bond}} \times 100$ = 240% of the average enthalpy of a σ bond.

N-N σ bonds are weaker than C-C σ bonds, while N-N π bonds are stronger than C-C π bonds. The relative energies of C-C σ and π bonds are similar, while N-N π bonds are much stronger than N-N σ bonds.

(c) N_2H_4, 14 valence e^-, 7 e^- pairs

H—N̈—N̈—H (each N also bonded to H)

4 electron domains around N, sp^3 hybridization

N_2H_2, 12 valence e^-, 6 e^- pairs

H—N̈=N̈—H

3 electron domains around N, sp^2 hybridization

N_2, 10 valence e^-, 5 e^- pairs

:N≡N:

2 electron domains around N, sp hybridization

(d) In the three types of N–N bonds, each N atom has a nonbonding or lone pair of electrons. The lone pair to bond pair repulsions are minimized going from 109° to 120° to 180° bond angles, making the π bonds stronger relative to the σ bond. In the three types of C–C bonds, no lone-pair to bond-pair repulsions exist, and the σ and π bonds have more similar energies.

9.87

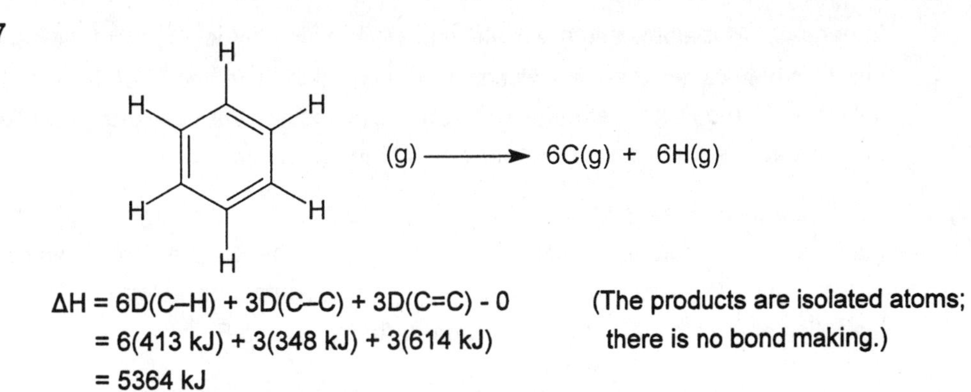

$\Delta H = 6D(C\text{–}H) + 3D(C\text{–}C) + 3D(C{=}C) - 0$ (The products are isolated atoms; there is no bond making.)

$= 6(413\text{ kJ}) + 3(348\text{ kJ}) + 3(614\text{ kJ})$

$= 5364\text{ kJ}$

According to Hess's Law:

$\Delta H^\circ = 6\Delta H_f^\circ\ C(g) + 6\Delta H_f^\circ\ H(g) - \Delta H_f^\circ\ C_6H_6(g)$

$= 6(718.4\text{ kJ}) + 6(217.94\text{ kJ}) - (+82.9\text{ kJ})$

$= 5535\text{ kJ}$

The difference in the two results, 171 kJ/mol C_6H_6 is due to the resonance stabilization in benzene. That is, because the π electrons are delocalized, the molecule has a lower overall energy than that predicted for the presence of 3 localized C–C and C=C bonds. Thus, the amount of energy actually required to decompose 1 mole of $C_6H_6(g)$, represented by the Hess's Law calculation, is greater than the sum of the localized bond enthalpies (not taking resonance into account) from the first calculation above.

9.88 (a) 1 eV = 96.485 kJ/mol

H_2: $15.4\text{ eV} \times \dfrac{96.485\text{ kJ/mol}}{1\text{ eV}} = 1486 = 1.49 \times 10^3\text{ kJ}$

N_2: $15.6\text{ eV} \times \dfrac{96.485\text{ kJ/mol}}{1\text{ eV}} = 1505 = 1.51 \times 10^3\text{ kJ}$

O_2: $12.1\text{ eV} \times \dfrac{96.485\text{ kJ/mol}}{1\text{ eV}} = 1167 = 1.17 \times 10^3\text{ kJ}$

F_2: $15.7\text{ eV} \times \dfrac{96.485\text{ kJ/mol}}{1\text{ eV}} = 1515 = 1.52 \times 10^3\text{ kJ}$

(b)

Z vs I_1 (Atoms)
Z vs I_1 (Molecules)
I_1 (KJ/mol)
1800
1700
1600
1500
1400
1300
1200
1100
0 2 4 6 8 10
Z

(c) In general, I_1 for atoms and molecules increases going across a row of the periodic chart. In both cases, there is a discontinuity at oxygen. The details of the trends are different. The deviation at O is larger for the molecules than the atoms, while the increase at F is much greater for the atoms than the molecules.

(d) According to Figures 9.34 and 9.41, H_2, N_2 and F_2 are diamagnetic and O_2 is paramagnetic. That is, ionization in H_2, N_2 and F_2 has to overcome spin-pairing energy, while ionization of O_2 removes an already unpaired electron. Thus, the ionization energy of O_2 is much less than I_1 for H_2, N_2 and F_2.

Despite differences in bond order, bond length and the bonding or antibonding nature of the HOMO in H_2, N_2 and F_2, the ionization energies for these molecules are very similar.

9.89 (a) $3d_{z^2}$

(b) Ignoring the donut of the d_{z^2} orbital

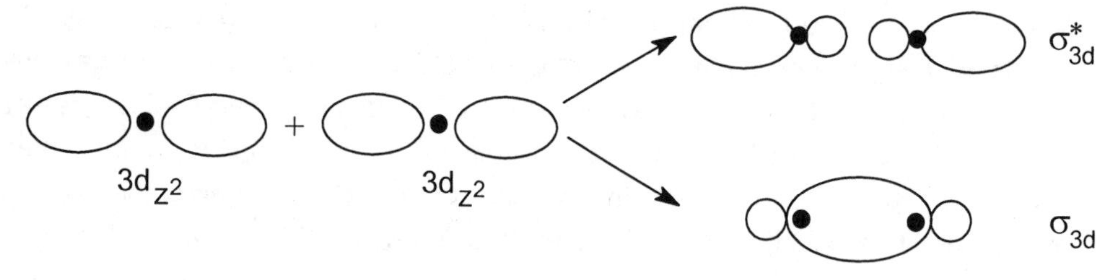

(c) Sc: [Ar] $4s^23d^1$ Omitting the core electrons, there are 6 e^- in the energy level diagram.

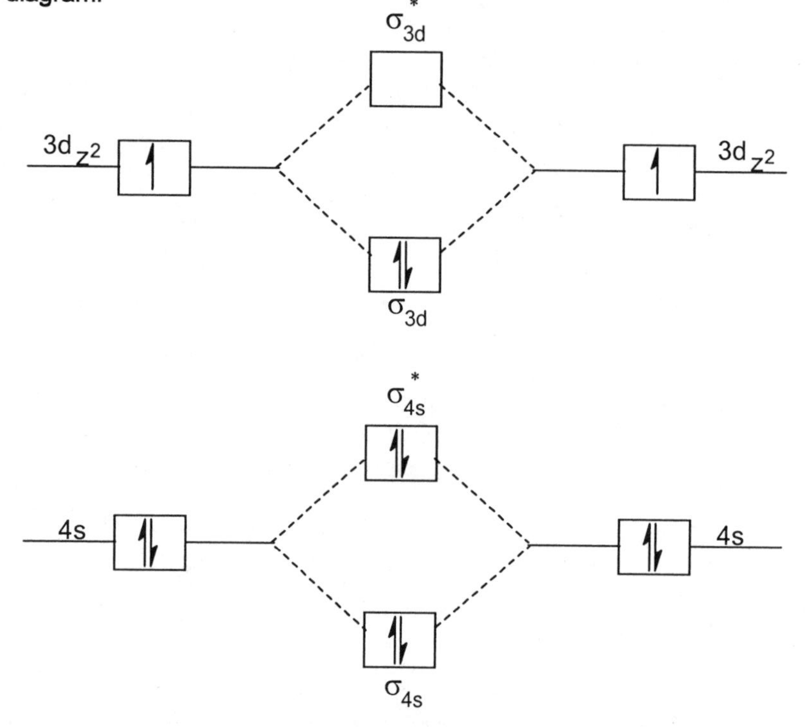

(d) The bond order in Sc_2 is 1/2 (4 - 2) = 1.0.

9.90 (a) Parentheses around the atomic weight indicate that the element is radioactive. The nuclei of radioactive elements spontaneously decay to form other, lighter nuclei. The difficulty with studying At is that it decays before a reliable set of physical and chemical properties can be measured.

(b) $1s^22s^22p^63s^23p^64s^23d^{10}4p^65s^24d^{10}5p^66s^24f^{14}5d^{10}6p^5$

(c) AtI is expected to have a slightly polar covalent bond, similar to other interhalogen molecules. Since At and I are adjacent in the halogen family, their electronegativities will be very similar, but not equal. When two atoms with similar electronegativities combine, the result is a covalent molecule with a small net dipole moment.

(d) 22 e^-, 11 e^- pairs; $[:\ddot{I}-\ddot{At}-\ddot{I}:]^-$

trigonal bipyramidal electron-domain geometry,
linear molecular geometry (In a trigonal bipyramid, placing nonbonding electron pairs in the equatorial plane minimizes repulsions.)

(e) By analogy to F_2 in Figure 9.41, the bond order should be 1 and the HOMO is π^*_{6p}.

10 Gases

Gas Characteristics; Pressure

10.1 In the gas phase molecules are far apart, while in the liquid they are touching.

(a) A gas is much less dense than a liquid because most of the volume of a gas is empty space.

(b) A gas is much more compressible because of the distance between molecules.

(c) Gaseous molecules are so far apart that there is no barrier to mixing, regardless of the identity of the molecule. All mixtures of gases are homogeneous. Liquid molecules are touching. In order to mix, they must displace one another. Similar molecules displace each other and form homogeneous mixtures. Very dissimilar molecules form heterogeneous mixtures.

10.2 (a) Because gas molecules are far apart and in constant motion, the gas expands to fill the container. Attractive forces hold liquid molecules together and the volume of the liquid does not change.

(b) H_2O and CCl_4 molecules are too dissimilar to displace each other and mix in the liquid state. All mixtures of gases are homogeneous. (See Solution 10.1 (c)).

(c) Because gas molecules are far apart, the mass present in 1 mL of a gas is very small. The mass of a gas present in 1 L is on the same order of magnitude as the mass of a liquid present in 1 mL.

10.3 (a) $F = m \times a$. Since both people have the same mass and both experience the acceleration of gravity, the forces they exert on the floor are exactly equal.

(b) $P = F / A$. The two forces are equal, but the person standing on one foot exerts this force over a smaller area. Thus, the person standing on one foot exerts a greater pressure on the floor.

10.4 The height of the mercury column in a barometer is a direct measure of atmospheric pressure. Atmospheric pressure decreases with increasing altitude (Section 18.1), so atmospheric pressure in Denver, at an altitude of nearly one mile, is measurably lower than atmospheric pressure in Los Angeles, near sea level. Thus, the mercury column in a Denver barometer should be lower than that in a Los Angeles barometer.

10.5 *Analyze.* Given: 760 mm column of Hg, densities of Hg and H_2O. Find: height of a column of H_2O at same pressure.

Plan. We must develop a relationship between pressure, height of a column of liquid, and density of the liquid. Relationships that might prove useful: P = F/A; F = m × a; m = d × V(density)(volume); V = A × height *Solve*:

$$P = \frac{F}{A} = \frac{m \times a}{A} = \frac{d \times V \times a}{A} = \frac{d \times A \times h \times a}{A} = d \times h \times a$$

(a) $P_{Hg} = P_{H_2O}$; Using the relationship derived above: $(d \times h \times a)_{H_2O} = (d \times h \times a)_{Hg}$

Since a, the acceleration due to gravity, is equal in both liquids,

$(d \times h)_{H_2O} = (d \times h)_{Hg}$

$1.00 \text{ g/mL} \times h_{H_2O} = 13.6 \text{ g/mL} \times 760 \text{ mm}$

$$h_{H_2O} = \frac{13.6 \text{ g/mL} \times 760 \text{ mm}}{1.00 \text{ g/mL}} = 1.034 \times 10^4 = 1.03 \times 10^4 \text{ mm} = 10.3 \text{ m}$$

(b) Pressure due to H_2O:

1 atm = 1.034×10^4 mm H_2O (from part (a))

$$36 \text{ ft } H_2O \times \frac{12 \text{ in}}{1 \text{ ft}} \times \frac{2.54 \text{ cm}}{1 \text{ in}} \times \frac{10 \text{ mm}}{1 \text{ cm}} \times \frac{1 \text{ atm}}{1.034 \times 10^4 \text{ mm}} = 1.061 = 1.1 \text{ atm}$$

$P_{total} = P_{atm} + P_{H_2O} = 0.95 \text{ atm} + 1.061 \text{ atm} = 2.011 = 2.0 \text{ atm}$

10.6 Using the relationship derived in Solution 10.5 for two liquids under the influence of gravity, $(d \times h)_{1id} = (d \times h)_{Hg}$. At 752 torr, the height of a Hg barometer is 752 mm.

$$\frac{1.20 \text{ g}}{1 \text{ mL}} \times h_{1id} = \frac{13.6 \text{ g}}{1 \text{ mL}} \times 760 \text{ mm}; \; h_{1id} = \frac{13.6 \text{ g/mL} \times 752 \text{ mm}}{1.20 \text{ g/mL}} = 8.52 \times 10^3 \text{ mm} = 8.52 \text{ m}$$

10.7 (a) The tube can have **any** cross-sectional area. (The height of the Hg column in a barometer is independent of the cross-sectional area. See the expression for pressure derived in Solution 10.5.)

(b) At equilibrium, the force of gravity per unit area acting on the mercury column at the level of the outside mercury is **not** equal to the force of gravity per unit area acting on the atmosphere. (F = ma; the acceleration due to gravity is equal for the two substances, but the mass of Hg for a given cross-sectional area is different than the mass of air for this same area.)

(c) The column of mercury is held up by the pressure of the atmosphere applied to the exterior pool of mercury.

10.8 The mercury would fill the tube completely; there would be no vacuum at the closed end. This is because atmospheric pressure will support a mercury column higher than 70 cm, while our tube is only 50 cm. No mercury flows from the tube into the dish and no vacuum forms at the top of the tube.

10.9 *Analyze/Plan.* Follow the logic in sample Exercise 10.1. *Solve:*

(a) $265 \text{ torr} \times \frac{1 \text{ atm}}{760 \text{ torr}} = 0.349 \text{ atm}$

(b) $265 \text{ torr} \times \frac{1 \text{ mm Hg}}{1 \text{ torr}} = 265 \text{ mm Hg}$

(c) $265 \text{ torr} \times \frac{1.01325 \times 10^5 \text{ Pa}}{760 \text{ torr}} = 3.53 \times 10^4 \text{ Pa}$

(d) $265 \text{ torr} \times \frac{1.01325 \times 10^5 \text{ Pa}}{760 \text{ torr}} \times \frac{1 \text{ bar}}{1 \times 10^5 \text{ Pa}} = 0.353 \text{ bar}$

10.10 (a) $2.44 \text{ atm} \times \frac{760 \text{ torr}}{1 \text{ atm}} = 1.85 \times 10^3 \text{ torr}$

(b) $682 \text{ torr} \times \frac{101.325 \text{ kPa}}{760 \text{ torr}} = 90.9 \text{ kPa}$

(c) $776 \text{ mm Hg} \times \frac{1 \text{ atm}}{760 \text{ mm Hg}} = 1.02 \text{ atm}$

(d) $1.456 \times 10^5 \text{ Pa} \times \frac{1 \text{ atm}}{1.01325 \times 10^5 \text{ Pa}} = 1.44 \text{ atm}$

(e) $3.44 \text{ atm} \times \frac{1.01325 \times 10^5 \text{ Pa}}{1 \text{ atm}} \times \frac{1 \text{ bar}}{1 \times 10^5 \text{ Pa}} = 3.49 \text{ bar}$

10.11 Analyze/Plan. Follow the logic in Sample Exercise 10.1. *Solve*:

(a) $30.45 \text{ in Hg} \times \frac{25.4 \text{ mm}}{1 \text{ in}} \times \frac{1 \text{ torr}}{1 \text{ mm Hg}} = 773.4 \text{ torr}$

[The result has 4 sig figs because 25.4 mm/in is considered to be an exact number. (Section 1.5)]

(b) The pressure in Chicago is greater than **standard atmospheric pressure**, 760 torr, so it makes sense to classify this weather system as a "high pressure system."

10.12 (a) $\frac{1.63105 \text{ Pa}}{1 \text{ Titan atm}} \times \frac{1 \text{ Earth atm}}{101{,}325 \text{ Pa}} = \frac{1.60972 \times 10^{-5} \text{ Earth atm}}{1 \text{ Titan atm}}$

(b) $\frac{90 \text{ Earth atm}}{1 \text{ Venus atm}} \times \frac{101.3 \text{ kPa}}{1 \text{ Earth atm}} = \frac{9.1 \times 10^3 \text{ kPa}}{1 \text{ Venus atm}}$

10.13 *Analyze*. Given: mass, area. Find: pressure. *Plan*. P = F/A = m × a/A; use this relationship, paying attention to units. *Solve*:

$$1 \text{ Pa} = \frac{1 \text{ N}}{\text{m}^2} = \frac{1 \text{ kg}\bullet\text{m}}{\text{s}^2} \times \frac{1}{\text{m}^2} = \frac{1 \text{ kg}}{\text{m}\bullet\text{s}^2}$$ Change mass to kg and area to m^2.

$$P = \frac{m \times a}{A} = \frac{125 \text{ lb}}{0.50 \text{ in}^2} \times \frac{9.81 \text{ m}}{1 \text{ s}^2} \times \frac{0.454 \text{ kg}}{1 \text{ lb}} \times \frac{39.4^2 \text{ in}^2}{1 \text{ m}^2} = 1.7 \times 10^6 \frac{\text{kg}}{\text{m}\bullet\text{s}^2}$$

$$= 1.7 \times 10^6 \text{ Pa} = 1.7 \times 10^3 \text{ kPa}$$

Check. $[1.25 \times 10 \times 0.5 \times (40)^2/0.5] \approx (125 \times 16{,}000) \approx 2.0 \times 10^6 \text{ Pa} \approx 2.0 \times 10^3 \text{ kPa}$. The units are correct.

10.14 $P = m \times a/A$; $1 \text{ Pa} = 1 \text{ kg/m} \bullet \text{s}^2$; $A = 2.2 \text{ cm} \times 30 \text{ cm} \times 2 = 132 \text{ cm}^2$

$$\frac{262 \text{ kg}}{132 \text{ cm}^2} \times \frac{9.81 \text{ m}}{\text{s}^2} \times \frac{(100)^2 \text{ cm}^2}{1 \text{ m}^2} \times \frac{1 \text{ Pa}\bullet\text{m}\bullet\text{s}^2}{\text{kg}} = 1.947 \times 10^5 = 1.9 \times 10^5 \text{ Pa}$$

10.15 *Analyze/Plan*. Follow the logic in Sample Exercise 10.2. *Solve*:

(i) The Hg level is lower in the open end than the closed end, so the gas pressure is less than atmospheric pressure.

$$P_{gas} = 0.975 \text{ atm} - \left(52 \text{ cm} \times \frac{1 \text{ atm}}{76 \text{ cm}}\right) = 0.29 \text{ atm}$$

(ii) The Hg level is higher in the open end, so the gas pressure is greater than atmospheric pressure.

$$P_{gas} = 0.975 \text{ atm} + \left(67 \text{ mm Hg} \times \frac{1 \text{ atm}}{760 \text{ mm Hg}}\right) = 1.063 \text{ atm}$$

(iii) This is a closed-end manometer so $P_{gas} = h$.

$$P_{gas} = 10.3 \text{ cm} \times \frac{1 \text{ atm}}{76 \text{ cm}} = 0.136 \text{ atm}$$

10.16 (a) The atmosphere is exerting 13.6 cm = 136 mm Hg (torr) more pressure than the gas.

$$P_{gas} = P_{atm} - 136 \text{ torr} = \left(1.05 \text{ atm} \times \frac{760 \text{ torr}}{1 \text{ atm}}\right) - 136 \text{ torr} = 662 \text{ torr}$$

(b) The gas is exerting 12 mm Hg (torr) more pressure than the atmosphere.

$$P_{gas} = P_{atm} + 12 \text{ torr} = \left(0.988 \text{ atm} \times \frac{760 \text{ torr}}{1 \text{ atm}}\right) + 12 \text{ torr} = 763 \text{ torr}$$

The Gas Laws

10.17 (a) $V_1/T_1 = V_2/T_2$

$V_1/300\ K = V_2/500\ K$

$V_2 = 5/3\ V_1$

(b) $P_1V_1 = P_2V_2$

$1\ atm \times V_1 = 2\ atm \times V_2$

$V_2 = 1/2\ V_1$

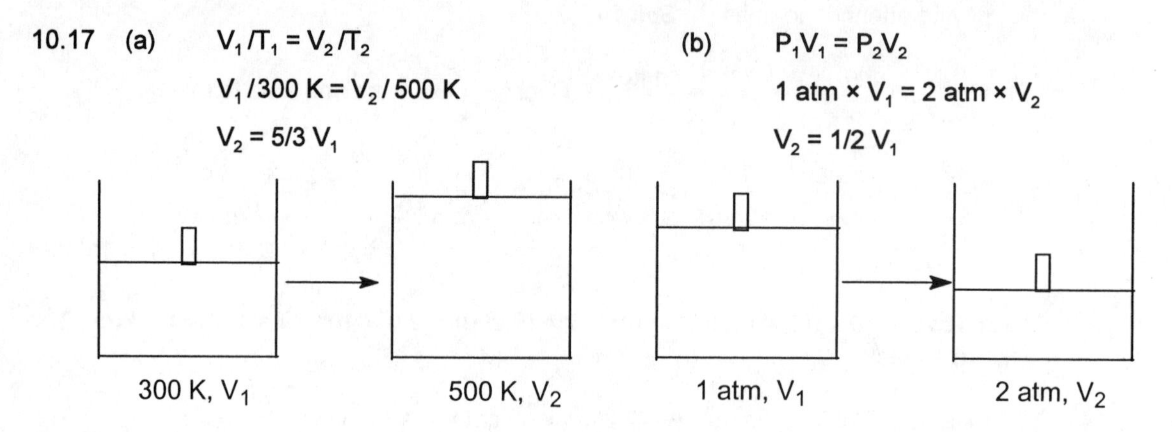

10.18 (a) P and V are inversely proportional at constant T. If the volume decreases by a factor of 3, the pressure increases by a factor of 3.

(b) P and T are directly proportional at constant V. If T decreases by a factor of 2, P also decreases by a factor of 2.

(c) P and n are directly proportional at constant V. If n decreases by a factor of 2, P also decreases by a factor of 2.

10.19 *Analyze.* Given: initial P, V, T. Find: final values of P, V, T for certain changes of condition. *Plan.* Select the appropriate Gas Law relationships from Section 10.3; solve for final conditions, paying attention to units. *Solve:*

(a) $P_1V_1 = P_2V_2$; the proportionality holds true for any pressure or volume units.

$P_1 = 748$ torr, $V_1 = 10.3$ L, $P_2 = 1.88$ atm

$$V_2 = \frac{P_1V_1}{P_2} = \frac{748\ \text{torr} \times 10.3\ \text{L}}{1.88\ \text{atm}} \times \frac{1\ \text{atm}}{760\ \text{torr}} = 5.39\ \text{L}$$

Check. As pressure increases, volume should decrease; our result agrees with this.

(b) $V_1/T_1 = V_2/T_2$; T must be in Kelvins for the relationship to be true.

$V_1 = 10.3$ L, $T_1 = 23°C = 296$ K, $T_2 = 165°C = 438$ K

$$V_2 = \frac{V_1T_2}{T_1} = \frac{10.3\ \text{L} \times 438\ \text{K}}{296\ \text{K}} = 15.2\ \text{L}$$

Check. As temperature increases, volume should increase; our result is consistent with this.

10.20 (a) $P_1V_1 = P_2V_2$; $P_1 = 0.988$ atm, $V_1 = 1248$ ft^3, $V_2 = 978$ ft^3

$$P_2 = \frac{P_1V_1}{V_2} = \frac{0.988 \text{ atm} \times 1248 \text{ ft}^3}{978 \text{ ft}^3} = 1.26 \text{ atm}$$

(b) $V_1/T_1 = V_2/T_2$; $V_1 = 1248$ ft^3 L, $V_2 = 1435$ ft^3, $T_1 = 28.0°C + 273.15 = 301.15$ K

$$T_2 = \frac{V_2T_1}{V_1} = \frac{1435 \text{ ft}^3 \times 301.15 \text{ K}}{1248 \text{ ft}^3} = 346.27 = 346.3 \text{ K}; \; 346.27 - 273.15 = 73.1°C$$

The result in K has 4 sig figs and 1 decimal place. The result in °C then has 1 decimal place (and 3 sig figs).

10.21 (a) Avogadro's hypothesis states that equal volumes of gases at the same temperature and pressure contain equal numbers of molecules. Since molecules react in the ratios of small whole numbers, it follows that the volumes of reacting gases (at the same temperature and pressure) are in the ratios of small whole numbers.

(b) Since the two gases are at the same temperature and pressure, the ratio of the numbers of atoms is the same as the ratio of volumes. There are 1.5 times as many Xe atoms as Ne atoms.

10.22 According to Avogadro's hypothesis, the mole ratios in the chemical equation will be volume ratios for the gases if they are at the same temperature and pressure.

$$N_2(g) + 3H_2(g) \rightarrow 2NH_3(g)$$

The volumes of H_2 and N_2 are in a stoichiometric $\frac{3.6 \text{ L}}{1.2 \text{ L}}$ or $\frac{3 \text{ vol } H_2}{1 \text{ vol } N_2}$ ratio, so either can be used to determine the volume of $NH_3(g)$ produced.

$$1.2 \text{ L } N_2 \times \frac{2 \text{ mol } NH_3}{1 \text{ mol } N_2} = 2.4 \text{ L } NH_3(g) \text{ produced.}$$

The Ideal-Gas Equation

(In *Solutions to Exercises*, the symbol for molar mass is $\mathcal{M}$.)

10.23 (a) PV = nRT; P in atmospheres, V in liters, n in moles, T in kelvins

(b) An ideal gas exhibits pressure, volume and temperature relationships which are described by the equation PV = nRT. (An ideal gas obeys the ideal-gas equation.)

10.24 (a) STP stands for standard temperature, 0°C (or 273 K), and standard pressure, 1 atm.

(b) $V = \frac{nRT}{P}$; $V = 1 \text{ mol} \times \frac{0.08206 \text{ L•atm}}{\text{K•mol}} \times \frac{273 \text{ K}}{1 \text{ atm}}$

V = 22.4 L for 1 mole of gas at STP

(c) 25°C + 273 = 298 K

$V = \frac{nRT}{P}$; $V = 1 \text{ mol} \times \frac{0.08206 \text{ L•atm}}{\text{K•mol}} \times \frac{298 \text{ K}}{1 \text{ atm}}$

V = 24.5 L for 1 mol of gas at 1 atm and 25°C

10.25 PV = nRT. At constant volume and temperature, P is directly proportional to n.

For samples with equal masses of gas, the gas with $\mathcal{M}$ = 30 will have twice as many moles of particles and twice the pressure. Thus, flask A contains the gas with $\mathcal{M}$ = 30 and flask B contains the gas with $\mathcal{M}$ = 60.

10.26 $n = g/\mathcal{M}$; $PV = nRT = gRT/\mathcal{M}$; $\mathcal{M} = gRT/PV$.

2-L flask: $\mathcal{M}$ = 4.8 T/2.0(X) = 2.4 T/X

3-L flask: $\mathcal{M}$ = 0.36 T/3.0 (0.1 X) = 1.2 T/X

The molar masses of the two gases are not equal. The gas in the 2-L flask has a molar mass that is twice as large as the gas in the 3-L flask.

10.27 *Analyze/Plan*. Follow the strategy for calculations involving many variables. *Solve*:

(a) n = 2.46 mol, P = 1.28 atm, T = -6°C = 267 K

$V = \frac{nRT}{P} = 2.46 \text{ mol} \times \frac{0.08206 \text{ L•atm}}{\text{K•mol}} \times \frac{267 \text{ K}}{1.28 \text{ atm}} = 42.1 \text{ L}$

(b) $n = 4.79 \times 10^{-2}$ mol, V = 135 mL = 0.135 L

$P = 720 \text{ torr} \times \frac{1 \text{ atm}}{760 \text{ torr}} = 0.9474 = 0.947 \text{ atm}$

$T = \frac{PV}{nR} = 0.9474 \text{ atm} \times \frac{0.135 \text{ L}}{0.0479 \text{ mol}} \times \frac{1 \text{ K•mol}}{0.08206 \text{ L•atm}} = 32.5 \text{ K}$

(c) $n = 5.52 \times 10^{-2}$ mol, V = 413 mL = 0.413 L, T = 88°C = 361 K

$P = \frac{nRT}{V}$; $= 0.0552 \text{ mol} \times \frac{0.08206 \text{ L•atm}}{\text{K•mol}} \times \frac{361 \text{ K}}{0.413 \text{ L}} = 3.96 \text{ atm}$

(d) V = 88.4 L, T = 54°C = 327 K,

$P = 9.84 \text{ kPa} \times \frac{1 \text{ atm}}{101.325 \text{ kPa}} = 0.09711 = 0.0971 \text{ atm}$

$n = \frac{PV}{RT} = 0.09711 \text{ atm} \times \frac{\text{K•mol}}{0.08206 \text{ L•atm}} \times \frac{88.4 \text{ L}}{327 \text{ K}} = 0.320 \text{ mol}$

10.28 (a) n = 0.215 mol, V = 338 mL = 0.338 L, T = 32°C = 305 K

$$P = \frac{nRT}{V} = 0.215 \text{ mol} \times \frac{0.08206 \text{ L•atm}}{\text{K•atm}} \times \frac{305 \text{ K}}{0.338 \text{ L}} = 15.9 \text{ atm}$$

(b) n = 0.0412 mol, V = 3.00 L, P = 1.05 atm

$$T = \frac{PV}{nR} = \frac{1.05 \text{ atm} \times 3.00 \text{ L}}{0.0412 \text{ mol}} \times \frac{\text{K•mol}}{0.08206 \text{ L•atm}} = 932 \text{ K}$$

(c) $V = 98.5 \text{ L}, \ T = 236 \text{ K}, \ P = 690 \text{ torr} \times \frac{1 \text{ atm}}{760 \text{ torr}} = 0.9079 = 0.908 \text{ atm}$

$$n = \frac{PV}{RT} = 0.9079 \text{ atm} \times \frac{\text{K•mol}}{0.08206 \text{ L•atm}} \times \frac{98.5 \text{ L}}{236 \text{ K}} = 4.62 \text{ mol}$$

(d) $n = 5.48 \times 10^{-3}$ mol, T = 55°C = 328 K

$$P = 3.87 \text{ kPa} \times \frac{1 \text{ atm}}{101.325 \text{ kPa}} = 0.03819 = 0.0382 \text{ atm}$$

$$V = \frac{nRT}{P} = 5.48 \times 10^{-3} \text{ mol} \times \frac{0.08206 \text{ L•atm}}{\text{K•mol}} \times \frac{328 \text{ K}}{0.03819 \text{ atm}} = 3.86 \text{ L}$$

10.29 *Analyze/Plan.* Follow the strategy for calculations involving many variables. *Solve:*

$n = g/\mathcal{M}$; PV = nRT; $PV = gRT/\mathcal{M}$; $g = \mathcal{M}PV/RT$

P = 1.0 atm, T = 23°C = 296 K, $V = 2.0 \times 10^5 \text{ m}^3$. Change m^3 to L, then calculate grams (or kg).

$$2.0 \times 10^5 \text{ m}^3 \times \frac{10^3 \text{ dm}^3}{1 \text{ m}^3} \times \frac{1 \text{ L}}{1 \text{ dm}^3} = 2.0 \times 10^8 \text{ L } H_2$$

$$g = \frac{2.02 \text{ g } H_2}{1 \text{ mol } H_2} \times \frac{\text{K•mol}}{0.08206 \text{ L•atm}} \times \frac{1.0 \text{ atm} \times 2.0 \times 10^8 \text{ L}}{296 \text{ K}} = 1.7 \times 10^7 \text{ g} = 1.7 \times 10^4 \text{ kg } H_2$$

10.30 Find the volume of the tube in cm^3; $1 \text{ cm}^3 = 1 \text{ mL}$.

r = d/2 = 4.5 cm/2 = 2.25 = 2.3 cm; $h = 5.3 \text{ m} = 5.3 \times 10^2 \text{ cm}$

$V = \pi r^2 h = 3.14159 \times (2.25 \text{ cm})^2 \times (5.3 \times 10^2 \text{ cm}) = 8.429 \times 10^3 \text{ cm}^3 = 8.4 \text{ L}$

$$PV = \frac{g}{\mathcal{M}} RT; \ g = \frac{\mathcal{M}PV}{RT}; \ P = 2.03 \text{ torr} \times \frac{1 \text{ atm}}{760 \text{ torr}} = 2.621 \times 10^{-3} = 2.62 \times 10^{-3} \text{ atm}$$

$$g = \frac{20.18 \text{ g Ne}}{1 \text{ mol Ne}} \times \frac{\text{K•mol}}{0.08206 \text{ L•atm}} \times \frac{2.67 \times 10^{-3} \text{ atm} \times 8.429 \text{ L}}{308 \text{ K}} = 0.018 \text{ g Ne}$$

10.31 *Analyze/Plan.* Follow the strategy for calculations involving many variables. *Solve:*

(a) $P = \frac{nRT}{V}$; $n = 0.29 \text{ kg } O_2 \times \frac{1000 \text{ g}}{1 \text{ kg}} \times \frac{1 \text{ mol } O_2}{32.00 \text{ g } O_2} = 9.0625 = 9.1 \text{ mol}$; V = 2.3 L;

T = 273 + 9°C = 282 K

$$P = \frac{9.0625 \text{ mol}}{2.3 \text{ L}} \times \frac{0.08206 \text{ L•atm}}{\text{K•mol}} \times 282 \text{ K} = 91 \text{ atm}$$

(b) $V = \frac{nRT}{P}; = \frac{9.0625 \text{ mol}}{0.95 \text{ atm}} \times \frac{0.08206 \text{ L} \bullet \text{atm}}{\text{K} \bullet \text{mol}} \times 299 \text{ K} = 2.3 \times 10^2 \text{ L}$

10.32 (a) $V = 0.456 \text{ L}, T = 23°\text{C} = 296 \text{ K}, n = 3.18 \text{ g } C_3H_8 \times \frac{1 \text{ mol } C_3H_8}{44.1 \text{ g } C_3H_8} = 0.07211$

$= 0.0721 \text{ mol}$

$P = \frac{nRT}{V} = 0.07211 \text{ mol} \times \frac{0.08206 \text{ L} \bullet \text{atm}}{\text{K} \bullet \text{mol}} \times \frac{296 \text{ K}}{0.456 \text{ L}} = 3.84 \text{ atm}$

(b) STP = 1.00 atm, 273 K

$V = \frac{nRT}{P} = 0.07211 \text{ mol} \times \frac{0.08206 \text{ L} \bullet \text{atm}}{\text{K} \bullet \text{mol}} \times \frac{273 \text{ K}}{1.00 \text{atm}} = 1.16154 = 1.62 \text{ L}$

(c) $°\text{C} = 5/9 (°\text{F} - 32°); \text{K} = °\text{C} + 273.15 = 5/9 (130°\text{F} - 32°) + 273.15 = 327.59 = 328 \text{ K}$

$P = \frac{nRT}{V} = 0.07211 \text{ mol} \times \frac{0.08206 \text{ L} \bullet \text{atm}}{\text{K} \bullet \text{mol}} \times \frac{327.59 \text{ K}}{0.456 \text{ L}} = 4.251 = 4.25 \text{ atm}$

10.33 *Analyze/Plan*. Follow the strategy for calculations involving many variables. *Solve*:

(a) $g = \frac{\mathcal{M} PV}{RT}; V = 9.22 \text{ L}, T = 24°\text{C} = 297 \text{ K}, P = 1124 \text{ torr} \times \frac{1 \text{ atm}}{760 \text{ torr}} = 1.4789$

$= 1.479 \text{ atm}$

$g = \frac{70.91 \text{ g } Cl_2}{1 \text{ mol } Cl_2} \times \frac{\text{K} \bullet \text{mol}}{0.08206 \text{ L} \bullet \text{atm}} \times \frac{1.4789 \text{ atm}}{297 \text{ K}} \times 9.22 \text{ L} = 39.7 \text{ g } Cl_2$

(b) $V_2 = \frac{P_1V_1T_2}{T_1P_2} = \frac{1124 \text{ torr} \times 9.22 \text{ L} \times 273 \text{ K}}{297 \text{ K} \times 760 \text{ torr}} = 12.5 \text{ L}$

(c) $T_2 = \frac{P_2V_2T_1}{P_1V_1} = \frac{876 \text{ torr} \times 15.00 \text{ L} \times 297 \text{ K}}{1124 \text{ torr} \times 9.22 \text{ L}} = 377 \text{ K}$

(d) $P_2 = \frac{P_1V_1T_2}{V_2T_1} = \frac{1124 \text{ torr} \times 9.22 \text{ L} \times 331 \text{ K}}{6.00 \text{ L} \times 297 \text{ K}} = 1.92 \times 10^3 \text{ torr} = 2.53 \text{ atm}$

10.34 (a) $g = \frac{\mathcal{M} PV}{RT}; V = 68.0 \text{ L}, T = 23°\text{C} = 296 \text{ K}, P = 15{,}900 \text{ kPa} \times \frac{1 \text{ atm}}{101.325} = 156.92$

$= 157 \text{ atm}$

$g = \frac{32.0 \text{ g } O_2}{1 \text{ mol } O_2} \times \frac{\text{K} \bullet \text{mol}}{0.08206 \text{ L} \bullet \text{atm}} \times \frac{156.92 \text{ atm}}{296 \text{ K}} \times 68.0 \text{ L} = 1.406 \times 10^4 \text{ g } O_2$

$= 14.1 \text{ kg } O_2$

(b) $V_2 = \frac{P_1V_1T_2}{T_1P_2} = \frac{15{,}900 \text{ kPa} \times 68.0 \text{ L} \times 273 \text{ K}}{296 \text{ K} \times 101.325 \text{ kPa}} = 9.84 \times 10^3 \text{ L}$

(c) $T_2 = \frac{P_2T_1}{P_1} = \frac{170 \text{ atm} \times 296 \text{ K}}{15{,}900 \text{ kPa}} \times \frac{101.325 \text{ kPa}}{1 \text{ atm}} = 321 \text{ K}$

(d) $P_2 = \frac{P_1V_1T_2}{V_2T_1} = \frac{15{,}900 \text{ kPa} \times 68.0 \text{ L} \times 297 \text{ K}}{52.6 \text{ L} \times 296 \text{ K}} = 20{,}625 = 2.06 \times 10^4 \text{ kPa}$

10.35 *Analyze.* Given: mass of cockroach, rate of O_2 consumption, temperature, percent O_2 in air, volume of air. Find: mol O_2 consumed per hour; mol O_2 in 1 quart of air; mol O_2 consumed in 48 hr.

(a) *Plan/Solve.* V of O_2 consumed = rate of consumption × mass × time. n = PV/RT.

$$5.2 \text{ g} \times 1 \text{ hr} \times \frac{0.8 \text{ mL } O_2}{1 \text{ g} \cdot \text{hr}} = 4.16 = 4 \text{ mL } O_2 \text{ consumed}$$

$$n = \frac{PV}{RT} = 1 \text{ atm} \times \frac{\text{K} \cdot \text{mol}}{0.08206 \text{ L} \cdot \text{atm}} \times \frac{0.00416 \text{ L}}{297 \text{ K}} = 1.71 \times 10^{-4} = 2 \times 10^{-4} \text{ mol } O_2$$

(b) *Plan/Solve.* qt air → L air → L O_2 available. mol O_2 available = PV/RT. mol O_2/hr (from part (a)) → total mol O_2 consumed. Compare O_2 available and O_2 consumed.

$$1 \text{ qt air} \times \frac{0.946 \text{ L}}{1 \text{ qt}} \times 0.21 \; O_2 \text{ in air} = 0.199 \text{ L } O_2 \text{ available}$$

$$n = 1 \text{ atm} \times \frac{\text{K} \cdot \text{mol}}{0.08206 \text{ L} \cdot \text{atm}} \times \frac{0.199 \text{ L}}{297 \text{ K}} = 8.16 \times 10^{-3} = 8 \times 10^{-3} \text{ mol } O_2 \text{ available}$$

$$\text{roach uses } \frac{1.71 \times 10^{-4} \text{ mol}}{1 \text{ hr}} \times 48 \text{ hr} = 8.19 \times 10^{-3} = 8 \times 10^{-3} \text{ mol } O_2 \text{ consumed}$$

Not only does the roach use 20% of the available O_2, it needs all the O_2 in the jar.

10.36 (a) $P = \dfrac{gRT}{\mathcal{M}V}$; mass = 1800×10^{-9} g = 1.8×10^{-6} g; V = 1 m^3 = 1×10^3 L

$$P = \frac{1.8 \times 10^{-6} \text{ g Hg} \times 1 \text{ mol Hg}}{200.6 \text{ g Hg}} \times \frac{0.08206 \text{ L} \cdot \text{atm}}{\text{K} \cdot \text{mol}} \times \frac{283 \text{ K}}{1 \times 10^3 \text{ L}} = 2.1 \times 10^{-10} \text{ atm}$$

(b)

$$\frac{1.8 \times 10^{-6} \text{ g Hg}}{1 \text{ m}^3} \times \frac{1 \text{ mol Hg}}{200.6 \text{ g Hg}} \times \frac{6.022 \times 10^{23} \text{ Hg atoms}}{1 \text{ mol Hg}} = 5.4 \times 10^{15} \text{ Hg atoms/m}^3$$

(c)

$$1600 \text{ km}^3 \times \frac{1000^3 \text{ m}^3}{1 \text{ km}^3} \times \frac{1.8 \times 10^{-6} \text{ g Hg}}{1 \text{ m}^3} = 2.9 \times 10^6 \text{ g Hg/day}$$

Further Applications of the Ideal-Gas Equation

10.37 (c) $Cl_2(g)$ is the most dense at 1.00 at and 298 K. Gas density is directly proportional to molar mass and pressure, and inversely proportional to temperature [Equation 10.10]. For gas samples at the same conditions, molar mass determines density. Of the three gases listed, Cl_2 has the largest molar mass.

10.38 (b) HCl(g) is least dense. For gases at the same conditions, density is directly proportional to molar mass, and HCl has the smallest molar mass.

10.39 (c) Because the helium atoms are of lower mass than the average air molecule, the helium gas is less dense than air. The balloon thus weighs less than the air displaced by its volume.

10.40 (b) Xe atoms have a higher mass than N_2 molecules. Because both gases at STP have the same number of molecules per unit volume, the Xe gas must be denser.

10.41 *Analyze/Plan.* Conditions (P, V, T) and amounts of gases are given. Rearrange the relationship $PV\mathcal{M} = gRT$ to obtain the desired of quantity, paying attention (as always!) to units. *Solve*:

(a) $d = \dfrac{\mathcal{M}P}{RT}$; $\mathcal{M} = 46.0$ g/mol; P = 0.970 atm, T = 35°C = 308 K

$$d = \frac{46.0 \text{ g } NO_2}{1 \text{ mol}} \times \frac{K \bullet mol}{0.08206 \text{ L} \bullet atm} \times \frac{0.970 \text{ atm}}{308 \text{ K}} = 1.77 \text{ g/L}$$

(b) $$\mathcal{M} = \frac{gRT}{PV} = \frac{2.50 \text{ g}}{0.875 \text{ L}} \times \frac{0.08206 \text{ L} \bullet atm}{K \bullet mol} \times \frac{308 \text{ K}}{685 \text{ torr}} \times \frac{760 \text{ torr}}{1 \text{ atm}} = 80.1 \text{ g/mol}$$

10.42 (a) $d = \dfrac{\mathcal{M}P}{RT}$; $\mathcal{M} = 146.1$ g/mol, T = 32°C = 305 K, P = 455 torr

$$d = \frac{146.1 \text{ g}}{1 \text{ mol}} \times \frac{K \bullet mol}{0.08206 \text{ L} \bullet atm} \times \frac{455 \text{ torr}}{305 \text{ K}} \times \frac{1 \text{ atm}}{760 \text{ torr}} = 3.49 \text{ g/L}$$

(b) $$\mathcal{M} = \frac{dRT}{P} = \frac{6.345 \text{ g}}{1 \text{ L}} \times \frac{0.08206 \text{ L} \bullet atm}{K \bullet mol} \times \frac{295 \text{ K}}{743 \text{ torr}} \times \frac{760 \text{ torr}}{1 \text{ atm}} = 157 \text{ g/mol}$$

10.43 *Analyze/Plan.* Given: mass, conditions (P, V, T) of unknown gas. Find: molar mass. $\mathcal{M} = gRT/PV$. *Solve*:

$$\mathcal{M} = \frac{gRT}{PV} = \frac{1.012 \text{ g}}{0.354 \text{ L}} \times \frac{0.08206 \text{ L} \bullet atm}{K \bullet atm} \times \frac{372 \text{ K}}{742 \text{ torr}} \times \frac{760 \text{ torr}}{1 \text{ atm}} = 89.4 \text{ g/mol}$$

10.44 $$\mathcal{M} = \frac{gRT}{PV} = \frac{0.963 \text{ g}}{0.418 \text{ L}} \times \frac{0.08206 \text{ L} \bullet atm}{K \bullet atm} \times \frac{373 \text{ K}}{752 \text{ torr}} \times \frac{760 \text{ torr}}{1 \text{ atm}} = 71.3 \text{ g/mol}$$

10.45 *Analyze/Plan.* Follow the logic in Sample Exercise 10.9. *Solve.*

$$\text{mol } O_2 = \frac{PV}{RT} = 3.5 \times 10^{-6} \text{ torr} \times \frac{1 \text{ atm}}{760 \text{ torr}} \times \frac{K \bullet mol}{0.08206 \text{ L} \bullet atm} \times \frac{0.382 \text{ L}}{300 \text{ K}} = 7.146 \times 10^{-11}$$

$$= 7.1 \times 10^{-11} \text{ mol } O_2$$

$$7.146 \times 10^{-11} \text{ mol } O_2 \times \frac{2 \text{ mol Mg}}{1 \text{ mol } O_2} \times \frac{24.3 \text{ g Mg}}{1 \text{ mol Mg}} = 3.5 \times 10^{-9} \text{ g Mg}$$

10.46 $n_{H_2} = \dfrac{P_{H_2}V}{RT}$; $P = 814 \text{ torr} \times \dfrac{1 \text{ atm}}{760 \text{ torr}} = 1.071 = 1.07 \text{ atm}$

$$n_{H_2} = 1.071 \text{ atm} \times \frac{K \bullet mol}{0.08206 \text{ L} \bullet atm} \times \frac{64.5 \text{ L}}{305 \text{ K}} = 2.760 = 2.76 \text{ mol } H_2$$

$$2.760 \text{ mol } H_2 \times \frac{1 \text{ mol } CaH_2}{2 \text{ mol } H_2} \times \frac{42.10 \text{ g } CaH_2}{1 \text{ mol } CaH_2} = 58.1 \text{ g } CaH_2$$

10.47 *Analyze/Plan*. kg H_2SO_4 → g H_2SO_4 → mol H_2SO_4 → mol NH_3 → V NH_3 *Solve*:

$$87 \text{ kg} \times \frac{1000 \text{ g}}{1 \text{ kg}} = 8.7 \times 10^4 \text{ g } H_2SO_4 \times \frac{1 \text{ mol}}{98.08 \text{ g}} = 887.03 = 887 \text{ mol } H_2SO_4$$

$$887.03 \text{ mol } H_2SO_4 \times \frac{2 \text{ mol } NH_3}{1 \text{ mol } H_2SO_4} = 1.774 \times 10^3 = 1.77 \times 10^3 \text{ mol } NH_3$$

$$V_{NH_3} = \frac{nRT}{P} = 1.774 \times 10^3 \text{ mol} \times \frac{0.08206 \text{ L} \bullet \text{atm}}{\text{K} \bullet \text{mol}} \times \frac{315 \text{ K}}{15.6 \text{ atm}} = 2.94 \times 10^3 \text{ L } NH_3$$

10.48 g glucose → mol glucose → mol CO_2 → V CO_2

$$24.5 \text{ g} \times \frac{1 \text{ mol glucose}}{180.1 \text{ g}} \times \frac{6 \text{ mol } CO_2}{1 \text{ mol glucose}} = 0.8162 = 0.816 \text{ mol } CO_2$$

$$V = \frac{nRT}{P} = 0.8162 \text{ mol} \times \frac{0.08206 \text{ L} \bullet \text{atm}}{\text{K} \bullet \text{mol}} \times \frac{310 \text{ K}}{0.970 \text{ atm}} = 21.4 \text{ L } CO_2$$

10.49 *Analyze/Plan*. The gas sample is a mixture of $H_2(g)$ and $H_2O(g)$. Find the partial pressure of $H_2(g)$ and then the moles of $H_2(g)$ and Zn(s). *Solve*:

$$P_t = 738 \text{ torr} = P_{H_2} + P_{H_2O}$$

From Appendix B, the vapor pressure of H_2O at 24°C = 22.38 torr

$$P_{H_2} = (738 \text{ torr} - 22.38 \text{ torr}) \times \frac{1 \text{ atm}}{760 \text{ torr}} = 0.9416 = 0.942 \text{ atm}$$

$$n_{H_2} = \frac{P_{H_2}V}{RT} = 0.9416 \text{ atm} \times \frac{\text{K} \bullet \text{mol}}{0.08206 \text{ L} \bullet \text{atm}} \times \frac{0.159 \text{ L}}{297 \text{ K}} = 0.006143 = 0.00614 \text{ mol } H_2$$

$$0.006143 \text{ mol } H_2 \times \frac{1 \text{ mol Zn}}{1 \text{ mol } H_2} \times \frac{65.39 \text{ g Zn}}{1 \text{ mol Zn}} = 0.402 \text{ g Zn}$$

10.50 The gas sample is a mixture of $C_2H_2(g)$ and $H_2O(g)$. Find the partial pressure of C_2H_2, then moles CaC_2 and C_2H_2.

$$P_t = 726 \text{ torr} = P_{C_2H_2} + P_{H_2O}. \quad P_{H_2O} \text{ at } 21°\text{C} = 18.65 \text{ torr}$$

$$P_{C_2H_2} = (748 \text{ torr} - 18.65 \text{ torr}) \times \frac{1 \text{ atm}}{760 \text{ torr}} = 0.9597 = 0.960 \text{ atm}$$

$$3.26 \text{ g } CaC_2 \times \frac{1 \text{ mol } CaC_2}{64.10 \text{ g}} \times \frac{1 \text{ mol } C_2H_2}{1 \text{ mol } CaC_2} = 0.050858 = 0.0509 \text{ mol } C_2H_2$$

$$V = 0.050858 \text{ mol} \times \frac{0.08206 \text{ L} \bullet \text{atm}}{\text{K} \bullet \text{mol}} \times \frac{294 \text{ K}}{0.9597 \text{ atm}} = 1.28 \text{ L}$$

Partial Pressures

10.51 (a) When the stopcock is opened, the volume occupied by $N_2(g)$ increases from 2.0 L to 5.0 L. At constant T, $P_1V_1 = P_2V_2$. 1.0 atm × 2.0 L = P_2 × 5.0 L; P_2 = 0.40 atm

(b) When the gases mix, the volume of $O_2(g)$ increases from 3.0 L to 5.0 L. At constant T, $P_1V_1 = P_2V_2$. 2.0 atm × 3.0 L = P_2 × 5.0 L; P_2 = 1.2 atm

(c) $P_T = P_{N_2} + P_{O_2}$ = 0.40 atm + 1.2 atm = 1.6 atm

10.52 (a) The partial pressure of gas A is **not affected** by the addition of gas C. The partial pressure of A depends only on moles of A, volume of container and conditions; none of these factors change when gas C is added.

(b) The total pressure in the vessel **increases** when gas C is added, because the total number of moles of gas increases.

(c) The mole fraction of gas B **decreases** when gas C is added. The moles of gas B stay the same, but the total moles increase, so the mole fraction of B (n_B/n_t) decreases.

10.53 *Analyze.* Given: amount, V, T of 3 gases. Find: P of each gas, total P.

Plan. P = nRT/V; $P_T = P_1 + P_2 + P_3 + \ldots$ *Solve*:

(a) $$P_{He} = \frac{nRT}{V} = 0.538 \text{ mol} \times \frac{0.08206 \text{ L}\bullet\text{atm}}{\text{K}\bullet\text{atm}} \times \frac{298 \text{ K}}{7.00 \text{ L}} = 1.88 \text{ atm}$$

$$P_{Ne} = \frac{nRT}{V} = 0.315 \text{ mol} \times \frac{0.08206 \text{ L}\bullet\text{atm}}{\text{K}\bullet\text{atm}} \times \frac{298 \text{ K}}{7.00 \text{ L}} = 1.10 \text{ atm}$$

$$P_{Ar} = \frac{nRT}{V} = 0.103 \text{ mol} \times \frac{0.08206 \text{ L}\bullet\text{atm}}{\text{K}\bullet\text{atm}} \times \frac{298 \text{ K}}{7.00 \text{ L}} = 0.360 \text{ atm}$$

(b) P_t = 1.88 atm + 1.10 atm + 0.360 atm = 3.34 atm

10.54 (a) $$3.15 \text{ g } CH_4 \times \frac{1 \text{ mol } CH_4}{16.04 \text{ g } CH_4} = 0.1964 = 0.196 \text{ mol } CH_4$$

$$P_{CH_4} = \frac{nRT}{V} = 0.1964 \text{ mol} \times \frac{0.08206 \text{ L}\bullet\text{atm}}{\text{K}\bullet\text{mol}} \times \frac{337 \text{ K}}{2.00 \text{ L}} = 2.715 = 2.72 \text{ atm}$$

$$3.15 \text{ g } C_2H_4 \times \frac{1 \text{ mol } C_2H_4}{28.05 \text{ g } C_2H_4} = 0.1123 = 0.112 \text{ mol } C_2H_4$$

$$P_{C_2H_4} = \frac{nRT}{V} = 0.1123 \text{ mol } C_2H_4 \times \frac{0.08206 \text{ L}\bullet\text{atm}}{\text{K}\bullet\text{mol}} \times \frac{337 \text{ K}}{2.00 \text{ L}} = 1.553 = 1.55 \text{ atm}$$

$$3.15 \text{ g } C_4H_{10} \times \frac{1 \text{ mol } C_4H_{10}}{58.12 \text{ g } C_4H_{10}} = 0.05420 = 0.0542 \text{ mol } C_4H_{10}$$

$$P_{C_4H_{10}} = \frac{nRT}{V} = 0.05420 \text{ mol} \times \frac{0.08206 \text{ L}\bullet\text{atm}}{\text{K}\bullet\text{mol}} \times \frac{337 \text{ K}}{2.00 \text{ L}} = 0.7494 = 0.749 \text{ atm}$$

(b) P_t = 2.715 atm + 1.553 atm + 0.749 atm = 5.017 = 5.02 atm

10.55 *Analyze/Plan*. The partial pressure of each component is equal to the mole fraction of that gas times the total pressure of the mixture. Find the mole fraction of each component and then its partial pressure. *Solve*:

n_t = 0.75 mol N_2 + 0.30 mol O_2 + 0.15 mol CO_2 = 1.20 mol

$$\chi_{N_2} = \frac{0.75}{1.20} = 0.625 = 0.63; \; P_{N_2} = 0.625 \times 1.56 \text{ atm} = 0.98 \text{ atm}$$

$$\chi_{O_2} = \frac{0.30}{1.20} = 0.250 = 0.25; \; P_{O_2} = 0.250 \times 1.56 \text{ atm} = 0.39 \text{ atm}$$

$$\chi_{CO_2} = \frac{0.15}{1.20} = 0.125 = 0.13; \; P_{CO_2} = 0.125 \times 1.56 \text{ atm} = 0.20 \text{ atm}$$

10.56 $$n_{N_2} = 12.47 \text{ g } N_2 \times \frac{1 \text{ mol}}{28.02 \text{ g}} = 0.4450 \text{ mol}; \; n_{H_2} = 1.98 \text{ g } H_2 \times \frac{1 \text{ mol}}{2.016 \text{ g}} = 0.982 \text{ mol}$$

$$n_{NH_3} = 8.15 \text{ g } NH_3 \times \frac{1 \text{ mol}}{17.03 \text{ g}} = 0.479 \text{ mol}; \; n_t = 0.4450 + 0.982 + 0.479 = 1.906 \text{ mol}$$

$$P_{N_2} = \frac{n_{N_2}}{n_t} \times P_t = \frac{0.4450}{1.906} \times 2.35 \text{ atm} = 0.549 \text{ atm}$$

$$P_{H_2} = \frac{0.982}{1.906} \times 2.35 \text{ atm} = 1.21 \text{ atm}; \; P_{NH_3} = \frac{0.479}{1.906} \times 2.35 \text{ atm} = 0.591 \text{ atm}$$

10.57 *Analyze/Plan*. Mole fraction = pressure fraction. Find the desired mole fraction of O_2 and change to mole percent. *Solve*:

$$\chi_{O_2} = \frac{P_{O_2}}{P_t} = \frac{0.21 \text{ atm}}{8.38 \text{ atm}} = 0.025; \text{ mole \%} = 0.025 \times 100 = 2.5\%$$

10.58 (a) $$n_{O_2} = 6.55 \text{ g } O_2 \times \frac{1 \text{ mol}}{32.00 \text{ g}} = 0.205 \text{ mol}; \; n_{N_2} = 4.92 \text{ g } N_2 \times \frac{1 \text{ mol}}{28.02 \text{ g}} = 0.176 \text{ mol}$$

$$n_{H_2} = 1.32 \text{ g } H_2 \times \frac{1 \text{ mol}}{2.016 \text{ g}} = 0.655 \text{ mol}; \; n_t = 0.205 + 0.176 + 0.655 = 1.036 \text{ mol}$$

$$\chi_{O_2} = \frac{n_{O_2}}{n_t} = \frac{0.205}{1.036} = 0.198; \; \chi_{N_2} = \frac{n_{N_2}}{n_t} = \frac{0.176}{1.036} = 0.170$$

$$\chi_{H_2} = \frac{0.655}{1.036} = 0.632$$

(b) $$P_{O_2} = n \times \frac{RT}{V}; \; P_{O_2} = 0.205 \text{ mol} \times \frac{0.08206 \text{ L} \cdot \text{atm}}{\text{K} \cdot \text{mol}} \times \frac{288 \text{ K}}{12.40 \text{ L}} = 0.391 \text{ atm}$$

$$P_{N_2} = 0.176 \text{ mol} \times \frac{0.08206 \text{ L} \cdot \text{atm}}{\text{K} \cdot \text{mol}} \times \frac{288 \text{ K}}{12.40 \text{ L}} = 0.335 \text{ atm}$$

$$P_{H_2} = 0.655 \text{ mol} \times \frac{0.08206 \text{ L} \cdot \text{atm}}{\text{K} \cdot \text{mol}} \times \frac{288 \text{ K}}{12.40 \text{ L}} = 1.25 \text{ atm}$$

10.59 *Analyze/Plan.* $N_2(g)$ and $O_2(g)$ undergo changes of conditions and are mixed. Calculate the new pressure of each gas and add them to obtain the total pressure of the mixture. $P_2 = P_1V_1T_2/V_2T_1$; $P_T = P_{N_2} + P_{O_2}$. *Solve:*

$$P_{N_2} = \frac{P_1V_1T_2}{V_2T_1} = \frac{3.80 \text{ atm} \times 1.00 \text{ L} \times 293 \text{ K}}{10.0 \text{ L} \times 299 \text{ K}} = 0.372 \text{ atm}$$

$$P_{O_2} = \frac{P_1V_1T_2}{V_2T_1} = \frac{4.75 \text{ atm} \times 5.00 \text{ L} \times 293 \text{ K}}{10.0 \text{ L} \times 299 \text{ K}} = 2.33 \text{ atm}$$

P_T = 0.372 atm + 2.33 atm = 2.70 atm

10.60 Calculate the pressure of the gas in the second vessel directly from mass and conditions using the ideal-gas equation.

(a) $$P_{SO_2} = \frac{gRT}{\mathcal{M}V} = \frac{5.25 \text{ g } SO_2}{64.07 \text{ g } SO_2/\text{mol}} \times \frac{0.08206 \text{ L} \bullet \text{atm}}{\text{K} \bullet \text{mol}} \times \frac{298 \text{ K}}{13.6 \text{ L}} = 0.14734$$

= 0.147 atm

(b) $$P_{N_2} = \frac{gRT}{\mathcal{M}V} = \frac{2.35 \text{ g } N_2}{28.01 \text{ g } N_2/\text{mol}} \times \frac{0.08206 \text{ L} \bullet \text{atm}}{\text{K} \bullet \text{mol}} \times \frac{298 \text{ K}}{13.6 \text{ L}} = 0.15086$$

= 0.151 atm

(c) $P_T = P_{SO_2} + P_{N_2}$ = 0.14734 atm + 0.15086 atm = 0.298 atm

Kinetic - Molecular Theory; Graham's Law

10.61 (a) Increase in temperature at constant volume, decrease in volume, increase in pressure (b) decrease in temperature (c) increase in volume (d) increase in temperature

10.62 (a) False. The average kinetic energy per molecule in a collection of gas molecules is the same for all gases at the same temperature. (b) True. (c) False. The molecules in a gas sample at a given temperature exhibit a distribution of kinetic energies. (d) True.

10.63 The fact that gases are readily compressible supports the assumption that most of the volume of a gas sample is empty space.

10.64 Newton's model provides no explanation of the effect of a change in temperature on the pressure of a gas at constant volume or on the volume of a gas at constant pressure. On the other hand, the assumption that the average kinetic energy of gas molecules increases with increasing temperature explains Charles Law, that an increase in temperature requires an increase in volume to maintain constant pressure.

10.65 *Analyze/Plan.* We have samples of two different gases at different pressures and temperatures. Compare the two samples by considering the postulates of the kinetic-molecular theory that pertain to the quantities in (a)-(d). *Solve:*

(a) $n \propto P/T$ (V/R is the same for A and B.) Since P is greater and T is smaller for vessel A, it has more molecules.

(b) Vessel A has more molecules but the molar mass of CO is smaller than the molar mass of SO_2, so we need to calculate the masses. Since volume is not specified, calculate g/L.

$$\frac{g_A}{V} = \frac{\mathcal{M}P}{RT} = \frac{28.01 \text{ g CO}}{1 \text{ mol CO}} \times \frac{\text{K} \bullet \text{mol}}{0.08206 \text{ L} \bullet \text{atm}} \times \frac{1 \text{ atm}}{273 \text{ K}} = 1.25 \text{ g CO/L}$$

$$\frac{g_B}{V} = \frac{\mathcal{M}P}{RT} = \frac{64.07 \text{ g } SO_2}{1 \text{ mol } SO_2} \times \frac{\text{K} \bullet \text{mol}}{0.08206 \text{ L} \bullet \text{atm}} \times \frac{0.5 \text{ atm}}{293 \text{ K}} = 1.33 \text{ g } SO_2\text{/L}$$

Vessel B has more mass.

(c) Vessel B is at a higher temperature so the average kinetic energy of its molecules is higher.

(d) The two factors that affect rms speed are temperature and molar mass. The molecules in vessel A have smaller molar mass but are at the lower temperature, so we must calculate the rms speeds.

Mathematically, according to Equation 10.24,

$$\frac{u_A}{u_B} = \sqrt{\frac{T_A / \mathcal{M}_A}{T_B / \mathcal{M}_B}} = \sqrt{\frac{273/28.01}{293/64.07}} = 1.46$$

The ratio is greater than 1; vessel A has the greater rms speed.

10.66 (a) They have the same number of molecules (equal volumes of gases at the same temperature and pressure contain equal numbers of molecules).

(b) N_2 is more dense because it has the larger molar mass. Since the volumes of the samples and the number of molecules are equal, the gas with the larger molar mass will have the greater density.

(c) The average kinetic energies are equal (statement 5, section 10.7).

(d) CH_4 will effuse faster. The lighter the gas molecules, the faster they will effuse (Graham's Law).

10.67 (a) *Plan*. The greater the molecular (and molar) mass, the smaller the rms speed of the molecules. *Solve*: In order of increasing speed (and decreasing molar mass):

$$CO_2 \approx N_2O < F_2 < HF < H_2$$

(b) *Plan*. Follow the logic of Sample Exercise 10.14. *Solve*:

$$u_{H_2} = \sqrt{\frac{3RT}{\mathcal{M}}} = \left(\frac{3 \times 8.314\ kg \bullet m^2/s^2 \bullet K \bullet mol \times 300\ K}{2.02 \times 10^{-3}\ kg/mol}\right)^{1/2} = 1.92 \times 10^3\ m/s$$

$$u_{CO_2} = \left(\frac{3 \times 8.314\ kg \bullet m^2/s^2 \bullet K \bullet mol \times 300\ K}{44.0 \times 10^{-3}\ kg/mol}\right)^{1/2} = 4.12 \times 10^2\ m/s$$

As expected, the lighter molecule moves at the greater speed.

10.68 (a) The larger the molar mass, the slower the average speed (at constant temperature). In order of increasing speed (and decreasing molar mass):
$HBr < NF_3 < SO_2 < CO < Ne$

(b)
$$u = \sqrt{\frac{3RT}{\mathcal{M}}} = \left(\frac{3 \times 8.314\ kg \bullet m^2/s^2 \bullet K \bullet mol \times 298\ K}{71.0 \times 10^{-3}\ kg/mol}\right)^{1/2} = 324\ m/s$$

10.69 *Plan*. The heavier the molecule, the slower the rate of effusion. Thus, the order for increasing rate of effusion is in the order of decreasing mass. *Solve*:

rate $^2H^{37}Cl$ < rate $^1H^{37}Cl$ < rate $^2H^{35}Cl$ < rate $^1H^{35}Cl$

10.70
$$\frac{\text{rate } ^{235}U}{\text{rate } ^{238}U} = \sqrt{\frac{238.05}{235.04}} = \sqrt{1.0128} = 1.0064$$

There is a slightly greater rate enhancement for $^{235}U(g)$ atoms than $^{235}UF_6(g)$ molecules (1.0043), because ^{235}U is a greater percentage (100%) of the mass of the diffusing particles than in $^{235}UF_6$ molecules. The masses of the isotopes were taken from *The Handbook of Chemistry and Physics.*

10.71 *Analyze*. Given: relative effusion rates of two gases at same temperature. Find: molecular formula of one of the gases. *Plan*. Use Graham's law to calculate the formula weight of arsenic (III) sulfide, and thus the molecular formula. *Solve*:

$$\frac{\text{rate (sulfide)}}{\text{rate (Ar)}} = \left[\frac{39.9}{\mathcal{M}(\text{sulfide})}\right]^{1/2} = 0.28$$

$\mathcal{M}$ (sulfide) = $(39.9 / 0.28)^2$ = 510 g/mol (two significant figures)

The empirical formula of arsenic(III) sulfide is As_2S_3, which has a formula mass of 246.1. Twice this is 490 g/mol, close to the value estimated from the effusion experiment. Thus, the formula of the vapor phase molecule is As_4S_6.

10.72 The time required is proportional to the reciprocal of the effusion rate.

$$\frac{\text{rate (X)}}{\text{rate } (O_2)} = \frac{105\ s}{31\ s} = \left[\frac{32\ g\ O_2}{\mathcal{M}_x}\right]^{1/2}; \quad \mathcal{M}_x = 32\ g\ O_2 \times \left[\frac{105}{31}\right]^2 = 370\ g/mol\ \text{(two sig figs)}$$

Nonideal-Gas Behavior

10.73 (a) Nonideal gas behavior is observed at very high pressures and/or low temperatures.

(b) The real volumes of gas molecules and attractive intermolecular forces between molecules cause gases to behave nonideally.

10.74 Ideal-gas behavior is most likely to occur at high temperature and low pressure, so the atmosphere on Mercury is more likely to obey the ideal gas law. The higher temperature on Mercury means that the kinetic energies of the molecules will be larger relative to intermolecular attractive forces. Further, the gravitational attractive forces on Mercury are lower because the planet has a much smaller mass. This means that for the same column mass of gas (Figure 10.1), atmospheric pressure on Mercury will be lower.

10.75 The ratio PV/RT is equal to the number of moles of molecules in an ideal-gas sample; this number should be a constant for all pressure, volume and temperature conditions. If the value of this ratio changes with increasing pressure, the gas sample is not behaving ideally (according to the ideal-gas equation).

10.76 (a) The initial drop in the value of PV/RT is due to attractive forces between molecules. These intermolecular forces cause the molecules to "stick" together and behave as if there are fewer net particles in the sample. At lower pressures, this is the dominant effect. At the same time, the real volume of gas molecules causes the amount of free space in the gas sample to be less than the container volume. Using the container volume to calculate PV/RT gives a value larger than that for an ideal gas (which assumes that the total container volume is free space). At higher pressures, this effect more than compensates for molecular attraction, and PV/RT is greater than 1.

(b) As the temperature of a gas increases, the average kinetic energy of the particles increases. The increased kinetic energy overcomes the attractive forces between molecules and keeps them separate.

10.77 *Plan.* The constants *a* and *b* are part of the correction terms in the van der Waals equation. The smaller the values of *a* and *b*, the smaller the corrections and the more ideal the gas. *Solve*:

Ar (a = 1.34, b = 0.0322) will behave more like an ideal gas than CO_2 (a = 3.59, b = 0.0427) at high pressures.

10.78 The constant *a* is a measure of the strength of intermolecular attractions among gas molecules; *b* is a measure of molecular volume. Both increase with increasing molecular mass and structural complexity.

10.79 *Analyze.* Conditions and amount of $CCl_4(g)$ are given. *Plan.* Use ideal-gas equation and van der Waals equation to calculate pressure of gas at these conditions. *Solve*:

(a) $P = 1.00 \text{ mol} \times \frac{0.08206 \text{ L}\cdot\text{atm}}{\text{K}\cdot\text{mol}} \times \frac{313 \text{ K}}{28.0 \text{ L}} = 0.917 \text{ atm}$

(b) $P = \frac{nRT}{V - nb} - \frac{an^2}{V^2} = \frac{1.00 \times 0.08206 \times 313}{28.0 - (1.00 \times 0.1383)} - \frac{20.4(1.00)^2}{(28.0)^2} = 0.896 \text{ atm}$

Check. The van der Waals result indicates that the real pressure will be less than the ideal pressure. That is, intermolecular forces reduce the effective number of particles and the real pressure. This is reasonable for 1 mole of gas at relatively low temperature and pressure.

10.80 (a) At STP, the molar volume = $1 \text{ mol} \times \frac{0.08206 \text{ L}\cdot\text{atm}}{\text{K}\cdot\text{mol}} \times \frac{273 \text{ K}}{1 \text{ atm}} = 22.4 \text{ L}$

Dividing the value for *b*, 0.0322 L/mol, by 4, we obtain 0.0080 L. Thus, the volume of the Ar atoms is (0.0080/22.4)100 = 0.036% of the total volume.

(b) At 100 atm pressure the molar volume is 0.224 L, and the volume of the Ar atoms is 3.6% of the total volume.

Additional Exercises

10.81 Over time, the gases will mix perfectly. Each bulb will contain 4 blue and 3 red atoms.

10.82 A mercury barometer with water trapped in its tip would not read the correct pressure. The standard relationship between height of an Hg column and atmospheric pressure (760 mm Hg = 1 atm) assumes that there is a vacuum in the closed end of the barometer and that gravity is the only downward force on the Hg column. Water at the top of the Hg column would establish a vapor pressure which would exert additional downward pressure and partially counterbalance the pressure of the atmosphere. The Hg column would read lower than the actual atmospheric pressure.

10.83 $P_1V_1 = P_2V_2$; $V_2 = P_1V_1/P_2$

$$V_2 = \frac{3.0 \text{ atm} \times 1.0 \text{ mm}^3}{695 \text{ torr}} \times \frac{760 \text{ torr}}{1 \text{ atm}} = 3.3 \text{ mm}^3$$

10.84 $P = \frac{nRT}{V}$; $n = 1.4 \times 10^{-5}$ mol, $V = 0.600$ L, $T = 23°C = 296$ K

$$P = 1.4 \times 10^{-5} \text{ mol} \times \frac{0.08206 \text{ L}\cdot\text{atm}}{\text{K}\cdot\text{mol}} \times \frac{296 \text{ K}}{0.600 \text{ L}} = 5.7 \times 10^{-4} \text{ atm} = 0.43 \text{ mm Hg}$$

10.85 (a) $n = \frac{PV}{RT} = 3.00 \text{ atm} \times \frac{\text{K}\cdot\text{mol}}{0.08206 \text{ L}\cdot\text{atm}} \times \frac{110 \text{ L}}{300 \text{ K}} = 13.4 \text{ mol } C_3H_8(g)$

(b) $\frac{0.590 \text{ g } C_3H_8 (l)}{1 \text{ mL}} \times 110 \times 10^3 \text{ mL} \times \frac{1 \text{ mol } C_3H_8}{44.094 \text{ g}} = 1.47 \times 10^3 \text{ mol } C_3H_8 (l)$

(c) Using C_3H_8 in a 110 L container as an example, the ratio of moles liquid to moles gas that can be stored in a certain volume is $\frac{1.47 \times 10^3 \text{ mol liquid}}{13.4 \text{ mol gas}} = 110.$

A container with a fixed volume holds many more moles (molecules) of $C_3H_8(l)$ because in the liquid phase the molecules are touching. In the gas phase, the molecules are far apart (statement 2, section 10.7) and many fewer molecules will fit in the container.

10.86 If the air in the room is at STP, the partial pressure of O_2 is 0.2095 × 1 atm = 0.2095 atm. Since the gases in air are perfectly mixed, the volume of O_2 is the volume of the room.

$$V = 10.0\text{ ft} \times 8.0\text{ ft} \times 8.0\text{ ft} \times \frac{(12)^3\text{ in}^3}{\text{ft}^3} \times \frac{(2.54)^3\text{cm}^3}{\text{in}^3} \times \frac{1\text{ L}}{1000\text{ cm}^3} = 1.812 \times 10^4 = 1.8 \times 10^4\text{ L}$$

$$g = \frac{\mathcal{M}PV}{RT} = \frac{32.00\text{ g }O_2}{\text{mol }O_2} \times \frac{\text{K}\bullet\text{mol}}{0.08026\text{ L}\bullet\text{atm}} \times \frac{0.2095\text{ atm} \times 1.812 \times 10^4\text{ L}}{273\text{ K}} = 5.4 \times 10^3\text{ g }O_2$$

10.87 Volume of laboratory = $54\text{ m}^2 \times 3.1\text{ m} \times \frac{1000\text{ L}}{1\text{ m}^3} = 1.674 \times 10^5 = 1.7 \times 10^5\text{ L}$

Calculate the **total** moles of gas in the laboratory at the conditions given.

$$n_t = \frac{PV}{RT} = 1.00\text{ atm} \times \frac{\text{K}\bullet\text{mol}}{0.08206\text{ L}\bullet\text{atm}} \times \frac{1.674 \times 10^5\text{ L}}{297\text{ K}} = 6.869 \times 10^3 = 6.9 \times 10^3\text{ mol gas}$$

An $Ni(CO)_4$ concentration of 1 part in 10^9 means 1 mol $Ni(CO)_4$ in 1×10^9 total moles of gas.

$$\frac{x\text{ mol Ni(CO)}_4}{6.869 \times 10^3\text{ mol gas}} = \frac{1}{10^9} = 6.869 \times 10^{-6}\text{ mol Ni(CO)}_4$$

$$6.869 \times 10^{-6}\text{ mol Ni(CO)}_4 \times \frac{170.74\text{ g Ni(CO)}_4}{1\text{ mol Ni(CO)}_4} = 1.2 \times 10^{-3}\text{ g Ni(CO)}_4$$

10.88 It is simplest to calculate the partial pressure of each gas as it expands into the total volume, then sum the partial pressures.

$P_2 = P_1V_1 / V_2$; P_{N_2} = 265 torr (1.0 L / 2.5 L) = 106 = 1.1×10^2 torr

P_{Ne} = 800 torr (1.0 L / 2.5 L) = 320 = 3.2×10^2 torr; P_{H_2} = 532 torr (0.5 L / 2.5 L) = 106 = 1.1×10^2 torr

$P_t = P_{N_2} + P_{Ne} + P_{H_2}$ = (106 + 320 + 106) torr = 532 = 5.3×10^2 torr

10.89 (a) $$n = \frac{PV}{RT} = 0.980\text{ atm} \times \frac{\text{K}\bullet\text{mol}}{0.08206\text{ L}\bullet\text{atm}} \times \frac{0.524\text{ L}}{347\text{ K}} = 0.018034 = 0.0180\text{ mol air}$$

$$\text{mol }O_2 = 0.018034\text{ mol air} \times \frac{0.2095\text{ mol }O_2}{1\text{ mol air}} = 0.003778 = 0.00378\text{ mol }O_2$$

(b) $C_8H_{18}(l) + 25/2\ O_2(g) \rightarrow 8CO_2(g) + 9H_2O(g)$

(The H_2O produced in an automobile engine is in the gaseous state.)

$$0.003778 \text{ mol } O_2 \times \frac{1 \text{ mol } C_8H_{18}}{12.5 \text{ mol } O_2} \times \frac{114.2 \text{ g } C_8H_{18}}{1 \text{ mol } C_8H_{18}} = 0.0345 \text{ g } C_8H_{18}$$

10.90 (a) $$5.00 \text{ g HCl} \times \frac{1 \text{ mol HCl}}{36.46 \text{ g HCl}} = 0.1371 = 0.137 \text{ mol HCl}$$

$$5.00 \text{ g } NH_3 \times \frac{1 \text{ mol } NH_3}{17.03 \text{ g } NH_3} = 0.2936 = 0.294 \text{ mol } NH_3$$

The gases react in a 1:1 mole ratio, HCl is the limiting reactant and is completely consumed. (0.2936 mol - 0.1371 mol) = 0.1565 = 0.157 mol NH_3 remain in the system. $NH_3(g)$ is the only gas remaining after reaction. V_t = 4.00 L

(b) $$P = \frac{nRT}{V} = 0.1565 \text{ mol} \times \frac{0.08206 \text{ L} \bullet \text{atm}}{\text{K} \bullet \text{mol}} \times \frac{298 \text{ K}}{4.00 \text{ L}} = 0.957 \text{ atm}$$

10.91 V and T are the same for He and O_2.

$P_{He}V = n_{He}RT$, $P_{He}/n_{He} = RT/V$; $P_{O_2}/n_{O_2} = RT/V$

$$\frac{P_{He}}{n_{He}} = \frac{P_{O_2}}{n_{O_2}} = n_{O_2} = \frac{P_{O_2} \times n_{He}}{P_{He}};\ n_{He} = 1.42 \text{ g He} \times \frac{1 \text{ mol He}}{4.003 \text{ g He}} = 0.3547 = 0.355 \text{ mol He}$$

$$n_{O_2} = \frac{158 \text{ torr}}{42.5 \text{ torr}} \times 0.355 \text{ mol} = 1.3188 = 1.32 \text{ mol } O_2;\ 1.3188 \text{ mol } O_2 \times \frac{32.00 \text{ g } O_2}{1 \text{ mol } O_2} = 42.2 \text{ g } O_2$$

10.92 $$\mathcal{M}_{avg} = \frac{dRT}{P} = \frac{1.104 \text{ g}}{1 \text{ L}} \times \frac{0.08206 \text{ L} \bullet \text{atm}}{\text{K} \bullet \text{mol}} \times \frac{300 \text{ K}}{435 \text{ torr}} \times \frac{760 \text{ torr}}{1 \text{ atm}}$$

$$= 47.48 = 47.5 \text{ g/mol}$$

χ = mole fraction O_2; 1 - χ = mole fraction Kr

47.48 g = χ(32.00) + (1-χ)(83.80)

36.3 = 51.8 χ; χ = 0.701; 70.1% O_2

10.93 Calculate the number of moles of Ar in the vessel:

n = (339.854 - 337.428)/39.948 = 0.060729 = 0.06073 mol

The total number of moles of the mixed gas is the same (Avogadro's Law). Thus, the average atomic weight is (339.076 - 337.428)/0.060729 = 27.137 = 27.14. Let the mole fraction of Ne be χ. Then,

χ (20.183) + (1 - χ) (39.948) = 27.137; 12.811 = 19.765 χ; χ = 0.6482

Neon is thus 64.82 mole percent of the mixture.

10.94 (a) The quantity d/P = $\mathcal{M}$/RT should be a constant at all pressures for an ideal gas. It is not, however, because of nonideal behavior. If we graph d/P vs P, the ratio should approach ideal behavior at low P. At P = 0, d/P = 2.2525. Using this value in the formula $\mathcal{M}$ = d/P × RT, $\mathcal{M}$ = 50.46 g/mol.

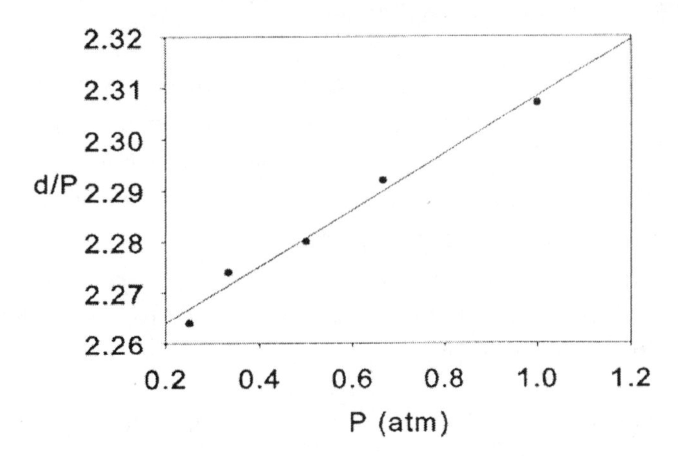

(b) The ratio d/P varies with pressure because of the finite volumes of gas molecules and attractive intermolecular forces.

10.95 Only item (b) is satisfactory. Item (c) would not have supported a column of Hg because it is open at both ends. The atmosphere would exert pressure on the top of the column as well as on the reservoir; the column would only be as high as the reservoir and the height would not change with changing pressure. Item (d) is not tall enough to support a nearly 760 mm Hg column. Items (a) and (e) are inappropriate for the same reason: they don't have a uniform cross-sectional area. The height of the Hg column is a direct measure of atmospheric pressure only if the cross-sectional area is constant over the entire tube.

10.96 In the expanded container volume, there are fewer collisions with the walls and with other gas particles. Attractive and repulsive forces which might change the speed of the particles become less important, and the particles maintain essentially the same average speed, kinetic energy and temperature after the expansion.

10.97

Kr (-50°C)
Kr (0°C)
Ar (0°C)
Fraction of Particles
Speed

10.98 (a) The effect of intermolecular attraction becomes more significant as a gas is compressed to a smaller volume at constant temperature. This compression causes the pressure, and thus the number of intermolecular collisions, to increase. Intermolecular attraction causes some of these collisions to be inelastic, which amplifies the deviation from ideal behavior.

(b) The effect of intermolecular attraction becomes less significant as the temperature of a gas is increased at constant volume. When the temperature of a gas is increased at constant volume, the pressure of the gas, the number of intermolecular collisions and the average kinetic energy of the gas particles increases. This higher average kinetic energy means that a larger fraction of the molecules has sufficient kinetic energy to overcome intermolecular attractions, even though there are more total collisions. This increases the fraction of elastic collisions, and the gas more closely obeys the ideal-gas equation.

10.99 (a) $$120.00 \text{ kg } N_2(g) \times \frac{1000 \text{ g}}{1 \text{ kg}} \times \frac{1 \text{ mol } N_2}{28.0135 \text{ g N}} = 4283.6 \text{ mol } N_2$$

$$P = \frac{nRT}{V} = 4283.6 \text{ mol} \times \frac{0.08206 \text{ L} \bullet \text{atm}}{\text{K} \bullet \text{mol}} \times \frac{553 \text{ K}}{1100.0 \text{ L}} = 176.72 = 177 \text{ atm}$$

(b) According to Equation 10.26,

$$P = \frac{nRT}{V - nb} - \frac{n^2 a}{V^2}$$

$$P = \frac{(4283.6 \text{ mol})(0.08206 \text{ L} \bullet \text{atm/K} \bullet \text{mol})(553 \text{ K})}{1100.0 \text{ L} - (4283.6 \text{ mol})(0.0391 \text{ L/mol})} - \frac{(4283.6 \text{ mol})^2 (1.39 \text{ L}^2 \bullet \text{atm/mol}^2)}{(1100.0 \text{ L})^2}$$

$$P = \frac{194{,}388 \text{ L} \bullet \text{atm}}{1100.0 \text{ L} - 167.5 \text{ L}} - 21.1 \text{ atm} = 208.5 \text{ atm} - 21.1 \text{ atm} = 187.4 \text{ atm}$$

(c) The pressure corrected for the real volume of the N_2 molecules is 208.5 atm, 31.8 atm higher than the ideal pressure of 176.7 atm. The 21.1 atm correction for intermolecular forces reduces the calculated pressure somewhat, but the "real" pressure is still higher than the ideal pressure. The correction for the real volume of molecules dominates. Even though the value of *b* is small, the number of moles of N_2 is large enough so that the molecular volume correction is larger than the attractive forces correction.

Integrative Exercises

10.100 (a) $$\mathcal{M} = \frac{gRT}{VP} = \frac{1.56 \text{ g}}{1.00 \text{ L}} \times \frac{0.08206 \text{ L} \bullet \text{atm}}{\text{K} \bullet \text{mol}} \times \frac{323 \text{ K}}{0.984 \text{ atm}} = 42.0 \text{ g/mol}$$

Assume 100 g cyclopropane

$$100 \text{ g} \times 0.857 \text{ C} = 85.7 \text{ g C} \times \frac{1 \text{ mol C}}{12.01 \text{ g}} = \frac{7.136 \text{ mol C}}{7.136} = 1 \text{ mol C}$$

$$100 \text{ g} \times 0.143 \text{ H} = 14.3 \text{ g H} \times \frac{1 \text{ mol H}}{1.008 \text{ g}} = \frac{14.19 \text{ mol H}}{7.136} = 2 \text{ mol H}$$

The empirical formula of cyclopropane is CH_2 and the empirical formula weight is 12 + 2 = 14 g. The ratio of molar mass to empirical formula weight, 42.0 g/14 g, is 3; therefore, there are three empirical formula units in one cyclopropane molecule. The molecular formula is 3 × (CH_2) = C_3H_6.

(b) Ar is a monoatomic gas. Cyclopropane molecules are larger and more structurally complex, even though the molar masses of Ar and C_3H_6 are similar. If both gases are at the same relatively low temperature, they have approximately the same average kinetic energy, and the same ability to overcome intermolecular attractions. We expect intermolecular attractions to be more significant for the more complex C_3H_6 molecules, and that C_3H_6 will deviate more from ideal behavior at the conditions listed. This conclusion is supported by the a values in Table 10.3. The a values for CH_4 and CO_2, more complex molecules than Ar atoms, are larger than the value for Ar. If the pressure is high enough for the volume correction in the van der Waals equation to dominate behavior, the larger C_3H_6 molecules definitely deviate more than Ar atoms from ideal behavior.

10.101 $n = \frac{PV}{RT} = 1.00 \text{ atm} \times \frac{\text{K}\cdot\text{mol}}{0.08206 \text{ L}\cdot\text{atm}} \times \frac{2.7 \times 10^{12} \text{ L}}{273 \text{ K}} = 1.205 \times 10^{11} = 1.2 \times 10^{11} \text{ mol } CH_4$

$CH_4(g) + 2O_2(g) \rightarrow CO_2(g) + 2H_2O(l) \qquad \Delta H° = -890.4 \text{ kJ}$

(At STP, H_2O is in the liquid state.)

$\Delta H^\circ_{rxn} = \Delta H^\circ_f CO_2(g) + 2\Delta H^\circ_f H_2O(l) - \Delta H^\circ_f CH_4(g) - \Delta H^\circ_f O_2(g)$

$\Delta H^\circ_{rxn} = -393.5 \text{ kJ} + 2(-285.83 \text{ kJ}) - (-74.8 \text{ kJ}) - 0 = -890.4 \text{ kJ}$

$\frac{-890.4 \text{ kJ}}{1 \text{ mol } CH_4} \times 1.205 \times 10^{11} \text{ mol } CH_4 = -1.073 \times 10^{14} = -1.1 \times 10^{14} \text{ kJ}$

The negative sign indicates heat evolved by the combustion reaction.

10.102 (a) *Analyze/Plan.* $AgF + S_8 \xrightarrow{\Delta}$ unknown gas

The gas probably contains the elements S and F in an unknown mole ratio. (Most compounds containing Ag are ionic and therefore solids.) Calculate the molar mass of the gas from its density at the given conditions. Determine the relative amount of F from the data on the reaction of the gas with water to produce HF. *Solve*:

$\mathcal{M} = \frac{dRT}{P} = \frac{0.803 \text{ g}}{\text{L}} \times \frac{0.08206 \text{ L}\cdot\text{atm}}{\text{K}\cdot\text{mol}} \times \frac{305 \text{ K}}{150 \text{ mm}} \times \frac{760 \text{ mm}}{1 \text{ atm}} = 101.83$

$= 102 \text{ g/mol}$

mol F in 480 mL sample: *M* × L = mol

0.081 *M* HF × 0.080 L = 0.00648 = 0.0065 mol HF = 0.0065 mol F in the sample

total mol gas in 480 mL sample: n = PV/RT

$n = 126 \text{ mm} \times \frac{1 \text{ atm}}{760 \text{ mm}} \times \frac{0.480 \text{ L}}{301 \text{ K}} \times \frac{\text{K}\cdot\text{mol}}{0.08206 \text{ L}\cdot\text{atm}} = 3.222 \times 10^{-3} = 3.22 \times 10^{-3} \text{ mol}$

mole ratio of S and F:

total g gas = 3.222×10^{-3} mol gas × 101.83 g/mol; = 0.32808 = 0.328 g gas

g F = 6.48×10^{-3} mol F × 18.998 g/mol F = 0.12311 = 0.123 g F

g S = 0.32808 - 0.12311 = 0.20497 = 0.205 g S

mol S = 0.205 g S/32.07 g/mol = 0.0639 mol S

0.00639 mol S/0.00648 mol F = 1:1 mole ratio of S:F

The empirical formula is SF; empirical FW = 32.07 + 19.00 = 51.07; since $\mathcal{M}$ = 102

The empirical formula is S_2F_2

Check. The empirical and molecular formula weights are in an integer ratio, so the result is reasonable.

(b) 26 valence e^-, 13 e^- pairs

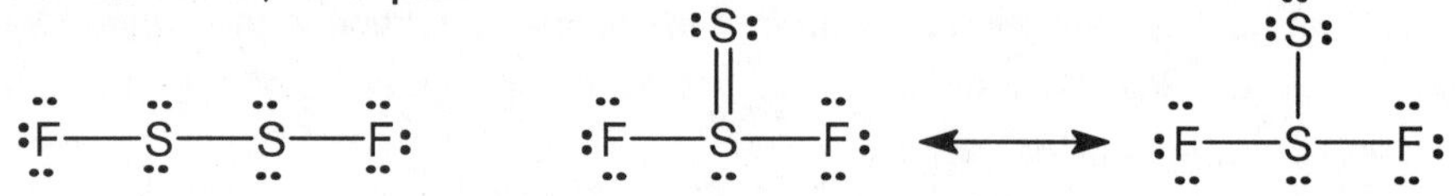

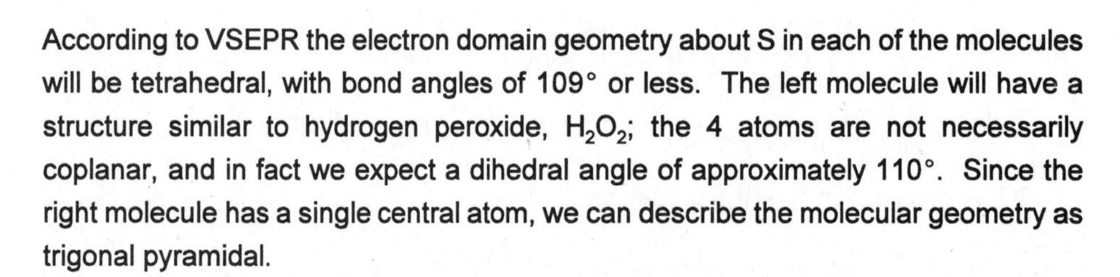

(c) According to VSEPR the electron domain geometry about S in each of the molecules will be tetrahedral, with bond angles of 109° or less. The left molecule will have a structure similar to hydrogen peroxide, H_2O_2; the 4 atoms are not necessarily coplanar, and in fact we expect a dihedral angle of approximately 110°. Since the right molecule has a single central atom, we can describe the molecular geometry as trigonal pyramidal.

From the given bond distances, we expect the single-bond covalent radii to be S = 1.02 Å, F = 0.72 Å. The simple conclusion is that the S–S distances will be ~2.04 Å and the S-F distances ~1.74 Å. In fact, the S–S distance in the left compound is 1.89 Å and in the right it is 1.86 Å. Clearly each of these bonds has some double bond character, as indicated by one of the resonance structures for the right molecule. The actual S-F distances are 1.63 Å (left) and 1.60 Å (right), also shorter than the predicted S-F single bond distance. One possible conclusion is that each of the bonds has some double bond character, that some of the "nonbonding" electron density is incorporated into a delocalized π-bonding network in the molecules.

10.103 (a) 19 e^-, 9.5 e^- pairs

:O—Cl—O:

Resonance structures can be drawn with the odd electron on O, but electronegativity considerations predict that it will be on Cl for most of the time.

(b) ClO_2 is very reactive because it is an odd-electron molecule. Adding an electron (reduction), both pairs the odd electron and completes the octet of Cl. Thus, ClO_2 has a strong tendency to gain an electron and be reduced.

10.104 (a) $ft^3\ CH_4 \rightarrow L\ CH_4 \rightarrow mol\ CH_4 \rightarrow mol\ CH_3OH \rightarrow g\ CH_3OH \rightarrow L\ CH_3OH$

$$10.7 \times 10^9\ ft\ CH_4 \times \frac{1\ yd^3}{3^3\ ft^3} \times \frac{1\ m^3}{(1.0936)^3\ yd^3} \times \frac{1\ L}{1 \times 10^{-3}\ m^3} = 3.03001 \times 10^{11}$$

$$= 3.03 \times 10^{11}\ L\ CH_4$$

$$n = \frac{PV}{RT} = \frac{3.03 \times 10^{11}\ L \times 1.00\ atm}{298\ K} \times \frac{K \bullet mol}{0.08206\ L \bullet atm} = 1.2391 \times 10^{10}$$

$$= 1.24 \times 10^{10}\ mol\ CH_4$$

1 mol CH_4 = 1 mol CH_3OH

$$1.2391 \times 10^{10}\ mol\ CH_3OH \times \frac{32.04\ g\ CH_3OH}{mol\ CH_3OH} \times \frac{1\ mL\ CH_3OH}{0.791\ g} \times \frac{1\ L}{1000\ mL}$$

$$= 5.0189 \times 10^8 = 5.02 \times 10^8\ L\ CH_3OH$$

(b) $CH_4(g) + 2O_2(g) \rightarrow CO_2(g) + 2H_2O(l)$ $\Delta H° = -890.4$ kJ (see Solution 10.101)

$$1.2391 \times 10^{10}\ mol\ CH_4 \times \frac{-890.4\ kJ}{1\ mol\ CH_4} = -1.10 \times 10^{13}\ kJ$$

$CH_3OH(l) + 3/2\ O_2(g) \rightarrow CO_2(g) + 2H_2O(l)$ $\Delta H° = -726.6$ kJ

$$\Delta H° = \Delta H_f^\circ\ CO_2(g) + 2\Delta H_f^\circ\ H_2O(l) - \Delta H_f^\circ\ CH_3OH(l) - 3/2\ \Delta H_f^\circ\ O_2(g)$$

$$= -393.5\ kJ + 2(-285.83\ kJ) - (-238.6\ kJ) - 0 = -726.6\ kJ$$

$$1.2391 \times 10^{10}\ mol\ CH_3OH \times \frac{-726.6\ kJ}{1\ mol\ CH_3OH} = -9.00 \times 10^{12}\ kJ$$

(c) Assume a volume of 1.00 L of each liquid. The densities stated in the Exercise should be 0.466 g/mL or 466 g/L for $CH_4(l)$ and 0.786 g/mL or 786 g/L for $CH_3OH(l)$. This is a slightly different density for $CH_3OH(l)$ than the value given in part (a).

$$1.00\ L\ CH_4(l) \times \frac{466\ g}{1\ L} \times \frac{1\ mol}{16.04\ g} \times \frac{-890.4\ kJ}{mol\ CH_4} = -2.59 \times 10^4\ kJ/L\ CH_4$$

$$1.00\ L\ CH_3OH \times \frac{786\ g}{1\ L} \times \frac{1\ mol}{32.04\ g} \times \frac{-726.6\ kJ}{mol\ CH_3OH} = -1.78 \times 10^4\ kJ/L\ CH_3OH$$

(Using 0.791 g/mL as the density of $CH_3OH(l)$ gives -1.79×10^4 kJ/L)

Clearly $CH_4(l)$ has the higher enthalpy of combustion per unit volume.

10.105 After reaction, the flask contains $IF_5(g)$ and whichever reactant is in excess. Determine the limiting reactant, which regulates the moles of IF_5 produced and moles of excess reactant.

$$I_2(s) + 5F_2(g) \rightarrow 2\ IF_5(g)$$

$$10.0\ g\ I_2 \times \frac{1\ mol\ I_2}{253.8\ g\ I_2} \times \frac{5\ mol\ F_2}{1\ mol\ I_2} = 0.1970 = 0.197\ mol\ F_2$$

$$10.0\ g\ F_2 \times \frac{1\ mol\ F_2}{38.00\ g\ F_2} = 0.2632 = 0.263\ mol\ F_2\ \text{available}$$

I_2 is the limiting reactant; F_2 is in excess.

0.263 mol F_2 available - 0.197 mol F_2 reacted = 0.066 mol F_2 remain.

$$10.0\ g\ I_2 \times \frac{1\ mol\ I_2}{253.8\ g\ I_2} \times \frac{2\ mol\ IF_5}{1\ mol\ I_2} = 0.0788\ mol\ IF_5\ \text{produced}$$

(a) $$P_{IF_5} = \frac{nRT}{V} = 0.0788\ mol \times \frac{0.08206\ L\bullet atm}{K\bullet mol} \times \frac{398\ K}{5.00\ L} = 0.515\ atm$$

(b) $$\chi_{IF_5} = \frac{mol\ IF_5}{mol\ IF_5 + mol\ F_2} = \frac{0.0788}{0.0788 + 0.066} = 0.544$$

10.106 (a) $MgCO_3(s) + 2HCl(aq) \rightarrow MgCl_2(aq) + H_2O(l) + CO_2(g)$

$CaCO_3(s) + 2HCl(aq) \rightarrow CaCl_2(aq) + H_2O(l) + CO_2(g)$

(b) $$n = \frac{PV}{RT} = 743\ torr \times \frac{1\ atm}{760\ torr} \times \frac{K\bullet mol}{0.08206\ L\bullet atm} \times \frac{1.72\ L}{301\ K}$$

$$= 0.06808 = 0.0681\ mol\ CO_2$$

(c) x = g $MgCO_3$, y = g $CaCO_3$, x + y = 6.53 g

mol $MgCO_3$ + mol $CaCO_3$ = mol CO_2 total

$$\frac{x}{84.32} + \frac{y}{100.09} = 0.06808;\ y = 6.53 - x$$

$$\frac{x}{84.32} + \frac{6.53 - x}{100.09} = 0.06808$$

100.09x - 84.32x + 84.32(6.53) = 0.06808 (84.32)(100.09)

15.77x + 550.610 = 574.549; x = 1.52 g $MgCO_3$

$$\text{mass \% } MgCO_3 = \frac{1.52\ g\ MgCO_3}{6.53\ g\ sample} \times 100 = 23.3\%$$

[By strict sig fig rules, the answer has 2 sig figs: 15.77x + 551 (3 digits from 6.53) = 575; 575 - 551 = 24 (no decimal places, 2 sig figs) leads to 1.5 g $MgCO_3$.]

11 Intermolecular Forces, Liquids and Solids

Kinetic-Molecular Theory

11.1 (a) solid $<$ liquid $<$ gas (b) gas $<$ liquid $<$ solid

11.2 Liquids are fluid and take the shape of their container. Solids are rigid, they have their own volume and do not flow. Solids are usually denser than liquids, indicating less empty space and a higher degree of order. Water is a notable exception.

11.3 Density is the ratio of the mass of a substance to the volume it occupies. For the same substance in different states, mass will be the same. In the liquid and solid states, the particles are touching and there is very little empty space, so the volumes occupied by a unit mass are very similar and the densities are similar. In the gas phase, the molecules are far apart, so a unit mass occupies a much greater volume than the liquid or solid, and the density of the gas phase is much less.

11.4 (a) The average distance between molecules is greater in the liquid state. Density is the ratio of the mass of a substance to the volume it occupies. For the same substance in different states, mass will be the same. The smaller the density, the greater the volume occupied, and the greater the distance between molecules. The liquid at 130° has the lower density (1.08 g/cm^3), so the average distance between molecules is greater.

(b) As the temperature of a substance increases, the average kinetic energy and speed of the molecules increases. At the melting point the molecules, on average, have enough kinetic energy to break away from the very orderly array that was present in the solid. As the translational motion of the molecules increases, the occupied volume increases and the density decreases. Thus, the density 1.266 g/cm^3 at 15°C is greater than the density 1.08 g/cm^3 at 130°C.

11.5 As the temperature of a substance is increased, the average kinetic energy of the particles increases. In a collection of particles (molecules), the state is determined by the strength of interparticle forces relative to the average kinetic energy of the particles. As the average kinetic energy increases, more particles are able to overcome intermolecular attractive forces and move to a less ordered state, from solid to liquid to gas.

11.6 (a) At constant temperature, the average kinetic energy of a collection of particles is constant. Compression brings particles closer together and increases the number of particle-particle collisions. With more collisions, the likelihood of intermolecular attractions causing the particles to coalesce (liquefy) is greater.

(b) In both liquids and solids, the component particles are touching. Most of the empty space in the sample volume has been eliminated and the sample is condensed.

Intermolecular Forces

11.7 (a) London-dispersion forces (b) dipole-dipole and London-dispersion forces
(c) dipole-dipole or in certain cases hydrogen bonding

11.8 (a) London-dispersion forces (b) dipole-dipole and London-dispersion forces
(c) hydrogen bonding

11.9 (a) Br_2 is a nonpolar covalent molecule, so only London-dispersion forces must be overcome to convert the liquid to a gas.

(b) CH_3OH is a polar covalent molecule that experiences London-dispersion, dipole-dipole and hydrogen-bonding (O–H bonds) forces. All of these forces must be overcome to convert the liquid to a gas.

(c) H_2S is a polar covalent molecule that experiences London-dispersion and dipole-dipole forces, so these must be overcome to change the liquid into a gas. (H–S bonds do not lead to hydrogen-bonding interactions.)

11.10 (a) CH_3OH experiences hydrogen bonding, but CH_3SH does not.

(b) Both gases are influenced by London-dispersion forces. The heavier the gas particles, the stronger the London-dispersion forces. The heavier Xe is a liquid at the specified conditions, while the lighter Ar is a gas.

(c) Both gases are influenced by London-dispersion forces. The larger, diatomic Cl_2 molecules are more polarizable, experience stronger dispersion forces and have the higher boiling point.

(d) Acetone and 2-methyl propane are molecules with similar molar masses and London-dispersion forces. Acetone also experiences dipole-dipole forces and has the higher boiling point.

11.11 (a) *Polarizability* is the ease with which the charge distribution (electron cloud) in a molecule can be distorted to produce a transient dipole.

(b) Te is most polarizable because its valence electrons are farthest from the nucleus and least tightly held.

(c) Polarizability increases as molecular size (and thus molecular weight) increases. In order of increasing polarizability: $CH_4 < SiH_4 < SiCl_4 < GeCl_4 < GeBr_4$

(d) The magnitude of London-dispersion forces and thus the boiling points of molecules increase as polarizability increases. The order of increasing boiling points is the order of increasing polarizability: $CH_4 < SiH_4 < SiCl_4 < GeCl_4 < GeBr_4$

11.12 (a) A more polarizable molecule can develop a larger transient dipole, increasing the strength of electrostatic attractions among polarized molecules.

(b) The noble gases are all monoatomic. Going down the column, the atomic radius and the size of the electron cloud increases. The larger the electron cloud, the more polarizable the atom, the stronger the London-disperson forces and the higher the boiling point.

(c) It is generally true that the greater the molecular weight, the stronger the dispersion forces experienced by a molecule. This is true because trends in molecular size and molecular weight are usually parallel.

(d) It is usually true that as the number of electrons in a molecule increases, the size of the molecule increases. Larger molecules tend to have diffuse electron clouds, which leads to greater polarizability. Thus, the statement that more electrons lead to increased dispersion forces (and greater polarizability) is correct.

11.13 *Analyze/Plan.* For molecules with similar structures, the strength of dispersion forces increases with molecular size (molecular weight and number of electrons in the molecule).

Solve: (a) H_2S (b) CO_2 (c) CCl_4

11.14 For molecules with similar structures, the strength of dispersion forces increases with molecular size (molecular weight and number of electrons in the molecule).

(a) Br_2 (b) $CH_3CH_2CH_2SH$

(c) $CH_3CH_2CH_2Cl$. These two molecules have the same molecular formula and molecular weight (C_3H_7Cl, molecular weight = 78.5 amu), so the shapes of the molecules determine which has the stronger dispersion forces. According to Figure 11.6, the cylindrical (not branched) molecule will have stronger dispersion forces.

11.15 Both hydrocarbons experience dispersion forces. Rod-like butane molecules can contact each other over the length of the molecule, while spherical 2-methylpropane molecules can only touch tangentially. The larger contact surface of butane produces greater polarizability and a higher boiling point.

11.16 Both molecules experience hydrogen bonding through their –OH groups and dispersion forces between their hydrocarbon portions. The position of the –OH group in isopropyl alcohol shields it somewhat from approach by other molecules and slightly decreases the extent of hydrogen bonding. Also, isopropyl alcohol is less rod-like (it has a shorter chain) than propyl alcohol, so dispersion forces are weaker. Since hydrogen bonding and dispersion forces are weaker in isopropyl alcohol, it has the lower boiling point.

11.17 Molecules with N–H, O–H and F–H bonds form hydrogen bonds with like molecules. $\mathbf{CH_3NH_2}$ and $\mathbf{CH_3OH}$ have N–H and O–H bonds, respectively. (CH_3F has C–F and C–H bonds, but no H–F bonds.)

11.18 (a) Replacing a hydroxyl hydrogen with a CH_3 group eliminates hydrogen bonding in that part of the molecule. This reduces the strength of intermolecular forces and leads to a (much) lower boiling point.

(b) $CH_3OCH_2CH_2OCH_3$ is a larger, more polarizable molecule with stronger London-dispersion forces and thus a higher boiling point.

11.19 (a) HF has the higher boiling point because hydrogen bonding is stronger than dipole-dipole forces.

(b) $CHBr_3$ has the higher boiling point because it has the higher molar mass which leads to greater polarizability and stronger dispersion forces.

(c) ICl has the higher boiling point because it is a polar molecule. For molecules with similar structures and molar masses, dipole-dipole forces are stronger than dispersion forces.

11.20 (a) C_6H_{14} – dispersion; C_8H_{18} – dispersion. C_8H_{18} has the higher boiling point due to greater molar mass and similar strength of forces.

(b) C_3H_8 – dispersion; CH_3OCH_3 – dipole-dipole and dispersion. CH_3OCH_3 has the higher boiling point due to stronger intermolecular forces and similar molar mass.

(c) CH_3OH – hydrogen bonding, dipole-dipole and dispersion; CH_3SH – dipole-dipole and dispersion. CH_3OH has the higher boiling point due to the influence of hydrogen bonding (Figure 11.7).

(d) NH_2NH_2 – hydrogen bonding, dipole-dipole and dispersion; CH_3CH_3 – dispersion. NH_2NH_2 has the higher boiling point due to much stronger intermolecular forces.

11.21 Surface tension (Section 11.3), high boiling point (relative to H_2S, H_2Se, H_2Te, Figure 11.7), high heat capacity per gram, high enthalpy of vaporization; the solid is less dense than the liquid; it is a liquid at room temperature despite its low molar mass.

11.22 (a) In the solid state, NH_3 molecules are arranged so as to form the maximum number of hydrogen bonds. At the melting point, the average kinetic energy of the molecules is large enough so that they are free to move relative to each other. As they move, old hydrogen bonds break and new ones form, but the strict relative order required for maximum hydrogen bonding is no longer present.

(b) In the liquid state, molecules are moving relative to one another while touching, which makes some hydrogen bonding possible. When molecules achieve enough kinetic energy to vaporize, the distance between them increases beyond the point where hydrogen bonds can form.

Viscosity and Surface Tension

11.23 (a) Viscosities and surface tensions of liquids both increase as intermolecular forces become stronger.

(b) As temperature increases, the average kinetic energy of the molecules increases and intermolecular attractions are more easily overcome. Surface tensions and viscosities decrease.

11.24 (a) *Cohesive* forces bind molecules to each other.
Adhesive forces bind molecules to surfaces.

(b) Viscosity and surface tension reflect attractive forces among molecules in a liquid. That is, they reflect cohesive forces.

(c) The shape of a meniscus depends on the strength of the cohesive forces within a liquid relative to the adhesive forces between the walls of the tube and the liquid. Adhesive forces between polar water molecules and silicates in glass (Figure 11.16) are even stronger than cohesive hydrogen-bonding forces among water molecules, so the meniscus is U-shaped (concave-upward).

(d) Water travels through the paper towel by capillary action. The stronger the adhesive forces between the cellulose fibers in the towel and the water molecules, the greater the capillary action. Also, the number and orientation of the capillaries in the towel affect the absorbency of the towel.

11.25 (a) $CHBr_3$ has a higher molar mass, is more polarizable and has stronger dispersion forces, so the surface tension is greater (see Solution 11.19(b)).

(b) As temperature increases, the viscosity of the oil decreases because the average kinetic energies of the molecules increase (Solution 11.23(b)).

(c) Adhesive forces between polar water and nonpolar car wax are weak, so the large surface tension of water draws the liquid into the shape with the smallest surface area, a sphere.

11.26 (a)

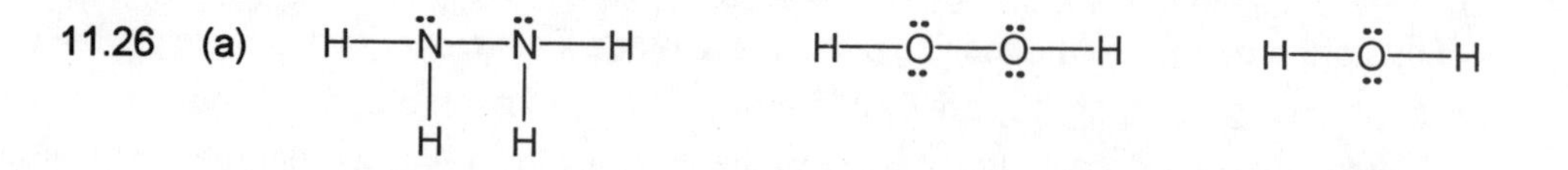

(b) All have bonds (N-H or O-H, respectively) capable of forming hydrogen bonds. Hydrogen bonding is the strongest intermolecular interaction between neutral molecules and leads to very strong cohesive forces in liquids. The stronger the cohesive forces in a liquid, the greater the surface tension.

Changes of State

11.27 Endothermic: melting (s → l), vaporization (l → g), sublimation (s → g)
Exothermic: condensation (g → l), freezing (l → s), deposition (g → s)

11.28 (a) Condensation, exothermic (b) sublimation, endothermic
(c) vaporization (evaporation), endothermic (d) freezing, exothermic

11.29 The heat energy required to increase the kinetic energy of molecules enough to melt the solid does not produce a large separation of molecules. The specific order is disrupted, but the molecules remain close together. On the other hand, when a liquid is vaporized, the intermolecular forces which maintain close molecular contacts must be overcome. Because molecules are being separated, the energy requirement is higher than for melting.

11.30 (a) Liquid ethyl chloride at room temperature is far above its boiling point. When the liquid contacts the metal surface, heat sufficient to vaporize the liquid is transferred from the metal to the ethyl chloride, and the heat content of the molecules increases. At constant atmospheric pressure, $\Delta H = q$, so the heat content and the enthalpy content of $C_2H_5Cl(g)$ is higher than that of $C_2H_5Cl(l)$.

(b) Attractive intermolecular forces hold the C_2H_5Cl molecules in close contact in the liquid phase. In order to overcome these attractive forces and maintain separation in the gas phase, the enthalpy content of the C_2H_5Cl molecules must increase when they change from the liquid to the gaseous state.

11.31 *Analyze.* the heat required to vaporize 50 g of H_2O equals the heat lost by the cooled water. *Plan.* Using the enthalpy of vaporization, calculate the heat required to vaporize 50 g of H_2O in this temperature range. Using the specific heat capacity of water, calculate the mass of water than can be cooled 13°C if this much heat is lost.

Solve. Evaporation of 50 g of water requires:

$$50 \text{ g } H_2O \times \frac{2.4 \text{ kJ}}{1 \text{ g } H_2O} = 1.2 \times 10^2 \text{ kJ or } 1.2 \times 10^5 \text{ J}$$

Cooling a certain amount of water by 13°C:

$$1.2 \times 10^5 \text{ J} \times \frac{1 \text{ g}\cdot\text{K}}{4.184 \text{ J}} \times \frac{1}{13°\text{C}} = 2206 = 2.2 \times 10^3 \text{ g } H_2O$$

Check. The units are correct. A surprisingly large mass of water (2200 g ≈ 2.2 L) can be cooled by this method.

11.32 Energy released when 100 g of H_2O is cooled from 18°C to 0°C:

$$\frac{4.184\ \text{J}}{\text{g}\bullet\text{K}} \times 100\ \text{g}\ H_2O \times 18°\text{C} = 7.53 \times 10^3\ \text{J} = 7.5\ \text{kJ}$$

Energy released when 100 g of H_2O is frozen (there is no change in temperature during a change of state):

$$\frac{334\ \text{J}}{\text{g}} \times 100\ \text{g}\ H_2O = 3.34 \times 10^4\ \text{J} = 33.4\ \text{kJ}$$

Total energy released = 7.53 kJ + 33.4 kJ = 40.93 = 40.9 kJ

Mass of freon that will absorb 40.9 kJ when vaporized:

$$40.93\ \text{kJ} \times \frac{1 \times 10^3\ \text{J}}{1\ \text{kJ}} \times \frac{1\ \text{g}\ CCl_2F_2}{289\ \text{J}} = 142\ \text{g}\ CCl_2F_2$$

11.33 *Analyze/Plan.* Follow the logic in Sample Exercise 11.4. *Solve*:

Heat the solid from -120°C to -114°C (153 K to 159 K), using the specific heat of the solid.

$$75.0\ \text{g}\ C_2H_5OH \times \frac{0.97\ \text{J}}{\text{g}\bullet\text{K}} \times 6\ \text{K} \times \frac{1\ \text{kJ}}{1000\ \text{J}} = 0.4365 = 0.4\ \text{kJ}$$

At -114°C (159 K), melt the solid, using its enthalpy of fusion.

$$75.0\ \text{g}\ C_2H_5OH \times \frac{1\ \text{mol}\ C_2H_5OH}{46.07\ \text{g}\ C_2H_5OH} \times \frac{5.02\ \text{kJ}}{1\ \text{mol}} = 8.172 = 8.17\ \text{kJ}$$

Heat the liquid from -114°C to 78°C (159 K to 351 K), using the specific heat of the liquid.

$$75.0\ \text{g}\ C_2H_5OH \times \frac{2.3\ \text{J}}{\text{g}\bullet\text{K}} \times 192\ \text{K} \times \frac{1\ \text{kJ}}{1000\ \text{J}} = 33.12 = 33\ \text{kJ}$$

At 78°C (351 K), vaporize the liquid, using its enthalpy of vaporization.

$$75.0\ \text{g}\ C_2H_5OH \times \frac{1\ \text{mol}\ C_2H_5OH}{46.07\ \text{g}\ C_2H_5OH} \times \frac{38.56\ \text{kJ}}{1\ \text{mol}} = 62.77 = 62.8\ \text{kJ}$$

The total energy required is 0.4365 kJ + 8.172 kJ + 33.12 kJ + 62.77 kJ = 104.50 = 105 kJ. (The result has zero decimal places, from 33 kJ required to heat the liquid.)

Check. The relative energies of the various steps are reasonable; vaporization is the largest.

11.34 Consider the process in steps, using the appropriate thermochemical constant.

Heat the liquid from 5.00°C to 47.6°C (278.00 K to 320.6 K), using the specific heat of the liquid.

$$25.0 \text{ g } C_2Co_3F_3 \times \frac{0.91 \text{ J}}{\text{g}\bullet\text{K}} \times 42.6 \text{ K} \times \frac{1 \text{ kJ}}{1000 \text{ J}} = 0.969 = 0.97 \text{ kJ}$$

Boil the liquid at 47.6°C (320.6 K), using the enthalpy of vaporization.

$$25.0 \text{ g } C_2Cl_3F_3 \times \frac{1 \text{ mol } C_2Cl_3F_3}{187.4 \text{ g } C_2Cl_3F_3} \times \frac{27.49 \text{ kJ}}{\text{mol}} = 3.667 = 3.67 \text{ kJ}$$

Heat the gas from 47.6°C to 82.00°C (320.6 K to 355.00 K), using the specific heat of the gas.

$$25.0 \text{ g } C_2Cl_3F_3 \times \frac{0.67 \text{ J}}{\text{g}\bullet\text{K}} \times 34.4 \text{ K} \times \frac{1 \text{ kJ}}{1000 \text{ J}} = 0.576 = 0.58 \text{ kJ}$$

The total energy required is 0.969 kJ + 3.667 kJ + 0.576 kJ = 5.21 kJ.

11.35 (a) The critical pressure is the pressure required to cause liquefaction at the critical temperature. The critical temperature is the highest temperature at which a gas can be liquefied, regardless of pressure.

(b) As the force of attraction between molecules increases, the critical temperature of the compound increases.

(c) The temperature of $N_2(l)$ is 77 K. All of the gases in Table 11.4 have critical temperatures higher than 77 K, so all of them can be liquefied at this temperature, given sufficient pressure.

11.36 (a) According to Solution 11.35(b), the higher the critical temperature, the stronger the intermolecular forces of a substance. Therefore, the strength of intermolecular forces decreases moving from left to right across the series and as molecular weight decreases.

(b) The molecules in this series experience London-dispersion forces and, except for CF_4, dipole-dipole forces. We expect the strength of dispersion forces to increase with increasing molecular weight, which agrees with the trends in critical temperature and pressure.

Vapor Pressure and Boiling Point

11.37 (a) No effect. (b) No effect.

(c) Vapor pressure decreases with increasing intermolecular attractive forces because fewer molecules have sufficient kinetic energy to overcome the attractive forces and escape to the vapor phase.

(d) Vapor pressure increases with increasing temperature because average kinetic energies of molecules increase.

11.38 (a) The pressure difference is 130 mm Hg and the gas is essentially 100% molecules of the substance in the vapor phase. When the vessel is evacuated, essentially all air is removed.

(b) The pressure difference is 1 atm. The gas is 130 mm Hg of the molecular vapor and the rest is air. The liquid vaporizes in contact with the atmosphere, so atmospheric pressure is maintained above the liquid, but the equilibrium gas composition reflects the amount of vapor necessary to maintain 130 mm pressure, plus enough air to maintain a total pressure of 1 atm.

(c) The pressure difference is 890 mm Hg (1 atm + 130 mm Hg) and the gas is a mixture of 130 mm vapor and 1 atm air. The initial air pressure in the flask is 1 atm and no air is allowed to escape. The gas in the flask is not in equilibrium with the atmosphere and the final pressure in the flask does not equal atmospheric pressure. After the liquid vaporizes, the total gas pressure is the result of 130 mm vapor and 1 atm air.

11.39 *Analyze/Plan.* Given the molecular formulae of several substances, determine the kind of intermolecular forces present, and rank the strength of these forces. The weaker the forces, the more volatile the substance. *Solve*:

$CBr_4 < CHBr_3 < CH_2Br_2 < CH_2Cl_2 < CH_3Cl < CH_4$

The weaker the intermolecular forces, the higher the vapor pressure, the more volatile the compound. The order of increasing volatility is the order of decreasing strength of intermolecular forces. By analogy to the boiling points of HCl and HBr (Section 11.2), the trend will be dominated by dispersion forces, even though four of the molecules ($CHBr_3$, CH_2Br_2, CH_2Cl_2 and CH_3Cl) are polar. Thus, the order of increasing volatility is the order of decreasing molar mass and decreasing strength of dispersion forces.

11.40 Both molecules are pyramidal, with a nonbonding electron pair on the central atom. Even though they are polar covalent, the intermolecular forces and physical properties are likely to be dominated by dispersion forces.

(a) PCl_3 (b) $AsCl_3$

(c) At the same temperature, the average kinetic energy of molecules of the two substances is equal.

(d) The strength of dispersion forces increases with increasing molecular weight, so $AsCl_3$ will experience stronger intermolecular forces. This is the basis of predictions in parts (a) and (b) above.

11.41 (a) The water in the two pans is at the same temperature, the boiling point of water at the atmospheric pressure of the room. During a phase change, the temperature of a system is constant. All energy gained from the surroundings is used to accomplish the transition, in this case to vaporize the liquid water. The pan of water that is boiling vigorously is gaining more energy and the liquid is being vaporized more quickly than in the other pan, but the temperature of the phase change is the same.

(b) Vapor pressure does not depend on either volume or surface area of the liquid. As long as the containers are at the same temperature, the vapor pressures of water in the two containers are the same.

11.42 (a) On a humid day, there are more gaseous water molecules in the air and more are recaptured by the surface of the liquid, making evaporation slower.

(b) At high altitude, atmospheric pressure is lower and water boils at a lower temperature. The eggs must be cooked longer at the lower temperature.

11.43 The boiling point is the temperature at which the vapor pressure of a liquid equals atmospheric pressure.

(a) The boiling point of diethyl ether at 400 torr is ~17°C, or, at 17°C, the vapor pressure of diethyl ether is 400 torr.

(b) At a pressure of 25 torr, water would boil at ~28°C, or, the vapor pressure of water at 28°C is 25 torr.

11.44 (a) A pressure of 1.2 atm corresponds to $1.4 \text{ atm} \times \frac{760 \text{ torr}}{1 \text{ atm}} = 912 = 9.1 \times 10^2$ torr. According to Appendix B, the vapor pressure of water reaches 912 torr somewhere between 104-106°C. By linear interpolation, the boiling point should be near

$$104°\text{C} + \left[\frac{(912 - 875)\text{torr}}{(938 - 875)\text{torr}} \times 2°\text{C}\right] = 105.2°\text{C} = 105°\text{C}$$

(b) The vapor pressure of ethyl alcohol at 70°C is approximately 510 torr. Thus, at 70°C ethyl alcohol would boil at an external pressure of 510 torr.

11.45 (a) From Appendix B, a vapor pressure of 340 torr is in the 70-80°C range. By linear interpolation,

$$\text{b.p.} = 70°\text{C} + \left[\frac{340 - 234}{355 - 234} \times 10°\text{C}\right] \approx 79°\text{C}$$

(b) According to Figure 11.22, the vapor pressure of diethyl ether at 12°C is approximately 325 torr. This is a substantial vapor pressure, but still less than the atmospheric pressure of 340 torr. In an open end manometer such as the one in Figure 10.3, the arm open to the atmosphere would be lower than the arm open to the container.

11.46 (a) The boiling point of a liquid is the temperature at which its vapor pressure equals atmospheric pressure. According to Appendix B, the vapor pressure of water is 680 torr at approximately 97°C.

(b) The temperature at which the vapor pressure of water is 752 mm Hg is almost 100°C, greater than the boiling temperature at 680 mm Hg. The average kinetic energy of the H_2O molecules at the boiling temperature in Chicago is greater than the average kinetic energy of the molecules at the boiling temperature in Reno. A liquid boils when its vapor pressure equals the external pressure acting on the liquid. If the external pressure acting on the liquid molecules is smaller (as is the atmospheric pressure in Reno), a smaller average kinetic energy is required for boiling (bubble formation within the liquid).

Phase Diagrams

11.47 The liquid/gas line of a phase diagram ends at the critical point, the temperature and pressure beyond which the gas and liquid phases are indistinguishable. At temperatures higher than the critical temperature, a gas cannot be liquefied, regardless of pressure.

11.48 (a) The triple point on a phase diagram represents the temperature at which the gas, liquid and solid phases are in equilibrium.

(b) No. A phase diagram represents a closed system, one where no matter can escape and no substance other than the one under consideration is present; air cannot be present in the system. Even if air is excluded, at 1 atm of external pressure, the triple point of water is inaccessible, regardless of temperature [see Sample Exercise 11.6(b)].

11.49 (a) The water vapor would condense to form a solid at a pressure of around 4 torr. At higher pressure, perhaps 5 atm or so, the solid would melt to form liquid water. This occurs because the melting point of ice, which is 0°C at 1 atm, decreases with increasing pressure.

(b) In thinking about this exercise, keep in mind that the **total** pressure is being maintained at a constant 0.50 atm. That pressure is composed of water vapor pressure and some other pressure, which could come from an inert gas. At 100°C and 0.50 atm, water is in the vapor phase. As it cools, the water vapor will condense to the liquid at the temperature where the vapor pressure of liquid water is 0.50 atm. From Appendix B, we see that condensation occurs at approximately 82°C. Further cooling of the liquid water results in freezing to the solid at approximately 0°C. The freezing point of water increases with decreasing pressure, so at 0.50 atm, the freezing temperature is very slightly above 0°C.

11.50 (a) Solid CO_2 sublimes to form $CO_2(g)$ at a temperature of about -60°C.

(b) Solid CO_2 melts to form $CO_2(l)$ at a temperature of about -50°C. The $CO_2(l)$ boils when the temperature reaches approximately -40°C.

11.51 (a)

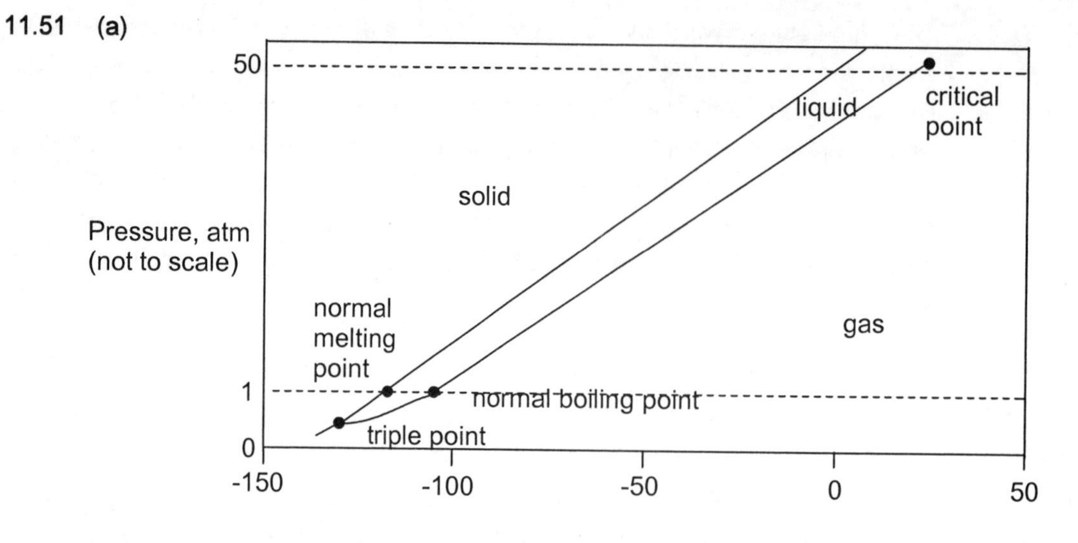

(b) The solid-liquid line on the phase diagram is normal and the melting point of Xe(s) increases with increasing pressure. This means that Xe(s) is denser than Xe(l).

(c) Cooling Xe(g) at 100 torr will cause deposition of the solid. A pressure of 100 torr is below the pressure of the triple point, so the gas will change directly to the solid upon cooling.

11.52 (a)

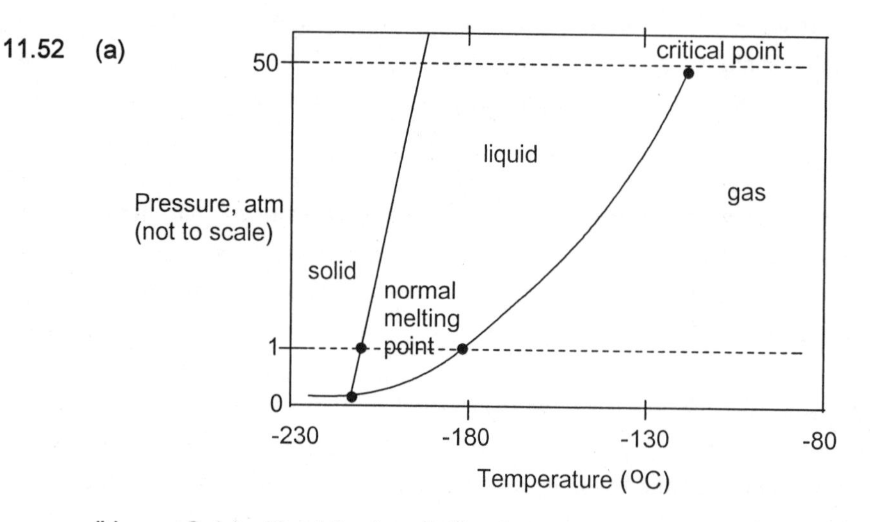

(b) O_2(s) will not float on O_2(l). O_2(s) is denser than O_2(l) because the solid-liquid line on the phase diagram is normal. That is, as pressure increases, the melting temperature increases. [Note that the solid-liquid line for O_2 is nearly vertical, indicating a small difference in the densities of O_2(s) and O_2(l)].

(c) O_2(s) will melt when heated at a pressure of 1 atm, since this is a much greater pressure than the pressure at the triple point.

Structures of Solids

11.53 In a crystalline solid, the component particles (atoms, ions or molecules) are arranged in an ordered repeating pattern. In an amorphous solid, there is no orderly structure. Quartz glass (Figure 11.30(b)) is an example of an amorphous solid. Paraffin wax is another example of an amorphous solid. Also, most plastics show no long range order and are amorphous overall, although they can show regions of order (see Section 12.2).

11.54 In amorphous silica (SiO_2) the regular structure of quartz is disrupted; the loose, disordered structure, Figure 11.30(b), has many vacant "pockets" throughout. There are fewer SiO_2 groups per volume in the amorphous solid; the packing is less efficient and less dense.

11.55 The unit cell is the building block of the crystal lattice. When repeated in three dimensions, it produces the crystalline solid. It is a parallelepiped with characteristic distances and angles. Unit cells can be primitive (lattice points only at the corners of the parallelepiped) or centered (lattice points at the corners and at the middle of faces or the middle of the parallelepiped).

11.56 Ca: Ca atoms occupy the 8 corners of the cube. 8 corners × 1/8 sphere/corner = 1 Ca atom

O: O atoms occupy the centers of the 6 faces of the cube.
6 faces × 1/2 atom/face = 3 O atoms

Ti: There is 1 Ti atom at the body center of the cube.

Formula: $CaTiO_3$

11.57 In a metallic solid such as gold, the atoms are held in their very orderly arrangement by metallic bonding, the result of valence electrons delocalized throughout the three-dimensional lattice. A large amount of kinetic energy is required to disrupt this delocalized bonding network and allow the atoms to translate relative to each, so the melting point of Au(s) is quite high. Xe atoms are held in a cubic close-packed arrangement by London-dispersion forces much weaker than metallic bonding. Very little kinetic energy is required for Xe atoms to overcome these forces and melt, so the melting point of Xe is quite low.

11.58 (a) Ti: 8 corners × 1/8 sphere/corner + [1 center × 1 sphere/center] = 2 Ti atoms

O: 4 faces × 1/2 sphere/face + [2 interior × 1 sphere/interior] = 4 O atoms

Formula: Ti_2O_4

(b) Rutile is an ionic solid; ion-ion forces among Ti^{4+} cations and O^{2-} anions are quite strong, owing to the magnitudes of the charges, and lead to the ordered structure.

11.59 *Analyze.* Given the cubic unit cell edge length and arrangement of Ir atoms, calculate the atomic radius and the density of the metal. *Plan.* There is space between the atoms along the unit cell edge, but they touch along the face diagonal. Use the geometry of the right equilateral triangle to calculate the atomic radius. From the definition of density and paying attention to units, calculate the density of Ir(s). *Solve:*

(a) The length of the face diagonal of a face-centered cubic unit cell is four times the radius of the atom and $\sqrt{2}$ times the unit cell dimension or edge length, usually designated *a* for cubic unit cells.

$$4\text{ r} = \sqrt{2}\ a;\ \text{r} = \sqrt{2}\ a/4 = \frac{\sqrt{2} \times 3.833\ \text{Å}}{4} = 1.3552 = 1.355\ \text{Å}$$

(b) The density of iridium is the mass of the unit cell contents divided by the unit cell volume. There are 4 Ir atoms in a face-centered cubic unit cell.

$$\frac{4\ \text{Ir atoms}}{(3.833 \times 10^{-8}\ \text{cm})^3} \times \frac{192.22\ \text{g Ir}}{6.022 \times 10^{23}\ \text{Ir atoms}} = 22.67\ \text{g/cm}^3$$

Check. The units of density are correct. Note that Ir is quite dense.

11.60 (a) 8 corners × 1/8 atom/corner + 6 faces × ½ atom/face = 4 atoms

(b) Each aluminum atom is in contact with 12 nearest neighbors, 6 in one plane, 3 above that plane, and 3 below. Its coordination number is thus 12.

(c) The length of the face diagonal of a face-centered cubic unit cell is four times the radius of the metal and $\sqrt{2}$ times the unit cell dimension (usually designated *a* for cubic cells).

$$4 \times 1.43\ \text{Å} = \sqrt{2} \times a;\quad a = \frac{4 \times 1.43\ \text{Å}}{\sqrt{2}} = 4.0447 = 4.04\ \text{Å} = 4.04 \times 10^{-8}\ \text{cm}$$

(d) The density of the metal is the mass of the unit cell contents divided by the volume of the unit cell.

$$\text{density} = \frac{4\ \text{Al atoms}}{(4.0447 \times 10^{-8}\ \text{cm})^3} \times \frac{26.98\ \text{g Al}}{6.022 \times 10^{23}\ \text{Al atoms}} = 2.71\ \text{g/cm}^3$$

11.61 *Analyze.* Given the atomic arrangement, length of the cubic unit cell edge and density of the solid, calculate the atomic weight of the element. *Plan.* If we calculate the mass of a single unit cell, and determine the number of atoms in one unit cell, we can calculate the mass of a single atom and of a mole of atoms. *Solve:*

The volume of the unit cell is $(2.86 \times 10^{-8}\ \text{cm})^3$. The mass of the unit cell is:

$$\frac{7.92\ \text{g}}{\text{cm}^3} \times \frac{(2.86 \times 10^{-8})^3\ \text{cm}^3}{\text{unit cell}} = 1.853 \times 10^{-22}\ \text{g/unit cell}$$

There are two atoms of the element present in the body-centered cubic unit cell. Thus the atomic weight is:

$$\frac{1.853 \times 10^{-22}\ \text{g}}{\text{unit cell}} \times \frac{1\ \text{unit cell}}{2\ \text{atoms}} \times \frac{6.022 \times 10^{23}\ \text{atoms}}{1\ \text{mol}} = 55.8\ \text{g/mol}$$

Check. The result is a reasonable atomic weight and the units are correct. The element could be iron.

11.62 Avogadro's number is the number of KCl formula units in 74.55 g of KCl.

$$74.55 \text{ g KCl} \times \frac{1 \text{ cm}^3}{1.984 \text{ g}} \times \frac{(1 \times 10^{10} \text{ pm})^3}{1 \text{ cm}^3} \times \frac{4 \text{ KCl units}}{628^3 \text{ pm}^3} = 6.07 \times 10^{23} \text{ KCl formula units}$$

11.63 (a) Each sphere is in contact with 12 nearest neighbors; its coordination number is thus 12.

(b) Each sphere has a coordination number of six.

(c) Each sphere has a coordination number of eight.

11.64 (a) Na^+, 6 (b) Zn^{2+}, 4 (c) Ca^{2+}, 8

11.65 *Analyze.* Given the atomic arrangement and density of the solid, calculate the unit cell edge length. *Plan.* Calculate the mass of a single unit cell and then use density to find the volume of a single unit cell. The edge length is the cube-root of the volume of a cubic cell. *Solve*:

There are four PbSe units in the unit cell. The unit cell edge is designated *a*.

$$8.27 \text{ g/cm}^3 = \frac{4 \text{ PbSe units}}{a^3} \times \frac{286.2 \text{ g}}{6.022 \times 10^{23} \text{ PbSe units}} \times \left(\frac{1 \text{ Å}}{1 \times 10^{-8} \text{ cm}}\right)^3$$

$a^3 = 229.87$ Å^3, $a = 6.13$ Å

11.66 In the face-centered cubic structure, there are four CaS units in the unit cell. Density is the mass of the unit cell contents divided by the unit cell volume (a^3).

$$\text{density} = \frac{4 \text{ CaS units}}{(5.689 \text{ Å})^3} \times \frac{72.144 \text{ g NiO}}{6.022 \times 10^{23} \text{ NiO units}} \times \left(\frac{1 \text{ Å}}{1 \times 10^{-8} \text{ cm}}\right)^3 = 2.603 \text{ g/cm}^3$$

11.67 (a) The U ions in UO_2 are represented by the smaller spheres in Figure 11.42(c). The chemical formula requires twice as many O^{2-} ions as U^{4+} ions. There are eight complete large spheres and four total (8 × 1/8 + 6 ×1/2) small spheres, so the small ones must represent U^{4+}. (It is probably true that O^{2-} has a physically larger radius than U^{4+}, but the elements' large separation on the periodic chart makes the relative radii difficult to estimate from trends.)

(b) According to Figure 11.42(c), there are four UO_2 units in the "fluorite" unit cell.

$$\frac{4 \text{ UO}_2 \text{ units}}{(5.468 \text{ Å})^3} \times \frac{270.03 \text{ g}}{6.022 \times 10^{23} \text{ UO}_2 \text{ units}} \times \left(\frac{1 \text{ Å}}{1 \times 10^{-8} \text{ cm}}\right)^3 = 10.97 \text{ g/cm}^3$$

11.68 (a) According to Figure 11.42(b), there are 4 HgS units in a unit cell with the zinc blende structure. [4 complete Hg^{2+} ions, 6(1/2) + 8(1/8) S^{2-} ions]

$$\text{density} = \frac{4 \text{ HgS units}}{(5.852 \text{ Å})^3} \times \frac{232.656 \text{ g}}{6.022 \times 10^{23} \text{ HgS units}} \times \left(\frac{1 \text{ Å}}{1 \times 10^{-8} \text{ cm}}\right)^3 = 7.711 \text{ g/cm}^3$$

(b) We expect Se^{2-} to have a larger ionic radius than S^{2-}, since Se is below S in the chalcogen family and both ions have the same charge. Thus, HgSe will occupy a larger volume and the unit cell edge will be longer.

(c) For HgSe:

$$\text{density} = \frac{4 \text{ HgSe units}}{(6.085 \text{ Å})^3} \times \frac{279.55 \text{ g HgSe}}{6.022 \times 10^{23} \text{ HgSe units}} \times \left(\frac{1 \text{ Å}}{1 \times 10^{-8} \text{ cm}}\right)^3 = 8.241 \text{ g/cm}^3$$

Even though HgSe has a larger unit cell volume than HgS, it also has a larger molar mass. The mass of Se is more than twice that of S, while the radius of Se^{2-} is only slightly larger than that of S (Figure 7.6). The greater mass of Se accounts for the greater density of HgSe.

Bonding in Solids

11.69 (a) Hydrogen bonding, dipole-dipole forces, London dispersion forces

(b) covalent chemical bonds (mainly)

(c) ionic bonds (mainly)

(d) metallic bonds

11.70 (a) ionic (b) metallic (c) covalent-network
(d) molecular (e) molecular (f) molecular

11.71 In molecular solids, relatively weak intermolecular forces (hydrogen bonding, dipole-dipole, dispersion) bind the molecules in the lattice, so relatively little energy is required to disrupt these forces. In covalent-network solids, covalent bonds join atoms into an extended network. Melting or deforming a covalent-network solid means breaking these covalent bonds, which requires a large amount of energy.

11.72 (a) metallic (b) molecular or metallic (physical properties of metals vary widely)
(c) covalent-network (d) covalent-network (e) ionic

11.73 Because of its relatively high melting point and properties as a conducting solution, the solid must be ionic.

11.74 According to Table 11.7, the solid could be either ionic with low water solubility or network covalent. Due to the extremely high sublimation temperature, it is probably covalent-network.

11.75 (a) B – nonmetallic element likely to adopt covalent-network lattice like C(s), versus weak dispersion forces in BF_3

(b) NaCl – ionic versus metallic bonding; Na(s) has only one valence electron and is a relatively soft, low-melting alkali metal.

(c) TiO_2 – due to higher charge on O^{2-} than Cl^-. [In fact, $TiCl_4$ is molecular, with a very low melting point, but this is difficult to predict.]

(d) MgF_2 – due to higher charge on Mg^{2+} than Na^+.

11.76 (a) Xe – greater atomic weight, stronger dispersion forces

(b) SiO_2 – covalent-network lattice versus weak dispersion forces

(c) KBr – strong ionic versus weak dispersion forces

(d) C_6Cl_6 – both are influenced by dispersion forces, C_6Cl_6 has the higher molar mass.

Additional Exercises

11.77 In intramolecular covalent or ionic bonding, the bonded atoms are closer together and the magnitudes of the interacting charges are greater. Intermolecular interactions are the result of weaker attractive forces between molecules with net partial or transient charges.

11.78 (a) Dipole-dipole attractions (polar covalent molecules): SO_2, IF, HBr

(b) Hydrogen bonding (O–H, N–H or F–H bonds): CH_3NH_2, HCOOH

11.79 (a) Correct

(b) The lower boiling liquid must experience less total intermolecular forces.

(c) If both liquids are structurally-similar nonpolar molecules, the lower boiling liquid has a lower molecular weight than the higher boiling liquid.

(d) Correct

(e) Correct

11.80 (a) The *cis* isomer has stronger dipole-dipole forces; the *trans* isomer is nonpolar. The higher boiling point of the *cis* isomer supports this conclusion.

(b) While boiling points are primarily a measure of strength of intermolecular forces, melting points are influenced by crystal packing efficiency as well as intermolecular forces. Since the nonpolar *trans* isomer with weaker intermolecular forces has the higher melting point, it must pack more efficiently.

11.81 (a) In dibromomethane, CH_2Br_2, the dispersion force contribution will be larger than for CH_2Cl_2, because bromine is more polarizable than the lighter element chlorine. At the same time, the dipole-dipole contribution for CH_2Cl_2 is greater than for CH_2Br_2 because CH_2Cl_2 has a larger dipole moment.

(b) Just the opposite comparisons apply to CH_2F_2, which is less polarizable and has a higher dipole moment than CH_2Cl_2.

11.82 (a) The ability to flow is determined by factors that hold molecules in the same position relative to one another. These include strength of intermolecular forces and structural features such as chain length or branching that cause molecules to become entangled.

(b) The tendency of a liquid to bead is caused by an imbalance of intermolecular forces at the surface of a liquid. If the liquid has no appreciable adhesive forces with the surface, the stronger the cohesive forces among liquid molecules, the greater the tendency to bead.

(c) The strength of intermolecular forces determines boiling point. Hydrogen bonding, dipole moment, molecular weight and shape are important molecular properties that contribute to the overall strength of intermolecular forces.

(d) The strength of intermolecular forces determines heat of vaporization. Hydrogen bonding, dipole moment, molecular weight and shape are important molecular properties that contribute to the overall strength of intermolecular forces.

11.83 (a) Decrease (b) increase (c) increase (d) increase
(e) increase (f) increase (g) increase

11.84 When a halogen atom (Cl or Br) is substituted for H in benzene, the molecule becomes polar. These molecules experience dispersion forces similar to those in benzene plus dipole-dipole forces, so they have higher boiling points than benzene. C_6H_5Br has a higher molar mass and is more polarizable than C_6H_5Cl so it has the higher boiling point. C_6H_5OH experiences hydrogen bonding, the strongest force between neutral molecules, so it has the highest boiling point.

11.85 (a) Propylamine experiences hydrogen bonding interactions while trimethylamine, with no N-H bonds, does not. Also, the rod-like shape of propylamine (see Solution 11.15) leads to stronger dispersion forces than in pyramidal trimethylamine. The stronger intermolecular forces in propylamine lead to the higher boiling point.

(b) Propylamine is most soluble in water by virtue of its $-NH_2$ group, which can act as both a donor and acceptor in hydrogen bonding. Trimethylamine is much more soluble than the structurally similar isobutane, because the pyramidal $\ddot{N}$ atom can act as a hydrogen bond acceptor. Trimethylamine is less soluble than propylamine because its participation in hydrogen bonding is less extensive.

11.86 The two O–H groups in ethylene glycol are involved in many hydrogen bonding interactions, leading to its high boiling point and viscosity, relative to pentane, which experiences only dispersion forces.

11.87 The more carbon atoms in the hydrocarbon, the longer the chain, the more polarizable the electron cloud, the higher the boiling point. A plot of the number of carbon atoms versus boiling point is shown below. For 8 C atoms, the boiling point is approximately 130°C.

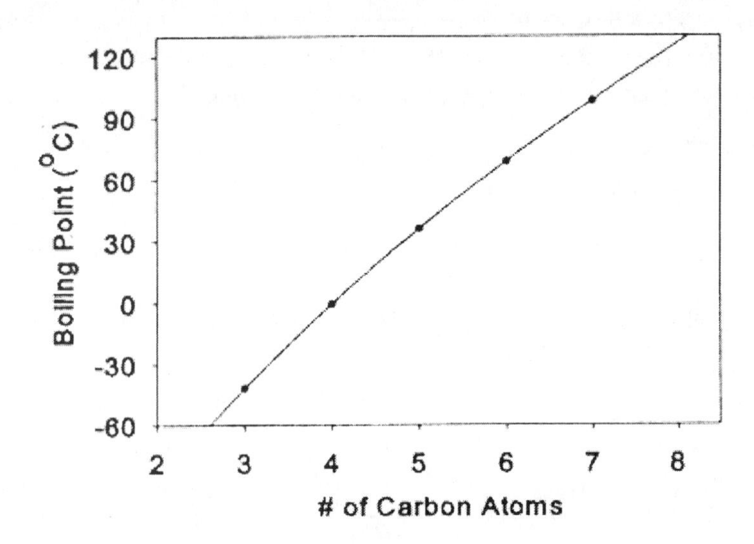

11.88 The vacuum pump reduces the pressure of the atmosphere (air + water vapor) above the water. Eventually, atmospheric pressure equals the vapor pressure of water and the water boils. Boiling is an endothermic process, and the temperature drops if the system is not able to absorb heat from the surroundings fast enough. As the temperature of the water decreases, the water freezes. (On a molecular level, the evaporation of water removes the molecules with the highest kinetic energies from the liquid. This decrease in average kinetic energy is what we experience as a temperature decrease.)

11.89 According to Figure 11.23, as the pressure of a gas above its critical temperature increases beyond critical pressure, the solubility of a solute increases. The solubility of the solute is essentially zero below critical pressure. Supercritical CO_2 at very high pressure dissolves caffeine and the solution leaves the extractor. The pressure reduction value reduces the pressure of CO_2 enough so that the caffeine becomes insoluble. The solid caffeine is deposited in the separator and the low pressure CO_2 gas is recycled.

11.90 (a) If the Clausius-Clapeyron equation is obeyed, a graph of ln P vs 1/T(K) should be linear. Here are the data in a form

T(K)	1/T	P(torr)	ln P
280.0	3.571×10^{-3}	32.42	3.479
300.0	3.333×10^{-3}	92.47	4.527
320.0	3.125×10^{-3}	225.1	5.417
330.0	3.030×10^{-3}	334.4	5.812
340.0	2.941×10^{-3}	482.9	6.180

6.5
6.0
5.5
5.0
4.5
4.0
3.5
ln P
0.0030 0.0032 0.0034 0.0036
1/T (K^{-1})

According to the graph, the Clausius-Clape equation is obeyed, to a first approximation

ΔH_{vap} = -slope × R; slope = $\dfrac{3.479 - 6.180}{(3.571 - 2.941) \times 10^{-3}} = -\dfrac{2.701}{0.630 \times 10^{-3}} = -4.29 \times 10^3$

$\Delta H_{vap} = -(-4.29 \times 10^3) \times 8.314$ J/K • mol = 35.7 kJ/mol

(b) The normal boiling point is the temperature at which the vapor pressure of the liquid equals atmospheric pressure, 760 torr. From the graph,

ln 760 = 6.63, 1/T for this vapor pressure = 2.828×10^{-3}; T = 353.6 K

11.91 (a) The Clausius-Clapeyron equation is $\ln P = \dfrac{-\Delta H_{vap}}{RT} + C$.

For two vapor pressures, P_1 and P_2, measured at corresponding temperatures T_1 and T_2, the relationship is

$$\ln P_1 - \ln P_2 = \left(\frac{-\Delta H_{vap}}{RT_1} + C\right) - \left(\frac{-\Delta H_{vap}}{RT_2} + C\right)$$

$$\ln P_1 - \ln P_2 = \frac{-\Delta H_{vap}}{R}\left(\frac{1}{T_1} - \frac{1}{T_2}\right) + C - C;\ \ln\frac{P_1}{P_2} = \frac{-\Delta H_{vap}}{R}\left(\frac{1}{T_1} - \frac{1}{T_2}\right)$$

(b) P_1 = 10.00 torr, T_1 = 716 K; P_2 = 400.0 torr, T_2 = 981 K

$$\ln\frac{10.00}{400.0} = \frac{-\Delta H_{vap}}{8.314\ \text{J/K}\bullet\text{mol}}\left(\frac{1}{716} - \frac{1}{981}\right)$$

$-3.6889\ (8.314$ J/K • mol) = $-\Delta H_{vap}(3.773 \times 10^{-4}$/K)

$\Delta H_{vap} = 8.129 \times 10^4 = 8.13 \times 10^4$ J/mol = 81.3 kJ/mol

(c) The normal boiling point of a liquid is the temperature at which the vapor pressure of the liquid is 760 torr.

P_1 = 400.0 torr, T_1 = 981 K; P_2 = 760 torr, T_2 = b.p. of potassium

$$\ln\left(\frac{400.0}{760.0}\right) = \frac{-8.129 \times 10^4\ \text{J/mol}}{8.314\ \text{J/K}\bullet\text{mol}}\left(\frac{1}{981\ \text{K}} - \frac{1}{T_2}\right)$$

$$\frac{-0.64185}{-9.7775 \times 10^3} = 1.0194 \times 10^{-3} - \frac{1}{T_2};\ \frac{1}{T_2} = 1.0194 \times 10^{-3} - 6.565 \times 10^{-5}$$

$$\frac{1}{T_2} = 9.5375 \times 10^{-4};\ \ T_2 = 1048\ \text{K}\ (775°\text{C})$$

(d) P_1 = VP of K(l) at 100°C, T_1 = 373 K; P_2 = 10.00 torr, T_2 = 716 K

$$\ln\frac{P_1}{10.00\ \text{torr}} = \frac{-8.129 \times 10^4\ \text{J/mol}}{8.314\ \text{J/K}\bullet\text{mol}}\left(\frac{1}{373} - \frac{1}{716}\right)$$

$$\ln\frac{P_1}{10.00\ \text{torr}} = \frac{-8.129 \times 10^4\ \text{J/mol}}{8.314\ \text{J/K}\bullet\text{mol}} \times 1.284 \times 10^{-3} = -12.5543$$

$$\frac{P_1}{10.00\ \text{torr}} = e^{-12.5543} = 3.530 \times 10^{-6};\ P_1 = 3.53 \times 10^{-5}\ \text{torr}$$

11.92 (a) The face diagonal of a face-centered cubic unit cell has length $a\sqrt{2}$ and also 4 r, where *a* is the cubic cell dimension and r is the atomic radius.

$4\,r = a\sqrt{2}$; $r = (a\sqrt{2})/4 = (4.078\ \text{Å})(\sqrt{2})/4 = 1.44179 = 1.442\ \text{Å}$

(b) Density is the mass of the unit cell contents divided by the unit cell volume (a^3). In a face-centered cubic unit cell, there are 4 Au atoms.

$$\text{density} = \frac{4\ \text{Au atoms}}{(4.078\ \text{Å})^3} \times \frac{196.97\ \text{g}}{6.0221 \times 10^{23}\ \text{Au atoms}} \times \left(\frac{1\ \text{Å}}{1 \times 10^{-8}\ \text{cm}}\right)^3 = 19.29\ \text{g/cm}^3$$

11.93 (a) 8 corners × 1/8 atom/corner = 1 atom

(b) 8 corners × 1/8 atom/corner + 1 center × 1atom/center = 2 atoms

(c) 8 corners × 1/8 atom/corner + 6 faces × 1/2 atom/face = 4 atoms

11.94 Physical data for the two compounds from the *Handbook of Chemistry and Physics*:

	$\mathcal{M}$	dipole moment	boiling point
CH_2Cl_2	85 g/mol	1.60 D	40.0°C
CH_3I	142 g/mol	1.62 D	42.4°C

(a) The two substances have very similar molecular structures; each is an unsymmetrical tetrahedron with a single central carbon atom and no hydrogen bonding. Since the structures are very similar, the magnitudes of the dipole-dipole forces should be similar. This is verified by their very similar dipole moments. The heavier compound, CH_3I, will have slightly stronger London dispersion forces. Since the nature and magnitude of the intermolecular forces in the two compounds are nearly the same, it is very difficult to predict which will be more volatile (or which will have the higher boiling point as in part (b)).

(b) Given the structural similarities discussed in part (a), one would expect the boiling points to be very similar, and they are. Based on its larger molar mass (and dipole-dipole forces being essentially equal) one might predict that CH_3I would have a slightly higher boiling point; this is verified by the known boiling points.

(c) According to Equation 11.1, $\ln P = \dfrac{-\Delta H_{vap}}{RT} + C$

A plot of ln P vs. 1/T for each compound is linear. Since the order of volatility changes with temperature for the two compounds, the two lines must cross at some temperature; the slopes of the two lines, ΔH_{vap} for the two compounds, and the y-intercepts, C, must be different.

(d)

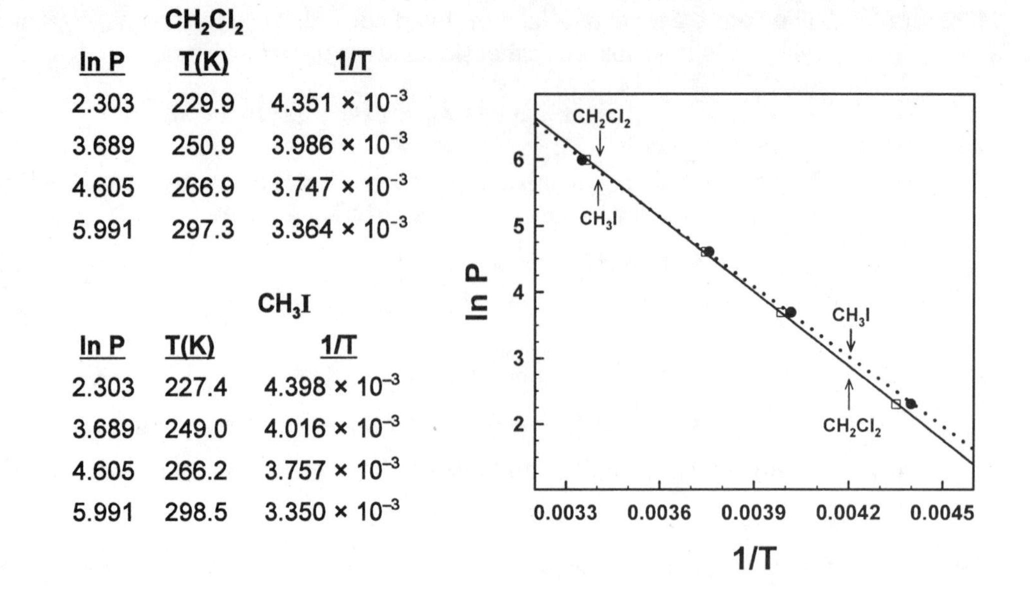

CH_2Cl_2

ln P	T(K)	1/T
2.303	229.9	4.351×10^{-3}
3.689	250.9	3.986×10^{-3}
4.605	266.9	3.747×10^{-3}
5.991	297.3	3.364×10^{-3}

CH_3I

ln P	T(K)	1/T
2.303	227.4	4.398×10^{-3}
3.689	249.0	4.016×10^{-3}
4.605	266.2	3.757×10^{-3}
5.991	298.5	3.350×10^{-3}

For CH_2Cl_2, $-\Delta H_{vap}/R = \text{slope} = \dfrac{(5.991 - 2.303)}{(3.364 \times 10^{-3} - 4.350 \times 10^{-3})} = \dfrac{-3.688}{0.987 \times 10^{-3}}$

$= -3.74 \times 10^3 = -\Delta H_{vap}/R$

$\Delta H_{vap} = 8.314\ (3.74 \times 10^3) = 3.107 \times 10^4$ J/mol = 31.1 kJ/mol

For CH_3I, $-\Delta H_{vap}/R = \text{slope} = \dfrac{(5.991 - 2.303)}{(3.350 \times 10^{-3} - 4.398 \times 10^{-3})} = \dfrac{-3.688}{1.048 \times 10^{-3}} = -3.519 \times 10^3$

$= -\Delta H_{vap}/R$

$\Delta H_{vap} = 8.314\ (3.519 \times 10^3) = 2.926 \times 10^4$ J/mol = 29.3 kJ/mol

11.95 The most effective diffraction of light by a grating occurs when the wavelength of light and the separation of the slits in the grating are similar. When X-rays are diffracted by a crystal, layers of atoms serve as the "slits." The most effective diffraction occurs when the distances between layers of atoms are similar to the wavelength of the X-rays. Typical interlayer distances in crystals range from 2 Å to 20 Å. Visible light, 400-700 nm or 4,000 to 7,000 Å, is too long to be diffracted effectively by crystals. Molybdenum X-rays of 0.71 Å are on the same order of magnitude as interlayer distances in crystals and are diffracted.

11.96 (a) Both diamond (d = 3.5 g/cm^3) and graphite (d = 2.3 g/cm^3) are covalent-network solids with efficient packing arrangements in the solid state; there is relatively little empty space in their respective crystal lattices. Diamond, with bonded C–C distances of 1.54 Å in all directions, is more dense than graphite, with shorter C–C distances within carbon sheets but longer 3.41 Å separations between sheets (Figure 11.40). Buckminsterfullerene has much more empty space, both inside each C_{60} "ball" and between balls, than either diamond or graphite, so its density will be considerably less than 2.3 g/cm^3.

(b) In a face-centered-cubic unit cell, there are 4 complete C_{60} units.

$$\frac{4\,C_{60}\text{ units}}{(14.2\text{ Å})^3} \times \frac{720.66\text{ g}}{6.022 \times 10^{23}\,C_{60}\text{ units}} \times \left(\frac{1\text{ Å}}{1 \times 10^{-8}\text{ cm}}\right)^3 = 1.67\text{ g/cm}^3$$

(1.67 g/cm^3 is the smallest density of the three allotropes, diamond, graphite and buckminsterfullerene.)

Integrative Exercises

11.97 (a) In Table 11.4, viscosity increases as the length of the carbon chain increases. Longer molecular chains become increasingly entangled, increasing resistance to flow.

(b) Whereas viscosity depends on molecular chain length in a critical way, surface tension depends on the strengths of intermolecular interactions between molecules. These dispersion forces do not increase as rapidly with increasing chain length and molecular weight as viscosity does.

(c) The –OH group in n-octyl alcohol gives rise to hydrogen bonding among molecules, which increases molecular entanglement and leads to greater viscosity and higher boiling point.

11.98 (a) 24 valence e^-, 12 e^- pairs

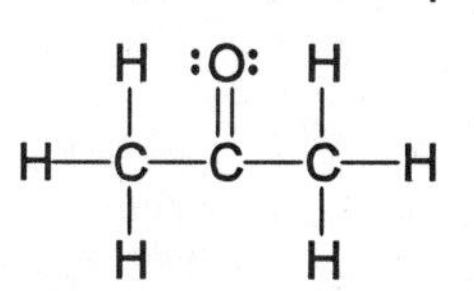

The geometry around the central C atom is trigonal planar, and around the two terminal C atoms, tetrahedral.

(b) Polar. The C=O bond is quite polar and the dipoles in the trigonal plane around the central C atom do not cancel.

(c) Dipole-dipole and London-dispersion forces

(d) Since the molecular weights of acetone and 1-propanol are similar, the strength of the London-dispersion forces in the two compounds is also similar. The big difference is that 1-propanol has hydrogen bonding, while acetone does not. These relatively strong attractive forces lead to the higher boiling point for 1-propanol.

11.99

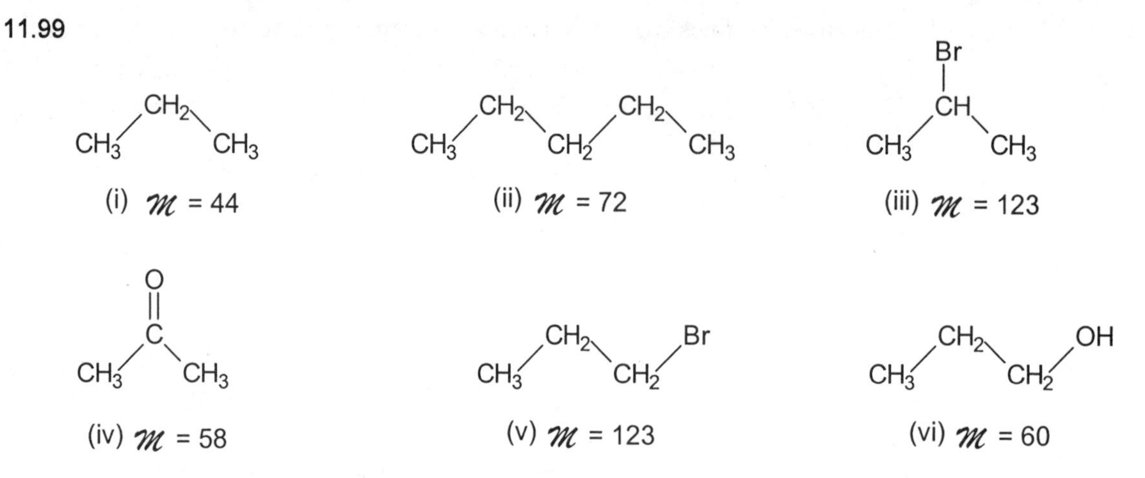

It is useful to draw the structural formulas because intermolecular forces are determined by the size and shape (structure) of molecules.

(a) *Molar mass*: compounds (i) and (ii) have similar rod-like structures; (ii) has a longer rod. The longer chain leads to greater molar mass, stronger London-dispersion forces and higher heat of vaporization.

(b) *Molecular shape*: compounds (iii) and (v) have the same chemical formula and molar mass but different molecular shapes (they are structural isomers). The more rod-like shape of (v) leads to more contact between molecules, stronger dispersion forces and higher heat of vaporization.

(c) *Molecular polarity*: rod-like hydrocarbons (i) and (ii) are essentially nonpolar, owing to free rotation about C-C σ bonds, while (iv) is quite polar, owing to the C=O group. (iv) has a smaller molar mass than (ii) but a larger heat of vaporization, which must be due to the presence of dipole-dipole forces in (iv). [Note that (iii) and (iv), with similar shape and molecular polarity, have very similar heats of vaporization.]

(d) *Hydrogen-bonding interactions*: molecules (v) and (vi) have similar structures, but (vi) has hydrogen bonding and (v) does not. Even though molar mass and thus dispersion forces are larger for (v), (vi) has the higher heat of vaporization. This must be due to hydrogen bonding interactions.

11.100 (a) In order for butane to be stored as a liquid at temperatures above its boiling point (-5°C), the pressure in the tank must be greater than atmospheric pressure. In terms of the phase diagram of butane, the pressure must be high enough so that, at tank conditions, the butane is "above" the gas-liquid line and in the liquid region of the diagram.

The pressure of a gas is described by the ideal gas law as $P = nRT/V$; pressure is directly proportional to moles of gas. The more moles of gas present in the tank the greater the pressure, until sufficient pressure is achieved for the gas to liquify. At the point where liquid and gas are in equilibrium and temperature is constant, liquid will vaporize or condense to maintain the equilibrium vapor pressure. That is, as long as some liquid is present, the gas pressure in the tank will be constant.

(b) If butane gas escapes the tank, butane liquid will vaporize (evaporate) to maintain the equilibrium vapor pressure. Vaporization is an endothermic process, so the butane will absorb heat from the surroundings. The temperature of the tank and the liquid butane will decrease.

(c) $155\ g\ C_4H_{10} \times \frac{1\ mol\ C_4H_{10}}{58.12\ g\ C_4H_{10}} \times \frac{21.3\ kJ}{mol} = 56.8\ kJ$

$$V = \frac{nRT}{P} = 155\ g \times \frac{1\ mol}{58.12\ g} \times \frac{0.08206\ L \bullet atm}{mol \bullet K} \times \frac{308\ K}{755\ torr} \times \frac{760\ torr}{1\ atm}$$

$= 67.851 = 67.9\ L$

11.101 *Plan*:

(i) Using thermochemical data from Appendix B, calculate the energy (enthalpy) required to melt and heat the H_2O.

(ii) Using Hess's Law, calculate the enthalpy of combustion, ΔH_{comb}, for C_3H_8.

(iii) Solve the stoichiometry problem.

(i) Heat $H_2O(s)$ from -14.0°C to 0.0°C; $2500\ g\ H_2O \times \frac{2.092\ J}{g \bullet {}^\circ C} \times 14.0\ {}^\circ C = 73.22$

$= 73.2\ kJ$

Melt $H_2O(s)$; $2500\ g\ H_2O \times \frac{6.008\ kJ}{mol\ H_2O} \times \frac{1\ mol\ H_2O}{18.02\ g\ H_2O} = 833.52 = 834.4\ kJ$

Heat $H_2O(l)$ from 0.0°C to 60.0°C; $2500\ g\ H_2O \times \frac{4.184\ J}{g \bullet {}^\circ C} \times 60.0\ {}^\circ C = 627.6$

$= 628\ kJ$

Total energy = 73.22 kJ + 833.52 kJ + 627.6 kJ = 1534.3 = 1.53×10^3 kJ

(The result has zero decimal places because 628 kJ has zero decimal places.)

(ii) $C_3H_8(g) + 5O_2(g) \rightarrow 3CO_2(g) + 4H_2O(l)$

Assume that one product is $H_2O(l)$, since this leads to a more negative ΔH_{comb} and fewer grams of $C_3H_8(g)$ required.

$$\Delta H_{comb} = 3\Delta H_f^\circ\ CO_2(g) + 4\Delta H_f^\circ\ H_2O(l) - \Delta H_f^\circ\ C_3H_8(g) - 5\Delta H_f^\circ\ O_2(g)$$

$$= 3(-393.5\ kJ) + 4(-285.83\ kJ) - (-103.85\ kJ) - 5(0) = -2219.97 = -2220\ kJ$$

(iii) $1.5343 \times 10^3\ kJ\ required \times \frac{1\ mol\ C_3H_8}{2219.97\ kJ} \times \frac{44.094\ g\ C_3H_8}{1\ mol\ C_3H_8} = 30.5\ g\ C_3H_8$

(1.53×10^3 kJ required has 3 sig figs and so does the result)

11.103 $P = \frac{nRT}{V} = \frac{g\ RT}{\mathcal{M}\ V}$; T = 273 + 26°C = 299 K; V = 5.00 L

$g\ C_6H_6(g) = 7.2146 - 5.1493 = 2.0653\ g\ C_6H_6(g)$

$$P\ (vapor) = \frac{2.0653\ g}{78.11\ g/mol} \times \frac{299\ K}{5.00\ L} \times \frac{0.08206\ L \bullet atm}{K \bullet mol} \times \frac{760\ torr}{1\ atm} = 98.6\ torr$$

11.104 *Plan*: relative humidity and v.p. of H_2O at 23°C → P_{H_2O}

ideal-gas law → mol $H_2O(g)$ → H_2O molecules

r.h. = (P_{H_2O} in air / v.p. of H_2O) × 100; From Appendix B, v.p. of H_2O at 23°C = 21.07 torr

P_{H_2O} in air = r.h. × v.p. of H_2O/100 = 45 × 21.07 torr/100 = 9.4815 = 9.5 torr

$$n = PV/RT;\ V = 14\ m \times 9.0\ m \times 8.6\ m \times \frac{1\ dm^3}{(0.1)^3\ m^3} \times \frac{1\ L}{dm^3} = 1.0836 \times 10^6 = 1.1 \times 10^6\ L$$

$$n = 9.4815\ torr \times \frac{1\ atm}{760\ torr} \times \frac{mol \cdot K}{0.08206\ L \cdot atm} \times \frac{1.0836 \times 10^6\ L}{296\ K} = 556.6 = 5.6 \times 10^2\ mol\ H_2O$$

$$556.6\ mol\ H_2O \times \frac{6.022 \times 10^{23}\ molecules}{1\ mol} = 3.4 \times 10^{26}\ H_2O\ molecules$$

11.105 Data are taken from the 74th edition of the *Handbook of Chemistry and Physics.*
T_m = melting point, T_b = boiling point

(a) W: T_m = 3410°C, T_b = 5660 °C; WF_6: T_m = 2.5°C, T_b = 17.5°C

W is a metal, with strong metallic bonding, and very high T_m and T_b. WF_6 is an octahedral, nonpolar molecule. Even though it has high molar mass, the spherical shape of the molecule prevents extensive molecular contacts. The resulting London-dispersion forces are very weak, which leads to the low T_m and T_b.

(b) SO_2: T_m = -72.7°C, T_b = -10°C; SF_4: T_m = -124°C, T_b = -40°C (sublimes)

Both SO_2 and SF_4 are polar covalent molecules with a nonbonding electron pair on the central S atom. The electron-domain geometry in SO_2 is trigonal planar and the molecule shape is bent. The electron-domain geometry in SF_4 is trigonal bipyramidal and the molecular shape is see-saw. Both are gases at ambient temperature and pressure. SF_4 has higher molar mass but lower melting and boiling points than SO_2. This indicates that dipole-dipole forces are more influential on the properties of these molecules and that SF_4 has a smaller dipole moment than SO_2.

(c) SiO_2: T_m = 1723°C, T_b = 2230°C, $\mathcal{M}$ = 60 g/mol

$SiCl_4$: T_m = -70°C, T_b = 57.57°C, $\mathcal{M}$ = 170 g/mol

SiO_2 is a covalent-network substance, with high T_m and T_b. Covalent bonds hold SiO_2 units in a rigid lattice, and high energy is required to break these bonds and melt or boil the substance. $SiCl_4$ is a tetrahedral, nonpolar molecule. Weak dispersion forces between the approximately spherical molecules result in predictably low T_m and T_b.

12 Modern Materials

Liquid Crystals

12.1 Both an ordinary liquid and a nematic liquid crystal phase are fluids; they are converted directly to the solid phase upon cooling. The nematic phase is cloudy and more viscous than an ordinary liquid. Upon heating, the nematic phase is converted to an ordinary liquid.

12.2 In an ordinary liquid, molecules are oriented randomly and their relative orientations are continuously changing. In liquid crystals, the molecules are aligned in at least one dimension. The relative orientations in the other two dimensions may change, but alignment in the oriented direction is maintained.

12.3 In the solid state, there is three-dimensional order; the relative orientation of the molecules is fixed and repeating in all three dimensions. Essentially no translational or rotational motion is allowed. When a substance changes to the nematic liquid-crystalline phase, the molecules remain aligned in one dimension (the long dimension of the molecule). Translational motion is allowed, but rotational motion is restricted. Transformation to the isotropic-liquid phase destroys the one-dimensional order. Free translational and rotational motion result in random molecular orientations that change continuously.

12.4 Reinitzer observed that cholesteryl benzoate has a phase that exhibits properties intermediate between those of the solid and liquid phases. This "liquid-crystalline" phase, formed by melting at 145°C, is opaque, changes color as the temperature increases, and becomes clear at 179°C.

12.5 The presence of polar groups or nonbonded electron pairs leads to relatively strong dipole-dipole interactions between molecules. These are a significant part of the orienting forces necessary for liquid crystal formation.

12.6 Because order is maintained in at least one dimension, the molecules in a liquid-crystalline phase are not totally free to change orientation. This makes the liquid-crystalline phase more resistant to flow, more viscous, than the isotropic liquid.

12.7 In the nematic phase, molecules are aligned in one dimension, the long dimension of the molecule. In a smectic phase (A or C), molecules are aligned in two dimensions. Not only are the long directions of the molecules aligned, but the ends are also aligned. The molecules are organized into layers; the height of the layer is related to the length of the molecule.

12.8 The "LCD molecule" is long relative to its thickness. It has C=C and C≡N groups that promote rigidity and polarizability along the length of the molecule. The C≡N group also provides dipole-dipole interactions that encourage alignment. Unlike the molecules in Figure 12.4, which contain planar phenyl rings, the LCD molecule contains nonaromatic, nonplanar six-membered rings. These rings could contribute to specific physical properties such as the liquid crystal temperature range that make this molecule particularly functional in LCD displays.

12.9 A nematic phase is composed of sheets of molecules aligned along their lengths, but with no additional order within the sheet or between sheets. A cholesteric phase also contains this kind of sheet, but with some ordering between sheets. In a cholesteric phase, there is a characterisitic angle between molecules in one sheet and those in an adjacent sheet. That is, one sheet of molecules is twisted at some characteristic angle relative to the next, producing a "screw" axis perpendicular to the sheets.

12.10 As the temperature of a substance increases, the average kinetic energy of the molecules increases. More molecules have sufficient kinetic energy to overcome intermolecular attractive forces, so overall ordering of the molecules decreases as temperature increases. Melting provides kinetic energy sufficient to disrupt alignment in one dimension in the solid, producing a smectic phase with ordering in two dimensions. Additional heating of the smectic phase provides kinetic energy sufficient to disrupt alignment in another dimension, producing a nematic phase with one-dimensional order.

Polymers

12.11 *n*-decane does not have a sufficiently high chain length or molecular mass to be considered a polymer.

12.12 Monomers are small molecules with low molecular mass that are joined together to form polymers. They are the repeating units of a polymer. Three monomers mentioned in this chapter are

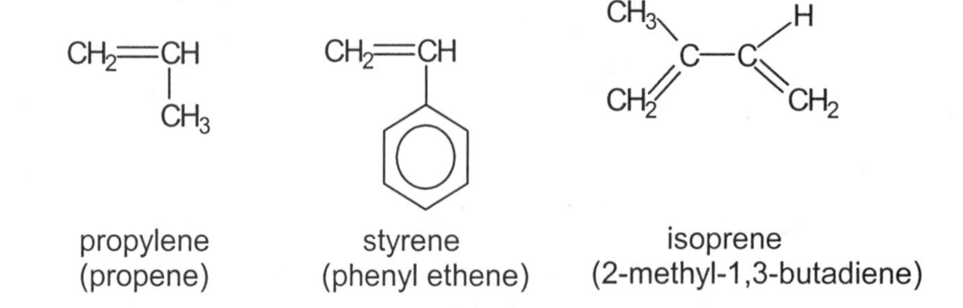

12.13 *Analyze.* Given two types of reactant molecules, we are asked to write a condensation reaction with an ester product. *Plan.* A condensation reaction occurs when two smaller molecules combine to form a larger molecule and a small molecule, often water. Consider the structures of the two reactants and how they could combine to join the larger fragments and split water. *Solve*:

A carboxylic acid contains the $-\overset{O}{\overset{\|}{C}}-OH$ functional group; an alcohol contains the –OH functional group. These can be arranged to form the $-\overset{O}{\overset{\|}{C}}-O-C-$ ester functional group and H_2O. Condensation reaction to form an ester:

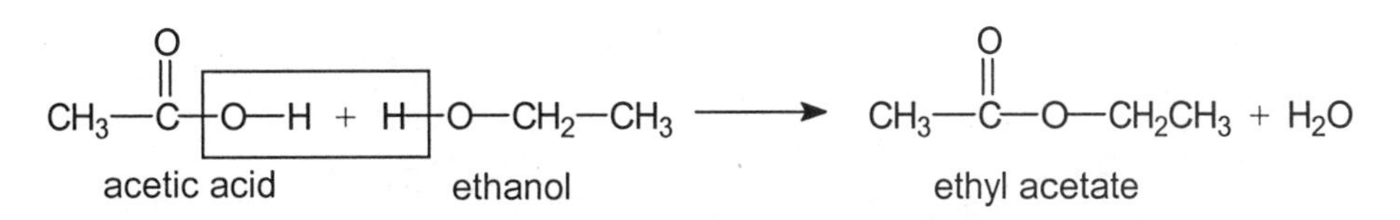

If a dicarboxylic acid (two –COOH groups, usually at opposite ends of the molecule) and a dialcohol (two –OH groups, usually at opposite ends of the molecule) are combined, there is the potential for propagation of the polymer chain at both ends of both monomers. Polyethylene terephthalate (Table 12.1) is an example of a polyester formed from the monomers ethylene glycol and terephthalic acid.

12.14

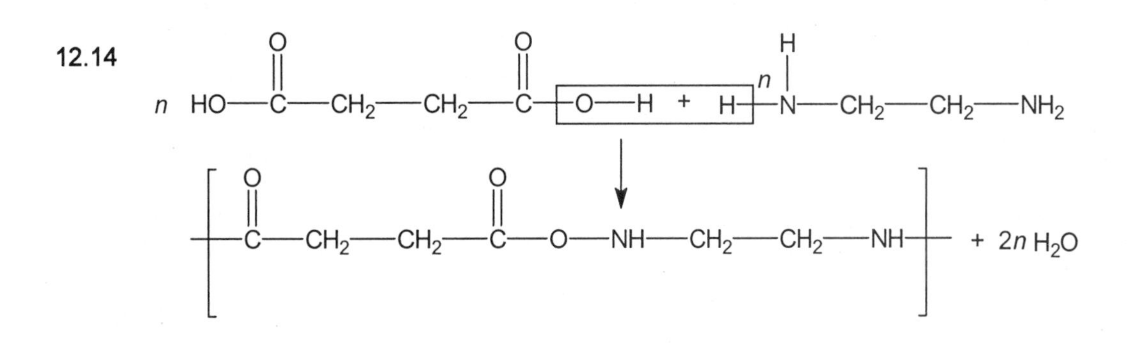

12.15 *Analyze/Plan.* Decide whether the given polymer is an addition or condensation polymer. Select the smallest repeat unit and deconstruct it into the monomer(s) with the specific functional group(s) that would form the stated polymer. *Solve*:

(a)

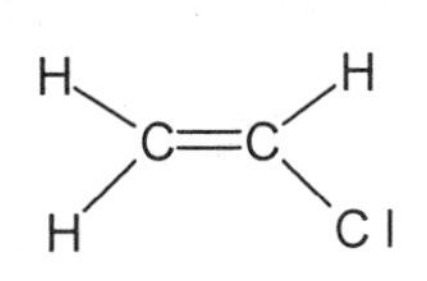

vinyl chloride (chloroethylene or chloroethene)

(b)

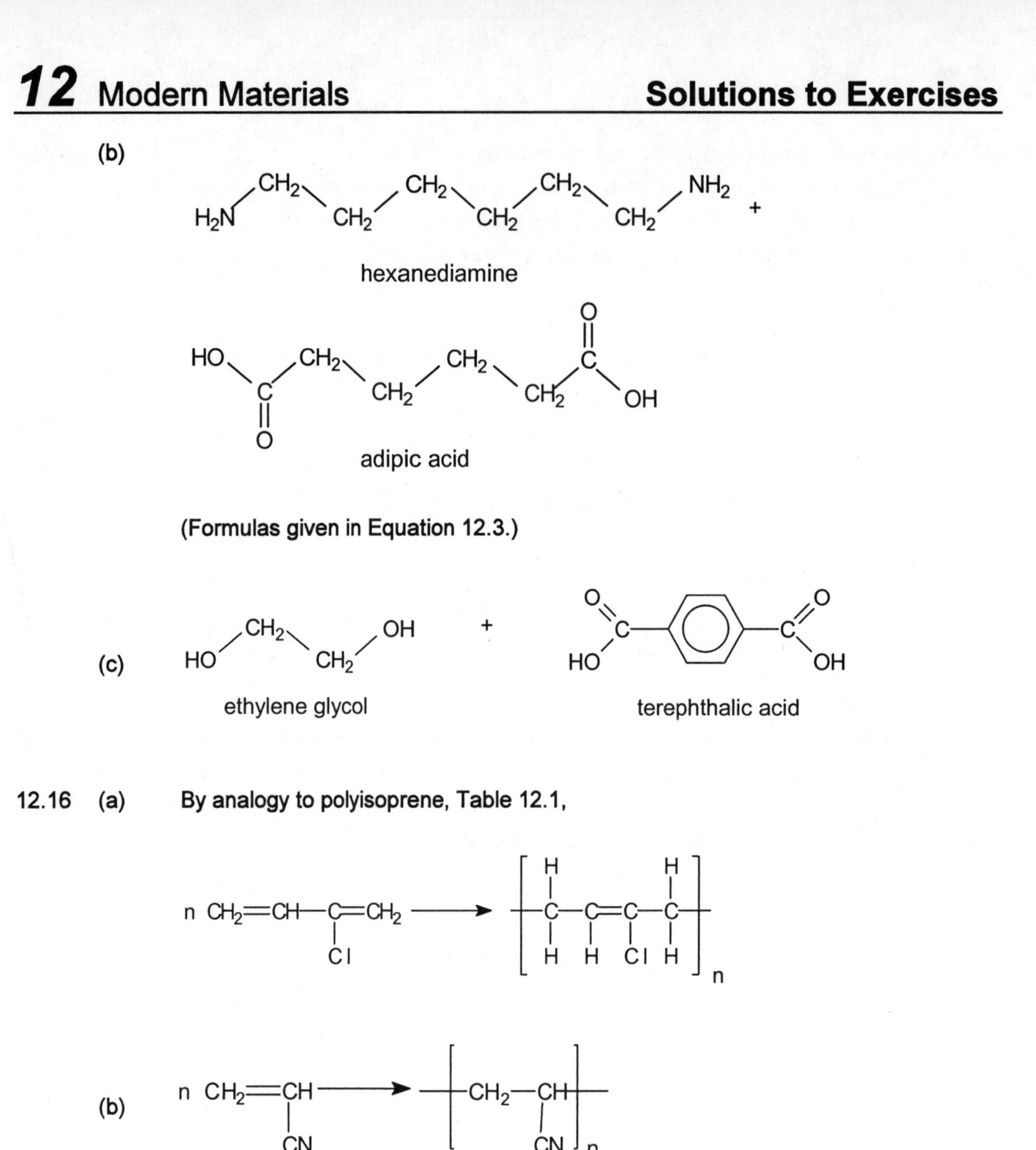

(Formulas given in Equation 12.3.)

(c)

12.16 (a) By analogy to polyisoprene, Table 12.1,

(b)

12.17 *Plan/Solve.* When nylon polymers are made, H_2O is produced as the C–N bonds are formed. Reversing this process (adding H_2O across the C–N bond), we see that the monomers used to produce Nomex™ are:

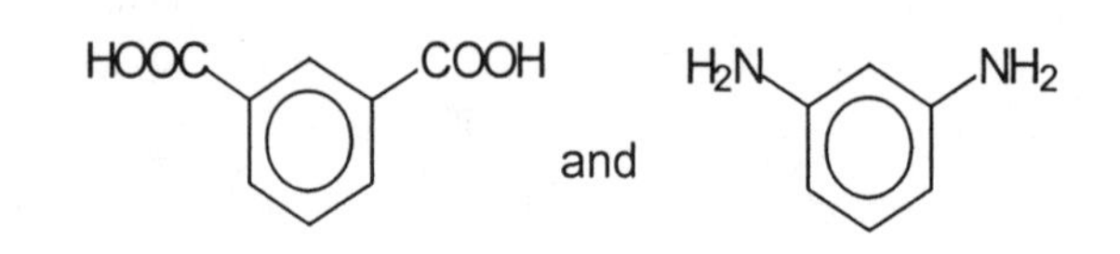

12.18

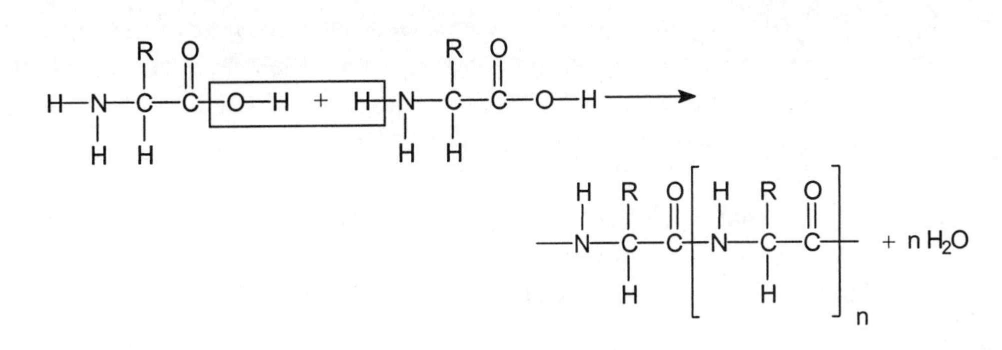

12.19 *Analyze/Plan.* Given the formula of a monomer, write the equation for condensation polymerization. The monomers are aligned so that the caroboxyl-end of one monomer joins the amine-end of another molecule. *Solve*:

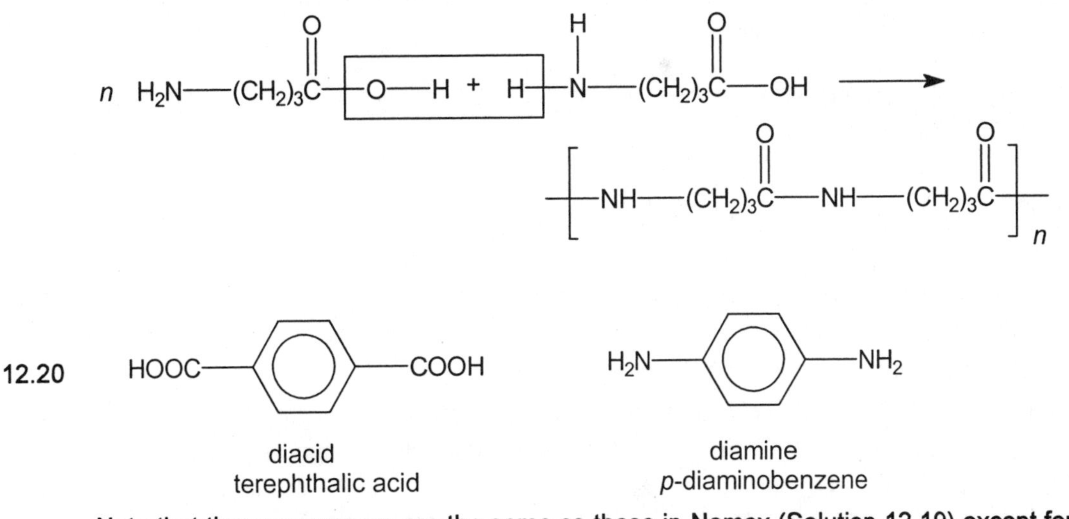

12.20

diacid
terephthalic acid

diamine
p-diaminobenzene

Note that these monomers are the same as those in Nomex (Solution 12.19) except for the orientation of the functional groups on the benzene rings. Clearly monomer structure strongly impacts polymer structure.

12.21 Most of a polymer backbone is composed of σ bonds. The geometry around individual atoms is tetrahedral with bond angles of 109°, so the polymer is not flat, and there is relatively free rotation around the σ bonds. The flexibility of the molecular chains causes flexibility of the bulk material. Flexibility is enhanced by molecular features that inhibit order, such as branching, and diminished by features that encourage order, such as crosslinking or delocalized π electron density.

Crosslinking is the formation of chemical bonds between polymer chains. It reduces flexibility of the molecular chains and increases the hardness of the material. Crosslinked polymers are less chemically reactive because of the links.

12.22 At the molecular level, the longer, unbranched chains of HDPE fit closer together and have more crystalline (ordered, aligned) regions than the shorter, branched chains of LDPE. Closer packing leads to higher density.

12.23 The function of the material (polymer) determines whether high molecular mass and high degree of crystallinity are desirable properties. If the material will be formed into containers or pipes, rigidity and structural strength are required. If the polymer will be used as a flexible wrapping or as a garment material, rigidity is an undesirable property.

12.24 (a) An *elastomer* is a polymer material that recovers its shape when released from a distorting force. A typical elastomeric polymer can be stretched to at least twice its original length and return to its original dimensions upon release.

(b) A *thermoplastic* material can be shaped and reshaped by application of heat and/or pressure.

(c) A *thermosetting plastic* can be shaped once, through chemical reaction in the shape-forming process, but cannot easily be reshaped, due to the presence of chemical bonds that crosslink the polymer chains.

(d) A *plasticizer* is a substance of relatively low molar mass added to a polymer material to soften it.

Biomaterials

12.25 Is the neoprene biocompatible: is the surface smooth enough and is the chemical composition appropriate so that there are no inflammatory reactions in the body? Does neoprene meet the physical requirements of a flexible lead: will it remain resistant to degradation by body fluids over a long time period; will it maintain elasticity over the same time period? Can neoprene be prepared in sufficiently pure form (free of trace amounts of monomer, catalyst, etc.) so that it can be classified as *medical grade*?

12.26 One structural characteristic of polymers that forms effective interfaces with biological systems is the presence of polar functional groups in the polymer backbone or as substituents. Polystyrene is a hydrocarbon; it has no polar functional groups and is a nonpolar substance. Polyurethane has polar carbon-oxygen, carbon-nitrogen and nitrogen-hydrogen functional groups. The N–H groups mean that it can act as a hydrogen-bond donor as well as an acceptor. In fact, the polyurethane backbone is very similar to the protein backbone shown in this section. We expect polyurethane to be the superior biointerface.

12.27 Current vascular-graft materials cannot be lined with cells similar to those in the native artery. The body detects that the graft is "foreign" and platelets attach to the inside surfaces, causing blood clots. The inside surfaces of the future vascular implants need to accommodate a lining of cells that do not attract or attach to platelets.

12.28 Surface roughness in synthetic heart valves causes *hemolysis*, the breakdown of red blood cells. The surface of the valve implant was probably not smooth enough.

12.29 In order for skin cells in a culture medium to develop into synthetic skin, a mechanical matrix must be present that holds the cells in contact with one another and allows them to differentiate. The matrix must be mechanically strong, biocompatible and biodegradable. It probably has polar functional groups that are capable of hydrogen bonding with biomolecules in the tissue cells.

12.30 Polystyrene is an essentially nonpolar hydrocarbon, while polyethyleneterephthalate (PET) contains polar ester groups, as well as nonpolar hydrocarbon portions. PET is more appropriate, because it provides polar ester groups $\left(\overset{\overset{\displaystyle O}{\|}}{C}—O—C\right)$ with hydrogen bonding capabilities where the cells can attach. Also, the ester linkages are susceptible to hydrolysis (the reverse of condensation); this renders the synthetic matrix biodegradable when employed in the body.

Ceramics

12.31 Ceramics are not readily recyclable because of their extremely high melting points and rigid ionic or covalent-network structures. According to Table 12.4, the melting points of ceramic materials are much higher than those of Al and Fe. This makes recycling ceramics technologically difficult and expensive. Crystalline ceramics have rigid, precise three-dimensional structures. If and when these materials can be melted, either covalent or ionic bonds are broken. The precise structures are usually not reformed upon cooling. Recyclable ceramics such as bottle glass are amorphous; there is no exact repeating structure that must be reformed after melting.

12.32 (a) The object that shatters is ceramic; the one that dents is metal.

(b) The two materials differ in their behavior because of their different solid state bonding characteristics. Ceramics are formed from inorganic materials linked by ionic or highly polar covalent bonds into three-dimensional bonding networks. During catastrophic failure (dropping 10 feet onto cement), the network structure prevents atoms from sliding over one-another and the ceramic shatters. A series of bond ruptures occurs, often along planes in the three-dimensional structure, leading to fragments with sharp edges.

Metallic bonding is characterized by delocalization of loosely-held valence electrons among metal atoms. Unlike the covalent or ionic network bonding in ceramics, metallic bonding is multidirectional. This allows metal atoms to slide over each other during deformation, resulting in dents and cracks rather than shattering.

12.33 Very small, uniformly sized and shaped particles are required for the production of a strong ceramic object by sintering. During sintering, the small ceramic particles are heated to a high temperature below the melting point of the solid. This high temperature initiates condensation reactions between atoms at the surfaces of the spheres; the spheres are then connected by chemical bonds between atoms in different spheres. The more uniform the particle size and the greater the total surface area of the solid, the more chemical bonds are formed, and the stronger the ceramic object.

12.34 Since Zr and Ti are in the same family, assume that the stoichiometry of the compounds in a sol-gel process will be the same for the two metals.

i. Alkoxide formation: oxidation-reduction reaction

$$Zr(s) + 4CH_3CH_2OH(l) \rightarrow \underset{\text{alkoxide}}{Zr(OCH_2CH_3)_4(s)} + 2H_2(g)$$

ii. Sol formation: metathesis reaction

$$Zr(OCH_2CH_3)_4(soln) + 4H_2O(l) \rightarrow \underset{\text{"precipitate" sol}}{Zr(OH)_4(s)} + \underset{\text{nonelectrolyte}}{4CH_3CH_2OH(l)}$$

$Zr(OCH_2CH_3)_4(s)$ is dissolved in an alcohol solvent and then reacted with water. In general, reaction with water is called *hydrolysis*. The alkoxide anions ($CH_3CH_2O^-$) combine with H^+ from H_2O to form the nonelectrolyte $CH_3CH_2OH(l)$, and Zr^{2+} cations combine with OH^- to form the $Zr(OH)_4$ solid. The product $Zr(OH)_4(s)$ is not a traditional coagulated precipitate, but a finely divided evenly dispersed collection of particles called a sol.

iii. Gel formation: condensation reaction

$$(OH)_3Zr–O–H(s) + H–O–Zr(OH)_3(s) \rightarrow \underset{\text{gel}}{(HO)_3Zr–O–Zr(OH)_3(s)} + H_2O(l)$$

Adjusting the acidity of the $Zr(OH)_4$ sol initiates condensation, the splitting-out of $H_2O(l)$ and formation of a zirconium-oxide network solid. The solid remains suspended in the solvent mixture and is called a gel.

iv. Processing: physical changes

The gel is heated to drive off solvent and the resulting solid consists of dry, uniform and finely divided ZrO_2 particles.

12.35 Concrete is a typically brittle ceramic that is susceptible to catastrophic fracture. Steel reinforcing rods are added to resist stress applied along the long direction of the rod. By analogy, the shape of the reinforcing material in the ceramic composite should be rod-like, with a length much greater than its diameter. This is the optimal shape because rods have great strength when the load or stress is applied parallel to the long direction of the rod. Rods can be oriented in many directions, so that the material (concrete or ceramic composite) is strengthened in all directions.

12.36 The ceramics are: MgO, soda lime glass, ZrB_2, Al_2O_3 and TaC. The criteria are a combination of chemical formula (with corresponding bonding characteristics) and Knoop values. Ceramics are ionic or covalent-network solids with fairly large hardness values. $CaCO_3$ is ionic, but carbonates are not one of the typical types of ceramics listed in Section 12.4. Ag and Cr are metals.

Hardness alone is not a sufficient criteria for ceramics. The range of hardness values for the ceramics in this group is large; Cr, a metal, lies in the middle of the range, as could other metals. Nonetheless, ceramics as a group are hard materials; hardness is a necessary, but not a sufficient condition for classification as a ceramic.

12.37 By analogy to the ZnS structure, the C atoms form a face-centered cubic array with Si atoms occupying **alternate** tetrahedral holes in the lattice. This means that the coordination numbers of both Si and C are 4; each Si is bound to 4 C atoms in a tetrahedral arrangement, and each C is bound to 4 Si atoms in a tetrahedral arrangement, producing an extended three-dimensional network. ZnS, an ionic solid, sublimes at 1185° and 1 atm pressure and melts at 1850° and 150 atm pressure. The considerably higher melting point of SiC, 2800° at 1 atm, indicates that SiC is probably not a purely ionic solid and that the Si–C bonding network has significant covalent character. This is reasonable, since the electronegativities of Si and C are similar (Figure 8.7). SiC is high-melting because a great deal of chemical energy is stored in the covalent Si–C bonds, and it is hard because the three-dimensional lattice resists any change that would weaken the Si–C bonding network.

12.38 The four ceramics in Table 12.4 are Al_2O_3, SiC, ZrO_2 and BeO. All are network solids with some ionic and some covalent bonding character. They have a higher modulus of elasticity, which is to say they are more rigid than the metals listed. The high melting points, hardness and rigidity of the these four solids are due to the rigid, highly ordered, three-dimensional network structure of the solids. These network structures resist deformation (hardness) and require a great deal of energy (high melting point) to break the chemical bonds that form the network.

Superconductivity

12.39 A superconducting material offers no resistance to the flow of electrical current; *superconductivity* is the frictionless flow of electrons. Superconductive materials could transmit electricity with no heat loss and therefore much greater efficiency than current carriers. Because of the Meisner effect, they are also potential materials for magnetically levitated trains.

12.40 A conductive metal such as Ag conducts electricity with a characteristic resistance to the flow of electrons given by Ohm's law, $E = IR$. A superconducting substance such as Nb_3Sn below its transition temperature conducts electricity with **no** resistance to the flow of electrons. Such a superconductor can transfer energy with no net loss, while a metallic conductor cannot be 100% efficient.

12.41 Below 39 K, MgB_2 conducts electricity with zero resistivity, the definition of a superconductor. Above 39 K, the material is not superconducting. The sharp drop in resistivity of MgB_2 near 39 K is the superconducting transition temperature, T_c.

12.42 (a) The superconducting transition temperature, T_c, is the temperature at which a material loses all resistance to the flow of electrical current, the temperature below which the material becomes superconducting.

(b) The temperature 77 K is significant because that is the temperature of liquid nitrogen, a readily-available, inexpensive and safe coolant. Materials with T_c temperatures above 77 K produce more financially viable devices than materials which must be cooled with liquid helium below 77 K to achieve superconductivity.

12.43 Because they are brittle ceramics, it is difficult to mold superconductors into useful shapes such as wires and these wires would be fragile at best. For presently known superconductors, the amount of current per cross-sectional area that can be carried by these wires is limited. The low temperatures required for superconductivity render today's superconducting ceramics impractical for widespread use.

12.44 Superconductivity acts on electrons that are conduction electrons. Thus you need to have electrons available to be conductors. Insulators have no electrons available for conduction.

Thin Films

12.45 Adhesion is due to attractive intermolecular forces. These include ion-dipole, dipole-dipole, dispersion and hydrogen bonding. Adhesive interactions will be strongest between substances with similar intermolecular forces, so the bonding characteristics (that determine intermolecular forces) of the thin film material should be matched to those of the substrate.

12.46 In general, a useful thin film should:

(a) be chemically stable in its working environment
(b) adhere to its substrate
(c) have a uniform thickness
(d) have an easily controllable composition
(e) be nearly free of imperfections

12.47 The coating in Figure 12.31 is a metallic film that reflects most of the incident sunlight. The exclusion of sunlight from the interior of the building reduces glare and cooling load. The opacity of the film provides privacy.

12.48 There are three major methods of producing thin films.

(i) In *vacuum deposition*, a substance is vaporized or evaporated by heating under vacuum and then deposited on the desired substrate. No chemical change occurs.

(ii) In *sputtering*, ions accelerated to high energies by applying a high voltage are allowed to strike the target material, knocking atoms from its surface. These target material atoms are further accelerated toward the substrate, forming a thin film. No net chemical change occurs, because the material in the film is the target material.

(iii) In *chemical vapor deposition*, two gas phase substances react at the substrate surface to form a stable product which is deposited as a thin film. This involves a net chemical change.

Additional Exercises

12.49 A dipole moment (permanent, partial charge separation) roughly parallel to the long dimension of the molecule would cause the molecules to reorient when an electric field is applied perpendicular to the usual direction of molecular orientation.

12.50

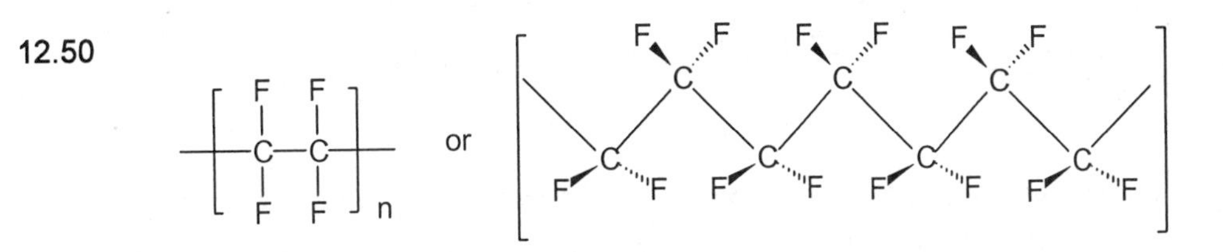

Teflon™ is formed by addition polymerization.

12.51 (a) Polymer (b) ceramic (c) ceramic (d) polymer (e) liquid crystal (an organic molecule with a characteristic long axis and the kinds of functional groups often found in compounds with liquid-crystalline phases (Figure 12.4); not enough repeating units to be a polymer)

12.52 At the temperature where a substance changes from the solid to the liquid-crystalline phase, kinetic energy sufficient to overcome most of the long range order in the solid has been supplied. A few van der Waals forces have sufficient attractive energy to impose the one-dimensional order characteristic of the liquid-crystalline state. Very little additional kinetic energy (and thus a relatively small increase in temperature) is required to overcome these aligning forces and produce an isotropic liquid.

12.53 Ceramics are usually three-dimensional network solids, whereas plastics most often consist of large, chain-like molecules (the chain may be branched) held loosely together by relatively weak van der Waals forces. Ceramics are rigid precisely because of the many strong bonding interactions intrinsic to the network. Once a crack forms, atoms near the defect are subject to great stress, and the crack is propagated. They are stable to high temperatures because tremendous kinetic energy (temperature) is required for an atom to break free from the bonding network. On the other hand, plastics are flexible because the molecules themselves are flexible (free rotation around the sigma bonds in the polymer chain), and it is easy for the molecules to move relative to one another (weak intermolecular forces). (However, recall that rigidity of the plastic increases as crosslinking of the polymer chain increases. The melamine-formaldehyde polymer in Figure 12.17 is a very rigid, brittle polymer.) Plastics are not thermally stable because their largely organic molecules are subject to oxidation and/or bond breaking at high temperatures.

12.54 In a liquid crystal display (Figure 12.7), the molecules must be free to rotate by 90°. The long directions of molecules remain aligned but any attractive forces between the ends of molecules are disrupted. At low Antarctic temperatures, the liquid crystalline phase is closer to its freezing point. The molecules have less kinetic energy due to temperature and the applied voltage may not be sufficient to overcome orienting forces among the ends of molecules. If some or all of the molecules do not rotate when the voltage is applied, the display will not function properly.

12.55 This phenomenon is similar to supercooling, Section 11.4. When the isotropic liquid is cooled below the liquid crystal-liquid transition temperature, the kinetic energy of the molecules has been decreased enough so that formation of the liquid crystalline phase is energetically favorable. However, the molecules may not be correctly organized so that long range ordering can take place.

12.56 Both metals and ceramics have highly ordered three-dimensional structures or lattices, but the bonding characteristics of these lattices are different. Many metals adopt close-packed structures where each metal has 12 nearest neighbors (Section 11.7). Since each metal atom has too few electrons to form a covalent (electron-pair) bond with each of its nearest neighbors, the bonding electrons are delocalized and free to move throughout the lattice. This provides a mechanism for stress relief without catastrophic fracture. Metal atoms or planes of metal atoms can slip relative to each other without significant damage to the metallic bonding network.

In the three dimensional lattices of ceramics, bonding electrons are localized. In ionic solids such as Al_2O_3 they are localized on the ions; in network covalent solids such as SiC (Solution 12.37), the electrons are localized in the covalent bonds between the atoms. In either case, atoms or ions cannot move without breaking chemical bonds. Localized (ionic or covalent) bonding in ceramics renders them harder than metals, which have delocalized bonding throughout the crystal lattice.

12.57 The degree of crystallinity of polyethylene increases with increasing molecular weight. Following the trends in Table 12.3, ultra high molecular weight polyethylene will be dense, mechanically strong and high melting. The important properties in this application are density and mechanical strength, so that the coating does not deform or wear down under load. Movement of the metal ball around the polymer surface should be essentially frictionless; adhesive forces between metal and polymer should be minimal. The high density polymer has no pockets of empty space, so it forms a smooth surface. Attractive interactions between the nonpolar hydrocarbon polymer and the metallic ball are essentially absent, facilitating frictionless movement of the ball.

12.58 (a)

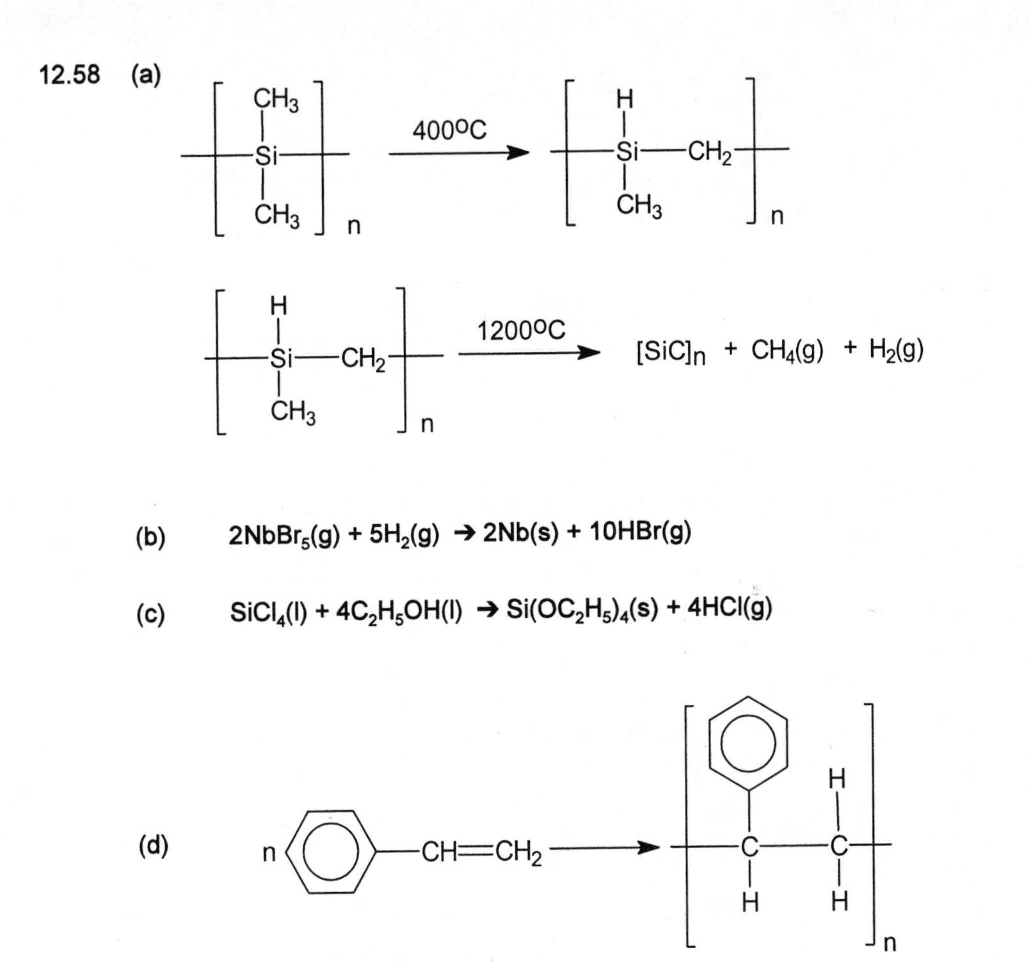

(b) $2NbBr_5(g) + 5H_2(g) \rightarrow 2Nb(s) + 10HBr(g)$

(c) $SiCl_4(l) + 4C_2H_5OH(l) \rightarrow Si(OC_2H_5)_4(s) + 4HCl(g)$

(d)

12.59

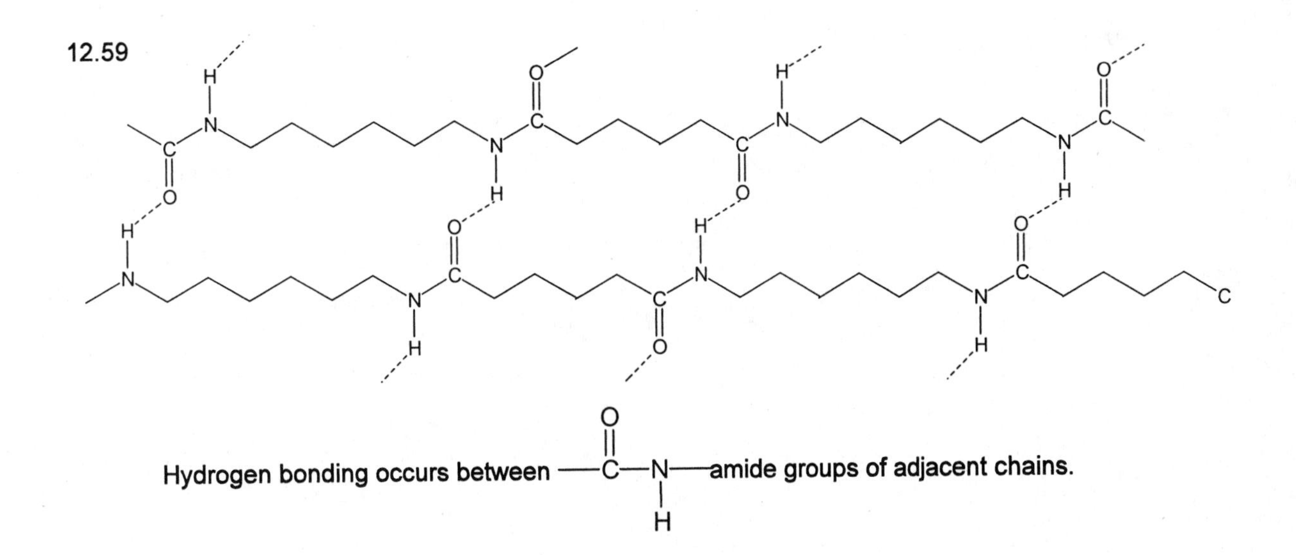

Hydrogen bonding occurs between —C(=O)—N(H)— amide groups of adjacent chains.

12.60

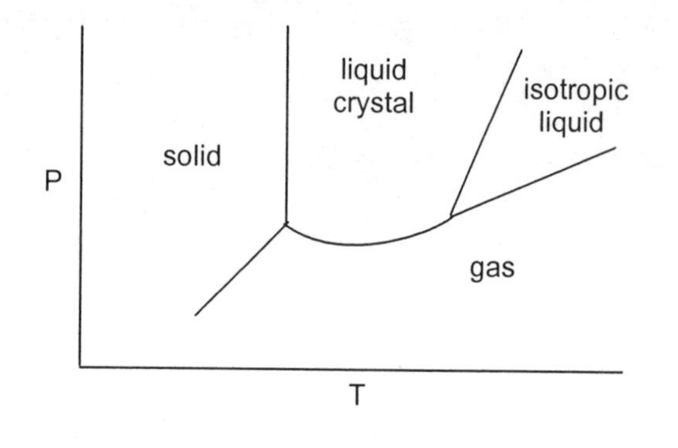

12.61 $TiCl_4(g) + 2SiH_4(g) \rightarrow TiSi_2(s) + 4HCl(g) + 2H_2(g)$

As a ceramic, $TiSi_2$ will have a three dimensional network structure similar to that of Si. (Si(s) has a diamond-like covalent-network structure, Figure 11.41). At the surface of the thin film there will be Ti atoms and Si atoms with incomplete valences that can and will chemically bond with Si atoms on the surface of the substrate. This kind of bonding would not be possible with a Cu thin film. Strong adherence to the surface is an essential component of thin film performance.

12.62 The formula of the compound deposited as a thin film is indicated by boldface type.

(a) $SiH_4(g) + 2H_2(g) + 2CO_2(g) \rightarrow \mathbf{SiO_2(s)} + 4H_2(g) + 2CO(g)$

(b) $TiCl_4(g) + 2H_2O(g) \rightarrow \mathbf{TiO_2(s)} + 4HCl(g)$

(c) $GeCl_4(g) \rightarrow \mathbf{Ge(s)} + 2Cl_2(g)$

The H_2 carrier gas dilutes the $GeCl_4(g)$ so the reaction occurs more evenly and at a controlled rate; it does not participate in the reaction.

Integrative Exercises

12.63 These compounds cannot be vaporized without destroying their chemical identities. Anions with names ending in "ite" or "ate" are oxyanions with nonmetallic elements as central atoms. Under the conditions of vacuum deposition, high temperatures and low pressures, these anions tend to chemically decompose to form gaseous nonmetal oxides. For example:

$$MnSO_4(s) \rightarrow MnO(s) + SO_3(g)$$

12.64 (a)

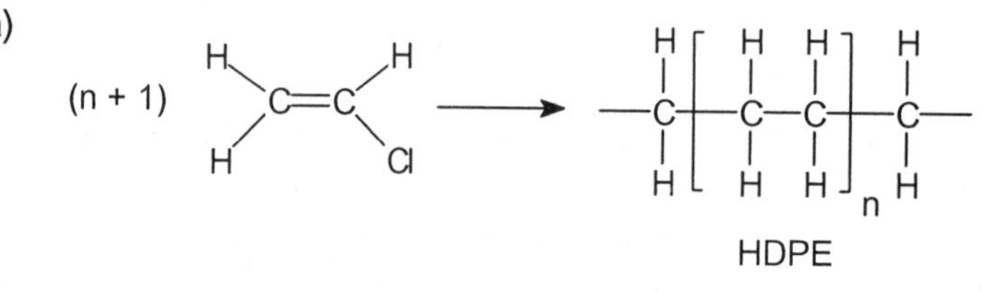

$$\Delta H = D(C{=}C) - 2D(C{-}C) = 614 - 2(348) = -82 \text{ kJ/mol } C_2H_4$$

(b)

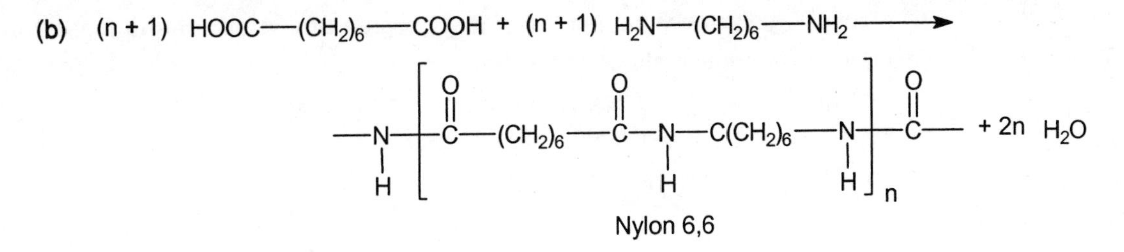

$\Delta H = 2D(C\text{-}O) + 2D(N\text{-}H) - 2D(C\text{-}N) - 2D(H\text{-}O)$

$\Delta H = 2(358) + 2(391) - 2(293) - 2(463) = -14$ kJ/mol

(This is -14 kJ/mol of either reactant.)

(c)

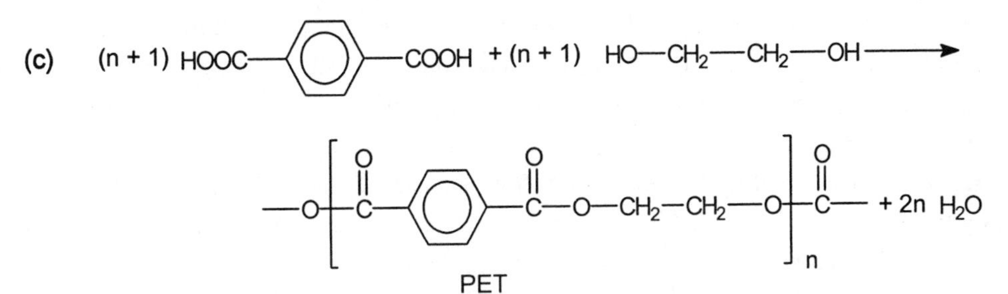

$\Delta H = 2D(C–O) + 2D(O–H) - 2D(C–O) - 2D(O–H) = 0$ kJ

12.65 (a) sp^3 hybrid orbitals at C, 109° bond angles around C

(b)

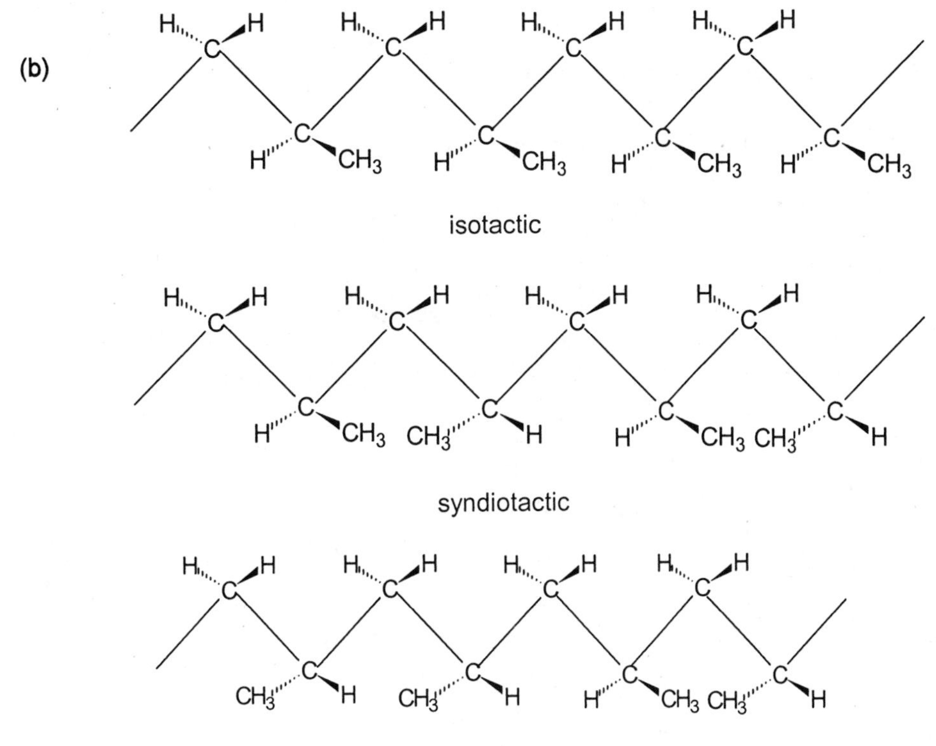

Isotactic polypropylene has the highest degree of crystallinity and highest melting point. The regular shape of the polymer backbone allows for close, orderly (almost zipper like) contact between chains. This maximizes dispersion forces between chains and produces higher order (crystallinity) and melting point. Atactic polypropylene has the least order and the lowest melting point.

(c) Cotton, with —C(OH)— groups and polyester, with —C(=O)—O—C groups, both participate in hydrogen bonding interactions with H_2O molecules. These are strong intermolecular forces that hold the "moisture" at the surface of the fabric next to the skin. Polypropylene has no strong interactions with water and capillary action "wicks" the moisture away from the skin.

12.66 If the expected oxidation states on Y and Ba are +3 and +2, respectively, the **average** oxidation state of Cu is +2 1/3. That is, two Cu ions are in the +2 state and one is in the +3 state. Y^{3+} and Ba^{2+} have the stable electron configurations of their nearest noble gases, while Cu has an incomplete d orbital set. Cu(II) is d^9, Cu(III) is d^8 and both have unpaired electrons. Although the mechanism by which the copper 3d electrons interact through the bridging oxygen atoms to form a superconducting state is still not clear, it is evident that the electronic structure of the copper ions is essential to the observed superconductivity.

12.67 (a) The average molar mass ($\mathcal{M}_{avg}$) of the gas is the average of the individual molar masses weighted for their relative abundances.

$$\mathcal{M}_{avg} = (40/41)(16.04) + (1/41)(2.016) = 15.70 \text{ g} = 16 \text{ g}$$

$$\text{gas density (g/L)} = \mathcal{M}_{avg}\ P/RT; \quad P = 90 \text{ torr} \times \frac{1 \text{ atm}}{760 \text{ torr}} = 0.1184 \text{ atm} = 0.12 \text{ atm}$$

$$\rho = 15.70 \text{ g/mol} \times 0.1184 \text{ atm} \times \frac{\text{K} \bullet \text{mol}}{0.08206 \text{ L} \bullet \text{atm}} \times \frac{1}{1123 \text{ K}} = 0.020 \text{ g/L}$$

(b) The density of diamond, C(s,d), is 3.51 g/mL. Calculate the volume of the diamond film, and then its mass.

$$V = 1 \text{ cm}^2 \times 2.5\ \mu\text{m} \times \frac{1 \times 10^{-6} \text{ m}}{1\ \mu\text{m}} \times \frac{100 \text{ cm}}{\text{m}} \times \frac{1 \text{ mL}}{1 \text{ cm}^3} = 2.5 \times 10^{-4} \text{ mL}$$

$V = 3 \times 10^{-4}$ mL (to 1 sig fig)

$$\frac{3.51 \text{ g C(s,d)}}{1 \text{ mL}} \times 2.5 \times 10^{-4} \text{ mL} = 8.775 \times 10^{-4} \text{ g} = 9 \times 10^{-4} \text{ g C (s, d)}$$

Calculate how many grams of gas are needed to deposit 9×10^{-4} g of film, and the volume of this amount of gas.

$$8.775 \times 10^{-4} \text{ g C (s, d)} \times \frac{15.70 \text{ g gas}}{12.01 \text{ g C(s,d)}} = 1.147 \times 10^{-3} \text{ g gas contains the amount of C(s,d) in the film.}$$

But the process is only 0.5% efficient.

$$\frac{1.147 \times 10^{-3} \text{ g gas needed}}{\text{total of gas}} \times 100 = 0.5; \quad \text{total g gas} = 0.2294 \text{ g} = 0.23 \text{ g gas}$$

$$V = \frac{\text{g RT}}{\mathcal{M}_{avg} \text{ P}} = \frac{0.2294 \text{ g}}{15.70 \text{ g/mol}} \times \frac{0.08206 \text{ L} \bullet \text{atm}}{\text{K} \bullet \text{mol}} \times \frac{1123 \text{ K}}{0.1184 \text{ atm}} = 11.37 \text{ L} = 11 \text{ L}$$

12.68 (a) The data (14.99%) has 4 sig figs, so use molar masses to 5 sig figs.

$$\text{mass \% O} = 14.99 = \frac{(8+x)15.999}{746.04 + (8+x)15.999} \times 100$$

rounded (to show sig figs)	**unrounded**
$(8+x)15.999$	$(8+x)15.999$
$= 0.1499 [746.04 + (8+x) 15.999]$	$= 0.1499 [746.04 + (8+x) 15.999]$
$127.99 + 15.999x$	$127.992 + 15.999x$
$= 0.1499(874.04 + 15.999x)$	$= 0.1499 (874.036 + 15.999x$
$15.999x - 2.398x$	$15.999x - 2.3983x$
$= 131.0 - 127.99$	$= 131.018 - 127.992$
$13.601x = 3.0; \; x = 0.22$	$13.6007x = 3.026; \; x = 0.2225$

(b) **Hg** and **Cu** both have more than one stable oxidation state. If different Cu ions (or Hg ions) in the solid lattice have different charges, then the average charge is a noninteger value. Ca and Ba are stable only in the +2 oxidation state; they are unlikely to have noninteger average charge.

(c) Ba^{2+} is largest; Cu^{2+} is smallest. For ions with the same charge, size decreases going up or across the periodic table. In the +2 state, Hg is smaller than Ba. If Hg has an average charge greater than 2+, it will be smaller yet. The same argument is true for Cu and Ca.

12.69 (a)

$$-\text{Si}(CH_3)_2-[\text{Si}(CH_3)_2]_n-\text{Si}(CH_3)_2-$$

Si-Si	226 kJ	← lowest
Si-C	301 kJ	
C-H	413 kJ	

(b) Si-Si bonds are weakest, so they are most likely to break upon heating.

(c) At this point, a C–H bond must be broken, because the product polymer has $-CH_2-$ in the backbone, while only terminal $-CH_3$ groups are present in the reactant polymer.

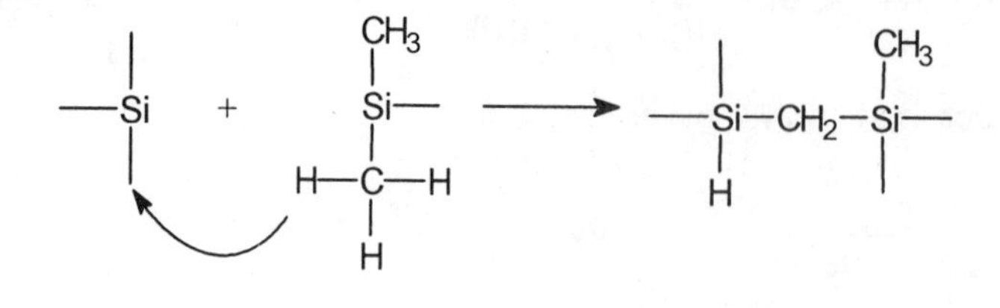

$$\Delta H = D(C\text{-}H) - D(Si\text{-}C) - D(Si\text{-}H) = 413 - 301 - 323 = -211 \text{ kJ/mol}$$

12.70 (a)

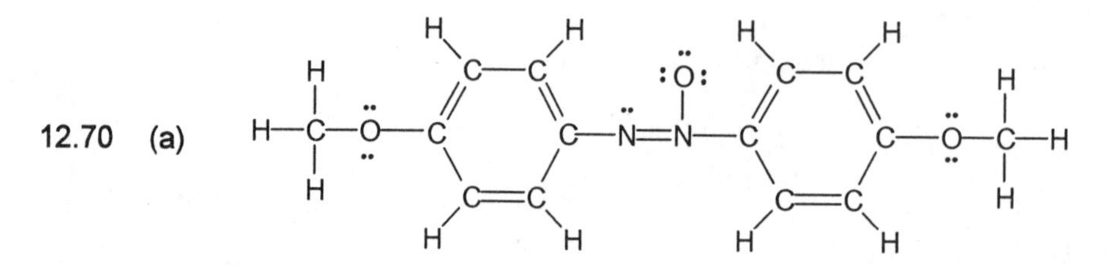

There are several other resonance structures involving alternate placement of the double bonds in the benzene rings.

(b) Both N atoms are surrounded by 3 VSEPR electron domains, so the hybridization at both atoms is sp^2. The bond angles around the N attached to O will be approximately 120°.

(c) When the $-OCH_3$ group is replaced by the $-CH_2CH_2CH_2CH_3$ group, a small rather compact group with some polarity is replaced by a larger, more flexible, nonpolar group. Thus, the molecules don't line up as well in the solid, and the melting point and liquid crystal temperature range are lower.

(d) The density decreases going from solid to nematic liquid crystal to isotropic liquid. In the nematic liquid crystal, most of the long range order of the solid state is lost. The molecules are moving relative to one another with their long axes more or less aligned. The result is more empty space and a lower density than the solid. There is a further small decrease in density when the last degree of order is lost and the substance becomes an isotropic liquid.

13 Properties of Solutions

The Solution Process

13.1 If the enthalpy released due to solute-solvent attractive forces (ΔH_3) is at least as large as the enthalpy required to separate the solute particles (ΔH_1), the overall enthalpy of solution (ΔH_{soln}) will be either slightly endothermic (owing to $+\Delta H_2$) or exothermic. Even if ΔH_{soln} is slightly endothermic, the increase in disorder due to mixing will cause a significant amount of solute to dissolve. If the magnitude of ΔH_3 is small relative to the magnitude of ΔH_1, ΔH_{soln} will be large and endothermic (energetically unfavorable) and not much solute will dissolve.

13.2 (a) For the same solute, NaCl, in different solvents, solute-solute interactions (ΔH_1) are the same. Because water experiences hydrogen bonding while benzene has only dispersion forces, solvent-solvent interactions (ΔH_2) are greater for water. On the other hand, solute-solvent interactions (ΔH_3) are much weaker between ionic NaCl and nonpolar benzene than between ionic NaCl and polar water. It is the large difference in ΔH_3 that causes NaCl to be soluble in water but not in benzene.

(b) Lattice energy is the main component of ΔH_1, the enthalpy required to separate solute particles. If ΔH_1, is too large, the dissolving process is prohibitively endothermic, and the substance is not very soluble.

(c) Ion-dipole forces between cations and water molecules and relatively small lattice energies (ion-ion forces between cations and anions) lead to strongly hydrated cations.

13.3 *Analyze/Plan*. Decide whether the solute and solvent in question are ionic, polar covalent or nonpolar covalent. Draw Lewis structures as needed. Then, state the appropriate type of solute-solvent interaction. *Solve*:

(a) CCl_4, nonpolar; benzene, nonpolar; dispersion forces

(b) $CaCl_2$, ionic; water, polar; ion-dipole forces

(c) propanol, polar with hydrogen bonding; water, polar with hydrogen bonding; hydrogen bonding

(d) HCl, polar; CH_3CN, polar; dipole-dipole forces

13.4 From weakest to strongest solvent-solute interactions:
(b), dispersion forces < (c), hydrogen bonding < (a), ion-dipole

13.5 (a) Lattice energy is the amount of energy required to completely separate a mole of solid ionic compound into its gaseous ions (Section 8.2). For ionic solutes, this corresponds to ΔH_1 (solute-solute interactions) in Equation 13.1.

(b) In Equation 13.1, ΔH_3 is always exothermic. Formation of attractive interactions, no matter how weak, always lowers the energy of the system, relative to the energy of the isolated particles.

13.6 Separation of solvent molecules, ΔH_2, will be smallest in this case, because hydrogen bonding is the weakest of the intermolecular forces involved. ΔH_1 involves breaking ionic bonds, and ΔH_3 involves formation of ion-dipole interactions, both stronger forces than hydrogen bonding.

13.7 (a) ΔH_{soln} is determined by the relative magnitudes of the "old" solute-solute (ΔH_1) and solvent-solvent (ΔH_2) interactions and the new solute-solvent interactions (ΔH_3); $\Delta H_{soln} = \Delta H_1 + \Delta H_2 + \Delta H_3$. Since the solute and solvent in this case experience very similar London dispersion forces, the energy required to separate them individually and the energy released when they are mixed are approximately equal.
$\Delta H_1 + \Delta H_2 \approx -\Delta H_3$. Thus, ΔH_{soln} is nearly zero.

(b) Mixing hexane and heptane produces a homogeneous solution from two pure substances, and the randomness of the system increases. Since no strong intermolecular forces prevent the molecules from mixing, they do so spontaneously due to the increase in disorder.

13.8 KBr is quite soluble in water because of the sizeable increase in disorder of the system (ordered KBr lattice → freely moving hydrated ions) associated with the dissolving process. An increase in disorder or randomness in a process tends to make that process spontaneous.

Saturated Solutions; Factors Affecting Solubility

13.9 (a) Supersaturated

(b) Add a seed crystal. Supersaturated solutions exist because not enough solute molecules are properly aligned for crystallization to occur. A seed crystal provides a nucleus of already aligned molecules, so that ordering of the dissolved particles is more facile.

13.10 (a) $$\frac{1.22 \text{ mol } MnSO_4 \cdot H_2O}{1 \text{ L soln}} \times \frac{169.0 \text{ g } MnSO_4 \cdot H_2O}{1 \text{ mol}} \times 0.100 \text{ L} = 20.6 \text{ g } MnSO_4 \cdot H_2O / 100 \text{ mL}$$

The 1.22 *M* solution is unsaturated.

(b) Add a known mass, say 5.0 g, of $MnSO_4 \cdot H_2O$ to the unknown solution. If the solid dissolves, the solution is unsaturated. If there is undissolved $MnSO_4 \cdot H_2O$, filter the solution and weigh the solid. If there is less than 5.0 g of solid, some of the added $MnSO_4 \cdot H_2O$ dissolved and the unknown solution is unsaturated. If there is exactly 5.0 g, no additional solid dissolved and the unknown is saturated. If there is more than 5.0 g, excess solute has precipitated and the solution is supersaturated.

13.11 *Analyze/Plan.* On Figure 13.17, find the solubility curve for the appropriate solute. Find the intersection of 40°C and 40 g solute on the graph. If this point is below the solubility curve, more solute can dissolve and the solution is unsaturated. If the intersection is on or above the curve, the solution is saturated. *Solve:*

(a) unsaturated (b) saturated (c) saturated (d) unsaturated

13.12 (a) at 30°C, $\frac{10 \text{ g } KClO_3}{100 \text{ g } H_2O} \times 250 \text{ g } H_2O = 25 \text{ g } KClO_3$

(b) $\frac{66 \text{ g } Pb(NO_3)_2}{100 \text{ g } H_2O} \times 250 \text{ g } H_2O = 165 = 1.7 \times 10^2 \text{ g } Pb(NO_3)_2$

(c) $\frac{3 \text{ g } Ce_2(SO_4)_3}{100 \text{ g } H_2O} \times 250 \text{ g } H_2O = 7.5 = 8 \text{ g } Ce_2(SO_4)_3$

13.13 The liquids water and glycerol form homogenous mixtures (solutions), regardless of the relative amounts of the two components. Glycerol has an –OH group on each C atom in the molecule. This structure facilitates strong hydrogen bonding similar to that in water. Like dissolves like and the two liquids are miscible in all proportions.

13.14 Immiscible means that oil and water do not mix homogeneously; they do not dissolve. Many substances are called 'oil', but they are typically nonpolar carbon-based molecules with fairly high molecular weights. As such, there are fairly strong dispersion forces among oil molecules. The properties of water are dominated by its strong hydrogen bonding. The dispersion-dipole interactions between water and oil are likely to be weak. Thus, ΔH_1 and ΔH_2 are large and positive, while ΔH_3 is small and negative. The net ΔH_{soln} is large and positive, and mixing does not occur.

13.15 For small *n* values, the dominant interactions among acid molecules will be hydrogen-bonding. As n increases, dispersion forces between carbon chains become more important and eventually dominate. Thus, as *n* increases, water solubility decreases and hexane solubility increases.

13.16 (a) Dispersion interactions among nonpolar $CH_3(CH_2)_{16}$–chains dominate the properties of stearic acid. It is more soluble in nonpolar CCl_4 than polar (hydrogen bonding) water, despite the presence of the –COOH group.

(b)

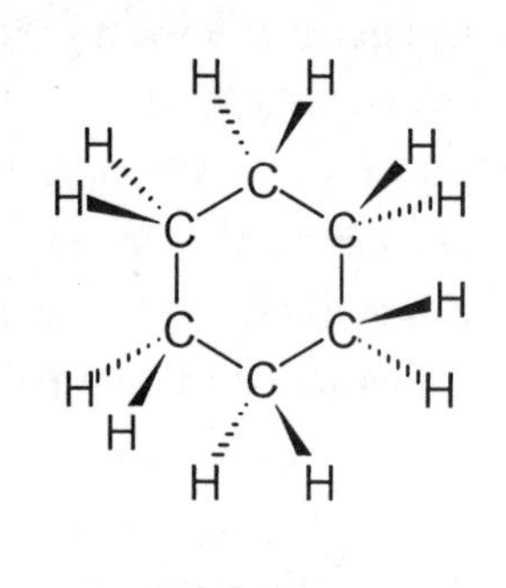

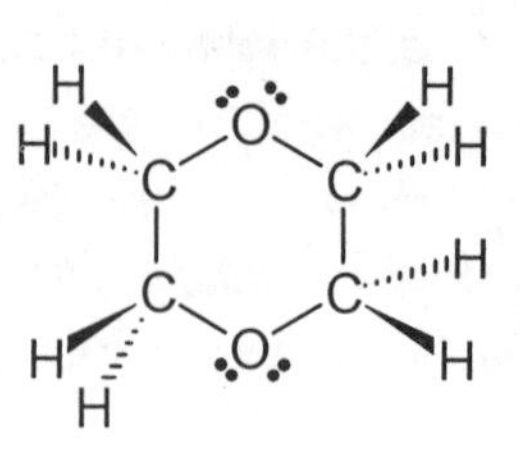

cyclohexane dioxane

Dioxane can act as a hydrogen bond acceptor, so it will be more soluble than cyclohexane in water.

13.17 *Analyze/Plan*. Water, H_2O, is a polar solvent that forms hydrogen bonds with other H_2O molecules. The more soluble solute in each case will have intermolecular interactions that are most similar to the hydrogen bonding in H_2O. *Solve*:

(a) Ionic $CaCl_2$ is more soluble because ion-dipole solute-solvent interactions are more similar to ionic solute-solute and hydrogen bonding solvent-solvent interactions than the weak dispersion forces between CCl_4 and H_2O.

(b) C_6H_5OH is more soluble because it is capable of hydrogen bonding. Nonpolar C_6H_6 is capable only of dispersion force interactions and does not have strong intermolecular interactions with polar (hydrogen bonding) H_2O.

13.18 Hexane is a nonpolar hydrocarbon that experiences dispersion forces with other C_6H_{14} molecules. Solutes that primarily experience dispersion forces will be more soluble in hexane.

(a) Cyclohexane, C_6H_{12}, is also a nonpolar hydrocarbon and will be more soluble in hexane. Glucose experiences hydrogen bonding with itself; these solute-solute interactions are less likely to be overcome by weak solute-solvent interactions.

(b) Propionic acid, which experiences hydrogen bonding, will be more soluble than sodium propionate, which experiences ion-ion forces. The hydrogen bonding is weaker (relatively speaking) and more likely to be overcome by dispersion forces with hexane.

(c) Ethyl chloride, which has a $-CH_2CH_3$ group capable of dispersion forces, will be more soluble. HCl experiences dipole-dipole forces less likely to be disrupted by dispersion forces with hexane.

13.19 (a) Carbonated beverages are stored with a partial pressure of $CO_2(g)$ greater than 1 atm above the liquid. A sealed container is required to maintain this CO_2 pressure.

(b) Since the solubility of gases increases with decreasing temperature, some $CO_2(g)$ will remain dissolved in the beverage if it is kept cool.

13.20 Pressure has an effect on O_2 solubility in water because, at constant temperature and volume, pressure is directly related to the amount of O_2 available to dissolve. The greater the partial pressure of O_2 above water, the more O_2 molecules are available for dissolution, and the more molecules that strike the surface of the liquid.

Pressure does not affect the amount or physical properties of NaCl, or ionic solids in general, so it has little influence on the dissolving of NaCl in water.

13.21 *Analyze/Plan.* Follow the logic in Sample Exercise 13.2. *Solve*:

$S_{He} = 3.7 \times 10^{-4}$ *M*/atm × 1.5 atm = 5.6×10^{-4} *M*

$S_{N_2} = 6.0 \times 10^{-4}$ *M*/atm × 1.5 atm = 9.0×10^{-4} *M*

13.22 $665 \text{ torr} \times \frac{1 \text{ atm}}{760 \text{ torr}} = 0.875 \text{ atm};\ P_{O_2} = \chi_{O_2}(P_t) = 0.21(0.875 \text{ atm}) = 0.1838 = 0.18 \text{ atm}$

$$S_{O_2} = kP_{O_2} = \frac{1.38 \times 10^{-3} \text{mol}}{\text{L} \bullet \text{atm}} \times 0.1838 \text{ atm} = 2.5 \times 10^{-4}\ M$$

Concentrations of Solutions

13.23 *Analyze/Plan.* Follow the logic in Sample Exercise 13.3. *Solve*:

(a) $$\text{mass \%} = \frac{\text{mass solute}}{\text{total mass solution}} \times 100 = \frac{11.7 \text{ g } Na_2SO_4}{11.7 \text{ g } Na_2SO_4 + 443 \text{ g } H_2O} \times 100 = 2.57\%$$

(b) $$\text{ppm} = \frac{\text{mass solute}}{\text{total mass solution}} \times 10^6;\ \frac{5.95 \text{ g Ag}}{1 \text{ ton ore}} \times \frac{1 \text{ ton}}{2000 \text{ lb}} \times \frac{1 \text{ lb}}{453.6 \text{ g}} \times 10^6$$

$= 6.56$ ppm

13.24 (a) $$\text{mass \%} = \frac{\text{mass solute}}{\text{total mass solution}} \times 100$$

$$\text{mass solute} = 0.045 \text{ mol } I_2 \times \frac{253.8 \text{ g } I_2}{1 \text{ mol } I_2} = 11.421 = 11 \text{ g } I_2$$

$$\text{mass \% } I_2 = \frac{11.421 \text{ g } I_2}{11.421 \text{ g } I_2 + 115 \text{ g } CCl_4} \times 100 = 9.034 = 9.0\% \ I_2$$

(b) $$\text{ppm} = \frac{\text{mass solute}}{\text{total mass solution}} \times 10^6 = \frac{0.0079 \text{ g } Sr^{2+}}{1 \times 10^3 \text{ g } H_2O} \times 10^6 = 7.9 \text{ ppm } Sr^{2+}$$

13.25 *Analyze/Plan.* Given masses of CH_3OH and H_2O, calculate moles of each component. (a) Mole fraction CH_3OH = (mol CH_3OH)/(total mol) (b) mass % CH_3OH = [(g CH_3OH)/(total mass)] × 100 (c) molality CH_3OH = (mol CH_3OH)/(kg H_2O). *Solve*:

(a) $7.5 \text{ g } CH_3OH \times \frac{1 \text{ mol } CH_3OH}{32.04 \text{ g } CH_3OH} = 0.234 = 0.23 \text{ mol } CH_3OH$

$245 \text{ g } H_2O \times \frac{1 \text{ mol } H_2O}{18.02 \text{ g } H_2O} = 13.60 = 13.6 \text{ mol } H_2O$

$\chi_{CH_3OH} = \frac{0.234}{0.234 + 13.60} = 0.0169 = 0.017$

(b) $\text{mass \% } CH_3OH = \frac{7.5 \text{ g } CH_3OH}{7.5 \text{ g } CH_3OH + 245 \text{ g } H_2O} \times 100 = 3.0\% \; CH_3OH$

(c) $m = \frac{0.234 \text{ mol } CH_3OH}{0.245 \text{ kg } H_2O} = 0.955 = 0.96 \; m \; CH_3OH$

13.26 (a) $\frac{25.5 \text{ g } C_6H_5OH}{94.11 \text{ g/mol}} = 0.2710 = 0.271 \text{ mol } C_6H_5OH$

$\frac{495 \text{ g } CH_3CH_2OH}{46.07 \text{ g/mol}} = 10.7445 = 10.7 \text{ mol } CH_3CH_2OH$

$\chi_{C_6H_5OH} = \frac{0.2710}{0.2710 + 10.7445} = 0.02460 = 0.0246$

(b) $\text{mass \%} = \frac{25.5 \text{ g } C_6H_5OH}{25.5 \text{ g } C_6H_5OH + 495 \text{ g } CH_3CH_2OH} \times 100 = 4.90\% \; C_6H_5OH$

(c) $m = \frac{0.2710 \text{ mol } C_6H_5OH}{0.495 \text{ kg } CH_3CH_2OH} = 0.54747 = 0.547 \; m \; C_6H_5OH$

13.27 *Analyze/Plan.* Given mass solute and volume solution, calculate mol solute, then molarity = mol solute/L solution. Or, for dilution, $M_c \times L_c = M_d \times L_d$ *Solve*:

(a) $M = \frac{\text{mol solute}}{\text{L soln}}; \; \frac{10.5 \text{ g } Mg(NO_3)_2}{0.2500 \text{ L soln}} \times \frac{1 \text{ mol } Mg(NO_3)_2}{148.3 \text{ g } Mg(NO_3)_2} = 0.283 \; M \; Mg(NO_3)_2$

(b) $\frac{22.4 \text{ g } LiClO_4 \cdot 3H_2O}{0.125 \text{ L soln}} \times \frac{1 \text{ mol } LiClO_4 \cdot 3H_2O}{160.4 \text{ g } LiClO_4 \cdot 3H_2O} = 1.12 \; M \; LiClO_4 \cdot 3H_2O$

(c) $M_c \times L_c = M_d \times L_d$; $3.50 \; M \; HNO_3 \times 0.0250 \text{ L} = ?M \; HNO_3 \times 0.250 \text{ L}$

250 mL of 0.350 M HNO_3

13.28 (a) $M = \frac{\text{mol solute}}{\text{L soln}}; \; \frac{15.0 \text{ g } Al_2(SO_4)_3}{0.350 \text{ L soln}} \times \frac{1 \text{ mol } Al_2(SO_4)_3}{342.2 \text{ g } Al_2(SO_4)_3} = 0.125 \; M \; Al_2(SO_4)_3$

(b) $\frac{5.25 \text{ g } Mn(NO_3)_2 \cdot 2H_2O}{0.175 \text{ L soln}} \times \frac{1 \text{ mol } Mn(NO_3)_2 \cdot 2H_2O}{215.0 \text{ g } Mn(NO_3)_2 \cdot 2H_2O} = 0.140 \; M \; Mn(NO_3)_2$

(c) $M_c \times L_c = M_d \times L_d$; 9.00 M H_2SO_4 × 0.0350 L = ?M H_2SO_4 × 0.500 L

500 mL of 0.630 M H_2SO_4

13.29 *Analyze/Plan*. Follow the logic in Sample Exercise 13.4. *Solve*:

(a) $$m = \frac{\text{mol solute}}{\text{kg solvent}};\ \frac{10.5 \text{ g } C_6H_6}{18.5 \text{ g } CCl_4} \times \frac{1 \text{ mol } C_6H_6}{78.11 \text{ g } C_6H_6} \times \frac{1000 \text{ g } CCl_4}{1 \text{ kg } CCl_4} = 7.27\ m\ C_6H_6.$$

(b) The density of H_2O = 0.997 g/mL = 0.997 kg/L.

$$\frac{4.15 \text{ g NaCl}}{0.250 \text{ L } H_2O} \times \frac{1 \text{ mol NaCl}}{58.44 \text{ g NaCl}} \times \frac{1 \text{ L } H_2O}{0.997 \text{ kg } H_2O} = 0.285\ m \text{ NaCl}$$

13.30 (a) $$16.0 \text{ mol } H_2O \times \frac{18.02 \text{ g } H_2O}{1 \text{ mol } H_2O} = 288.3 \text{ g } H_2O = 0.288 \text{ kg } H_2O$$

$$m = \frac{1.50 \text{ mol KCl}}{0.2883 \text{ kg } H_2O} = 5.2026 = 5.20\ m \text{ KCl}$$

(b) $$m = \frac{\text{mol solute}}{\text{kg solvent}};\ \text{mol } S_8 = m \times \text{kg } C_{10}H_8 = 0.12\ m \times 0.1000 \text{ kg } C_{10}H_8 = 0.012 \text{ mol}$$

$$0.012 \text{ mol } S_8 \times \frac{256.5 \text{ g } S_8}{1 \text{ mol } S_8} = 3.078 = 3.1 \text{ g } S_8$$

13.31 *Analyze/Plan*. Assume 1 L of solution. Density gives the total mass of 1 L of solution. The g H_2SO_4/L are also given in the problem. Mass % = mass solute/total mass solution. Calculate mass solvent from mass solution and mass solute. Calculate moles solute and solvent and use the appropriate definitions to calculate mole fraction, molality and molarity. *Solve*:

(a) $$\frac{571.6 \text{ g } H_2SO_4}{1 \text{ L soln}} \times \frac{1 \text{ L soln}}{1329 \text{ g soln}} = 0.430098 \text{ g } H_2SO_4\text{/g soln}$$

mass percent is thus 0.4301 × 100 = 43.01% H_2SO_4

(b) In a liter of solution there are 1329 - 571.6 = 757.4 = 757 g H_2O.

$$\frac{571.6 \text{ g } H_2SO_4}{98.09 \text{ g/mol}} = 5.827 \text{ mol } H_2SO_4;\ \frac{757.4 \text{ g } H_2O}{18.02 \text{ g/mol}} = 42.03 = 42.0 \text{ mol } H_2O$$

$$\chi_{H_2SO_4} = \frac{5.827}{42.03 + 5.827} = 0.122$$

(The result has 3 sig figs because 42.0 mol H_2O limits the denominator to 3 sig figs.)

(c) $$\text{molality} = \frac{5.827 \text{ mol } H_2SO_4}{0.7574 \text{ kg } H_2O} = 7.693 = 7.69\ m\ H_2SO_4$$

(d) $$\text{molarity} = \frac{5.827 \text{ mol } H_2SO_4}{1 \text{ L soln}} = 5.827\ M\ H_2SO_4$$

13.32 (a) $$\text{mass \%} = \frac{\text{mass } C_6H_8O_6}{\text{total mass solution}} \times 100;$$

$$\frac{80.5 \text{ g } C_6H_8O_6}{80.5 \text{ g } C_6H_8O_6 + 210 \text{ g } H_2O} \times 100 = 27.71 = 27.7\% \; C_6H_8O_6$$

(b) $$\text{mol } C_6H_8O_6 = \frac{80.5 \text{ g } C_6H_8O_6}{176.1 \text{ g/mol}} = 0.4571 = 0.457 \text{ mol } C_6H_8O_6$$

$$\text{mol } H_2O = \frac{210 \text{ g } H_2O}{18.02 \text{ g/mol}} = 11.654 = 11.7 \text{ mol } H_2O$$

$$\chi_{C_6H_8O_6} = \frac{0.4571 \text{ mol } C_6H_8O_6}{0.4571 \text{ mol } C_6H_8O_6 + 11.654 \text{ mol } H_2O} = 0.0377$$

(c) $$m = \frac{0.4571 \text{ mol } C_6H_8O_6}{0.210 \text{ kg } H_2O} = 2.18 \; m \; C_6H_8O_6$$

(d) $$M = \frac{\text{mol } C_6H_8O_6}{\text{L solution}}; \; 290.5 \text{ g soln} \times \frac{1 \text{ mL}}{1.22 \text{ g}} \times \frac{1 \text{ L}}{1000 \text{ mL}} = 0.2381 = 0.238 \text{ L}$$

$$M = \frac{0.4571 \text{ mol } C_6H_8O_6}{0.2381 \text{ L soln}} = 1.92 \; M \; C_6H_8O_6$$

13.33 *Analyze/Plan.* Given: 90.0 mL of $CH_3CN(l)$, 0.786 g/mL; 15.0 mL CH_3OH, 0.791 g/mL. Use the density and volume of each component to calculate mass and then moles of each component. Use the definitions to calculate mole fraction, molality and molarity. *Solve:*

(a) $$\text{mol } CH_3CN = \frac{0.786 \text{ g}}{1 \text{ mL}} \times 90.0 \text{ mL} \times \frac{1 \text{ mol } CH_3CN}{41.05 \text{ g } CH_3CN} = 1.7233 = 1.72 \text{ mol}$$

$$\text{mol } CH_3OH = \frac{0.791 \text{ g}}{1 \text{ mL}} \times 15.0 \text{ mL} \times \frac{1 \text{ mol } CH_3OH}{32.04 \text{ g } CH_3OH} = 0.3703 = 0.370 \text{ mol}$$

$$\chi_{CH_3OH} = \frac{0.3703 \text{ mol } CH_3OH}{1.7233 \text{ mol } CH_3CN + 0.3703 \text{ mol } CH_3OH} = 0.177$$

(b) Assuming CH_3OH is the solute and CH_3CN is the solvent,

$$90.0 \text{ mL } CH_3CN \times \frac{0.786 \text{ g}}{1 \text{ mL}} \times \frac{1 \text{ kg}}{1000 \text{ g}} = 0.07074 = 0.0707 \text{ kg } CH_3CN$$

$$m_{CH_3OH} = \frac{0.3703 \text{ mol } CH_3OH}{0.7074 \text{ kg } CH_3CN} = 5.2347 = 5.23 \; m \; CH_3OH$$

(c) The total volume of the solution is 105.0 mL, assuming volumes are additive.

$$M = \frac{0.3703 \text{ mol } CH_3OH}{0.1050 \text{ L solution}} = 3.53 \; M \; CH_3OH$$

13.34 Given: 10.0 g C_4H_4S, 1.065 g/mL; 105.0 mL C_7H_8, 0.897 g/mL

(a) $$\text{mol } C_4H_4S = 10.0 \text{ g } C_4H_4S \times \frac{1 \text{ mol } C_4H_4S}{84.15 \text{ g } C_4H_4S} = 0.1188 = 0.119 \text{ mol } C_4H_4S$$

$$\text{mol } C_7H_8 = \frac{0.867 \text{ g}}{1 \text{ mL}} \times 250.0 \text{ mL} \times \frac{1 \text{ mol } C_7H_8}{92.14 \text{ g } C_7H_8} = 2.352 = 2.35 \text{ mol}$$

$$\chi_{C_4H_4S} = \frac{0.1188 \text{ mol } C_4H_4S}{0.1188 \text{ mol } C_4H_4S + 2.352 \text{ mol } C_7H_8} = 0.04809 = 0.0481$$

(b) $$m_{C_4H_4S} = \frac{\text{mol } C_4H_4S}{\text{kg } C_7H_8}; \; 250.0 \text{ mL} \times \frac{0.867 \text{ g}}{1 \text{ mL}} \times \frac{1 \text{ kg}}{1000 \text{ g}} = 0.2168 = 0.217 \text{ kg } C_7H_8$$

$$m_{C_4H_4S} = \frac{0.1188 \text{ mol } C_4H_4S}{0.2168 \text{ kg } C_7H_8} = 0.548 \; m \; C_4H_4S$$

(c) $$10.0 \text{ g } C_4H_4S \times \frac{1 \text{ mL}}{1.065 \text{ g}} = 9.390 = 9.39 \text{ mL } C_4H_4S;$$

$$V_{soln} = 9.39 \text{ mL } C_4H_4S + 250.0 \text{ mL } C_7H_8 = 259.4 \text{ mL}$$

$$M_{C_4H_4S} = \frac{0.1188 \text{ mol } C_4H_4S}{0.2594 \text{ L soln}} = 0.458 \; M \; C_4H_4S$$

13.35 *Analyze/Plan.* Given concentration and volume of solution use definitions of the appropriate concentration units to calculate amount of solute; change amount to moles if needed. *Solve:*

(a) $$\text{mol} = M \times \text{L}; \quad \frac{0.250 \text{ mol } CaBr_2}{1 \text{ L soln}} \times 0.255 \text{ L} = 6.38 \times 10^{-2} \text{ mol } CaBr_2$$

(b) Assume that for dilute aqueous solutions, the mass of the solvent is the mass of solution. Use proportions to get mol KCl.

$$\frac{0.150 \text{ mol KCl}}{1 \text{ kg } H_2O} = \frac{x \text{ mol KCl}}{0.0500 \text{ kg } H_2O}; \; x = 7.50 \times 10^{-3} \text{ mol KCl}$$

(c) Use proportions to get mass of glucose, then change to mol glucose.

$$\frac{2.50 \text{ g } C_6H_{12}O_6}{100 \text{ g soln}} = \frac{x \text{ g } C_6H_{12}O_6}{50.0 \text{ g soln}}; \; x = 1.25 \text{ g } C_6H_{12}O_6$$

$$1.25 \text{ g } C_6H_{12}O_6 \times \frac{1 \text{ mol } C_6H_{12}O_6}{180.2 \text{ g } C_6H_{12}O_6} = 6.94 \times 10^{-3} \text{ mol } C_6H_{12}O_6$$

13.36 (a) $$\frac{1.50 \text{ mol } HNO_3}{1 \text{ L soln}} \times 0.245 \text{ L} = 0.3675 = 0.368 \text{ mol } HNO_3$$

(b) Assume that for dilute aqueous solutions, the mass of the solvent is the mass of solution.

$$\frac{0.125 \text{ mol NaCl}}{1 \text{ kg } H_2O} = \frac{x \text{ mol}}{50.0 \times 10^{-6} \text{ kg}}; \; x = 6.25 \times 10^{-5} \text{ mol NaCl}$$

(c) $\dfrac{1.50 \text{ g } C_{12}H_{22}O_{11}}{100 \text{ g soln}} = \dfrac{x \text{ g } C_{12}H_{22}O_{11}}{75.0 \text{ g soln}}$; $x = 1.125 = 1.13 \text{ g } C_{12}H_{22}O_{11}$

$$1.125 \text{ g } C_{12}H_{22}O_{11} \times \frac{1 \text{ mol } C_{12}H_{22}O_{11}}{342.3 \text{ g } C_{12}H_{22}O_{11}} = 3.287 \times 10^{-3} = 3.29 \times 10^{-3} \text{ mol } C_{12}H_{22}O_{11}$$

13.37 *Analyze/Plan.* When preparing solution, we must know amount of solute and solvent. Use the appropriate concentration definition to calculate amount of solute. If this amount is in moles, use molar mass to get grams; use mass in grams directly. Amount of solvent can be expressed as total volume or mass of solution. Combine mass solute and solvent to produce required amount (mass or volume) of solution. *Solute*:

(a) mol = $M \times$ L; $\dfrac{1.50 \times 10^{-2} \text{ mol KBr}}{1 \text{ L soln}} \times 0.75 \text{ L} \times \dfrac{119.0 \text{ g KBr}}{1 \text{ mol KBr}} = 1.3 \text{ g KBr}$

Weigh out 1.5 g KBr, dissolve in water, dilute with stirring to 0.75 L (750 mL).

(b) Mass of solution is required, but density is not specified. Use molality to calculate mass fraction, and then the masses of solute and solvent needed for 125 g of solution.

$$\frac{0.180 \text{ mol KBr}}{1000 \text{ g } H_2O} \times \frac{119.0 \text{ g KBr}}{1 \text{ mol KBr}} = 21.42 = 21.4 \text{ g KBr/kg } H_2O$$

Thus, mass fraction $= \dfrac{21.42 \text{ g KBr}}{1000 + 21.42} = 0.02097 = 0.0210$

In 125 g of the 0.180 *m* solution, there are

$$(125 \text{ g soln}) \times \frac{0.02097 \text{ g KBr}}{1 \text{ g soln}} = 2.621 = 2.62 \text{ g KBr}$$

Weigh out 2.62 g KBr, dissolve it in 125 - 2.62 = 122.38 = 122 g H_2O to make exactly 125 g of 0.180 *m* solution.

(c) Using solution density, calculate the total mass of 1.85 L of solution, and from the mass % of KBr, the mass of KBr required.

$$1.85 \text{ L soln} \times \frac{1000 \text{ mL}}{1 \text{ L}} \times \frac{1.10 \text{ g soln}}{1 \text{ mL}} = 2035 = 2.04 \times 10^3 \text{ g soln}$$

0.120 (2035 g soln) = 244.2 = 244 g KBr

Dissolve 244 g KBr in water, dilute with stirring to 1.85 L.

(d) Calculate moles KBr needed to precipitate 16.0 g AgBr. $AgNO_3$ is present in excess.

$$16.0 \text{ g AgBr} \times \frac{1 \text{ mol AgBr}}{187.8 \text{ g AgBr}} \times \frac{1 \text{ mol KBr}}{1 \text{ mol AgBr}} = 0.08520 = 0.0852 \text{ mol KBr}$$

$$0.0852 \text{ mol KBr} \times \frac{1 \text{ L soln}}{0.150 \text{ mol KBr}} = 0.568 \text{ L soln}$$

Weigh out 0.0852 mol KBr (10.1 g KBr), dissolve it in a small amount of water and dilute to 0.568 L.

13.38 (a)

$$\frac{0.110 \text{ mol } (NH_4)_2SO_4}{1 \text{ L soln}} \times 1.50 \text{ L} \times \frac{132.2 \text{ g } (NH_4)_2SO_4}{1 \text{ mol } (NH_4)_2SO_4} = 21.81 = 21.8 \text{ g } (NH_4)_2SO_4$$

Weigh 21.8 g $(NH_4)_2SO_4$, dissolve in a small amount of water, continue adding water with thorough mixing up to a total solution volume of 1.50 L.

(b) Determine the mass fraction of Na_2CO_3 in the solution:

$$\frac{0.65 \text{ mol } Na_2CO_3}{1000 \text{ g } H_2O} \times \frac{106.0 \text{ g } Na_2CO_3}{1 \text{ mol } Na_2CO_3} = 68.9 \text{ g} = \frac{69 \text{ g } Na_2CO_3}{1000 \text{ g } H_2O}$$

$$\text{mass fraction} = \frac{68.9 \text{ g } Na_2CO_3}{1000 \text{ g } H_2O + 68.9 \text{ g } Na_2CO_3} = 0.06446 = 0.064$$

In 120 g of solution, there are 0.06446(120) = 7.735 = 7.7 g Na_2CO_3.

Weigh out 7.7 g Na_2CO_3 and dissolve it in 120 - 7.7 = 112.3 g H_2O to make exactly 120 g of solution.

(112.3 g H_2O/0.997 g H_2O/mL @ 25° = 112.6 mL H_2O)

(c)

$$1.20 \text{ L} \times \frac{1000 \text{ mL}}{1 \text{ L}} \times \frac{1.16 \text{ g}}{1 \text{ mL}} = 1392 \text{ g solution; } 0.150(1392 \text{ g soln}) = 209 \text{ g } Pb(NO_3)_2$$

Weigh 209 g $Pb(NO_3)$ and add (1392 - 209) = 1183 g H_2O to make exactly (1392 = 1.39 × 10^3) g or 1.20 L of solution.

(1183 g H_2O/0.997 g/mL @ 25°C = 1187 mL H_2O)

(d) Calculate the mol HCl necessary to neutralize 5.5 g $Ba(OH)_2$.

$Ba(OH)_2(s) + 2HCl(aq) \rightarrow BaCl_2(aq) + 2H_2O(l)$

$$5.5 \text{ g } Ba(OH)_2 \times \frac{1 \text{ mol } Ba(OH)_2}{171 \text{ g } Ba(OH)_2} \times \frac{2 \text{ mol HCl}}{1 \text{ mol } Ba(OH)_2} = 0.0643 = 0.064 \text{ mol HCl}$$

$$M = \frac{\text{mol}}{\text{L}};\ \text{L} = \frac{\text{mol}}{M} = \frac{0.0643 \text{ mol HCl}}{0.50\ M \text{ HCl}} = 0.1287 = 0.13 \text{ L} = 130 \text{ mL}$$

130 mL of 0.50 *M* HCl are needed.

$M_c \times L_c = M_d \times L_d$; $6.0\ M \times L_c = 0.50\ M \times 0.1287$ L; $L_c = 0.01072$ L = 11 mL

Using a pipette, measure exactly 11 mL of 6.0 *M* HCl and dilute with water to a total volume of 130 mL.

13.39 *Analyze/Plan.* Assume 1.00 L of solution. Calculate mass of 1 L of solution using density. Calculate mass of NH_3 using mass %, then mol NH_3 in 1.00 L. *Solve:*

$$1.00 \text{ L soln} \times \frac{1000 \text{ mL}}{1 \text{ L}} \times \frac{0.90 \text{ g soln}}{1 \text{ mL soln}} = 9.0 \times 10^2 \text{ g soln/L}$$

$$\frac{900 \text{ g soln}}{1.00 \text{ L soln}} \times \frac{28 \text{ g } NH_3}{100 \text{ g soln}} \times \frac{1 \text{ mol } NH_3}{17.03 \text{ g } NH_3} = 14.80 = 15 \text{ mol } NH_3\text{/L soln} = 15\ M\ NH_3$$

13.40 Assume a solution volume of 1.00 L. Calculate the mass of 1.00 L of solution and the mass of HNO_3 in 1.00 L of solution.

$$1.00 \text{ L} \times \frac{1000 \text{ mL}}{1 \text{ L}} \times \frac{1.42 \text{ g soln}}{\text{mL soln}} = 1.42 \times 10^3 \text{ g soln}$$

$$16\ M = \frac{16 \text{ mol } HNO_3}{1 \text{ L soln}} \times \frac{63.02 \text{ g } HNO_3}{1 \text{ mol } HNO_3} = 1008 = 1.0 \times 10^3 \text{ g } HNO_3$$

$$\text{mass \%} = \frac{1008 \text{ g } HNO_3}{1.42 \times 10^3 \text{ g soln}} \times 100 = 71\% \ HNO_3$$

13.41 *Analyze/Plan.*

$$\chi_{C_3H_6(OH)_2} = 0.100 = \frac{\text{mol } C_3H_6(OH)_2}{\text{mol } C_3H_6(OH)_2 + \text{mol } H_2O}$$

$0.100 [\text{mol } C_3H_6(OH)_2 + \text{mol } H_2O] = \text{mol } C_3H_6(OH)_2$

$0.100 \text{ mol } H_2O = 0.900 \text{ mol } C_3H_6(OH)_2$; $\text{mol } H_2O = 9[\text{mol } C_3H_6(OH)_2]$

The solution has nine times as many moles of H_2O as moles of $C_3H_6(OH)_2$. Assume 1.00 mol $C_3H_6(OH)_2$ and 9.00 mol H_2O. Calculate mass of $C_3H_6(OH)_2$ and H_2O. Use definitions to calculate mass % and molality. *Solve:*

(a) $76.09 = 76.1 \text{ g } C_3H_6(OH)_2$; $9.00 \text{ mol } H_2O \times 18.02 \text{ g } H_2O\text{/mol} = 162.18 = 162 \text{ g } H_2O$

$$\text{mass \%} = \frac{76.09 \text{ g } C_3H_6(OH)_2}{76.09 \text{ g } C_3H_6(OH)_2 + 162.18 \text{ g } H_2O} \times 100 = 31.9\% \ C_3H_6(OH)_2 \text{ by mass}$$

(b) $$m = \frac{\text{mol } C_3H_6(OH)_2}{\text{kg } H_2O}; \quad \frac{1.00 \text{ mol } C_3H_6(OH)_2}{0.16218 \text{ kg } H_2O} = 6.166 = 6.17\ m\ C_3H_6(OH)_2$$

13.42 (a) $$\frac{0.0750 \text{ mol } C_8H_{10}N_4O_2}{1 \text{ kg } CHCl_3} \times \frac{194.2 \text{ g } C_8H_{10}N_4O_2}{1 \text{ mol } C_8H_{10}N_4O_2} = 14.565$$

$$= 14.6 \text{ g } C_8H_{10}N_4O_2 \text{ / kg } CHCl_3$$

$$\frac{14.565 \text{ g } C_8H_{10}N_4O_2}{14.565 \text{ g } C_8H_{10}N_4O_2 + 1000.00 \text{ g } CHCl_3} \times 100 = 1.436 = 1.44\% \ C_8H_{10}N_4O_2 \text{ by mass}$$

(b) $$1000 \text{ g } CHCl_3 \times \frac{1 \text{ mol } CHCl_3}{119.4 \text{ } CHCl_3} = 8.375 = 8.38 \text{ mol } CHCl_3$$

$$\chi_{C_8H_{10}N_4O_2} = \frac{0.0750}{0.0750 + 8.375} = 0.00888$$

Colligative Properties

13.43 freezing point depression, $\Delta T_f = K_f(m)$; boiling point elevation, $\Delta T_b = K_b(m)$;

osmotic pressure, $\pi = M\,RT$; vapor pressure lowering, $P_A = \chi_A P_A^\circ$

13.44 (a) decrease (b) decrease (c) increase (d) increase

13.45 (a) An *ideal solution* is a solution that obeys Raoult's Law.

(b) *Analyze/Plan.* Calculate the vapor pressure predicted by Raoult's law and compare it to the experimental vapor pressure. Assume ethylene glycol (eg) is the solute. *Solve:*

$\chi_{H_2O} = \chi_{eg} = 0.500$; $P_A = \chi_A P_A^\circ = 0.500(149)$ mm Hg = 74.5 mm Hg

The experimental vapor pressure (P_A), 67 mm Hg, is less than the value predicted by Raoult's law for an ideal solution. The solution is not ideal.

Check. An ethylene glycol-water solution has extensive hydrogen bonding, which causes deviation from ideal behavior. We expect the experimental vapor pressure to be less than the ideal value and it is.

13.46 The vapor pressure over the sucrose solution is higher than the vapor pressure over the glucose solution. Since sucrose has a greater molar mass, 10 g of sucrose contains fewer particles than 10 g of glucose. The solution that contains fewer particles, the sucrose solution, will have the higher vapor pressure.

13.47 (a) *Analyze/Plan.* H_2O vapor pressure will be determined by the mole fraction of H_2O in the solution. The vapor pressure of pure H_2O at 338 K (65°C) = 187.5 torr. *Solve:*

$$\frac{15.0 \text{ g } C_{12}H_{22}O_{11}}{342.3 \text{ g/mol}} = 0.04382 = 0.0438 \text{ mol}; \quad \frac{100.0 \text{ g } H_2O}{18.02 \text{ g/mol}} = 5.5494 = 5.549 \text{ mol}$$

$$P_{H_2O} = \chi_{H_2O}\, P^\circ_{H_2O} = \frac{5.5494 \text{ mol } H_2O}{5.5494 + 0.04382} \times 187.5 \text{ torr} = 186.0 \text{ torr}$$

(b) *Analyze/Plan.* For this problem, it will be convenient to express Raoult's law in terms of the lowering of the vapor pressure of the solvent, ΔP_A.

$\Delta P_A = P_A^\circ - \chi_A P_A^\circ = P_A^\circ(1 - \chi_A)$. $1 - \chi_A = \chi_B$, the mole fraction of the *solute* particles $\Delta P_A^\circ = \chi_B P_A^\circ$; the vapor pressure of the solvent (A) is lowered according to the mole fraction of solute (B) particles present. *Solve:*

$$P_{H_2O} \text{ at } 40°\text{C} = 55.3 \text{ torr}; \quad \frac{500 \text{ g } H_2O}{18.02 \text{ g/mol}} = 27.747 = 27.7 \text{ mol } H_2O$$

$$\chi_{C_3H_8O_2} = \frac{4.60 \text{ torr}}{55.3 \text{ torr}} = \frac{y \text{ mol } C_3H_8O_2}{y \text{ mol } C_3H_8O_2 + 27.747 \text{ mol } H_2O} = 0.08318 = 0.0832$$

$$0.08318 = \frac{y}{y + 27.747}; \; 0.08318\,y + 2.308 = y; \; 0.9168\,y = 2.308,$$

$$y = 2.517 = 2.52 \text{ mol } C_3H_8O_2$$

This result has 3 sig figs because (27.7 × 0.0832 = 2.31) has 3 sig figs.

$$2.517 \text{ mol } C_3H_8O_2 \times \frac{76.09 \text{ g } C_3H_8O_2}{\text{mol } C_3H_8O_2} = 191.52 = 192 \text{ g } C_3H_8O_2$$

13.48 (a) H_2O vapor pressure will be determined by the mole fraction of H_2O in the solution. The vapor pressure of pure H_2O at 343 K (70°C) = 233.7 torr.

$$\frac{35.0 \text{ g } C_3H_8O_3}{92.10 \text{ g/mol}} = 0.3800 = 0.380 \text{ mol}; \quad \frac{125 \text{ g } H_2O}{18.02 \text{ g/mol}} = 6.937 = 6.94 \text{ mol}$$

$$P_{H_2O} = \frac{6.937 \text{ mol } H_2O}{6.937 + 0.380} \times 233.7 \text{ torr} = 221.6 = 222 \text{ torr}$$

(b) Calculate χ_B by vapor pressure lowering; $\chi_B = \Delta P_A / P_A°$ (see Solution 13.47(b)). Given moles solvent, calculate moles solute from the definition of mole fraction.

$$\chi_{C_2H_6O_2} = \frac{10.0 \text{ torr}}{100 \text{ torr}} = 0.100$$

$$\frac{1.00 \times 10^3 \text{ g } C_2H_5OH}{46.07 \text{ g/mol}} = 21.71 = 21.7 \text{ mol } C_2H_5OH; \text{ let } y = \text{mol } C_2H_6O_2$$

$$\chi_{C_2H_6O_2} = \frac{y \text{ mol } C_2H_6O_2}{y \text{ mol } C_2H_6O_2 + 21.71 \text{ mol } C_2H_5OH} = 0.100 = \frac{y}{y + 21.71}$$

$$0.100\,y + 2.171 = y; \; 0.900\,y = 2.171; \; y = 2.412 = 2.41 \text{ mol } C_2H_6O_2$$

$$2.412 \text{ mol } C_2H_6O_2 \times \frac{62.07 \text{ g}}{1 \text{ mol}} = 150 \text{ g } C_2H_6O_2$$

13.49 *Analyze/Plan*. At 63.5°C, $P^{\circ}_{H_2O}$ = 175 torr, P°_{Eth} = 400 torr. Let G = the mass of H_2O and/or C_2H_5OH. *Solve*:

(a)

$$\chi_{Eth} = \frac{\dfrac{G}{46.07 \text{ g } C_2H_5OH}}{\dfrac{G}{46.07 \text{ g } C_2H_5OH} + \dfrac{G}{18.02 \text{ g } H_2O}}$$

Multiplying top and bottom of the right side of the equation by 1/G gives:

$$\chi_{Eth} = \frac{1/46.07}{1/46.07 + 1/18.02} = \frac{0.02171}{0.02171 + 0.05549} = 0.2812$$

(b) $P_T = P_{Eth} + P_{H_2O}$; $P_{Eth} = \chi_{Eth} \times P^{\circ}_{Eth}$; $P_{H_2O} = \chi_{H_2O} P^{\circ}_{H_2O}$

$\chi_{Eth} = 0.2812$, P_{Eth} = 0.2812 (400 torr) = 112.48 = 112 torr

χ_{H_2O} = 1 - 0.2812 = 0.7188; P_{H_2O} = 0.7188(175 torr) = 125.8 =126 torr

P_T = 112.5 torr + 125.8 torr = 238.3 = 238 torr

(c) χ_{Eth} in vapor $= \frac{P_{Eth}}{P_{total}} = \frac{112.5 \text{ torr}}{238.3 \text{ torr}} = 0.4721 = 0.472$

13.50 (a) Since C_6H_6 and C_7H_8 form an ideal solution, we can use Raoult's Law. Since both components are volatile, both contribute to the total vapor pressure of 35 torr.

$P_T = P_{C_6H_6} + P_{C_7H_8}$; $P_{C_6H_6} = \chi_{C_6H_6} P^o_{C_6H_6}$; $P_{C_7H_8} = \chi_{C_7H_8} P^o_{C_7H_8}$

$\chi_{C_7H_8} = 1 - \chi_{C_6H_6}$; $P_T = \chi_{C_6H_6} P^o_{C_6H_6} + (1 - \chi_{C_6H_6}) P^o_{C_7H_8}$

35 torr $= \chi_{C_6H_6}(75 \text{ torr}) + (1 - \chi_{C_6H_6})22$ torr

13 torr = 53 torr $(\chi_{C_6H_6})$; $\chi_{C_6H_6} = \frac{13 \text{ torr}}{53 \text{ torr}} = 0.2453 = 0.25$; $\chi_{C_7H_8} = 0.7547 = 0.75$

(b) $P_{C_6H_6} = 0.2453(75 \text{ torr}) = 18.4$ torr; $P_{C_7H_8} = 0.7547(22 \text{ torr}) = 16.6$ torr

In the vapor, $\chi_{C_6H_6} = \frac{P_{C_6H_6}}{P_T} = \frac{18.4 \text{ torr}}{35 \text{ torr}} = 0.53$; $\chi_{C_7H_8} = 0.47$

13.51 (a) Because NaCl is a soluble ionic compound and a strong electrolyte, there are 2 mol dissolved particles for every 1 mol of NaCl solute. $C_6H_{12}O_6$ is a molecular solute, so there is 1 mol of dissolved particles per mol solute. Boiling point elevation is directly related to total moles of dissolved particles; 0.10 *m* NaCl has more dissolved particles so its boiling point is higher than 0.10 *m* $C_6H_{12}O_6$.

(b) *Analyze/Plan.* $\Delta T = K_b m$; K_b for H_2O is 0.51 °C/*m* (Table 13.4) *Solve*:

0.10 *m* NaCl: $\Delta T = \frac{0.51 °C}{m} \times 0.20\ m = 0.102$ °C; $T_b = 100.0 + 0.102 = 100.1$°C

0.10 *m* $C_6H_{12}O_6$: $\Delta T = \frac{0.51 °C}{m} \times 0.10\ m = 0.051$ °C; $T_b = 100.0 + 0.051 = 100.1$°C

Check. Because K_b for H_2O is so small, there is little real difference in the boiling points of the two solutions.

13.52 *Analyze/Plan.* ΔT_b depends on mol dissolved particles. Assume 100 g of each solution, calculate mol solute and mol dissolved particles. Glucose and sucrose are molecular solutes, but $NaNO_3$ dissociates into 2 mol particles per mol solute. *Solve*:

10% by mass means 10 g solute in 100 g solution. If we have 10 g of each solute, the one with the smallest molar mass will have the largest mol solute. The molar masses are: glucose, 180.2 g/mol; sucrose, 342.3 g/mol; $NaNO_3$, 85.0 g/mol. $NaNO_3$ has most mol solute, and twice as many dissolved particles, so it will have the highest boiling point. Sucrose has least mol solute and lowest boiling point. Glucose is intermediate.

In order of increasing boiling point: 10% sucrose < 10% glucose < 10% $NaNO_3$.

13.53 0.030 *m* phenol > 0.040 *m* glycerin = 0.020 *m* KBr. Phenol is very slightly ionized in water, but not enough to match the number of particles in a 0.040 *m* glycerin solution. The KBr solution is 0.040 *m* in particles, so it has the same freezing point as 0.040 glycerin, which is a nonelectrolyte.

13.54 For dilute aqueous solutions such as these, *M* is essentially equal to *m*. For the purposes of comparison, assume we can use *M*.

The more solute particles, the higher the boiling point. Since LiBr and $Zn(NO_3)_2$ are electrolytes, the particle concentrations in these solutions are 0.10 *M* and 0.15 *M*, respectively (although ion-ion attractive forces may decrease the real concentrations somewhat). Thus, the order of increasing boiling points is:

0.050 *M* LiBr < 0.120 *M* glucose < 0.050 *M* $Zn(NO_3)_2$

13.55 *Analyze/Plan.* $\Delta T = K(m)$; first, calculate the **molality** of each solution

(a) 0.35 *m*

(b) $14.2 \text{ mol } CHCl_3 \times \frac{119.4 \text{ g } CHCl_3}{\text{mol } CHCl_3} = 1.6955 = 1.70 \text{ kg};$

$$\frac{1.58 \text{ mol } C_{10}H_8}{1.6955 \text{ kg } CHCl_3} = 0.9319 = 0.932\ m$$

(c) $5.13 \text{ g KBr} \times \frac{1 \text{ mol KBr}}{119.0 \text{ g KBr}} \times \frac{2 \text{ mol particles}}{1 \text{ mol KBr}} = 0.08622 = 0.0862 \text{ mol particles}$

$$6.85 \text{ g } C_6H_{12}O_6 \times \frac{1 \text{ mol } C_6H_{12}O_6}{180.2 \text{ g } C_6H_{12}O_6} = 0.03801 = 0.0380 \text{ mol particles}$$

$$m = \frac{(0.08622 + 0.03801) \text{ mol particles}}{0.255 \text{ kg } H_2O} = 0.48718 = 0.487\ m$$

Solve: Then, f.p. = $T_f - K_f(m)$; b.p. = $T_b + K_b(m)$; T in °C

	m	T_f	$-K_f(m)$	f.p.	T_b	$+K_b(m)$	b.p.
(a)	0.35	-114.6	-1.99(0.35) = -0.70	-115.3	78.4	1.22(0.35) = 0.43	78.8
(b)	0.932	-63.5	-4.68(0.932) = -4.36	-67.9	61.2	3.63(0.932) = 3.38	64.6
(c)	0.487	0.0	-1.86(0.487) = -0.906	-0.91	100.0	0.52(0.487) = 0.25	100.3

13.56 $\Delta T = K(m)$; first calculate the **molality** of the solute particles.

(a) 0.40 *m*

(b) $\frac{20.0 \text{ g } C_{10}H_{22}}{0.455 \text{ kg } CHCl_3} \times \frac{1 \text{ mol } C_{10}H_{22}}{142.3 \text{ g } C_{10}H_{22}} = 0.3089 = 0.309\ m$

(c) $m = \frac{0.45 \text{ mol eg} + 2(0.15) \text{ mol KBr}}{0.150 \text{ kg } H_2O} = \frac{0.75 \text{ mol particles}}{0.150 \text{ kg } H_2O} = 5.0\ m$

Then, f.p. = $T_f - K_f(m)$; b.p. = $T_b + K_b(m)$; T in °C

	m	T_f	$-K_f(m)$	f.p.	T_b	$+K_b(m)$	b.p.
(a)	0.40	-114.6	-1.99(0.40) = -0.80	-115.4	78.4	1.22(0.40) = 0.49	78.9
(b)	0.309	-63.5	-4.68(0.309) = -1.45	-65.0	61.2	3.63(0.309) = 1.12	62.3
(c)	5.0	0.0	-1.86(5.0) = -9.3	-9.3	100.0	0.51(5.0) = 2.6	102.6

13.57 *Analyze/Plan.* $\pi = M\,RT$; T = 25°C + 273 = 298 K; M = mol $C_9H_8O_4$/L soln *Solve*:

$$M = \frac{50.0 \text{ mg } C_9H_8O_4}{0.250 \text{ L}} \times \frac{1 \text{ g}}{1000 \text{ mg}} \times \frac{1 \text{ mol } C_9H_8O_4}{180.2 \text{ g } C_9H_8O_4} = 1.1099 \times 10^{-3} = 1.11 \times 10^{-3}\ M$$

$$\pi = \frac{1.1099 \times 10^{-3} \text{ mol}}{\text{L}} \times \frac{0.08206 \text{ L}\bullet\text{atm}}{\text{K}\bullet\text{mol}} \times 298 \text{ K} = 0.02714 = 0.0271 \text{ atm}$$

13.58 $\pi = MRT$; T = 20°C + 273 = 293 K

$$M \text{ (of ions)} = \frac{\text{mol NaCl} \times 2}{\text{L soln}} = \frac{3.4 \text{ g NaCl}}{1 \text{ L soln}} \times \frac{1 \text{ mol NaCl}}{58.4 \text{ g NaCl}} \times \frac{2 \text{ mol ions}}{1 \text{ mol NaCl}} = 0.116 = 0.12\ M$$

$$\pi = \frac{0.116 \text{ mol}}{\text{L}} \times \frac{0.08206 \text{ L}\bullet\text{atm}}{\text{K}\bullet\text{mol}} \times 293 \text{ K} = 2.8 \text{ atm}$$

13.59 *Analyze/Plan.* Follow the logic in Sample Exercise 13.11. *Solve*:

$$\Delta T_b = K_b m;\quad m = \frac{\Delta T_b}{K_b} = \frac{+0.49}{5.02} = 0.0976 = 0.098\ m \text{ adrenaline}$$

$$m = \frac{\text{mol adrenaline}}{\text{kg } CCl_4} = \frac{\text{g adrenaline}}{\mathcal{M} \text{ adrenaline} \times \text{kg CCl4}}$$

$$\mathcal{M} \text{ adrenaline} = \frac{\text{g adrenaline}}{m \times \text{kg } CCl_4} = \frac{0.64 \text{ g adrenaline}}{0.0976\ m \times 0.0360 \text{ kg } CCl_4} = 1.8 \times 10^2 \text{ g/mol adrenaline}$$

13.60 $\Delta T_f = 5.5 - 4.1 = 1.4;\quad m = \dfrac{\Delta T_f}{K_f} = \dfrac{1.4}{5.12} = 0.273 = 0.27\ m$

$$\mathcal{M} \text{ lauryl alcohol} = \frac{\text{g lauryl alcohol}}{m \times \text{kg } C_6H_6} = \frac{5.00 \text{ g lauryl alcohol}}{0.273 \times 0.100 \text{ kg } C_6H_6}$$

$$= 1.8 \times 10^2 \text{ g/mol lauryl alcohol}$$

13.61 *Analyze/Plan.* Follow the logic in Sample Exercise 13.12. *Solve*:

$\pi = MRT$; $M = \dfrac{\pi}{RT}$; T = 25°C + 273 = 298 K

$$M = 0.953 \text{ torr} \times \frac{1 \text{ atm}}{760 \text{ torr}} \times \frac{\text{K}\bullet\text{mol}}{0.08206 \text{ L}\bullet\text{atm}} \times \frac{1}{298 \text{ K}} = 5.128 \times 10^{-5} = 5.13 \times 10^{-5}\ M$$

$$\text{mol} = M \times \text{L} = 5.128 \times 10^{-5} \times 0.210 \text{ L} = 1.077 \times 10^{-5} = 1.08 \times 10^{-5} \text{ mol lysozyme}$$

$$\mathcal{M} = \frac{\text{g}}{\text{mol}} = \frac{0.150 \text{ g}}{1.077 \times 10^{-5} \text{ mol}} = 1.39 \times 10^4 \text{ g/mol lysozyme}$$

13.62 $M = \pi/RT = \dfrac{0.605 \text{ atm}}{298 \text{ K}} \times \dfrac{\text{mol}\bullet\text{K}}{0.08206 \text{ L}\bullet\text{atm}} = 0.02474 = 0.0247\ M$

$$\mathcal{M} = \frac{\text{g}}{M \times \text{L}} = \frac{2.35 \text{ g}}{0.02474\ M \times 0.250 \text{ L}} = 380 \text{ g/mol}$$

13.63 (a) *Analyze/Plan.* $i = \pi$ (measured) / π (calculated for a nonelectrolyte);

π (calculated) = MRT. *Solve*:

$$\pi \text{ (calculated)} = 0.010 \frac{\text{mol}}{\text{L}} \times \frac{0.08206 \text{ L} \bullet \text{atm}}{\text{mol} \bullet \text{K}} \times 298 \text{ K} = 0.2445 = 0.24 \text{ atm}$$

i = 0.674 atm/0.2445 atm = 2.756 = 2.76

(b) The van't Hoff factor is the effective number of particles per mole of solute. The closer the measured i value is to a theoretical integer value, the more ideal the solution. Ion-pairing and other interparticle attractive forces reduce the effective number of particles in solution and reduce the measured value of i. The more concentrated the solution, the greater the ion-pairing and the smaller the measured value of i.

13.64 If these were ideal solutions, they would have equal ion concentrations and equal ΔT_f values. Data in Table 13.5 indicates that the van't Hoff factors (i) for both salts are less than the ideal values. For 0.030 m NaCl, i is between 1.87 and 1.94, about 1.92. For 0.020 m K_2SO_4, i is between 2.32 and 2.70, about 2.62. From Equation 13.14,

ΔT_f (measured) = i × ΔT_f (calculated for nonelectrolyte)

NaCl: ΔT_f (measured) =1.92 × 0.030 m × 1.86 °C/m = 0.11 °C

K_2SO_4: ΔT_f (measured) = 2.62 × 0.020 m × 1.86°C/m = 0.097 °C

0.030 m NaCl would have the larger ΔT_f.

The deviations from ideal behavior are due to ion-pairing in the two electrolyte solutions. K_2SO_4 has more extensive ion pairing and a larger deviation from ideality because of the higher charge on SO_4^{2-} relative to Cl^-.

Colloids

13.65 (a) In the gaseous state, the particles are far apart and intermolecular attractive forces are small. When two gases combine, all terms in Equation 13.1 are essentially zero and the mixture is always homogeneous.

(b) The outline of a light beam passing through a colloid is visible, whereas light passing through a true solution is invisible unless collected on a screen. This is the Tyndall effect. To determine whether Faraday's (or anyone's) apparently homogeneous dispersion is a true solution or a colloid, shine a beam of light on it and see if the light is scattered.

13.66 (a) Suspensions are classified as solutions or colloids according to the size of the dispersed particles. Solute particles have diameters less than 10 Å. Clearly a protein with a molecular mass of 30,000 amu will be longer than 10 Å. The aqueous suspensions are colloids because of the size of protein molecules.

(b) Emulsion. An emulsifying agent is one that aids in the formation of an emulsion. It usually has a polar part and a nonpolar part, to facilitate mixing of immiscible liquids with very different molecular polarities.

13.67 (a) hydrophobic (b) hydrophilic (c) hydrophobic

13.68 (a) When the colloid particle mass becomes large enough so that gravitational and interparticle attractive forces are greater than the kinetic energies of the particles, settling and aggregation can occur.

(b) Hydrophobic colloids do not attract a sheath of water molecules around them and thus tend to aggregate from aqueous solution. They can be stabilized as colloids by adsorbing charges on their surfaces. The charged particles interact with solvent water, stabilizing the colloid.

(c) Charges on colloid particles can stabilize them against aggregation. Particles carrying like charges repel one another and are thus prevented from aggregating and settling out.

13.69 Colloid particles are stabilized by attractive intermolecular forces with the dispersing medium (solvent) and do not coalesce because of electrostatic repulsions between groups at the surface of the dispersed particles. Colloids can be coagulated by heating (more collisions, greater chance that particles will coalesce); hydrophilic colloids can be coagulated by adding electrolytes, which neutralize surface charges allowing the colloid particles to collide more freely.

13.70 (a) The nonpolar hydrophobic tails of soap particles (the hydrocarbon chain of stearate ions) establish attractive intermolecular dispersion forces with the nonpolar oil molecules, while the charged hydrophilic head of the soap particles interacts with H_2O to keep the oil molecules suspended. (This is the mechanism by which laundry detergents remove greasy dirt from clothes.)

(b) Electrolytes from the acid neutralize surface charges of the suspended particles in milk, causing the colloid to coagulate.

Additional Exercises

13.71 The outer periphery of the BHT molecule is mostly hydrocarbon-like groups, such as $-CH_3$. The one –OH group is rather buried inside, and probably does little to enhance solubility in water. Thus, BHT is more likely to be soluble in the nonpolar hydrocarbon hexane, C_6H_{14}, than in polar water.

13.72 In this equilibrium system, molecules move from the surface of the solid into solution, while molecules in solution are deposited on the surface of the solid. As molecules leave the surface of the small particles of powder, the reverse process preferentially deposits other molecules on the surface of a single crystal. Eventually, all molecules that were present in the 50 g of powder are deposited on the surface of a 50 g crystal; this can only happen if the dissolution and deposition processes are ongoing.

13.73 Assume that the density of the solution is 1.00 g/mL.

(a) $4 \text{ ppm } O_2 = \dfrac{4 \text{ mg } O_2}{1 \text{ kg soln}} = \dfrac{4 \times 10^{-3} \text{ g } O_2}{1 \text{ L soln}} \times \dfrac{1 \text{ mol } O_2}{32.0 \text{ g } O_2} = 1.25 \times 10^{-4} = 1 \times 10^{-4}\ M$

(b) $C_{O_2} = kP_{O_2};\ P_{O_2} = C_{O_2}/k = \dfrac{1.25 \times 10^{-4} \text{ mol}}{\text{L}} \times \dfrac{\text{L}\bullet\text{atm}}{1.71 \times 10^{-3} \text{ mol}} = 0.0731 = 0.07 \text{ atm}$

$0.0731 \text{ atm} \times \dfrac{760 \text{ mm Hg}}{1 \text{ atm}} = 55.6 = 60 \text{ mm Hg}$

13.74 $P_{Rn} = \chi_{Rn}P_{total};\ P_{Rn} = 3.5 \times 10^{-6} (32 \text{ atm}) = 1.12 \times 10^{-4} = 1.1 \times 10^{-4} \text{ atm}$

$S_{Rn} = k\, P_{Rn};\ S_{Rn} = \dfrac{7.27 \times 10^{-3}\ M}{1 \text{ atm}} \times 1.12 \times 10^{-4} \text{ atm} = 8.1 \times 10^{-7}\ M$

13.75 0.10% by mass means 0.10 g glucose/100 g blood.

(a) $\text{ppm glucose} = \dfrac{\text{g glucose}}{\text{g solution}} \times 10^6 = \dfrac{0.10 \text{ g glucose}}{100 \text{ g blood}} \times 10^6 = 1000 \text{ ppm glucose}$

(b) m = mol glucose/kg solvent. Assume that the mixture of nonglucose components is the 'solvent'.

mass solvent = 100 g blood - 0.10 g glucose = 99.9 g solvent = 0.0999 kg solvent

$\text{mol glucose} = 0.10 \text{ g} \times \dfrac{1 \text{ mol}}{180.2 \text{ g } C_6H_{12}O_6} = 5.55 \times 10^{-4} = 5.6 \times 10^{-4} \text{ mol glucose}$

$m = \dfrac{5.55 \times 10^{-4} \text{ mol glucose}}{0.0999 \text{ kg solvent}} = 5.6 \times 10^{-3}\ m \text{ glucose}$

In order to calculate molarity, volume solution must be known. The density of blood is needed to relate mass and volume.

13.76 $15 \text{ ppm KBr} = \dfrac{15 \text{ mg KBr}}{1 \text{ kg soln}} = \dfrac{15 \times 10^{-3} \text{ g KBr}}{1 \text{ L soln}} \times \dfrac{1 \text{ mol KBr}}{119.0 \text{ g KBr}} = 1.26 \times 10^{-4} = 1.3 \times 10^{-4}\ M$

$12 \text{ ppm KCl} = \dfrac{12 \text{ mg KBr}}{1 \text{ kg soln}} = \dfrac{12 \times 10^{-3} \text{ g KCl}}{1 \text{ L soln}} \times \dfrac{1 \text{ mol KCl}}{74.55 \text{ g KCl}} = 1.61 \times 10^{-4} = 1.6 \times 10^{-4}\ M$

A solution that is 12 ppm KCl has the higher molarity of K^+ ions.

13.77 Assume 1.00 L of solution. 1000 mL soln × 0.945 g/mL = 945 g solution

$\dfrac{945 \text{ g soln}}{1.0 \text{ L soln}} \times \dfrac{32.0 \text{ g } C_3H_7OH}{100 \text{ g soln}} \times \dfrac{1 \text{ mol } C_3H_7OH}{60.09 \text{ g } C_3H_7OH} = 5.032 = 5.03\ M$

$945 \text{ g soln} \times \dfrac{32.0 \text{ g } C_3H_7OH}{100 \text{ g soln}} = 302.4 = 302 \text{ g } C_3H_7OH;\ 945 \text{ g} - 302.4 \text{ g} = 642.6 = 643 \text{ g } H_2O$

$\dfrac{302.4 \text{ g } C_3H_7OH}{0.6426 \text{ kg } H_2O} \times \dfrac{1 \text{ mol } C_3H_7OH}{60.09 \text{ g } C_3H_7OH} = 7.831 = 7.83\ m$

13.78 (a) $\frac{1.80 \text{ mol LiBr}}{1 \text{ L soln}} \times \frac{86.85 \text{ g LiBr}}{1 \text{ mol LiBr}} = 156.3 = 156 \text{ g LiBr}$

1 L soln = 826 g soln; g CH_3CN = 826 - 156.3 = 669.7 = 670 g CH_3CN

$m \text{ LiBr} = \frac{1.80 \text{ mol LiBr}}{0.6697 \text{ kg } CH_3CN} = 2.69\ m$

(b) $\frac{669.7 \text{ g } CH_3CN}{41.05 \text{ g/mol}} = 16.31 = 16.3 \text{ mol } CH_3CN;\ \chi_{LiBr} = \frac{1.80}{1.80 + 16.31} = 0.0994$

(c) $\text{mass \%} = \frac{669.7 \text{ g } CH_3CN}{826 \text{ g soln}} \times 100 = 81.1\% \ CH_3CN$

13.79 (a) $m = \frac{\text{mol Na(s)}}{\text{kg Hg(l)}};\ 1.0 \text{ cm}^3 \text{ Na(s)} \times \frac{0.97 \text{ g}}{1 \text{ cm}^3} \times \frac{1 \text{ mol}}{23.0 \text{ g Na}} = 0.0422 = 0.042 \text{ mol Na}$

$20.0 \text{ cm}^3 \text{ Hg(l)} \times \frac{13.6 \text{ g}}{1 \text{ cm}^3} \times \frac{1 \text{ kg}}{1000 \text{ g}} = 0.272 \text{ kg Hg(l)};\ m = \frac{0.0422 \text{ mol Na}}{0.272 \text{ kg Hg(l)}}$

$= 0.155 = 0.16\ m \text{ Na}$

(b) $M = \frac{\text{mol Na(s)}}{\text{L soln}} = \frac{0.0422 \text{ mol Na}}{0.021 \text{ L soln}} = 2.01 = 2.0\ M \text{ Na}$

(c) Clearly, molality and molarity are not the same for this amalgam. Only in the instance that one kg solvent and the mass of one liter solution are nearly equal do the two concentration units have similar values. In this example, one kg Hg has a volume much less than one liter.

13.80 Mole fraction ethylalcohol, $\chi_{C_2H_5OH} = \frac{P_{C_2H_5OH}}{P^{o}_{C_2H_5OH}} = \frac{8 \text{ torr}}{100 \text{ torr}} = 0.08$

$\frac{620 \times 10^3 \text{ g } C_{24}H_{50}}{338.6 \text{ g/mol}} = 1.83 \times 10^3 \text{ mol } C_{24}H_{50};$ let y = mol C_2H_5OH

$\chi_{C_2H_5OH} = 0.08 = \frac{y}{y + 1.83 \times 10^3};\ 0.92 y = 146.4;\ y = 1.6 \times 10^2 \text{ mol } C_2H_5OH$

(Strictly speaking, y should have 1 sig fig because 0.08 has 1 sig fig, but this severely limits the calculation.)

$1.6 \times 10^2 \text{ mol } C_2H_5OH \times \frac{46 \text{ g } C_2H_5OH}{1 \text{ mol}} = 7.4 \times 10^3 \text{ g or } 7.4 \text{ kg } C_2H_5OH$

13.81 The solvent vapor pressure over each solution is determined by the total particle concentrations present in the solutions. When the particle concentrations are equal, the vapor pressures will be equal and equilibrium established. The particle concentration of the nonelectrolyte is just 0.060 M, the ion concentration of the NaCl is 2 × 0.040 M = 0.080 M. Solvent will diffuse from the less concentrated nonelectrolyte solution. Let x = volume of solvent transferred.

$$\frac{0.060\ M \times 20.0\ \text{mL}}{(20 - x)\ \text{mL}} = \frac{0.080\ M \times 20.0\ \text{mL}}{(20 + x)\ \text{mL}};\ 1.2(20.0 + x) = 1.6(20.0 - x)$$

$2.8\ x = 8$, $x = 2.86 = 3$ mL transferred

13.82 (a) 0.100 m K_2SO_4 is 0.300 m in particles. H_2O is the solvent.

$\Delta T_f = K_f m = -1.86(0.300) = -0.558$; $T_f = 0.0 - 0.558 = -0.558°C = -0.6°C$

(b) ΔT_f (nonelectrolyte) $= -1.86(0.100) = -0.186$; $T_f = 0.0 - 0.186 = -0.186°C = -0.2°C$

T_f (measured) $= i \times T_f$ (nonelectrolyte)

From Table 13.5, i for 0.100 m $K_2SO_4 = 2.32$

T_f (measured) $= 2.32(-0.186°C) = -0.432°C = -0.4°C$

13.83 Assume the radiator solution is prepared by mixing 1.00 L of each of the liquids. To calculate the freezing and boiling points, we need the molality of this solution. Assume H_2O is the solvent.

$$m = \frac{\text{mol } C_2H_6O_2}{\text{kg } H_2O};\ \text{kg } H_2O = 1.00\ \text{L} \times \frac{1000\ \text{mL}}{1\ \text{L}} \times \frac{1.00\ \text{g}}{\text{mL}} = 1.00\ \text{kg}$$

$$\text{mol } C_2H_6O_2 = 1.00\ \text{L} \times \frac{1000\ \text{mL}}{1\ \text{L}} \times \frac{1.12\ \text{g}}{1\ \text{mL}} \times \frac{1\ \text{mol } C_2H_6O_2}{62.07\ \text{g } C_2H_6O_2} = 18.04 = 18.0\ \text{mol } C_2H_6O_2$$

$$m = \frac{18.04\ \text{mol } C_2H_6O_2}{1.00\ \text{kg } H_2O} = 18.04\ m$$

$\Delta T_f = K_f m = -1.86(18.04) = -33.6°C$; f.p. $= 0.0 - 33.6 = -33.6°C$

$\Delta T_b = K_b m = 0.52(18.04) = 9.4°C$; b.p. $= 100.0 + 9.4 = +109.4°C$

13.84 The compound with the larger i value is the stronger electrolyte.

$$i = \frac{\Delta T_f\ (\text{measured})}{\Delta T_f\ (\text{calculated})}$$ The idealized value is 3 for both salts.

$$Hg(NO_3)_2:\ m = \frac{10.0\ \text{g } Hg(NO_3)_2}{1.00\ \text{kg } H_2O} \times \frac{1\ \text{mol } Hg(NO_3)_2}{324.6\ \text{g } Hg(NO_3)_2} = 0.0308\ m$$

ΔT_f (nonelectrolyte) $= -1.86(0.0308) = -0.0573°C$

$$i = \frac{-0.162°C}{-0.0573°C} = 2.83$$

$$HgCl_2:\ m = \frac{10.0\ \text{g } HgCl_2}{1.00\ \text{kg } H_2O} \times \frac{1\ \text{mol } HgCl_2}{271.5\ \text{g } HgCl_2} = 0.0368\ m$$

ΔT_f (nonelectrolyte) $= -1.86(0.0368) = -0.0685°C$

$$i = \frac{-0.0685}{-0.0685} = 1.00$$

With an i value of 2.83, $Hg(NO_3)_2$ is almost completely dissociated into ions; with an i value of 1.00, the $HgCl_2$ behaves essentially like a nonelectrolyte. Clearly, $Hg(NO_3)_2$ is the stronger electrolyte.

13.85 (a) $K_b = \frac{\Delta T_b}{m}$; $\Delta T_b = 47.46°C - 46.30°C = 1.16°C$

$$m = \frac{\text{mol solute}}{\text{kg } CS_2} = \frac{0.250 \text{ mol}}{400.0 \text{ mL } CS_2} \times \frac{1 \text{ mL } CS_2}{1.261 \text{ g } CS_2} \times \frac{1000 \text{ g}}{1 \text{ kg}} = 0.4956 = 0.496\ m$$

$$K_b = \frac{1.16°C}{0.4956\ m} = 2.34°C/m$$

(b) $$m = \frac{\Delta T_b}{K_b} = \frac{(47.08 - 46.30)°C}{2.34°C/m} = 0.333 = 0.33\ m$$

$$m = \frac{\text{mol unknown}}{\text{kg } CS_2};\quad m \times \text{kg } CS_2 = \frac{\text{g unknown}}{\mathcal{M} \text{ unknown}};\quad \mathcal{M} = \frac{\text{g unknown}}{m \times \text{kg } CS_2}$$

$$50.0 \text{ mL } CS_2 \times \frac{1.261 \text{ g } CS_2}{1 \text{ mL}} \times \frac{1 \text{ kg}}{1000 \text{ g}} = 0.06305 = 0.0631 \text{ kg } CS_2$$

$$\mathcal{M} = \frac{5.39 \text{ g unknown}}{0.333\ m \times 0.06305 \text{ kg } CS_2} = 257 = 2.6 \times 10^2 \text{ g/mol}$$

13.86 (a) Assume 1000 g of solution. $1000 \text{ g soln} \times \frac{1 \text{ mL}}{1.22 \text{ g}} = 819.7 = 820 \text{ mL}$

$$1000 \text{ g soln} \times \frac{40.0 \text{ g KSCN}}{100 \text{ g soln}} = 400 \text{ g KSCN}; 1000 \text{ g soln} - 400 \text{ g KSCN} = 600 \text{ g } H_2O$$

$$\frac{400 \text{ g KSCN}}{97.19 \text{ g/mol}} = 4.116 = 4.12 \text{ mol KSCN};\quad \frac{600 \text{ g } H_2O}{18.02 \text{ g/mol}} = 33.30 = 33.3 \text{ mol } H_2O$$

$$\chi_{KSCN} = \frac{4.116}{4.116 + 33.30} = 0.110;\quad m = \frac{4.116 \text{ mol KSCN}}{0.600 \text{ kg } H_2O} = 6.86\ m$$

$$M = \frac{4.116 \text{ mol KSCN}}{0.8197 \text{ L}} = 5.02\ M$$

(b) If there are 4.12 mol of KSCN, there are 8.24 moles of ions. There are then 33.3 mol H_2O/8.24 mol ions $\approx$ 4 mol H_2O for each mol of ions, or 4 water molecules for each ion. This is too few water molecules to completely hydrate the anions and cations in the solution.

For a solution that is this concentrated, one would expect significant ion-pairing, because the ions are not completely surrounded and separated by H_2O molecules. Because of ion-pairing, the effective number of particles will be less than that indicated by m and M, so the observed colligative properties will be significantly different from those predicted by formulas for ideal solutions. The observed freezing point will be higher, the boiling point lower and the osmotic pressure lower than predicted.

13.87 $M = \frac{\pi}{RT} = \frac{6.41 \text{ atm}}{298 \text{ K}} \times \frac{\text{K} \cdot \text{mol}}{0.08206 \text{ L} \cdot \text{atm}} = 0.2621 = 0.262 \, M$

There are 0.262 mol of **particles** per liter of solution and 0.131 mol particles in 500 mL. Each mole of sucrose provides 1 mol of particles and each mole of NaCl provides 2.

Let x = g $C_{12}H_{22}O_{11}$, 15.0 - x = g NaCl

$$\frac{x \text{ g } C_{12}H_{22}O_{11}}{342.3 \text{ g } C_{12}H_{22}O_{11}} + \frac{2(15.0 - x)}{58.44 \text{ g NaCl}} = 0.131$$

$58.44x + 2(342.3)(15.0 - x) = 0.131(58.44)(342.3)$

$58.44x - 684.6x + 10{,}269 = 2{,}621$

$-626.2x = -7648$; x = 12.2 g $C_{12}H_{22}O_{11}$

$$\frac{12.2 \text{ g } C_{12}H_{22}O_{11}}{15.0 \text{ g mixture}} \times 100 = 81.3\% \text{ sucrose, } 18.7\% \text{ NaCl}$$

13.88 $M = \frac{\pi}{RT} = \frac{57.1 \text{ torr}}{298 \text{ K}} \times \frac{1 \text{ atm}}{760 \text{ torr}} \times \frac{\text{K} \cdot \text{mol}}{0.08206 \text{ L} \cdot \text{atm}} = 3.072 \times 10^{-3} = 3.07 \times 10^{-3} \, M$

$$\frac{0.036 \text{ g solute}}{100 \text{ g } H_2O} \times \frac{1000 \text{ g } H_2O}{1 \text{ kg } H_2O} = 0.36 \text{ g solute/kg } H_2O$$

Assuming molarity and molality are the same in this dilute solution, we can then say 0.36 g solute = 3.072×10^{-3} mol; $\mathcal{M}$ = 117 g/mol. Because the salt is completely ionized, the formula weight of the lithium salt is **twice** this calculated value, or **234 g/mol**. The organic portion, $C_nH_{2n-1}O_2^-$, has a formula weight of 234-7 = 227 g. Subtracting 32 for the oxygens, and adding 1 to make the formula C_nH_{2n}, we have C_nH_{2n}, $\mathcal{M}$ = 196 g/mol. Since each CH_2 unit has a mass of 14, n = 196/14 = 14. Thus the formula for our salt is $LiC_{14}H_{27}O_2$.

Integrative Exercises

13.89 Since these are very dilute solutions, assume that the density of the solution ≈ the density of H_2O ≈ 1.0 g/mL at 25°C. Then, 100 g solution = 100 g H_2O = 0.100 kg H_2O.

(a) CF_4: $\frac{0.0015 \text{ g } CF_4}{0.100 \text{ kg } H_2O} \times \frac{1 \text{ mol } CF_4}{88.00 \text{ g } CF_4} = 1.7 \times 10^{-4} \, m$

$CClF_3$: $\frac{0.009 \text{ g } CClF_3}{0.100 \text{ kg } H_2O} \times \frac{1 \text{ mol } CClF_3}{104.46 \text{ g } CClF_3} = 8.6 \times 10^{-4} \, m = 9 \times 10^{-4} \, m$

CCl_2F_2: $\frac{0.028 \text{ g } CCl_2F_2}{0.100 \text{ kg } H_2O} \times \frac{1 \text{ mol } CCl_2F_2}{120.9 \text{ g } CCl_2F_2} = 2.3 \times 10^{-3} \, m$

$CHClF_2$: $\frac{0.30 \text{ g } CHClF_2}{0.100 \text{ kg } H_2O} \times \frac{1 \text{ mol } CHClF_2}{86.47 \text{ g } CHClF_2} = 3.5 \times 10^{-2} \, m$

(b) $m = \frac{\text{mol solute}}{\text{kg solvent}}$; $M = \frac{\text{mol solute}}{\text{L solution}}$

Molality and molarity are numerically similar when kilograms solvent and liters solution are nearly equal. This is true when solutions are dilute, so that the density of the solution is essentially the density of the solvent, and when the density of the solvent is nearly 1 g/mL. That is, for dilute aqueous solutions such as the ones in this problem, $M \approx m$.

(c) Water is a polar solvent; the solubility of solutes increases as their polarity increases. All the fluorocarbons listed have tetrahedral molecular structures. CF_4, a symmetrical tetrahedron, is nonpolar and has the lowest solubility. As more different atoms are bound to the central carbon, the electron density distribution in the molecule becomes less symmetrical and the molecular polarity increases. The most polar fluorocarbon, $CHClF_2$, has the greatest solubility in H_2O. It may act as a weak hydrogen bond acceptor for water.

(d) $S_g = k P_g$. Assume $M = m$ for $CHClF_2$. P_g = 1 atm

$$k = \frac{S_g}{P_g} = \frac{M}{P_g}; \quad k = \frac{3.5 \times 10^{-2}\, M}{1.0 \text{ atm}} = 3.5 \times 10^{-2} \text{ mol/L} \bullet \text{atm}$$

This value is greater than the Henry's law constant for $N_2(g)$, because $N_2(g)$ is nonpolar and of lower molecular mass than $CHClF_2$. In fact, the Henry's law constant for nonpolar CF_4, 1.7×10^{-4} mol/L • atm is similar to the value for N_2, 6.8×10^{-4} mol L • atm.

13.90 $$\frac{0.015 \text{ g } N_2}{1 \text{ L blood}} \times \frac{1 \text{ mol } N_2}{28.01 \text{ g } N_2} = 5.355 \times 10^{-4} = 5.4 \times 10^{-4} \text{ mol } N_2\text{/L blood}$$

At 100 ft, the partial pressure of N_2 in air is 0.78 (4.0 atm) = 3.12 atm. This is just four times the partial pressure of N_2 at 1.0 atm air pressure. According to Henry's law, $S_g = kP_g$, a 4-fold increase in P_g results n a 4-fold increase in S_g, the solubility of the gas. Thus, the solubility of N_2 at 100 ft is $4(5.355 \times 10^{-4}\, M) = 2.142 \times 10^{-3} = 2.1 \times 10^{-3}\, M$. If the diver suddenly surfaces, the amount of N_2/L blood released is the difference in the solubilities at the two depths: $(2.142 \times 10^{-3}$ mol/L - 5.355×10^{-4} mol/L) = $1.607 \times 10^{-3} = 1.6 \times 10^{-3}$ mol N_2/L blood.

At surface conditions of 1.0 atm external pressure and 37°C = 310 K,

$$V = \frac{nRT}{P} = 1.607 \times 10^{-3} \text{ mol} \times \frac{310 \text{ K}}{1.0 \text{ atm}} \times \frac{0.08206 \text{ L} \bullet \text{atm}}{\text{mol} \bullet \text{K}} = 0.041 \text{ L}$$

That is, 41 mL of tiny N_2 bubbles are released from each L of blood.

13.91 The stronger the intermolecular forces, the higher the heat (enthalpy) of vaporization.

(a) None of the substances are capable of hydrogen bonding in the pure liquid, and they have similar molar masses. All intermolecular forces are van der Waals forces, dipole-dipole and dispersion forces. In decreasing order of strength of forces:

acetonitrile > acetone > acetaldehyde > ethylene oxide > cyclopropane > propane

The first four compounds have dipole-dipole and dispersion forces, the last two only dispersion forces.

(b) The order of solubility in hexane should be the reverse of the order above. The least polar substance, propane, will be most soluble in hexane. Ethanol, CH_3CH_2OH, is capable of hydrogen bonding with the four polar compounds. Thus, acetonitrile, acetaldehyde, acetone and ethylene oxide should be more soluble than propane and cyclopropane, but without further information we cannot distinguish among the polar molecules.

13.92 For ionic solids, the exothermic part of the solution process is step (3), surrounding the separated ions by solvent molecules. The released energy comes from the attractive interaction of the solvent with the separated ions. In the hydrates, one water molecule is already associated with the cation, reducing the total energy released during solvation.

13.93 (a)

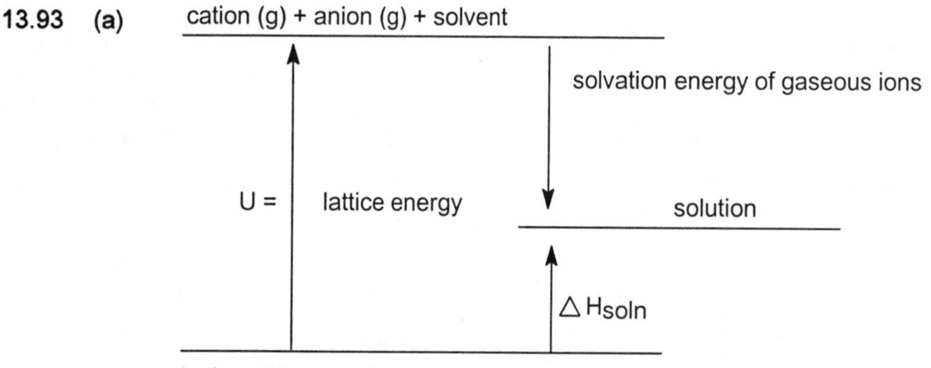

(b) If the lattice energy (U) of the ionic solid (ion-ion forces) is too large relative to the solvation energy of the gaseous ions (ion-dipole forces), ΔH_{soln} will be too large and positive (endothermic) for solution to occur. This is the case for solutes like NaBr.

Lattice energy is inversely related to the distance between ions, so salts with large cations like $(CH_3)_4N^+$ have smaller lattice energies than salts with simple cations like Na^+. The smaller lattice energy of $(CH_4)_3NBr$ causes it to be more soluble in nonaqueous polar solvents. Also, the $-CH_3$ groups in the large cation are capable of dispersion interactions with the $-CH_3$ (or other nonpolar groups) of the solvent molecules. This produces a more negative solvation energy for the salts with large cations.

Overall, for salts with larger cations, U is smaller (less positive), the solvation energy of the gaseous ions is more negative, and ΔH_{soln} is less endothermic. These salts are more soluble in polar nonaqueous solvents.

13.94 (a) $Zn(s) + H_2SO_4(aq) \rightarrow ZnSO_4(aq) + H_2(g)$

$$2.050 \text{ g Zn} \times \frac{1 \text{ mol Zn}}{65.39 \text{ g Zn}} = 0.03135 \text{ mol Zn}$$

$$1.00 \, M \, H_2SO_4 \times 0.0150 \text{ L} = 0.0150 \text{ mol } H_2SO_4$$

Since Zn and H_2SO_4 react in a 1:1 mole ratio, H_2SO_4 is the limiting reactant; 0.0150 mol of $H_2(g)$ are produced.

(b) $$P = \frac{nRT}{V} = \frac{0.0150 \text{ mol}}{0.122 \text{ L}} \times \frac{0.08206 \text{ L} \bullet \text{atm}}{\text{mol} \bullet \text{K}} \times 298 \text{ K} = 3.0066 = 3.01 \text{ atm}$$

(c) $$S_{H_2} = kP_{H_2} = \frac{7.8 \times 10^{-4} \text{ mol}}{\text{L} \bullet \text{atm}} \times 3.0066 \text{ atm} = 0.002345 = 2.3 \times 10^{-3} \, M$$

$$\frac{0.002345 \text{ mol } H_2}{\text{L soln}} \times 0.0150 \text{ L} = 3.518 \times 10^{-5} = 3.5 \times 10^{-5} \text{ mol dissolved } H_2$$

$$\frac{3.5 \times 10^{-5} \text{ mol dissolved } H_2}{0.0150 \text{ mol } H_2 \text{ produced}} \times 100 = 0.23\% \text{ dissolved } H_2$$

This is approximately 2.3 ppt; for every 10,000 H_2 molecules, 23 are dissolved. It was reasonable to ignore dissolved $H_2(g)$ in part (b).

13.95 (a) $$\frac{1.3 \times 10^{-3} \text{ mol } CH_4}{\text{L soln}} \times 4.0 \text{ L} = 5.2 \times 10^{-3} \text{ mol } CH_4$$

$$V = \frac{nRT}{P} = \frac{5.2 \times 10^{-3} \text{ mol} \times 273 \text{ K}}{1.0 \text{ atm}} \times \frac{0.08206 \text{ L} \bullet \text{atm}}{\text{K} \bullet \text{mol}} = 0.12 \text{ L}$$

(b) All three hydrocarbons are nonpolar; they have zero net dipole moment. In CH_4 and C_2H_6, the C atoms are tetrahedral and all bonds are σ bonds. C_2H_6 has a higher molar mass than CH_4, which leads to stronger dispersion forces and greater water solubility. In C_2H_4, the C atoms are trigonal planar and the π electron cloud is symmetric above and below the plane that contains all the atoms. The π cloud in C_2H_4 is an area of concentrated electron density that experiences attractive forces with the positive ends of H_2O molecules. These forces increase the solubility of C_2H_4 relative to the other hydrocarbons.

(c) The molecules have similar molar masses. NO is most soluble because it is polar. The triple bond in N_2 is shorter than the double bond in O_2. It is more difficult for H_2O molecules to surround the smaller N_2 molecules, so they are less soluble than O_2 molecules.

(d) H_2S and SO_2 are polar molecules capable of hydrogen bonding with water. Hydrogen bonding is the strongest force between neutral molecules and causes the much greater solubility. H_2S is weakly acidic in water. SO_2 reacts with water to form H_2SO_3, a weak acid. The large solubility of SO_2 is a sure sign that a chemical process has occurred.

(e) N_2 and C_2H_4. N_2 is too small to be easily hydrated so C_2H_4 is more soluble in H_2O.

NO (31) and C_2H_6 (30). The structures of these two molecules are very different, yet they have similar solubilities. NO is slightly polar, but too small to be easily hydrated. The larger C_2H_6 is nonpolar, but more polarizable (stronger dispersion forces).

NO (31) and O_2 (32). The slightly polar NO is more soluble than the slightly larger (longer O=O bond than N $\equiv$ O bond) but nonpolar O_2.

13.96 *Plan.* Assume 100 g sample. Calculate empirical formula from mass % data. Calculate molar mass from osmotic pressure. Deduce molecular formula.

$$61.00 \text{ g C} \times \frac{1 \text{ mol C}}{12.01 \text{ g C}} = 5.079 \text{ mol C; } 5.079/0.8467 = 6.0$$

$$6.83 \text{ g H} \times \frac{1 \text{ mol H}}{1.008 \text{ g H}} = 6.776 \text{ mol H; } 6.776/0.8467 = 8.0$$

$$11.86 \text{ g N} \times \frac{1 \text{ mol N}}{14.007 \text{ g N}} = 0.8467 \text{ mol N; } 0.8467/0.8467 = 1$$

$$20.32 \text{ g O} \times \frac{1 \text{ mol O}}{16.00 \text{ g O}} = 1.270 \text{ mol O; } 1.270/0.8467 = 1.5$$

Multiplying by 2 to obtain an integer ratio, the empirical formula is $C_{12}H_{16}N_2O_3$. The formula weight is 236.3 g.

$$\pi = M\text{RT; } M = \text{mol/L; mol} = \text{g}/\mathcal{M}\text{; } M = \frac{\text{g}}{\mathcal{M} \times \text{L}}\text{; } \pi = \frac{\text{g}}{\mathcal{M} \times \text{L}} \times \text{RT; } \mathcal{M} = \frac{\text{g RT}}{\pi \times \text{L}}$$

$$\mathcal{M} = \frac{2.505 \times 10^{-3} \text{ g}}{0.01000 \text{ L}} \times \frac{0.08206 \text{ L}\bullet\text{atm}}{\text{mol}\bullet\text{K}} \times \frac{298 \text{ K}}{19.7 \text{ torr}} \times \frac{760 \text{ torr}}{1 \text{ atm}} = 236 \text{ g/mol}$$

Since the formula weight and molar mass are equal, the empirical and molecular formula is $C_{12}H_{16}N_2O_3$.

13.97 (a)

$$\Delta T_f = K_f m = K_f \times \frac{\text{mol } C_7H_6O_2}{\text{kg } C_6H_6} = K_f \times \frac{\text{g } C_7H_6O_2}{\text{kg } C_6H_6 \times \mathcal{M}\, C_7H_6O_2}$$

$$\mathcal{M} = \frac{K_f \times \text{g } C_7H_6O_2}{\Delta T_f \times \text{kg } C_6H_6} = \frac{5.12 \times 0.55}{0.360 \times 0.032} = 244.4 = 2.4 \times 10^2 \text{ g/mol}$$

(b) The formula weight of $C_7H_6O_2$ is 122 g/mol. The experimental molar mass is twice this value, indicating that benzoic acid is associated into dimers in benzene solution. This is reasonable, since the carboxyl group, -COOH, is capable of strong hydrogen bonding with itself. Many carboxylic acids exist as dimers in solution. The structure of benzoic acid dimer in benzene solution is:

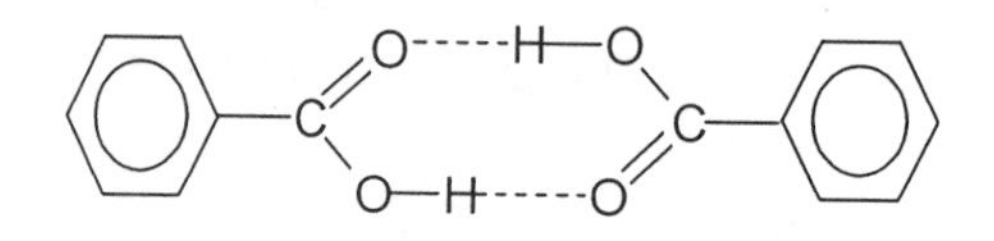

13.98 $\chi_{CHCl_3} = \chi_{C_3H_6O} = 0.500$

(a) For an ideal solution, Raoult's Law is obeyed.

$P_T = P_{CHCl_3} + P_{C_3H_6O}$; $P_{CHCl_3} = 0.5(300 \text{ torr}) = 150 \text{ torr}$

$P_{C_3H_6O} = 0.5(360 \text{ torr}) = 180 \text{ torr}$; $P_T = 150 \text{ torr} + 180 \text{ torr} = 330 \text{ torr}$

(b) The real solution has a lower vapor pressure, 250 torr, than an ideal solution of the same composition, 330 torr. Thus, fewer molecules escape to the vapor phase from the liquid. This means that fewer molecules have sufficient kinetic energy to overcome intermolecular attractions. Clearly, even weak hydrogen bonds such as this one are stronger attractive forces than dipole-dipole or dispersion forces. These hydrogen bonds prevent molecules from escaping to the vapor phase and result in a lower than ideal vapor pressure for the solution. There is essentially no hydrogen bonding in the individual liquids.

(c) According to Coulomb's law, electrostatic attractive forces lead to an overall lowering of the energy of the system. Thus, when the two liquids mix and hydrogen bonds are formed, the energy of the system is decreased and $\Delta H_{soln} < 0$; the solution process is exothermic.

14 Chemical Kinetics

Reaction Rates

14.1 (a) *Reaction rate* is the change in the amount of products or reactants in a given amount of time; it is the speed of a chemical reaction.

(b) Rates depend on concentration of reactants, surface area of reactants, temperature and presence of catalyst.

(c) The stoichiometry of the reaction (mole ratios of reactants and products) must be known to relate rate of disappearance of reactants to rate of appearance of products.

14.2 (a) M /s

(b) The hotter the oven, the faster the cake bakes. Milk sours faster in hot weather than cool weather.

(c) The *average rate* is the rate over a period of time, while the *instantaneous rate* is the rate at a particular time.

14.3 *Analyze/Plan*. Given mol A at a series of times in minutes calculate mol B produced, molarity of A at each time, change in M of A at each 10 min interval, and ΔM A/s. For this reaction, mol B produced equals mol A consumed. M of A or [A] = mol A/0.100 L. The average rate of disappearance of A for each 10 minute interval is

$$\frac{\Delta[A]}{s} = \frac{[A]_0 - [A]_1}{10\ \text{min}} \times \frac{1\ \text{min}}{60\ \text{s}}$$

Solve:

Time(min)	Mol A	(a) Mol B	[A]	Δ [A]	(b) Rate (Δ [A]/s)
0	0.065	0.000	0.65		
10	0.051	0.014	0.51	-0.14	2.3×10^{-4}
20	0.042	0.023	0.42	-0.09	1.5×10^{-4}
30	0.036	0.029	0.36	-0.06	1.0×10^{-4}
40	0.031	0.034	0.31	-0.05	0.8×10^{-4}

(c) $$\frac{\Delta M_B}{\Delta t} = \frac{(0.029 - 0.014)\ \text{mol}/0.100\ \text{L}}{(30 - 10)\ \text{min}} \times \frac{1\ \text{min}}{60\ \text{s}} = 1.25 \times 10^{-4} = 1.3 \times 10^{-4}\ M/\text{s}$$

14.4

Time(s)	Mol A	(a) Mol B	Δ Mol A	(b) Rate (Δ mol A/s)
0	0.100	0.000		
40	0.067	0.033	0.033	8.3×10^{-4}
80	0.045	0.055	-0.022	5.5×10^{-4}
120	0.030	0.070	-0.015	3.8×10^{-4}
160	0.020	0.080	-0.010	2.5×10^{-4}

(c) The volume of the container must be known to report the rate in units of concentration (mol/L) per time.

14.5 *Analyze/Plan.* Follow the logic in Sample Exercise 14.1. *Solve:*

Time (sec)	Time Interval (sec)	Concentration (*M*)	Δ *M*	Rate (*M* /s)
0		0.0165		
2,000	2,000	0.0110	-0.0055	28×10^{-7}
5,000	3,000	0.00591	-0.0051	17×10^{-7}
8,000	3,000	0.00314	-0.00277	9.23×10^{-7}
12,000	4,000	0.00137	-0.00177	4.43×10^{-7}
15,000	3,000	0.00074	-0.00063	2.1×10^{-7}

14.6

Time (min)	Time Interval (min)	Concentration (*M*)	Δ *M*	Rate (*M* /s)
0.0		1.85		
54.0	54.0	1.58	-0.27	8.3×10^{-5}
107.0	53.0	1.36	-0.22	6.9×10^{-5}
215.0	108	1.02	-0.34	5.3×10^{-5}
430.0	215	0.580	-0.44	3.4×10^{-5}

14.7 From the slopes of the lines in the figure at right, the rates are -1.2×10^{-6} *M* /s at 5000 s, -5.8×10^{-7} *M* /s at 8000 s.

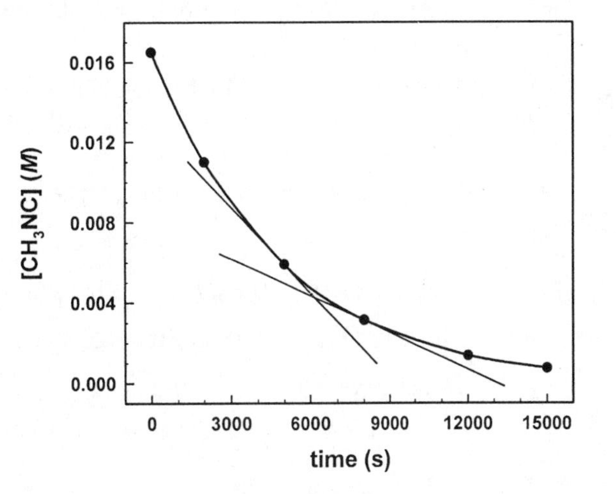

14.8 From the slopes of the lines in the figure at right, the rates are: at t = 75.0 min, -4.2×10^{-3} *M*/min or -7.0×10^{-5} *M*/s; at 250 min, -2.1×10^{-3} *M*/min or -3.5×10^{-5} M/s.

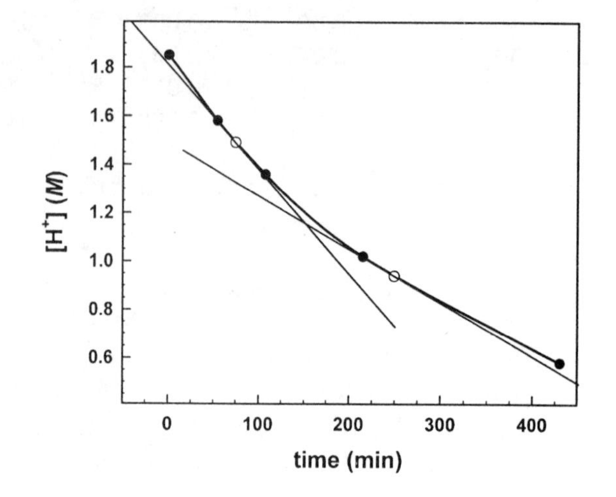

14.9 *Analyze/Plan*. Follow the logic in Sample Exercise 14.3. *Solve*:

(a) $-\Delta[H_2O_2]/\Delta t = \Delta[H_2]/\Delta t = \Delta[O_2]/\Delta t$

(b) $-\Delta[N_2O]/2\Delta t = \Delta[N_2]/2\Delta t = \Delta[O_2]/\Delta t$
$-\Delta[N_2O]/\Delta t = \Delta[N_2]/\Delta t = 2\Delta[O_2]/\Delta t$

(c) $-\Delta[N_2]/\Delta t = \Delta[NH_3]/2\Delta t$; $-\Delta[H_2]/3\Delta t = \Delta[NH_3]/2\Delta t$
$-2\Delta[N_2]/\Delta t = \Delta[NH_3]/\Delta t$; $-\Delta[H_2]/\Delta t = 3\Delta[NH_3]/2\Delta t$

14.10 (a) rate $= -\Delta[HBr]/2\Delta t = \Delta[H_2]/\Delta t = \Delta[Br_2]/\Delta t$

(b) rate $= -\Delta[SO_2]/2\Delta t = -\Delta[O_2]/\Delta t = \Delta[SO_3]/2\Delta t$

(c) rate $= -\Delta[NO]/2\Delta t = -\Delta[H_2]/2\Delta t = \Delta[N_2]/\Delta t = \Delta[H_2O]/2\Delta t$

14.11 *Analyze/Plan*. Use Equation 14.4 to relate the rate of disappearance of reactants to the rate of appearance of products. Use this relationship to calculate desired quantities. *Solve*:

(a) $\Delta[H_2O]/2\Delta t = -\Delta[H_2]/2\Delta t = -\Delta[O_2]/\Delta t$

H_2 is burning, $-\Delta[H_2]/\Delta t = 0.85$ mol/s

O_2 is consumed, $-\Delta[O_2]/\Delta t = -\Delta[H_2]/2\Delta t = 0.85$ mol/s/2 = 0.43 mol/s

H_2O is produced, $+\Delta[H_2O]/\Delta t = -\Delta[H_2]/\Delta t = 0.85$ mol/s

(b) The change in total pressure is the sum of the changes of each partial pressure. NO and Cl_2 are disappearing and NOCl is appearing.

$-\Delta P_{NO}/\Delta t = -23$ torr/min

$-\Delta P_{Cl_2}/\Delta t = \Delta P_{NO}/2\Delta t = -12$ torr/min

$+\Delta P_{NOCl}/\Delta t = -\Delta P_{NO}/\Delta t = +23$ torr/min

$\Delta P_T/\Delta t = -23$ torr/min - 12 torr/min + 23 torr/min = -12 torr/min

14.12 (a) $-\Delta[C_2H_4]/\Delta t = \Delta[CO_2]/2\Delta t = \Delta[H_2O]/2\Delta t$

$-2\Delta[C_2H_4]/\Delta t = \Delta[CO_2]/\Delta t = \Delta[H_2O]/\Delta t$

C_2H_4 is burning, $-\Delta[C_2H_4]/\Delta t = 0.23$ *M*/s

CO_2 and H_2O are produced, at twice the rate that C_2H_4 is consumed.

$\Delta[CO_2]/\Delta t = \Delta[H_2O]/\Delta t = 2(0.23)$ *M*/s = 0.46 *M*/s

(b) In this reaction, pressure is a measure of concentration.

$-\Delta[N_2H_4]/\Delta t = -\Delta[H_2]/\Delta t = \Delta[NH_3]/2\Delta t$

N_2H_4 is consumed, $-\Delta[N_2H_4]/\Delta t$ = 45 torr/hr

H_2 is consumed, $-\Delta[H_2]/\Delta t$ = 45 torr/hr

NH_3 is produced at twice the rate that N_2H_4 and H_2 are consumed,

$\Delta[NH_3]/\Delta t = -2\Delta[N_2H_4]/\Delta t$ = 2(45) torr/hr = 90 torr/hr

$\Delta P_T/\Delta t$ = (+90 torr/hr - 45 torr/hr - 45 torr/hr) = 0 torr/hr

Rate Laws

14.13 (a) If [A] doubles, the rate will increase by a factor of four; the rate constant, k, is unchanged. Rate is proportional to $[A]^2$, so when the value of [A] doubles, rate changes by 2^2 or 4. The rate constant, k, is the proportionality constant that does not change (unless the temperature changes).

(b) The reaction is second order in A, first order in B, and third order overall.

(c) Units of k = $\frac{M/s}{M^3} = M^{-2}\,s^{-1}$

14.14 (a) If [A] is doubled, there will be no change in the rate or the rate constant. The overall rate is unchanged because [A] does not appear in the rate law; the rate constant changes only with a change in temperature.

(b) The reaction is zero order in A, second order in B and second order overall.

(c) Units of k = $\frac{M/s}{M^2} = M^{-1}\,s^{-1}$

14.15 *Analyze/Plan*. Follow the logic in Sample Exercise 14.6. *Solve*:

(a) rate = $k[N_2O_5] = 4.82 \times 10^{-3}\,s^{-1}\,[N_2O_5]$

(b) rate = $4.82 \times 10^{-3}\,s^{-1}\,(0.0240\,M) = 1.16 \times 10^{-4}\,M/s$

(c) rate = $4.82 \times 10^{-3}\,s^{-1}\,(0.0480\,M) = 2.31 \times 10^{-4}\,M/s$

When the concentration of N_2O_5 doubles, the rate of the reaction doubles.

14.16 (a) rate = $k[H_2][NO]^2$

(b) rate = $(6.0 \times 10^4\,M^{-2}\,s^{-1})(0.050\,M)^2(0.010\,M) = 1.5\,M/s$

(c) rate = $(6.0 \times 10^4\,M^{-2}\,s^{-1})(0.10\,M)^2(0.010\,M) = 6.0\,M/s$

(Note that doubling [NO] causes a quadrupling in rate.)

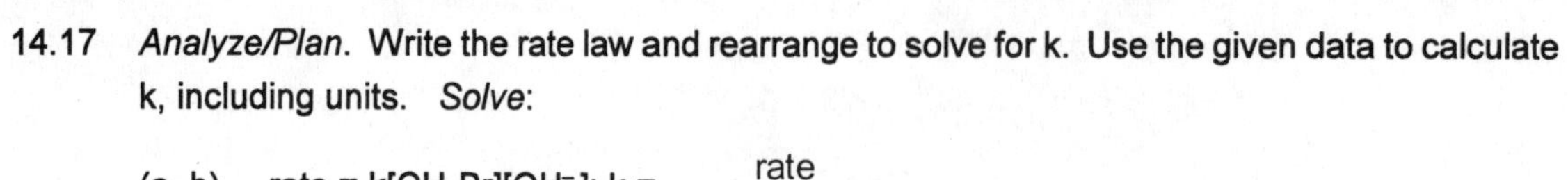

14.17 *Analyze/Plan*. Write the rate law and rearrange to solve for k. Use the given data to calculate k, including units. *Solve*:

(a, b) $\text{rate} = k[CH_3Br][OH^-]$; $k = \dfrac{\text{rate}}{[CH_3Br][OH^-]}$

at 298 K, $k = \dfrac{0.0432\ M/s}{(5.0 \times 10^{-3}\ M)(0.050\ M)} = 1.7 \times 10^2\ M^{-1}\,s^{-1}$

(c) Since the rate law is first order in $[OH^-]$, if $[OH^-]$ is tripled, the rate triples.

14.18 (a, b) $\text{rate} = k[C_2H_5Br][OH^-]$; $k = \dfrac{\text{rate}}{[C_2H_5Br][OH^-]}$

at 298 K, $k = \dfrac{1.7 \times 10^{-7}\ M/s}{[0.0477\ M][0.100\ M]} = 3.6 \times 10^{-5}\ M^{-1}\,s^{-1}$

(c) Adding an equal volume of ethyl alcohol reduces both $[C_2H_5Br]$ and $[OH^-]$ by a factor of two. new rate = (1/2)(1/2) = 1/4 of old rate

14.19 *Analyze/Plan*. Substitute relative values into the rate law and solve for x. *Solve*:

(a) $\text{rate} = [A]^x$; $3 = [3]^x$; $x = 1$

(b) $8 = [2]^x$, $x = 3$

(c) $1 = [3]^x$; $x = 0$ (The rate does not depend on [A].)

14.20 (a) Experiments 1 and 2 show the effect of a change in [A] on the rate while holding [B] constant. Experiment 3 must show the effect of changing [B] while holding [A] constant. The only restrictions on Experiment 3 are that [B] cannot be 1.0 *M* and [A] must be either 1.0 *M* or 2.0 *M*. Two (of the many) possibilities are [A] = 1.0 M, [B] = 2.0 *M* or [A] = 2.0 *M*, [B] = 2.0 *M*.

(b) Based on the observations, the rate law is: $\text{rate} = k[A][C]^2$. If all concentrations are half of their original values, $\text{rate} = k(1/2)(1/2)^2 = k(1/8)$; the rate is 1/8 of the original rate.

14.21 *Analyze/Plan*. Follow the logic in Sample Exercise 14.6. *Solve*:

(a) From the data given, when $[OCl^-]$ doubles, rate doubles. When $[I^-]$ doubles, rate doubles. The reaction is first order in both $[OCl^-]$ and $[I^-]$. $\text{rate} = [OCl^-][I^-]$

(b) Using the first set of data:

$$k = \frac{\text{rate}}{[OCl^-][I^-]} = \frac{1.36 \times 10^4\ M/s}{(1.5 \times 10^{-3} M)(1.5 \times 10^{-3} M)} = 6.0444 \times 10^9 = 6.04 \times 10^9\ M^{-1}\,s^{-1}$$

(c) $\text{rate} = \dfrac{6.044 \times 10^9}{M \bullet s}(1.0 \times 10^{-3}\ M)(5.0 \times 10^{-4}\ M) = 3.02 \times 10^3\ M/s$

14.22 (a) From the data given, when $[ClO_2]$ increases by a factor of 3 (Expt. 2 to Expt. 1), the rate increases by a factor of 9. When $[OH^-]$ increases by a factor of 3 (Expt. 2 to Expt. 3), the rate increases by a factor of 3. The reaction in second order is $[ClO_2]$ and first order is $[OH^-]$. rate = $k[ClO_2]^2[OH^-]$.

(b) Using data from Expt 2:

$$k = \frac{\text{rate}}{[ClO_2]^2[OH^-]} = \frac{0.00276\ M/s}{(0.020\ M)^2(0.030\ M)} = 2.3 \times 10^2\ M^{-2}\ s^{-1}$$

(c) rate = $2.3 \times 10^2\ M^{-2}\ s^{-1}\ (0.010\ M)^2(0.015\ M) = 3.45 \times 10^{-4} = 3.5 \times 10^{-4}\ M/s$

14.23 *Analyze/Plan.* Follow the logic in Sample Exercise 4.6 to deduce the rate law. Rearrange the rate law to solve for k and deduce units. Calculate a k value for each set of concentrations and then average the three values. *Solve:*

(a) Doubling [NO] while holding $[O_2]$ constant increases the rate by a factor of 4 (experiments 1 and 3). Reducing $[O_2]$ by a factor of 2 while holding [NO] constant reduces the rate by a factor of 2 (experiments 2 and 3). The rate is second order in [NO] and first order in $[O_2]$. rate = $k[NO]^2[O_2]$

(b, c) From experiment 1: $k_1 = \dfrac{1.41 \times 10^{-2}\ M/s}{(0.0126\ M)^2(0.0125\ M)} = 7105 = 7.11 \times 10^3\ M^{-2}\ s^{-1}$

$k_2 = 0.113/(0.0252)^2(0.0250) = 7118 = 7.12 \times 10^3\ M^{-2}\ s^{-1}$

$k_3 = 5.64 \times 10^{-2}/(0.0252)^2(0.125) = 7105 = 7.11 \times 10^3\ M^{-2}\ s^{-1}$

$k_{avg} = (7105 + 7118 + 7105)/3 = 7109 = 7.11 \times 10^3\ M^{-2}\ s^{-1}$

14.24 (a) Doubling $[NH_3]$ while holding $[BF_3]$ constant doubles the rate (experiments 1 and 2). Doubling $[BF_3]$ while holding $[NH_3]$ constant doubles the rate (experiments 4 and 5). Thus, the reaction is first order in both BF_3 and NH_3; rate = $k[BF_3][NH_3]$.

(b) The reaction is second order overall.

(c) From experiment 1: $k = \dfrac{0.2130\ M/s}{(0.250\ M)(0.250\ M)} = 3.41\ M^{-1}\ s^{-1}$

(Any of the five sets of initial concentrations and rates could be used to calculate the rate constant k. The average of these 5 values is $k_{avg} = 3.41\ M^{-1}\ s^{-1}$)

14.25 *Analyze/Plan.* Follow the logic in Sample Exercise 4.6 to deduce the rate law. Rearrange the rate law to solve for k and deduce units. Calculate a k value for each set of concentrations and then average the three values. *Solve:*

(a) Increasing [NO] by a factor of 2.5 while holding $[Br_2]$ constant (experiments 1 and 2) increases the rate by a factor 6.25 or $(2.5)^2$. Increasing $[Br_2]$ by a factor of 2.5 while holding [NO] constant increases the rate by a factor of 2.5. The rate law for the appearance of NOBr is: rate = $\Delta[NOBr]/\Delta t = k[NO]^2[Br_2]$.

(b) From experiment 1: $k_1 = \dfrac{24\ M/s}{(0.10\ M)^2(0.20\ M)} = 1.20 \times 10^4 = 1.2 \times 10^4\ M^{-2}\,s^{-1}$

$k_2 = 150/(0.25)^2(0.20) = 1.20 \times 10^4 = 1.2 \times 10^4\ M^{-2}\,s^{-1}$

$k_3 = 60/(0.10)^2(0.50) = 1.20 \times 10^4 = 1.2 \times 10^4\ M^{-2}\,s^{-1}$

$k_4 = 735/(0.35)^2(0.50) = 1.2 \times 10^4 = 1.2 \times 10^4\ M^{-2}\,s^{-1}$

$k_{avg} = (1.2 \times 10^4 + 1.2 \times 10^4 + 1.2 \times 10^4 + 1.2 \times 10^4)/4 = 1.2 \times 10^4\ M^{-2}\,s^{-1}$

(c) Use the reaction stoichiometry and Equation 14.4 to relate the designated rates. $\Delta[NOBr]/2\Delta t = -\Delta[Br_2]/\Delta t$; the rate of disappearance of Br_2 is half the rate of appearance of NOBr.

(d) Note that the data is given in terms of appearance of NOBr.

$$\frac{-\Delta[Br_2]}{\Delta t} = \frac{k[NO]^2[Br_2]}{2} = \frac{1.2 \times 10^4}{2\ M^2 s} \times (0.075\ M)^2 \times (0.25\ M) = 8.4\ M/s$$

14.26 (a) Increasing $[S_2O_8^{2-}]$ by a factor of 1.5 while holding $[I^-]$ constant increases the rate by a factor of 1.5 (Experiments 1 and 2). Doubling $[S_2O_8^{2-}]$ and increasing $[I^-]$ by a factor of 1.5 triples the rate ($2 \times 1.5 = 3$, experiments 1 and 3). Thus the reaction is first order in both $[S_2O_8^{2-}]$ and $[I^-]$; rate = $k[S_2O_8^{2-}][I^-]$.

(b) $k = \text{rate}/[S_2O_8^{2-}][I^-]$

$k_1 = 2.6 \times 10^{-6}\ M/s\ /\ (0.018\ M)(0.036\ M) = 4.01 \times 10^{-3} = 4.0 \times 10^{-3}\ M^{-1}\,s^{-1}$

$k_2 = 3.9 \times 10^{-6}/(0.027)(0.036) = 4.01 \times 10^{-3} = 4.01 \times 10^{-3} = 4.0 \times 10^{-3}\ M^{-1}\,s^{-1}$

$k_3 = 7.8 \times 10^{-6}/(0.036)(0.054) = 4.01 \times 10^{-3} = 4.01 \times 10^{-3} = 4.0 \times 10^{-3}\ M^{-1}\,s^{-1}$

$k_4 = 1.4 \times 10^{-5}/(0.050)(0.072) = 3.89 \times 10^{-3} = 3.9 \times 10^{-3}\ M^{-1}\,s^{-1}$

$k_{avg} = 3.98 \times 10^{-3} = 4.0 \times 10^{-3}\ M^{-1}\,s^{-1}$

(c) $-\Delta[S_2O_8^{2-}]/\Delta t = -\Delta[I^-]/3\Delta t$; the rate of disappearance of $S_2O_8^{2-}$ is one-third the rate of disappearance of I^-.

(d) Note that the data is given in terms of disappearance of $S_2O_8^{2-}$.

$$\frac{-\Delta[I^-]}{\Delta t} = \frac{-3\Delta[S_2O_8^{2-}]}{\Delta t} = 3(3.98 \times 10^{-3}\ M^{-1}\,s^{-1})(0.015\ M)(0.040\ M) = 7.2 \times 10^{-6}\ M/s$$

Change of Concentration with Time

14.27 (a) $[A]_0$ is the molar concentration of reactant A at time zero, the initial concentration of A. $[A]_t$ is the molar concentration of reactant A at time t. $t_{1/2}$ is the time required to reduce $[A]_0$ by a factor of 2, the time when $[A]_t = [A]_0/2$. k is the rate constant for a particular reaction. k is independent of reactant concentration but varies with reaction temperature.

(b) A graph of ln[A] vs time yields a straight line for a first-order reaction.

14.28 (a) A graph of $1/[A]$ vs time yields a straight line for a second-order reaction.

(b) The half-life of a first-order reaction is independent of $[A]_0$, $t_{1/2} = 0.693/k$. Whereas, the half-life of a second-order reaction does depend on $[A]_0$, $t_{1/2} = 1/k[A]_0$.

14.29 *Analyze/Plan.* The half-life of a first-order reaction depends only on the rate constant, $t_{1/2} = 0.693/k$. Use this relationship to calculate k for a given $t_{1/2}$, and, at a different temperature, $t_{1/2}$ given k. *Solve*:

(a) $t_{1/2} = 2.3 \times 10^5$ s; $t_{1/2} = 0.693/k$, $k = 0.693/t_{1/2}$

$k = 0.693/2.3 \times 10^5 \text{ s} = 3.0 \times 10^{-6} \text{ s}^{-1}$

(b) $k = 2.2 \times 10^{-5} \text{ s}^{-1}$. $t_{1/2} = 0.693/2.2 \times 10^{-5} \text{ s}^{-1} = 3.15 \times 10^4 = 3.2 \times 10^4$ s

14.30 (a) $t_{1/2} = 0.693/k = 0.693/7.0 \times 10^{-4} \text{ s}^{-1} = 990 = 9.9 \times 10^2$ s

(b) $$k = \frac{0.693}{t_{1/2}} = \frac{0.693}{56.3 \text{ min}} \times \frac{1 \text{ min}}{60 \text{ s}} = 2.052 \times 10^{-4} = 2.05 \times 10^{-4} \text{ s}^{-1}$$

14.31 *Analyze/Plan.* Follow the logic in Sample Exercise 14.7. In this reaction, pressure is a measure of concentration. In (a) we are given k, $[A]_0$, t and asked to find $[A]_t$, using Equation 14.13, the integrated form of the first-order rate law. In (b), $[A_t] = 0.1[A_0]$, find t. *Solve*:

(a) $\ln P_t = -kt + \ln P_0$; $P_0 = 375$ torr; t = 65 s

$\ln P_{65} = -4.5 \times 10^{-2} \text{ s}^{-1}(65) + \ln(375) = -2.925 + 5.927 = 3.002$

$P_{65} = 20.12 = 20$ torr

(b) $P_t = 0.10\ P_0$; $\ln(P_t/P_0) = -kt$

$\ln(0.10\ P_0/P_0) = -kt$, $\ln(0.10) = -kt$; $-\ln(0.10)/k = t$

$t = -(-2.303)/4.5 \times 10^{-2} \text{ s}^{-1} = 51.2 = 51$ s

Check. From part (a), the pressure at 65 s is 20 torr, $P_t \sim 0.05\ P_0$. In part (b) we calculate the time where $P_t = 0.10\ P_0$ to be 51 s. This time should be smaller than 65 s, and it is. Data and results in the two parts are consistent.

14.32 (a) Using Equation 14.13 for a first order reaction: $\ln[A]_t = -kt + \ln[A]_0$

2.5 min = 150 s; $[N_2O_5]_0 = (0.0250 \text{ mol}/2.0 \text{ L}) = 0.0125 = 0.013\ M$

$\ln[N_2O_5]_{150} = -(6.82 \times 10^{-3} \text{ s}^{-1})(150 \text{ s}) + \ln(0.0125)$

$\ln[N_2O_5]_{150} = -1.0230 + (-4.3820) = -5.4050 = -5.41$

$[N_2O_5]_{150} = 4.494 \times 10^{-3} = 4.5 \times 10^{-3}\ M$; mol $N_2O_5 = 4.494 \times 10^{-3}\ M \times 2.0$ L

$= 9.0 \times 10^{-3}$ mol

(b) $[N_2O_5]_t = 0.010 \text{ mol}/2.0 \text{ L} = 0.0050\ M$; $[N_2O_5]_0 = 0.0125\ M$

$\ln(0.0050) = -(6.82 \times 10^{-3} \text{ s}^{-1})(t) + \ln(0.0125)$

$$t = \frac{-[\ln(0.0050) - \ln(0.0125)]}{(6.82 \times 10^{-3} \text{s}^{-1})} = 134.35 = 1.3 \times 10^2 \text{ s} \times \frac{1 \text{ min}}{60 \text{ s}} = 2.24 = 2.2 \text{ min}$$

(c) $t_{1/2} = 0.693/k = 0.693/6.82 \times 10^{-3} s^{-1} = 101.6 = 102$ s or 1.69 min

14.33 *Analyze/Plan.* Given reaction order, various values for t, P_t, find the rate constant for the reaction at this temperature. For a first-order reaction, a graph of lnP vs t is linear with as slope of -k. *Solve:*

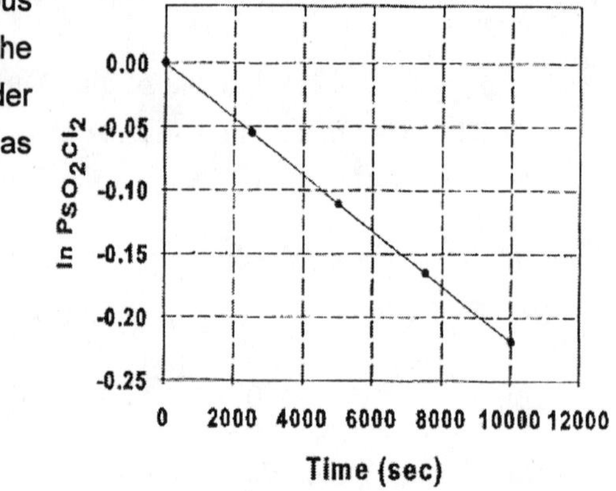

t(s)	$P_{SO_2Cl_2}$	$\ln P_{SO_2Cl_2}$
0	1.000	0
2500	0.947	-0.0545
5000	0.895	-0.111
7500	0.848	-0.165
10000	0.803	-0.219

Graph ln $P_{SO_2Cl_2}$ vs. time. (Pressure is a satisfactory unit for a gas, since the concentration in moles/liter is proportional to P.) The graph is linear with slope $-2.19 \times 10^{-5} s^{-1}$ as shown on the figure. The rate constant $k = -\text{slope} = 2.19 \times 10^{-5} s^{-1}$.

14.34

t(s)	P_{CH_3NC}	$\ln P_{CH_3NC}$
0	502	6.219
2000	335	5.814
5000	180	5.193
8000	95.5	4.559
12000	41.7	3.731
15000	22.4	3.109

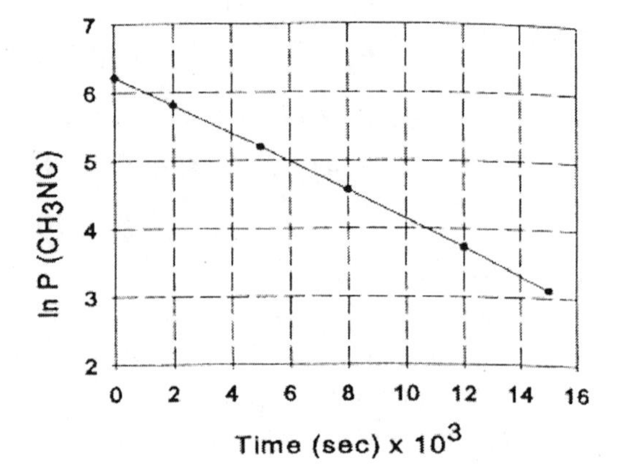

A graph of ln P vs t is linear with a slope of $-2.08 \times 10^{-4} s^{-1}$. The rate constant $k = -\text{slope} = 2.08 \times 10^{-4} s^{-1}$. Half-life = $t_{1/2} = 0.693/k = 3.33 \times 10^3$ s.

14.35 *Analyze/Plan.* Given: mol A, t. Change mol to *M* at various times. Make both first- and second-order plots to see which is linear. *Solve:*

(a)

time(min)	mol A	[A] (*M*)	ln[A]	1/mol A
0	0.065	0.65	-0.43	1.5
10	0.051	0.51	-0.67	2.0
20	0.042	0.42	-0.87	2.4
30	0.036	0.36	-1.02	2.8
40	0.031	0.31	-1.17	3.2

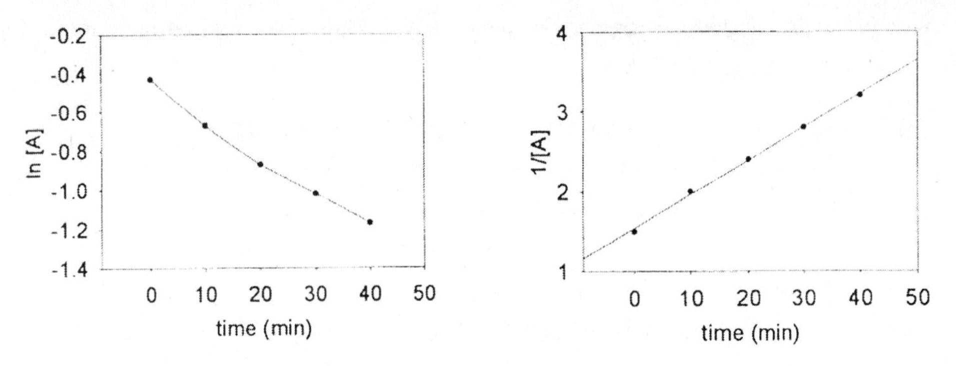

The plot of 1/[A] vs time is linear, so the reaction is second-order in [A].

(b) For a second-order reaction, a plot of 1/[A] vs. t is linear with slope k.
$k = \text{slope} = (3.2 - 2.0)\ M^{-1} / 30\ \text{min} = 0.040\ M^{-1}\ \text{min}^{-1}$
(The best fit to the line yields slope = $0.042\ M^{-1}\text{min}^{-1}$.)

(c) $t_{1/2} = 1/k[A]_0 = 1/(0.040\ M^{-1}\ \text{min}^{-1})(0.65\ M) = 38.46 = 38$ min
(Using the "best-fit" slope, $t_{1/2}$ = 37 min.)

14.36 (a) Make both first- and second-order plots to see which is linear. Moles is a satisfactory concentration unit, since volume is constant.

time(s)	mol A	ln (mol A)	1/mol A
0	0.1000	-2.303	10.00
40	0.067	-2.70	14.9
80	0.045	-3.10	22.2
120	0.030	-3.51	33.3
160	0.020	-3.91	50.0

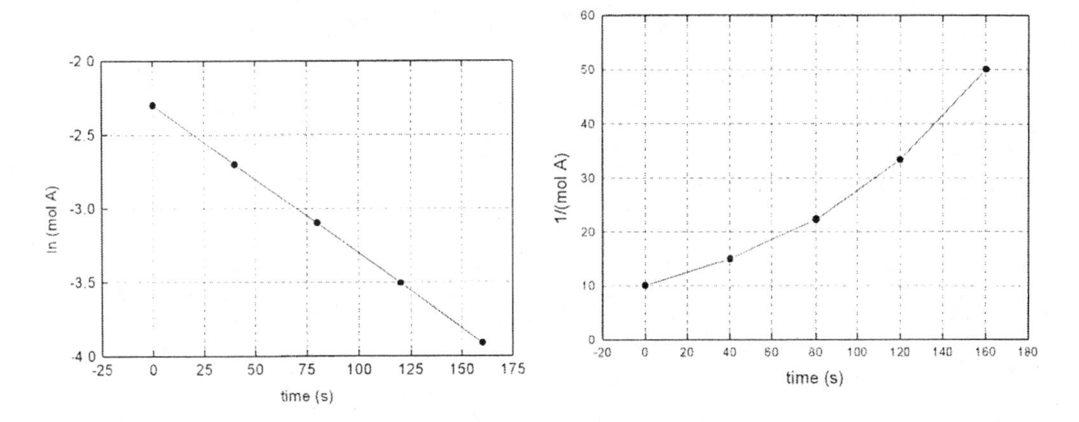

The plot of ln (mol A) vs time is linear, so the reaction is first-order in A.

(b) $k = -\text{slope} = -[-3.91 - (-2.70)]/120 = 0.010083 = 0.0101\ \text{s}^{-1}$
(The best fit to this line yields the same value for the slope, $0.01006 = 0.0101\ \text{s}^{-1}$)

(c) $t_{1/2} = 0.693/k = 0.693/0.010083\ \text{s}^{-1} = 68.7$ s

14.37 Analyze/Plan. Follow the logic in Solution 14.35. Make both first and second order plots to see which is linear. *Solve:*

(a)

time(s)	$[NO_2](M)$	$\ln[NO_2]$	$1/[NO_2]$
0.0	0.100	-2.303	10.0
5.0	0.017	-4.08	59
10.0	0.0090	-4.71	110
15.0	0.0062	-5.08	160
20.0	0.0047	-5.36	210

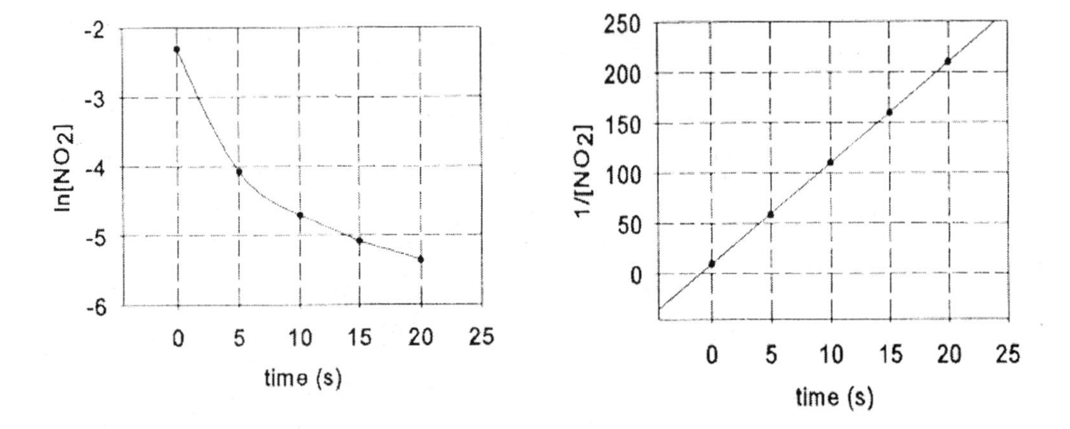

The plot of $1/[NO_2]$ vs time is linear, so the reaction is second order in NO_2.

(b) The slope of the line is (210 - 59) M^{-1} / 15.0 s = 10.07 = 10 $M^{-1}s^{-1}$ = k. (The slope of the best-fit line is 10.02 = 10 $M^{-1}s^{-1}$.)

14.38 (a) Make both first- and second-order plots to see which is linear.

time(min)	$[C_{12}H_{22}O_{11}]$ (M)	$\ln[C_{12}H_{22}O_{11}]$	$1/[C_{12}H_{22}O_{11}]$
0	0.316	-1.152	3.16
39	0.274	-1.295	3.65
80	0.238	-1.435	4.20
140	0.190	-1.661	5.26
210	0.146	-1.924	6.85

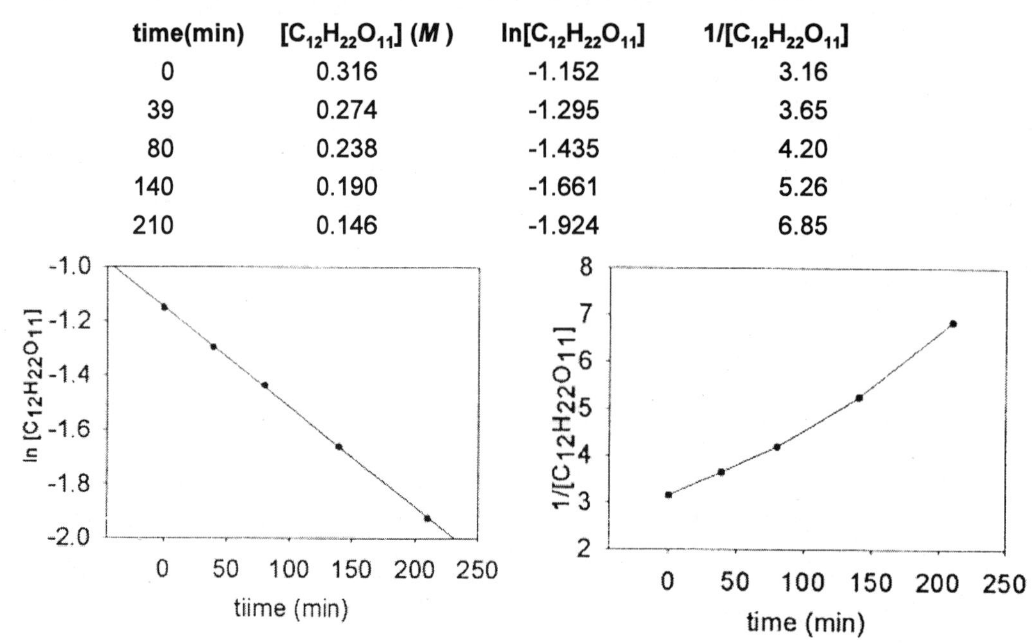

The plot of ln $[C_{12}H_{22}O_{11}]$ in linear, so the reaction is first order in $C_{12}H_{22}O_{11}$.

(b) $k = -\text{slope} = -[-1.924 - (-1.295)] / 171 \text{ min} = 3.68 \times 10^{-3} \text{ min}^{-1}$

(The slope of the best-fit line is $-3.67 \times 10^{-3} \text{ min}^{-1}$.)

Temperature and Rate

14.39 (a) The central idea of the *collision model* is that molecules must collide to react.

(b) The energy of the collision and the orientation of the molecules when they collide determine whether a reaction will occur.

(c) According to the Kinetic Molecular Theory (Chapter 10), the higher the temperature, the greater the speed and kinetic energy of the molecules. Therefore, at a higher temperature, there are more total collisions and each collision is more energetic.

14.40 (a) In order for isomerization of methyl isonitrile to acetonitrile ($CH_3N \equiv C \rightarrow CH_3C \equiv N$) to occur, $CH_3N \equiv C$ molecules must collide with each other. The more $CH_3N \equiv C$ molecules present, the more collisions and the faster the rate. The higher the temperature of the sample, the more collisions and the faster the rate.

(b) No. Not only must collisions of A and B be sufficiently energetic, A and B must collide in the correct orientation for the activated complex to form.

(c) The kinetic molecular theory tells us that at some temperature T, there will be a distribution of molecular speeds and kinetic energies, and that the average kinetic energy of the sample is proportional to temperature. That is, as temperature of the sample increases, the average speed and kinetic energy of the molecules increases. At higher temperatures, there will be more molecular collisions (owing to greater speeds) and more energetic collisions (owing to greater kinetic energies). Overall there will be more collisions that have sufficient energy to form an activated complex, and the reaction rate will be greater.

14.41 *Analyze/Plan.* Given the temperature and energy, use Equation 14.18 to calculate the fraction of Ar atoms that have at least this energy. *Solve*:

$f = e^{-E_a/RT}$ $E_a = 12.5 \text{ kJ/mol} = 1.25 \times 10^4 \text{ J/mol}$; $T = 400 \text{ K } (127°\text{C})$

$$-E_a/RT = -\frac{1.25 \times 10^4 \text{ J/mol}}{400 \text{ K}} \times \frac{\text{mol} \cdot \text{K}}{8.314 \text{ J}} = -3.7587 = -3.76$$

$f = e^{-3.7587} = 2.33 \times 10^{-2}$

At 400 K, approximately 1 out of 43 molecules has this kinetic energy.

14.42 (a) $f = e^{-E_a/RT}$ $E_a = 160 \text{ kJ/mol} = 1.60 \times 10^5 \text{ J/mol}$, $T = 500 \text{ K}$

$$-E_a/RT = -\frac{1.60 \times 10^5 \text{ J/mol}}{500 \text{ K}} \times \frac{\text{mol} \cdot \text{K}}{8.314 \text{ J}} = -38.489 = -38.5$$

$f = e^{-38.489} = 1.924 \times 10^{-17} = 1.92 \times 10^{-17}$

(b) $-E_a/RT = -\frac{1.60 \times 10^5 \text{ J/mol}}{510 \text{ K}} \times \frac{\text{mol} \bullet \text{K}}{8.314 \text{ J}} = -37.735 = -37.7$

$f = e^{-37.735} = 4.093 \times 10^{-17} = 4.09 \times 10^{-17}$

$\frac{\text{f at 510 K}}{\text{f at 500 K}} = \frac{4.09 \times 10^{-17}}{1.92 \times 10^{-17}} = 2.13$

An increase of 10 K means that 2.13 times more molecules have this energy.

14.43 *Analyze/Plan.* Use the definitions of activation energy ($E_{max} - E_{react}$) and ΔE ($E_{prod} - E_{react}$) to sketch the graphs and calculate E_a for the reverse reaction. *Solve:*

(a)

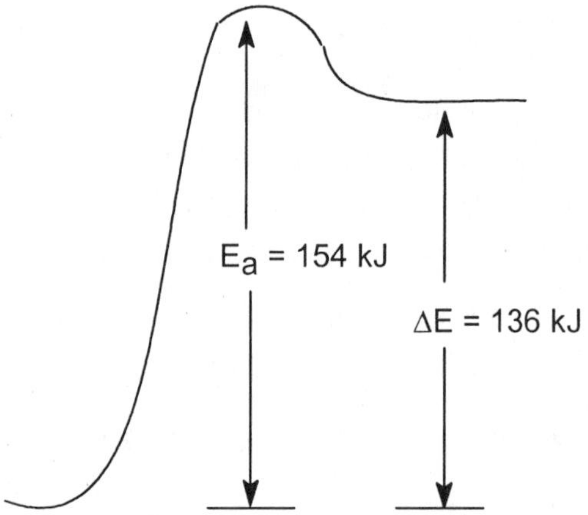

(b) E_a(reverse) = 18 kJ/mol

14.44 (a)

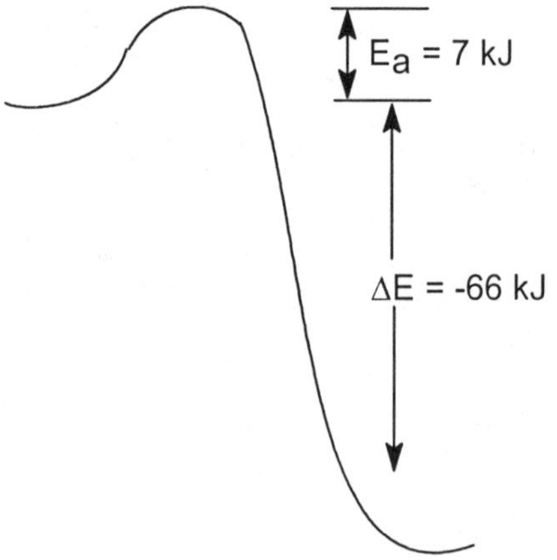

(b) E_a(reverse) = 73 kJ

14.45 Assuming all collision factors (A) to be the same, reaction rate depends only on E_a; it is independent of ΔE. Based on the magnitude of E_a, reaction (b) is fastest and reaction (c) is slowest.

14.46 E_a for the reverse reactions is:

(a) 45 - (-25) = 70 kJ (b) 35 - (-10) = 45 kJ (c) 55 - 10 = 45 kJ

Based on the magnitude of E_a, the reverse of reactions (b) and (c) occur at the same rate, which is faster than the reverse of reaction (a).

14.47 *Analyze/Plan.* Given k_1, at T_1, calculate k_2 at T_2. Change T to Kelvins, then use the Ahrrenius equation [14.21] to calculate k_2. *Solve*:

$T_1 = 20°C + 273 = 293$ K; $T_2 = 60°C + 273 = 333$ K; $k_1 = 2.75 \times 10^{-2} s^{-1}$

(a) $$\ln\left(\frac{k_1}{k_2}\right) = \frac{E_a}{R}\left(\frac{1}{333} - \frac{1}{293}\right) = \frac{75.5 \times 10^3 \text{ J/mol}}{8.314 \text{ J/mol}}(-4.100 \times 10^{-4})$$

$$\ln(k_1/k_2) = -3.7229 = -3.72;\ k_1/k_2 = 0.0242 = 0.024;\ k_2 = \frac{0.0275 \text{ s}^{-1}}{0.0242} = 1.1 \text{ s}^{-1}$$

(b) $$\ln\left(\frac{k_1}{k_2}\right) = \frac{105 \times 10^3 \text{ J/mol}}{8.314 \text{ J/mol}}\left(\frac{1}{333} - \frac{1}{293}\right) = -5.1776 = -5.18$$

$$k_1/k_2 = 5.642 \times 10^{-3} = 5.6 \times 10^{-3};\ k_2 = \frac{0.0275 \text{ s}^{-1}}{5.642 \times 10^{-3}} = 4.9 \text{ s}^{-1}$$

14.48 $T_1 = 30°C + 273 = 303$ K; $T_2 = 60°C + 273 = 333$ K

For reaction A, $E_a = 45.5$ kJ/mol $= 4.55 \times 10^4$ J/mol

$$\ln\left(\frac{k_{30}}{k_{60}}\right) = \frac{E_a}{R}\left[\frac{1}{T_2} - \frac{1}{T_1}\right] = \frac{4.55 \times 10^4 \text{ J/mol}}{8.314 \text{ J/mol}}\left(\frac{1}{333} - \frac{1}{303}\right) = -1.6272 = -1.63$$

$$\left(\frac{k_{A30}}{k_{A60}}\right) = e^{-1.6272} = 0.1965;\ k_{A30} = 0.1965\ k_{A60}$$

For reaction B, $E_a = 25.2$ kJ/mol $= 2.52 \times 10^4$ J/mol

$$\ln\left(\frac{k_{30}}{k_{60}}\right) = \frac{2.52 \times 10^4 \text{ J/mol}}{8.314 \text{ J/mol}}\left(\frac{1}{333} - \frac{1}{303}\right) = -0.9012 = -0.901$$

$$\left(\frac{k_{B30}}{k_{B60}}\right) = e^{-0.9012} = 0.4061;\ k_{B30} = 0.4061\ k_{B60}$$

$k_{A30} = k_{B30}$; $0.1965\ k_{A60} = 0.4061\ k_{B60}$

$k_{A60} / k_{B60} = 0.4061 / 0.1965 = 2.067 = 2.07$

14.49 *Analyze/Plan*. Follow the logic in Sample Exercise 14.11. *Solve*:

k	ln k	T(K)	1/T(× 10^3)
0.0521	-2.955	288	3.47
0.101	-2.293	298	3.36
0.184	-1.693	308	3.25
0.332	-1.103	318	3.14

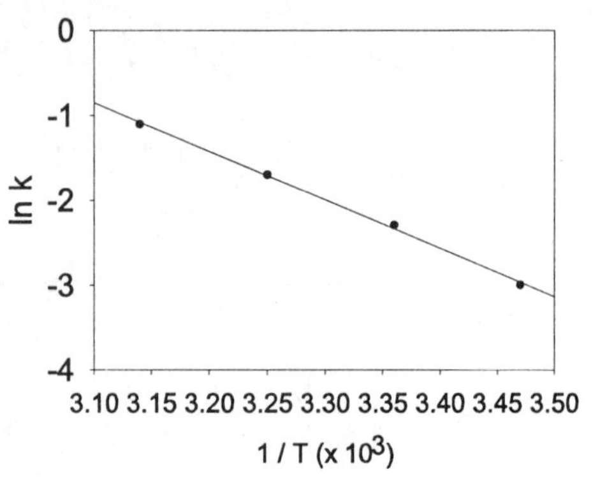

The slope, -5.71×10^3, equals $-E_a/R$. Thus, $E_a = 5.71 \times 10^3 \times 8.314$ J/mol = 47.5 kJ/mol.

14.50

k	ln k	T(K)	1/T(× 10^3)
0.028	-3.58	600	1.67
0.22	-1.51	650	1.54
1.3	0.26	700	1.43
6.0	1.79	750	1.33
23	3.14	800	1.25

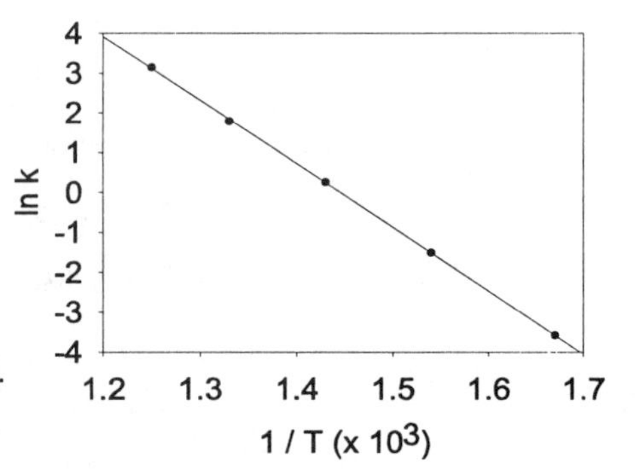

Using the relationship $\ln k = \ln A - E_a/RT$, the slope, $-15.94 \times 10^3 = -16 \times 10^3$, is $-E_a/R$. $E_a = 15.94 \times 10^3 \times 8.314$ J/mol $= 1.3 \times 10^2$ kJ/mol. To calculate A, we will use the rate data at 700 K. From the equation given above, 0.262 = $\ln A - 15.94 \times 10^3/700$; $\ln A = 0.262 + 22.771$. $A = 1.0 \times 10^{10}$.

14.51 *Analyze/Plan*. Given E_a, find the ratio of rates for a reaction at two temperatures. Assuming initial concentrations are the same at the two temperatures, the ratio of rates will be the ratio of rate constants, k_1/k_2. Use Equation [14.21] to calculate this ratio. *Solve*:

$T_1 = 50°C + 273 = 323$ K; $T_2 = 0°C + 273 = 273$ K

$$\ln\left(\frac{k_1}{k_2}\right) = \frac{E_a}{R}\left[\frac{1}{T_2} - \frac{1}{T_1}\right] = \frac{65.7 \text{ kJ/mol}}{8.314 \text{ J/mol}} \times \frac{1000 \text{ J}}{1 \text{ kJ}}\left[\frac{1}{273} - \frac{1}{323}\right]$$

$\ln(k_1/k_2) = 7.902 \times 10^3 (5.670 \times 10^{-4}) = 4.481 = 4.48$; $k_1/k_2 = 88.3 = 88$

The reaction will occur 88 times faster at 50°C, assuming equal initial concentrations.

14.52 (a) $T_1 = 77°F$; °C = 5/9 (°F - 32) = 5/9 (77 - 32) = 25°C = 298 K

$T_2 = 59°F$; °C = 5/9 (59-32) = 15°C = 288 K; $k_1/k_2 = 6$

$$\ln\left(\frac{k_1}{k_2}\right) = \frac{E_a}{R}\left[\frac{1}{T_2} - \frac{1}{T_1}\right];\ \ln(6) = \frac{E_a}{8.314 \text{ J/mol}}\left[\frac{1}{288} - \frac{1}{298}\right]$$

$$E_a = \frac{\ln(6)(8.314 \text{ J/mol})}{1.165 \times 10^{-4}} = 1.28 \times 10^5 \text{ J} = 1.3 \times 10^2 \text{ kJ/mol}$$

T_1 = 77°F = 25°C = 298 K; T_2 = 41°F = 5°C = 278 K, k_1/k_2 = 40

$$\ln(40) = \frac{E_a}{8.314 \text{ J/mol}}\left[\frac{1}{278} - \frac{1}{298}\right];\ E_a = \frac{\ln(40)(8.314 \text{ J/mol})}{2.414 \times 10^{-4}}$$

$E_a = 1.27 \times 10^5$ J = 1.3×10^2 kJ/mol

The values are amazingly consistent, considering the precision of the data.

(b) For a first order reaction, $t_{1/2} = 0.693/k$, $k = 0.693/t_{1/2}$

k_1 at 298 K = 0.693/2.7 yr = 0.257 = 0.26 yr^{-1}

T_1 = 298 K, T_2 = 273 - 15°C = 258 K

$$\ln\left(\frac{0.257}{k_2}\right) = \frac{1.27 \times 10^5 \text{ J}}{8.314 \text{ J/mol}}\left[\frac{1}{258} - \frac{1}{298}\right] = 7.9497$$

$0.257/k_2 = e^{7.9497} = 2.835 \times 10^3$; $k_2 = 0.257/2.835 \times 10^3 = 9.066 \times 10^{-5} = 9.1 \times 10^{-5}\ yr^{-1}$

$t_{1/2} = 0.693/k = 0.693/9.066 \times 10^{-5} = 7.64 \times 10^3$ yr = 7.6×10^3 yr

Reaction Mechanisms

14.53 (a) An *elementary step* is a process that occurs in a single event; the order is given by the coefficients in the balanced equation for the step.

(b) A *unimolecular* elementary step involves only one reactant molecule; the activated complex is derived from a single molecule. A *bimolecular* elementary step involves two reactant molecules in the activated complex and the overall process.

(c) A *reaction mechanism* is a series of elementary steps that describe how an overall reaction occurs and explain the experimentally determined rate law.

14.54 (a) The *molecularity* of a process indicates the number of molecules that participate as reactants in the process. A unimolecular process has one reactant molecule, a bimolecular process has two reactant molecules and a termolecular process has three reactant molecules.

(b) Termolecular processes are rare because it is highly unlikely that three molecules will simultaneously collide with the correct energy and orientation to form an activated complex.

(c) An *intermediate* is a substance that is produced and then consumed during a chemical reaction. It does not appear in the balanced equation for the overall reaction.

14.55 *Analyze/Plan.* Elementary processes occur as a single step, so the molecularity is determined by the number of reactant molecules; the rate law reflects reactant stoichiometry.
Solve:

(a) unimolecular, rate = $k[Cl_2]$

(b) bimolecular, rate = $k[OCl^-][H_2O]$

(c) bimolecular, rate = $k[NO][Cl_2]$

14.56 (a) bimolecular, rate = $k[NO]^2$

(b) unimolecular, rate = $k[C_3H_6]$

(c) unimolecular, rate = $k[SO_3]$

14.57 This is the profile of a two-step mechanism, A → B and B → C. There is one intermediate, B. Because there are two energy maxima, there are two transition states. The B → C step is faster, because its activation energy is smaller. The reaction is exothermic because the energy of the products is lower than the energy of the reactants.

14.58 Two intermediates, three transition states, the C → D step is fastest, the overall reaction is endothermic.

14.59 (a)

$$H_2(g) + ICl(g) \rightarrow HI(g) + HCl(g)$$
$$HI(g) + ICl(g) \rightarrow I_2(g) + HCl(g)$$

$$H_2(g) + 2ICl(g) \rightarrow I_2(g) + 2HCl(g)$$

(b) Intermediates are produced and consumed during reaction. HI is the intermediate.

(c) Follow the logic in Sample Exercise 14.13.
First step: rate = $k[H_2][ICl]$
Second step: rate = $k[HI][ICl]$

(d) The slow step determines the rate law for the overall reaction. If the first step is slow, the observed rate law is: rate = $k[H_2][HCl]$.

14.60 (a)

$$NO(g) + NO(g) \rightarrow N_2O_2(g)$$
$$N_2O_2(g) + H_2(g) \rightarrow N_2O(g) + H_2O(g)$$

$$2NO(g) + N_2O_2(g) + H_2(g) \rightarrow N_2O_2(g) + N_2O(g) + H_2O(g)$$
$$2NO(g) + H_2(g) \rightarrow N_2O(g) + H_2O(g)$$

(b) First step: $-\Delta[NO]/\Delta t = k[NO]\,[NO] = k[NO]^2$
Second step: $-\Delta[H_2]/\Delta t = k[H_2][N_2O_2]$

(c) N_2O_2 is the intermediate; it is produced in the first step and consumed in the second.

(d) Since $[H_2]$ appears in the rate law, the second step must be slow relative to the first.

14.61 *Analyze/Plan.* Given a proposed mechanism and an observed rate law, determine which step is rate determining. *Solve*:

(a) If the first step is slow, the observed rate law is the rate law for this step.
rate = k[NO][Cl_2]

(b) Since the observed rate law is second-order in [NO], the second step must be slow relative to the first step; the second step is rate determining.

14.62 (a)

i. $HBr + O_2 \rightarrow HOOBr$

ii. $HOOBr + HBr \rightarrow 2HOBr$

iii. $2HOBr + 2HBr \rightarrow 2H_2O + 2Br_2$

$4HBr + O_2 \rightarrow 2H_2O + 2Br_2$

(b) The observed rate law is: rate = k[HBr][O_2], the rate law for the first elementary step. The first step must be rate-determining.

(c) HOOBr and HOBr are both intermediates; HOOBr is produced in i and consumed in ii and HOBr is produced in ii and consumed in iii.

(d) Since the first step is rate-determining, it is possible that neither of the intermediates accumulates enough to be detected. This does not disprove the mechanism, but indicates that steps ii and iii are very fast, relative to step i.

Catalysis

14.63 (a) A catalyst increases the rate of reaction by decreasing the activation energy, E_a, or increasing the frequency factor A. Lowering the activation energy is more common and more dramatic.

(b) A *homogeneous catalyst* is in the same phase as the reactants; a *heterogeneous catalyst* is in a different phase and is usually a solid.

14.64 (a) The smaller the particle size of a solid catalyst, the greater the surface area. The greater the surface area, the more active sites and the greater the increase in reaction rate.

(b) *Adsorption* is the binding of reactants onto the surface of the heterogeneous catalyst. It is usually the first step in the catalyzed reaction.

14.65 (a)

$2[NO_2(g) + SO_2(g) \rightarrow NO(g) + SO_3(g)]$

$2NO(g) + O_2(g) \rightarrow 2NO_2(g)$

$2SO_2(g) + O_2(g) \rightarrow 2SO_3(g)$

(b) $NO_2(g)$ is a catalyst because it is consumed and then reproduced in the reaction sequence. (NO(g) is an intermediate because it is produced and then consumed.)

(c) Since NO_2 is in the same state as the other reactants, this is homogeneous catalysis.

14.66 (a)

$$2[NO(g) + N_2O(g) \rightarrow N_2(g) + NO_2(g)]$$
$$2NO_2(g) \rightarrow 2NO(g) + O_2(g)$$

$$2N_2O(g) \rightarrow 2N_2(g) + O_2(g)$$

(b) An intermediate is produced and then consumed during the course of the reaction. A catalyst is consumed and then reproduced. In other words, the catalyst is present when the reaction sequence begins and after the last step is completed. In this reaction, NO is the catalyst and NO_2 is an intermediate.

(c) No. The proposed mechanism cannot be ruled out, based on the behavior of NO_2. NO_2 functions as an intermediate; it is produced and then consumed during the reaction. That there is no measurable build-up of NO_2 indicates the first step is slow relative to the second; as soon as NO_2 is produced by the slow first step, it is consumed by the faster second step.

14.67 Use of chemically stable supports such as alumina and silica makes it possible to obtain very large surface areas per unit mass of the precious metal catalyst. This is so because the metal can be deposited in a very thin, even monomolecular, layer on the surface of the support.

14.68 (a) Catalytic converters are heterogeneous catalysts that adsorb gaseous CO and hydrocarbons and speed up their oxidation to $CO_2(g)$ and $H_2O(g)$. They also adsorb nitrogen oxides, NO_x, and speed up their reduction to $N_2(g)$ and $O_2(g)$. If a catalytic converter is working effectively, the exhaust gas should have very small amounts of the undesirable gases CO, $(NO)_x$ and hydrocarbons.

(b) The high temperatures could increase the rate of the desired catalytic reactions given in part (a). It could also increase the rate of undesirable reactions such as corrosion, which decrease the lifetime of the catalytic converter.

(c) The rate flow of exhaust gases over the converter will determine the rate of adsorption of CO, $(NO)_x$ and hydrocarbons onto the catalyst and thus the rate of conversion to desired products. Too fast an exhaust flow leads to less than maximum adsorption. A very slow flow leads to back pressure and potential damage to the exhaust system. Clearly the flow rate must be adjusted to balance chemical and mechanical efficiency of the catalytic converter.

14.69 As illustrated in Figure 14.21, the two C–H bonds that exist on each carbon of the ethylene molecule before adsorption are retained in the process in which a D atom is added to each C (assuming we use D_2 rather than H_2). To put two deuteriums on a single carbon, it is necessary that one of the already existing C–H bonds in ethylene be broken while the molecule is adsorbed, so the H atom moves off as an adsorbed atom, and is replaced by a D. This requires a larger activation energy than simply adsorbing C_2H_4 and adding one D atom to each carbon.

14.70 Just as the π electrons in C_2H_4 are attracted to the surface of a hydrogenation catalyst, the nonbonding electron density on S causes compounds of S to be attracted to these same surfaces. Strong interactions could cause the sulfur compounds to be permanently attached to the surface, blocking active sites and reducing adsorption of alkenes for hydrogenation.

14.71 (a) Living organisms operate efficiently in a very narrow temperature range; heating to increase reaction rate is not an option. Therefore, the role of enzymes as homogeneous catalysts that speed up desirable reactions without heating and undesirable side-effects is crucial for biological systems.

(b) *catalase*: $2H_2O_2 \rightarrow 2H_2O + O_2$; *nitrogenase*: $N_2 \rightarrow 2NH_3$ (nitrogen fixation)

14.72 The individual structure of each enzyme molecule leads to a unique coiling and folding pattern. The resulting shape and electronic properties of the active site in each enzyme leads to its substrate specificity.

14.73 *Analyze/Plan.* Let k = the rate constant for the uncatalyzed reaction,
k_c = the rate constant for the catalyzed reaction

According to Equation [14.22], $\ln k = -E_a/RT + \ln A$

Subtracting ln k from ln k_c,

$$\ln k_c - \ln k = -\left[\frac{55\ \text{kJ/mol}}{RT} + \ln A\right] - \left[-\frac{95\ \text{kJ/mol}}{RT} + \ln A\right].$$ *Solve:*

(a) $RT = 8.314\ \text{J/K}\bullet\text{mol} \times 298\ \text{K} \times 1\ \text{kJ}/1000\ \text{J} = 2.478\ \text{kJ/mol}$; ln A is the same for both reactions.

$$\ln(k_c/k) = \frac{95\ \text{kJ/mol} - 55\ \text{kJ/mol}}{2.478\ \text{kJ/mol}};\quad k_c/k = 1.0 \times 10^7$$

The catalyzed reaction is approximately 10,000,000 (ten million) times faster at 25°C.

(b) $RT = 8.314\ \text{J/K}\bullet\text{mol} \times 398\ \text{K} \times 1\ \text{kJ}/1000\ \text{J} = 3.309\ \text{kJ/mol}$

$$\ln(k_c/k) = \frac{40\ \text{kJ/mol}}{3.309\ \text{kJ/mol}};\quad k_c/k = 1.8 \times 10^5$$

The catalyzed reaction is 180,000 times faster at 125°C.

14.74 Let k and E_a equal the rate constant and activation energy for the uncatalyzed reaction. Let k_c and E_{ac} equal the rate constant and activation energy of the catalyzed reaction. A is the same for the uncatalyzed and catalyzed reactions. $k_c/k = 1 \times 10^5$, T = 37°C = 310 K.

According to Equation [14.20], $\ln k = E_a/RT + \ln A$. Subtracting ln k from ln k_c

$$\ln k_c - \ln k = \left[\frac{-E_{ac}}{RT}\right] + \ln A - \left[\frac{-E_a}{RT}\right] - \ln A$$

$$\ln(k_c/k) = \frac{E_a - E_{ac}}{RT}; \quad E_a - E_{ac} = RT \ln(k_c/k)$$

$$E_a - E_{ac} = \frac{8.314 \text{ J}}{\text{K} \bullet \text{mol}} \times 310 \text{ K} \times \ln(1 \times 10^5) = 2.966 \times 10^4 \text{ J} = 29.7 \text{ kJ}$$

The enzyme must lower the activation energy by 29.7 kJ in order to achieve a 1×10^5 -fold increase in reaction rate.

Additional Exercises

14.75 $\text{rate} = \frac{-\Delta[H_2S]}{\Delta t} = \frac{\Delta[Cl^-]}{2\Delta t} = k[H_2S][Cl_2]$

$$\frac{-\Delta[H_2S]}{\Delta t} = (3.5 \times 10^{-2}\ M^{-1}s^{-1})(1.6 \times 10^{-4}\ M)\ (0.070\ M) = 3.92 \times 10^{-7} = 3.9 \times 10^{-7}\ M/s$$

$$\frac{\Delta[Cl^-]}{\Delta t} = \frac{2\Delta[H_2S]}{\Delta t} = 2(3.92 \times 10^{-7}\ M/s) = 7.8 \times 10^{-7}\ M/s$$

14.76 (a) $\text{rate} = \frac{-\Delta[NO]}{2\Delta t} = \frac{-\Delta[O_2]}{\Delta t} = \frac{9.3 \times 10^{-5}\ M/s}{2} = 4.7 \times 10^{-5}\ M/s$

(b,c) $\text{rate} = k[NO]^2[O_2]$; $k = \text{rate}/[NO]^2[O_2]$

$$k = \frac{9.3 \times 10^{-5}\ M/s}{(0.040\ M)^2(0.035\ M)} = 1.7\ M^{-2}s^{-1}$$

(d) Since the reaction is second order in NO, if the [NO] is increased by a factor of 1.8, the rate would increase by a factor of 1.8^2, or (3.24) = 3.2.

14.77 (a) $\text{rate} = k[I^-][OCl^-]/[OH^-]$

(b) Since the reaction is first order in $[I^-]$, tripling $[I^-]$ triples the rate.

(c) Since the rate is inversely proportional to $[OH^-]$, doubling $[OH^-]$ cuts the rate in half.

14.78 (a) The rate increases by a factor of nine when $[C_2O_4^{2-}]$ triples (compare experiments 1 and 2). The rate doubles when $[HgCl_2]$ doubles (compare experiments 2 and 3). The rate law is apparently: $\text{rate} = k[HgCl_2][C_2O_4^{2-}]^2$

(b) $k = \frac{\text{rate}}{[HgCl_2][C_2O_4^{2-}]^2}$ Using the data for Experiment 1,

$$k = \frac{(3.2 \times 10^{-5}\ M/s)}{[0.164\ M][0.15\ M]^2} = 8.672 \times 10^{-3} = 8.7 \times 10^{-3}\ M^{-2}\ s^{-1}$$

(c) $\text{rate} = (8.672 \times 10^{-3}\ M^{-2}s^{-1})(0.12\ M)(0.10\ M)^2 = 1.0 \times 10^{-5}\ M/s$

14.79 The units of rate are M/s. The reaction must be second order overall if the units of the rate constant are M^{-1} s^{-1}. If rate = k$[NO_2]^x$, then the cumulative units of $[NO_2]^x$ must be M^2, and x = 2.

If $[NO_2]_0$ = 0.100 M and $[NO_2]_t$ = 0.025 M, use the integrated form of the second order rate equation, $\frac{1}{[A]_t} = kt + \frac{1}{[A]_0}$, Equation 14.14, to solve for t.

$$\frac{1}{0.025\,M} = 0.63\ M^{-1}\ s^{-1}\ (t) + \frac{1}{0.100\,M};\quad \frac{(40 - 10)\,M^{-1}}{0.63\,M^{-1}\,s^{-1}} = t = 47.62 = 48\ s.$$

14.80 (a) $k = (8.56 \times 10^{-5}\ M/s)/(0.200\ M) = 4.28 \times 10^{-4}\ s^{-1}$

(b) $\ln\ [urea] = -(4.28 \times 10^{-4} s^{-1} \times 5.00 \times 10^3\ s) + \ln\ (0.500)$

$\ln\ [urea] = -2.14 - 0.693 = -2.833 = -2.83$; $[urea] = 0.0588 = 0.059\ M$

(c) $t_{1/2} = 0.693/k = 0.693/4.28 \times 10^{-4}\ s^{-1} = 1.62 \times 10^3\ s$

14.81 (a) A = abc, Equation 14.5. A = 0.605, a = $5.60 \times 10^3\ cm^{-1}\ M^{-1}$, b = 1.00 cm

$$c = \frac{A}{ab} = \frac{0.605}{(5.60 \times 10^3\ cm^{-1}\ M^{-1})(1.00\ cm)} = 1.080 \times 10^{-4} = 1.08 \times 10^{-4}\ M$$

(b) Calculate $[c]_t$ using Beer's law. We calculated $[c]_0$ in part (a). Use Equation [14.13] to calculate k.

$$A_{30} = abc_{30};\ c_{30} = \frac{A_{30}}{ab} = \frac{0.250}{(5.60 \times 10^3\ cm^{-1}\ M^{-1})(1.00\ cm)} = 4.464 \times 10^{-5}\ M$$

$$\ln[c]_t = -kt + \ln[c]_0;\quad \frac{\ln[c]_0 - \ln[c]_t}{t} = k;\ t = 30\ min \times \frac{60\ s}{min} = 1800\ s$$

$k = \ln(1.080 \times 10^{-4}) - \ln(4.464 \times 10^{-5})\,/\,1800\ s = 4.910 \times 10^{-4} = 4.91 \times 10^{-4}\ s^{-1}$

(c) For a first order reaction, $t_{1/2} = 0.693/k$.

$t_{1/2} = 0.693/4.910 \times 10^{-4}\ s^{-1} = 1.411 \times 10^3 = 1.41 \times 10^3\ s = 23.5\ min$

(d) A_t = 0.100; calculate c_t using Beer's law, then t from the first order integrated rate equation.

$$c_t = \frac{A}{ab} = \frac{0.100}{(5.60 \times 10^3\ cm^{-1}\ M^{-1})(1.00\ cm)} = 1.786 \times 10^{-5} = 1.79 \times 10^{-5}\ M$$

$$t = \frac{\ln[c]_0 - \ln[c]_t}{k} = \frac{\ln(1.080 \times 10^{-4}) - \ln(1.786 \times 10^{-5})}{4.910 \times 10^{-4}\ s^{-1}}$$

$t = 3.666 \times 10^3 = 3.67 \times 10^3\ s = 61.1\ min$

14.82

Time (s)	$[C_5H_6]$ (*M*)	$\ln[C_5H_6]$	$1/[C_5H_6]$
0	0.0400	-3.219	25.0
50	0.0300	-3.507	33.3
100	0.0240	-3.730	41.7
150	0.0200	-3.912	50.0
200	0.0174	-4.051	57.5

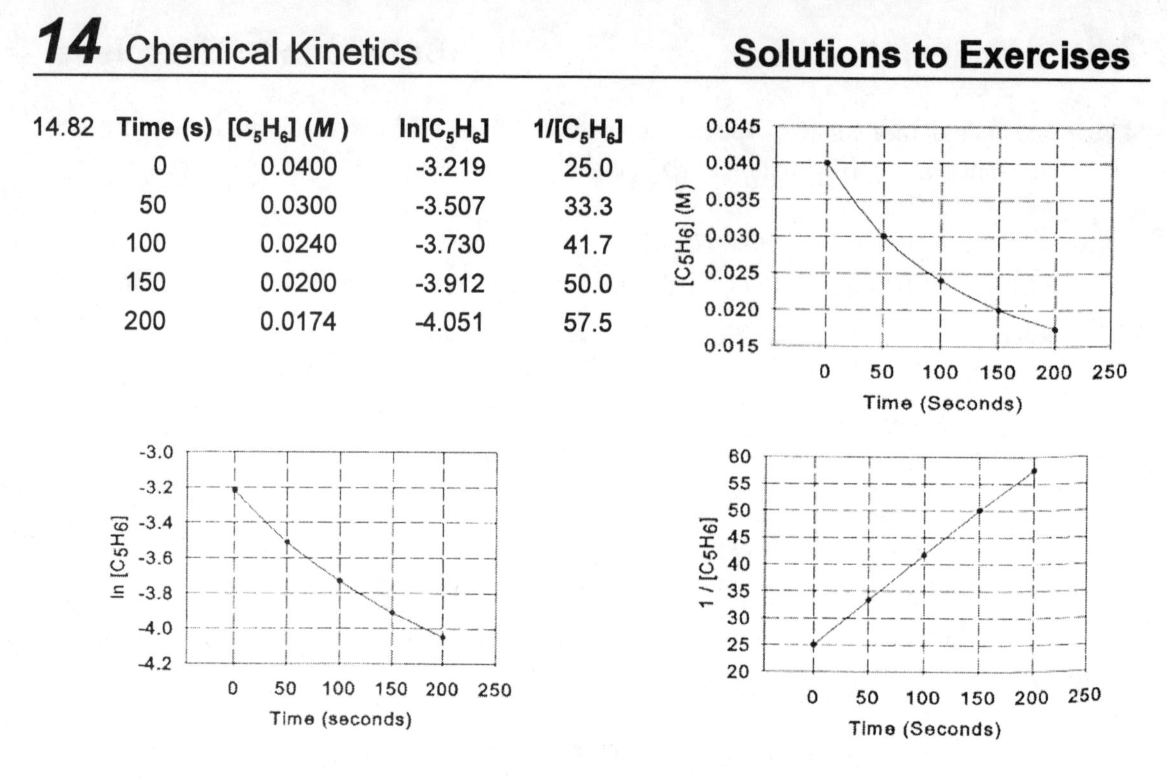

The plot of $1/[C_5H_6]$ vs time is linear and the reaction is second order.

The slope of this line is k. k = slope = $(50.0 - 25.0)\ M^{-1} / (150-0)\text{s} = 0.167\ M^{-1}\ \text{s}^{-1}$

(The best-fit slope and k value is $0.163\ M^{-1}\ \text{s}^{-1}$.)

14.83 (a) No. The value of A, which is related to frequency and effectiveness of collisions, can be different for each reaction and k is proportional to A.

(b) From Equation 14.21, reactions with different variations of k with respect to temperature have different activation energies, E_a. The fact that k for the two reactions is the same at a certain temperature is coincidental. The reaction with the higher rate at 35°C has the larger activation energy, because it was able to use the increase in energy more effectively.

14.84

ln k	1/T
-24.17	3.33×10^{-3}
-20.72	3.13×10^{-3}
-17.32	2.94×10^{-3}
-15.24	2.82×10^{-3}

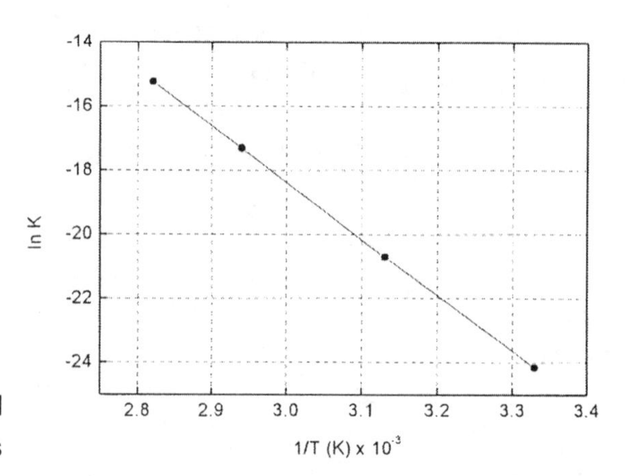

The calculated slope is -1.751×10^4. The activation energy E_a, equals $-(\text{slope}) \times (8.314\ \text{J/mol})$. Thus, $E_a = 1.8 \times 10^4 (8.314) = 1.5 \times 10^5$ J/mol $= 1.5 \times 10^2$ kJ/mol. (The best-fit slope is $-1.76 \times 10^4 = -1.8 \times 10^4$ and the value of E_a is 1.5×10^2 kJ/mol.)

14.85 (a) rate = $k[H_2O_2][I^-]$

(b) $2H_2O_2(aq) \rightarrow 2H_2O(l) + O_2(g)$

(c) $IO^-(aq)$ is the intermediate.

14.86 (a)

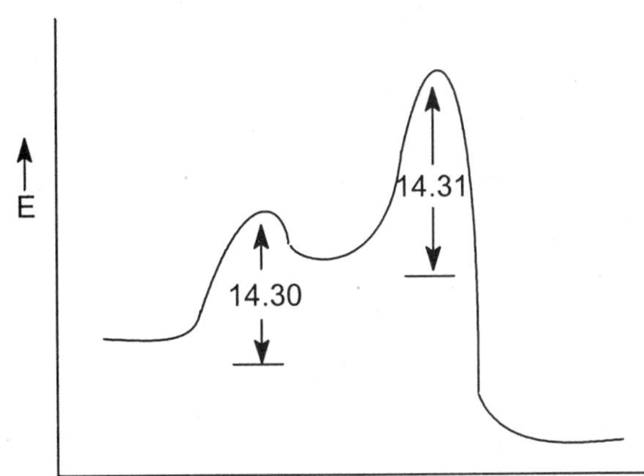

14.30 = E_a for reaction [14.30]

14.31 = E_a for reaction [14.31]

(b) The fact that Br_2 builds up during the reaction tells us that the appearance of Br_2 (reaction [14.30]) is faster than the disappearance of Br_2 (reaction [14.31]). This is the reason that E_a of [14.31] in the energy profile above is larger than E_a of [14.30].

14.87 (a)

$$Cl_2(g) \rightleftharpoons 2Cl(g)$$
$$Cl(g) + CHCl_3(g) \rightarrow HCl(g) + CCl_3(g)$$
$$Cl(g) + CCl_3(g) \rightarrow CCl_4(g)$$

$$Cl_2(g) + 2Cl(g) + CHCl_3(g) + CCl_3(g) \rightarrow 2Cl(g) + HCl(g) + CCl_3(g) + CCl_4(g)$$
$$Cl_2(g) + CHCl_3(g) \rightarrow HCl(g) + CCl_4(g)$$

(b) $Cl(g)$, $CCl_3(g)$

(c) Step 1 - unimolecular, Step 2 - bimolecular, Step 3 - bimolecular

(d) Step 2, the slow step, is rate determining.

(e) If Step 2 is rate determining, rate = $k_2[CHCl_3][Cl]$. Cl is an intermediate formed in Step 1, an equilibrium. By definition, the rates of the forward and reverse processes are equal; $k_1[Cl_2] = k_{-1}[Cl]^2$. Solving for [Cl] in terms of $[Cl_2]$,

$$[Cl]^2 = \frac{k_1}{k_{-1}}[Cl_2]; \quad [Cl] = \left(\frac{k_1}{k_{-1}}[Cl_2]\right)^{1/2}$$

Substituting into the overall rate law

$$\text{rate} = k_2\left(\frac{k_1}{k_{-1}}\right)^{1/2}[CHCl_3][Cl_2]^{1/2} = k[CHCl_3][Cl_2]^{1/2} \text{ (The overall order is 3/2.)}$$

14.88 (a) $(CH_3)_3AuPH_3 \rightarrow C_2H_6 + (CH_3)AuPH_3$

(b) $(CH_3)_3Au$, $(CH_3)Au$ and PH_3 are intermediates.

(c) Step 1 is unimolecular, Step 2 is unimolecular, Step 3 is bimolecular.

(d) Step 2, the slow step, is rate determining.

(e) If Step 2 is rate determining, rate = $k_2[(CH_3)_3Au]$.

$(CH_3)_3Au$ is an intermediate formed in Step 1, an equilibrium. By definition, the rates of the forward and reverse processes in Step 1 are equal:

$k_1[(CH_3)_3AuPH_3] = k_{-1}[(CH_3)_3Au][PH_3]$; solving for $[(CH_3)_3Au]$,

$$[(CH_3)_3Au] = \frac{k_1[(CH_3)_3AuPH_3]}{k_{-1}[PH_3]}$$

Substituting into the rate law

$$\text{rate} = \left(\frac{k_2k_1}{k_{-1}}\right)\frac{[(CH_3)_3AuPH_3]}{[PH_3]} = \frac{k[(CH_3)_3AuPH_3]}{[PH_3]}$$

(f) The rate is inversely proportional to $[PH_3]$, so adding PH_3 to the $(CH_3)_3AuPH_3$ solution would decrease the rate of the reaction.

14.89 Enzyme: carbonic anhydrase; substrate: carbonic acid (H_2CO_3); turnover number: 1×10^7 molecules/s.

14.90 (a) The fact that the rate doubles with a doubling of the concentration of sugar tells us that the fraction of enzyme tied up in the form of an enzyme-substrate complex is small. A doubling of the substrate concentration leads to a doubling of the concentration of enzyme-substrate complex, because most of the enzyme molecules are available to bind substrates.

(b) The behavior of inositol suggests that it acts as a competitor with sucrose for binding at the active sites of the enzyme system. Such a competition results in a lower effective concentration of active sites for binding of sucrose, and thus results in a lower reaction rate.

Integrative Exercises

14.91 *Analyze/Plan.* $2N_2O_5 \rightarrow 4NO_2 + O_2$ rate = $k[N_2O_5] = 1.0 \times 10^{-5}\ s^{-1}[N_2O_5]$
Use the integrated rate law for a first-order reaction, Equation 14.13, to calculate $k[N_2O_5]$ at 20.0 hr. Build a stoichiometry table to determine mol O_2 produced in 20.0 hr. Assuming that $O_2(g)$ is insoluble in chloroform, calculate the pressure of O_2 in the 10.0 L container. *Solve:*

$$20.0\ \text{hr} \times \frac{60\ \text{min}}{1\ \text{hr}} \times \frac{60\ \text{s}}{1\ \text{min}} = 7.20 \times 10^4\ \text{s};\ [N_2O_5]_0 = 0.600\ M$$

$\ln[A]_t - \ln[A]_0 = -kt$; $\ln[N_2O_5]_t = -kt + \ln[N_2O_5]_0$

$\ln [N_2O_5]_t = -1.0 \times 10^{-5}\ s^{-1} (7.20 \times 10^4\ s) + \ln(0.600) = -0.720 - 0.511 = -1.231$

$[N_2O_5]_t = e^{-1.231} = 0.292\ M$

N_2O_5 was present initially as 1.00 L of 0.600 *M* solution.

mol N_2O_5 = *M* × L = 0.600 mol N_2O_5 initial, 0.292 mol N_2O_5 at 20.0 hr

	$2N_2O_5$ →	$4NO_2$ +	O_2
t = 0	0.600 mol	0	0
change	-0.308 mol	0.616 mol	0.154 mol
t = 20 hr	0.292 mol	0.616 mol	0.154 mol

[Note that the reaction stoichiometry is applied to the 'change' line.]

PV = nRT; P = nRT/V; V = 10.0 L, T = 45°C = 318 K, n = 0.154 mol

$$P = 0.154\ \text{mol} \times \frac{318\ \text{K}}{10.0\ \text{L}} \times \frac{0.08206\ \text{L}\bullet\text{atm}}{\text{mol}\bullet\text{K}} = 0.402\ \text{atm}$$

14.92 (a) $\ln k = -E_a/RT + \ln A$; E_a = 86.8 kJ/mol = 8.68×10^4 J/mol;

T = 35°C + 273 = 308 K; A = $2.10 \times 10^{11}\ M^{-1}\ s^{-1}$

$$\ln k = \frac{-8.68 \times 10^4\ \text{J/mol}}{308\ \text{K}} \times \frac{\text{mol}\bullet\text{K}}{8.314\ \text{J}} + \ln(2.10 \times 10^{11}\ M^{-1}\ s^{-1})$$

$\ln k = -33.8968 + 26.0704 = -7.8264$; $k = 3.99 \times 10^{-4}\ M^{-1}\ s^{-1}$

(b)

$$\frac{0.335\ \text{g KOH}}{0.250\ \text{L soln}} \times \frac{1\ \text{mol KOH}}{56.1\ \text{g KOH}} = 0.02389 = 0.0239\ M\ \text{KOH}$$

$$\frac{1.453\ \text{g}\ C_2H_5I}{0.250\ \text{L soln}} \times \frac{1\ \text{mol}\ C_2H_5I}{156.0\ \text{g}\ C_2H_5I} = 0.03726 = 0.0373\ M\ C_2H_5I$$

If equal volumes of the two solutions are mixed, the initial concentrations in the reaction mixture are 0.01194 *M* KOH and 0.01863 M C_2H_5I. Assuming the reaction is first order in each reactant:

rate = $k[C_2H_5I][OH^-] = 3.99 \times 10^{-4}\ M^{-1}\ s^{-1} (0.01194\ M)(0.01863\ M) = 8.88 \times 10^{-8}\ M/s$

(c) Since C_2H_5I and OH^- react in a 1 : 1 mole ratio and equal volumes of the solutions are mixed, the reactant with the smaller concentration, KOH, is the limiting reactant.

14.93 (a) Use an apparatus such as the one pictured in Figure 10.3 (c) (an open-end manometer), a clock, a rule and a constant temperature bath. Since P = (n/V)RT, $\Delta P/\Delta t$ at constant temperature is an acceptable measure of reaction rate.

Load the flask with HCl(aq) and read the height of the Hg in both arms of the manometer. Quickly add Zn(s) to the flask and record time = 0 when the Zn(s) contacts the acid. Record the height of the Hg in one arm of the manometer at convenient time intervals such as 5 sec. (The decrease in the short arm will be the same as the increase in the tall arm). Calculate the pressure of $H_2(g)$ at each time.

(b) Keep the amount of Zn(s) constant and vary the concentration of HCl(aq) to determine the reaction order for H^+ and Cl^-. Keep the concentration of HCl(aq) constant and vary the amount of Zn(s) to determine the order for Zn(s). Combine this information to write the rate law.

(c) $-\Delta[H^+]/2\Delta t = \Delta[H_2]/\Delta t$; $-\Delta[H^+]/\Delta t = 2\Delta[H_2]/\Delta t$

$[H_2]$ = mol H_2/L H_2 = n/V; $[H_2]$ = P (in atm)/RT

Then, the rate of disappearance of H^+ is twice the rate of appearance of $H_2(g)$.

(d) By changing the temperature of the constant temperature bath, measure the rate data at several (at least three) temperatures and calculate the rate constant k at these temperatures. Plot ln k vs 1/T. The slope of the line is $-E_a/R$ and E_a = -slope (R).

(e) Measure rate data at constant temperature, HCl concentration and mass of Zn(s), varying only the form of the Zn(s). Compare the rate of reaction for metal strips and granules.

14.94 (a) $\ln k = -E_a/RT + \ln A$, Equation 14.20. E_a = 6.3 kJ/mol = 6.3×10^3 J/mol

T = 100°C + 273 = 373 K

$$\ln k = \frac{-6.3 \times 10^3 \text{ J/mol}}{8.314 \text{ J/K} \cdot \text{mol} \times 373 \text{ K}} + \ln(6.0 \times 10^8\ M^{-1}\ s^{-1})$$

$\ln k = -2.032 + 20.212 = 18.181 = 18.2$; $k = 7.87 \times 10^7 = 8 \times 10^7\ M^{-1}\ s^{-1}$

(b) NO, 11 valence e^-, 5.5 e^- pair (Assume the less electronegative N atom will be electron deficient.)

:N̊=Ö:

ONF, 18 valence e^-, 9 e^- pr

:Ö=N̈—F̤: ⟷ (:Ö—N̈=F̤)

The resonance form on the right is a very minor contributor to the true bonding picture, due to high formal charges and the unlikely double bond involving F.

(c) ONF has trigonal planar electron domain geometry, which leads to a "bent" structure with a bond angle of approximately 120°.

(d) [O=N···F—F]

(e) The electron deficient NO molecule is attracted to electron-rich F_2, so the driving force for formation of the transition state is greater than simple random collisions.

14.95 (a) $\Delta H^{\circ}_{rxn} = 2\Delta H^{\circ}_{f}\ H_2O(g) + 2\Delta H^{\circ}_{f}\ Br_2(g) - 4\Delta H^{\circ}_{f}\ HBr(g) - \Delta H^{\circ}_{f}\ O_2(g)$

$\Delta H^{\circ}_{rxn} = 2(-241.82) + 2(30.71) - 4(-36.23) - (0) = -277.30$ kJ

(b) Since the rate of the uncatalyzed reaction is very slow at room temperature, the magnitude of the activation energy for the rate-determining first step must be quite large. At room temperature, the reactant molecules have a distribution of kinetic energies (Chapter 10), but very few molecules even at the high end of the distribution have sufficient energy to form an activated complex. E_a for this step must be much greater than 3/2 RT, the average kinetic energy of the sample.

(c) 20 e^-, 10 e^- pr

H—Ö—Ö—Br:

The intermediate resembles hydrogen peroxide, H_2O_2.

14.96 In the lock and key model of enzyme action, the active site is the specific location in the enzyme where reaction takes place. The precise geometry (size and shape) of the active site both accommodates and activates the substrate (reactant). Proteins are large biopolymers, with the same structural flexibility as synthetic polymers (Chapter 12). The three-dimensional shape of the protein in solution, including the geometry of the active site, is determined by many intermolecular forces of varying strengths.

Changes in temperature change the kinetic energy of the various groups on the enzyme and their tendency to form intermolecular associations or break free from them. Thus, changing the temperature changes the overall shape of the protein and specifically the shape of the active site. At the operating temperature of the enzyme, the competition between kinetic energy driving groups apart and intermolecular attraction pulling them together forms an active site that is optimum for a specific substrate. At temperatures above the temperature of maximum activity, sufficient kinetic energy has been imparted so that the forces driving groups apart win the competition, and the three-dimensional structure of the enzyme is destroyed. This is the process of *denaturation*. The activity of the enzyme is destroyed because the active site has collapsed. The protein or enzyme is denatured, because it is no longer capable of its "natural" activity.

14.97 (a) If the reaction proceeds in a single elementary step, the coefficients in the balanced equation are the reaction orders for the respective reactants.

rate $= k[Ce^{4+}]^2\ [Tl^+]$

(b) If the uncatalyzed reaction occurs in a single step, it is *termolecular*. The activated complex requires collision of three particles with the correct energy and orientation for reaction. The probability of an effective three-particle collision is low and the rate is slow.

(c) The first step is rate-determining.

(d) The ability of Mn to adopt every oxidation state from +2 to +7 makes it especially suitable to catalyze this (and many other) reactions.

14.98 (a) D(Cl–Cl) = 242 kJ/mol Cl_2

$$\frac{242 \text{ kJ}}{\text{mol } Cl_2} \times \frac{1000 \text{ J}}{\text{kJ}} \times \frac{1 \text{ mol}}{6.022 \times 10^{23} \text{ molecules}} = 4.019 \times 10^{-19} = 4.02 \times 10^{-19} \text{ J}$$

$$\lambda = hc/E = \frac{6.626 \times 10^{-34} \text{ J}\bullet\text{s} \times 2.998 \times 10^{8} \text{ m/s}}{4.019 \times 10^{-19} \text{ J}} = 4.94 \times 10^{-7} \text{ m}$$

This wavelength, 494 nm, is in the visible portion of the spectrum.

(b)

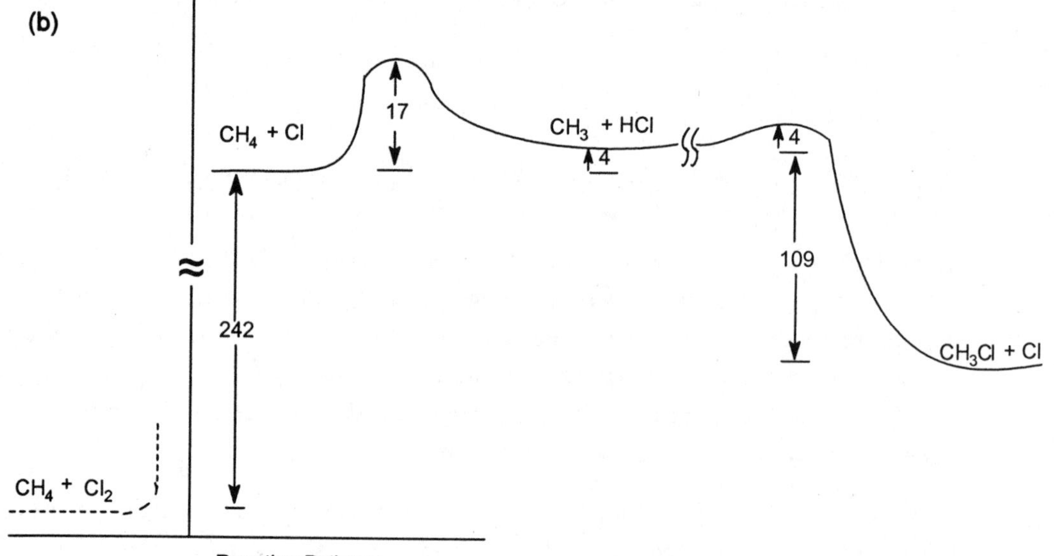

(c) Since D(Cl-Cl) is 242 kJ/mol, $CH_4(g) + Cl_2(g)$ should be about 242 kJ below the starting point on the diagram. For the reaction $CH_4(g) + Cl_2(g) \rightarrow CH_3(g) + HCl(g) + Cl(g)$, E_a is 242 + 17 = 259 kJ. (From bond dissociation enthalpies, ΔH for the overall reaction $CH_4(g) + Cl_2(g) \rightarrow CH_3Cl(g) + Cl(g)$ is -104 kJ, so the graph above is simply a sketch of the relative energies of some of the steps in the process.)

(d) CH_3, 7 valence e^-, odd electron species

```
     •
  H—C—H
     |
     H
```

(e) This sequence is called a chain reaction because Cl• radicals are regenerated in Reaction 4, perpetuating the reaction. Absence of Cl• terminates the reaction, so Cl•+Cl• → Cl_2 is a termination step.

15 Chemical Equilibrium

The Concept of Equilibrium; Equilibrium - Constant Expressions

15.1 Yes. The first box is pure reactant A. As the reaction proceeds, some A changes to B. In the fourth and fifth boxes, the relative amounts (concentrations) of A and B are constant. Although the reaction is ongoing the rates of A → B and B → A are equal, and the relative amounts of A and B are constant.

15.2 (a) At equilibrium the forward and reverse reactions proceed at equal rates. Reactants are continually transformed into products, but products are also transformed into reactants at the same rate, so the net concentrations of reactants and products are constant at equilibrium.

(b) At equilibrium, it is the forward and reverse rates, not the rate constants, that are equal. It is true that the ratio of the two rate constants is constant at equilibrium, but this ratio is not constrained to have a value of 1.

(c) At equilibrium, the net concentrations of reactants and products are *constant* (see part (a)), but not necessarily equal. The relative concentrations of reactants and products at equilibrium are determined by their initial concentrations and the value of the equilibrium constant.

15.3 *Analyze/Plan.* Given the forward and reverse rate constants, calculate the equilibrium constant using Equation [15.5]. At equilibrium, the rates of the forward and reverse reactions are equal. Write the rate laws for the forward and reverse reactions and use their equality to answer part (b). *Solve*:

(a) $K_{eq} = \frac{k_f}{k_r}$, Equation [15.5]; $K_{eq} = \frac{4.2 \times 10^{-3}\ s^{-1}}{1.5 \times 10^{-1}\ s^{-1}} = 2.8 \times 10^{-2}$

(b) $rate_f = rate_r$; $k_f[A] = k_r[B]$

Since $k_f < k_r$, in order for the two rates to be equal, [A] must be greater than [B].

15.4 (a) $K_{eq} = \frac{[C][D]}{[A][B]}$; if K_{eq} is large, the numerator of the K_{eq} expression is much greater than the denominator and products will predominate at equilibrium.

(b) $K_{eq} = k_f/k_r$; if K_{eq} is large, k_f is larger than k_r and the forward reaction has the greater rate constant.

15.5 (a) The *law of mass action* expresses the relationship between the concentrations of reactants and products at equilibrium for any reaction. The law of mass action is a generic equilibrium expression.

$$K_{eq} = \frac{P_{NOBr_2}}{P_{NO} \times P_{Br_2}}$$

(b) The *equilibrium-constant expression* is an algebraic equation where the variables are the equilibrium concentrations of the reactants and products for a specific chemical reaction. The *equilibrium constant* is a number; it is the ratio calculated from the equilibrium expression for a particular chemical reaction. For any reaction, there are an infinite number of sets of equilibrium concentrations, depending on initial concentrations, but there is only one equilibrium constant.

(c) Introduce a known quantity of $NOBr_2(g)$ into a vessel of known volume at constant (known) temperature. After equilibrium has been established, measure the total pressure in the flask. Using an equilibrium table, such as the one in Sample Exercise 15.8, calculate equilibrium pressures and concentrations of $NO(g)$, $Br_2(g)$ and $NOBr_2(g)$ and calculate K_{eq}.

15.6 (a) Yes. The algebraic form of the law of mass action depends only on the coefficients of a chemical equation, not on the reaction mechanism.

(b) $N_2(g) + 3H_2(g) \rightleftharpoons 2NH_3(g)$. The *Haber process* is the primary industrial method of nitrogen fixation, that is, of converting $N_2(g)$ into usable forms. The major use of $NH_3(g)$ from the Haber process is for fertilizer.

(c) $$K_{eq} = \frac{P_{NH_3}^2}{P_{N_2} \times P_{H_2}^3}$$

15.7 *Analyze/Plan*. Follow the logic in Sample Exercises 15.1 and 15.5. *Solve*:

(a) $K_{eq} = \dfrac{P_{N_2O} \times P_{NO_2}}{P_{NO}^3}$ (b) $K_{eq} = \dfrac{P_{CS_2} \times P_{H_2}^4}{P_{CH_4} \times P_{H_2S}^2}$

(c) $K_{eq} = \dfrac{P_{CO}^4}{P_{Ni(CO)_4}}$ (d) $K_{eq} = \dfrac{[H^+][F^-]}{[HF]}$ e) $K_{eq} = \dfrac{[Ag^+]^2}{[Zn^{2+}]}$

homogeneous: (a), (b), (d); heterogeneous: (c), (e)

15.8 (a) $K_{eq} = \dfrac{P_{NO}^2}{P_{N_2} \times P_{O_2}}$ (b) $K_{eq} = \dfrac{1}{P_{Cl_2}^2}$ (c) $K_{eq} = \dfrac{P_{C_2H_6}^2 \times P_{O_2}}{P_{C_2H_4}^2 \times P_{H_2O}^2}$

(d) $K_{eq} = \dfrac{[Co^{2+}] \times P_{H_2}}{[H^+]^2}$ (e) $K_{eq} = \dfrac{[NH_4^+][OH^-]}{[NH_3]}$

homogeneous: (a), (c), (e); heterogeneous: (b), (d)

Note: Categorizing reaction (e) as homogeneous or heterogeneous is problematic. It contains a pure liquid, H_2O, that doesn't appear in the equilibrium expression. However, an aqueous solution of these components is perfectly mixed, with no phase barriers or distinct domains. I have listed it as homogeneous.

15.9 *Analyze/Plan*. Follow the logic in Sample exercise 15.2. *Solve*:

(a) mostly reactants ($K_{eq} << 1$)

(b) mostly products ($K_{eq} >> 1$)

15.10 (a) equilibrium lies to right, favoring products ($K_{eq} >> 1$)

(b) equilibrium lies to left, favoring reactants ($K_{eq} << 1$)

15.11 *Analyze/Plan*. Follow the logic in Sample Exercise 15.3. *Solve*:

(a) $2SO_2(g) + O_2(g) \rightleftharpoons 2SO_3(g)$ is the reverse of the reaction given.

$K_{eq}' = (K_{eq})^{-1} = 1/2.4 \times 10^{-3} = 4.2 \times 10^2$

(b) Since $K_{eq} < 1$ (when SO_3 is the reactant) and $K_{eq}' > 1$ (when SO_3 is the product), the equilibrium favors SO_3 at this temperature.

15.12 (a) $2NOBr(g) \rightleftharpoons 2NO(g) + Br_2(g)$ is the reverse of the reaction given.

$K_{eq}' = (K_{eq})^{-1} = 1/1.3 \times 10^{-2} = 76.92 = 77$

(b) $K_{eq} < 1$ when NOBr is the product, and $K_{eq}' > 1$ when NOBr is the reactant. At this temperature, the equilibrium favors NO and Br_2.

15.13 *Analyze/Plan*. Follow the logic in Sample Exercise 15.3. *Solve*:

(a) $K_{eq}' = 1/K_{eq} = 1/0.112 = 8.93$

(b) $K_{eq}' = 1/K_{eq}^2 = (0.112)^2 = 1.25 \times 10^{-2}$

(c) $K_{eq}' = 1/K_{eq}^2 = 1/(0.112)^2 = 79.7$

15.14 (a) $K_{eq}' = 1/K_{eq} = 1/0.0752 = 13.3$

(b) $K_{eq}' = (K_{eq})^{1/2} = (0.0752)^{1/2} = 0.274$

(c) $K_{eq}' = 1/K_{eq}^{1/2} = 1/(0.0752)^{1/2} = 3.65$

15.15 *Analyze/Plan*. Follow the logic in Sample Exercise 15.3. *Solve*:

(a) $K_{eq}' = (K_{eq})^{1/2} = (1.1 \times 10^{-2})^{1/2} = 0.1049 = 0.10$

(b) $K_{eq}' = (K_{eq})^2 = (1.1 \times 10^{-2})^2 = 1.2 \times 10^{-4}$

(c) $K_{eq}' = 1/(K_{eq})^2 = 1/(1.1 \times 10^{-2})^2 = 8.3 \times 10^3$

15.16 According to Hess's law, Section 5.6:

$$A(aq) + B(aq) \rightleftharpoons C(aq) \qquad K_1 = 1.9 \times 10^{-4}$$

$$C(aq) + D(aq) \rightleftharpoons E(aq) + A(aq) \qquad K_2 = 8.5 \times 10^{2}$$

$$A(aq) + B(aq) + C(aq) + D(aq) \rightleftharpoons A(aq) + C(aq) + E(aq)$$

$$B(aq) + D(aq) \rightleftharpoons E(aq) \qquad K_{eq} = K_1 \times K_2 = 0.16$$

$K_{eq} = (1.9 \times 10^{-4})(8.5 \times 10^{2}) = 0.162 = 0.16$

15.17 *Analyze/Plan.* Follow the logic in Sample Exercise 15.5. *Solve:*

(a) $K_{eq} = \dfrac{[Hg]^4 P_{O_2}}{[Hg_2O]^2}$

(b) The molar concentration, the ratio of moles of a substance to volume occupied by the substance, is a constant for pure solids and liquids.

(c) $K_{eq} = P_{O_2}$

15.18 (a) $K_{eq} = \dfrac{[Na_2SO_3]}{[Na_2O]P_{SO_2}}$

(b) The molar concentration, the ratio of moles of a substance to volume occupied by the substance, is a constant for pure solids and liquids.

(c) $K_{eq} = 1/P_{SO_2}$

Calculating Equilibrium Constants

15.19 *Analyze/Plan.* Follow the logic in Sample Exercise 15.7. *Solve:*

$$K_{eq} = \frac{P_{H_2} \times P_{I_2}}{P_{HI}^2} = \frac{(0.0274)(0.0274)}{(0.202)^2} = 1.84 \times 10^{-2}$$

15.20 $K_{eq} = P_{CH_3OH}/P_{CO} \times P_{H_2}^2$; $P = nRT/V$; RT/V is the same for all gases.

$$RT/V = \frac{0.08206\ L \bullet atm}{mol \bullet K} \times \frac{500\ K}{2.00\ L} = 20.515 = 20.5\ atm/mol$$

$$K_{eq} = \frac{0.0406\ mol\,(20.5\ atm/mol)}{0.170\ mol\,(20.5\ atm/mol) \times [0.302\ mol\,(20.5\ atm/mol)]^2}$$

$$= \frac{0.0406}{0.170[0.302(20.515)]^2} = 0.006222 = 6.22 \times 10^{-3}$$

15.21 *Analyze/Plan*. Follow the logic in Sample Exercise 15.7. *Solve*:

$2NO(g) + Cl_2(g) \rightleftharpoons 2NOCl(g)$

$$K_{eq} = \frac{P^2_{NOCl}}{P^2_{NO} \times P_{Cl_2}} = \frac{(0.28)^2}{(0.095)^2(0.171)} = 50.80 = 51$$

15.22 (a) $$K_{eq} = \frac{P_{PCl_5}}{P_{PCl_3} \times P_{Cl_2}} = \frac{1.30 \text{ atm}}{0.124 \text{ atm} \times 0.157 \text{ atm}} = 66.8$$

(b) Since $K_{eq} > 1$, products (the numerator of the K_{eq} expression) are favored over reactants (the denonimator of the K_{eq} expression).

15.23 *Analyze/Plan*. Follow the logic in Sample Exercise 15.8, using pressure rather than concentration as a measure of amount. Change mol of substances present initially to pressures and construct an equilibrium table. *Solve*:

(a) $P = nRT/V$; $$P_{NO} = 0.10 \text{ mol} \times \frac{300 \text{ K}}{1.0 \text{ L}} \times \frac{0.08206 \text{ L} \cdot \text{atm}}{\text{K} \cdot \text{mol}} = 2.462 = 2.5 \text{ atm}$$

$$P_{H_2} = 0.050 \text{ mol} \times \frac{300 \text{ K}}{1.0 \text{ L}} \times \frac{0.08206 \text{ L} \cdot \text{atm}}{\text{K} \cdot \text{mol}} = 1.231 = 1.2 \text{ atm};$$

$P_{H_2O} = P_{NO} = 2.462 = 2.5$ atm

First calculate the change in P_{NO}, 2.462 - 1.53 = 0.932 = 0.9 atm. From the stoichiometry of the reaction, calculate the change in the other pressures. Finally, calculate the equilibrium pressures.

	$2NO(g)$ +	$2H_2(g)$ $\rightleftharpoons$	$N_2(g)$ +	$2H_2O(g)$
initial	2.462 atm	1.231 atm	0 atm	2.462 atm
change	-0.932 atm	-0.932 atm	+0.466 atm	+0.932 atm
equil.	1.53 atm	0.299 atm	0.466 atm	3.394 atm

Strictly speaking, the change in P_{NO} has one decimal place and thus one sig fig. This limits equilibrium pressures to one decimal place, and K_{eq} to one sig fig. We compute the extra figures and then round.

(b) $$K_{eq} = \frac{P_{N_2} \times P^2_{H_2O}}{P^2_{NO} \times P^2_{H_2}} = \frac{(0.466)(3.394)^2}{(1.53)^2(0.299)^2} = \frac{(0.5)(3.4)^2}{(1.5)^2(0.3)^2} = 25.65 = 3 \times 10^1$$

15.24 (a) Calculate the initial partial pressures of $H_2(g)$ and $Br_2(g)$ and the equilibrium partial pressure of $H_2(g)$. $P = gRT/\mathcal{M}V$.

$$P_{H_2} = \frac{1.374 \text{ g}}{2.016 \text{ g/mol}} \times \frac{0.08206 \text{ L} \cdot \text{atm}}{\text{mol} \cdot \text{K}} \times \frac{700 \text{ K}}{2.00 \text{ L}} = 19.5747 = 19.57 \text{ atm}$$

$$P_{Br_2} = \frac{70.31 \text{ g}}{159.8 \text{ g/mol}} \times \frac{0.08206 \text{ L} \cdot \text{atm}}{\text{mol} \cdot \text{K}} \times \frac{700 \text{ K}}{2.00 \text{ L}} = 12.6369 = 12.64 \text{ atm}$$

$$P_{H_2} = \frac{0.566\text{ g}}{2.016\text{ g/mol}} \times \frac{0.08206\text{ L}\bullet\text{atm}}{\text{mol}\bullet\text{K}} \times \frac{700\text{ K}}{2.00\text{ L}} = 8.0635 = 8.06\text{ atm}$$

	$H_2(g)$	+ $Br_2(g)$	$\rightleftharpoons$ $2HBr(g)$
initial	19.57atm	12.64 atm	0
change	-11.51 atm	-11.51 atm	+2(11.51 atm)
equil.	8.06 atm	1.13 atm	23.02 atm

The change in P_{H_2} is (19.57 atm - 8.06 atm = 11.51 atm). The changes in P_{Br_2} and P_{HBr} are set by stoichiometry, resulting in the equilibrium concentrations shown in the table.

(b) $$K_{eq} = \frac{P_{HBr}^2}{P_{H_2} \times P_{Br_2}} = \frac{(23.02)^2}{(8.06)(1.13)} = 58.2$$

15.25 *Analyze/Plan*. Follow the logic in Sample Exercise 15.8 and Solution 15.23. *Solve*:

(a) $$P = nRT/V;\ P_{CO_2} = 0.2000\text{ mol} \times \frac{500\text{ K}}{2.000\text{L}} \times \frac{0.08206\text{ L}\bullet\text{atm}}{\text{K}\bullet\text{mol}} = 4.1030 = 4.10\text{ atm}$$

$$P_{H_2} = 0.1000\text{ mol} \times \frac{500\text{ K}}{2.000\text{L}} \times \frac{0.08206\text{ L}\bullet\text{atm}}{\text{K}\bullet\text{mol}} = 2.0515 = 2.05\text{ atm}$$

$$P_{H_2O} = 0.1600 \times \frac{500\text{ K}}{2.000\text{L}} \times \frac{0.08206\text{ L}\bullet\text{atm}}{\text{K}\bullet\text{mol}} = 3.2824 = 3.28\text{ atm}$$

The change in P_{H_2O} is 3.51 - 3.28 = 0.2276 = 0.23 atm. From the reaction stoichiometry, calculate the change in the other pressures, and the equilibrium pressures.

	$CO_2(g)$ +	$H_2(g)$	$\rightleftharpoons$	$CO(g)$ +	$H_2O(g)$
initial	4.10 atm	2.05 atm		0 atm	3.28 atm
change	-0.23 atm	-0.23 atm		+0.23	+0.23 atm
equil	3.87 atm	1.82 atm		0.23 atm	3.51 atm

(b) $$K_{eq} = \frac{P_{CO} \times P_{H_2O}}{P_{CO_2} \times P_{H_2}} = \frac{(0.23)(3.51)}{(3.87)(1.82)} = 0.1146 = 0.11$$

Without intermediate rounding, equilibrium pressures are $P_{H_2O} = 3.51$, $P_{CO} = 0.2276$, $P_{H_2} = 1.8239$, $P_{CO_2} = 3.8754$ and $K_{eq} = 0.1130 = 0.11$, in good agreement with the value above.

15.26 (a)

	$N_2O_4(g)$	$\rightleftharpoons$	$2NO_2(g)$
initial	1.500 atm		1.000 atm
change	+0.244 atm		-0.488 atm
equil	1.744 atm		0.512 atm

The change in P_{NO_2} is (1.000 - 0.512) = -0.488 atm, so the change in $P_{N_2O_4}$ is +(0.488/2) = +0.244 atm.

(b) $K_{eq} = \frac{P^2_{NO_2}}{P_{N_2O_4}} = \frac{(0.512)^2}{(1.744)} = 0.1503 = 0.150$

Applications of Equilibrium Constants

15.27 (a) A *reaction quotient* is the result of the law of mass action for a general set of concentrations, whereas the equilibrium constant requires equilibrium concentrations.

(b) In the direction of more products, to the right.

(c) If Q = K, the system is at equilibrium; the concentrations used to calculate Q must be equilibrium concentrations.

15.28 (a) If the value of Q equals the value of K, the system is at equilibrium.

(b) In the direction of less products (more reactants), to the left.

(c) Q = 0 if the concentration of any product is zero.

15.29 *Analyze/Plan.* follow the logic in Sample Exercise 15.9. We are given partial pressures so we can calculate Q directly and decide on the direction to equilibrium.

Solve: $K_{eq} = \frac{P_{CO} \times P_{Cl_2}}{P_{COCl_2}} = 6.71 \times 10^{-9}$ at 100°C

(a) $Q = \frac{(1.01 \times 10^{-4})(2.03 \times 10^{-4})}{(6.12 \times 10^{-2})} = 3.35 \times 10^{-7}$; $Q > K_{eq}$

The reaction will proceed to the left to attain equilibrium.

(b) $Q = \frac{(3.37 \times 10^{-6})(6.89 \times 10^{-5})}{(1.38)} = 1.68 \times 10^{-10}$; $Q < K_{eq}$

The reaction will proceed to the right to attain equilibrium.

(c) $Q = \frac{(4.53 \times 10^{-5})^2}{(3.06 \times 10^{-1})} = 6.71 \times 10^{-9}$, $Q = K_{eq}$

The reaction is at equilibrium.

15.30 Calculate the reaction quotient in each case, compare with $K_{eq} = \frac{P^2_{NH_3}}{P_{N_2} \times P^3_{H_2}} = 4.51 \times 10^{-5}$

(a) $Q = \frac{(105)^2}{(35)(495)^3} = 2.6 \times 10^{-6}$

Since $Q < K_{eq}$, the reaction will shift to the right to attain equilibrium.

(b) $Q = \frac{(35)^2}{(0)(595)^3} = \infty$

Since $Q > K_{eq}$, reaction must shift to the left to attain equilibrium. There must be **some** N_2 present to attain equilibrium. In this example, the only source of N_2 is the decomposition of NH_3.

(c) $Q = \frac{(26)^2}{(42)^3(202)} = 4.52 \times 10^{-5}$; $Q = K_{eq}$ Reaction is at equilibrium.

(d) $Q = \frac{(105)^2}{(5.0)(55)^3} = 1.3 \times 10^{-2}$; $Q > K_{eq}$

The reaction will proceed to the left to attain equilibrium.

15.31 *Analyze/Plan*. Follow the logic in Sample Exercise 15.10. *Solve*:

$$K_{eq} = \frac{P_{SO_2} \times P_{Cl_2}}{P_{SO_2Cl_2}};\ P_{Cl_2} = \frac{K_{eq} \times P_{SO_2Cl_2}}{P_{SO_2}} = \frac{(2.39)(3.31)}{(1.59)} = 4.9754 = 4.98 \text{ atm}$$

Check. $K_{eq} = \frac{(1.59)(4.98)}{(3.31)} = 2.39$. Our values are self-consistent.

15.32 $K_{eq} = \frac{P^2_{SO_3}}{P^2_{SO_2} \times P_{O_2}}$; $P_{SO_3} = \left(K_{eq} \times P^2_{SO_2} \times P_{O_2}\right)^{1/2} = [(0.345)(0.165)^2(0.755)]^{1/2} = 0.0842$ atm

15.33 *Analyze/Plan*. Follow the logic in Sample Exercise 15.10. In each case, change given masses to pressures, solve for the equilibrium pressure of the desired component, and calculate mass of that substance present at equilibrium. *Solve*:

(a) $K_{eq} = \frac{P^2_{Br}}{P_{Br_2}} = 0.133$; $P = \frac{gRT}{\mathcal{M}V}$; $T = 273 + 1285°C = 1558$ K

$$P_{Br_2} = \frac{0.245 \text{ g}}{159.8 \text{ g/mol}} \times \frac{1558 \text{ K}}{0.200 \text{ L}} \times \frac{0.08206 \text{ L}\cdot\text{atm}}{\text{K}\cdot\text{mol}} = 0.9801 = 0.980 \text{ atm}$$

$$P_{Br} = (K_{eq} \times P_{Br_2})^{1/2} = (0.133 \times 0.9801)^{1/2} = 0.36104 = 0.361 \text{ atm}$$

$$g_{Br} = \frac{\mathcal{M}_{Br}P_{Br}V}{RT} = \frac{79.904 \text{ g/mol} \times 0.36104 \text{ atm} \times 0.200 \text{ L}}{1558 \text{ K} \times 0.08206 \text{ L}\cdot\text{atm/mol}\cdot\text{K}} = 0.04513 = 0.0451 \text{ g Br}$$

(b) $PV = nRT$; $P = gRT/(\mathcal{M} \times V)$

$$P_{H_2} = \frac{0.056 \text{ g } H_2}{2.016 \text{ g/mol}} \times \frac{0.08206 \text{ L}\cdot\text{atm}}{\text{K}\cdot\text{mol}} \times \frac{700 \text{ K}}{2.000 \text{ L}} = 0.7978 = 0.80 \text{ atm}$$

$$P_{I_2} = \frac{4.36 \text{ g } I_2}{253.8 \text{ g/mol}} \times \frac{0.08206 \text{ L}\cdot\text{atm}}{\text{K}\cdot\text{mol}} \times \frac{700 \text{ K}}{2.000 \text{ L}} = 0.4934 = 0.494 \text{ atm}$$

$$K_p = 55.3 = ;\ \frac{P^2_{HI}}{P_{H_2} \times_{I2}} \quad P_{HI} = [55.3\,(P_{H_2})(P_{I_2})]^{1/2} = [55.3(0.7978)(0.4934)]^{1/2}$$
$$= 4.666 = 4.7 \text{ atm}$$

$$g_{HI} = \frac{\mathcal{M}_{HI}P_{HI}V}{RT} = \frac{128.0 \text{ g HI}}{\text{mol HI}} \times \frac{\text{K}\cdot\text{mol}}{0.08206 \text{ L}\cdot\text{atm}} \times \frac{4.666 \text{ atm} \times 2.000 \text{ L}}{700 \text{ K}}$$
$$= 20.79 = 21 \text{ g HI}$$

15.34 (a) $K_{eq} = P_I^2 / P_{I_2}$. $P_I = \frac{gRT}{\mathcal{M}V} = \frac{3.22 \times 10^{-2}\ g}{126.9\ g/mol} \times \frac{0.08206\ L \cdot atm}{mol \cdot K} \times \frac{800\ K}{10.0\ L}$

$= 1.666 \times 10^{-3} = 1.67 \times 10^{-3}$ atm

$P_{I_2} = P_I^2/K_{eq} = (1.666 \times 10^{-3})^2 / 2.04 \times 10^{-3} = 1.360 \times 10^{-3} = 1.36 \times 10^{-3}$ atm

$$g_{I_2} = \frac{\mathcal{M}PV}{RT} = \frac{253.8\ g\ I_2}{mol\ I_2}\ \frac{K \cdot mol}{0.08206\ L \cdot atm} \times \frac{1.360 \times 10^{-3}\ atm \times 10.0\ L}{800\ K}$$

$= 0.0526$ g I_2

(b) $PV = nRT$; $P = \frac{gRT}{\mathcal{M}V}$

$$P_{SO_3} = \frac{2.65\ g\ SO_3}{80.06\ g/mol} \times \frac{0.08206\ L \cdot atm}{K \cdot mol} \times \frac{700\ K}{2.00\ L} = 0.9507 = 0.951\ atm$$

$$P_{O_2} = \frac{1.08\ g\ O_2}{32.00\ g/mol} \times \frac{0.08206\ L \cdot atm}{K \cdot mol} \times \frac{700\ K}{2.00\ L} = 0.9693 = 0.969\ atm$$

$$K_{eq} = 3.0 \times 10^4 = \frac{P_{SO_3}^2}{P_{SO_2}^2 \times P_{O_2}};\ P_{SO_2} = \left[P_{SO_3}^2 / (K_{eq})(P_{O_2})\right]^{1/2}$$

$P_{SO_2} = [(0.9507)^2/(3.0 \times 10^4)(0.9693)]^{1/2} = 5.575 \times 10^{-3} = 5.58 \times 10^{-3}$ atm

$$g\ SO_2 = \frac{\mathcal{M}PV}{RT} = \frac{64.06\ g\ SO_2}{mol\ SO_2} \times \frac{K \cdot mol}{0.08206\ L \cdot atm} \times \frac{5.575 \times 10^{-3}\ atm \times 2.00\ L}{700\ K}$$

$= 0.01243 = 0.0124$ g SO_2

15.35 *Analyze/Plan*. Follow the logic in Sample Exercise 15.11. Since pressure of NO is given directly, we can construct the equilibrium table straight away. *Solve*:

$$2NO(g) \rightleftharpoons N_2(g) + O_2(g) \qquad K_{eq} = \frac{P_{N_2} \times P_{O_2}}{P_{NO}^2} = 2.4 \times 10^{-3}$$

	2NO(g)	N_2(g)	O_2(g)
initial	37.3 atm	0	0
change	-2x	+x	+x
equil.	37.3 - 2x	+x	+x

$$2.4 \times 10^3 = \frac{x^2}{(37.3-2x)^2};\ (2.4 \times 10^3)^{1/2} = \frac{x}{37.3 - 2x}$$

$x = (2.4 \times 10^3)^{1/2} (37.3 - 2x)$; $x = 1827.3 - 97.98x$; $98.98x = 1827.3$, $x = 18.46 = 18$ atm

$P_{N_2} = P_{O_2} = 18$ atm; $P_{NO} = 37.3 - 2(18.46) = 0.377 = 0.4$ atm

Strictly speaking, 18 atm has 2 sig figs and no decimal places, so P_{NO} should have no decimal places, or $P_{NO} = 0$ atm. This result is not very useful.

Check. $K_{eq} = \frac{(18)^2}{(0.4)^2} = 2 \times 10^3$; more sig figs are needed for closer agreement.

15.36 $P_{Br_2} = P_{Cl_2} = \frac{nRT}{V} = 0.30 \text{ mol} \times \frac{0.08206 \text{ L} \bullet \text{atm}}{\text{K} \bullet \text{mol}} \times \frac{400 \text{ K}}{1.0 \text{ L}} = 9.847 = 9.8 \text{ atm}$

	$Br_2(g)$	+ $Cl_2(g)$	$\rightleftharpoons$ $2BrCl(g)$
initial	9.8 atm	9.8 atm	0
change	-x	-x	+2x
equil.	(9.8-x) atm	(9.8-x) atm	+2x

$K_{eq} = \frac{P^2_{BrCl}}{P_{Br_2} \times P_{Cl_2}} = 7.0$

$7.0 = \frac{(2x)^2}{(9.8-x)^2}$ (Assuming x is small leads to x = 13 atm, P_{BrCl} = 26 atm. Clearly x is not small compared to 9.8 atm.)

$(7.0)^{1/2} = \frac{2x}{9.847-x}$, 2.646(9.847-x) = 2x, 26.056 = 4.646x, x = 5.608 = 5.6 atm

P_{BrCl} = 2x = 11.216 = 11 atm

15.37 *Analyze/Plan*. Write the K_{eq} expression, substitute the stated pressure relationship and solve for P_{Br_2}. *Solve*:

$$K_{eq} = \frac{P^2_{NO} \times P_{Br_2}}{P^2_{NOBr}}$$

When $P_{NOBr} = P_{NO}$, these terms cancel and $P_{Br_2} = K_{eq}$ = 0.416 atm. This is true for all cases where $P_{NOBr} = P_{NO}$.

15.38 $K_{eq} = P_{H_2S} \times P_{NH_3} = 7.0 \times 10^{-2}$. The equilibrium pressures of H_2S and NH_3 will be equal; call this quantity y. Then, $y^2 = 7.0 \times 10^{-2}$, y = 0.2646 = 0.26 atm.

15.39 (a) *Analyze/Plan*. If only $PH_3BCl_3(s)$ is present initially, the equation requires that the equilibrium partial pressures of $PH_3(g)$ and $BCl_3(g)$ are equal. Write the K_{eq} expression and solve for $x = P_{PH_3} = P_{BCl_3}$. *Solve*:

$K_{eq} = P_{PH_3} \times P_{BCl_3}$; $5.42 \times 10^{-2} = x^2$; x = 0.2328 = 0.233 atm PH_3 and BCl_3

(b) Since the mole ratios are 1:1:1, mol $PH_3BCl_3(s)$ required = mol PH_3 or BCl_3 produced.

$$n_{PH_3} = P_{PH_3} V/RT = \frac{0.2328 \text{ atm} \times 0.500 \text{ L}}{353 \text{ K} \times 0.08206 \text{ L} \bullet \text{atm/K} \bullet \text{mol}}$$

$$= 4.018 \times 10^{-3} = 4.02 \times 10^{-3} \text{ mol } PH_3$$

$$4.018 \times 10^{-3} \text{ mol } PH_3 = 4.018 \times 10^{-3} \text{ mol } PH_3BCl_3 \times \frac{151.2 \text{ g } PH_3BCl_3}{1 \text{ mol } PH_3BCl_3}$$

$$= 0.6076 = 0.608 \text{ g } PH_3BCl_3$$

In fact, some $PH_3BCl_3(s)$ must remain for the system to be in equilibrium, so a bit more than 0.608 g PH_3BCl_3 is needed.

15.40 (a) $CaSO_4(s) \rightleftharpoons Ca^{2+}(aq) + SO_4^{2-}(aq)$ $K_{eq} = [Ca^{2+}][SO_4^{2-}] = 2.4 \times 10^{-5}$

At equilibrium, $[Ca^{2+}] = [SO_4^{2-}] = x$

$K_{eq} = 2.4 \times 10^{-5} = x^2$; $x = 4.9 \times 10^{-3}$ *M* Ca^{2+} and SO_4^{2-}

(b) A saturated solution of $CaSO_4(aq)$ is 4.9×10^{-3} *M*.
3.0 L of this solution contain:

$$4.9 \times 10^{-3} \frac{\text{mol}}{\text{L}} \times 3.0\,\text{L} \times \frac{136.14\text{ g CaSO}_4}{\text{mol}} = 2.001 = 2.0\text{ g CaSO}_4$$

A bit more than 2.0 g $CaSO_4$ is needed in order to have some undissolved $CaSO_4(s)$ in equilibrium with 3.0 L of saturated solution.

15.41 *Analyze/Plan.* Follow the approach in Solution 15.35. Calculate P_{IBr} from mol IBr and construct the equilibrium table.

$$K_{eq} = \frac{P_{IBr}^2}{P_{I_2} \times P_{Br_2}} = 280;\ P_{IBr} = \frac{nRT}{V} = \frac{0.500\text{ mol}}{1.000\text{ L}} \times 423\text{ K} \times 0.08206\ \frac{\text{L}\bullet\text{atm}}{\text{K}\bullet\text{mol}}$$

$$= 17.356 = 17.4\text{ atm}$$

Since no I_2 or Br_2 were present initially, the amounts present at equilibrium are produced by the reverse reaction and stoichiometrically equal. Let these amounts equal x. The amount of HBr that reacts is then 2x. Substitute the equilibrium pressures (in terms of x) into the equilibrium expression and solve for x.

	I_2	+ Br_2	$\rightleftharpoons$ 2IBr
initial	0 atm	0 atm	17.356
change	+x atm	+x atm	-2x
equil.	x atm	x atm	(17.356 - 2x) atm

$K_{eq} = 280 = \frac{(17.356 - 2x)^2}{x^2}$; taking the square root of both sides

$16.733 = \frac{17.356 - 2x}{x}$; $16.733x + 2x = 17.356$; $18.733x = 17.356$

$x = 0.92647 = 0.926$ atm; $P_{I_2} = 0.926$ atm; $P_{Br_2} = 0.926$ atm

$P_{IBr} = 17.356 - 2x = 17.356 - 1.853 = 15.503 = 15.5$ atm

Check. $\frac{(15.5)^2}{(0.926)^2} = 280$. Our values are self-consistent.

15.42 $CaCrO_4(s) \rightleftharpoons Ca^{2+}(aq) + CrO_4^{2-}(aq)$ $K_{eq} = [Ca^{2+}][CrO_4^{2-}] = 7.1 \times 10^{-4}$

At equilibrium, $[Ca^{2+}] = [CrO_4^{2-}] = x$

$K_{eq} = 7.1 \times 10^{-4} = x^2$, $x = 0.0266 = 0.027$ *M* Ca^{2+} and CrO_4^{2-}

LeChâtelier's Principle

15.43 *Analyze/Plan.* Follow the logic in Sample Exercise 15.12. *Solve:*

(a) Shift equilibrium to the right; more $SO_3(g)$ is formed, the amount of $SO_2(g)$ decreases.

(b) Heating an exothermic reaction decreases the value of K. More SO_2 and O_2 will form, the amount of SO_3 will decrease.

(c) Since, $\Delta n = -1$, a change in volume will affect the equilibrium position and favor the side with more moles of gas. The amounts of SO_2 and O_2 increase and the amount of SO_3 decreases.

(d) No effect. Speeds up the forward and reverse reactions equally.

(e) No effect. Does not appear in the equilibrium expression.

(f) Shift equilibrium to the right; amounts of SO_2 and O_2 decrease.

15.44 (a) Increase (b) increase (c) decrease (d) no effect (e) no effect
(f) no effect

15.45 *Analyze/Plan.* Given certain changes to a reaction system, determine the effect on K_{eq}, if any. Only changes in temperature cause changes to the value of K_{eq}. *Solve:*

(a) No effect (b) no effect (c) increase equilibrium constant (d) no effect

15.46 (a) The reaction must be endothermic ($+\Delta H$) if heating increases the fraction of products.

(b) There must be more moles of gas in the products if increasing the volume of the vessel increases the fraction of products.

15.47 *Analyze/Plan.* Use Hess's Law, $\Delta H° = \Sigma\Delta H_f^\circ$ products - $\Sigma\Delta H_f^\circ$ reactants, to calculate $\Delta H°$. According to the sign of $\Delta H°$, describe the effect of temperature on the value of K_{eq}. According to the value of Δn, describe the effect of changes to container volume. *Solve:*

(a) $\Delta H° = \Delta H_f^\circ\ NO_2(g) + \Delta H_f^\circ\ N_2O(g) - 3\Delta H_f^\circ\ NO(g)$

$\Delta H° = 33.84\ kJ + 81.6\ kJ - 3(90.37\ kJ) = -155.7\ kJ$

(b) The reaction is exothermic ($-\Delta H°$), so the equilibrium constant will decrease with increasing temperature.

(c) Δn does not equal zero, so a change in volume at constant temperature will affect the fraction of products in the equilibrium mixture. An increase in container volume would favor reactants, while a decrease in volume would favor products.

15.48 (a) $\Delta H° = \Delta H_f^\circ\ CH_3OH(g) - \Delta H_f^\circ\ CO(g) - 2\Delta H_f^\circ\ H_2(g)$

$= -201.2\ kJ - (-110.5\ kJ) - 0\ kJ$

$= -90.7\ kJ$

(b) The reaction is exothermic; an increase in temperature would decrease the value of K_{eq} and decrease the yield. A *low temperature* is needed to maximize yield.

(c) Increasing total pressure would increase the partial pressure of each gas, shifting the equilibrium toward products. The extent of conversion to CH_3OH increases as the total pressure increases.

Additional Exercises

15.49 (a) Since both the forward and reverse processes are elementary steps, we can write the rate laws directly from the chemical equation.

$$\text{rate}_f = k_f\,[CO][Cl_2] = \text{rate}_r = k_r\,[COCl][Cl]$$

$$\frac{k_f}{k_r} = \frac{[COCl]\,[Cl]}{[CO]\,[Cl_2]} = K$$

$$K_{eq} = \frac{k_f}{k_r} = \frac{1.4 \times 10^{-28}\ M^{-1}\,s^{-1}}{9.3 \times 10^{10}\ M^{-1}\,s^{-1}} = 1.5 \times 10^{-39}$$

For a homogeneous equilibrium in the gas phase, we usually write K_{eq} in terms of partial pressures. In this exercise, concentrations are more convenient because the rate constants are expressed in terms of molarity. For this reaction, the value of K_{eq} is the same regardless of how K_{eq} is expressed, because there is no change in the moles of gas in going from reactants to products.

(b) Since the K is quite small, reactants are much more plentiful than products at equilibrium.

15.50 $$K_{eq} = \frac{P_{CO} \times P_{H_2}^3}{P_{CH_4} \times P_{H_2O}};\ P = \frac{g\,RT}{\mathcal{M}\,V};\ T = 1000\ K$$

$$P_{CO} = \frac{8.62\ g}{28.01\ g/mol} \times \frac{0.08206\ L\bullet atm}{mol\bullet K} \bullet \frac{1000\ K}{5.00\ L} = 5.0507 = 5.05\ atm$$

$$P_{H_2} = \frac{2.60\ g}{2.016\ g/mol} \times \frac{0.08206\ L\bullet atm}{mol\bullet K} \times \frac{1000\ K}{5.00\ L} = 21.1663 = 21.2\ atm$$

$$P_{CH_4} = \frac{43.0\ g}{16.04\ g/mol} \times \frac{0.08206\ L\bullet atm}{mol\bullet K} \bullet \frac{1000\ K}{5.00\ L} = 43.9973 = 44.0\ atm$$

$$P_{H_2O} = \frac{48.4\ g}{18.02\ g/mol} \times \frac{0.08206\ L\bullet atm}{mol\bullet K} \bullet \frac{1000\ K}{5.00\ L} = 44.0811 = 44.1\ atm$$

$$K_{eq} = \frac{(5.0507)(21.1663)^3}{(43.9973)(44.0811)} = 24.6949 = 24.7$$

15.51 $$K_{eq} = \frac{P_{SO_2} \times P_{Cl_2}}{P_{SO_2Cl_2}};\quad P_{SOCl_2} = \frac{nRT}{V} = \frac{2.00\ mol}{2.00\ L} \times 303\ K \times \frac{0.08206\ L\bullet atm}{K\bullet mol}$$
$$= 24.864 = 24.9\ atm$$

The change in $P_{SOCl_2} = x = 0.56(24.864) = 13.924 = 14$ atm

	$SO_2Cl_2(g)$ $\rightleftharpoons$	$SO_2(g)$	+ $Cl_2(g)$
initial	24.9 atm	0	0
change	-14 atm	+14 atm	+14 atm
equil.	11 atm	+14 atm	+14 atm

$$K_{eq} = \frac{(13.924)^2}{10.940} = 17.72 = 18$$

15.52 (a) $H_2(g) + S(s) \rightleftharpoons H_2S(g)$ $\qquad K_{eq} = P_{H_2S} / P_{H_2}$

(b) Calculate the partial pressures of H_2S and H_2. $P = \frac{gRT}{\mathcal{M}V}$; T = 90°C + 273 = 363 K

$$P_{H_2S} = \frac{0.46\text{ g}}{34.1\text{ g/mol}} \times \frac{0.08206\text{ L}\bullet\text{atm}}{\text{K}\bullet\text{mol}} \times \frac{363\text{ K}}{1.0\text{ L}} = 0.4018 = 0.40\text{ atm}$$

$$P_{H_2} = \frac{0.40\text{ g}}{2.02\text{ g/mol}} \times \frac{0.08206\text{ L}\bullet\text{atm}}{\text{K}\bullet\text{mol}} \times \frac{363\text{ K}}{1.0\text{ L}} = 5.899 = 5.9\text{ atm}$$

K_{eq} = 0.4018/5.899 = 0.06811 = 0.068

(c) Since S is a pure solid, its 'partial pressure' doesn't change during the reaction, so P_S does not appear in the equilibrium expression.

15.53 (a) $K_{eq} = \frac{P_{Br_2} \times P_{NO}^2}{P_{NOBr}^2}$; $P = \frac{gRT}{PV}$; T = 100°C + 273 = 373

$$P_{Br_2} = \frac{4.19\text{ g}}{159.8\text{ g/mol}} \times \frac{0.08206\text{ L}\bullet\text{atm}}{\text{K}\bullet\text{mol}} \times \frac{373}{5.0\text{ L}} = 0.16051 = 0.161\text{ atm}$$

$$P_{NO} = \frac{3.08\text{ g}}{30.01\text{ g/mol}} \times \frac{0.08206\text{ L}\bullet\text{atm}}{\text{K}\bullet\text{mol}} \times \frac{373}{5.0\text{ L}} = 0.62828 = 0.628\text{ atm}$$

$$P_{NOBr} = \frac{3.22\text{ g NOBr}}{109.9\text{ g/mol}} \times \frac{0.08206\text{ L}\bullet\text{atm}}{\text{K}\bullet\text{mol}} \times \frac{373}{5.0\text{ L}} = 0.17936 = 0.179\text{ atm}$$

$$K_{eq} = \frac{(0.16051)(0.62828)^2}{(0.17936)^2} = 1.9695 = 1.97$$

(b) $P_t = P_{Br_2} + P_{NO} + P_{NOBr}$ = 0.16051 + 0.62828 + 0.17936 = 0.96815 = 0.968 atm

15.54 (a)

	$A(g)$ $\rightleftharpoons$	$2B(g)$
initial	0.55 atm	0
change	-0.19 atm	+0.38 atm
equil.	0.36 atm	0.38 atm

$P_t = P_A + P_B$ = 0.36 atm + 0.38 atm = 0.74 atm

(b) $K_{eq} = \frac{(P_B)^2}{P_A} = \frac{(0.38)^2}{0.36} = 0.4011 = 0.40$

15.55 (a) $K_{eq} = \frac{P_{NH_3}^2}{P_{N_2} \times P_{H_2}^3} = 4.34 \times 10^{-3}$; T = 300°C + 273 = 573 K

$P_{NH_3} = \frac{gRT}{\mathcal{M} \times V} = \frac{1.05\text{ g}}{17.03\text{ g/mol}} \times \frac{0.08206\text{ L}\cdot\text{atm}}{\text{K}\cdot\text{mol}} \times \frac{573\text{ K}}{1.00\text{ L}} = 2.899 = 2.90\text{ atm}$

	$N_2(g)$	+ $3H_2(g)$	$\rightleftharpoons$ $2NH_3(g)$
initial	0 atm	0 atm	?
change	x	3x	-2x
equil.	x atm	3x atm	2.899 atm

(Remember, only the change line reflects the stoichiometry of the reaction.)

$K_{eq} = \frac{(2.899)^2}{(x)(3x)^3} = 4.34 \times 10^{-3}$; $27x^4 = \frac{(2.899)^2}{4.34 \times 10^{-3}}$; $x^4 = 71.725$

x = 2.910 = 2.91 atm = P_{N_2}; P_{H_2} = 3x = 8.730 = 8.73 atm

$g_{N_2} = \frac{\mathcal{M} \times PV}{RT} = \frac{28.02\text{ g }N_2}{\text{mol }N_2} \times \frac{\text{K}\cdot\text{mol}}{0.08206\text{ L}\cdot\text{atm}} \times \frac{2.910\text{ atm} \times 1.000\text{ L}}{573\text{ K}} = 1.73\text{ g }N_2$

$g_{H_2} = \frac{2.016\text{ g }H_2}{\text{mol }H_2} \times \frac{\text{K}\cdot\text{mol}}{0.08206\text{ L}\cdot\text{atm}} \times \frac{8.730\text{ atm} \times 1.00\text{ L}}{573\text{ K}} = 0.374\text{ g }N_2$

(b) The initial P_{NH_3} = 2.899 atm + 2(2.910 atm) = 8.719 = 8.72 atm

$g_{NH_3} = = \frac{17.03\text{ g }NH_3}{\text{mol }NH_3} \times \frac{\text{K}\cdot\text{mol}}{0.08206\text{ L}\cdot\text{atm}} \times \frac{8.719\text{ atm} \times 1.000\text{ L}}{573\text{ K}} = 3.16\text{ g }NH_3$

(c) $P_t = P_{N_2} + P_{H_2} + P_{NH_3}$ = 2.910 atm + 8.730 atm + 8.719 atm = 20.36 atm

15.56 $P_{IBr} = \frac{nRT}{V} = 0.025\text{ mol} \times \frac{0.08206\text{ L}\cdot\text{atm}}{\text{mol}\cdot\text{K}} \times \frac{423\text{ K}}{2.0\text{ L}} = 0.4339 = 0.43\text{ atm}$

	2IBr	$\rightleftharpoons$ I_2	+ Br_2
initial	0.43 atm	0	0
change	-2x	x	x
equil.	(0.43-2x) atm	x	x

$K_{eq} = 8.5 \times 10^{-3} = \frac{P_{I_2} \times P_{Br_2}}{P_{IBr}} = \frac{x^2}{(0.43-2x)^2}$; Taking the square root of both sides

$\frac{x}{0.4339-2x} = (8.5 \times 10^{-3})^{1/2} = 0.0922$; x = 0.0922(0.4339 - 2x)

x + 0.184x = 0.04001; 1.184x = 0.04001, x = 0.03379 = 0.034 atm

P_{IBr} at equilibrium = 0.04339 - 2(0.03379) = 0.3663 = 0.37 atm

15.57 $P_{BCl_3} = \frac{nRT}{V} = \frac{0.0128 \text{ mol} \times 333 \text{ K}}{0.500 \text{ L}} \times \frac{0.08206 \text{ L} \cdot \text{atm}}{\text{mol} \cdot \text{K}} = 0.69955 = 0.700 \text{ atm}$

PH_3BCl_3 is a solid and its concentration is taken as a constant, C.

	PH_3BCl_3 $\rightleftharpoons$	PH_3 +	BCl_3
initial	C	0 atm	0.700 atm
change		+x atm	+x atm
equil.	C	+x atm	0.700 atm

$K_{eq} = P_{PH_3} \times P_{BCl_3} = 0.052(0.700+x)$; $x^2 + 0.700 - 0.052 = 0$

$$x = \frac{-0.700 \pm [(0.700)^2 - 4(-0.052)]^{1/2}}{2} = 0.06773 = 6.8 \times 10^{-2} \text{ atm } PH_3$$

Check: (0.068)(0.700 + 0.068) = 0.052; the solution is correct to two significant figures.

15.58 $K_{eq} = P_{NH_3} \times P_{H_2S}$; $P_t = 0.614$ atm

If the equilibrium amounts of NH_3 and H_2S are due solely to the decomposition of $NH_4HS(s)$, the equilibrium pressures of the two gases are equal, and each is 1/2 of the total pressure.

$P_{NH_3} = P_{H_2S} = 0.614 \text{ atm}/2 = 0.307 \text{ atm}$

$K_{eq} = (0.307)^2 = 0.0943$

15.59 Initial $P_{SO_3} = \frac{gRT}{\mathcal{M} V} = \frac{0.831 \text{ g}}{80.07 \text{ g/mol}} \times \frac{0.08206 \text{ L} \cdot \text{atm}}{\text{K} \cdot \text{mol}} \times \frac{1100 \text{ K}}{1.00 \text{ L}} = 0.9368 = 0.937 \text{ atm}$

	$2SO_3$ $\rightleftharpoons$	$2SO_2$ +	O_2
initial	0.9368 atm	0	0
change	-2x	+2x	+x
equil.	0.9368-2x	2x	x
[equil.]	0.2104 atm	0.7264 atm	0.3632 atm

$P_t = (0.9368-2x) + 2x + x$; $0.9368 + x = 1.300$ atm; $x = 1.300 - 0.9368 = 0.3632 = 0.363$ atm

$$K_{eq} = \frac{P^2_{SO_2} \times P_{O_2}}{P^2_{SO_3}} = \frac{(0.7264)^2(0.3632)}{(0.2104)^2} = 4.33$$

15.60 In general, the reaction quotient is of the form $Q = \frac{P^2_{NOCl}}{P^2_{NO} \times P_{Cl_2}}$.

(a) $Q = \frac{(0.11)^2}{(0.15)^2(0.31)} = 1.7$

$Q > K_{eq}$. Therefore, the reaction will shift toward reactants, to the left, in moving toward equilibrium.

(b) $Q = \frac{(0.050)^2}{(0.12)^2(0.10)} = 1.7$

$Q > K_{eq}$. Therefore, the reaction will shift toward reactants, to the left, in moving toward equilibrium.

(c) $Q = \frac{(5.10 \times 10^{-3})^2}{(0.15)^2(0.20)} = 5.8 \times 10^{-3}$

$Q < K_{eq}$. Therefore, the reaction mixture will shift in the direction of more product, to the right, in moving toward equilibrium.

15.61 $K_{eq} = P_{CO_2} = 0.0108$; $P_{CO_2} = \frac{gRT}{\mathcal{M}V}$; T = 900°C + 273 = 1173 K

In each case, calculate P_{CO_2} and determine the position of the equilibrium.

(a) $P_{CO_2} = \frac{4.25\ g}{44.01\ g/mol} \times \frac{0.08206\ L \bullet atm}{mol \bullet K} \times \frac{1173\ K}{10.0\ L} = 0.92954 = 0.930\ atm$

$Q = 0.930 > K_{eq}$. The reaction proceeds to the left to achieve equilibrium and the amount of $CaCO_3(s)$ increases.

(b) 5.66 g CO_2 means $P_{CO_2} > 0.930$ atm; $Q > 0.930 >> K_{eq}$, the amount of $CaCO_3$ increases.

(c) 6.48 g CO_2 means $P_{CO_2} > 0.930$ atm; $Q > 0.930 >> K_{eq}$, the amount of $CaCO_3$ increases.

15.62 (a) $K_{eq} = \frac{P_{Ni(CO)_4}}{P_{CO}^4}$

(b) Increasing the temperature to 200°C favors the reverse process (decomposition of $Ni(CO)_4(g)$) and thus the value of K_{eq} is smaller at the higher temperature. This is the behavior expected from an **exothermic** reaction (heat is a product).

(c) At the temperature of the exhaust pipe, the $Ni(CO)_4$ product is a gas and is carried into the atmosphere with other exhaust gases. Thus, equilibrium is never established (we do not have a closed system) and the reaction proceeds to the right as $Ni(CO)_4$ product is removed.

15.63 $K_{eq} = \frac{P_{CO_2}}{P_{CO}} = 6.0 \times 10^2$

If P_{CO} is 150 torr, P_{CO_2} can never exceed 760 - 150 = 610 torr. Then Q = 610/150 = 4.1. Since this is far less than K, the reaction will shift in the direction of more product. Reduction will therefore occur.

15.64 (a)

	$CCl_4(g)$	$\rightleftharpoons$	C(s) + $2Cl_2(g)$
initial	2.00 atm		0 atm
change	-x atm		+2x atm
equil.	(2.00-x) atm		2x atm

$$K_{eq} = 0.76 = \frac{P^2_{Cl_2}}{P_{CCl_4}} = \frac{(2x)^2}{(2.00-x)}$$

$1.52 - 0.76x = 4x^2$; $4x^2 + 0.76x - 1.52 = 0$

Using the quadratic formula, a = 4, b = 0.76, c = -1.52

$$x = \frac{-0.76 \pm \sqrt{(0.76)^2 - 4(4)(-1.52)}}{2(4)} = \frac{-0.76 + 4.99}{8} = 0.5287 = 0.53 \text{ atm}$$

$$\text{Fraction } CCl_4 \text{ reacted} = \frac{x \text{ atm}}{2.00 \text{ atm}} = \frac{0.53}{2.00} = 0.264 = 26\%$$

(b) $P_{Cl_2} = 2x = 2(0.5287) = 1.06$ atm

$P_{CCl_4} = 2.00 - x = 2.00 - 0.5287 = 1.47$ atm

15.65 (a)

$$Q = \frac{P_{PCl_5}}{P_{PCl_3} \times P_{Cl_2}} = \frac{(0.20)}{(0.50)(0.50)} = 0.80$$

0.80 (Q) > 0.0870 (K), the reaction proceeds to the left.

(b)

	$PCl_3(g)$	+	$Cl_2(g)$	$\rightleftharpoons$	$PCl_5(g)$
initial	0.50 atm		0.50 atm		0.20 atm
change	+x atm		+x atm		-x atm
equil.	(0.50 + x) atm		(0.50 + x) atm		(0.20 - x) atm

(Since the reaction proceeds to the left, P_{PCl_5} must decrease and P_{PCl_3} and P_{Cl_2} must increase.)

$$K_{eq} = 0.0870 = \frac{(0.20 - x)}{(0.50 + x)(0.50 + x)}; \quad 0.0870 = \frac{(0.20 - x)}{(0.250 + 1.00x + x^2)}$$

$0.0870(0.250 + 1.00x + x^2) = 0.20 - x$; $-0.17825 + 1.0870x + 0.0870x^2 = 0$

$$x = \frac{-1.0870 \pm \sqrt{(1.0870)^2 - 4(0.0870)(-0.17825)}}{2(0.0870)} = \frac{-1.0870 + 1.1152}{0.174} = 0.162$$

$P_{PCl_3} = (0.50 + 0.162)$ atm = 0.662 $P_{Cl_2} = (0.50 + 0.162)$ atm = 0.662 atm

$P_{PCl_5} = (0.20 - 0.162)$ atm = 0.038 atm

To 2 decimal places, the pressures are 0.66, 0.66 and 0.04 atm, respectively. When substituting into the K_{eq} expression, pressures to 3 decimal places yield a result much closer to 0.0870.

(c) Increasing the volume of the container favors the process where more moles of gas are produced, so the reverse reaction is favored and the equilibrium shifts to the left; the mole fraction of Cl_2 increases.

(d) For an exothermic reaction, increasing the temperature decreases the value of K; more reactants and fewer products are present at equilibrium and the mole fraction of Cl_2 increases.

15.66 *Analyze/Plan.* Equilibrium pressures of H_2, I_2, HI → K_{eq} → equilibrium table → new equilibrium concentrations. *Solve*:

$$P = \frac{nRT}{V};\ \frac{RT}{V} = \frac{0.08206\ \text{L}\bullet\text{atm}}{\text{mol}\bullet\text{K}} \times \frac{731\ \text{K}}{5.00\ \text{L}} = 11.997 = 12.0\ \text{atm/mol}$$

$$P_{H_2} = P_{I_2} = 0.112\ \text{mol} \times 11.997\ \frac{\text{atm}}{\text{mol}} = 1.344 = 1.34\ \text{atm}$$

$$P_{HI} = 0.775\ \text{mol} \times 11.997\ \frac{\text{atm}}{\text{mol}} = 9.298 = 9.30\ \text{atm}$$

$$H_2(g) + I_2(g) \rightleftharpoons 2HI(g);\ K_{eq} = \frac{P_{HI}^2}{P_{H_2} \times P_{I_2}} = \frac{(9.298)^2}{(1.344)^2} = 47.861 = 47.9$$

$$P_{HI}\ (\text{added}) = 0.100\ \text{mol} \times \frac{11.997\ \text{atm}}{\text{mol}} = 1.1997 = 1.20\ \text{atm}$$

	$H_2(g)$ +	$I_2(g)$ $\rightleftharpoons$	$2HI(g)$
initial	1.34 atm	1.34 atm	9.30 atm + 1.20 atm
change	+x atm	+x atm	-2x atm
equil.	(1.34+x) atm	(1.34+x) atm	(10.50-2x) atm

$K_{eq} = 47.86 = \dfrac{(10.50 - 2x)^2}{(1.34 + x)^2}$. Take the square root of both sides:

$$6.918 = \frac{10.50 - 2x}{1.34 + x};\ 9.270 + 6.918\,x = 10.50 - 2x;\ 8.918\,x = 1.230,\ x = 0.1379 = 0.140$$

$$P_{H_2} = P_{I_2} = 1.34 + 0.140 = 1.48\ \text{atm};\ P_{HI} = 10.50 - 2(0.140) = 10.22\ \text{atm}$$

Check: $\dfrac{(10.22)^2}{(1.48)^2} = 47.68 = 47.7$

15.67 (a) Since the volume of the vessel = 1.00 L, mol = *M*. The reaction will proceed to the left to establish equilibrium.

	A(g)	+ 2B(g)	$\rightleftharpoons$ 2C(g)
initial	0 *M*	0 *M*	1.00 *M*
change	+x *M*	+2x *M*	-2x *M*
equil.	x *M*	2x *M*	(1.00 - 2x) *M*

At equilibrium, [C] = (1.00 - 2x) *M*, [B] = 2x *M*.

(b) x must be less than 0.50 *M* (so that [C], 1.00 -2x, is not less than zero).

(c) $K_{eq} = \dfrac{[C]^2}{[A][B]^2}$; $\dfrac{(1.00-2x)^2}{(x)(2x)^2} = 0.25$

$1.00 - 4x + 4x^2 = 0.25(4x)^3$; $x^3 - 4x^2 + 4x - 1 = 0$

(d)

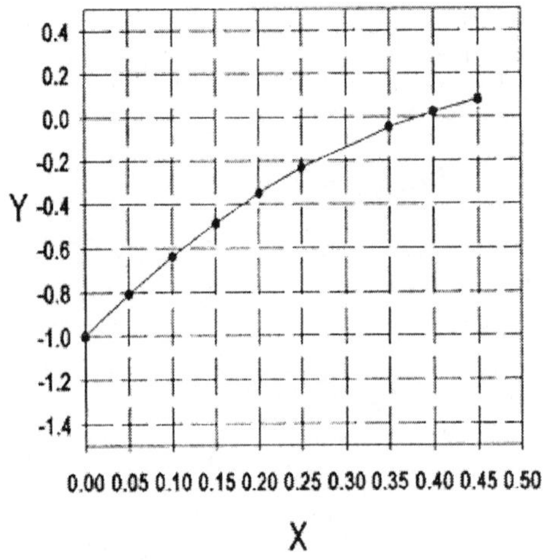

x	y
0.0	-1.000
0.05	-0.810
0.10	-0.639
0.15	-0.487
0.20	-0.352
0.25	-0.234
0.35	-0.047
0.40	+0.024
0.45	+0.081
~0.383	0.00

(e) From the plot, x ≈ 0.383 *M*

[A] = x = 0.383 *M*; [B] = 2x = 0.766 *M*

[C] = 1.00 - 2x = 0.234 *M*

Using the K_{eq} expression as a check:

$K_{eq} = 0.25$; $\dfrac{(0.234)^2}{(0.383)(0.766)^2} = 0.24$; the estimated values are reasonable.

15.68 $K_{eq} = \dfrac{P_{O_2} \times P_{CO}^2}{P_{CO_2}^2} = 1 \times 10^{-13}$; $P_{O_2} = (0.03)(1\text{ atm}) = 0.03\text{ atm}$

$P_{CO} = (0.002)(1\text{ atm}) = 0.002\text{ atm}$; $P_{CO_2} = (0.12)(1\text{ atm}) = 0.12\text{ atm}$

$Q = \dfrac{(0.03)(0.002)^2}{(0.12)^2} = 8.3 \times 10^{-6} = 8 \times 10^{-6}$

Since Q > K_{eq}, the system will shift to the left to attain equilibrium. Thus a catalyst that promoted the attainment of equilibrium would result in a lower CO content in the exhaust.

15.69 The patent claim is false. A catalyst does not alter the position of equilibrium in a system, only the rate of approach to the equilibrium condition.

Integrative Exercises

15.70 (a) (i) $K_{eq} = [Na^+]/[Ag^+]$ (ii) $K_{eq} = [Hg^{2+}]^3 / [Al^{3+}]^2$

(iii) $K_{eq} = [Zn^{2+}]\ P_{H_2} / [H^+]^2$

(b) According to Table 4.5, the activity series of the metals, a metal can be oxidized by any metal cation below it on the table.

(i) Ag^+ is far below Na, so the reaction will proceed to the right and K_{eq} will be large.

(ii) Al^{3+} is above Hg, so the reaction will not proceed to the right and K_{eq} will be small.

(iii) H^+ is below Zn, so the reaction will proceed to the right and K_{eq} will be large.

(c) $K_{eq} < 1$ for this reaction, so Fe^{2+} (and thus Fe) is above Cd on the table. In other words, Cd is below Fe. The value of K_{eq}, 0.06, is small but not extremely small, so Cd will be only a few rows below Fe.

15.71 (a) $AgCl(s) \rightarrow Ag^+(aq) + Cl^-(aq)$

(b) $K_{eq} = [Ag^+][Cl^-]$

(c) Using thermodynamic data from Appendix C, calculate ΔH for the reaction in part (a).

$$\Delta H^\circ = \Delta H_f^\circ\ Ag^+(aq) + \Delta H_f^\circ\ Cl^-(aq) - \Delta H_f^\circ\ AgCl(s)$$

$$\Delta H^\circ = +105.90\ kJ - 167.2\ kJ - (-127.0\ kJ) = +65.7\ kJ$$

The reaction is endothermic (heat is a reactant), so the solubility of AgCl(s) in $H_2O(l)$ will increase with increasing temperature.

15.72 (a) At equilibrium, the forward and reverse reactions occur at **equal** rates.

(b) One expects the reactants to be favored at equilibrium since they are lower in energy.

(c) A catalyst lowers the activation energy for both the forward and reverse reactions; the "hill" would be lower.

(d) Since the activation energy is lowered for both processes, the new rates would be equal and the ratio of the rate constants, k_f / k_r, would remain unchanged.

(e) Since the reaction is endothermic (the energy of the reactants is lower than that of the products, ΔE is positive), the value of K should increase with increasing temperature.

15.73 Consider the energy profile for an exothermic reaction.

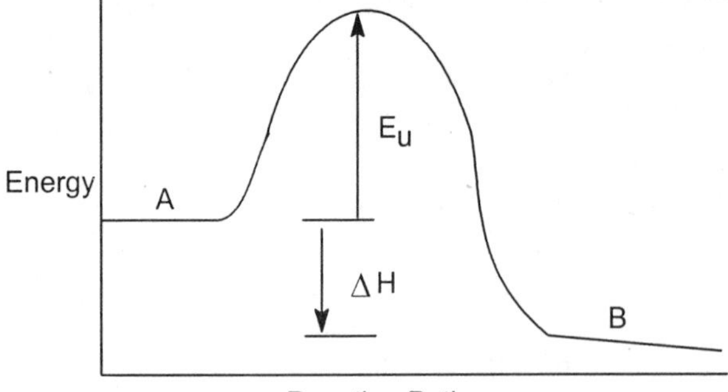

The activation energy in the forward direction, E_{af}, equals E_u, and the activation energy in the reverse reaction, E_{ar}, equals $E_u - \Delta H$. (The same is true for an endothermic reaction because the sign of ΔH is the positive and $E_{ar} < E_{af}$). For the reaction in question,

$$K = \frac{k_f}{k_r} = \frac{A_f e^{-E_{af}/RT}}{A_r e^{-E_{ar}/RT}}$$

Since the ln form of the Arrhenius equation is easier to manipulate, we will consider ln K.

$$\ln K = \ln\left(\frac{k_f}{k_r}\right) = \ln k_f - \ln k_r = \frac{-E_{af}}{RT} + \ln A - \left[\frac{-E_{ar}}{RT} + \ln A_r\right]$$

Substituting E_u for E_{af} and $(E_u - \Delta H)$ for E_{ar}

$$\ln K = \frac{-E_u}{RT} + \ln A_f - \left[\frac{-(E_u - \Delta H)}{RT} + \ln Ar\right]; \ \ln K = \frac{-E_u + (E_u - \Delta H)}{RT} + \ln A_f - \ln A_r$$

$$\ln K = \frac{-\Delta H}{RT} + \ln\frac{A_f}{A_r}$$

For the catalyzed reaction, $E_c < E_u$ and $E_{af} = E_c$, $E_{ar} = E_c - \Delta H$. The catalyst does not change the value of ΔH.

$$\ln K_c = \frac{-E_c + (E_c - \Delta H)}{RT} + \ln A_f - \ln A_r$$

$$\ln K_c = \frac{-\Delta H}{RT} + \frac{\ln A_f}{A_r}$$

Thus, assuming A_f and A_r are not changed by the catalyst, $\ln K = \ln K_c$ and $K = K_c$.

15.74 (a) $$P = \frac{gRT}{\mathcal{M}V} = \frac{0.300 \text{ g } H_2S}{34.08 \text{ g/mol } H_2S} \times \frac{298 \text{ K}}{5.00 \text{ L}} \times \frac{0.08206 \text{ L} \bullet \text{atm}}{\text{mol} \bullet \text{K}} = 0.043053$$
$$= 0.0431 \text{ atm}$$

(b) $K_{eq} = P_{NH_3} \times P_{H_2S}$. Before solid is added, $Q = P_{NH_3} \times P_{H_2S} = 0 \times 0.0431 = 0$. $Q < K$ and the reaction will proceed to the right. However, no $NH_4SH(s)$ is present to produce $NH_3(g)$, so the reaction cannot proceed.

(c)

	$NH_4HS(s) \rightleftharpoons$	$NH_3(g)$	+ $H_2S(g)$
initial		0 atm	0.043053 atm
change		+x atm	+x atm
equil.		+x atm	(0.043053+x) atm

Since Q < K initially (part (a)), P_{H_2S} must increase along with P_{NH_3} until equilibrium is established.

$K_{eq} = P_{NH_3} \times P_{H_2S}$; $0.120 = (x)(0.043053 + x)$; $0 = x^2 + 0.043053\,x - 0.120$

Solve for x using the quadratic formula.

$$x = \frac{-0.043053 \pm \sqrt{(0.043053)^2 - 4(1)(-0.120)}}{2}; \quad x = 0.3256 = 0.326 \text{ atm}$$

$P_{NH_3} = 0.326$ atm; $P_{H_2S} = (0.043053 + 0.3256)$ atm $= 0.3686 = 0.369$ atm

(d) $$\chi_{H_2S} = \frac{P_{H_2S}}{P_t} = \frac{0.3686 \text{ atm}}{(0.3256 + 0.3686) \text{ atm}} = 0.531$$

(e) The minimum amount of $NH_4HS(s)$ required is slightly greater than the number of moles NH_3 present at equilibrium. We can calculate the mol NH_3 present at equilibrium using the ideal-gas equation.

$$n_{NH_3} = \frac{P_{NH_3}V}{RT} = 0.3256 \text{ atm} \times \frac{\text{K} \cdot \text{mol}}{0.08206 \text{ L} \cdot \text{atm}} \times \frac{5.00 \text{ L}}{298 \text{ K}} = 0.06657$$

$$= 0.0666 \text{ mol } NH_3$$

$$0.06657 \text{ mol } H_2S \times \frac{1 \text{ mol } NH_4SH}{1 \text{ mol } H_2S} \times \frac{51.12 \text{ g } NH_4SH}{1 \text{ mol } NH_4SH} = 3.40 \text{ g } NH_4SH$$

The minimum amount is slightly greater than 3.40 g NH_4HS.

15.75 $K_{eq} = P_{CO}^2 / P_{CO_2}$. mole % = pressure %. Since the total pressure is 1 atm, mol %/100 = mol fraction = partial pressure.

Temp (K)	P_{CO_2} **(atm)**	P_{CO} **(atm)**	K_{eq}
1123	0.0623	0.9377	14.1
1223	0.0132	0.9868	78.8
1323	0.0037	0.9963	2.7×10^2
1473	0.0006	0.9994	1.7×10^3 (2×10^3 to 1 sig fig)

Because K_{eq} grows larger with increasing temperature, the reaction must be endothermic in the forward direction.

15.76 (a) $H_2O(l) \rightleftharpoons H_2O(g)$; $K_{eq} = P_{H_2O}$

(b) At 30°C, the vapor pressure of $H_2O(l)$ is 31.82 torr. $K_{eq} = P_{H_2O}$ = 31.82 torr

K_{eq} = 31.82 torr × 1 atm/760 torr = 0.041868 = 0.04187 atm

(c) From part (b), the value of K_{eq} is the vapor pressure of the liquid at that temperature. By definition, vapor pressure = atmospheric pressure = 1 atm at the normal boiling point. K_{eq} = 1 atm

15.77 (a)

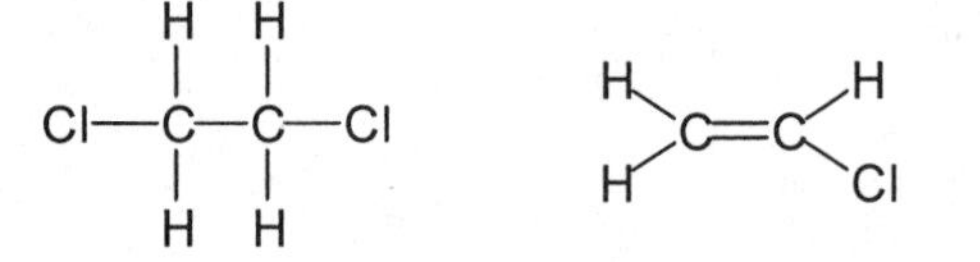

C-C B.O. = 1 C=C B.O. = 2

(b) ΔH = D(bond breaking) - D (bond making)

E1: ΔH = D(C=C) + D(Cl-Cl) - D(C-C) - 2D (C-Cl)

ΔH = 614 + 242 - 348 - 2(328) = -148 kJ

E2: ΔH = D(C-C) + D(C-H) + D(C-Cl) - D(C=C) - D(H-Cl)

= 348 + 413 + 328 - 614 - 431 = +44 kJ

(c) E1 is exothermic with Δn = -1. The yield of $C_2H_4Cl_2(g)$ would decrease with increasing temperature and with increasing container volume.

(d) E2 is endothermic with Δn = 1. The yield of C_2H_3Cl would increase with increasing temperature and with increasing container volume.

(e) The boiling points of the reactants and products are: C_2H_4, -103.7°C; Cl_2, -34.6°C, $C_2H_4Cl_2$, +83.5°C; C_2H_3Cl, -13.4°C; HCl, -84.9°C.

Because the products of E1 and E2 are optimized by different conditions, carry out the two equilibria in separate reactors. Since E1 is exothermic and Δn = -1, the reactor on the left should be as small and cold as possible to maximize yield of $C_2H_4Cl_2$. At temperatures below 83.5°C, $C_2H_4Cl_2$ will condense to the liquid and it can be easily transferred to the second (right) reactor.

Since E_2 is endothermic, the reactor on the right should be as large and hot as possible to optimize production of C_2H_3Cl. The outlet stream will be a mixture of $C_2H_3Cl(g)$, $C_2H_4Cl_2(g)$ and HCl(g). This mixture could be run through a heat exchanger to condense and subsequently recycle $C_2H_4Cl_2(l)$. Since HCl has a lower boiling point than C_2H_3Cl, it cannot be removed by condensation. The $C_2H_3Cl(g)$ / HCl(g) mixture could be bubbled through a basic aqueous solution such as $NaHCO_3(aq)$ or NaOH(aq) to remove HCl(g), leaving pure $C_2H_3Cl(g)$.

Other details of reactor design such as the use of catalysts to speed up these reactions, the exact costs and benefits of heat exchange, recycling unreacted components and separation and recovery of products are issues best resolved by chemical engineers.

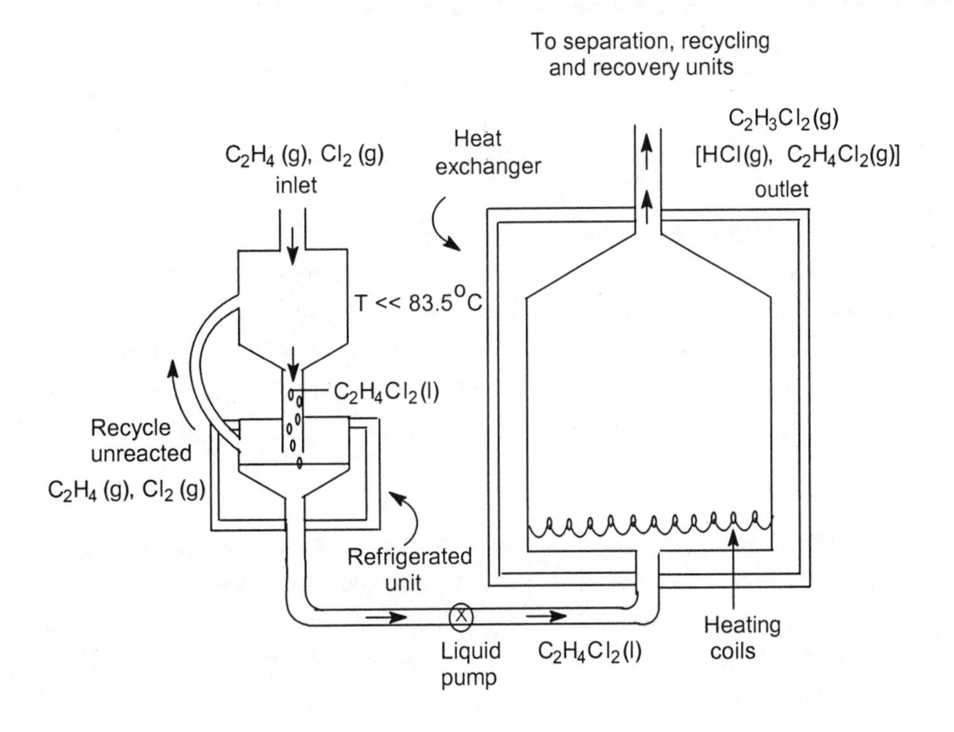

16 Acid-Base Equilibria

Arrhenius and Brønsted-Lowry Acids and Bases

16.1 Solutions of HCl and H_2SO_4 taste sour, turn litmus paper red (are acidic), neutralize solutions of bases, react with active metals to form $H_2(g)$ and conduct electricity. The two solutions have these properties in common because both solutes are strong acids. That is, they both ionize completely in H_2O to form $H^+(aq)$ and an anion. (The first ionization step for H_2SO_4 is complete, but the second is not.) The presence of ions enables the solutions to conduct electricity; the presence of $H^+(aq)$ in excess of 1×10^{-7} *M* accounts for all the other listed properties.

16.2 When NaOH dissolves in water, it completely dissociates to form $Na^+(aq)$ and $OH^-(aq)$. CaO is the oxide of a metal; it dissolves in water according to the following process: $CaO(s) + H_2O(l) \rightarrow Ca^{2+}(aq) + 2OH^-(aq)$. Thus, the properties of both solutions are dominated by the presence of $OH^-(aq)$. Both solutions taste bitter, turn litmus paper blue (are basic), neutralize solutions of acids and conduct electricity.

16.3 (a) According to the Arrhenius definition, an acid when dissolved in water increases $[H^+]$. According to the Brønsted-Lowry definition, an acid is capable of donating H^+, regardless of physical state. The Arrhenius definition of an acid is confined to an aqueous solution; the Brønsted-Lowry definition applies to any physical state.

(b) $HCl(g) + NH_3(g) \rightarrow NH_4^+Cl^-(s)$ HCl is the B-L (Brønsted-Lowry) acid; it donates an H^+ to NH_3 to form NH_4^+. NH_3 is the B-L base; it accepts the H^+ from HCl.

16.4 (a) According to the Arrhenius definition, a base when dissolved in water increases $[OH^-]$. According to the Brønsted-Lowry theory, a base is an H^+ acceptor regardless of physical state. A Brønsted-Lowry base is not limited to aqueous solution and need not contain OH^- or produce it in aqueous solution.

(b) $NH_3(g) + H_2O(l) \rightleftharpoons NH_4^+(aq) + OH^-(aq)$ When NH_3 dissolves in water, it accepts H^+ from H_2O (B-L definition). In doing so, OH^- is produced (Arrhenius definition).

Note that the OH^- produced was originally part of the H_2O molecule, not part of the NH_3 molecule.

16.5 *Analyze/Plan*. Follow the logic in Sample Exercise 16.1. A conjugate base has one less H^+ than its conjugate acid. *Solve*:

(a) HSO_3^- (b) $C_2H_3O_2^-$ (c) $HAsO_4^{2-}$ (d) NH_3

16.6 A conjugate acid has one more H^+ than its conjugate base.

(a) $H_2AsO_4^-$ (b) $CH_3NH_3^+$ (c) HSO_4^- (d) H_3PO_4

16.7 *Analyze/Plan.* Use the definitions of B-L acids and bases, and conjugate acids and bases to make the designations. Evaluate the changes going from reactant to product to inform your choices. *Solve*:

	B-L acid	+	B-L base	⇌	Conjugate acid	+	Conjugate base
(a)	$NH_4^+(aq)$		$CN^-(aq)$		$HCN(aq)$		$NH_3(aq)$
(b)	$H_2O(l)$		$(CH_3)_3N(aq)$		$(CH_3)_3NH^+(aq)$		$OH^-(aq)$
(c)	$HCHO_2(aq)$		$PO_4^{3-}(aq)$		$HPO_4^{2-}(aq)$		$CHO_2^-(aq)$

16.8

	B-L acid	+	B-L base	⇌	Conjugate acid	+	Conjugate base
(a)	$H_2O(l)$		$CHO_2^-(aq)$		$HCHO_2(aq)$		$OH^-(aq)$
(b)	$HSO_4^-(aq)$		$HCO_3^-(aq)$		$H_2CO_3(aq)$		$SO_4^{2-}(aq)$
(c)	$H_3O^+(aq)$		$HSO_3^-(aq)$		$H_2SO_3(aq)$		$H_2O(l)$

16.9 *Analyze/Plan.* Follow the logic in Sample Exercise 16.2. *Solve*:

(a) Acid: $HC_2O_4^-(aq) + H_2O(l) \rightleftharpoons C_2O_4^{2-}(aq) + H_3O^+(aq)$

B-L acid B-L base conj. base conj. acid

Base: $HC_2O_4^-(aq) + H_2O(l) \rightleftharpoons H_2C_2O_4(aq) + OH^-(aq)$

B-L base B-L acid conj. acid conj. base

(b) $H_2C_2O_4$ is the conjugate acid of $HC_2O_4^-$.

$C_2O_4^{2-}$ is the conjugate base of $HC_2O_4^-$.

16.10 (a) $H_2C_6H_7O_5^-(aq) + H_2O(l) \rightleftharpoons H_3C_6H_7O_5(aq) + OH^-(aq)$

(b) $H_2C_6H_7O_5^-(aq) + H_2O(l) \rightleftharpoons HC_6H_7O_5^{2-}(aq) + H_3O^+(aq)$

(c) $H_3C_6H_7O_5$ is the conjugate acid of $H_2C_6H_7O_5^-$

$HC_6H_7O_5^{2-}$ is the conjugate base of $H_2C_6H_7O_5^-$

16.11 *Analyze/Plan.* Based on the chemical formula, decide whether the acid is strong, weak or negligible. Is it one of the known seven strong acids (Section 16.5)? Also check Figure 16.4. Remove a single H and decrease the particle charge by one to write the formula of the conjugate base. *Solve*:

(a) weak, NO_2^- (b) strong, HSO_4^- (c) weak, PO_4^{3-}

(d) negligible, CH_3^- (e) weak, CH_3NH_2

16.12 (a) weak, $HC_2H_3O_2$ (b) weak, H_2CO_3 (c) strong, OH^-

(d) negligible, HCl (e) weak, NH_4^+

16.13 *Analyze/Plan.* Given chemical formula, determine strength of acids and bases by checking the known strongs (Section 16.5). Recall the paradigm 'The **stronger** the acid, the **weaker** its conjugate base, and vice versa!' *Solve:*

(a) HBr. It is one of the seven strong acids (Section 16.5).

(b) F^-. HCl is a stronger acid than HF, so F^- is the stronger conjugate base.

16.14 (a) HNO_3. It is one of the seven strong acids (Section 16.5). Also, in a series of oxyacids with the same central atom (N), the acid with more O atoms is stronger (Section 16.10).

(b) NH_3. When NH_3 and H_2O are combined, as in $NH_3(aq)$, NH_3 acts as the B-L base, accepting H^+ from H_2O. NH_3 has the greater tendency to accept H^+. For binary hydrides, base strength increases going to the left across a row of the periodic chart (Section 16.10).

16.15 *Analyze/Plan.* Acid-base equilibria favor formation of the weaker acid and base. Compare the relative strengths of the substances acting as acids on opposite sides of the equation. (Bases can also be compared; the conclusion should be the same.) *Solve:*

	Base	+	Acid	$\rightleftharpoons$	Conjugate acid	+	Conjugate base
(a)	F^- (aq)	+	HCO_3^-(aq)		HF(aq)	+	CO_3^{2-}(aq)

HF is a stronger acid than HCO_3^-, so the equilibrium lies to the left.

(b)	O^{2-}(aq)	+	H_2O(l)	$\rightleftharpoons$	OH^-(aq)	+	OH^-(aq)

H_2O is a stronger acid than OH^-, so the equilibrium lies to the right.

(c)	HS^-(aq)	+	$HC_2H_3O_2$(aq)	$\rightleftharpoons$	H_2S(aq)	+	$C_2H_3O_2^-$(aq)

$HC_2H_3O_2$ is a stronger acid than H_2S, so the equilibrium lies to the right.

16.16 Acid-base equilibria favor formation of the weaker acid and base. Compare the relative strengths of the substances acting as acids on opposite sides of the equation. (Bases can also be compared; the conclusion should be the same.)

	Base	+	Acid	$\rightleftharpoons$	Conjugate acid	+	Conjugate base
(a)	Cl^-(aq)	+	H_3O^+(aq)		HCl(aq)	+	H_2O(l)

HCl is a stronger acid than H_3O^+ (Figure 16.4) , so the equilibrium lies to the left.

(b)	H_2O(l)	+	HNO_2(aq)	$\rightleftharpoons$	H_3O^+(aq)	+	NO_2^-(aq)

H_3O^+ is a stronger acid than HNO_2, so the equilibrium lies to the left.

(c)	NO_3^-(aq)	+	H_2O(l)	$\rightleftharpoons$	HNO_3(aq)	+	OH^-(aq)

HNO_3 is a stronger acid than H_2O (Solution 16.14), so the equilibrium lies to the right.

Autoionization of Water

16.17 (a) *Autoionization* is the ionization of a neutral molecule (in the absence of any other reactant) into an anion and a cation. The equilibrium expression for the autoionization of water is $H_2O(l) \rightleftharpoons H^+(aq) + OH^-(aq)$.

(b) Pure water is a poor conductor of electricity because it contains very few ions. Ions, mobile charged particles, are required for the conduction of electricity in liquids.

(c) If a solution is *acidic*, it contains more H^+ than OH^- ($[H^+] > [OH^-]$).

16.18 (a) $H_2O(l) \rightleftharpoons H^+(aq) + OH^-(aq)$

(b) $K_w = [H^+][OH^-]$. The $[H_2O(l)]$ is omitted because water is a pure liquid. The molarity (mol/L) of pure solids or liquids does not change as equilibrium is established, so it is usually omitted from equilibrium expressions.

(c) If a solution is *basic*, it contains more OH^- than H^+ ($[OH^-] > [H^+]$).

16.19 *Analyze/Plan*. Follow the logic in Sample Exercise 16.5. In pure water at 25°C, $[H^+] = [OH^-] = 1 \times 10^{-7}$ *M*. If $[H^+] > 1 \times 10^{-7}$ *M*, the solution is acidic; if $[H^+] < 1 \times 10^{-7}$ *M*, the solution is basic. *Solve*:

(a) $[H^+] = \dfrac{K_w}{[OH^-]} = \dfrac{1.0 \times 10^{-14}}{5 \times 10^{-5}\,M} = \mathbf{2 \times 10^{-10}\,M} < 1 \times 10^{-7}\,M$; basic

(b) $[H^+] = \dfrac{K_w}{[OH^-]} = \dfrac{1.0 \times 10^{-14}}{3.2 \times 10^{-9}\,M} = \mathbf{3.1 \times 10^{-6}\,M} > 1 \times 10^{-7}\,M$; acidic

(c) $[OH^-] = 100[H^+]$; $K_w = [H^+] \times 100[H^+] = 100[H^+]^2$;

$[H^+] = (K_w/100)^{1/2} = \mathbf{1.0 \times 10^{-8}\,M} < 1 \times 10^{-7}\,M$; basic

16.20 In pure water at 25°C, $[H^+] = [OH^-] = 1 \times 10^{-7}$ *M*. If $[OH^-] > 1 \times 10^{-7}$ *M*, the solution is basic; if $[OH^-] < 1 \times 10^{-7}$ *M*, the solution is acidic.

(a) $[OH^-] = \dfrac{K_w}{[H^+]} = \dfrac{1.0 \times 10^{-14}}{4.1 \times 10^{-3}\,M} = \mathbf{2.4 \times 10^{-12}\,M} < 1 \times 10^{-7}\,M$; acidic

(b) $[OH^-] = \dfrac{K_w}{[H^+]} = \dfrac{1.0 \times 10^{-14}}{3.5 \times 10^{-9}\,M} = \mathbf{2.9 \times 10^{-6}\,M} < 1 \times 10^{-7}\,M$; basic

(c) $[H^+] = 10[OH^-]$; $K_w = 10[OH^-][OH^-] = 10[OH^-]^2$

$[OH^-] = (K_w/10)^{1/2} = \mathbf{3.2 \times 10^{-8}\,M} < 1 \times 10^{-7}\,M$; acidic

16.21 *Analyze/Plan*. Follow the logic in Sample Exercise 16.4. Note that the value of the equilibrium constant (in this case, K_w) changes with temperature. *Solve*:

At 0°C, $K_w = 1.2 \times 10^{-15} = [H^+][OH^-]$.

In pure water, $[H^+] = [OH^-]$; $2.4 \times 10^{-14} = [H^+]^2$; $[H^+] = (1.2 \times 10^{-15})^{1/2}$

$[H^+] = [OH^-] = 3.5 \times 10^{-8}$ *M*

16.22 $K_w = [D^+][OD^-]$; for pure D_2O, $[D^+] = [OD^-]$; $8.9 \times 10^{-16} = [D^+]^2$; $[D^+] = [OD^-] = 3.0 \times 10^{-8}$ *M*

The pH Scale

16.23 *Analyze/Plan.* A change of one pH unit (in either direction) is:

$\Delta pH = pH_2 - pH_1 = -(\log[H^+]_2 - \log [H^+]_1) = -\log \dfrac{[H^+]_2}{[H^+]_1} = \pm 1$. The antilog of +1 is 10; the antilog of -1 is 1×10^{-1}. Thus, a ΔpH of one unit represents an increase or decrease in $[H^+]$ by a factor of 10. *Solve:*

(a) $\Delta pH = \pm 2.00$ is a change of $10^{2.00}$; $[H^+]$ changes by a factor of 100.

(b) $\Delta pH = \pm 0.5$ is a change of $10^{0.50}$; $[H^+]$ changes by a factor of 3.2.

16.24 $[H^+]_A = 500\,[H^+]_B$ From Solution 16.23, $\Delta pH = -\log \dfrac{[H^+]_B}{[H^+]_A}$

$$\Delta pH = -\log \frac{[H^+]_B}{500\,[H^+]_B} = -\log\left(\frac{1}{500}\right) = 2.70$$

The pH of solution A is 2.70 pH units lower than the pH of solution B, because $[H^+]_A$ is 500 times greater than $[H^+]_B$. The greater $[H^+]$, the lower the pH of the solution.

16.25 (a) $K_w = [H^+][OH^-]$. If NaOH is added to water, it dissociates into $Na^+(aq)$ and $OH^-(aq)$. This increases $[OH^-]$ and necessarily decreases $[H^+]$. When $[H^+]$ decreases, pH increases.

(b) 0.00003 M = 3×10^{-5} *M*. On figure 16.5, this is $[H^+] > 1 \times 10^{-5}$ but $< 1 \times 10^{-4}$. The pH is between 4 and 5. by calculation:

$pH = -\log [H^+] = -\log (3 \times 10^{-5}) = 4.5$ If pH < 7, the solution is acidic.

(c) pH = 7.8 is between pH 7 and pH 8 on Figure 16.5, closer to pH = 8. A pH = 8, $[H^+] = 1 \times 10^{-8}$; at pH = 7, $[H^+] = 1 \times 10^{-7} = 10 \times 10^{-8}$. A good estimate is 3×10^{-8} *M* H^+.

By calculation: $[H^+] = 10^{-pOH} = 10^{-7.8} = 2 \times 10^{-8}$ *M*

At pH = 7, $[OH^-] = 1 \times 10^{-7}$; at pH = 8, $[OH^-] = 1 \times 10^{-6} = 10 \times 10^{-7}$.

Since pH = 7.8 is closer to pH = 8, we estimate 7×10^{-7} *M* OH^-.

By calculation: pOH = 14.0 - 7.8 = 6.2

$[OH^-] = 10^{-pOH} = 10^{-6.2} = 6 \times 10^{-7}$ *M* OH^-

16.26 (a) $K_w = [H^+][OH^-]$. If HNO_3 is added to water, it ionizes to form $H^+(aq)$ and $NO_3^-(aq)$. This increases $[H^+]$ and necessarily decreases $[OH^-]$. When $[H^+]$ increases, pH decreases.

(b) On Figure 16.5, 1.4×10^{-2} *M* OH^- is between pH = 12 (1×10^{-2} *M* OH^-) and pH 13 (1×10^{-2} *M* OH^-), slightly higher than pH = 12, so we estimate pH = 12.1. By calculation:

$[H^+] = K_w / [OH^-] = 10 \times 10^{-14} / 1.4 \times 10^{-2}\ M = 7.1 \times 10^{-13}\ M$

pH = -log (7.1×10^{-13}) = 12.15. If pH > 7, the solution is basic.

(c) pH = 6.6 is midway between pH 6 and pH 7 on Figure 16.5.

At pH = 7, $[H^+] = 1 \times 10^{-7}$; at pH = 6, $[H^+] = 1 \times 10^{-6} = 10 \times 10^{-7}$.

A reasonable estimate is 5×10^{-7} *M* H^+. By calculation:

pH = 6.6, $[H^+] = 10^{-pH} = 10^{-6.6} \times 10^{-7} = 3 \times 10^{-7}$

At pH = 6 $[OH^-] = 1 \times 10^{-7}$; at pH = 7, $[OH^-] = 1 \times 10^{-7} = 1 \times 10^{-8}$.

A reasonable estimate 5×10^{-7} M OH^-. By calculation:

pOH = 14.0 - 6.6 = 7.4; $[OH^-] = 10^{-pOH} = 10^{-7.4} = 4 \times 10^{-8}$ *M* OH^-.

16.27 *Analyze/Plan*. At 25°C, $[H^+][OH^-] = 1 \times 10^{-14}$; pH = pOH = 14. Use these relationships to complete the table. If pH < 7, the solution is acidic; if pH > 7, the solution is basic. *Solve*:

$[H^+]$	$[OH^-]$	pH	pOH	acidic or basic
7.5×10^{-3} *M*	1.3×10^{-12} *M*	2.12	11.88	acidic
2.8×10^{-5} *M*	3.6×10^{-10} *M*	4.56	9.44	acidic
5.6×10^{-9} *M*	1.8×10^{-6} *M*	8.25	5.75	basic
5.0×10^{-9} *M*	2.0×10^{-6} *M*	8.30	5.70	basic

16.28

pH	pOH	$[H^+]$	$[OH^-]$	acidic or basic
6.21	7.79	6.2×10^{-7} *M*	1.6×10^{-8} *M*	acidic
3.87	10.13	1.3×10^{-4} *M*	7.4×10^{-11} *M*	acidic
2.46	11.54	3.5×10^{-3} *M*	2.9×10^{-12} *M*	acidic
10.75	3.25	1.8×10^{-11} *M*	5.6×10^{-4} *M*	basic

16.29 *Analyze/Plan*. Given pH and a new value of the equilibrium constant K_w, calculate equilibrium concentrations of $H^+(aq)$ and $OH^-(aq)$. The definition of pH remains pH = $-\log[H^+]$. *Solve*:

$pH = 7.40; [H^+] = 10^{-pH} = 10^{-7.40} = 4.0 \times 10^{-8}$ *M*

$K_w = 2.4 \times 10^{-14} = [H^+][OH^-]; [OH^-] = 2.4 \times 10^{-14} / [H^+]$

$[OH^-] = 2.4 \times 10^{-14} / 4.0 \times 10^{-8} = 6.0 \times 10^{-7}$ *M*

Alternately, pH + pOH = pK_w. At 37°C, pH + pOH = $-\log(2.4 \times 10^{-14})$

pH + pOH = 13.62; pOH = 13.62 - 7.40 = 6.22

$[OH^-] = 10^{-pOH} = 10^{-6.22} = 6.0 \times 10^{-7}$ *M*

16.30 The pH ranges from 5.2-5.6; pOH ranges from (14.9-5.2 =) 8.8 to (14.0-5.6 =) 8.4.

$[H^+] = 10^{-pH}$, $[OH^-] = 10^{-pOH}$

$[H^+] = 10^{-5.2} = 6.31 \times 10^{-6} = 6 \times 10^{-6}$ *M*; $[H^+] = 10^{-5.6} = 2.51 \times 10^{-6} = 3 \times 10^{-6}$ *M*

The range of $[H^+]$ is 6×10^{-6} *M* to 3×10^{-6} *M*.

$[OH^-] = 10^{-8.8} = 1.58 \times 10^{-9} = 2 \times 10^{-9}$ *M*; $[OH^-] = 10^{-8.4} = 3.98 \times 10^{-9} = 4 \times 10^{-9}$ *M*.

The range of $[OH^-]$ is 2×10^{-9} *M* to 4×10^{-9} *M*.

[The pH has one decimal place, so concentrations are reported to one sig fig.]

Strong Acids and Bases

16.31 (a) A *strong* acid is completely ionized in aqueous solution; a strong acid is a strong electrolyte.

(b) For a strong acid such as HCl, $[H^+]$ = initial acid concentration. $[H^+] = 0.500$ *M*

(c) HCl, HBr, HI

16.32 (a) A *strong* base is completely dissociated in aqueous solution; a strong base is a strong electrolyte.

(b) $Sr(OH)_2$ is a soluble strong base.

$Sr(OH)_2(aq) \rightarrow Sr^{2+}(aq) + 2OH^-(aq)$

0.125 *M* $Sr(OH)_2(aq)$ = 0.250 *M* OH^-

(c) Base strength should not be confused with solubility. Base strength describes the tendency of a dissolved molecule (formula unit for ionic compounds such as $Mg(OH)_2$) to dissociate into cations and hydroxide ions. $Mg(OH)_2$ is a strong base because each $Mg(OH)_2$ unit that dissolves also dissociates into $Mg^{2+}(aq)$ and $OH^-(aq)$. $Mg(OH)_2$ is not very soluble, so relatively few $Mg(OH)_2$ units dissolve when the solid compound is added to water.

16.33 *Analyze/Plan*. Follow the logic in Sample Exercise 16.8. Strong acids are completely ionized, so $[H^+]$ = original acid concentration, and pH = $-\log[H^+]$. For the solutions obtained by dilution, use the 'dilution' formula, $M_1V_1 = M_2V_2$, to calculate molarity of the acid. *Solve:*

(a) 8.5×10^{-3} *M* HBr = 8.5×10^{-3} *M* H^+; pH = $-\log(8.5 \times 10^{-3}) = 2.07$

(b) $\frac{1.52 \text{ g } HNO_3}{0.575 \text{ L soln}} \times \frac{1 \text{ mol } HNO_3}{63.02 \text{ g } HNO_3} = 0.041947 = 0.0419 \; M \; HNO_3$

$[H^+] = 0.0419 \; M$; pH = -log (0.041947) = 1.377

(c) $M_c \times V_c = M_d \times V_d$; $0.250 \; M \times 0.00500 \text{ L} = ? \; M \times 0.0500 \text{ L}$

$$M_d = \frac{0.250 \; M \times 0.00500 \text{ L}}{0.0500 \text{ L}} = 0.0250 \; M \text{ HCl}$$

$[H^+] = 0.0250 \; M$; pH = -log (0.0250) = 1.602

(d) $$[H^+]_{total} = \frac{\text{mol } H^+ \text{ from HBr} + \text{mol } H^+ \text{ from HCl}}{\text{total L solution}}$$

$$[H^+]_{total} = \frac{(0.100 \; M \text{ HBr} \times 0.0100 \text{ L}) + (0.200 \; M \times 0.0200 \text{ L})}{0.0300 \text{ L}}$$

$$[H^+]_{total} = \frac{1.00 \times 10^{-3} \text{ mol } H^+ + 4.00 \times 10^{-3} \text{ mol } H^+}{0.0300 \text{ L}} = 0.1667 = 0.167 \; M$$

pH = -log (0.1667 M) = 0.778

16.34 For a strong acid, which is completely ionized, $[H^+]$ = the initial acid concentration.

(a) $0.0575 \; M \; HNO_3 = 0.0575 \; M \; H^+$; pH = -log (0.0575) = 1.24

(b) $\frac{0.723 \text{ g } HClO_4}{2.00 \text{ L soln}} \times \frac{1 \text{ mol } HClO_4}{100.5 \text{ g } HClO_4} = 3.597 \times 10^{-3} = 3.60 \times 10^{-3} \; M \; HClO_4$

$[H^+] = 3.60 \times 10^{-3} \; M$; pH = -log (3.60×10^{-3}) = 2.444

(c) $M_c \times V_c = M_d \times V_d$

$1.00 \; M \text{ HCl} \times 5.00 \text{ mL HCl} = M_d \text{ HCl} \times 750 \text{ mL HCl}$

$$M_d \text{ HCl} = \frac{1.00 \; M \times 5.00 \text{ mL}}{750 \text{ mL}} = 6.67 \times 10^{-3} \; M \text{ HCl} = 6.67 \times 10^{-3} \; M \; H^+$$

pH = -log (6.67×10^{-3}) = 2.176

(d) $$[H^+]_{total} = \frac{\text{mol } H^+ \text{ from HCl} + \text{mol } H^+ \text{ from HI}}{\text{total L solution}}; \quad \text{mol} = M \times \text{L}$$

$$[H^+]_{total} = \frac{(0.020 \; M \text{ HCl} \times 0.0500 \text{ L}) + (0.010 \; M \text{ HI} \times 0.125 \text{ L})}{0.175 \text{ L}}$$

$$[H^+]_{total} = \frac{1.0 \times 10^{-3} \text{ mol } H^+ + 1.25 \times 10^{-3} \text{ mol } H^+}{0.175 \text{ L}} = 0.01286 = 0.013 \; M$$

pH = -log (0.01286) = 1.89

16.35 *Analyze/Plan.* Follow the logic in Sample Exercise 16.9. Strong bases dissociate completely upon dissolving. pOH = -log[OH^-]; pH = 14 - pOH.

(a) Pay attention to the formula of the base to get [OH^-]. *Solve*:

$[OH^-] = 2[Sr(OH)_2] = 2(1.5 \times 10^{-3}\ M) = 3.0 \times 10^{-3}\ M\ OH^-$ (see Exercise 16.32(b))

pOH = -log (3.0 × 10^{-3}) = 2.52; pH = 14 - pOH = 11.48

(b) mol/LiOH = g LiOH/molar mass LiOH. [OH^-] = [LiOH]. *Solve*:

$$\frac{2.250 \text{ g LiOH}}{0.2500 \text{ L soln}} \times \frac{1 \text{ mol LiOH}}{23.948 \text{ g LiOH}} = 0.37581 = 0.3758\ M \text{ LiOH} = [OH^-]$$

pOH = -log (0.37581) = 0.4250; pH = 14 - pOH = 13.5750

(c) Use the dilution formula to get the [NaOH] = [OH^-]. *Solve*:

$M_c \times V_c = M_d \times V_d$; 0.175 M × 0.00100 L = ? M × 2.00 L

$$M_d = \frac{0.0175\ M \times 0.00100 \text{ L}}{2.00 \text{ L}} = 8.75 \times 10^{-5}\ M \text{ NaOH} = [OH^-]$$

pOH = -log (8.75 × 10^{-5}) = 4.058; pH = 14 - pOH = 9.942

(d) Consider total mol OH^- from KOH and $Ca(OH)_2$, as well as total solution volume. *Solve*:

$$[OH^-]_{total} = \frac{\text{mol } OH^- \text{ from KOH} + \text{mol } OH^- \text{ from } Ca(OH)_2}{\text{total L soln}}$$

$$[OH^-]_{total} = \frac{(0.105\ M \times 0.00500 \text{ L}) + 2(9.5 \times 10^{-2} \times 0.0150 \text{ L})}{0.0200 \text{ L}}$$

$$[OH^-]_{total} = \frac{0.525 \times 10^{-3} \text{ mol } OH^- + 2.85 \times 10^{-3} \text{ mol } OH^-}{0.0200 \text{ L}} = 0.16875 = 0.17\ M$$

pOH = -log (0.16875) = 0.77; pH = 14 - pOH = 13.23

(9.5 × 10^{-2} M has 2 sig figs, so the [OH^-] has 2 sig figs and pH and pOH have 2 decimal places.)

16.36 For a strong base, which is completely dissociated, [OH^-] = the initial base concentration. Then, pOH = -log [OH^-] and pH = 14 - pOH.

(a) 0.0050 M KOH = 0.0050 M OH^-; pOH = -log (0.0050) = 2.30;

pH = 14 - 1.30 = 11.70

(b) $$\frac{2.055 \text{ g KOH}}{0.5000 \text{ L}} \times \frac{1 \text{ mol KOH}}{56.106 \text{ g KOH}} = 0.073254 = 0.07325\ M = [OH^-]$$

pOH = -log (0.073254) = 1.1352; pH = 14 - pOH = 12.8648

(c) $M_c \times V_c = M_d \times V_d$

$0.250\ M\ Ca(OH)_2 \times 10.0\ mL = M_d\ Ca(OH)_2 \times 500\ mL$

$$M_d\ Ca(OH)_2 = \frac{0.250\ M\ Ca(OH)_2 \times 10.0\ mL}{500.0\ mL} = 0.00500\ M\ Ca(OH)_2$$

$Ca(OH)_2(aq) \rightarrow Ca^{2+}(aq) + 2OH^-(aq)$

$[OH^-] = 2[Ca(OH)_2] = 2(5.00 \times 10^{-3}\ M) = 0.0100\ M$

pOH = -log (0.0100) = 2.000; pH = 14 - pOH = 12.00

(d) $$[OH^-]_{total} = \frac{\text{mol } OH^- \text{ from NaOH} + \text{mol } OH^- \text{ from } Ba(OH)_2}{\text{total L solution}}$$

$$\frac{(7.5 \times 10^{-3}\ M \times 0.0300\ L) + 2(0.015\ M \times 0.0100\ L)}{0.0400\ L}$$

$$[OH^-]_{total} = \frac{2.25 \times 10^{-4}\ \text{mol } OH^- + 3.0 \times 10^{-4}\ \text{mol } OH^-}{0.0400\ L} = 0.0131 = 0.013\ M\ OH^-$$

pOH = -log (0.0131) = 1.88; pH = 14 - pOH = 12.12

16.37 *Analyze/Plan.* pH → pOH → $[OH^-]$ = [NaOH]. *Solve*:

pOH = 14 - pH = 14.00 - 11.50 = 2.50

pOH = 2.50 = $-\log[OH^-]$; $[OH^-] = 10^{-2.50} = 3.2 \times 10^{-3}\ M$

$[OH^-] = [NaOH] = 3.2 \times 10^{-3}\ M$

16.38 pOH = 14 - pH = 14.00 - 12.00 = 2.00

pOH = 2.00 = $-\log[OH^-]$; $[OH^-] = 10^{-2.00} = 1.0 \times 10^{-2}\ M$

$[OH^-] = 2[Ca(OH)_2]$; $[Ca(OH)_2] = [OH^-] / 2 = 1.0 \times 10^{-2} / 2 = 5.0 \times 10^{-3}\ M$

16.39 *Analyze/Plan.* $NaH(aq) \rightarrow Na^+(aq) + H^-(aq)$

$H^-(aq) + H_2O(l) \rightarrow H_2(g) + OH^-(aq)$

Thus, initial $[NaH] = [OH^-]$; $[NaH] = g\ NaH/[\mathcal{M}(NaH) \times V]$. *Solve*:

$$[NaH] = \frac{\text{mol NaH}}{\text{L solution}} = 15.00\ g\ NaH \times \frac{1\ \text{mol NaH}}{24.00\ g\ NaH} \times \frac{1}{2.50\ L} = 0.250\ M$$

$[OH^-] = 0.250\ M$; pOH = -log (0.250) = 0.602, pH = 14 - pOH = 13.400

16.40 Upon dissolving, Li_2O dissociates to form Li^+ and O^{2-}. According to Equation 16.19, O^{2-} is completely protonated in aqueous solution.

Thus, initial $[Li_2O] = [O_2^-]$; $[OH^-] = 2[O^{2-}] = 2[Li_2O]$

$$[Li_2O] = \frac{\text{mol } Li_2O}{\text{L solution}} = 2.50 \text{ g } Li_2O \times \frac{1 \text{ mol } Li_2O}{29.88 \text{ g } Li_2O} \times \frac{1}{1.20 \text{ L}} = 0.06972 = 0.697\ M$$

$[OH^-] = 0.1394 = 0.139\ M$; pOH = 0.856 pH = 14.00 - pOH = 13.144

Weak Acids

16.41 *Analyze/Plan.* Remember that K_{eq} = [products]/[reactants]. If $H_2O(l)$ appears in the equilibrium reaction, it will **not** appear in the K_a expression, because it is a pure liquid. *Solve:*

(a) $HBrO_2(aq) \rightleftharpoons H^+(aq) + BrO_2^-(aq)$; $K_a = \dfrac{[H^+][BrO_2^-]}{[HBrO_2]}$

$HBrO_2(aq) + H_2O(l) \rightleftharpoons H_3O^+(aq) + BrO_2^-(aq)$; $K_a = \dfrac{[H_3O^+][BrO_2^-]}{[HBrO_2]}$

(b) $HC_3H_5O_2(aq) \rightleftharpoons H^+(aq) + C_3H_5O_2^-(aq)$; $K_a = \dfrac{[H^+][C_3H_5O_2^-]}{[HC_3H_5O_2]}$

$HC_3H_5O_2(aq) + H_2O(l) \rightleftharpoons H_3O^+(aq) + C_3H_5O_2^-(aq)$; $K_a = \dfrac{[H_3O^+][C_3H_5O_2^-]}{[HC_3H_5O_2]}$

16.42 (a) $HC_6H_5O(aq) \rightleftharpoons H^+(aq) + C_6H_5O^-(aq)$; $K_a = \dfrac{[H^+][C_6H_5O^-]}{[HC_6H_5O]}$

$HC_6H_5O(aq) + H_2O(l) \rightleftharpoons H_3O^+(aq) + C_6H_5O^-(aq)$; $K_a = \dfrac{[H_3O^+][C_6H_5O^-]}{[HC_6H_5O]}$

(b) $HCO_3^-(aq) \rightleftharpoons H^+(aq) + CO_3^{2-}(aq)$; $K_a = \dfrac{[H^+][CO_3^{2-}]}{[HCO_3^-]}$

$HCO_3^-(aq) + H_2O(aq) \rightleftharpoons H_3O^+(aq) + CO_3^{2-}(aq)$; $K_a = \dfrac{[H_3O^+][CO_3^{2-}]}{[HCO_3^-]}$

16.43 *Analyze/Plan.* Follow the logic in Sample Exercise 16.10. *Solve:*

$HC_3H_5O_3(aq) \rightleftharpoons H^+(aq) + C_3H_5O_3^-(aq)$; $K_a = \dfrac{[H^+][C_3H_5O_3^-]}{[HC_3H_5O_3]}$

$[H^+] = [C_3H_5O_3^-] = 10^{-2.44} = 3.63 \times 10^{-3} = 3.6 \times 10^{-3}\ M$

$[HC_3H_5O_3] = 0.10 - 3.63 \times 10^{-3} = 0.0964 = 0.096\ M$

$$K_a = \frac{(3.63 \times 10^{-3})^2}{(0.0964)} = 1.4 \times 10^{-4}$$

16.44 $HC_8H_7O_2(aq) \rightleftharpoons H^+(aq) + C_8H_7O_2^-(aq)$; $K_a = \dfrac{[H^+][C_8H_7O_2^-]}{[HC_8H_7O_2]}$

$[H^+] = [C_8H_7O_2^-] = 10^{-2.68} = 2.09 \times 10^{-3} = 2.1 \times 10^{-3}\ M$

$[HC_8H_7O_2] = 0.085 - 2.09 \times 10^{-3} = 0.0829 = 0.083\ M$

$$K_a = \frac{(2.09 \times 10^{-3})^2}{0.0829} = 5.3 \times 10^{-5}$$

16.45 *Analyze/Plan.* Write the equilibrium reaction and the K_a expression. Use % ionization to get equilibrium concentration of $[H^+]$, and by stoichiometry, $[X^-]$ and $[HX]$. Calculate K_a. *Solve:*

$[H^+] = 0.094 \times [HX]_{initial} = 0.0188 = 0.019\ M$

	$HX(aq)$	$\rightleftharpoons$	$H^+(aq)$	+	$X^-(aq)$
initial	0.200 M		0		0
equil.	(0.200 - 0.019) M		0.019 M		0.019 M

$$K_a = \frac{[H^+][X^-]}{[HX]} = \frac{(0.0188)^2}{0.181} = 2.0 \times 10^{-3}$$

16.46 $[H^+] = 0.110 \times [CH_2ClCOOH]_{initial} = 0.0110\ M$

	$CH_2ClCOOH(aq)$	$\rightleftharpoons$	$H^+(aq)$	+	$CH_2ClCOO^-(aq)$
initial	0.100 M		0		0
equil.	0.089M		0.0110 M		0.0110 M

$$K_a = \frac{[H^+][CH_2ClCOO^-]}{[CH_2ClCOOH]} = \frac{(0.0110)^2}{0.089} = 1.4 \times 10^{-3}$$

16.47 *Analyze/Plan.* Write the equilibrium reaction and the K_a expression.
$[H^+] = 10^{-pH} = [C_2H_3O_2^-]$ $\quad [HC_2H_3O_2] = x - [H^+]$.
Substitute into the K_a expression and solve for x. *Solve:*

$[H^+] = 10^{-pH} = 10^{-2.90} = 1.26 \times 10^{-3} = 1.3 \times 10^{-3}\ M$

$$K_a = 1.8 \times 10^{-5} = \frac{[H^+][C_2H_3O_2^-]}{[HC_2H_3O_2]} = \frac{(1.26 \times 10^{-3})^2}{(x - 1.26 \times 10^{-3})}$$

$1.8 \times 10^{-5} (x - 1.26 \times 10^{-3}) = (1.26 \times 10^{-3})^2;$

$1.8 \times 10^{-5}\ x = 1.585 \times 10^{-6} + 2.266 \times 10^{-8} = 1.608 \times 10^{-6};$

$x = 0.08931 = 0.089\ M\ HC_2H_3O_2$

16.48 $[H^+] = 10^{-pH} = 10^{-2.70} = 1.995 \times 10^{-3} = 2.0 \times 10^{-3}\ M$

$$K_a = 6.8 \times 10^{-4} = \frac{[H^+][F^-]}{[HF]} = \frac{(1.995 \times 10^{-3})^2}{x - 1.995 \times 10^{-3}}$$

$6.8 \times 10^{-4}(x - 1.995 \times 10^{-3}) = (1.995 \times 10^{-3})^2;$

$6.8 \times 10^{-4} x = 1.357 \times 10^{-6} + 3.981 \times 10^{-6} = 5.338 \times 10^{-6}$

$x = 7.85 \times 10^{-3} = 7.9 \times 10^{-3}\ M$ HF

mol = $M \times L = 7.85 \times 10^{-3}\ M \times 0.500\ L = 3.925 \times 10^{-3} = 3.9 \times 10^{-3}$ mol HF

16.49 *Analyze/Plan*. Follow the logic in Sample Exercise 16.11. Write K_a, construct the equilibrium table, solve for x = $[H^+]$, then get equilibrium $[C_7H_5O_2^-]$ and $[HC_7H_5O_2]$ by substituting $[H^+]$ for x. *Solve*:

	$HC_7H_5O_2(aq)$	$\rightleftharpoons$	$H^+(aq)$	+	$C_7H_5O_2^-(aq)$
initial	0.050 *M*		0		0
equil.	(0.050 - x) *M*		x *M*		x *M*

$$K_a = \frac{[H^+][C_7H_5O_2^-]}{[HC_7H_5O_2]} = \frac{x^2}{(0.050 - x)} \approx \frac{x^2}{0.050} = 6.3 \times 10^{-5}$$

$x^2 = 0.050\ (6.3 \times 10^{-5})$; $x = 1.8 \times 10^{-3}\ M = [H^+] = [H_3O^+] = [C_7H_5O_2^-]$

$[HC_7H_5O_2] = 0.050 - 0.0018 = 0.048\ M$

Check. $\frac{1.8 \times 10^{-3}\ M\ H^+}{0.050\ M\ HC_7H_5O_2} \times 100 = 3.6\%$ ionization; the assumption is valid

16.50

	$HClO(aq)$	$\rightleftharpoons$	$H^+(aq)$	+	$ClO^-(aq)$
initial	0.0075 *M*		0		0
equil.	(0.0075 - x) *M*		x *M*		x *M*

$$K_a = \frac{[H^+][ClO^-]}{[HClO]} = \frac{x^2}{(0.0075 - x)} \approx \frac{x^2}{0.0075} = 3.0 \times 10^{-8}$$

$x^2 = 0.0075\ (3.0 \times 10^{-8})$; $x = 1.5 \times 10^{-5}\ M = [H^+] = [H_3O^+] = [ClO^-]$

$[HClO] = 7.5 \times 10^{-3} - 1.5 \times 10^{-5} = 7.485 \times 10^{-3} = 7.5 \times 10^{-3}\ M$

Check. $\frac{4.7 \times 10^{-5}\ M\ H^+}{0.0075\ M\ HClO} \times 100 = 0.20\%$ ionization; the assumption is valid

16.51 *Analyze/Plan*. Follow the logic in Sample Exercise 16.11.

(a) *Solve*:

	$HC_3H_5O_2(aq)$	$\rightleftharpoons$	$H^+(aq)$	+	$C_3H_5O_2^-(aq)$
initial	0.095 *M*		0		0
equil	(0.095 - x) *M*		x *M*		x *M*

$$K_a = \frac{[H^+][C_3H_5O_2^-]}{[HC_3H_5O_2]} = \frac{x^2}{(0.095 - x)} \approx \frac{x^2}{0.095} = 1.3 \times 10^{-5}$$

$x^2 = 0.095(1.3 \times 10^{-5})$; $x = 1.111 \times 10^{-3} = 1.1 \times 10^{-3}\ M\ H^+$; pH = 2.95

Check. $\frac{1.1 \times 10^{-3}\ M\ H^+}{0.095\ M\ HC_3H_5O_2} \times 100 = 1.2\%$ ionization; the assumption is valid

(b) *Solve*:

$$K_a = \frac{[H^+][CrO_4^{2-}]}{[HCrO_4^-]} = \frac{x^2}{(0.100 - x)} \approx \frac{x^2}{0.100} = 3.0 \times 10^{-7}$$

$x^2 = 0.100(3.0 \times 10^{-7})$; $x = 1.732 \times 10^{-4} = 1.7 \times 10^{-4}$ M H^+

pH = -log(1.732×10^{-4}) = 3.7614 = 3.76

Check. $\frac{1.7 \times 10^{-4}\ M\ H^+}{0.100\ M\ HCrO_4^-} \times 100 = 0.17\%$ ionization; the assumption is valid

(c) Follow the logic in Sample Exercise 16.14. pOH = -log[OH^-]. pH = 14 - pOH

Solve:

	$C_5H_5N(aq) + H_2O(l)$	$\rightleftharpoons$	$C_5H_5NH^+(aq)$ +	OH^-
initial	0.120 *M*		0	0
equil	(0.120 - x) *M*		x *M*	x *M*

$$K_b = \frac{[C_5H_5NH^+][OH^-]}{[C_5H_5N]} = \frac{x^2}{(0.120 - x)} \approx \frac{x^2}{0.120} = 1.7 \times 10^{-9}$$

$x^2 = 0.120(1.7 \times 10^{-9})$; $x = 1.428 \times 10^{-5} = 1.4 \times 10^{-5}$ *M* OH^-; pH = 9.15

Check. $\frac{1.4 \times 10^{-5}\ M\ OH^-}{0.120\ M\ C_5H_5N} \times 100 = 0.011\%$ ionization; the assumption is valid

16.52 (a)

	$HOCl(aq)$	$\rightleftharpoons$	$H^+(aq)$ +	$OCl^-(aq)$
initial	0.125 *M*		0	0
equil	(0.125 - x) *M*		x *M*	x *M*

$$K_a = \frac{[H^+][OCl^-]}{[HOCl]} = \frac{x^2}{(0.125 - x)} \approx \frac{x^2}{0.125} = 3.0 \times 10^{-8}$$

$x^2 = 0.125\ (3.0 \times 10^{-8})$; $x = [H^+] = 6.1 \times 10^{-5}$ *M*, pH = 4.21

Check. $\frac{6.1 \times 10^{-5}\ M\ H^+}{0.125\ M\ HOCl} \times 100 = 0.049\%$ ionization; the assumption is valid

(b) $$K_a = \frac{[H^+][C_6H_5O^-]}{[C_6H_5OH]} = \frac{x^2}{(0.0085 - x)} \approx \frac{x^2}{0.0085} = 1.3 \times 10^{-10}$$

$x^2 = 0.0085\ (1.3 \times 10^{-10})$; $x = [H^+] = 1.1 \times 10^{-6}$ *M*, pH = 5.98

Check. Clearly 1.1×10^{-6} *M* H^+ is small compared to 8.5×10^{-3} *M* C_6H_5OH, and the assumption is valid.

(c) $HONH_2(aq) + H_2O(l) \rightleftharpoons HONH_3^+(aq) + OH^-(aq)$

	$HONH_2(aq)$	$HONH_3^+(aq)$	$OH^-(aq)$
initial	0.095 M	0	0
equil	(0.095 - x) M	x M	x M

$$K_b = \frac{[HONH_3^+][OH^-]}{[HONH_2]} = \frac{x^2}{(0.095 - x)} \approx \frac{x^2}{0.095} = 1.1 \times 10^{-8}$$

$x^2 = 0.095\ (1.1 \times 10^{-8})$; $x = [OH^-] = 3.2 \times 10^{-5}\ M$, pH = 9.51

Check. $\frac{3.2 \times 10^{-5}\ M\ OH^-}{0.095\ M\ HONH_2} \times 100 = 0.034\%$ ionization; the assumption is valid

16.53 *Analyze/Plan.* $K_a = 10^{-pK_a}$. Follow the logic in Sample Exercise 16.11. *Solve:*

Let $[H^+] = [NC_7H_4SO_3^-] = z$. K_a = antilog (-2.32) = $4.79 \times 10^{-3} = 4.8 \times 10^{-3}$

$\frac{z^2}{0.10 - z} = 4.79 \times 10^{-3}$. Since K_a is relatively large, solve the quadratic.

$z^2 = 4.79 \times 10^{-3} z - 4.79 \times 10^{-4} = 0$

$$z = \frac{-4.79 \times 10^{-3} \pm \sqrt{(4.79 \times 10^{-3})^2 - 4(1)(-4.79 \times 10^{-4})}}{2(1)} = \frac{-4.79 \times 10^{-3} \pm \sqrt{1.937 \times 10^{-3}}}{2}$$

$z = 1.96 \times 10^{-2} = 2.0 \times 10^{-2}\ M\ H^+$; pH = -log (1.96×10^{-2}) = 1.71

16.54 Calculate the initial concentration of $HC_9H_7O_4$.

$$2 \text{ tablets} \times \frac{500 \text{ mg}}{\text{tablet}} \times \frac{1 \text{ g}}{1000 \text{ mg}} \times \frac{1 \text{ mol } HC_9H_7O_4}{180.2 \text{ g } HC_9H_7O_4} = 0.005549 = 0.00555 \text{ mol } HC_9H_7O_4$$

$$\frac{0.005549 \text{ mol } HC_9H_7O_4}{0.250 \text{ L}} = 0.002220 = 0.0222\ M\ HC_9H_7O_4$$

$HC_9H_7O_4(aq) \rightleftharpoons C_9H_7O_4^-(aq) + H^+(aq)$

	$HC_9H_7O_4(aq)$	$C_9H_7O_4^-(aq)$	$H^+(aq)$
initial	0.0222 M	0 M	0 M
equil	(0.0222 - x)	x M	x M

$$K_a = 3.3 \times 10^{-4} = \frac{[H^+][C_7H_9O_4^-]}{[HC_7H_9O_4]} = \frac{x^2}{(0.00222 - x)}$$

Assuming x is small compared to 0.0222,

$x^2 = 0.00222\ (3.3 \times 10^{-4})$; $x = [H^+] = 2.7 \times 10^{-3}\ M$

$\frac{2.7 \times 10^{-3}\ M\ H^+}{0.0222\ M\ HC_9H_7O_4} \times 100 = 12\%$ ionization; the assumption is not valid

Using the quadratic formula, $x^2 + 3.3 \times 10^{-4} x - 7.325 \times 10^{-6} = 0$

$$x = \frac{-3.3 \times 10^{-4} \pm \sqrt{(3.3 \times 10^{-4})^2 - 4(1)(-7.325 \times 10^{-6})}}{2(1)} = \frac{-3.3 \times 10^{-4} \pm \sqrt{2.941 \times 10^{-5}}}{2}$$

$x = 2.547 \times 10^{-3} = 2.5 \times 10^{-3}$ *M* H^+; pH = $-\log(2.547 \times 10^{-3}) = 2.594 = 2.59$

16.55 *Analyze/Plan.* Follow the logic in Sample Exercise 16.12. *Solve:*

(a)

	$HN_3(aq)$	$\rightleftharpoons$	$H^+(aq)$	+	$N_3^-(aq)$
initial	0.400 *M*		0		0
equil	(0.400 - x) *M*		x *M*		x *M*

$$K_a = \frac{[H^+][N_3^-]}{[HN_3]} = 1.9 \times 10^{-5};\ \frac{x^2}{(0.400 - x)} \approx \frac{x^2}{0.400} = 1.9 \times 10^{-5}$$

$$x = 0.00276 = 2.8 \times 10^{-3}\ M = [H^+];\ \%\ \text{ionization} = \frac{2.76 \times 10^{-3}}{0.400} \times 100 = 0.69\%$$

(b) $1.9 \times 10^{-5} \approx \dfrac{x^2}{0.100}$; $x = 0.00138 = 1.4 \times 10^{-3}\ M\ H^+$

$$\%\ \text{ionization} = \frac{1.38 \times 10^{-3}\ M\ H^+}{0.100\ M\ HN_3} \times 100 = 1.4\%$$

(c) $1.9 \times 10^{-5} \approx \dfrac{x^2}{0.0400}$; $x = 8.72 \times 10^{-4} = 8.7 \times 10^{-4}\ M\ H^+$

$$\%\ \text{ionization} = \frac{8.72 \times 10^{-4}\ M\ H^+}{0.0400\ M\ HN_3} \times 100 = 2.2\%$$

Check. Notice that a tenfold dilution [part (a) versus part (c)] leads to a slightly more than threefold increase in percent ionization.

16.56 (a) $HCrO_4^-(aq) \rightleftharpoons H^+(aq) + CrO_4^{2-}(aq)$

$$K_a = 3.0 \times 10^{-7} = \frac{[H^+][CrO_4^{2-}]}{[HCrO_4^-]} = \frac{x^2}{0.250 - x}$$

$x^2 = 0.250\ (3.0 \times 10^{-7})$; $x = 2.74 \times 10^{-4} = 2.7 \times 10^{-4}\ M\ H^+$

$$\%\ \text{ionization} = \frac{2.74 \times 10^{-4}\ M\ H^+}{0.250\ M\ HCrO_4^-} \times 100 = 0.11\%$$

(b) $\dfrac{x^2}{0.0800} \approx 3.0 \times 10^{-7}$; $x = 1.549 \times 10^{-4} = 1.5 \times 10^{-4}\ M\ H^+$

$$\%\ \text{ionization} = \frac{1.55 \times 10^{-4}\ M\ H^+}{0.0800\ M\ HCrO_4^-} \times 100 = 0.19\%$$

(c) $\dfrac{x^2}{0.0200} \approx 3.0 \times 10^{-7}$; $x = 7.746 = 7.7 \times 10^{-5}\ M\ H^+$

$$\%\ \text{ionization} = \frac{7.75 \times 10^{-5}\ M\ H^+}{0.0200\ M\ HCrO_4^-} \times 100 = 0.39\%$$

16.57 *Analyze/Plan.* Let the weak acid be HX. $HX(aq) \rightleftharpoons H^+(aq) + X^-(aq)$. Solve the K_a expression symbolically for $[H^+]$ in terms of [HX]. Substitute into the formula for % ionization, $([H^+]/[HX]) \times 100$. *Solve*:

$$K_a = \frac{[H^+][X^-]}{[HX]};\ [H^+] = [X^-] = y;\ K_a = \frac{y^2}{[HX] - y};\ \text{assume that \% ionization is small}$$

$$K_a = \frac{y^2}{[HX]};\ y = K_a^{1/2}[HX]^{1/2}$$

$$\%\text{ ionization} = \frac{y}{[HX]} \times 100 = \frac{K_a^{1/2}[HX]^{1/2}}{[HX]} \times 100 = \frac{K_a^{1/2}}{[HX]^{1/2}} \times 100$$

That is, percent ionization varies inversely as the square root of concentration HX.

16.58 $HX(aq) \rightleftharpoons H^+(aq) + X^-(aq);\ K_a = \frac{[H^+][X^-]}{[HX]}$

$[H^+] = [X^-]$; assume the % ionization is small; $K_a = \frac{[H^+]^2}{[HX]};\ [H^+] = K_a^{1/2}[HX]^{1/2}$

$pH = -\log K_a^{1/2}[HX]^{1/2} = -\log K_a^{1/2} - \log [HX]^{1/2};\ pH = -1/2 \log K_a - 1/2 \log [HX]$

This is the equation of a straight line, where the intercept is $-1/2 \log K_a$, the slope is -1/2, and the independent variable is log [HX].

16.59 Analyze/Plan. Follow the logic in Sample Exercise 16.13. Citric acid is a triprotic acid with three K_a values that do not differ by more than 10^3. We must consider all three steps. Also, $C_6H_5O_7^{3-}$ is only produced in step 3. *Solve*:

$$H_3C_6H_5O_7(aq) \rightleftharpoons H^+(aq) + H_2C_6H_5O_7^-(aq) \qquad K_{a1} = 7.4 \times 10^{-4}$$
$$H_2C_6H_5O_7^-(aq) \rightleftharpoons H^+(aq) + HC_6H_5O_7^{2-}(aq) \qquad K_{a2} = 1.7 \times 10^{-5}$$
$$HC_6H_5O_7^{2-}(aq) \rightleftharpoons H^+(aq) + C_6H_5O_7^{3-}(aq) \qquad K_{a3} = 4.0 \times 10^{-7}$$

To calculate the pH of a 0.050 *M* solution, assume initially that only the first ionization is important:

	$H_3C_6H_5O_7(aq)$	$\rightleftharpoons$	$H^+(aq)$	+	$H_2C_6H_5O_7^-(aq)$
initial	0.050 *M*		0		0
equil.	(0.050 - x) *M*		x *M*		x *M*

$$K_{a1} = \frac{[H^+][H_2C_6H_5O_7^-]}{[H_3C_6H_5O_7]} = \frac{x^2}{(0.050 - x)} = 7.4 \times 10^{-4}$$

$x^2 = (0.050 - x)(7.4 \times 10^{-4});\ x^2 \approx (0.050)(7.4 \times 10^{-4});\ x = 0.00608 = 6.1 \times 10^{-3}\ M$

Since this value for x is rather large in relation to 0.050, a better approximation for x can be obtained by substituting this first estimate into the expression for x^2, then solving again for x:

$$x^2 = (0.050 - x)(7.4 \times 10^{-4}) = (0.050 - 6.08 \times 10^{-3})(7.4 \times 10^{-4})$$
$$x^2 = 3.2 \times 10^{-5};\ x = 5.7 \times 10^{-3}\ M$$

(This is the same result obtained from the quadratic formula.)

The correction to the value of x, though not large, is significant. Does the second ionization produce a significant additional concentration of H^+?

	$H_2C_6H_5O_7^-$ (aq)	$\rightleftharpoons$	H^+(aq)	+	$HC_6H_5O_7^{2-}$ (aq)
initial	5.7×10^{-5} *M*		5.7×10^{-3} *M*		0
equil.	$(5.7 \times 10^{-3} - y)$		$(5.7 \times 10^{-3} + y)$		y

$$K_{a2} = \frac{[H^+][HC_6H_5O_7^{2-}]}{[H_2C_6H_5O_7^-]} = 1.7 \times 10^{-5}; \quad \frac{(5.7 \times 10^{-3} + y)(y)}{(5.7 \times 10^{-3} - y)} = 1.7 \times 10^{-5}$$

Assume that y is small relative to 5.7×10^{-3}; that is, that additional ionization of $H_2C_6H_5O_7^-$ is small, then

$$\frac{(5.7 \times 10^{-3})y}{(5.7 \times 10^{-3})} = 1.7 \times 10^{-5}\ M; \quad y = 1.7 \times 10^{-5}\ M$$

This value is indeed small compared to 5.7×10^{-3} *M*; $[H^+]$ and pH are determined by the first ionization step. If we were only interested in pH, we could stop here. However, to calculate $[C_6H_5O_7^{3-}]$, we must consider the third ionization, with adjusted
$[H^+] = 5.7 \times 10^{-3} + 1.7 \times 10^{-5} = 5.72 \times 10^{-5}\ M\ (= 5.7 \times 10^{-5})$

	$HC_6H_5O_7^{2-}$	$\rightleftharpoons$	H^+(aq)	+	$HC_6H_5O_7^{3-}$(aq)
initial	1.75×10^{-5} *M*		5.72×10^{-3} *M*		0
equil.	$1.7 \times 10^{-5} - z$		$5.72 \times 10^{-3} + z$		z

$$K_{a3} = \frac{[H^+][C_6H_5O_7^{3-}]}{[HC_6H_5O_7^{2-}]} = \frac{(5.72 \times 10^{-3} + z)(z)}{(1.7 \times 10^{-5} - z)} = 4.0 \times 10^{-7}$$

Assume z is small relative to 5.72×10^{-3}, but not relative to 1.7×10^{-5}.

$(4.0 \times 10^{-7})(1.7 \times 10^{-5} - z) = 5.72 \times 10^{-3}\ z$; $6.8 \times 10^{-12} - 4.0 \times 10^{-7}\ z = 5.72 \times 10^{-3}\ z$;

$6.8 \times 10^{-12} = 5.72 \times 10^{-3}\ z + 4.0 \times 10^{-7}\ z = 5.72 \times 10^{-3}\ z$; $z = 1.19 \times 10^{-9} = 1.2 \times 10^{-9}\ M$

$[C_6H_5O_7^{3-}] = 1.2 \times 10^{-9}\ M$; $[H^+] = 5.72 \times 10^{-3}\ M + 1.2 \times 10^{-9}\ M = 5.72 \times 10^{-3}\ M$

$pH = -\log(5.72 \times 10^{-3}) = 2.24$

Note that neither the second nor third ionizations contributed significantly to $[H^+]$ and pH.

16.60 $H_2C_4H_4O_6(aq) \rightleftharpoons H^+(aq) + HC_4H_4O_6^-(aq) \quad K_{a1} = 1.0 \times 10^{-3}$
$HC_4H_4O_6^-(aq \rightleftharpoons H^+(aq) + C_4H_4O_6^{2-}(aq) \quad K_{a2} = 4.6 \times 10^{-5}$

Begin by calculating the $[H^+]$ from the first ionization. The equilibrium concentrations are $[H^+] = [HC_4H_4O_6^-] = x$, $[H_2C_4H_4O_6] = 0.25 - x$.

$$K_{a1} = \frac{[H^+][HC_4H_4O_6^-]}{[H_2C_4H_4O_6]} = \frac{x^2}{0.25 - x}; \; x^2 + 1.0 \times 10^{-3} x - 2.5 \times 10^{-4} = 0$$

Using the quadratic formula, $x = 1.532 \times 10^{-2} = 0.015$ *M* H^+ from the first ionization. Next calculate the H^+ contribution from the second ionization.

	$HC_4H_4O_6^-(aq)$	$\rightleftharpoons$	$H^+(aq)$	+	$C_4H_4O_6^{2-}(aq)$
initial	0.015		0.015		0
equil.	(0.015 - y)		(0.015 + y)		y

$$K_{a2} = \frac{(0.015 + y)(y)}{(0.015 - y)} = 4.6 \times 10^{-5}$$; assuming y is small compared to 0.015,

$y = 4.6 \times 10^{-5}$ *M* $HC_4H_4O_6^{2-}(aq)$

This assumption is reasonable, since 4.6×10^{-5} is only 0.3% of 0.015. $[H^+] = 0.015$ *M* (first ionization) + 4.6×10^{-5} (second ionization). Since 4.6×10^{-5} is 0.3% of 0.015 *M*, it can be safely ignored when calculating total $[H^+]$. Thus, pH = -log(0.01532) = 1.18148 = 1.181.

Assumptions:

1) The ionization can be treated as a series of steps (valid by Hess's Law).

2) The extent of ionization in the second step (y) is small relative to that from the first step (valid for this acid and initial concentration). This assumption was used twice, to calculate the value of y from K_{a2} and to calculate total $[H^+]$ and pH.

Weak Bases

16.61 All Brønsted-Lowry bases contain at least one nonbonded (lone) pair of electrons to attract H^+.

16.62 Organic amines (neutral molecules with nonbonded pairs on N atoms) and anions that are the conjugate bases of weak acids function as weak bases.

16.63 Analyze/Plan. Remember that K_{eq} = [products]/[reactants]. If $H_2O(l)$ appears in the equilibrium reaction, it will **not** appear in the K_b expression, because it is a pure liquid. *Solve:*

(a) $(CH_3)_2NH(aq) + H_2O(l) \rightleftharpoons (CH_3)_2NH_2^+(aq) + OH^-(aq)$; $K_b = \frac{[(CH_3)_2NH_2^+][OH^-]}{[(CH_3)_2NH]}$

(b) $CO_3^{2-}(aq) + H_2O(l) \rightleftharpoons HCO_3^-(aq) + OH^-(aq)$; $K_b = \frac{[HCO_3^-][OH^-]}{[CO_3^{2-}]}$

(c) $CHO_2^-(aq) + H_2O(l) \rightleftharpoons HCHO_2(aq) + OH^-(aq)$; $K_b = \frac{[HCHO_2][OH^-]}{[CHO_2^-]}$

16.64 (a) $C_3H_7NH_2(aq) + H_2O(l) \rightleftharpoons C_3H_7NH_3^+(aq) + OH^-(aq)$; $K_b = \frac{[C_3H_7NH_3^+][OH^-]}{[C_3H_7NH_2]}$

(b) $HPO_4^{2-}(aq) + H_2O(l) \rightleftharpoons H_2PO_4^-(aq) + OH^-(aq)$; $K_b = \frac{[H_2PO_4^-][OH^-]}{[HPO_4^{2-}]}$

(c) $C_6H_5CO_2^-(aq) + H_2O(l) \rightleftharpoons C_6H_5CO_2H(aq) + OH^-(aq)$; $K_b = \frac{[C_6H_5CO_2H][OH^-]}{[C_6H_5CO_2^-]}$

16.65 *Analyze/Plan.* Follow the logic in Sample Exercise 16.14. *Solve:*

$$C_2H_5NH_2(aq) + H_2O(l) \rightleftharpoons C_2H_5NH_3^+(aq) + OH^-(aq)$$

	$C_2H_5NH_2$	$C_2H_5NH_3^+$	OH^-
initial	0.075 *M*	0	0
equil.	(0.075 - x) *M*	x *M*	x *M*

$$K_b = \frac{[C_2H_5NH_3^+][OH^-]}{[C_2H_5NH_2]} = \frac{(x)(x)}{(0.075 - x)} \approx \frac{x^2}{0.075} = 6.4 \times 10^{-4}$$

$x^2 = 0.075\ (6.4 \times 10^{-4})$; $x = [OH^-] = 6.9 \times 10^{-3}\ M$; pH = 11.84

Check. $\frac{6.9 \times 10^{-3}\ M\ OH^-}{0.075\ M\ C_2H_5NH_2} \times 100 = 9.2\%$ ionization; the assumption is **not** valid

To obtain a more precise result, the K_b expression is rewritten in standard quadratic form and solved via the quadratic formula.

$$\frac{x^2}{0.075 - x} = 6.4 \times 10^{-4};\ x^2 + 6.4 \times 10^{-4}\,x - 4.8 \times 10^{-5} = 0$$

$$x = \frac{b \pm \sqrt{b^2 - 4ac}}{2a} = \frac{-6.4 \times 10^{-4} \pm \sqrt{(6.4 \times 10^{-4})^2 - 4(1)(-4.8 \times 10^{-5})}}{2}$$

$x = 6.61 \times 10^{-3} = 6.6 \times 10^{-3}\ M\ OH^-$; pOH = 2.18, pH = 14.00 - pOH = 11.82

Note that the pH values obtained using the two algebraic techniques are very similar.

16.66

$$BrO^-(aq) + H_2O(l) \rightleftharpoons HOBr(aq) + OH^-(aq)$$

	BrO^-	HOBr	OH^-
initial	1.15 *M*	0	0
equil.	(1.15 - x) *M*	x *M*	x *M*

$$K_b = \frac{[HOBr][OH^-]}{[BrO^-]} = \frac{x^2}{1.15 - x} \approx \frac{x^2}{1.15} = 4.0 \times 10^{-6}$$

$x^2 = 1.15\ (4.0 \times 10^{-6})$; $x = [OH^-] = 2.14 \times 10^{-3} = 2.1 \times 10^{-3}\ M$; pH = 11.33

Check. $\frac{2.1 \times 10^{-3}\ M\ OH^-}{1.15\ M\ C_3H_5O_2^-} \times 100 = 0.19\%$ hydrolysis; the assumption is valid

16.67 *Analyze/Plan.* Given pH and initial concentration of base, calculate all equilibrium concentrations. pH → pOH → $[OH^-]$ at equilibrium. Construct the equilibrium table and calculate other equilibrium concentrations. Substitute into the K_b expression and calculate K_b. *Solve:*

(a) $[OH^-] = 10^{-pOH}$; pOH = 14 - pH = 14.00 - 11.33 = 2.67

$[OH^-] = 10^{-2.67} = 2.138 \times 10^{-3} = 2.1 \times 10^{-3}$ *M*

	$C_{10}H_{15}ON(aq) + H_2O(l)$	$\rightleftharpoons$	$C_{10}H_{15}ONH^+(aq)$	+	$OH^-(aq)$
initial	0.035 *M*		0		0
equil.	0.033 *M*		2.1×10^{-3} *M*		2.1×10^{-3} *M*

(b) $$K_b = \frac{[C_{10}H_{15}ONH^+][OH^-]}{[C_{10}H_{15}ON]} = \frac{(2.138 \times 10^{-3})^2}{(0.03286)} = 1.4 \times 10^{-4}$$

16.68 (a) pOH = 14.00 - 9.95 = 4.05; $[OH^-] = 10^{-4.05} = 8.91 \times 10^{-5} = 8.9 \times 10^{-5}$ *M*

	$C_{18}H_{21}NO_3(aq) + H_2O(l)$	$\rightleftharpoons$	$C_{18}H_{21}NO_3H^+(aq)$	+	$OH^-(aq)$
initial	0.0050 *M*		0		0
equil.	$(0.0050 - 8.9 \times 10^{-5})$		8.9×10^{-5} *M*		8.9×10^{-5} *M*

$$K_b = \frac{C_{18}H_{21}NO_3H^+][OH^-]}{[C_{18}H_{21}NO_3]} = \frac{(8.91 \times 10^{-5})^2}{(0.0050 - 8.91 \times 10^{-5})} = 1.62 \times 10^{-6} = 1.6 \times 10^{-6}$$

(b) $pK_b = -\log(K_b) = -\log(1.62 \times 10^{-6}) = 5.79$

The K_a - K_b Relationship; Acid-Base Properties of Salts

16.69 (a) For a conjugate acid/conjugate base pair such as $C_6H_5OH/C_6H_5O^-$, K_b for the conjugate base is always K_w / K_a for the conjugate acid. K_b for the conjugate base can always be calculated from K_a for the conjugate acid, so a separate list of K_b values is not necessary.

(b) $K_b = K_w / K_a = 1.0 \times 10^{-14} / 1.3 \times 10^{-10} = 7.7 \times 10^{-5}$

(c) K_b for phenolate $(7.7 \times 10^{-5}) > K_b$ for ammonia (1.8×10^{-5}).
Phenolate is a stronger base than NH_3.

16.70 (a) We need K_a for the conjugate acid of CO_3^{2-}, K_a for HCO_3^-. K_a for HCO_3^- is K_{a2}.

(b) $K_b = K_w / K_a = 1.0 \times 10^{-14} / 5.6 \times 10^{-11} = 1.8 \times 10^{-4}$

(c) K_b for CO_3^{2-} $(1.8 \times 10^{-4}) > K_b$ for NH_3 (1.8×10^{-5}).
CO_3^{2-} is a stronger base than NH_3.

16.71 *Analyze/Plan.* Given K_a, determine relative strengths of the acids and their conjugate bases. The greater the magnitude of K_a, the stronger the acid and the weaker the conjugate base. K_b (conjugate base) = K_w/K_a. *Solve:*

(a) Acetic acid is stronger, because it has the larger K_a value.

(b) Hypochlorite ion is the stronger base because the weaker acid, hypochlorous acid, has the stronger conjugate base.

(c) K_b for $C_2H_3O_2^-$ = K_w/K_a for $HC_2H_3O_2$ = $1.0 \times 10^{-14}/1.8 \times 10^{-5} = 5.6 \times 10^{-10}$

K_b for ClO^- = K_w/K_a for HClO = $1 \times 10^{-14}/3.0 \times 10^{-8} = 3.3 \times 10^{-7}$

Note that K_b for ClO^- is greater than K_b for $C_2H_3O_2^-$.

16.72 (a) Ammonia is the stronger base because it has the larger K_b value.

(b) Hydroxylammonium is the stronger acid because the weaker base, hydroxylamine, has the stronger conjugate acid.

(c) K_a for NH_4^+ = K_w/K_b for NH_3 = $1.0 \times 10^{-14}/1.8 \times 10^{-5} = 5.6 \times 10^{-10}$

K_a for $HONH_3^+$ = K_w/K_b for $HONH_2$ = $1.0 \times 10^{-14}/1.1 \times 10^{-8} = 9.1 \times 10^{-7}$

Note that K_a for $HONH_3^+$ is larger than K_a for NH_4^+.

16.73 *Analyze.* When the solute in an aqueous solution is a salt, evaluate the acid/base properties of the component ions.

(a) *Plan.* NaCN is a soluble salt and thus a strong electrolyte. When it is dissolved in H_2O, it dissociates completely into Na^+ and CN^-. [NaCN] = $[Na^+]$ = $[CN^-]$ = 0.10 M. Na^+ is the conjugate acid of the strong base NaOH and thus does not influence the pH of the solution. CN^-, on the other hand, is the conjugate base of the weak acid HCN and **does** influence the pH of the solution. Like any other weak base, it hydrolyzes water to produce OH^-(aq). Solve the equilibrium problem to determine $[OH^-]$. *Solve*:

$$CN^-(aq) + H_2O(l) \rightleftharpoons HCN(aq) + OH^-(aq)$$

	CN^-	HCN	OH^-
initial	0.10 *M*	0	0
equil.	(0.10 - x) *M*	x *M*	x *M*

$$K_b \text{ for } CN^- = \frac{[HCN][OH^-]}{[CN^-]} = \frac{K_w}{K_a \text{ for HCN}} = \frac{1 \times 10^{-14}}{4.9 \times 10^{-10}} = 2.04 \times 10^{-5} = 2.0 \times 10^{-5}$$

$2.04 \times 10^{-5} = \dfrac{(x)(x)}{(0.10 - x)}$; assume the percent of CN^- that hydrolyzes is small

$x^2 = 0.10\,(2.04 \times 10^{-5})$; $x = [OH^-] = 0.00143 = 1.4 \times 10^{-3}$ *M*

pOH = 2.85; pH = 14 - 2.85 = 11.15

(b) *Plan.* $Na_2CO_3(aq) \rightarrow 2Na^+(aq) + CO_3^{2-}(aq)$

CO_3^{2-} is the conjugate base of HCO_3^- and its hydrolysis reaction will determine the $[OH^-]$ and pH of the solution (see similar explanation for NaCN in part (a)). We will assume the process $HCO_3^-(aq) + H_2O(l) \rightleftharpoons H_2CO_3(aq) + OH^-$ will not add significantly to the $[OH^-]$ in solution because $[HCO_3^-(aq)]$ is so small. Solve the equilibrium problem for $[OH^-]$. *Solve:*

$$CO_3^{2-}(aq) + H_2O(l) \rightleftharpoons HCO_3^-(aq) + OH^-(aq)$$

initial	0.080 *M*	0	0
equil.	(0.080 - x) *M*	x	x

$$K_b = \frac{[HCO_3^-][OH^-]}{[CO_3^{2-}]} = \frac{K_w}{K_a \text{ for } HCO_3^-} = \frac{1.0 \times 10^{-14}}{5.6 \times 10^{-11}} = 1.79 \times 10^{-4} = 1.8 \times 10^{-4}$$

$$1.8 \times 10^{-4} = \frac{x^2}{(0.080 - x)}; \; x^2 = 0.080\,(1.79 \times 10^{-4}); \; x = 0.00378 = 3.8 \times 10^{-3}\, M\, OH^-$$

(Assume x is small compared to 0.080); pOH = 2.42; pH = 14 - 2.42 = 11.58

Check. $\frac{3.8 \times 10^{-3}\, M\, OH^-}{0.080\, M\, CO_3^{2-}} \times 100 = 4.75\%$ hydrolysis; the assumption is valid

(c) *Plan.* For the two salts present, Na^+ and Ca^{2+} are negligible acids. NO_2^- is the conjugate base of HNO_2 and will determine the pH of the solution. *Solve:*

Calculate total $[NO_2^-]$ present initially.

$[NO_2^-]_{total}$ = $[NO_2^-]$ from $NaNO_2$ + $[NO_2^-]$ from $Ca(NO_2)_2$

$[NO_2^-]_{total}$ = 0.10 *M* + 2(0.20 M) = 0.50 *M*

The hydrolysis equilibrium is:

$$NO_2^-(aq) + H_2O(l) \rightleftharpoons HNO_2 + OH^-(aq)$$

initial	0.50 *M*	0	0
equil.	(0.50 - x) *M*	x *M*	x *M*

$$K_b = \frac{[HNO_2][OH^-]}{[NO_2^-]} = \frac{K_w}{K_a \text{ for } HNO_2} = \frac{1.0 \times 10^{-14}}{4.5 \times 10^{-4}} = 2.22 \times 10^{-11} = 2.2 \times 10^{-11}$$

$$2.2 \times 10^{-11} = \frac{x^2}{(0.50 - x)} \approx \frac{x^2}{0.50}; \; x^2 = 0.50\,(2.22 \times 10^{-11})$$

$x = 3.33 \times 10^{-6} = 3.3 \times 10^{-6}\, M\, OH^-$; pOH = 5.48; pH = 14 - 5.48 = 8.52

16.74 (a) Proceeding as in Solution 16.73(a):

$$F^-(aq) + H_2O(l) \rightleftharpoons HF(aq) + OH^-(aq)$$

initial	0.036 *M*	0 *M*	0 *M*
equil	(0.036 - x) *M*	x *M*	x *M*

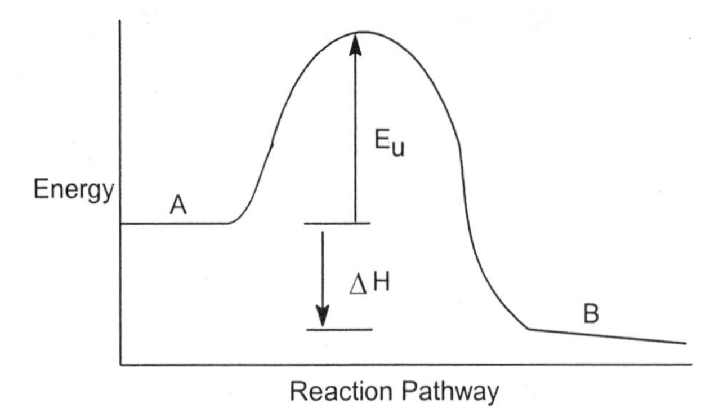

17.38 (a) $PbBr_2(s) \rightleftharpoons Pb^{2+}(aq) + 2Br^-(aq)$

$K_{sp} = [Pb^{2+}][Br^-]^2$; $[Pb^{2+}] = 1.0 \times 10^{-2}\ M$, $[Br^-] = 2.0 \times 10^{-2}\ M$

$K_{sp} = (1.0 \times 10^{-2}\ M)(2.0 \times 10^{-2}\ M)^2 = 4.0 \times 10^{-6}$

(b) $AgIO_3(s) \rightleftharpoons Ag^+(aq) + IO_3^-(aq)$; $K_{sp} = [Ag^+][IO_3^-]$

$$[Ag^+] = [IO_3^-] = \frac{0.0490 \text{ g } AgIO_3}{1.00 \text{ L soln}} \times \frac{1 \text{ mol } AgIO_3}{282.8 \text{ g } AgIO_3} = 1.733 \times 10^{-4} = 1.73 \times 10^{-4}\ M$$

$K_{sp} = (1.733 \times 10^{-4}\ M)(1.733 \times 10^{-4}\ M) = 3.00 \times 10^{-8}$

(c) $Cu(OH)_2(s) \rightleftharpoons Cu^{2+}(aq) + 2OH^-(aq)$; $K_{sp} = [Cu^{2+}][OH^-]^2$

$[Cu^{2+}] = x$, $[OH^-] = 2x$; $K_{sp} = 4.8 \times 10^{-20} = (x)(2x)^2$

$4.8 \times 10^{-20} = 4x^3$; $x = [Cu^{2+}] = 2.290 \times 10^{-7} = 2.3 \times 10^{-7}\ M$

$$\frac{2.290 \times 10^{-7} \text{ mol } Cu(OH)_2}{1 \text{ L}} \times \frac{97.56 \text{ g } Cu(OH)_2}{1 \text{ mol } Cu(OH)_2} = 2.2 \times 10^{-5} \text{ g } Cu(OH)_2$$

However, $[OH^-]$ from $Cu(OH)_2 = 4.58 \times 10^{-7}$ M; this is similar to $[OH^-]$ from the autoionization of water.

$K_w = [H^+][OH^-]$; $[H^+] = y$, $[OH^-] = (4.58 \times 10^{-7} + y)$

$1.0 \times 10^{-14} = y(4.58 \times 10^{-7} + y)$; $y^2 + 4.58 \times 10^{-7}\ y - 1.0 \times 10^{-14}$

$$y = \frac{-4.58 \times 10^{-7} \pm \sqrt{(4.58 \times 10^{-7})^2 - 4(1)(-1.0 \times 10^{-14})}}{2}; \quad y = 2.09 \times 10^{-8}$$

$[OH^-]_{total} = 4.58 \times 10^{-7}\ M + 0.209 \times 10^{-7}\ M = 4.79 \times 10^{-7}\ M$

Recalculating $[Cu^{2+}]$ and thus molar solubility of $Cu(OH)_2(s)$:

$4.8 \times 10^{-20} = x(4.79 \times 10^{-7})^2$; $x = 2.09 \times 10^{-7}\ M\ Cu^{2+}$

$$\frac{2.09 \times 10^{-7} \text{ mol } Cu(OH)_2(s)}{1 \text{ L}} \times \frac{97.56 \text{ g } Cu(OH)_2}{1 \text{ mol } Cu(OH)_2} = 2.0 \times 10^{-5} \text{ g } Cu(OH)_2$$

Note that the presence of OH^- as a common ion decreases the water solubility of $Cu(OH)_2$.

17.39 *Analyze/Plan.* Given gram solubility of a compound, calculate K_{sp}. Write the dissociation equilibrium and K_{sp} expression. change gram solubility to molarity of the individual ions, taking the stoichiometry of the compound into account. Calculate K_{sp}. *Solve:*

$CaC_2O_4(s) \rightleftharpoons Ca^{2+}(aq) + C_2O_4^{2-}(aq)$; $K_{sp} = [Ca^{2+}][C_2O_4^{2-}]$

$$[Ca^{2+}] = [C_2O_4^{2-}] = \frac{0.0061 \text{ g } CaC_2O_4}{1.00 \text{ L soln}} \times \frac{1 \text{ mol } CaC_2O_4}{128.1 \text{ g } CaC_2O_4} = 4.76 \times 10^{-5} = 4.8 \times 10^{-5}\ M$$

$K_{sp} = (4.76 \times 10^{-5}\ M)(4.76 \times 10^{-5}\ M) = 2.3 \times 10^{-9}$

17.40 $PbI_2(s) \rightleftharpoons Pb^{2+}(aq) + 2I^-(aq)$; $K_{sp} = [Pb^{2+}][I^-]^2$

$$[Pb^{2+}] = \frac{0.54 \text{ g } PbI_2}{1.00 \text{ L soln}} \times \frac{1 \text{ mol } PbI_2}{461.0 \text{ g } PbI_2} = 1.17 \times 10^{-3} = 1.2 \times 10^{-3}\ M$$

$[I^-] = 2[Pb^{2+}]$; $K_{sp} = [Pb^{2+}](2[Pb^{2+}])^2 = 4[Pb^{2+}]^3 = 4(1.17 \times 10^{-3})^3 = 6.4 \times 10^{-9}$

17.41 *Analyze/Plan*. Follow the logic in Sample Exercises 17.11 and 17.12. *Solve*:

(a) $AgBr(s) \rightleftharpoons Ag^+(aq) + Br^-(aq)$; $K_{sp} = [Ag^+][Br^-] = 5.0 \times 10^{-13}$

molar solubility = x = $[Ag^+] = [Br^-]$; $K_{sp} = x^2$

$x = (5.0 \times 10^{-13})^{1/2}$; $x = 7.1 \times 10^{-7}$ mol AgBr/L

(b) Molar solubility = x = $[Br^-]$; $[Ag^+] = 0.030\ M + x$

$K_{sp} = (0.030 + x)(x) \approx 0.030(x)$

$5.0 \times 10^{-13} = 0.030(x)$; $x = 1.7 \times 10^{-11}$ mol AgBr/L

(c) Molar solubility = x = $[Ag^+]$

There are two sources of Br^-: NaBr(0.10 *M*) and AgBr(x *M*)

$K_{sp} = (x)(0.10 + x)$; Assuming x is small compared to 0.10 *M*

$5.0 \times 10^{-13} = 0.10\ (x)$; $x \approx 5.0 \times 10^{-12}$ mol AgBr/L

17.42 $LaF_3(s) \rightleftharpoons La^{3+}(aq) + 3F^-(aq)$; $Ksp = [La^{3+}][F^-]^3$

(a) molar solubility = x = $[La^{3+}]$; $[F^-] = 3x$

$K_{sp} = 2 \times 10^{-19}\ (x)(3x)3$; $2 \times 10^{-19} = 27\ x^4$; $x = (7.41 \times 10^{-21})^{1/4}$, $x = 9.28 \times 10^{-6}$
$= 9 \times 10^{-6}$ M La^{3+}

$$\frac{9.28 \times 10^{-6} \text{ mol } LaF_3}{1 \text{ L}} \times \frac{195.9 \text{ g } LaF_3}{1 \text{ mol}} = 1.82 \times 10^{-3} = 2 \text{ g } LaF_3/L$$

(b) molar solubility = x = $[La^{3+}]$

There are two sources of F^-: KF(0.025 *M*) and LaF_3 (3 x *M*)

$K_{sp} = (x)(0.25 + 3x)^3$; assume x is small compared to 0.025 M.

$2 \times 10^{-19} = (0.025)^3\ x$; $x = 2 \times 10^{-19}/1.563 \times 10^{-5} = 1.28 \times 10^{-14}$ M La^{3+}

$$\frac{1.28 \times 10^{14} \text{ mol } LaF_3}{1 \text{ L}} \times \frac{195.9 \text{ g } LaF_3}{1 \text{ mol}} = 2.51 \times 10^{-12} = 3 \times 10^{-12} \text{ g } LaF_3/L$$
$= 3$ pg LaF_3/L

(c) molar solubllity = x, $[F^-] = 3x$, $[La^{3+}] = 0.150\ M + x$

$K_{sp} = (0.150 + x)(3x)^3$; assume x is small compared to 0.150 *M*.

$2 \times 10^{-19} = (0.150)(27\ x^3) = 4.05\ x^3$; $x = (4.94 \times 10^{-20})^{1/3} = 3.67 \times 10^{-7} = 4 \times 10^{-7}\ M$

$$\frac{3.67 \times 10^{-7} \text{ mol } LaF_3}{1 \text{ L}} \times \frac{195.9 \text{ g } LaF_3}{1 \text{ mol}} = 7.19 \times 10^{-5} = 7 \times 10^{-5} \text{ g } LaF_3/L$$
$= 70 \times 10^{-6} = 70$ μg/L $= 70$ ppb LaF_3

17.43 *Analyze/Plan.* We are asked to calculate the solubility of a slightly-soluble hydroxide salt at various pH values. This is a common ion problem; pH tells us not only $[H^+]$ but also $[OH^-]$, which is an ion common to the salt. Use pH to calculate $[OH^-]$, then proceed as in Sample Exercise 17.12. *Solve*:

$Mn(OH)_2(s) \rightleftharpoons Mn^{2+}(aq) + 2OH^-(aq)$; $K_{sp} = 1.6 \times 10^{-13}$

Since $[OH^-]$ is set by the pH of the solution, the solubility of $Mn(OH)_2$ is just $[Mn^{2+}]$.

(a) pH = 7.0, pOH = 14 - pH = 7.0, $[OH^-] = 10^{-pOH} = 1.0 \times 10^{-7}$ *M*

$$K_{sp} = 1.6 \times 10^{-13} = [Mn^{2+}](1.0 \times 10^{-7})^2;\ [Mn^{2+}] = \frac{1.6 \times 10^{-13}}{1.0 \times 10^{-14}} = 16\ M$$

$$\frac{16 \text{ mol } Mn(OH)_2}{1 \text{ L}} \times \frac{88.95 \text{ g } Mn(OH)_2}{1 \text{ mol } Mn(OH)_2} = 1423 = 1.4 \times 10^3 \text{ g } Mn(OH)_2/L$$

Check: Note that the solubility of $Mn(OH)_2$ in pure water is 3.6×10^{-5} *M*, and the pH of the resulting solution is 9.0. The relatively low pH of a solution buffered to pH 7.0 actually increases the solubility of $Mn(OH)_2$.

(b) pH = 9.5, pOH = 4.5, $[OH^-] = 3.16 \times 10^{-5} = 3.2 \times 10^{-5}$ *M*

$$K_{sp} = 1.6 \times 10^{-13} = [Mn^{2+}](3.16 \times 10^{-5})^2;\ [Mn^{2+}] = \frac{1.6 \times 10^{-13}}{1.0 \times 10^{-9}} = 1.6 \times 10^{-4}\ M$$

1.6×10^{-4} *M* $Mn(OH)_2 \times 88.95$ g/mol = 0.0142 = 0.014 g/L

(c) pH = 11.8, pOH = 2.2, $[OH^-] = 6.31 \times 10^{-3} = 6.3 \times 10^{-3}$ *M*

$$K_{sp} = 1.6 \times 10^{-13} = [Mn^{2+}](6.31 \times 10^{-3})^2;\ [Mn^{2+}] = \frac{1.6 \times 10^{-13}}{3.98 \times 10^{-5}} = 4.0 \times 10^{-9}\ M$$

4.02×10^{-9} *M* $Mn(OH)_2 \times 88.95$ g/mol = $3.575 \times 10^{-7} = 3.6 \times 10^{-7}$ g/L

17.44 $Fe(OH)_2(s) \rightleftharpoons Fe^{2+}(aq) + 2OH^-(aq)$; $K_{sp} = 8.0 \times 10^{-16}$

Since the $[OH^-]$ is set by the pH of the solution, the solubility of $Fe(OH)_2$ is just $[Fe^{2+}]$.

(a). pH = 7.0, pOH = 14 - pH = 7.0, $[OH^-] = 10^{-pOH} = 1.0 \times 10^{-7}$ *M*

$$K_{sp} = 7.9 \times 10^{-16} = [Fe^{2+}](1.0 \times 10^{-7})^2;\ [Fe^{2+}] = \frac{7.9 \times 10^{-16}}{1.0 \times 10^{-14}} = 7.9 \times 10^{-2}\ M$$

Check: In pure water, $[OH^-]$ from $Fe(OH)_2$ is similar to (OH^-) from the autoionization of water, resulting in a cubic equation for $[Fe^{2+}]$. The solubility of $Fe(OH)_2$ at a buffered pH = 7.0 is actually greater than the solubility in pure water.

(b) pH = 10.0, pOH = 4.0, $[OH^-] = 1.0 \times 10^{-4}$

$$K_{sp} = 7.9 \times 10^{-16} = [Fe^{2+}][1.0 \times 10^{-4}]^2;\ [Fe^{2+}] = \frac{7.9 \times 10^{-16}}{1.0 \times 8^{-10}} = 7.9 \times 10^{-8}\ M$$

(c) pH = 12.0, pOH = 2.0, $[OH^-] = 1.0 \times 10^{-2}$

$$K_{sp} = 7.9 \times 10^{-16} = [Fe^{2+}][1.0 \times 10^{-2}]^2;\ [Fe^{2+}] = \frac{7.9 \times 10^{-16}}{1.0 \times 10^{-4}} = 7.9 \times 10^{-12}\ M$$

17.45 *Analyze/Plan.* Follow the logic in Sample Exercise 17.13. *Solve:*

If the anion of the salt is the conjugate base of a weak acid, it will combine with H^+, reducing the concentration of the free anion in solution, thereby causing more salt to dissolve. More soluble in acid: (a) $ZnCO_3$ (b) ZnS (d) AgCN (e) $Ba_3(PO_4)_2$

17.46 If the anion in the slightly soluble salt is the conjugate base of a strong acid, there will be no reaction.

(a) $MnS(s) + 2H^+(aq) \rightarrow H_2S(aq) + Mn^{2+}(aq)$

(b) $PbF_2(s) + 2H^+(aq) \rightarrow 2HF(aq) + Pb^{2+}(aq)$

(c) $AuCl_3(s) + H^+(aq) \rightarrow$ no reaction

(d) $Hg_2C_2O_4(s) + 2H^+(aq) \rightarrow H_2C_2O_4(aq) + Hg_2^{2+}(aq)$

(e) $CuBr(s) + H^+(aq) \rightarrow$ no reaction

17.47 *Analyze/Plan.* Follow the logic in Sample Exercise 17.14. *Solve:*

The formation equilibrium is

$$Cu^{2+}(aq) + 4NH_3(aq) \rightleftharpoons Cu(NH_3)_4^{2+}(aq) \quad K_f = \frac{[Cu(NH_3)_4^{2+}]}{[Cu^{2+}][NH_3]^4} = 5 \times 10^{12}$$

Assuming that nearly all the Cu^{2+} is in the form $Cu(NH_3)_4^{2+}$

$[Cu(NH_3)_4^{2+}] = 1 \times 10^{-3}\ M$; $[Cu^{2+}] = x$; $[NH_3] = 0.10\ M$

$$5 \times 10^{12} = \frac{(1 \times 10^{-3})}{x(0.10)^4};\ x = 2 \times 10^{-12}\ M = [Cu^{2+}]$$

17.48 $NiC_2O_4(s) \rightleftharpoons Ni^{2+}(aq) + C_2O_4^{2-}(aq)$; $K_{sp} = [Ni^{2+}][C_2O_4^{2-}] = 4 \times 10^{-10}$

When the salt has just dissolved, $[C_2O_4^{2-}]$ will be 0.020 *M*. Thus $[Ni^{2+}]$ must be less than $4 \times 10^{-10} / 0.020 = 2 \times 10^{-8}\ M$. To achieve this low $[Ni^{2+}]$ we must complex the Ni^{2+} ion with NH_3: $Ni^{2+}(aq) + 6NH_3(aq) \rightleftharpoons Ni(NH_3)_6^{2+}(aq)$. Essentially all Ni(II) is in the form of the complex, so $[Ni(NH_3)_6^{2+}] = 0.020$. Find K_f for $Ni(NH_3)_6^{2+}$ in Table 17.1.

$$K_f = \frac{[Ni(NH_3)_6^{2+}]}{[Ni^{2+}][NH_3]^6} = \frac{(0.020)}{(2 \times 10^{-8})[NH_3]^6} = 5.5 \times 10^8;\ [NH_3]^6 = 1.82 \times 10^{-3};$$

$$[NH_3] = 0.349 = 0.3\ M$$

17.49 *Analyze/Plan.* We are asked to calculate K_{eq} for a particular reaction, making use of pertinent K_{sp} and K_f values from Appendix D and Table 17.1. Write the dissociation equilibrium for AgI and the formation reaction for $Ag(CN)_2^-$. Use algebra to manipulate these equations and their associated equilibrium constants to obtain the desired reaction and its equilibrium constant.

Solve:

$$AgI(s) \rightleftharpoons Ag^{+}(aq) + I^{-}(aq)$$

$$Ag^{+}(aq) + 2CN^{-}(aq \rightleftharpoons Ag(CN)_2^{-}(aq)$$

$$AgI(s) + 2CN^{-}(aq) \rightleftharpoons Ag(CN)_2^{-}(aq) + I^{-}(aq)$$

$$K = K_{sp} \times K_f = [Ag^{+}][I^{-}] \times \frac{[Ag(CN)_2^{-}]}{[Ag^{+}][CN^{-}]^2} = (8.3 \times 10^{-17})(1 \times 10^{21}) = 8 \times 10^4$$

17.50

$Ag_2S(s) \rightleftharpoons 2Ag^{+}(aq) + S^{2-}(aq)$	K_{sp}
$S^{2-}(aq) + 2H^{+}(aq) \rightleftharpoons H_2S(aq)$	$1/(K_{a1} \times K_{a2})$
$2[Ag^{+}(aq) + 2Cl^{-}(aq) \rightleftharpoons AgCl_2^{-}(aq)]$	K_f^2

Add: $Ag_2S(s) + 2H^{+}(aq) + 4Cl^{-}(aq) \rightarrow 2AgCl_2^{-}(aq) + H_2S(aq)$

$$K = \frac{K_{sp} \times K_f^2}{K_{a1} \times K_{a2}} = \frac{(6 \times 10^{-51})(1.1 \times 10^5)^2}{(9.5 \times 10^{-8})(1 \times 10^{-19})} = 7.64 \times 10^{-15} = 8 \times 10^{-15}$$

Precipitation; Qualitative Analysis

17.51 *Analyze/Plan*. Follow the logic in Sample Exercise 17.15. Precipitation conditions: will Q (see Chapter 15) exceed K_{sp} for the compound? *Solve*:

(a) In base, Ca^{2+} can form $Ca(OH)_2(s)$.

$Ca(OH)_2(s) \rightleftharpoons Ca^{2+}(aq) + 2OH^{-}(aq)$; $K_{sp} = [Ca^{2+}][OH^{-}]^2$

$Q = [Ca^{2+}][OH^{-}]^2$; $[Ca^{2+}] = 0.050\ M$; pOH = 6; $[OH^{-}] = 10^{-6} = 1 \times 10^{-6}\ M$

$Q = (0.050)(1 \times 10^{-6})^2 = 5 \times 10^{-14}$; $K_{sp} = 6.5 \times 10^{-6}$ (Appendix D)

$Q < K_{sp}$, no $Ca(OH)_2$ precipitates.

(b) $Ag_2SO_4(s) \rightleftharpoons 2Ag^{+}(aq) + SO_4^{2-}(aq)$; $K_{sp} = [Ag^{+}]^2[SO_4^{2-}]$

$$[Ag^{+}] = \frac{0.050\ M \times 100\ mL}{110\ mL} = 4.545 \times 10^{-2} = 4.5 \times 10^{-2}\ M$$

$$[SO_4^{2-}] = \frac{0.050\ M \times 10\ mL}{110\ mL} = 4.545 \times 10^{-3} = 4.5 \times 10^{-3}\ M$$

$Q = (4.545 \times 10^{-2})^2(4.545 \times 10^{-3}) = 9.4 \times 10^{-6}$; $K_{sp} = 1.5 \times 10^{-5}$

$Q < K_{sp}$, no Ag_2SO_4 precipitates.

17.52 (a) $Co(OH)_2(s) \rightleftharpoons Co^{2+}(aq) + 2OH^{-}(aq)$; $K_{sp} = [Co^{2+}][OH^{-}]^2 = 1.3 \times 10^{-15}$

pH = 8.5; pOH = 14 - 8.5 = 5.5; $[OH^{-}] = 10^{-5.5} = 3.16 \times 10^{-6} = 3 \times 10^{-6}\ M$

$(Q = (0.020)(3.16 \times 10^{-6})^2 = 2 \times 10^{-13}$; $Q > K_{sp}$, $Co(OH)_2$ will precipitate

(b) $AgIO_3(s) \rightleftharpoons Ag^+(aq) + IO_3^-(aq);\ K_{sp} = [Ag^+][IO_3^-] = 3.1 \times 10^{-8}$

$$[Ag^+] = \frac{0.010\ M\ Ag^+ \times 0.100\ L}{0.110\ L} = 9.09 \times 10^{-3} = 9.1 \times 10^{-3}\ M$$

$$[IO_3^-] = \frac{0.015\ M\ IO_3^- \times 0.010\ L}{0.110\ L} = 1.36 \times 10^{-3} = 1.4 \times 10^{-3}\ M$$

$Q = (9.09 \times 10^{-3})(1.36 \times 10^{-3}) = 1.2 \times 10^{-5}$; $Q > K_{sp}$, $AgIO_3$ will precipitate

17.53 *Analyze/Plan.* We are asked which ion will precipitate first from a solution containing $Pb^{2+}(aq)$ and $Ag^+(aq)$ when $I^-(aq)$ is added. Follow the logic in Sample Exercise 17.16.

$Mn(OH)_2(s) \rightleftharpoons Mn^{2+}(aq) + 2OH^-(aq);\ K_{sp} = [Mn^{2+}][OH^-]^2 = 1.6 \times 10^{-13}$

At equilibrium, $[Mn^{2+}][OH^-]^2 = 1.6 \times 10^{-13}$. Change $[Mn^{2+}]$ to mol/L and solve for $[OH^-]$.
Solve:

$$\frac{1\ \mu g\ Mn^{2+}}{1\ L} \times \frac{1 \times 10^{-6}\ g}{1\ \mu g} \times \frac{1\ mol\ Mn^{2+}}{54.94\ g\ Mn^{2+}} = 1.82 \times 10^{-8} = 2 \times 10^{-8}\ M\ Mn^{2+}$$

$1.6 \times 10^{-13} = (1.82 \times 10^{-8})[OH^-]^2$; $[OH^-]^2 = 8.79 \times 10^{-6}$; $[OH^-] = 2.96 \times 10^{-3} = 3 \times 10^{-3}\ M$

pOH = 2.53; pH = 14 - 2.53 = 11.47 = 11.5

17.54 $AgCl(s) \rightleftharpoons Ag^+(aq) + Cl^-(aq);\ K_{sp} = [Ag^+][Cl^-] = 1.8 \times 10^{-10}$

$$[Ag^+] = \frac{0.10\ M \times 0.2\ mL}{20\ mL} = 1 \times 10^{-3}\ M;\quad [Cl^-] = \frac{1.8 \times 10^{-10}}{1 \times 10^{-3}\ M} = 1.8 \times 10^{-7} = 2 \times 10^{-7}\ M$$

$$\frac{1.8 \times 10^{-7}\ mol\ Cl^-}{1\ L} \times \frac{35.45\ g\ Cl^-}{1\ mol\ Cl^-} \times 0.020\ L = 1.28 \times 10^{-7}\ g\ Cl^- = 1 \times 10^{-7}\ g\ Cl^-$$

17.55 *Analyze/Plan.* We are asked which ion will precipitate first from a solution containing $Pb^{2+}(aq)$ and $Ag^+(aq)$ when $I^-(aq)$ is added. Follow the logic in Sample Exercise 17.16. Calculate $[I^-]$ needed to initiate precipitation of each ion. The cation that requires lower $[I^-]$ will precipitate first. *Solve*:

$$Ag^+:\ K_{sp} = [Ag^+][I^-];\ 8.3 \times 10^{-17} = (2.0 \times 10^{-4})[I^-];\ [I^-] = \frac{8.3 \times 10^{-17}}{2.0 \times 10^{-4}} = 4.2 \times 10^{-13}\ M\ I^-$$

$$Pb^{2+}:\ K_{sp} = [Pb^{2+}][I^-]^2;\ 7.9 \times 10^{-9} = (1.5 \times 10^{-3})[I^-]^2;\ [I^-] = \left(\frac{7.9 \times 10^{-9}}{1.5 \times 10^{-3}}\right)^{1/2} = 2.3 \times 10^{-3}\ M\ I^-$$

AgI will precipitate first, at $[I^-] = 4.2 \times 10^{-13}\ M$.

17.56 (a) Precipitation will begin when $Q = K_{sp}$.

$BaSO_4$: $K_{sp} = [Ba^{2+}][SO_4^{2-}] = 1.1 \times 10^{-10}$

$1.1 \times 10^{-10} = (0.015)[SO_4^{2-}]$; $[SO_4^{2-}] = 7.3 \times 10^{-9}\ M$

$SrSO_4$: $K_{sp} = [Sr^{2+}][SO_4^{2-}] = 3.2 \times 10^{-7}$

$3.2 \times 10^{-7} = (0.015)[SO_4^{2-}]$; $[SO_4^{2-}] = 2.1 \times 10^{-5}$ *M*

The $[SO_4^{2-}]$ necessary to begin precipitation is the smaller of the two values, 7.3×10^{-9} *M* SO_4^{2-}.

(b) Ba^{2+} precipitates first, because it requires the smaller $[SO_4^{2-}]$.

(c) Sr^{2+} will begin to precipitate when $[SO_4^{2-}]$ in solution (not bound in $BaSO_4$) reaches 2.1×10^{-5} *M*.

17.57 *Analyze/Plan*. Use Figure 17.22 and the description of the five qualitative analysis "groups" in Section 17.7 to analyze the given data. *Solve*:

The first two experiments eliminate Group 1 and 2 ions (Figure 17.22). The fact that no insoluble phosphates form in the filtrate from the third experiment rules out Group 4 ions. The ions which might be in the sample are those of Group 3, that is, Al^{3+}, Fe^{2+}, Zn^{2+}, Cr^{3+}, Ni^{2+}, Co^{2+}, or Mn^{2+}, and those of Group 5, NH_4^+, Na^+ or K^+.

17.58 Initial solubility in water rules out CdS and HgO. Formation of a precipitate on addition of HCl indicates the presence of $Pb(NO_3)_2$ (formation of $PbCl_2$). Formation of a precipitate on addition of H_2S at pH 1 probably indicates $Cd(NO_3)_2$ (formation of CdS). (This test can be misleading because enough Pb^{2+} can remain in solution after filtering $PbCl_2$ to lead to visible precipitation of PbS.) Absence of a precipitate on addition of H_2S at pH 8 indicates that $ZnSO_4$ is not present. The yellow flame test indicates presence of Na^+. In summary, $Pb(NO_3)_2$ and Na_2SO_4 are definitely present, $Cd(NO_3)_2$ is probably present, and CdS, HgO and $ZnSO_4$ are definitely absent.

17.59 *Analyze/Plan*. We are asked to devise a procedure to separate various pairs of ions in aqueous solutions. In each case, refer to Figure 17.22 to find a set of conditions where the solubility of the two ions differs. Construct a procedure to generate these conditions. *Solve*:

(a) Cd^{2+} is in Gp. 2, but Zn^{2+} is not. Make the solution acidic using 0.5 *M* HCl; saturate with H_2S. CdS will precipitate, ZnS will not.

(b) $Cr(OH)_3$ is amphoteric but $Fe(OH)_3$ is not. Add excess base; $Fe(OH)_3(s)$ precipitates, but Cr^{3+} forms the soluble complex $Cr(OH)_4^-$.

(c) Mg^{2+} is a member of Gp. 4, but K^+ is not. Add $(NH_4)_2HPO_4$; Mg^{2+} precipitates as $MgNH_4PO_4$, K^+ remains in solution.

(d) Ag^+ is a member of Gp. 1, but Mn^{2+} is not. Add 6 *M* HCl, precipitate Ag^+ as AgCl(s).

17.60 (a) Make the solution slightly basic and saturate with H_2S; CdS will precipitate, Na^+ remains in solution.

(b) Make the solution acidic, saturate with H_2S; CuS will precipitate, Mg^{2+} remains in solution.

(c) Add HCl, $PbCl_2$ precipitates. (It is best to carry out the reaction in an ice-water bath to reduce the solubility of $PbCl_2$.)

(d) Add dilute HCl; AgCl precipitates, Hg^{2+} remains in solution.

17.61 (a) Because phosphoric acid is a weak acid, the concentration of free $PO_4^{3-}(aq)$ in an aqueous phosphate solution is low except in strongly basic media. In less basic media, the solubility product of the phosphates that one wishes to precipitate is not exceeded.

(b) K_{sp} for those cations in Group 3 is much larger. Thus, to exceed K_{sp} a higher $[S^{2-}]$ is required. This is achieved by making the solution more basic.

(c) They should all redissolve in strongly acidic solution, e.g., in 12 *M* HCl (all the chlorides of Group 3 metals are soluble).

17.62 The addition of $(NH_4)_2HPO_4$ could result in precipitation of salts from metal ions of the other groups. The $(NH_4)_2HPO_4$ will render the solution basic, so metal hydroxides could form as well as insoluble phosphates. It is essential to separate the metal ions of a group from other metal ions before carrying out the specific tests for that group.

Additional Exercises

17.63 The equilibrium of interest is

$$HC_5H_3O_3(aq) \rightleftharpoons H^+(aq) + C_5H_3O_3^-(aq);\ K_a = 6.76 \times 10^{-4} = \frac{[H^+][C_5H_3O_3^-]}{[HC_5H_3O_3]}$$

Begin by calculating $[HC_5H_3O_3]$ and $[C_5H_3O_3^-]$ for each case.

(a) $$\frac{35.0\text{ g }HC_5H_3O_3}{0.250\text{ L soln}} \times \frac{1\text{ mol }HC_5H_3O_3}{112.1\text{ g }HC_5H_3O_3} = 1.249 = 1.25\ M\ HC_5H_3O_3$$

$$\frac{30.0\text{ g }NaC_5H_3O_3}{0.250\text{ L soln}} \times \frac{1\text{ mol }NaC_5H_3O_3}{134.1\text{ g }NaC_5H_3O_3} = 0.8949 = 0.895\ M\ C_5H_3O_3^-$$

$$[H^+] = \frac{K_a[HC_5H_3O_3]}{[C_5H_3O_3^-]} = \frac{6.76 \times 10^{-4}(1.249 - x)}{(0.8949 + x)} \approx \frac{6.76 \times 10^{-4}(1.249)}{(0.8949)}$$

$[H^+] = 9.43 \times 10^{-4}\ M$, pH = 3.025

(b) For dilution, $M_1V_1 = M_2V_2$

$$[HC_5H_3O_3] = \frac{0.250\ M \times 30.0\text{ mL}}{125\text{ mL}} = 0.0600\ M$$

$$[C_5H_3O_3^-] = \frac{0.220\ M \times 20.0\text{ mL}}{125\text{ mL}} = 0.0352\ M$$

$$[H^+] \approx \frac{6.76 \times 10^{-4}(0.0600)}{0.0352} = 1.15 \times 10^{-3}\ M,\ \text{pH} = 2.938$$

(yes, $[H^+]$ is < 5% of 0.0352 *M*)

(c) $0.0850\ M \times 0.500\ L = 0.0425$ mol $HC_5H_3O_3$

$1.65\ M \times 0.0500\ L = 0.0825$ mol NaOH

	$HC_5H_3O_3(aq)$	+ NaOH(aq)	→ $NaC_5H_3O_3(aq)$	+ $H_2O(l)$
initial	0.0425 mol	0.0825 mol		
reaction	-0.0425 mol	-0.0425mol	+0.0425 mol	
after	0 mol	0.0400 mol	0.0425 mol	

The strong base NaOH dominates the pH; the contribution of $C_5H_3O_3^-$ is negligible. This combination would be "after the equivalence point" of a titration. The total volume is 0.550 L.

$$[OH^-] = \frac{0.0400\ \text{mol}}{0.550\ \text{L}} = 0.0727\ M;\ \ pOH = 1.138,\ pH = 12.862$$

17.64 From Equation [17.9] we see that when the acid and base forms are present in equal concentrations the pH = pK_a. Thus $pK_a = 7.80$.

17.65 $K_a = \dfrac{[H^+][In^-]}{[HIn]}$; at pH = 4.68, [HIn] = $[In^-]$; $[H^+] = K_a$; pH = pK_a = 4.68

17.66 (a) $HA(aq) + B(aq) \rightleftharpoons HB^+(aq) + A^-(aq) \quad K_{eq} = \dfrac{[HB^+][A^-]}{[HA][B]}$

(b) Note that the solution is slightly basic because B is a stronger base than HA is an acid. (Or, equivalently, that A^- is a stronger base than HB^+ is an acid.) Thus, a little of the A^- is used up in reaction: $A^-(aq) + H_2O(l) \rightleftharpoons HA(aq) + OH^-(aq)$. Since pH is not very far from neutral, it is reasonable to assume that the reaction in part (a) has gone far to the right, and that $[A^-] \approx [HB^+]$ and $[HA] \approx [B]$. Then

$$K_a = \frac{[A^-][H^+]}{[HA]} = 8.0 \times 10^{-5};\ \text{when pH} = 9.2,\ [H^+] = 6.31 \times 10^{-10} = 6 \times 10^{-10}\ M$$

$$\frac{[A^-]}{[HA]} = 8.0 \times 10^{-5} / 6.31 \times 10^{-10} = 1.268 \times 10^5 = 1 \times 10^5$$

From the assumptions above, $\dfrac{[A^-]}{[HA]} = \dfrac{[HB^+]}{[B]}$, so $K_{eq} \approx \dfrac{[A^-]^2}{[HA]^2} = 1.608 \times 10^{10}$

$= 2 \times 10^{10}$

(c) K_b for the reaction $B(aq) + H_2O(l) \rightleftharpoons BH^+(aq) + OH^-(aq)$ can be calculated by noting that the equilibrium constant for the reaction in part (a) can be written as $K_{eq} = K_a\,(HA) \times K_b\,(B) / K_w$. (You should prove this to yourself.) Then,

$$K_b\,(B) = \frac{K_{eq} \times K_w}{K_a\,(HA)} = \frac{(1.608 \times 10^{10})(1.0 \times 10^{-14})}{8.0 \times 10^{-5}} = 2.010 = 2$$

K_b (B) is larger than K_a (HA), as it must be if the solution is basic.

17.67 (a) $K_a = \dfrac{[H^+][CHO_2^-]}{[HCHO_2]}$; $[H^+] = \dfrac{K_a[HCHO_2]}{[CHO_2^-]}$

Buffer A: $[HCHO_2] = [CHO_2^-] = \dfrac{1.00 \text{ mol}}{1.00 \text{ L}} = 1.00\ M$

$[H^+] = \dfrac{1.8 \times 10^{-4}(1.00\ M)}{(1.00\ M)} = 1.8 \times 10^{-4}\ M$, pH = 3.74

Buffer B: $[HCHO_2] = [CHO_2^-] = \dfrac{0.010 \text{ mol}}{1.00 \text{ L}} = 0.010\ M$

$[H^+] = \dfrac{1.8 \times 10^{-4}(0.010\ M)}{(0.010\ M)} = 1.8 \times 10^{-4}\ M$, pH = 3.74

The pH of a buffer is determined by the identity of the conjugate acid/conjugate base pair (that is, the relevant K_a value) and the **ratio of concentrations** of the conjugate acid and conjugate base. The absolute concentrations of the components is not relevant. The pH values of the two buffers are equal because they both contain $HCHO_2$ and $NaCHO_2$ and the $[HCHO_2] / [CHO_2^-]$ **ratio** is the same in both solutions.

(b) Buffer capacity is determined by the absolute amount of conjugate acid and conjugate base available to absorb strong acid (H^+) or strong base (OH^-) that is added to the buffer. Buffer A has the greater capacity because it contains the greater absolute concentrations of $HCHO_2$ and CHO_2^-.

(c) Buffer A: $CHO_2^- + HCl \rightarrow HCHO_2 + Cl^-$

	CHO_2^-	HCl	$HCHO_2$
	1.00 mol	0.001 mol	1.00 mol
	0.999 mol	0	1.001 mol

$[H^+] = \dfrac{1.8 \times 10^{-4}(1.001)}{(0.999)} = 1.8 \times 10^{-4}\ M$, pH = 3.74

(In a buffer calculation, volumes cancel and we can substitute moles directly into the K_a expression.)

Buffer B: $CHO_2^- + HCl \rightarrow HCHO_2 + Cl^-$

	CHO_2^-	HCl	$HCHO_2$
	0.010 mol	0.001 mol	0.010 mol
	0.009 mol	0	0.011 mol

$[H^+] = \dfrac{1.8 \times 10^{-4}(0.011)}{(0.009)} = 2.2 \times 10^{-4}\ M$, pH = 3.66

(d) Buffer A: $1.00\ M$ HCl × 0.010 L = 0.010 mol H^+ added

mol $HCHO_2$ = 1.00 + 0.010 = 1.01 mol

mol CHO_2^- = 1.00 - 0.010 = 0.99 mol

$[H^+] = \dfrac{1.8 \times 10^{-4}(1.01)}{(0.99)} = 1.8 \times 10^{-4}\ M$, pH = 3.74

Buffer B: mol $HCHO_2$ = 0.010 + 0.010 = 0.020 mol = 0.020 M

mol CHO_2^- = 0.010 - 0.010 = 0.000 mol

The solution is no longer a buffer; the only source of CHO_2^- is the dissociation of $HCHO_2$.

$$K_a = \frac{[H^+][CHO_2^-]}{[HCHO_2]} = \frac{x^2}{(0.020-x)\ M}$$

The extent of ionization is greater than 5%; from the quadratic formula, $x = [H^+] = 1.8 \times 10^{-3}$, pH = 2.74.

(e) Adding 10 mL of 1.00 *M* HCl to buffer B exceeded its capacity, while the pH of buffer A was unaffected. This is quantitative confirmation that buffer A has a significantly greater capacity than buffer B. In fact, 1.0 L of 1.0 *M* HCl would be required to exceed the capacity of buffer A. Buffer A, with 100 times more $HCHO_2$ and CHO_2^- has 100 times the capacity of buffer B.

17.68 The pH of a buffer is centered around pK_a for its conjugate acid. From Table D.1, hypobromous acid, hypochlorous acid or phenol have pK_a values near 8.6. For the bases in Table D.2, pK_a for the conjugate acids = 14 - pK_b. 14 - pK_b = 8.6; pK_b = 5.4, $K_b = 10^{-5.4}$ $= 4 \times 10^{-6}$. Select two bases with K_b values near 4×10^{-6}. Ammonia and hydrazine have K_b values closest to 4×10^{-6}, and hydroxylamine would probably also work. We will select the two acid-base pairs with pK_a values close to 8.6, $HBrO/BrO^-$ and $H_2NNH_3^+/H_2NNH_2$. In both cases, consider the dissociation of equilibrium for the conjugate acid. For the $HOBr/OBr^-$ buffer:

$HA(aq) \rightleftharpoons H^+(aq) + A^-(aq)$; $HOBr(aq) \rightleftharpoons H^+(aq) + A^-(aq)$

$$K_a = \frac{[H^+][A^-]}{[HA]};\ [H^+] = \frac{K_a[HA]}{[A^-]};\ \frac{[HA]}{[A^-]} = \frac{[H^+]}{K_a};\ \frac{[HOBr]}{[OBr^-]} = \frac{[H^+]}{K_a}$$

$[H^+] = 10^{-8.6} = 2.512 \times 10^{-9}$; $K_a = 2.5 \times 10^{-9}$

$\frac{[HOBr]}{[OBr^-]} = \frac{2.512 \times 10^{-9}}{2.5 \times 10^{-9}} = 1.005 = 1.005 = 1$. Because K_a for HOBr is very close to $[H^+]$ for the buffer, the $[HA/[A^-]$ ratio is essentially 1. The acid and its conjugate base should be present in the same concentrations. This condition can be easily reached by starting with x moles of the acid and adding x/2 moles $OH^-(aq)$ to neutralize half of the acid. The absolute number of moles of HOBr and OBr^- aren't important, just their ratio.

For the hydrazine buffer, $H_2NNH_3^+/H_2NNH_2$, $BH^+(aq) \rightleftharpoons B(aq) + H^+(aq)$;

$H_2NNH_3^+(aq) \rightleftharpoons H_2NNH_2(aq) + H^+(aq)$

$$K_a = \frac{[B][H^+]}{[BH^+]};\ \frac{[BH^+]}{[B]} = \frac{[H^+]}{K_a};\ [H^+] = 2.512 \times 10^{-9} = 3 \times 10^{-9}$$

For $H_2NNH_3^+$, $K_a = K_w/K_b\ (H_2NNH_2) = 1.0 \times 10^{-14} / 1.3 \times 10^{-6} = 7.692 \times 10^{-9} = 7.7 \times 10^{-9}$

$$\frac{[H_2NNH_3^+]}{[H_2NNH_2]} = \frac{2.512 \times 10^{-9}}{7.692 \times 10^{-9}} = 0.3265 \approx 0.3$$

The ratio of $[H_2NNH_3^+]$ to $[H_2NNH_2]$ is 0.3 to 1 or 1 to 3.1. The concentration of base in the buffer is roughly three times the concentration of the conjugate acid. To prepare this buffer, begin with x moles of H_2NNH_2 and add x/4 moles $H^+(aq)$ to neutralize one-fourth of the base. The ratio $[BH^+]/[B]$ is then 0.25/0.75x or 1 to 3.

17.69 $\frac{0.20 \text{ mol } HC_2H_3O_2}{1 \text{ L soln}} \times 0.750 \text{ L} = 0.150 = 0.15 \text{ mol } HC_2H_3O_2$

$$0.15 \text{ mol } HC_2H_3O_2 \times \frac{60.05 \text{ g } HC_2H_3O_2}{1 \text{ mol } HC_2H_3O_2} \times \frac{1 \text{ g gl acetic acid}}{0.99 \text{ g } HC_2H_3O_2} \times \frac{1.00 \text{ mL gl acetic acid}}{1.05 \text{ g gl acetic acid}}$$

$= 8.7$ mL glacial acetic acid

At pH 4.50, $[H^+] = 10^{-4.50} = 3.16 \times 10^{-5} = 3.2 \times 10^{-5}$ *M*; this is small compared to 0.20 *M* $HC_2H_3O_2$.

$$K_a = \frac{(3.16 \times 10^{-5})[C_2H_3O_2^-]}{0.20} = 1.8 \times 10^{-5}; \quad [C_2H_3O_2^-] = 0.114 = 0.11\ M$$

$$\frac{0.114 \text{ mol } NaC_2H_3O_2}{1 \text{ L soln}} \times 0.750 \text{ L} \times \frac{82.03 \text{ g } NaC_2H_3O_2}{1 \text{ mol } NaC_2H_3O_2} = 7.004 = 7.0 \text{ g } NaC_2H_3O_2$$

17.70 (a) For a monoprotic acid (one H^+ per mole of acid), at the equivalence point moles OH^- added = moles H^+ originally present

$M_B \times V_B$ = g acid/molar mass

$$\mathcal{M} = \frac{\text{g acid}}{M_B \times V_B} = \frac{0.2140 \text{ g}}{0.0950\ M \times 0.0274 \text{ L}} = 82.21 = 82.2 \text{ g/mol}$$

(b) initial mol HA = $\frac{0.2140 \text{ g}}{82.21 \text{ g/mol}} = 2.603 \times 10^{-3} = 2.60 \times 10^{-3}$ mol HA

mol OH^- added to pH 6.50 = 0.0950 *M* × 0.0150 L = $1.425 \times 10^{-3} = 1.43 \times 10^{-3}$ mol OH^-

	HA(aq)	+	NaOH(aq)	→	NaA(aq) + H_2O
before rx	2.603×10^{-3} mol		1.425×10^{-3} mol		0
change	-1.425×10^{-3} mol		-1.425×10^{-3} mol		$+1.425 \times 10^{-3}$ mol
after rx	1.178×10^{-3} mol		0		1.425×10^{-3} mol

$$[HA] = \frac{1.178 \times 10^{-3} \text{ mol}}{0.0400 \text{ L}} = 0.02945 = 0.0295\ M$$

$$[A^-] = \frac{1.425 \times 10^{-3} \text{ mol}}{0.0400 \text{ L}} = 0.03563 = 0.0356\ M; \quad [H^+] = 10^{-6.50} = 3.162 \times 10^{-7}$$

$= 3.2 \times 10^{-7}$ *M*

The mixture after reaction (a buffer) can be described by the acid dissociation equilibrium

	HA(aq)	$\rightleftharpoons$	$H^+(aq)$	+	$A^-(aq)$
initial	0.0295 *M*		0		0.0356 *M*
equil	(0.0295 - 3.2 × 10^{-7} *M*)		3.2 × 10^{-7} *M*		(0.0356 + 3.2 × 10^{-7}) *M*

$$K_a = \frac{[H^+][A^-]}{[HA]} \approx \frac{(3.162 \times 10^{-7})(0.03563)}{(0.02945)} = 3.8 \times 10^{-7}$$

(Although we have carried 3 figures through the calculation to avoid rounding errors, the data dictate an answer with 2 significant figures.)

17.71 At the equivalence point of a titration, moles strong base added equals moles weak acid initially present. $M_B \times V_B$ = mol base added = mol acid initial

At the half-way point, the volume of base is one-half of the volume required to reach the equivalence point, and the moles base delivered equals one-half of the mol acid initially present. This means that one-half of the weak acid HA is converted to the conjugate base A^-. If exactly half of the acid reacts, mol HA = mol A^- and [HA] = $[A^-]$ at the half-way point.

From Equation 17.9, $pH = pK_a + \log\frac{[\text{conj. base}]}{[\text{conj. acid}]} = pK_a + \log\frac{[A^-]}{[HA^-]}$.

If $[A^-]/[HA] = 1$, log(1) = 0 and pH = pK_a of the weak acid being titrated.

17.72 (a) $$\frac{0.4885 \text{ g KHP}}{0.100 \text{ L}} \times \frac{1 \text{ mol KHP}}{204.2 \text{ g KHP}} = 0.02392 = 0.0239\ M\ P^{2-} \text{ at the equivalence point}$$

The pH at the equivalence point is determined by the hydrolysis of P^{2-}.

$P^{2-}(aq) + H_2O(l) \rightleftharpoons HP^-(aq) + OH^-(aq)$

$$K_b = \frac{[HP^-][OH^-]}{[P^{2-}]} = \frac{K_w}{K_a \text{ for } HP^-} = \frac{1.0 \times 10^{-14}}{3.1 \times 10^{-6}} = 3.23 \times 10^{-9} = 3.2 \times 10^{-9}$$

$$3.23 \times 10^{-9} = \frac{x^2}{(0.02392 - x)} \approx \frac{x^2}{0.02392}; \ x = [OH^-] = 8.8 \times 10^{-6}\ M$$

pH = 14 - 5.06 = 8.94. From Figure 16.7, either phenolphthalein (pH 8.2 - 10.0) or thymol blue (pH 8.0 - 9.6) could be used to detect the equivalence point.

Phenolphthalein is usually the indicator of choice because the colorless to pink change is easier to see.

(b) $$0.4885 \text{ g KHP} \times \frac{1 \text{ mol KHP}}{204.2 \text{ g KHP}} \times \frac{1 \text{ mol NaOH}}{1 \text{ mol KHP}} \times \frac{1}{0.03855 \text{ L NaOH}} = 0.06206\ M \text{ NaOH}$$

17.73 (a) Initially, the solution is 0.100 *M* in CO_3^{2-}.

$CO_3^{2-}(aq) + H_2O(l) \rightleftharpoons HCO_3^-(aq) + OH^-(aq)$

$$K_b = \frac{[HCO_3^-][OH^-]}{[CO_3^{2-}]} = \frac{K_w}{K_a[HCO_3^-]} = 1.79 \times 10^{-4} = 1.8 \times 10^{-4}$$

Proceeding in the usual way for a weak base, calculate $[OH^-] = 4.23 \times 10^{-3}$

$= 4.2 \times 10^{-3}$ *M*, pH = 11.63.

(b) It will require 40.00 mL of 0.100 *M* HCl to reach the first equivalence point, at which point HCO_3^- is the predominant species.

(c) An additional 40.00 mL are required to react with HCO_3^- to form H_2CO_3, the predominant species at the second equivalence point.

(d) At the second equivalence point there is a 0.0333 *M* solution of H_2CO_3. By the usual procedure for a weak acid:

$$H_2CO_3(aq) \rightleftharpoons H^+(aq) + HCO_3^-(aq)$$

$$K_a = \frac{[H^+][HCO_3^-]}{[H_2CO_3]} = 4.3 \times 10^{-7}; \quad \frac{(x)^2}{(0.0333 - x)} \approx \frac{x^2}{0.0333} \approx 4.3 \times 10^{-7}$$

$x = 1.20 \times 10^{-4} = 1.2 \times 10^{-4}$ *M* H^+; pH = 3.92

17.74 The reaction involved is $HA(aq) + OH^-(aq) \rightleftharpoons A^-(aq) + H_2O(l)$. We thus have 0.080 mol A^- and 0.12 mol HA in a total volume of 1.0 L, so the "initial" molarities of A^- and HA are 0.080 *M* and 0.12 *M*, respectively. The weak acid equilibrium of interest is

$$HA(aq) \rightleftharpoons H^+(aq) + A^-(aq)$$

(a) $K_a = \frac{[H^+][A^-]}{[HA]}$; $[H^+] = 10^{-4.80} = 1.58 \times 10^{-5} = 1.6 \times 10^{-5}$ *M*

Assuming $[H^+]$ is small compared to [HA] and $[A^-]$,

$$K_a \approx \frac{(1.58 \times 10^{-5})(0.080)}{(0.12)} = 1.06 \times 10^{-5} = 1.1 \times 10^{-5}, \quad pK_a = 4.98$$

(b) At pH = 5.00, $[H^+] = 1.0 \times 10^{-5}$ *M*. Let b = extra moles NaOH.

[HA] = 0.12 - b, $[A^-]$ = 0.080 + b

$$1.06 \times 10^{-5} \approx \frac{(1.0 \times 10^{-5})(0.080 + b)}{(0.12 - b)}; \quad 2.06 \times 10^{-5}\, b = 4.72 \times 10^{-7};$$

b = 0.023 mol NaOH

17.75 Assume that H_3PO_4 will react with NaOH in a stepwise fashion. (This is not unreasonable, since the three K_a values for H_3PO_4 are significantly different.)

	$H_3PO_4(aq)$ +	$NaOH(aq)$ →	$H_2PO_4^-(aq)$ + $Na^+(aq)$ + $H_2O(l)$
before	0.20 mol	0.30 mol	0 mol
after	0 mol	0.10 mol	0.20 mol

	$H_2PO_4^-(aq)$ +	$NaOH(aq)$ →	$HPO_4^{2-}(aq)$ + $Na^+(aq)$ + $H_2O(l)$
before	0.20 mol	0.10 mol	0.25 mol
after	0.10 mol	0	0.35 mol

Thus, after all NaOH has reacted, the resulting 1.00 L solution is a buffer containing 0.10 mol $H_2PO_4^-$ and 0.35 mol HPO_4^{2-}. $H_2PO_4^-(aq) \rightleftharpoons H^+(aq) + HPO_4^{2-}(aq)$

$$K_a = 6.2 \times 10^{-8} = \frac{[HPO_4^{2-}][H^+]}{[H_2PO_4^-]};\ [H^+] = \frac{6.2 \times 10^{-8}\,(0.10\,M)}{0.35\,M} = 1.77 \times 10^{-8} = 1.8 \times 10^{-8}\,M;$$

pH = 7.75

17.76 The simplest way to obtain the desired buffer mixture would be to titrate the H_3PO_4 solution with NaOH, using a pH meter, until the pH had risen to 7.20. From the values of K_a for phosphoric acid, we can guess that the equilibrium of importance will be the second acid dissociation.

$$H_2PO_4^-(aq) \rightleftharpoons HPO_4^{2-}(aq) + H^+(aq);\ K_a = 6.2 \times 10^{-8},\ pK_a = 7.21$$

Using Equation [17.9]

$$7.20 = 7.21 + \log\frac{[HPO_4^{2-}]}{[H_2PO_4^-]};\ \frac{[HPO_4^{2-}]}{[H_2PO_4^-]} = 0.98$$

That is, at pH 7.20 we have added enough 1.0 *M* NaOH to the solution to have produced approximately equal concentrations of the two anions $H_2PO_4^-$ and HPO_4^{2-}. If we had begun with 100 mL of the 1.0 *M* H_3PO_4 solution we would have added about 150 mL of 1.0 *M* NaOH, so the total volume of solution would be 250 mL. The total moles phosphate present = 1.0 *M* × 0.100 L = 0.10 mol, so mol $H_2PO_4^-$ = mol HPO_4^{2-} = 0.050 mol. The concentration of each component is 0.050 mol/0.25 L = 0.20 *M*.

17.77 The pH of a buffer system is centered around pK_a for the conjugate acid component. For a diprotic acid, two conjugate acid/conjugate base pairs are possible.

$$H_2X(aq) \rightleftharpoons H^+(aq) + HX^-(aq);\ K_{a1} = 2 \times 10^{-2};\quad pK_{a1} = 1.70$$

$$HX^-(aq) \rightleftharpoons H^+(aq) + X^{2-}(aq);\ K_{a2} = 5.0 \times 10^{-7};\ pK_{a2} = 6.30$$

Clearly HX^- / X^{2-} is the more appropriate combination for preparing a buffer with pH = 6.50. The $[H^+]$ in this buffer = $10^{-6.50} = 3.16 \times 10^{-7} = 3.2 \times 10^{-7}$ *M*. Using the K_{a2} expression to calculate the $[X^{2-}] / [HX^-]$ ratio:

$$K_{a2} = \frac{[H^+][X^{2-}]}{[HX^-]};\ \frac{K_{a2}}{[H^+]} = \frac{[X^{2-}]}{[HX^-]} = \frac{5.0 \times 10^{-7}}{3.16 \times 10^{-7}} = 1.58 = 1.6$$

Since X^{2-} and HX^- are present in the same solution, the ratio of concentrations is also a ratio of moles.

$$\frac{[X^{2-}]}{[HX^-]} = \left(\frac{\text{mol } X^{2-}/\text{L soln}}{\text{mol } HX^-/\text{L soln}}\right) = \frac{\text{mol } X^{2-}}{\text{mol } HX^-} = 1.58;\ \text{mol } X^{2-} = (1.58)\text{ mol } HX^-$$

In the 1.0 L of 1.0 *M* H_2X, there is 1.0 mol of X^{2-} containing material.
Thus, mol HX^- + 1.58 (mol HX^-) = 1.0 mol. 2.58 (mol HX^-) = 1.0;
mol HX^- = 1.0 / 2.58 = 0.39 mol HX^-; mol X^{2-} = 1.0 - 0.39 = 0.61 mol X^{2-}.

Thus enough 1.0 *M* NaOH must be added to produce 0.39 mol HX^- and 0.61 mol X^{2-}.
Considering the neutralization in a step-wise fashion (see discussion of titrations of polyprotic acids in Section 17.3).

	$H_2X(aq)$	+ $NaOH(aq)$	→ $HX^-(aq)$ + $H_2O(l)$
before	1.0 mol	1 mol	0
after	0	0	1.0 mol

	$HX^-(aq)$	+ $NaOH(aq)$	→ $X^{2-}(aq)$ + $H_2O(l)$
before	1.0		0.61
change	-0.61	-0.61	+0.61
after	0.39	0	0.61

Starting with 1.0 mol of H_2X, 1.0 mol of NaOH is added to completely convert it to 1.0 mol of HX^-. Of that 1.0 mol of HX^-, 0.61 mol must be converted to 0.61 mol X^{2-}. The total moles of NaOH added is (1.00 + 0.61) = 1.61 mol NaOH.

$$\text{L NaOH} = \frac{\text{mol NaOH}}{M\ \text{NaOH}} = \frac{1.61\ \text{mol}}{1.0\ M} = 1.6\ \text{L of } 1.0\ M\ \text{NaOH}$$

17.78 $C_3H_5O_3^-$ will be formed by reaction of $HC_3H_5O_3$ with NaOH.
0.1000 *M* × 0.02500 L = 2.500 × 10^{-3} mol $HC_3H_5O_3$; b = mol NaOH needed

	$HC_3H_5O_3$	+ NaOH	→ $C_3H_5O_3^-$ + H_2O + Na^+
initial	2.500 × 10^{-3}	b mol	
rx	-b mol	-b mol	+b mol
after rx	2.500 × 10^{-3} - b mol	0	b mol

$$K_a = \frac{[H^+][C_3H_5O_3^-]}{[HC_3H_5O_3]};\quad K_a = 1.4 \times 10^{-4};\quad [H^+] = 10^{-pH} = 10^{-3.75} = 1.778 \times 10^{-4} = 1.8 \times 10^{-4}\ M$$

Since solution volume is the same for $HC_3H_5O_3$ and $C_3H_5O_3^-$, we can use moles in the equation for $[H^+]$.

$$K_a = 1.4 \times 10^{-4} = \frac{1.778 \times 10^{-4}\ (b)}{(2.500 \times 10^{-3} - b)};\quad 0.7874\ (2.500 \times 10^{-3} - b) = b,\ 1.969 \times 10^{-3} = 1.7874\ b,$$

b = 1.10 × 10^{-3} = 1.1 × 10^{-3} mol OH^-

(The precision of K_a dictates that the result has 2 sig figs.)

Substituting this result into the K_a expression gives $[H^+]$ = 1.8 × 10^{-4}. This checks and confirms our result.

Calculate volume NaOH required from M = mol/L.

$$1.10 \times 10^{-3} \text{ mol OH}^- \times \frac{1 \text{ L}}{1.000 \text{ mol}} \times \frac{1 \text{ µL}}{1 \times 10^{-6} \text{ L}} = 1.1 \times 10^3 \text{ µL (1.1 mL)}$$

17.79 (a) CdS: 8.0×10^{-28}; CuS: 6×10^{-37} CdS has greater molar solubility.

(b) $PbCO_3$: 7.4×10^{-14}; $BaCrO_4$: 2.1×10^{-10} $BaCrO_4$ has greater molar solubility.

(c) Since the stoichiometry of the two complexes is not the same, K_{sp} values can't be compared directly; molar solubilities must be calculated from K_{sp} values.

$Ni(OH)_2$: $K_{sp} = 6.0 \times 10^{-16} = [Ni^{2+}][OH^-]^2$; $[Ni^{2+}] = x$, $[OH^-] = 2x$

$6.0 \times 10^{-16} = (x)(2x)^2 = 4x^3$; $x = 5.3 \times 10^{-6}$ M Ni^{2+}

Note that $[OH^-]$ from the autoionization of water is less than 1% of $[OH^-]$ from $Ni(OH)_2$ and can be neglected.

$NiCO_3$: $K_{sp} = 1.3 \times 10^{-7} = [Ni^{2+}][CO_3^{2-}]$; $[Ni^{2+}] = [CO_3^{2-}] = x$

$1.3 \times 10^{-7} = x^2$; $x = 3.6 \times 10^{-4}$ M Ni^{2+}

$NiCO_3$ has greater molar solubility than $Ni(OH)_2$, but the values are much closer than expected from inspection of K_{sp} values alone.

(d) Again, molar solubilities must be calculated for comparison.

Ag_2SO_4: $K_{sp} = 1.5 \times 10^{-5} = [Ag^+]^2[SO_4^{2-}]$; $[SO_4^{2-}] = x$, $[Ag^+] = 2x$

$1.5 \times 10^{-5} = (2x)^2(x) = 4x^3$; $x = 1.6 \times 10^{-2}$ M SO_4^{2-}

AgI: $K_{sp} = 8.3 \times 10^{-17} = [Ag^+][I^-]$; $[Ag^+] = [I^-] = x$

$8.3 \times 10^{-17} = x^2$; $x = 9.1 \times 10^{-9}$ M Ag^+

Ag_2SO_4 has greater molar solubility than AgI.

17.80 pH = 10.38; pOH = 14.00 - 10.38 = 3.62; $[OH^-] = 10^{-3.62}$

$[OH^-] = 2.40 \times 10^{-4} = 2.4 \times 10^{-4}$ M; $[Mg^{2+}] = 0.5[OH^-] = 1.20 \times 10^{-4} = 1.2 \times 10^{-4}$ M

$K_{sp} = [Mg^{2+}][OH^-]^2 \approx (1.20 \times 10^{-4})(2.40 \times 10^{-4})^2 \approx 6.9 \times 10^{-12}$

17.81 After precipitation, the solution in contact with $CaF_2(s)$ is saturated. The $[Ca^{2+}]$ calculated from the K_{sp} expression gives an upper limit of $[Ca^{2+}]$ remaining in solution.

$K_{sp} = [Ca^{2+}][F^-]^2 = 3.9 \times 10^{-11}$; $[F^-] = 0.20$ M

$3.9 \times 10^{-11} = [Ca^{2+}](0.20)^2$; $[Ca^{2+}] = 9.75 \times 10^{-10} = 9.8 \times 10^{-10}$ M

17.82 $K_{sp} = [Ba^{2+}][MnO_4^-]^2 = 2.5 \times 10^{-10}$

$[MnO_4^-]^2 = 2.5 \times 10^{-10} / 2.0 \times 10^{-8} = 0.0125$; $[MnO_4^-] = \sqrt{0.0125} = 0.11$ M

17.83 $[Ca^{2+}][CO_3^{2-}] = 4.5 \times 10^{-9}$; $[Fe^{2+}][CO_3^{2-}] = 2.1 \times 10^{-11}$

Since $[CO_3^{2-}]$ is the same for both equilibria:

$$[CO_3^{2-}] = \frac{4.5 \times 10^{-9}}{[Ca^{2+}]} = \frac{2.1 \times 10^{-11}}{[Fe^{2+}]}; \text{ rearranging } \frac{[Ca^{2+}]}{[Fe^{2+}]} = \frac{4.5 \times 10^{-9}}{2.1 \times 10^{-11}} = 214 = 2.1 \times 10^2$$

17.84 $PbSO_4(s) \rightleftharpoons Pb^{2+}(aq) + SO_4^{2-}(aq)$; $K_{sp} = 6.3 \times 10^{-7} = [Pb^{2+}][SO_4^{2-}]$

$SrSO_4(s) \rightleftharpoons Sr^{2+}(aq) + SO_4^{2-}(aq)$; $K_{sp} = 3.2 \times 10^{-7} = [Sr^{2+}][SO_4^{2-}]$

Let $x = [Pb^{2+}]$, $y = [Sr^{2+}]$, $x + y = [SO_4^{2-}]$

$$\frac{x(x+y)}{y(x+y)} = \frac{6.3 \times 10^{-7}}{3.2 \times 10^{-7}}; \frac{x}{y} = 1.9688 = 2.0; \; x = 1.969\,y = 2.0\,y$$

$y(1.969\,y + y) = 3.2 \times 10^{-7}$; $2.969\,y^2 = 3.2 \times 10^{-7}$; $y = 3.283 \times 10^{-4} = 3.3 \times 10^{-4}$

$x = 1.969\,y$; $x = 1.969(3.283 \times 10^{-4}) = 6.464 \times 10^{-4} = 6.5 \times 10^{-4}$

$[Pb^{2+}] = 6.5 \times 10^{-4}\,M$, $[Sr^{2+}] = 3.3 \times 10^{-4}\,M$, $[SO_4^{2-}] = (3.283 + 6.464) \times 10^{-4} = 9.7 \times 10^{-4}\,M$

17.85 $MgC_2O_4(s) \rightleftharpoons Mg^{2+}(aq) + C_2O_4^{2-}(aq)$

$K_{sp} = [Mg^{2+}][C_2O_4^{2-}] = 8.6 \times 10^{-5}$

If $[Mg^{2+}]$ is to be $3.0 \times 10^{-2}\,M$, $[C_2O_4^{2-}] = 8.6 \times 10^{-5} / 3.0 \times 10^{-2} = 2.87 \times 10^{-3} = 2.9 \times 10^{-3}\,M$

The oxalate ion undergoes hydrolysis:

$C_2O_4^{2-}(aq) + H_2O(l) \rightleftharpoons HC_2O_4^-(aq) + OH^-(aq)$

$$K_b = \frac{[HC_2O_4^-][OH^-]}{[C_2O_4^{2-}]} = 1.0 \times 10^{-14} / 6.4 \times 10^{-5} = 1.56 \times 10^{-10} = 1.6 \times 10^{-10}$$

$[Mg^{2+}] = 3.0 \times 10^{-2}\,M$, $[C_2O_4^{2-}] = 2.87 \times 10^{-3} = 2.9 \times 10^{-3}\,M$

$[HC_2O_4^-] = (3.0 \times 10^{-2} - 2.87 \times 10^{-3})\,M = 2.71 \times 10^{-2} = 2.7 \times 10^{-2}\,M$

$$[OH^-] = 1.56 \times 10^{-10} \times \frac{[C_2O_4^{2-}]}{[HC_2O_4^-]} = 1.56 \times 10^{-10} \times \frac{(2.87 \times 10^{-3})}{(2.71 \times 10^{-2})} = 1.652 \times 10^{-11}$$

$[OH^-] = 1.7 \times 10^{-11}$; pOH = 10.78, pH = 3.22

17.86 The student failed to account for the hydrolysis of the AsO_4^{3-} ion. If there were no hydrolysis, $[Mg^{2+}]$ would indeed be 1.5 times that of $[AsO_4^{3-}]$. However, as the reaction $AsO_4^{3-}(aq) + H_2O(l) \rightleftharpoons HAsO_4^{2-}(aq) + OH^-(aq)$ proceeds, the ion product $[Mg^{2+}]^3[AsO_4^{3-}]^2$ falls below the value for K_{sp}. More $Mg_3(AsO_4)_2$ dissolves, more hydrolysis occurs, and so on, until an equilibrium is reached. At this point $[Mg^{2+}]$ in solution is much greater than 1.5 times free $[AsO_4^{3-}]$. However, it is exactly 1.5 times the **total** concentration of all arsenic-containing species. That is, $[Mg^{2+}] = 1.5\,([AsO_4^{3-}] + [HAsO_4^{2-}] + [H_2AsO_4^-] + [H_3AsO_4])$

17.87

$$Zn(OH)_2(s) \rightleftharpoons Zn^{2+}(aq) + 2OH^-(aq) \qquad K_{sp} = 3.0 \times 10^{-16}$$
$$Zn^{2+}(aq) + 4OH^-(aq) \rightleftharpoons Zn(OH)_4^{2-}(aq) \qquad K_f = 4.6 \times 10^{17}$$

$$Zn(OH)_2(s) + 2OH^-(aq) \rightleftharpoons Zn(OH)_4^{2-}(aq) \qquad K = K_{sp} \times K_f = 138 = 1.4 \times 10^2$$

$$K = 138 = 1.4 \times 10^2 = \frac{[Zn(OH)_4^{2-}]}{[OH^-]^2}$$

If 0.015 mol $Zn(OH)_2$ dissolves, 0.015 mol $Zn(OH)_4^{2-}$ should be present at equilibrium.

$$[OH^-]^2 = \frac{(0.015)}{138};\ [OH^-] = 1.043 \times 10^{-2}\ M \quad [OH^-] \geq 1.0 \times 10^{-2}\ M \text{ or pH} \geq 12.02$$

Integrative Exercises

17.88 (a) Complete ionic:

$$H^+(aq) + Cl^-(aq) + Na^+(aq) + CHO_2^-(aq) \rightarrow HCHO_2(aq) + Na^+(aq) + Cl^-(aq)$$

Na^+ and Cl^- are spectator ions.

Net ionic: $H^+(aq) + CHO_2^-(aq) \rightleftharpoons HCHO_2(aq)$

(b) The net ionic equation in part (a) is the reverse of the dissociation of $HCHO_2$.

$$K = \frac{1}{K_a} = \frac{1}{1.8 \times 10^{-4}} = 5.55 \times 10^3 = 5.6 \times 10^3$$

(c) For Na^+ and Cl^-, this is just a dilution problem.

$M_1V_1 = M_2V_2$; V_2 is 50.0 mL + 50.0 mL = 100.0 mL

$$Cl^-: \frac{0.15\ M \times 50.0\ \text{mL}}{100.0\ \text{mL}} = 0.075\ M;\quad Na^+: \frac{0.15\ M \times 50.0\ \text{mL}}{100.0\ \text{mL}} = 0.075\ M$$

H^+ and CHO_2^- react to form $HCHO_2$. Since K >> 1, the reaction essentially goes to completion.

$0.15\ M \times 0.0500$ mL $= 7.5 \times 10^{-3}$ mol H^+

$0.15\ M \times 0.0500$ mL $= 7.5 \times 10^{-3}$ mol CHO_2^-

$= 7.5 \times 10^{-3}$ mol $HCHO_2$

Solve the weak acid problem to determine $[H^+]$, $[CHO_2^-]$ and $[HCHO_2]$ at equilibrium.

$$K_a = \frac{[H^+][CHO_2^-]}{[HCHO_2]};\ [H^+] = [CHO_2^-] = x\ M;\ [HCHO_2] = \frac{(7.5 \times 10^{-3} - x)\ \text{mol}}{0.100\ \text{L}} = (0.075 - x)\ M$$

$$1.8 \times 10^{-4} = \frac{x^2}{(0.075 - x)} \approx \frac{x^2}{0.075};\ x = 3.7 \times 10^{-3}\ M \quad H^+ \text{ and } HCHO_2^-$$

$$[HCHO_2] = (0.075 - 0.0037) = 0.071\ M$$

$$\frac{[H^+]}{[HNO_2]} \times 100 = \frac{3.7 \times 10^{-3}}{0.075} \times 100 = 4.9\% \text{ dissociation}$$

In summary:

$[Na^+] = [Cl^-] = 0.075\ M$, $[HCHO_2] = 0.071\ M$, $[H^+] = [CHO_2^-] = 0.0037\ M$

17.89 (a) For a monoprotic acid (one H^+ per mole of acid), at the equivalence point

moles OH^- added = moles H^+ originally present

$$M_B \times V_B = \text{g acid/molar mass}$$

$$\mathcal{M} = \frac{\text{g acid}}{M_B \times V_B} = \frac{0.1044\ \text{g}}{0.0500\ M \times 0.0220\ \text{L}} = 94.48 = 94.5\ \text{g/mol}$$

(b) 11.05 mL is exactly half-way to the equivalence point (22.10 mL). When half of the unknown acid is neutralized, $[HA] = [A^-]$, $[H^+] = K_a$ and pH = pK_a.

$K_a = 10^{-4.89} = 1.3 \times 10^{-5}$

(c) From Appendix D, Table D.1, acids with K_a values close to 1.3×10^{-5} are

name	K_a	formula	molar mass
propionic	1.3×10^{-5}	$HC_3H_5O_2$	74.1
butanoic	1.5×10^{-5}	$HC_4H_7O_2$	88.1
acetic	1.8×10^{-5}	$HC_2H_3O_2$	60.1
hydroazoic	1.9×10^{-5}	HN_3	43.0

Of these, butanoic has the closest match for K_a and molar mass, but the agreement is not good.

17.90 $$n = \frac{PV}{RT} = 735\ \text{torr} \times \frac{1\ \text{atm}}{760\ \text{torr}} \times \frac{7.5\ \text{L}}{295\ \text{K}} \times \frac{\text{K}\bullet\text{mol}}{0.08206\ \text{L}\bullet\text{atm}} = 0.300 = 0.30\ \text{mol } NH_3$$

$0.40\ M \times 0.50\ \text{L} = 0.20$ mol HCl

	$HCl(aq)$ +	$NH_3(g)$ →	$NH_4^+(aq)$ +	$Cl^-(aq)$
before	0.20 mol	0.30 mol		
after	0	0.10 mol	0.20 mol	0.20 mol

The solution will be a buffer because of the substantial concentrations of NH_3 and NH_4^+ present. Use K_a for NH_4^+ to describe the equilibrium.

	$NH_4^+(aq)$ ⇌	$NH_3(aq)$ +	$H^+(aq)$
equil.	0.20 - x	0.10 + x	x

$$K_a = \frac{1.0 \times 10^{-14}}{1.8 \times 10^{-5}} = 5.56 \times 10^{-10} = 5.6 \times 10^{-10}\ ;\ K_a = \frac{[NH_3][H^+]}{[NH_4^+]};\ [H^+] = \frac{K_a[NH_4^+]}{[NH_3]}$$

Since this expression contains a ratio of concentrations, volume will cancel and we can substitute moles directly. Assume x is small compared to 0.10 and 0.20.

$$[H^+] = \frac{5.56 \times 10^{-10}\ (0.20)}{(0.10)} = 1.111 \times 10^{-9} = 1.1 \times 10^{-9}\ M,\ \text{pH} = 8.95$$

17.91 Calculate the initial *M* of aspirin in the stomach and solve the equilibrium problem to find equilibrium concentrations of $C_8H_7O_2COOH$ and $C_8H_7O_2COO^-$. At pH = 2, $[H^+] = 1 \times 10^{-2}$.

$$\frac{325\text{ mg}}{\text{tablet}} \times 2\text{ tablets} \times \frac{1\text{ g}}{1000\text{ mg}} \times \frac{1\text{ mol } C_8H_7O_2COOH}{180.2\text{ g } C_8H_7O_2COOH} \times \frac{1}{1\text{ L}} = 3.61 \times 10^{-3} = 4 \times 10^{-3}\ M$$

$$C_8H_7O_2COOH(aq) \rightleftharpoons C_8H_7O_2COO^-(aq) + H^+(aq)$$

	$C_8H_7O_2COOH(aq)$	$C_8H_7O_2COO^-(aq)$	$H^+(aq)$
initial	$3.61 \times 10^{-3}\ M$	0	$1 \times 10^{-2}\ M$
equil	$(3.61 \times 10^{-3} - x)\ M$	$x\ M$	$(1 \times 10^{-2} + x)\ M$

$$K_a = 3 \times 10^{-5} = \frac{[H^+][C_8H_7O_2COO^-]}{[C_8H_7O_2COOH]} = \frac{(0.01 + x)(x)}{(3.61 \times 10^{-3} - x)} \approx \frac{0.01x}{3.61 \times 10^{-3}}$$

$x = [C_8H_7O_2COO^-] = 1.08 \times 10^{-5} = 1 \times 10^{-5}\ M$

$$\%\text{ ionization} = \frac{1.08 \times 10^{-5}\ M\ C_8H_7O_2COO^-}{3.61 \times 10^{-3}\ M\ C_8H_7O_2COOH} \times 100 = 0.3\%$$

(% ionization is small, so the assumption was valid.)

% aspirin molecules = 100.0% - 0.3% = 99.7% molecules

17.92 According to Equation 13.4, $S_g = kP_g$

$$S_{CO_2} = 3.1 \times 10^{-2} \frac{\text{mol}}{\text{L} \cdot \text{atm}} \times 1.10\text{ atm} = 0.0341 = \frac{0.034\text{ mol}}{\text{L}} = 0.034\ M\ CO_2$$

$CO_2(g) + H_2O(l) \rightarrow H_2CO_3(aq)$; $0.0341\ M\ CO_2 = 0.0341\ M\ H_2CO_3$

Consider the stepwise dissociation of $H_2CO_3(aq)$.

$$H_2CO_3(aq) \rightleftharpoons H^+(aq) + HCO_3^-(aq)$$

	$H_2CO_3(aq)$	$H^+(aq)$	$HCO_3^-(aq)$
initial	$0.0341\ M$	0	0
equil.	$(0.0341 - x)\ M$	x	x

$$K_{a1} = \frac{[H^+][HCO_3^-]}{[H_2CO_3]} = \frac{x^2}{(0.0341 - x)} \approx \frac{x^2}{0.0341} \approx 4.3 \times 10^{-7}$$

$x^2 = 1.47 \times 10^{-8}$; $x = 1.2 \times 10^{-4}\ M\ H^+$; pH = 3.92

$K_{a2} = 5.6 \times 10^{-11}$; assume the second ionization does not contribute significantly to $[H^+]$.

17.93 $\pi = MRT$, $M = \dfrac{\pi}{RT} = \dfrac{21\text{ torr}}{298\text{ K}} \times \dfrac{1\text{ atm}}{760\text{ torr}} \times \dfrac{\text{K} \cdot \text{mol}}{0.08206\text{ L} \cdot \text{atm}} = 1.13 \times 10^{-3} = 1.1\ M$

$SrSO_4(s) \rightleftharpoons Sr^{2+}(aq) + SO_4^{2-}(aq)$; $K_{sp} = [Sr^{2+}][SO_4^{2-}]$

The total particle concentration is $1.13 \times 10^{-3}\ M$. Each mole of $SrSO_4$ that dissolves produces 2 mol of ions, so $[Sr^{2+}] = [SO_4^{2-}] = 1.13 \times 10^{-3}\ M / 2 = 5.65 \times 10^{-4} = 5.7 \times 10^{-4}\ M$.

$K_{sp} = (5.65 \times 10^{-4})^2 = 3.2 \times 10^{-7}$

17.94 For very dilute aqueous solutions, assume the solution density is 1 g/mL.

$$\text{ppb} = \frac{\text{g solute}}{10^9 \text{ g solution}} = \frac{1 \times 10^{-6} \text{ g solute}}{1 \times 10^3 \text{ g solution}} = \frac{\mu\text{g solute}}{\text{L solution}}$$

(a) $K_{sp} = [Ag^+][Cl^-] = 1.8 \times 10^{-10}$; $[Ag^+] = (1.8 \times 10^{-10})^{1/2} = 1.34 \times 10^{-5} = 1.3 \times 10^{-5}$ *M*

$$\frac{1.34 \times 10^{-5} \text{ mol Ag}^+}{\text{L}} \times \frac{107.9 \text{ g Ag}^+}{1 \text{ mol Ag}^+} \times \frac{1 \mu\text{g}}{1 \times 10^{-6} \text{ g}} = \frac{1.4 \times 10^3 \mu\text{g Ag}^+}{\text{L}}$$

$$= 1.4 \times 10^3 \text{ ppb} = 1.4 \text{ ppm}$$

(b) $K_{sp} = [Ag^+][Br^-] = 5.0 \times 10^{-13}$; $[Ag^+] = (5.0 \times 10^{-13})^{1/2} = 7.07 \times 10^{-7} = 7.1 \times 10^{-7}$ *M*

$$\frac{7.07 \times 10^{-7} \text{ mol Ag}^+}{\text{L}} \times \frac{107.9 \text{ g Ag}^+}{1 \text{ mol Ag}^+} \times \frac{1 \mu\text{g}}{1 \times 10^{-6} \text{ g}} = 76 \text{ ppb}$$

(c) $K_{sp} = [Ag^+][I^-] = 8.3 \times 10^{-17}$; $[Ag^+] = (8.3 \times 10^{-17})^{1/2} = 9.11 \times 10^{-9} = 9.1 \times 10^{-9}$ *M*

$$\frac{9.11 \times 10^{-9} \text{ mol Ag}^+}{\text{L}} \times \frac{107.9 \text{ g Ag}^+}{1 \text{ mol Ag}^+} \times \frac{1 \mu\text{g}}{1 \times 10^{-6} \text{ g}} = 0.98 \text{ ppb}$$

AgBr(s) would maintain $[Ag^+]$ in the correct range.

17.95 To determine precipitation conditions, we must know K_{sp} for CaF_2(s) and calculate Q under the specified conditions. $K_{sp} = 3.9 \times 10^{-11} = [Ca^{2+}][F^-]^2$

$[Ca^{2+}]$ and $[F^-]$: The term 1 ppb means 1 part per billion or 1 g solute per billion g solution. Assuming that the density of this very dilute solution is the density of water:

$$1 \text{ ppb} = \frac{1 \text{ g solute}}{1 \times 10^9 \text{ g solution}} \times \frac{1 \text{ g solution}}{1 \text{ mL solution}} \times \frac{1 \times 10^3 \text{ mL}}{1 \text{ L}} = \frac{1 \times 10^{-6} \text{ g solute}}{1 \text{ L solution}}$$

$$\frac{1 \times 10^{-6} \text{ g solute}}{1 \text{ L solution}} \times \frac{1 \mu\text{g}}{1 \times 10^{-6} \text{ g}} = 1 \mu\text{g} / 1 \text{ L}$$

$$8 \text{ ppb } Ca^{2+} \times \frac{1 \mu\text{g}}{1 \text{ L}} = \frac{8 \mu\text{g } Ca^{2+}}{1 \text{ L}} = \frac{8 \times 10^{-6} \text{ g } Ca^{2+}}{1 \text{ L}} \times \frac{1 \text{ mol } Ca^{2+}}{40 \text{ g}} = 2 \times 10^{-7} \ M \ Ca^{2+}$$

$$1 \text{ ppb } F^- \times \frac{1 \mu\text{g}}{1 \text{ L}} = \frac{1 \mu\text{g } F^-}{1 \text{ L}} = \frac{1 \times 10^{-6} \text{ g } F^-}{1 \text{ L}} \times \frac{1 \text{ mol } F^-}{19.0 \text{ g}} = 5 \times 10^{-8} \ M \ F^-$$

$Q = [Ca^{2+}][F^-]^2 = (2 \times 10^{-7})(5 \times 10^{-8})^2 = 5 \times 10^{-22}$

$5 \times 10^{-22} < 3.9 \times 10^{-11}$, $Q < K_{sp}$, no CaF_2 will precipitate

18 Chemistry of the Environment

Earth's Atmosphere

18.1 (a) The temperature profile of the atmosphere (Figure 18.1) is the basis of its division into regions. The center of each peak or trough in the temperature profile corresponds to a new region.

(b) Troposphere, 0-12 km; stratosphere, 12-50 km; mesosphere, 50-85 km; thermosphere, 85-110 km.

18.2 (a) The tropopause is the boundary between the troposphere and the stratosphere, the first minimum in the temperature profile of the atmosphere.

(b) Boundaries between regions of the atmosphere are at maxima and minima (peaks and valleys) in the atmospheric temperature profile. For example, in the troposphere, temperature decreases with altitude, while in the stratosphere, it increases with altitude. The temperature minimum is the tropopause boundary.

(c) Atmospheric pressure in the troposphere ranges from 1.0 atm to 0.4 atm, while pressure in the stratosphere ranges from 0.4 atm to 0.001 atm. Gas density (g/L) is directly proportional to pressure. The much lower density of the stratosphere means it has the smaller mass, despite having a larger volume than the troposphere.

18.3 *Analyze/Plan.* Given O_3 concentration in ppm, calculate partial pressure. Use the definition of ppm to get mol fraction O_3. For gases mole fraction = pressure fraction; $P_{O_3} = \chi_{O_3} \cdot P_{atm}$.

$$0.37 \text{ ppm } O_3 = \frac{0.37 \text{ mol } O_3}{1 \times 10^6 \text{ mol air}} = 3.7 \times 10^{-7} = \chi_{O_3}$$

Solve: $P_{O_3} = \chi_{O_3} \cdot P_{atm} = 3.7 \times 10^{-7} (650 \text{ torr}) = 2.4 \times 10^{-4}$ torr

18.4 $P_{Ar} = \chi_{Ar} \cdot P_{atm}$; $P_{Ar} = 0.00934 (98.6 \text{ kPa}) = 0.921 \text{ kPa}$; $0.921 \text{ kPa} \times \dfrac{760 \text{ torr}}{101.325 \text{ kPa}} = 6.91 \text{ torr}$

$P_{CO_2} = \chi_{CO_2} \cdot P_{atm}$; $P_{CO_2} = 0.000355 (98.6 \text{ kPa}) = 0.0350 \text{ kPa}$; $0.0350 \text{ kPa} \times \dfrac{760 \text{ torr}}{101.325 \text{ kPa}}$

$= 0.263$ torr

18.5 *Analyze/Plan.* Given CO concentration in ppm, calculate number of CO molecules in 1.0 L air at given conditions. ppm CO → χ_{O_3} → atm CO → mol CO → molecules CO Use the ideal gas law to change atm CO to mol CO, then Avogadro's number to get molecules. *Solve*:

$$6.0 \text{ ppm CO} = \frac{6.0 \text{ mol CO}}{1 \times 10^6 \text{ mol air}} = 6.0 \times 10^{-6} = \chi_{CO}$$

$$P_{CO} = \chi_{CO} \bullet P_{atm} = 6.0 \times 10^{-6} \times 745 \text{ torr} \times \frac{1 \text{ atm}}{760 \text{ torr}} = 5.88 \times 10^{-6} = 5.9 \times 10^{-6} \text{ atm}$$

$$n_{CO} = \frac{P_{CO}V}{RT} = \frac{5.88 \times 10^{-6} \text{ atm} \times 1.0 \text{ L}}{290 \text{ K}} \times \frac{\text{K}\bullet\text{mol}}{0.08206 \text{ L}\bullet\text{atm}} = 2.4715 \times 10^{-7} = 2.5 \times 10^{-7} \text{ mol CO}$$

$$2.4715 \times 10^{-7} \text{ mol CO} \times \frac{6.022 \times 10^{23} \text{ molecules}}{\text{mol}} = 1.488 \times 10^{17} = 1.5 \times 10^{17} \text{ molecules CO}$$

18.6 (a) ppm Ne = mol Ne/1 × 10^6 mol air; $\chi_{Ne} = 1.818 \times 10^{-5}$ mol Ne/mol air

$$\frac{1.818 \times 10^{-5} \text{ mol Ne}}{1 \text{ mol air}} = \frac{x \text{ mol Ne}}{1 \times 10^6 \text{ mol air}}; \; x = 18.18 \text{ ppm Ne}$$

(b) $$P_{Ne} = \chi_{Ne} \bullet P_{atm} = 1.818 \times 10^{-5} \times 743 \text{ torr} \times \frac{1 \text{ atm}}{760 \text{ torr}} = 1.7773 \times 10^{-5}$$

$$= 1.78 \times 10^{-5} \text{ atm}$$

T = 295°C + 273 = 568 K

$$\frac{n_{Ne}}{V} = \frac{P_{Ne}V}{RT} = \frac{1.7773 \times 10^{-5} \text{ atm} \times 1.0 \text{ L}}{568 \text{ K}} \times \frac{\text{K}\bullet\text{mol}}{0.08206 \text{ L}\bullet\text{atm}} = 3.8132 \times 10^{-7}$$

$$= 3.813 \times 10^{-7} \text{ mol/L}$$

$$\frac{3.8132 \times 10^{-7} \text{ mol Ne}}{\text{L}} \times \frac{6.022 \times 10^{23} \text{ atoms}}{\text{mol}} = 2.296 \times 10^{17} = 2.30 \times 10^{17} \text{ Ne atoms/L}$$

The Upper Atmosphere; Ozone

18.7 *Analyze/Plan.* Given bond dissociation energy in kJ/mol, calculate the wavelength of a single photon that will rupture a C--Br bond. kJ/mol → J/molecule. λ = hc/E. (λ = hc/E describes the energy/wavelength relationship of a single photon.) *Solve*:

$$\frac{210 \times 10^3 \text{ J}}{1 \text{ mol}} \times \frac{1 \text{ mol}}{6.022 \times 10^{23} \text{ molecules}} = 3.487 \times 10^{-19} = 3.49 \times 10^{-19} \text{ J/molecule}$$

λ = c/ν We also have that E = hν, so ν = E/h. Thus,

$$\lambda = \frac{hc}{E} = \frac{(6.626 \times 10^{-34} \text{ J}\bullet\text{sec})(3.00 \times 10^8 \text{ m/sec})}{3.487 \times 10^{-19} \text{ J}} = 5.70 \times 10^{-7} \text{ m} = 570 \text{ nm}$$

18.8 $\frac{339 \times 10^3 \text{ J}}{1 \text{ mol}} \times \frac{1 \text{ mol}}{6.022 \times 10^{23} \text{ molecules}} = 5.6294 \times 10^{-19} = 5.63 \times 10^{-19}$ J/molecule

$$\lambda = \frac{hc}{E} = \frac{(6.626 \times 10^{-34} \text{ J} \bullet \text{sec})(3.00 \times 10^8 \text{ m/sec})}{5.6294 \times 10^{-19} \text{ J}} = 3.53 \times 10^{-7} \text{ m} = 353 \text{ nm}$$

$$\frac{293 \times 10^3 \text{ J}}{1 \text{ mol}} \times \frac{1 \text{ mol}}{6.022 \times 10^{23} \text{ molecules}} = 4.8655 \times 10^{-19} = 4.87 \times 10^{-19} \text{ J/molecule}$$

$$\lambda = \frac{(6.626 \times 10^{-34} \text{ J} \bullet \text{sec})(3.00 \times 10^8 \text{ m/sec})}{4.8655 \times 10^{-19} \text{ J}} = 4.09 \times 10^{-7} \text{ m} = 409 \text{ nm}$$

Photons of wavelengths longer than 409 nm cannot cause rupture of the C--Cl bond in either CF_3Cl or CCl_4. Photons with wavelengths between 409 and 353 nm can cause C–Cl bond rupture in CCl_4, but not in CF_3Cl.

18.9 Photoionization of O_2 requires 1205 kJ/mol. Photodissociation requires only 495 kJ/mol . At lower elevations, solar radiation with wavelengths corresponding to 1205 kJ/mol or shorter has already been absorbed, while the longer wavelength radiation has passed through relatively well. Below 90 km, the increased concentration of O_2 and the availability of longer wavelength radiation cause the photodissociation process to dominate.

18.10 The bond dissociation energy of N_2, 941 kJ/mol, is much higher than that of O_2, 495 kJ/mol. Photons with a wavelength short enough to photodissociate N_2 are not as abundant as the ultraviolet photons which lead to photodissociation of O_2. Also, N_2 does not absorb these photons as readily as O_2 so even if a short-wavelength photon is available, it may not be absorbed by an N_2 molecule.

18.11 (a) The highest rate of ozone, O_3, formation occurs at about 50 km, near the stratopause. The formation of ozone is an exothermic process as M* carries excess energy away from the O_3 molecule. The heat energy from the formation of O_3 causes the temperature to be higher near the stratopause than the lower altitude tropopause.

(b) The first step in the formation of O_3 is the photodissociation of O_2 to form two O atoms. Then, an O atom and an O_2 molecule collide to form O_3^*, a species with excess energy. If no other collisions occur, O_3^* spontaneously decomposes. If a carrier molecule such as N_2 or O_2 collides with O_3^* and removes the excess energy, O_3 is formed. It is the energy carried by M* that contributes to the temperature maximum at 50 km altitude.

18.12 (a) At 120 km and beyond, oxygen atoms are primarily generated by photodissociation of O_2 molecules: $O_2(g) + h\nu \rightarrow 2O(g)$

(b) Oxygen atoms exist longer at 120 km because there are fewer particles (atoms and molecules) at this altitude and thus fewer collisions and subsequent reactions that consume O atoms.

(c) Ozone is the primary absorber of high energy ultraviolet radiation in the 200-310 nm range. If this radiation were not absorbed in the stratosphere, plants and animals at the earth's surface would be seriously and adversely affected.

18.13 A *hydrofluorocarbon* is a compound that contains hydrogen, fluorine and carbon; it contains hydrogen in place of chlorine. HFCs are potentially less harmful than CFCs because photodissociation does not produce Cl atoms, which catalyze the destruction of ozone.

18.14 32 e^-, 16 e^- pr

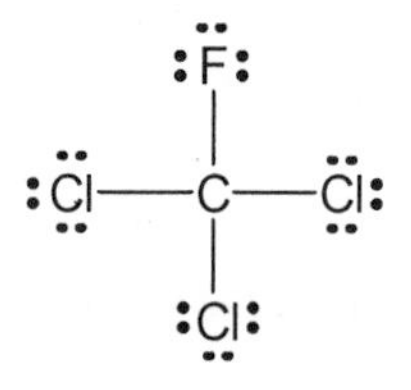

It contains C–Cl bonds that can be cleaved by UV light in the stratosphere to produce Cl atoms. It is chlorine in atomic form that catalyzes the destruction of stratospheric ozone.

CFC–11, $CFCl_3$, is chemically inert and resists decomposition in the troposphere, so that it eventually reaches the stratosphere in molecular form.

18.15 (a) In order to catalyze ozone depletion, the halogen must be present as single halogen atoms. These halogen atoms are produced in the stratosphere by photodissociation of a carbon-halogen bond. According to Table 8.4, the average C–F average bond dissociation energy is 485 kJ/mol, while that of C-Cl is 328 kJ/mol. The C–F bond requires more energy for dissociation and is not readily cleaved by the available wavelengths of UV light.

(b) Chlorine is present as chlorine atoms and chlorine oxide molecules, Cl and ClO.

18.16 Yes. Assuming $CFBr_3$ reaches the stratosphere intact, it contains C–Br bonds that are even more susceptible to cleavage by UV light than C–Cl bonds. According to Table 8.4, the average C–Br bond dissociation energy is 273 kJ/mol, compared to 328 kJ/mol for C–Cl bonds. Once in atomic form, Br atoms catalyze the destruction of ozone by a mechanism similar to that of Cl atoms.

Chemistry of the Troposphere

18.17 (a) CO binds with hemoglobin in the blood to block O_2 transport to the cells; people with CO poisoning suffocate from lack of O_2.

(b) SO_2 is corrosive to lung tissue and contributes to higher levels of respiratory disease and shorter life expectancy, especially for people with other respiratory problems such as asthma. It also is a major source of acid rain, which damages forests and wildlife in natural waters.

(c) O_3 is extremely reactive and toxic because of its ability to form free radicals upon reaction with organic molecules in the body. It is particularly dangerous for asthma suffers, exercisers and the elderly. O_3 can also react with organic compounds in polluted air to form peroxyacylnitrates, which cause eye irritation and breathing difficulties.

18.18 CO in unpolluted air is typically 0.05 ppm, whereas in urban air CO is about 10 ppm. A major source is automobile exhaust. SO_2 is less than 0.01 ppm in unpolluted air and on the order of 0.08 ppm in urban air. A major source is coal and oil-burning power plants, but there is also some SO_2 in auto exhaust. NO is about 0.01 ppm in unpolluted air and about 0.05 ppm in urban air. It comes mainly from auto exhaust.

18.19 (a) Methane, CH_4, arises from decomposition of organic matter by certain microorganisms; it also escapes from underground gas deposits.

(b) SO_2 is released in volcanic gases, and also is produced by bacterial action on decomposing vegetable and animal matter.

(c) Nitric oxide, NO, results from oxidation of decomposing organic matter, and is formed in lightning flashes.

(d) CO is a possible product of some vegetable matter decay.

18.20 All oxides of nonmetals produce acid solutions when dissolved in water. Sulfur oxides are produced naturally during volcanic eruptions and carbon oxides are products of combustion and metabolism. These dissolved gases cause rainwater to be naturally acidic.

18.21 (a) Acid rain is primarily $H_2SO_4(aq)$.

$H_2SO_4(aq) + CaCO_3(s) \rightarrow CaSO_4(s) + H_2O(l) + CO_2(g)$

(b) The $CaSO_4(s)$ would be much less reactive with acidic solution, since it would require a strongly acidic solution to shift the relevant equilibrium to the right.

$CaSO_4(s) + 2H^+(aq) \rightleftharpoons Ca^{2+}(aq) + 2HSO_4^-(aq)$

Note, however, that $CaSO_4(s)$ is brittle and easily dislodged; it provides none of the structural strength of limestone.

18.22 (a) $Fe(s) + H_2SO_4(aq) \rightarrow Fe^{2+}(aq) + SO_4^{2-}(aq) + H_2(g)$

(b) No. Silver is a 'noble' metal. It is relatively resistant to oxidation, and much more resistant than iron. In Table 4.5, The Activity Series of Metals in Aqueous Solution, Ag is much, much lower than Fe and it is below hydrogen, while Fe is above hydrogen. This means that Fe is susceptible to oxidation by acid, while Ag is not.

18.23 *Analyze/Plan*. Given wavelength of a photon, place it in the electromagnetic spectrum, calculate its energy in kJ/mol, and compare it to an average bond dissociation energy. Use Figure 6.4; E(J/photon) = hc/λ. J/photon → kJ/mol. *Solve*:

(a) Ultraviolet (Figure 6.4)

(b) $$E_{photon} = hc/\lambda = \frac{6.626 \times 10^{-34}\ J \bullet s \times 3.00 \times 10^{8}\ m/s}{335 \times 10^{-9}\ m} = 5.934 \times 10^{-19}$$

$$= 5.93 \times 10^{-19}\ J/photon$$

$$\frac{5.934 \times 10^{-19}\ J}{1\ photon} \times \frac{6.022 \times 10^{23}\ photons}{1\ mol} \times \frac{1\ kJ}{1000\ J} = 357\ kJ/mol$$

(c) The average C–H bond energy from Table 8.4 is 413 kJ/mol. The energy calculated in part (b), 357 kJ/mol, is the energy required to break 1 mol of C–H bonds in formaldehyde, CH_2O. The C–H bond energy in CH_2O must be less than the "average" C–H bond energy.

18.24 (a) Visible (Figure 6.4)

(b) $$E_{photon} = hc/\lambda = \frac{6.626 \times 10^{-34}\ J \bullet s \times 3.00 \times 10^{8}\ m/s}{420 \times 10^{-9}\ m} = 4.733 \times 10^{-19}$$

$$= 4.73 \times 10^{-19}\ J/photon$$

$$\frac{4.733 \times 10^{-19}\ J}{1\ photon} \times \frac{6.022 \times 10^{23}\ photons}{1\ mol} \times \frac{1\ kJ}{1000\ J} = 285\ kJ/mol$$

18.25 Most of the energy entering the atmosphere from the sun is in the form of visible radiation, while most of the energy leaving the earth is in the form of infrared radiation. CO_2 is transparent to the incoming visible radiation, but absorbs the outgoing infrared radiation.

18.26 (a) A *greenhouse gas* absorbs energy in the 10,000 - 30,000 nm or infrared region. It absorbs wavelengths of radiation emitted by earth and returns it as heat. A non-greenhouse gas is transparent to radiation in this wavelength range.

(b) Ar(g) is monatomic, while CH_4(g) contains 4 C–H bonds. Infrared radiation has insufficient energy to cause electron transitions or bond cleavage, but it has an appropriate amount of energy to cause molecular deformations, bond stretching and angle bending. Monatomic gases such as Ar cannot "use" infrared radiation and are transparent to it.

The World Ocean

18.27 *Analyze/Plan*. Given salinity and density, calculate molarity. A salinity of 5.3 denotes that there are 5.3 g of dry salt per kg of water. *Solve*:

$$\frac{5.3\ g\ NaCl}{1\ kg\ soln} \times \frac{1.03\ kg\ soln}{1\ L\ soln} \times \frac{1\ mol\ NaCl}{58.44\ g\ NaCl} \times \frac{1\ mol\ Na^{+}}{1\ mol\ NaCl} = 0.0934 = 0.093\ M\ Na^{+}$$

18.28 If the phosphorous is present as phosphate, there is a 1:1 ratio between the molarity of phosphorus and molarity of phosphate. Thus, we can calculate the molarity based on the given mass of P.

$$\frac{0.07\text{ g P}}{1\times10^6\text{ g H}_2\text{O}} \times \frac{1\text{ mol P}}{31\text{ g P}} \times \frac{1\text{ mol PO}_4^{3-}}{1\text{ mol P}} \times \frac{1\times10^3\text{ g H}_2\text{O}}{1\text{ L H}_2\text{O}} = 2.26\times10^{-6}$$

$$= 2.3\times10^{-6}\ M\ \text{PO}_4^{3-}$$

18.29 *Analyze/Plan.* g $Mg(OH)_2$ → mol $Mg(OH)_2$ → mol ratio → mol CaO → g CaO. *Solve:*

$$5.0\times10^6\text{ g Mg(OH)}_2 \times \frac{1\text{ mol Mg(OH)}_2}{58.3\text{ g Mg(OH)}_2} \times \frac{1\text{ mol CaO}}{1\text{ mol Mg(OH)}_2} \times \frac{56.1\text{ g CaO}}{1\text{ mol CaO}} = 4.8\times10^6\text{ g CaO}$$

18.30 $$1\times10^{11}\text{ g Br} \times \frac{1\times10^3\text{ g H}_2\text{O}}{0.067\text{ g Br}} \times \frac{1\text{ L H}_2\text{O}}{1\times10^3\text{ g H}_2\text{O}} = 1.5\times10^{12}\text{ L H}_2\text{O}$$

Because the process is only 10% efficient, ten times this much, or 1.5×10^{13} L H_2O, must be processed.

18.31 *Analyze/Plan.* Given the concentration difference between the two solutions ($\Delta M = 0.22 - 0.01 = 0.21\ M$) and temperature, calculate the minimum pressure for reverse osmosis. Use the relationship $\pi = MRT$ from Section 13.5. This is the pressure required to halt osmosis from the more dilute (0.01 M) to the more concentrated (0.22 M) solution. Slightly more pressure will initiate reverse osmosis. *Solve*:

$$\pi = \Delta MRT = \frac{0.21\text{ mol}}{\text{L}} \times \frac{0.08206\text{ L}\bullet\text{atm}}{\text{mol}\bullet\text{K}} \times 298\text{ K} = 5.135 = 5.1\text{ atm}$$

The minimum pressure required to initiate reverse osmosis is greater than 5.1 atm.

18.32 Calculate the total ion concentration of sea water by summing the molarities given in Table 18.6. Then use $\pi = \Delta MRT$ to calculate pressure.

$$M_{total} = 0.55 + 0.47 + 0.028 + 0.054 + 0.010 + 0.010 + 2.3\times10^{-3} + 8.3\times10^{-4}$$
$$+ 4.3\times10^{-4} + 9.1\times10^{-5} + 7.0\times10^{-5} = 1.1257 = 1.13\ M$$

$$\pi = \frac{(1.1257 - 0.02)\text{ mol}}{\text{L}} \times \frac{0.08206\text{ L}\times\text{atm}}{\text{mol}\bullet\text{K}} \times 305\text{ K} = 27.674 = 27.7\text{ atm}$$

Check. The largest numbers in the molarity sum have 2 decimal places, so M_{total} has 2 decimal places and 3 sig figs. ΔM also has 2 decimal places and 3 sig figs so the calculated pressure has 3 sig figs. Units are correct.

Freshwater

18.33 *Analyze/Plan.* Under aerobic conditions, excess oxyen is present and decomposition leads to oxidized products, the element in its maximum oxidation state combined with oxygen. Under anaerobic conditions, little or no oxygen is present so decomposition leads to reduced products, the element in its minimum oxidation state combined with hydrogen. *Solve*:

(a) CO_2, HCO_3^-, H_2O, SO_4^{2-}, NO_3^-, HPO_4^{2-}, $H_2PO_4^-$.

(b) $CH_4(g)$, $H_2S(g)$, $NH_3(g)$, $PH_3(g)$

18.34 (a) Decomposition of organic matter by aerobic bacteria depletes dissolved O_2. A low dissolved oxygen concentration indicates the presence of organic pollutants.

(b) According to Section 13.3, the solubility of $O_2(g)$ (or any gas) in water decreases with increasing temperature.

18.35 *Analyze/Plan*. Given the balanced equation, calculate the amount of one reactant required to react exactly with a certain amount of the other reactants. Solve the stoichiometry problem. g $C_{18}H_{29}O_3S^-$ → mol → mol ratio → mol O_2 → g O_2. *Solve*:

$$1.0\text{ g } C_{18}H_{29}O_3S^- \times \frac{1\text{ mol } C_{18}H_{29}O_3S^-}{325\text{ g } C_{18}H_{29}O_3S^-} \times \frac{51\text{ mol } O_2}{2\text{ mol } C_{18}H_{29}O_3S^-} \times \frac{32.0\text{ g } O_2}{1\text{ mol } O_2} = 2.5\text{ g } O_2$$

Notice that the mass of O_2 required is 2.5 times greater than the mass of biodegradable material.

18.36 $$85{,}000\text{ persons} \times \frac{59\text{ g } O_2}{1\text{ person}} \times \frac{1 \times 10^6\text{ g } H_2O}{9\text{ g } O_2} \times \frac{1\text{ L } H_2O}{1 \times 10^3\text{ g } H_2O} = 5.6 \times 10^8\text{ L } H_2O$$

18.37 *Analyze/Plan*. Slaked lime is $Ca(OH)_2(s)$. The reaction is metathesis. *Solve*:

$Mg^{2+}(aq) + Ca(OH)_2(s) \rightarrow Mg(OH)_2(s) + Ca^{2+}(aq)$

The excess $Ca^{2+}(aq)$ is removed as $CaCO_3$ by naturally occurring bicarbonate or added Na_2CO_3.

18.38 (a) Ca^{2+}, Mg^{2+}, Fe^{2+}

(b) Divalent cations (ions with +2 changes) contribute to water hardness. These ions react with soap to form scum on surfaces or leave undesirable deposits on surfaces, particularly inside pipes, upon heating.

18.39 *Analyze/Plan*. Given $[Ca^{2+}]$ and $[HCO_3^-]$ calculate mole $Ca(OH)_2$ and Na_2CO_3 needed to remove the Ca^{2+} and HCO_3^-. Consider the chemical equations and reaction stoichiometry in the stepwise process. *Solve*:

$Ca(OH)_2$ is added to remove Ca^{2+} as $CaCO_3(s)$, and Na_2CO_3 removes the remaining Ca^{2+}. $Ca^{2+}(aq) + 2HCO_3^-(aq) + [Ca^{2+}(aq) + 2OH^-(aq)] \rightarrow 2CaCO_3(s) + 2H_2O(l)$. One mole $Ca(OH)_2$ is needed for each 2 moles of $HCO_3^-(aq)$ present. If there are 7.0×10^{-4} mol $HCO_3^-(aq)$ per liter, we must add 3.5×10^{-4} mol $Ca(OH)_2$ per liter, or a total of 0.35 mol $Ca(OH)_2$ for 10^3 L. This reaction removes 3.5×10^{-4} mol of the original Ca^{2+} from each liter of solution, leaving 1.5×10^{-4} *M* $Ca^{2+}(aq)$. To remove this $Ca^{2+}(aq)$, we add 1.5×10^{-4} mol Na_2CO_3 per liter, or a total of 0.15 mol Na_2CO_3, forming $CaCO_3(s)$.

18.40 $Ca(OH)_2$ is added to remove Ca^{2+} as $CaCO_3(s)$, and Na_2CO_3 removes the remaining Ca^{2+}. $Ca^{2+}(aq) + 2HCO_3^-(aq) + [Ca^{2+}(aq) + 2OH^-(aq)] \rightarrow 2CaCO_3(s) + 2H_2O(l)$. One mole $Ca(OH)_2$ is needed for each 2 moles of $HCO_3^-(aq)$ present.

$$5.0 \times 10^7 \text{ L } H_2O \times \frac{1.7 \times 10^{-3} \text{ mol } HCO_3^-}{1 \text{ L } H_2O} \times \frac{1 \text{ mol } Ca(OH)_2}{2 \text{ mol } HCO_3^-} \times \frac{74 \text{ g } Ca(OH)_2}{1 \text{ mol } Ca(OH)_2}$$

$$= 3.1 \times 10^6 \text{ g } Ca(OH)_2$$

Half of the native HCO_3^- precipitates the added Ca^{2+} so this operation reduces the Ca^{2+} concentration from 5.7×10^{-3} *M* to $(5.7 \times 10^{-3} - 8.5 \times 10^{-4})$ *M* = 4.85×10^{-3} *M*. Next we must add sufficient Na_2CO_3 to further reduce $[Ca^{2+}]$ to 1.1×10^{-3} *M*. We thus need to reduce $[Ca^{2+}]$ by $(4.85 \times 10^{-3} - 1.1 \times 10^{-3})$ *M* = 3.75×10^{-3} *M*. $Ca^{2+}(aq) + CO_3^{-2}(aq) \rightarrow CaCO_3(s)$.

$$5.0 \times 10^7 \text{ L } H_2O \times \frac{3.75 \times 10^{-3} \text{ mol } Ca^{2+}}{1 \text{ L } H_2O} \times \frac{1 \text{ mol } Na_2CO_3}{1 \text{ mol } Ca^{2+}} \times \frac{106 \text{ g } Na_2CO_3}{1 \text{ mol } Na_2CO_3}$$

$$= 2.0 \times 10^7 \text{ g } Na_2CO_3$$

18.41 A slightly basic solution contains a small excess of $OH^-(aq)$. $Al_2(SO_4)_3$ reacts with OH^- to form $Al(OH)_3(s)$, a gelatinous precipitate that occludes fine particles and bacteria present in the water. The $Al(OH)_3(s)$ settles slowly, removing the undesirable particulate matter.

$Al_2(SO_4)_3(s) + 6OH^-(aq) \rightarrow 2Al(OH)_3(s) + 3SO_4^{2-}(aq)$

18.42 $4FeSO_4(aq) + O_2(aq) + 2H_2O(l) \rightarrow 4Fe^{3+}(aq) + 4OH^-(aq) + 4SO_4^{2-}(aq)$

SO_4^{2-} is a spectator, so the net ionic equation is $4Fe^{2+}(aq) + 2H_2O(l) \rightarrow 4Fe^{3+}(aq) + 4OH^-(aq)$.

$Fe^{3+}(aq) + 3HCO_3^-(aq) \rightarrow Fe(OH)_3(s) + 3CO_2(g)$

In this reaction, Fe^{3+} acts as a Lewis acid, and HCO_3^- acts as a Lewis base.

Green Chemistry

18.43 Production of any form of energy requires a fuel and generates waste products. A more energy-efficient device or process uses less energy, which requires less fuel and generates fewer waste-products. Automobile fuel efficiency is a clear example. The better the gas mileage of a vehicle, the less gasoline per mile is burned and the fewer waste products (unburned by hydrocarbons, CO, NO_x, etc.) are generated. Electrical appliances are a less obvious but equally pertinent example. Electricity is generated at a power plant, which uses coal, oil or nuclear fuel. The less electricity required by an appliance, the less waste per appliance use that is generated at the power plant.

18.44 Catalysts increase the rate of a reaction by lowering activation energy, E_a. For an uncatalyzed reaction that requires extreme temperatures and pressures to generate product at a viable rate, finding a suitable catalyst reduces the required temperature and/or pressure, which reduces the amount of energy used to run the process. A catalyst can also increase rate of production, which would reduce the net time and thus energy required to generate a certain amount of product.

18.45 One use of phosgene, Cl—C(=O)—Cl, is as the source of the —C(=O)— carbonyl group in a condensation polymerization. The small-molecule that is 'condensed' is HCl, a corrosive strong acid and source of Cl^- that definitely requires treatment. Use of dimethylcarbonate in place of phosgene as a carbonyl source condenses methanol, CH_3OH, rather than HCl. Methanol is much less toxic than HCl, and it has the potential to be a second useful product, rather than waste.

18.46 (a) cyclohexanone ($C_6H_{10}O$: ring of C=O and five CH_2) + H—O—O—H ⟶ ε-caprolactone (seven-membered ring: C=O, O, five CH_2) + H_2O

(b)

- It is better to prevent waste than to treat it. The alternative process eliminates production of 3-chlorobenzoic acid by-product, chlorine-containing waste that must be treated.
- Produce as little, nontoxic waste as possible. The by-product of the alternative process is nontoxic water. The low molar mass of water means that a small amount of 'waste' is generated.
- Chemical processes should be efficient. The alternative process is catalyzed, which could mean that the process will be more energy efficient than the Baeyer-Villiger reaction (see Solution 18.44).
- Raw materials should be renewable. The catalyst can be recovered from the reaction mixture and reused. We don't have information about solvents or other auxiliary substances.

Additional Exercises

18.47 (a) *Acid rain* is rain with a larger $[H^+]$ and thus a lower pH than expected. The additional H^+ is produced by the dissolution of sulfur and nitrogen oxides such as $SO_3(g)$ and $NO_2(g)$ in rain droplets to form sulfuric and nitric acid, $H_2SO_4(aq)$ and $HNO_3(aq)$.

(b) The *greenhouse effect* is warming of the atmosphere caused by heat trapping gases such as $CO_2(g)$ and $H_2O(g)$. That is, these gases absorb infrared or "heat" radiation emitted from the earth's surface and serve to maintain a relatively constant temperature on the surface. A significant increase in the amount of atmospheric CO_2

(from burning fossil fuels and other sources) could cause a corresponding increase in the average surface temperature and drastically change the global climate.

(c) *Photochemical smog* is an unpleasant collection of atmospheric pollutants initiated by photochemical dissociation of NO_2 to form NO and O atoms. The major components are NO(g), NO_2(g), CO(g) and unburned hydrocarbons, all produced by automobile engines, and O_3(g), ozone.

(d) The *ozone hole* is a region of depleted O_3 in the stratosphere over Antarctica. It is caused by reactions between O_3 and Cl atoms originating from chlorofluorocarbons (CFC's), CF_xCl_{4-x}. Depletion of the ozone layer would allow damaging ultraviolet radiation disruptive to the plant and animal life in our ecosystem to reach earth.

18.48 $\mathcal{M}_{avg}$ at the surface = 40.0(0.17) + 16.0(0.38) + 32.0(0.45) = 27.28 = 27 g/mol.

Next, calculate the percentage composition at 200 km. The fractions can be "normalized" by saying that the 0.45 fraction of O_2 is converted into **two** 0.45 fractions of O atoms, then dividing by the total fractions, 0.17 + 0.38 + 0.45 + 0.45 = 1.45:

$$\mathcal{M}_{avg} = \frac{40.0(0.17) + 16.0(0.38) + 16.0(0.90)}{1.45} = 18.81 = 19 \text{ g/mol}$$

18.49 Stratospheric ozone is formed and destroyed in a cycle of chemical reactions. The decomposition of O_3 to O_2 and O produces oxygen atoms, an essential ingredient for the production of ozone. While single O_3 molecules exist for only a few seconds, new O_3 molecules are constantly reformed. This cyclic process ensures a finite concentration of O_3 in the stratosphere available to absorb ultraviolet radiation. (This explanation assumes that the cycle is not disrupted by outside agents such as CFCs.)

18.50

$$2[Cl(g) + O_3(g) \rightarrow ClO(g) + O_2(g)] \qquad [18.7]$$

$$2Cl(g) + 2O_3(g) \rightarrow 2ClO(g) + 2O_2(g)$$

$$2ClO(g) \rightarrow O_2(g) + 2Cl(g) \qquad [18.9]$$

$$2Cl(g) + 2O_3(g) + 2ClO(g) \rightarrow 2ClO(g) + 3O_2(g) + 2Cl(g)$$

$$2O_3(g) \xrightarrow{Cl} 3O_2(g) \qquad [18.10]$$

Note that Cl(g) fits the definition of a catalyst in this reaction.

18.51 (a) The production of Cl atoms in the stratosphere is the result of the photodissociation of a C–Cl bond in the chlorofluorocarbon molecule.

$$CF_2Cl_2(g) \xrightarrow{h\nu} CF_2Cl(g) + Cl(g)$$

According to Table 8.4, the bond dissociation energy of a C–Br bond is 276 kJ/mol, while the value for a C–Cl bond is 328 kJ/mol. Photodissociation of $CBrF_3$ to form Br atoms requires less energy than the production of Cl atoms and should occur readily in the stratosphere.

(b) $CBrF_3(g) \xrightarrow{hv} CF_3(g) + Br(g)$

$Br(g) + O_3(g) \longrightarrow BrO(g) + O_2(g)$

Also, under certain conditions

$BrO(g) + BrO(g) \longrightarrow Br_2O_2(g)$

$Br_2O_2(g) + hv \longrightarrow O_2(g) + 2Br(g)$

18.52 (a) HNO_3 is a major component in acid rain.

(b) While it removes CO, the reaction produces NO_2. The photodissociation of NO_2 to form O atoms is the first step in the formation of tropospheric ozone and photochemical smog.

(c) Again, NO_2 is the initiator of photochemical smog. Also, methoxyl radical, OCH_3, is a reactive species capable of initiating other undesirable reactions.

18.53 From section 18.4:

$N_2(g) + O_2(g) \rightleftharpoons 2NO(g)$ $\Delta H = +180.8$ kJ (1)

$2\,NO(g) + O_2(g) \rightleftharpoons 2NO_2(g)$ $\Delta H = -113.1$ kJ (2)

In an endothermic reaction, heat is a reactant. As the temperature of the reaction increases, the addition of heat favors formation of products and the value of K increases. The reverse is true for exothermic reactions; as temperature increases, the value of K decreases. Thus, K for reaction (1), which is endothermic, increases with increasing temperature and K for reaction (2), which is exothermic, decreases with increasing temperature.

18.54 Oxygen is present in the atmosphere to the extent of 209,000 parts per million. If CO binds 210 times more effectively than O_2, then the **effective** concentration of CO is 210 × 112 ppm = 23,520 = 23,500 ppm. The fraction of carboxyhemoglobin in the blood leaving the lungs is thus $\dfrac{23{,}520}{23{,}520 + 209{,}000} = 0.101$. Thus, 10.1 percent of the blood is in the form of carboxyhemoglobin, 89.9 percent as the O_2-bound oxyhemoglobin.

18.55 (a) $CH_4(g) + 2O_2(g) \rightarrow CO_2(g) + 2H_2O(g)$

(b) $2CH_4(g) + 3O_2(g) \rightarrow 2CO(g) + 4H_2O(g)$

(c) vol CH_4 → vol O_2 → volume air ($\chi_{O_2} = 0.20948$)

Equal volumes of gases at the same temperature and pressure contain equal numbers of moles (Avogadro's law). If 2 moles of O_2 are required for 1 mole of CH_4, 2.0 L of pure O_2 are needed to burn 1.0 L of CH_4.

$$\text{vol } O_2 = \chi_{O_2} \times \text{vol}_{air} = \frac{\text{vol } O_2}{\chi_{O_2}} = \frac{2.0\text{ L}}{0.20948} = 9.5\text{ L air}$$

(d) $2NO_2(g) + 2CH_4(g) \xrightarrow{\text{catalyst}} N_2(g) + 2C(s) + 4H_2O(l)$

$2NO(g) + CH_4(g) \xrightarrow{\text{catalyst}} N_2(g) + C(s) + 2H_2O(l)$

NO_x contains N in an oxidized form and CH_4 contains C in a reduced form, so the reaction is likely to generate a reduced form of N, N_2, and a more oxidized form of C. The carbon-containing products could also be CO(g) or CO_2(g), depending on the relative amounts of NO_x.

18.56 (a) According to Section 13.3, the solubility of gases in water decreases with increasing temperature. Thus, the solubility of CO_2(g) in the ocean would decrease if the temperature of the ocean increased.

(b) If the solubility of CO_2(g) in the ocean decreased because of global warming, more CO_2(g) would be released into the atmosphere, perpetuating a cycle of increasing temperature and concomitant release of CO_2(g) from the ocean.

18.57 Most of the 390 watts/m^2 radiated from Earth's surface is in the infrared region of the spectrum. Tropospheric gases, particularly H_2O(g) and CO_2(g), absorb much of this radiation and prevent it from escaping into space (Figure 18.10). The energy absorbed by these so-called "greenhouse gases" warms the atmosphere close to Earth's surface and makes the planet liveable.

18.58 (a) $NO(g) + h\nu \rightarrow N(g) + O(g)$

(b) $NO(g) + h\nu \rightarrow NO^+(g) + e^-$

(c) $NO(g) + O_3(g) \rightarrow NO_2(g) + O_2(g)$

(d) $3NO_2(g) + H_2O(l) \rightarrow 2HNO_3(aq) + NO(g)$

18.59 (a) CO_3^{2-} is a relatively strong Brønsted base and produces OH^- in aqueous solution according to the hydrolysis reaction:

$$CO_3^{2-}(aq) + H_2O(l) \rightleftharpoons HCO_3^-(aq) + OH^-(aq), \quad K_b = 1.8 \times 10^{-4}$$

If $[OH^-(aq)]$ is sufficient to exceed K_{sp} for $Mg(OH)_2$, the solid will precipitate.

(b)
$$\frac{125 \text{ mg } Mg^{2+}}{1 \text{ kg soln}} \times \frac{1 \text{ g } Mg^{2+}}{1000 \text{ mg } Mg^{2+}} \times \frac{1.00 \text{ kg soln}}{1.00 \text{ L soln}} \times \frac{1 \text{ mol } Mg^{2+}}{24.305 \text{ g } Mg^{2+}} = 5.143 \times 10^{-3}$$

$$= 5.14 \times 10^{-3}\ M\ Mg^{2+}$$

$$\frac{4.0 \text{ g } Na_2CO_3}{1.0 \text{ L soln}} \times \frac{1 \text{ mol } CO_3^{2-}}{106.0 \text{ g } Na_2CO_3} = 0.03774 = 0.038\ M\ CO_3^{2-}$$

$$K_b = 1.8 \times 10^{-4} = \frac{[HCO_3^-][OH^-]}{[CO_3^{2-}]} \approx \frac{x^2}{0.03774}; \quad x = [OH^-] = 2.606 \times 10^{-3}$$

$$= 2.6 \times 10^{-3}\ M$$

(This represents 6.9% hydrolysis, but the result will not be significantly different using the quadratic formula.)

$Q = [Mg^{2+}][OH^-]^2 = (5.143 \times 10^{-3})(2.606 \times 10^{-3})^2 = 3.5 \times 10^{-8}$

K_{sp} for $Mg(OH)_2 = 1.6 \times 10^{-12}$; $Q > K_{sp}$, so $Mg(OH)_2$ will precipitate.

18.60 Chlorine is more convenient, because it can be shipped to the treatment sight as condensed gas and introduced in a measured way. It requires no new capital expense, because the chlorination equipment is in place. However, Cl_2 is expensive to produce and ship because it is corrosive and toxic. It is a safety hazard for workers. Its use in water treatment produces trihalomethanes, THM's, which pose a small cancer risk when dissolved in water. Perhaps more importantly, THM's are greenhouse gases and contribute to the destruction of stratospheric ozone.

18.61
- Use of chlorine as an oxidizing agent generates toxic chlorine-containing waste that must be treated.
- Cl_2 is not a renewable feed stock.
- Cl_2 is produced by electrolysis of molten NaCl. This is an energy-intensive process that reduces the net energy-efficiency of pulp bleaching.
- Cl_2 is an auxiliary substance and is certainly not innocuous.

18.62 Because NO has an odd electron, like Cl(g), it could act as a catalyst for decomposition of ozone in the stratosphere. The increased destruction of ozone by NO would result in less absorption of short wavelength UV radiation now being screened out primarily by the ozone. Radiation in this wavelength range is known to be harmful to humans; it causes skin cancer. There is evidence that many plants don't tolerate it very well either, though more research is needed to test this idea.

In Chapter 22 the oxidation of NO to NO_2 by oxygen is described. On dissolving in water, NO_2 disproportionates into NO_3^- (aq) and NO(g). Thus, over time the NO in the troposphere will be converted into NO_3^- which is in turn incorporated into soils.

Integrative Exercises

18.63 (a) $$0.021 \text{ ppm } NO_2 = \frac{0.021 \text{ mol } NO_2}{1 \times 10^6 \text{ mol air}} = 2.1 \times 10^{-8} = \chi_{NO_2}$$

$$P_{NO_2} = \chi_{NO_2} \cdot P_{atm} = 2.1 \times 10^{-8} (745 \text{ torr}) = 1.565 \times 10^{-5} = 1.6 \times 10^{-5} \text{ torr}$$

(b) $$n = \frac{PV}{RT}; \text{ molecules} = n \times \frac{6.022 \times 10^{23} \text{ molecules}}{\text{mol}} = \frac{PV}{RT} \times \frac{6.022 \times 10^{23} \text{ molecules}}{\text{mol}}$$

$$V = 15 \text{ ft} \times 14 \text{ ft} \times 8 \text{ ft} \times \frac{12^3 \text{ in}^3}{\text{ft}^3} \times \frac{2.54^3 \text{ cm}^3}{\text{in}^3} \times \frac{1 \text{ L}}{1000 \text{ cm}^3} = 4.757 \times 10^4 = 5 \times 10^4 \text{ L}$$

$$1.565 \times 10^{-5} \text{ torr} \times \frac{1 \text{ atm}}{760 \text{ torr}} \times \frac{4.757 \times 10^4 \text{ L}}{293 \text{ K}} \times \frac{\text{K} \cdot \text{mol}}{0.08206 \text{ L} \cdot \text{atm}}$$

$$\times \frac{6.022 \times 10^{23} \text{ molecules}}{\text{mol}} = 2.453 \times 10^{19} = 2 \times 10^{19} \text{ molecules}$$

18.64 (a) $8{,}376{,}726 \text{ tons coal} \times \frac{83 \text{ ton C}}{100 \text{ ton coal}} \times \frac{44.01 \text{ ton } CO_2}{12.01 \text{ ton C}} = 2.5 \times 10^7 \text{ ton } CO_2$

$8{,}376{,}726 \text{ tons coal} \times \frac{2.5 \text{ ton S}}{100 \text{ ton coal}} \times \frac{64.07 \text{ ton } SO_2}{32.07 \text{ ton S}} = 4.18 \times 10^5$

(b) $CaO(s) + SO_2(g) \rightarrow CaSO_3(s)$

$4.18 \times 10^5 \text{ ton } SO_2 \times \frac{55 \text{ ton } SO_2 \text{ removed}}{100 \text{ ton } SO_2 \text{ produced}} \times \frac{120.15 \text{ ton } CaSO_3}{64.07 \text{ ton } SO_2}$

$= 4.3 \times 10^5 \text{ ton } CaSO_3$

18.65 *Course sand* is removed by course sand filtration. *Finely divided particles* and some *bacteria* are removed by precipitation with aluminum hydroxide. Remaining *harmful bacteria* are removed by ozonation. *Trihalomethanes* are removed by either aeration or activated carbon filtration; use of activated carbon might be preferred because it does not involve release of TCMs into the atmosphere. *Dissolved organic substances* are oxidized (and rendered less harmful, but not removed) by both aeration and ozonation. Dissolved *nitrates* and *phosphates* are not removed by any of these processes, but are rendered less harmful by adequate aeration.

18.66 (a) $\Delta H = 2D(O–H) - D(O–H) = D(O–H) = 463 \text{ kJ/mol}$

$\frac{463 \text{ kJ}}{\text{mol } H_2O} \times \frac{1 \text{ mol } H_2O}{6.022 \times 10^{23} \text{ molecules}} \times \frac{1000 \text{ J}}{\text{kJ}} = 7.688 \times 10^{-19}$

$= 7.69 \times 10^{-19} \text{ J/}H_2O \text{ molecule}$

$\lambda = \frac{hc}{\Delta E} = \frac{6.626 \times 10^{-34} \text{ J} \cdot \text{sec} \times 2.998 \times 10^8 \text{ m/s}}{7.688 \times 10^{-19} \text{ J}} = 2.58 \times 10^{-7} \text{ m} = 258 \text{ nm}$

This wavelength is in the UV region of the spectrum, close to the visible.

(b)

$OH(g) + O_3(g) \rightarrow HO_2(g) + O_2(g)$

$HO_2(g) + O(g) \rightarrow OH(g) + O_2(g)$

$OH(g) + O_3(g) + HO_2(g) + O(g) \rightarrow HO_2(g) + 2O_2(g) + OH(g)$

$O_3(g) + O(g) \rightarrow 2O_2(g)$

OH(g) is the catalyst in this overall reaction, another pathway for the destruction of ozone.

18.67 (i) $ClO(g) + O_3(g) \rightarrow ClO_2(g) + O_2(g)$

$\Delta H_i = \Delta H^\circ_f \, ClO_2(g) + \Delta H^\circ_f \, O_2(g) - \Delta H^\circ_f \, ClO(g) - \Delta H^\circ_f \, O_3(g)$

$\Delta H_i = 102 + 0 - 101 - (142.3) = -141 \text{ kJ}$

(ii) $ClO_2(g) + O(g) \rightarrow ClO(g) + O_2(g)$

$\Delta H_{ii} = \Delta H^\circ_f \, ClO(g) + \Delta H^\circ_f \, O_2(g) - \Delta H^\circ_f \, ClO_2(g) + \Delta H^\circ_f \, O(g)$

$\Delta H_{ii} = 101 + 0 - 102 - (247.5) = -249 \text{ kJ}$

(overall) $ClO(g) + O_3(g) + ClO_2(g) + O(g) \rightarrow ClO_2(g) + O_2(g) + ClO(g) + O_2(g)$

$O_3(g) + O(g) \rightarrow 2O_2(g)$

$\Delta H = \Delta H_i + \Delta H_{ii} = -141 \text{ kJ} + (-249) \text{ kJ} = -390 \text{ kJ}$

Because the enthalpies of both (i) and (ii) are distinctly exothermic, it is possible that the ClO - ClO_2 pair could be a catalyst for the destruction of ozone.

18.68 (a) Assume the density of water at 20°C is the same as at 25°C.

$$1.00 \text{ gal} \times \frac{4 \text{ qt}}{1 \text{ gal}} \times \frac{1 \text{ L}}{1.057 \text{ qt}} \times \frac{1000 \text{ mL}}{1 \text{ L}} \times \frac{0.99707 \text{ g } H_2O}{1 \text{ mL}} = 3773$$

$$= 3.77 \times 10^3 \text{ g } H_2O$$

The $H_2O(l)$ must be heated from 20°C to 100°C and then vaporized at 100°C.

$$3.773 \times 10^3 \text{ g } H_2O \times \frac{4.184 \text{ J}}{\text{g °C}} \times 80 \text{ °C} \times \frac{1 \text{ kJ}}{1000 \text{ J}} = 1263 = 1.3 \times 10^3 \text{ kJ}$$

$$3.773 \times 10^3 \text{ g } H_2O \times \frac{1 \text{ mol } H_2O}{18.02 \text{ g } H_2O} \times \frac{40.67 \text{ kJ}}{\text{mol } H_2O} = 8516 = 8.52 \times 10^3 \text{ kJ}$$

energy = 1263 kJ + 8516 kJ = 9779 = 9.8×10^3 kJ/gal H_2O

(b) According to Solution 5.6, 1 kwh = 3.6×10^6 J.

$$\frac{9779 \text{ kJ}}{\text{gal } H_2O} \times \frac{1000 \text{ J}}{\text{kJ}} \times \frac{1 \text{ kwh}}{3.6 \times 10^6 \text{ J}} \times \frac{\$0.085}{\text{kwh}} = \$0.23/\text{gal}$$

(c) $\frac{\$0.23}{\$1.26} \times 100 = 18\%$ of the total cost is energy

18.69 (a) A rate constant of $M^{-1}s^{-1}$ is indicative of a reaction that is second order overall. For the reaction given, the rate law is probably rate = $k[O][O_3]$. (Although rate = $k[O]^2$ or $k[O_3]^2$ are possibilities, it is difficult to envision a mechanism consistent with either one that would result in two molecules of O_2 being produced.)

(b) Yes. Most atmospheric processes are initiated by collision. One could imagine an activated complex of four O atoms collapsing to form two O_2 molecules. Also, the rate constant is large, which is less likely for a multistep process. The reaction is analogous to the destruction of O_3 by Cl atoms (Equation 18.7), which is also second order with a large rate constant.

(c) According to the Arrhenius equation, $k = Ae^{-E_a/RT}$. Thus, the larger the value of k, the smaller the activation energy, E_a. The value of the rate constant for this reaction is large, so the activation energy is small.

(d) $\Delta H^\circ_f = 2\Delta H^\circ_f \; O_2(g) - \Delta H^\circ_f \; O(g) - \Delta H^\circ_f \; O_3(g)$

$\Delta H^\circ_f = 0 - 247.5 \text{ kJ} - 142.3 \text{ kJ} = -389.8 \text{ kJ}$

The reaction is exothermic, so energy is released; the reaction would raise the temperature of the stratosphere.

18.70 From the composition of air at sea-level (Table 18.1) calculate the partial pressures of $N_2(g)$ and $O_2(g)$ in the original sample.

$P_x = \chi_x \cdot P_T$; $P_{N_2} = 0.78084\ (1.0\text{ atm}) = 0.78\text{ atm}$; $P_{O_2} = 0.20948\ (1.0\text{ atm}) = 0.21\text{ atm}$

	$N_2(g)$	+	$O_2(g)$	$\rightleftharpoons$	$2NO(g)$
initial	0.78 atm		0.21 atm		0
charge	-x		-x		+2x
equil	(0.78-x) atm		(0.21-x) atm		2x atm

$$K_p = \frac{P_{NO}^2}{P_{N_2} \times P_{O_2}} = \frac{(2x)^2}{(0.78-x)(0.21-x)} = \frac{4x^2}{0.164 - 0.99x + x^2} = \frac{4x^2}{0.164 - 0.99x + x^2} = 0.05$$

$0.05\ (0.164 - 0.99x + x^2) = 4x^2$; $0 = 3.95x^2 + 0.05x - 0.0082$

Using the quadratic formula, $x = \frac{-b \pm \sqrt{b^2 - 4ac}}{2a} = \frac{-0.05 \pm \sqrt{(0.05)^2 - 4(3.95)(-0.0082)}}{2(3.95)}$

$$x = \frac{-0.05 \pm \sqrt{0.0025 + 0.1296}}{7.90} = \frac{-0.05 \pm 0.363}{7.90}$$

The negative result is meaningless; $x = 0.04$ atm; $P_{NO} = 2x = 0.08$ atm

Assuming that the total pressure of the gaseous mixture at equilibrium is still 1.0 atm,

$\chi_{NO} = P_{NO} / P_T = 0.08\text{ atm}/1.0\text{ atm} = 0.08$

ppm for gases $= \chi \times 10^6$ (see Section 18.1)

$ppm_{CO} = 0.08 \times 10^6 = 8 \times 10^4$ ppm

18.71 (a) 17 e^-, 8.5 e^- pairs

$\ddot{O}{=}\dot{N}{-}\ddot{O}: \longleftrightarrow :\ddot{O}{-}\dot{N}{=}\ddot{O}$

Owing to its lower electronegativity, N is more likely to be electron deficient and to accommodate the odd electron.

(b) The fact that NO_2 is an electron deficient molecule indicates that it will be highly reactive. Dimerization results in formation of a N–N single bond which completes the octet of both N atoms. NO_2 and N_2O_4 exist in equilibrium in a closed system. The reaction is exothermic, Equation [22.65]. In an urban environment, NO_2 is produced from hot automobile combustion. At these temperatures, equilibrium favors the monomer because the reaction is exothermic.

(c) $2NO_2(g) + 4CO(g) \rightarrow N_2(g) + 4CO_2(g)$

$NO_2(g) + CO(g) \rightarrow NO(g) + CO_2(g)$

NO_2 is an oxidizing agent and CO is a reducing agent, so we expect products to contain N in a more reduced form, NO or N_2, and C in a more oxidized form, CO_2.

(d) No. Because it is an odd-electron molecule, NO_2 is very reactive. We expect it to undergo chemical reactions or photodissociate before it can migrate to the stratosphere. The expected half-life of an NO_2 molecule is short.

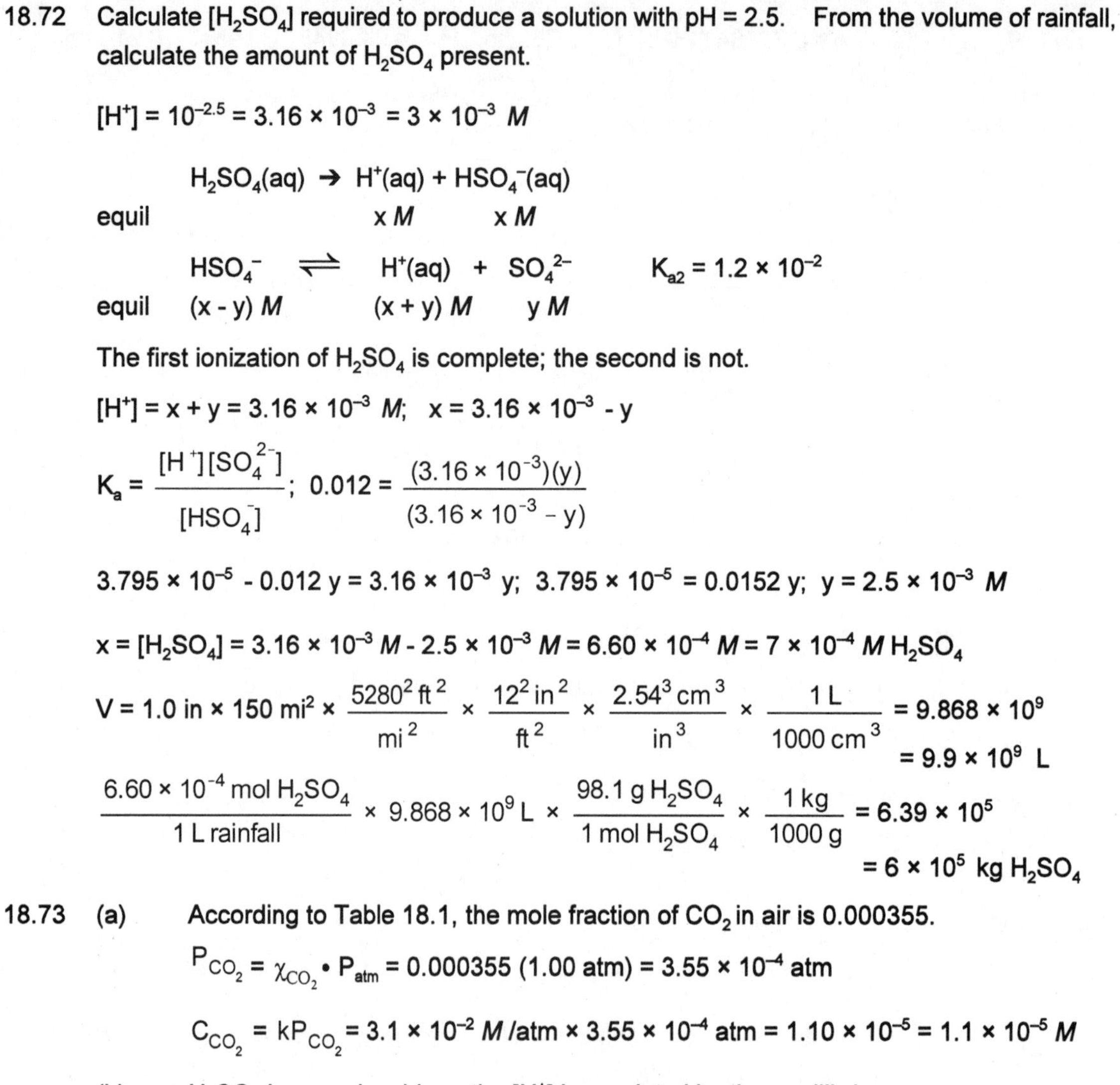

18.72 Calculate $[H_2SO_4]$ required to produce a solution with pH = 2.5. From the volume of rainfall, calculate the amount of H_2SO_4 present.

$[H^+] = 10^{-2.5} = 3.16 \times 10^{-3} = 3 \times 10^{-3}$ *M*

$$H_2SO_4(aq) \rightarrow H^+(aq) + HSO_4^-(aq)$$

equil $\quad x\ M \quad x\ M$

$$HSO_4^- \rightleftharpoons H^+(aq) + SO_4^{2-} \qquad K_{a2} = 1.2 \times 10^{-2}$$

equil $\quad (x - y)\ M \quad (x + y)\ M \quad y\ M$

The first ionization of H_2SO_4 is complete; the second is not.

$[H^+] = x + y = 3.16 \times 10^{-3}\ M; \quad x = 3.16 \times 10^{-3} - y$

$$K_a = \frac{[H^+][SO_4^{2-}]}{[HSO_4^-]}; \quad 0.012 = \frac{(3.16 \times 10^{-3})(y)}{(3.16 \times 10^{-3} - y)}$$

$3.795 \times 10^{-5} - 0.012\ y = 3.16 \times 10^{-3}\ y; \quad 3.795 \times 10^{-5} = 0.0152\ y; \quad y = 2.5 \times 10^{-3}\ M$

$x = [H_2SO_4] = 3.16 \times 10^{-3}\ M - 2.5 \times 10^{-3}\ M = 6.60 \times 10^{-4}\ M = 7 \times 10^{-4}\ M\ H_2SO_4$

$$V = 1.0\ \text{in} \times 150\ \text{mi}^2 \times \frac{5280^2\ \text{ft}^2}{\text{mi}^2} \times \frac{12^2\ \text{in}^2}{\text{ft}^2} \times \frac{2.54^3\ \text{cm}^3}{\text{in}^3} \times \frac{1\ \text{L}}{1000\ \text{cm}^3} = 9.868 \times 10^9$$
$$= 9.9 \times 10^9\ \text{L}$$

$$\frac{6.60 \times 10^{-4}\ \text{mol}\ H_2SO_4}{1\ \text{L rainfall}} \times 9.868 \times 10^9\ \text{L} \times \frac{98.1\ \text{g}\ H_2SO_4}{1\ \text{mol}\ H_2SO_4} \times \frac{1\ \text{kg}}{1000\ \text{g}} = 6.39 \times 10^5$$
$$= 6 \times 10^5\ \text{kg}\ H_2SO_4$$

18.73 (a) According to Table 18.1, the mole fraction of CO_2 in air is 0.000355.

$P_{CO_2} = \chi_{CO_2} \bullet P_{atm} = 0.000355\ (1.00\ \text{atm}) = 3.55 \times 10^{-4}$ atm

$C_{CO_2} = kP_{CO_2} = 3.1 \times 10^{-2}\ M/\text{atm} \times 3.55 \times 10^{-4}\ \text{atm} = 1.10 \times 10^{-5} = 1.1 \times 10^{-5}\ M$

(b) H_2CO_3 is a weak acid, so the $[H^+]$ is regulated by the equilibria:

$$H_2CO_3(aq) \rightleftharpoons H^+(aq) + HCO_3^-(aq) \qquad K_{a1} = 4.3 \times 10^{-7}$$

$$HCO_3^-(aq) \rightleftharpoons H^+(aq) + CO_3^{2-}(aq) \qquad K_{a2} = 5.6 \times 10^{-11}$$

Since the value of K_{a2} is small compared to K_{a1}, we will assume that most of the $H^+(aq)$ is produced by the first dissociation.

$$K_{a1} = 4.3 \times 10^{-7} = \frac{[H^+][HCO_3^-]}{[H_2CO_3]}; \quad [H^+] = [HCO_3^-] = x,\ [H_2CO_3] = 1.1 \times 10^{-5} - x$$

Since K_{a1} and $[H_2CO_3]$ have similar values, we cannot assume x is small compared to 1.1×10^{-5}.

$$4.3 \times 10^{-7} = \frac{x^2}{(1.1 \times 10^{-5} - x)}; \quad 4.73 \times 10^{-12} - 4.3 \times 10^{-7}\ x = x^2$$

$0 = x^2 + 4.3 \times 10^{-7} - 4.73 \times 10^{-12}$

$$x = \frac{-4.3 \times 10^{-7} \pm \sqrt{(4.3 \times 10^{-7})^2 - 4(1)(-4.73 \times 10^{-12})}}{2(1)}$$

$$x = \frac{-4.3 \times 10^{-7} \pm \sqrt{1.85 \times 10^{-13} + 1.89 \times 10^{-11}}}{2} = \frac{-4.3 \times 10^{-7} \pm 4.37 \times 10^{-6}}{2}$$

The negative result is meaningless; $x = 1.97 \times 10^{-6} = 2.0 \times 10^{-6}$ M H^+; pH = 5.71 Since this $[H^+]$ is quite small, the $[H^+]$ from the autoionization of water might be significant. Calculation shows that for $[H^+] = 2.0 \times 10^{-6}$ M from H_2CO_3, $[H^+]$ from $H_2O = 5.2 \times 10^{-9}$ M, which we can ignore.

18.74 (a) $Al(OH)_3(s) \rightleftharpoons Al^{3+}(aq) + 3OH^-(aq)$ $K_{sp} = 1.3 \times 10^{-33} = [Al^{3+}][OH^-]^3$

This is a precipitation conditions problem. At what $[OH^-]$ (we can get pH from $[OH^-]$) will $Q = 1.3 \times 10^{-33}$, the requirement for the onset of precipitation?

$Q = 1.3 \times 10^{-33} = [Al^{3+}][OH^-]^3$. Find the molar concentration of $Al_2(SO_4)_3$ and thus $[Al^{3+}]$.

$$\frac{2.0 \text{ lb } Al_2(SO_4)_3}{1000 \text{ gal } H_2O} \times \frac{453.6 \text{ g}}{1 \text{ lb}} \times \frac{1 \text{ mol } Al_2(SO_4)_3}{342.2 \text{ g } Al_2(SO_4)_3} \times \frac{1 \text{ gal}}{4 \text{ qt}} \times \frac{1 \text{ qt}}{0.946 \text{ L}}$$

$$= 7.01 \times 10^{-4}\ M\ Al_2(SO_4)_3 = 1.40 \times 10^{-3}\ M\ Al^{3+}$$

$Q = 1.3 \times 10^{-33} = (1.4 \times 10^{-3})[OH^-]^3$; $[OH^-]^3 = 9.28 \times 10^{-31}$

$[OH^-] = 9.753 \times 10^{-11} = 9.8 \times 10^{-11}$ M; pOH = 10.01; pH = 14 - 10.01 = 3.99

(b) $CaO(s) + H_2O(l) \rightarrow Ca^{2+}(aq) + 2OH^-(aq)$; $[OH^-] = 9.753 \times 10^{-11}$ mol/L

$$\text{mol } OH^- = \frac{9.753 \times 10^{-11} \text{ mol}}{1 \text{ L}} \times 1000 \text{ gal} \times \frac{4 \text{ qt}}{1 \text{ gal}} \times \frac{0.946 \text{ L}}{1 \text{ qt}} = 3.69 \times 10^{-7}$$

$$= 3.7 \times 10^{-7} \text{ mol } OH^-$$

$$3.69 \times 10^{-7} \text{ mol } OH^- \times \frac{1 \text{ mol CaO}}{2 \text{ mol } OH^-} \times \frac{56.1 \text{ g CaO}}{1 \text{ mol CaO}} \times \frac{1 \text{ lb}}{453.6 \text{ g}} = 2.3 \times 10^{-8} \text{ lb CaO}$$

This is a **very** small amount of CaO, about 10 mg.

19 Chemical Thermodynamics

Spontaneous Processes

19.1 *Analyze/Plan*. Follow the logic in Sample Exercise 19.1. *Solve*:

(a) Nonspontaneous; -5°C is below the melting point of ice, so melting does not happen without continuous intervention.

(b) Spontaneous; sugar is soluble in water, and even more soluble in hot coffee.

(c) Spontaneous; N_2 molecules are stable relative to isolated N atoms.

(d) Spontaneous; the filings organize in a magnetic field without intervention.

(e) Nonspontaneous; CO_2 and H_2O are in contact continuously at atmospheric conditions in nature and do not form CH_4 and O_2.

19.2 (a) Spontaneous; a gas, in this case perfume vapor, expands to fill its container, the room.

(b) Nonspontaneous; a mixture cannot be separated without outside intervention.

(c) Nonspontaneous; an inflated balloon doesn't burst without external stress, such as a pin pric, a squeeze or adding more gas.

(d) Spontaneous; see Figure 8.2.

(e) Spontaneous; the very polar HCl molecules readily dissolve in water to form concentrated HCl(aq).

19.3 (a) NH_4NO_3(s) dissolves in water, as in a chemical cold pack. Naphthalene (moth balls) sublimes at room temperature.

(b) Melting of a solid is spontaneous above its melting point but nonspontaneous below its melting point.

19.4 Berthelot's suggestion is incorrect. Some examples of nonexothermic spontaneous processes are expansion of certain pressurized gases, dissolving of one liquid in another, and dissolving of many salts in water.

19.5 *Analyze/Plan.* Define the system and surroundings. Use the appropriate definition to answer the specific questions. *Solve:*

(a) Water is the system. Heat must be added to the system to evaporate the water. The process is endothermic.

(b) At 1 atm, the reaction is spontaneous at temperatures above 100°C.

(c) At 1 atm, the reaction is nonspontaneous at temperatures below 100°C.

(d) The two phases are in equilibrium at 100°C.

19.6 (a) Exothermic. If melting requires heat and is endothermic, freezing must be exothermic.

(b) At 1 atm (indicated by the term 'normal' freezing point), the freezing of 1-propanol is spontaneous at temperatures below -127°C.

(c) At 1 atm, the freezing of 1-propanol is nonspontaneous at temperatures above -127°C.

(d) At 1 atm and -127°C, the normal freezing point of 1-propanol, the solid and liquid phases are in equilibrium. That is, at the freezing point, 1-propanol molecules escape to the liquid phase at the same rate as liquid 1-propanol solidifies, assuming no heat is exchanged between 1-propanol and the surroundings.

19.7 *Analyze/Plan.* Define the system and surroundings. Use the appropriate definition to answer the specific questions. *Solve:*

(a) For a *reversible* process, the forward and reverse changes occur by the same path. There is only one reversible pathway for a specified set of conditions. Work can only be realized from a reversible process.

(b) If a system is returned to its original state via a reversible path, the surroundings are also returned to their original state. That is, there is no net change in the surroundings.

(c) The vaporization of water to steam is reversible if it occurs at the boiling temperature of water for a specified external (atmospheric) pressure. This is the temperature and pressure at which the two phases are in equilibrium.

19.8 (a) A process is *irreversible* if the system cannot be returned to its original state by the same path that the forward process took place. No work is done by the system on the surroundings during an irreversible process.

(b) Since the system returned to its initial state via a different path (different q_r and w_r than q_f and w_f), there is a net change in the surroundings.

(c) The condensation of a liquid will be irreversible if it occurs at any temperature other than the boiling point of the liquid, at a specified pressure.

19.9 No. ΔE is a state function. $\Delta E = q + w$; q and w are not state functions. Their values do depend on path, but their sum, ΔE, does not.

19.10 (a) $\Delta E (1 \rightarrow 2) = -\Delta E (2 \rightarrow 1)$

(b) We can say nothing about the values of q and w because we have no information about the paths.

(c) If the changes of state are reversible, the two paths are the same and $w (1 \rightarrow 2) = -w (2 \rightarrow 1)$. This is the maximum realizable work from this system.

19.11 *Analyze/Plan.* Define the system and surroundings. Use the appropriate definition to answer the specific questions. *Solve:*

We know that melting is a process that increases the energy of the system, even though there is no change in temperature. ΔE is not zero for the process.

19.12 (a) The detonation of an explosive is definitely spontaneous, once it is initiated.

(b) At constant atmospheric pressure, $\Delta H = q$. The detonation reaction is highly exothermic, so the value of q is large and negative.

(c) The sign (and magnitude) of w depend on the path of the process, the exact details of how the detonation is carried out. It seems clear, however, that work will be done by the system on the surroundings in almost all circumstances (buildings collapse, earth and air are moved), so the sign of w is probably negative.

(d) $\Delta E = q + w$. If q and w are both negative, then the sign of ΔE is negative, regardless of the magnitudes of q and w.

Entropy and the Second Law of Thermodynamics

19.13 *Analyze/Plan.* Review the definitions of isothermal and spontaneous. $w = -P_{ext}\Delta V$. *Solve:*

(a) Yes, the process is spontaneous.

(b) $w = -P_{ext}\Delta V$. Since the gas expands into a vacuum, $P_{ext} = 0$ and $w = 0$.

(c) The driving force for this expansion is the increase in the possible arrangements of the molecules, the increase in disorder of the system.

19.14 In the general case, heat gained or lost by a system is path dependent; the value of q depends on how the change occurred, as well as the initial and final states. For a reversible process, the path is specified; it is the reversible path. There is only one reversible pathway for a process. Therefore, q_{rev} depends only on the initial and final states of the system, which is the definition of a state function.

19.15 *Analyze/Plan*. Consider the discussion in Section 19.2 and Figures 19.5 and 19.6. *Solve*:

(a) Each of the 4 molecules can be in either the left or the right bulb. Thus, there are $(2)^4 = 16$ total arrangements.

(b) Only one arrangement has all 4 molecules in the right-hand flask.

(c) The gas will spontaneously adopt the state with maximum disorder, the state with the most possible arrangements for the molecules.

19.16 (a) The two arrangements are equally probable. This is because the identity of the molecules is specified.

(b) The answer in part (a) is consistent with the idea that gas molecules will arrange themselves equally between the two flasks. There are many more **equivalent** arrangements of equally distributed molecules [like arrangement (a)] that there are asymmetric 7:1 arrangements like (b). This means that equal distribution is more likely, but any two specific arrangements are equally probable.

19.17 (a) Entropy is the order or randomness of a system.

(b) ΔS is negative if order increases.

(c) No. ΔS is a state function, so it is independent of path.

19.18 (a) When a liquid freezes, the entropy of the system decreases.

(b) ΔS is negative.

(c) Entropy being a state function means that ΔS is independent of the path of the process.

19.19 *Analyze/Plan*. Consider the conditions that lead to an increase in entropy: more mol gas in products than reactants, increase in volume of sample and, therefore, number of possible arrangements, more motional freedom of molecules, etc. *Solve*:

(a) More gaseous particles means more possible arrangements and greater disorder; ΔS is positive.

(b) ΔS is positive for Exercise 19.2 (a) and (c). Both processes represent an increase in volume and possible arrangements for the sample. (In (e), even though HCl(aq) is a mixture, there are fewer moles of gas in the product, so ΔS is not positive.)

19.20 (a) Solids are much more ordered than gases, so ΔS is negative.

(b) ΔS is positive for Exercise 19.1 (a), (b) and (e). (At room temperature and 1 atm pressure, H_2O is a liquid, so there are more moles of gas in the products in part (e) and $\Delta S > 0$.)

19.21 *Analyze/Plan.* Consider the conditions that lead to an increase in entropy: more mol gas in products than reactants, increase in volume of sample and, therefore, number of possible arrangements, more motional freedom of molecules, etc. *Solve:*

S increases in (a), (b) and (c); S decreases in (d).

19.22 More disorder is associated with forming a gas because of the large volume expansion and increased motional freedom associated with a gas.

19.23 (a) $CH_3OH(l) \rightarrow CH_3OH(g)$, entropy increases, more mol gas in products, greater motional freedom.

(b) $$\Delta S = \frac{\Delta H}{T} = \frac{71.8 \text{ kJ}}{\text{mol } CH_3OH(l)} \times 1.00 \text{ mol } CH_3OH(l) \times \frac{1}{(273.15 + 64.7)K} \times \frac{1000 \text{ J}}{1 \text{ kJ}}$$
$$= 213 \text{ J/K}$$

19.24 (a) $Cs(l) \rightarrow Cs(s)$, ΔS is negative

(b) $$\Delta H = 15.0 \text{ g Cs} \times \frac{1 \text{ mol Cs}}{132.9 \text{ g Cs}} \times \frac{2.09 \text{ kJ}}{\text{mol Cs}} = 0.2359 = 0.236$$

$$\Delta S = \frac{\Delta H}{T} = 7.2359 \text{ kJ} \times \frac{1000 \text{ J}}{1 \text{ kJ}} \times \frac{1}{(273.15 + 28.4)K} = 0.782 \text{ J/K}$$

19.25 (a) For a spontaneous process, the entropy of the universe increases; for a reversible process, the entropy of the universe does not change.

(b) In a reversible process, $\Delta S_{system} + \Delta S_{surroundings} = 0$. If ΔS_{system} is positive, $\Delta S_{surroundings}$ must be negative.

(c) Since $\Delta S_{universe}$ must be positive for a spontaneous process, $\Delta S_{surroundings}$ must be greater than -42 J/K.

19.26 (a) For a spontaneous process, $\Delta S_{universe} > 0$. For a reversible process, $\Delta S_{universe} = 0$.

(b) $\Delta S_{surroundings}$ is positive and greater than the magnitude of the decrease in ΔS_{system}.

(c) $\Delta S_{system} = 78$ J/K.

Molecular Interpretation of Entropy

19.27 (a) The entropy of a pure crystalline substance at absolute zero is zero.

(b) In *translational* motion, the entire molecule moves in a single direction; in *rotational* motion, the molecule rotates or spins around a fixed axis. *Vibrational* motion is reciprocating motion. The bonds within a molecule stretch and bend, but the average position of the atoms does not change.

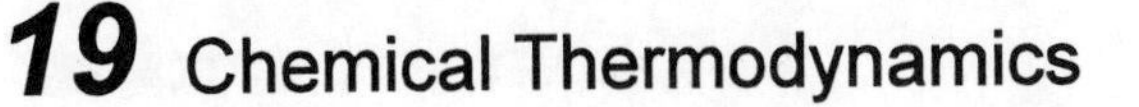

(c)

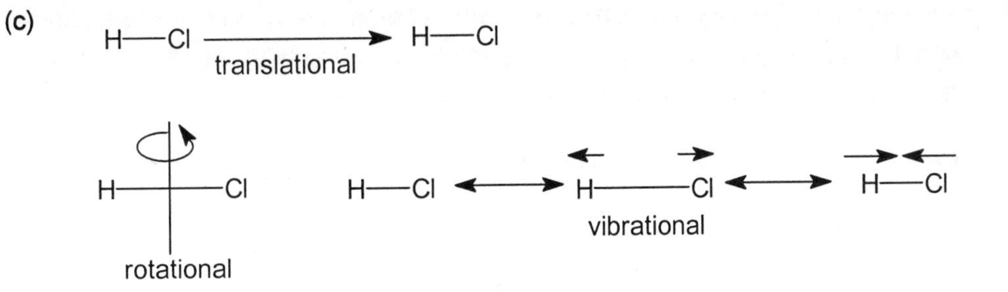

19.28 (a) A value of zero for entropy requires that the substance is a perfectly crystalline (not disordered or amorphous, see Chapter 11) solid at a temperature of 0 K.

(b) Since CO_2 has more than one atom, the thermal energy can be distributed as translational, vibrational or rotational motion.

(c)

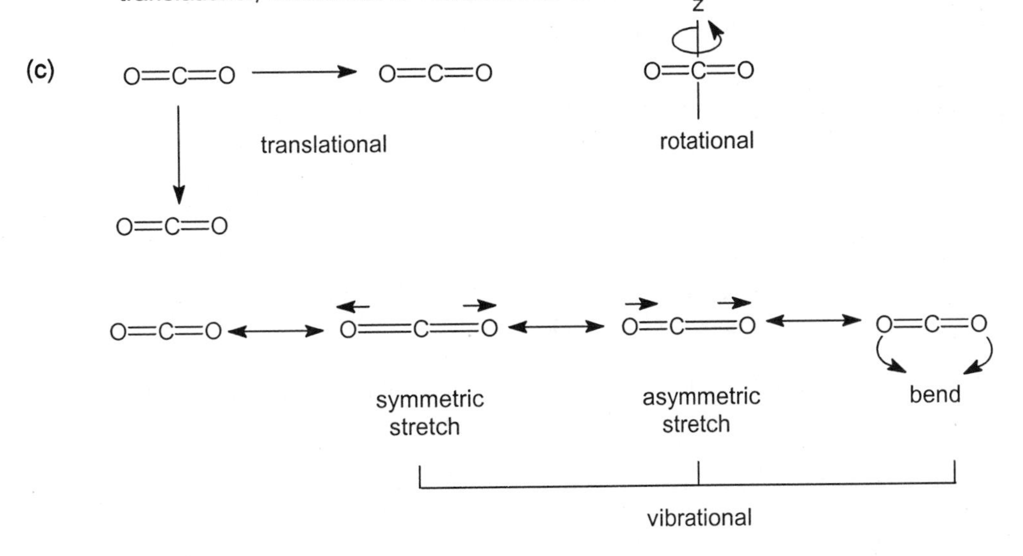

19.29 *Analyze/Plan*. Consider the factors that lead to higher entropy: more mol gas in products than reactants, increase in volume of sample and, therefore, number of possible arrangements, more motional freedom of molecules, etc. *Solve*:

(a) Ar(g) (gases have higher entropy due primarily to much larger volume)

(b) He(g) at 1.5 atm (larger volume and more motional freedom)

(c) 1 mol of Ne(g) in 15.0 L (larger volume provides more motional freedom)

(d) CO_2(g) (more motional freedom)

19.30 (a) 1 mol of As_4(g) at 300°C, 0.01 atm (As_4 has more massive atoms in a comparable system at the same temperature.)

(b) 1 mol H_2O(g) at 100°C, 1 atm (larger volume occupied by H_2O(g))

(c) 0.5 mol CH_4(g) at 298 K, 20-L volume (more complex molecule, more rotational and vibrational degrees of freedom)

(d) 100 g of Na_2SO_4(aq) at 30°C (more motional freedom in aqueous solution)

19.31 *Analyze/Plan.* Consider the markers of an increase in entropy for a chemical reaction: liquids or solutions formed from solids, gases formed from either solids or liquids, increase in moles gas during reaction. *Solve:*

(a) ΔS negative (moles of gas decrease)

(b) ΔS positive (gas produced, increased disorder)

(c) ΔS negative (moles of gas decrease)

(d) ΔS positive (moles of gas increase)

19.32 (a) $Fe(l) \rightarrow Fe(s)$; ΔS is negative (less motional freedom)

(b) $2Li(s) + Cl_2(g) \rightarrow 2LiCl$; ΔS is negative (moles of gas decrease)

(c) $Zn(s) + 2HCl(aq) \rightarrow ZnCl_2(aq) + H_2(g)$; ΔS is positive (moles of gas increase)

(d) $AgNO_3(aq) + KBr(aq) \rightarrow AgBr(s) + KNO_3(aq)$; ΔS is negative (less motional freedom)

19.33 *Analyze/Plan.* Consider the conditions that lead to an increase in entropy: more mol gas in products than reactants, increase in volume of sample and, therefore, number of possible arrangements, more motional freedom of molecules, etc. *Solve:*

(a) Sc(s), 34.6 J/mol•K; Sc(g), 174.7 J/mol•K. In general, the gas phase of a substance has a larger S° than the solid phase because of the greater volume and motional freedom of the molecules.

(b) $NH_3(g)$, 192.5 J/mol•K; $NH_3(aq)$, 111.3 J/mol•K. Molecules in the gas phase have more motional freedom than molecules in solution.

(c) 1 mol of $P_4(g)$, 280 J/K; 2 mol of $P_2(g)$, 2(218.1) = 436.2 J/K. More particles have a greater number of arrangements.

(d) C(diamond), 2.43 J/mol•K; C(graphite) 5.69 J/mol•K. Diamond is a network covalent solid with each C atom tetrahedrally bound to four other C atoms. Graphite consists of sheets of fused planar 6-membered rings with each C atom bound in a trigonal planar arrangement to three other C atoms. The internal entropy in graphite is greater because there is translational freedom among the planar sheets of C atoms while there is very little vibrational freedom within the network covalent diamond lattice.

19.34 (a) CuO(s), 42.59 J/mol•K ; $Cu_2O(s)$, 92.36 J/mol•K. Molecules in the solid state have only vibrational motion available to them. The more complex Cu_2O molecule has more vibrational degrees of freedom and a larger standard entropy.

(b) 1 mol $N_2O_4(g)$, 304.3 J/K; 2 mol $NO_2(g)$, 2(240.45) = 480.90 J/K. More particles have a greater number of arrangements.

(c) $CH_3OH(g)$, 237.6 J/mol•K; $CH_3OH(l)$, 126.8 J/mol•K. Molecules in the gas phase occupy a larger volume and have more motional freedom than molecules in the liquid state.

(d) 1 mol $PbCO_3(s)$, 131.0 J/K; 1 mol PbO(s) + 1 mol $CO_2(g)$, (68.70 + 213.6) = 282.3 J/K. The second member of the pair has more total particles and half of them are in the gas phase for greater total motional freedom. Note that 1 mol of $PbCO_3(s)$ has greater entropy than 1 mol of PbO(s), because of the additional ways to store energy in the more complex CO_3^{2-} anion.

19.35 *Analyze/Plan*. Consider the molecular interpretation of entropy. *Solve*:

Hydrocarbon	S° (J/mol•K)
$CH_4(g)$	186.3
$C_2H_6(g)$	229.5
$C_3H_8(g)$	269.9
$C_4H_{10}(g)$	310.0

As the number of C atoms increases, the S° of the hydrocarbon increases. The increased structural complexity means more motional degrees of freedom for each molecule.

19.36 For elements with similar structures, the heavier the atoms, the lower the vibrational frequencies at a given temperature. This means that more vibrations can be accessed at a particular temperature resulting in a greater absolute entropy for the heavier elements.

19.37 *Analyze/Plan*. Follow the logic in Sample Exercise 19.7. *Solve*:

(a) $\Delta S° = S°\ C_2H_6(g) - S°\ C_2H_4(g) - S°\ H_2(g)$

$= 229.5 - 219.4 - 130.58 = -120.5$ J/K

ΔS° is negative because there are fewer moles of gas in the products.

(b) $\Delta S° = 2S°\ NO_2(g) - \Delta S°\ N_2O_4(g) = 2(240.45) - 304.3 = +176.6$ J/K

ΔS° is positive because there are more moles of gas in the products.

(c) $\Delta S° = \Delta S°\ BeO(s) + \Delta S°\ H_2O(g) - \Delta S°\ Be(OH)_2(s)$

$= 13.77 + 188.83 - 50.21 = +152.39$ J/K

ΔS° is positive because the product contains more total particles and more moles of gas.

(d) $\Delta S° = 2S°\ CO_2(g) + 4S°\ H_2O(g) - 2S°\ CH_3OH(g) - 3S°\ O_2(g)$

$= 2(213.6) + 4(188.83) - 2(237.6) - 3(205.0) = +92.3$ J/K

ΔS° is positive because the product contains more total particles and more moles of gas.

19.38 (a) $\Delta S° = 2S°\ NH_3(g) - S°\ N_2H_4(g) - S°\ H_2(g)$

$= 2(192.5) - 238.5 - 130.58 = +15.92$ J/K

ΔS° is small because there are the same number of moles of gas in the products as in reactants. The slight increase is due to the relatively small S° value of $H_2(g)$, which has fewer degrees of freedom than molecules with more than two atoms.

(b) $\Delta S° = 2S°\ AlCl_3(s) - S°\ Al(s) - 3S°\ Cl_2(g)$
$= 2(109.3) - 28.32 - 3(222.96) = -478.6$ J/K

$\Delta S°$ is negative because the products contain fewer (no) moles of gas.

(c) $\Delta S° = S°\ MgCl_2(s) + 2S°H_2O(l) - S°\ Mg(OH)_2(s) - 2S°\ HCl(g)$
$= 89.6 + 2(69.91) - 63.24 - 2(186.69) = -207.20$ J/K

$\Delta S°$ is negative because the products contain fewer (no) moles of gas.

(d) $\Delta S° = S°\ C_2H_6(g) + S°\ H_2(g) - 2S°\ CH_4(g)$
$= 229.5 + 130.58 - 2(186.3) = -12.5$ J/K

$\Delta S°$ is very small because there are the same number of moles of gas in the products and reactants. The slight decrease is related to the relatively small S° value for $H_2(g)$, which has fewer degrees of freedom than molecules with more than two atoms.

Gibbs Free Energy

19.39 (a) $\Delta G = \Delta H - T\Delta S$

(b) If ΔG is positive, the process is nonspontaneous, but the reverse process is spontaneous.

(c) There is no relationship between ΔG and rate of reaction. A spontaneous reaction, one with a $-\Delta G$, may occur at a very slow rate. For example: $2H_2(g) + O_2(g) \rightarrow 2H_2O(g)$, $\Delta G = -457$ kJ is very slow if not initiated by a spark.

19.40 (a) The *standard* free energy change, $\Delta G°$, represents the free energy change for the process when all reactants and products are in their standard states. When any or all reactants or products are not in their standard states, the free energy is represented simply as ΔG. The value for ΔG thus depends on the specific states of all reactants and products.

(b) When $\Delta G = 0$, the system is at equilibrium.

(c) The sign and magnitude of ΔG give no information about rate; we cannot predict whether the reaction will occur rapidly.

19.41 *Analyze/Plan.* Consider the definitions of ΔH, $\Delta S°$ and $\Delta G°$, along with sign conventions. $\Delta G° = \Delta H° - T\Delta S°$ *Solve:*

(a) $\Delta H°$ is negative; the reaction is exothermic.

(b) $\Delta S°$ is negative; the reaction leads to decrease in disorder (increase in order) of the system.

(c) $\Delta G° = \Delta H° - T\Delta S° = -35.4\ \text{kJ} - 298\ \text{K}\ (-0.0855\ \text{kJ/K}) = -9.921 = -9.9$ kJ

(d) At 298 K, $\Delta G°$ is negative. If all reactants and products are present in their standard states, the reaction is spontaneous at this temperature.

19.42 (a) ΔH° is negative; the reaction is exothermic.

(b) ΔS° is positive; the reaction leads to an increase in disorder.

(c) $\Delta G^\circ = \Delta H^\circ - T\Delta S^\circ = -19.5 \text{ kJ} - 298 \text{ K } (0.0427 \text{ kJ/K}) = -32.225 = -32.2 \text{ kJ}$

(d) At 298 K, ΔG° is negative. If all reactants and products are present in their standard states, the reaction is spontaneous at this temperature.

19.43 *Analyze/Plan*. Calculate ΔH° according to Equation 5.31, ΔS° by Equation 19.8 and ΔG° by Equation 19.13. Then use ΔH° and ΔS° to calculate ΔG° using Equation 19.20, $\Delta G^\circ = \Delta H^\circ - T\Delta S^\circ$. *Solve*:

(a) $\Delta H^\circ = 2(-268.61) - [0 + 0] = -537.22 \text{ kJ}$

$\Delta S^\circ = 2(173.51) - [130.58 + 202.7] = 13.74 = 13.7 \text{ J/K}$

$\Delta G^\circ = 2(-270.70) - [0 + 0] = -541.40 \text{ kJ}$

$\Delta G^\circ = -537.22 \text{ kJ} - 298(0.01374) \text{ kJ} = -541.31 \text{ kJ}$

(b) $\Delta H^\circ = -106.7 - [0 + 2(0)] = -106.7 \text{ kJ}$

$\Delta S^\circ = 309.4 - [5.69 + 2(222.96)] = -142.21 = -142.2 \text{ J/K}$

$\Delta G^\circ = -64.0 - [0 + 2(0)] = -64.0 \text{ kJ}$

$\Delta G^\circ = -106.7 \text{ kJ} - 298(-0.14221) \text{ kJ} = -64.3 \text{ kJ}$

(c) $\Delta H^\circ = 2(-542.2) - [2(-288.07) + 0] = -508.26 = -508.3 \text{ kJ}$

$\Delta S^\circ = 2(325) - [2(311.7) + 205.0] = -178.4 = -178 \text{ J/K}$

$\Delta G^\circ = 2(-502.5) - [2(-269.6) + 0] = -465.8 \text{ kJ}$

$\Delta G^\circ = -508.26 \text{ kJ} - 298(-0.1784) \text{ kJ} = -455.097 = -455.1 \text{ kJ}$

(The discrepancy in ΔG° values is due to experimental uncertainties in the tabulated thermodynamic data.)

(d) $\Delta H^\circ = -84.68 + 2(-241.82) - [2(-201.2) + 0] = -165.92 = -165.9 \text{ kJ}$

$\Delta S^\circ = 229.5 + 2(188.83) - [2(237.6) + 130.58] = 1.38 = 1.4 \text{ J/K}$

$\Delta G^\circ = -32.89 + 2(-228.57) - [2(-161.9) + 0] = -166.23 = -166.2 \text{ kJ}$

$\Delta G^\circ = -165.92 \text{ kJ} - 298(0.00138) \text{ kJ} = -166.33 = -166.3 \text{ kJ}$

19.44 (a) $\Delta H^\circ = -305.3 - [0 + 0] = -305.3 \text{ kJ}$

$\Delta S^\circ = 97.65 - [29.9 + 222.96] = -155.21 = -155.2 \text{ J/K}$

$\Delta G^\circ = -259.0 - [0 + 0] = -259.0 \text{ kJ}$

$\Delta G^\circ = -305.3 \text{ kJ} - 298(-0.15521) \text{ kJ} = -259.047 = -259.0 \text{ kJ}$

(b) $\Delta H^\circ = -635.5 + (-393.5) - (-1207.1) = 178.1 \text{ kJ}$

$\Delta S^\circ = 39.75 + 213.6 - (92.88) = 160.47 = 160.5 \text{ J/K}$

$\Delta G^\circ = -604.17 + (-394.4) - (-1128.76) = 130.19 = 130.2 \text{ kJ}$

$\Delta G^\circ = 178.1 \text{ kJ} - 298(0.16047) \text{ kJ} = 130.28 = 130.3 \text{ kJ}$

(c) $\Delta H° = 4(-1288.3) - [-2940.1 + 6(-285.83)] = -498.12 = -498.1$ kJ

$\Delta S° = 4(158.2) - [228.9 + 6(69.91)] = -15.56 = -15.6$ J/K

$\Delta G° = 4(-1142.6) - [-2675.2 + 6(-237.13)] = -472.42 = -472.4$ kJ

$\Delta G° = -498.12$ kJ $- 298(-0.01556)$ kJ $= -493.48 = -493.5$ kJ

(The discrepancy in $\Delta G°$ values is due to experimental uncertainties in the tabulated thermodynamic data.)

(d) $\Delta H° = 2(-393.5) + 4(-285.83) - [2(-238.6) + 3(0)] = -1453.1$ kJ

$\Delta S° = 2(213.6) + 4(69.91) - [2(126.8) + 3(205.0)] = -161.76 = -161.8$ J/K

$\Delta G° = 2(-394.4) + 4(-237.13) - [2(-166.23) + 3(0)] = -1404.86 = -1404.9$ kJ

$\Delta G° = -1453.2$ kJ $- 298(-0.16176)$ kJ $= -1404.996 = -1405.0$ kJ

19.45 *Analyze/Plan.* Follow the logic in Sample Exercise 19.8. *Solve:*

(a) $\Delta G° = 2\Delta G°\ SO_3(g) - [2\Delta G°\ SO_2(g) + \Delta G°\ O_2(g)]$

$= 2(-370.4) - [2(-300.4) + 0] = -140.0$ kJ, spontaneous

(b) $\Delta G° = 3\Delta G°\ NO(g) - [\Delta G°\ NO_2(g) + \Delta G°\ N_2O(g)]$

$= 3(86.71) - [51.84 + 103.59] = +104.70$ kJ, nonspontaneous

(c) $\Delta G° = 4\Delta G°\ FeCl_3(s) + 3\Delta G°\ O_2(g) - [6\Delta G°\ Cl_2(g) + 2\Delta G°\ Fe_2O_3(s)]$

$= 4(-334) + 3(0) - [6(0) + 2(-740.98)] = +146$ kJ, nonspontaneous

(d) $\Delta G° = \Delta G°\ S(s) + 2\Delta G°\ H_2O(g) - [\Delta G°\ SO_2(g) + 2\Delta G°\ H_2(g)]$

$= 0 + 2(-228.57) - [(-300.4) + 2(0)] = -156.7$ kJ, spontaneous

19.46 (a) $\Delta G° = 2\Delta G°\ HCl(g) - [\Delta G°\ H_2(g) + \Delta G°\ Cl_2(g)]$

$= 2(-95.27$ kJ$) - 0 - 0 = -190.5$ kJ, spontaneous

(b) $\Delta G° = \Delta G°\ MgO(s) + 2\Delta G°\ HCl(g) - [\Delta G°\ MgCl_2(s) + \Delta G°\ H_2O(l)]$

$= -569.6 + 2(-95.27) - [-592.1 + (-237.13)] = +69.1$ kJ, nonspontaneous

(c) $\Delta G° = \Delta G°\ N_2H_4(g) + \Delta G°\ H_2(g) - 2\Delta G°\ NH_3(g)$

$= 159.4 + 0 - 2(-16.66) = +192.7$ kJ, nonspontaneous

(d) $\Delta G° = 2\Delta G°\ NO(g) + \Delta G°\ Cl_2(g) - 2\Delta G°\ NOCl(g)$

$= 2(86.71) + 0 - 2(66.3) = +40.8$ kJ, nonspontaneous

19.47 *Analyze/Plan.* Follow the logic in Sample Exercise 19.9(a). *Solve:*

(a) $C_6H_{12}(l) + 9O_2(g) \rightarrow 6CO_2(g) + 6H_2O(l)$

(b) Because there are fewer moles of gas in the products, $\Delta S°$ is negative, which makes $-T\Delta S$ positive. $\Delta G°$ is less negative (more positive) than $\Delta H°$.

19.48 (a) $\Delta G°$ should be less negative than $\Delta H°$. Products contain fewer moles of gas, so $\Delta S°$ is negative. $\Delta G° = \Delta H° - T\Delta S°$; $-T\Delta S°$ is positive so $\Delta G°$ is less negative than $\Delta H°$.

(b) We can estimate $\Delta S°$ using a similar reaction and then use $\Delta G° = \Delta H° - T\Delta S°$(estimate) to get a ballpark figure. There are no sulfite salts listed in Appendix C, so use a reaction such as $CO_2(g) + CaO(s) \rightarrow CaCO_3(s)$ or $CO_2(g) + BaO(s) \rightarrow BaCO_3(s)$. Or calculate both $\Delta S°$ values and use the average as your estimate.

19.49 *Analyze/Plan*. Based on the signs of ΔH and ΔS for a particular reaction, assign a category from Table 19.4 to each reaction. *Solve*:

(a) ΔG is negative at low temperatures, positive at high temperatures. That is, the reaction proceeds in the forward direction spontaneously at lower temperatures but spontaneously reverses at higher temperatures.

(b) ΔG is positive at all temperatures. The reaction is nonspontaneous in the forward direction at all temperatures.

(c) ΔG is positive at low temperatures, negative at high temperatures. That is, the reaction will proceed spontaneously in the forward direction at high temperature.

19.50 $\Delta G° = \Delta H° - T\Delta S°$

(a) $\Delta G° = -844 \text{ kJ} - 298 \text{ K}(-0.165 \text{ kJ/K}) = -795 \text{ kJ}$, spontaneous

(b) $\Delta G° = +572 \text{ kJ} - 298 \text{ K}(0.179 \text{ kJ/K}) = +519 \text{ kJ}$, nonspontaneous

To be spontaneous, ΔG must be negative ($\Delta G < 0$).

Thus, $\Delta H° - T\Delta S° < 0$; $\Delta H° < T\Delta S°$; $T > \Delta H°/\Delta S°$; $T > \dfrac{572 \text{ kJ}}{0.179 \text{ kJ/K}} = 3{,}200 \text{ K}$

19.51 *Analyze/Plan*. We are told that the reaction is spontaneous and endothermic, and asked to estimate the sign and magnitude of ΔS. If a reaction is spontaneous, $\Delta G < 0$. Use this information with Equation 19.20 to solve the problem. *Solve*:

At 450 K, $\Delta G < 0$; $\Delta G = \Delta H - T\Delta S < 0$

$34.5 \text{ kJ} - 450 \text{ K} (\Delta S) < 0$; $34.5 \text{ kJ} < 450 \text{ K} (\Delta S)$; $\Delta S > 34.5 \text{ kJ}/450 \text{ K}$

$\Delta S > 0.0767 \text{ kJ/K}$ or $\Delta S > +76.7 \text{ J/K}$

19.52 At -25°C or 248 K, $\Delta G > 0$. $\Delta G = \Delta H - T\Delta S > 0$

$\Delta H - 248 \text{ K } (95 \text{ J/K}) > 0$; $\Delta H > +2.4 \times 10^4 \text{ J}$; $\Delta H > +24 \text{ kJ}$

19.53 *Analyze/Plan*. Follow the logic in Sample Exercise 19.11. Use Equation 19.20 to calculate T when $\Delta G = 0$. Use Table 19.4 to determine whether the reaction is spontaneous or non-spontaneous above this temperature. *Solve*:

(a) $\Delta G = \Delta H - T\Delta S$; $0 = -32 \text{ kJ} - T(-98 \text{ J/K})$; $32 \times 10^3 \text{ J} = T(98 \text{ J/K})$

$T = 32 \times 10^3 \text{ J}/(98 \text{ J/K}) = 326.5 = 330 \text{ K}$

(b) Nonspontaneous. The sign of ΔS is negative, so as T increases, ΔG becomes more positive.

19.54 ΔG is negative when $T\Delta S > \Delta H$ or $T > \Delta H/\Delta S$.

$\Delta H° = \Delta H°\ CH_4(g) + \Delta H°\ CO_2(g) - \Delta H°\ CH_3COOH(l)$

$= -74.8 + (-393.5) - (-487.0) = +18.7$ kJ

$\Delta S° = S°\ CH_4(g) + S°\ CO_2(g) - S°\ CH_3COOH(l) = +186.3 + 213.6 - 159.8 = +240.1$ J/K

$$T > \frac{18.7\ \text{kJ}}{0.2401\ \text{kJ/K}} = 77.9\ \text{K}$$

The reaction is spontaneous above 77.9 K (-195°C).

19.55 *Analyze/Plan.* Given a chemical equation and thermodynamic data (values of ΔH_f°, ΔG_f° and $S°$) for reactants and products, predict the variation of $\Delta G°$ with temperature and calculate $\Delta G°$ at 800 K and 1000 K. Use Equations 5.31 and 19.8 to calculate $\Delta H°$ and $\Delta S°$, respectively; use these values to calculate $\Delta G°$ at various temperatures, using Equation 19.20. The signs of $\Delta H°$ and $\Delta S°$ determine the variation of $\Delta G°$ with temperature. *Solve:*

(a) Calculate $\Delta H°$ and $\Delta S°$ to determine the sign of $T\Delta S°$.

$\Delta H° = 3\Delta H°\ NO(g) - \Delta H°\ NO_2(g) - \Delta H°\ N_2O(g)$

$= 3(90.37) - 33.84 - 81.6 = 155.7$ kJ

$\Delta S° = 3S°\ NO(g) - S°\ NO_2(g) - S°\ N_2O(g)$

$= 3(210.62) - 240.45 - 220.0 = 171.4$ J/K

$\Delta G° = \Delta H° - T\Delta S°$. Since $\Delta S°$ is positive, $-T\Delta S°$ becomes more negative as T increases and $\Delta G°$ becomes more negative.

(b) $\Delta G° = \Delta H° - T\Delta S° = 155.7\text{kJ} - (800\ \text{K})(0.1714\ \text{kJ/K})$

$\Delta G° = 155.7\ \text{kJ} - 137\ \text{kJ} = 19$ kJ

Since $\Delta G°$ is positive at 800 K, the reaction is not spontaneous at this temperature.

(c) $\Delta G° = 155.7\ \text{kJ} - (1000\ \text{K})(0.1714\ \text{kJ/K}) = 155.7\ \text{kJ} - 171.4\ \text{kJ} = -15.7$ kJ

$\Delta G°$ is negative at 1000 K and the reaction is spontaneous at this temperature.

19.56 (a) $\Delta H° = \Delta H_f^\circ\ CH_3OH(g) - \Delta H_f^\circ\ CH_4(g) - 1/2\ \Delta H_f^\circ\ O_2(g)$

$= -201.2 - (-74.8) - 0 = -126.4$ kJ

$\Delta S° = S°CH_3OH(g) - S°CH_4(g) - 1/2\ S°O_2(g)$

$= 237.6 - 186.3 - 1/2(205.0) = -51.2\ \text{J/K} = -0.0512$ kJ/K

(b) $\Delta G° = \Delta H° - T\Delta S°$. $-T\Delta S°$ is positive, so $\Delta G°$ becomes more positive as temperature increases.

(c) $\Delta G° = \Delta H° - T\Delta S° = -126.4\ \text{kJ} - 298\ \text{K}(-0.0512\ \text{kJ/K}) = -111.1$ kJ

The reaction is spontaneous at 298 K because $\Delta G°$ is negative at this temperature. In this case, $\Delta G°$ could have been calculated from ΔG_f° values in Appendix C, since these values are tabulated at 298 K.

(d) The reaction is at equilibrium when $\Delta G° = 0$.

$\Delta G° = \Delta H° - T\Delta S° = 0$. $\Delta H° = T\Delta S°$, $T = \Delta H°/\Delta S°$

$T = -126.4$ kJ/-0.0512 kJ/K = 2469 = 2470 K.

This temperature is so high that the reactants and products are likely to decompose. At standard conditions, equilibrium is functionally unattainable for this reaction.

19.57 *Analyze/Plan.* Follow the logic in Sample Exercise 19.11. *Solve:*

(a) $\Delta S^{\circ}_{vap} = \Delta H^{\circ}_{vap}/T_b$; $T_b = \Delta H^{\circ}_{vap}/\Delta S^{\circ}_{vap}$

$\Delta H^{\circ}_{vap} = \Delta H°\ C_6H_6(g) - \Delta H°\ C_6H_6(l) = 82.9 - 49.0 = 33.9$ kJ

$\Delta S^{\circ}_{vap} = S°\ C_6H_6(g) - S°\ C_6H_6(l) = 269.2 - 172.8 = 96.4$ J/K

$T_b = 33.9 \times 10^3$ J/96.4 J/K = 351.66 = 352 K = 79°C

(b) From the *Handbook of Chemistry and Physics*, 74th Edition, $T_b = 80.1$°C. The values are remarkably close; the small difference is due to deviation from ideal behavior by $C_6H_6(g)$ and experimental uncertainty in the boiling point measurement and the thermodynamic data.

19.58 (a) As in Sample Exercise 19.11, $T_{sub} = \Delta H^{\circ}_{sub}/\Delta S^{\circ}_{sub}$

Use Data from Appendix C to calculate ΔH°_{sub} and ΔS°_{sub} for $I_2(s)$.

$I_2(s) \rightarrow I_2(l)$ melting

$I_2(l) \rightarrow I_2(g)$ boiling

$I_2(s) \rightarrow I_2(g)$ sublimation

$\Delta H^{\circ}_{sub} = \Delta H^{\circ}_f\ I_2(g) - \Delta H^{\circ}_f\ I_2(s) = 62.25 - 0 = 62.25$ kJ

$\Delta S^{\circ}_{sub} = S°\ I_2(g) - S°\ I_2(s) = 260.57 - 116.73 = 143.84$ J/K = 0.14384 kJ/K

$$T_{sub} = \frac{\Delta H°_{sub}}{\Delta S°_{sub}} = \frac{62.25\ \text{kJ}}{0.14384\ \text{kJ/K}} = 432.8\ \text{K} = 159.6°\text{C}$$

(b) T_m for $I_2(s)$ = 386.85 K = 113.7°C; T_b = 457.4 K = 184.3°C (from Webelements, 2002)

(c) The boiling point of I_2 is closer to the sublimation temperature. Both boiling and sublimation begin with molecules in a condensed phase (little space between molecules) and end in the gas phase (large intermolecular distances). Separation of the molecules is the main phenomenon that determines both ΔH and ΔS, so it is not surprising that the ratio of $\Delta H/\Delta S$ is similar for sublimation and boiling.

19.59 *Analyze/Plan.* We are asked to write a balanced equation for the combustion of acetylene, calculate $\Delta H°$ for this reaction and calculate maximum useful work possible by the system. Combustion is combination with O_2 to produce CO_2 and H_2O. Calculate $\Delta H°$ using data from Appendix C and Equation 5.31. The maximum obtainable work is ΔG (Equation 19.19), which can be calculated from data in Appendix C and Equation 19.13. *Solve:*

(a) $C_2H_2(g) + 5/2\ O_2(g) \rightarrow 2CO_2(g) + H_2O(l)$

(b) $\Delta H° = 2\Delta H°\ CO_2(g) + \Delta H°\ H_2O(l) - \Delta H°\ C_2H_2(g) - 5/2\Delta H°\ O_2(g)$

$= 2(-393.5) - 285.83 - 226.7 - 5/2(0) = -1299.5$ kJ produced/mol C_2H_2 burned

(c) $w_{max} = \Delta G° = 2\Delta G°\ CO_2(g) + \Delta G°\ H_2O(l) - \Delta G°\ C_2H_2(g) - 5/2\ \Delta G°\ O_2(g)$

$= 2(-394.4) - 237.13 - 209.2 - 5/2(0) = -1235.1$ kJ

The negative sign indicates that the system does work on the surroundings; the system can accomplish a maximum of 1235.1 kJ of work on its surroundings.

19.60 (a) $C_2H_4(g) + 3O_2(g) \rightarrow 2CO_2(g) + 2H_2O(l)$

$\Delta H° = 2\Delta H_f°\ CO_2(g) + 2\Delta H_f°\ H_2O(l) - \Delta H_f°\ C_2H_4(g) - 3\Delta H_f°\ O_2(g)$

$= 2(-393.5) + 2(-285.83) - 52.30 - 3(0) = -1410.96 = -1411.0$ kJ/mol C_2H_4 burned

(b) $w_{max} = \Delta G° = 2\Delta G_f°\ CO_2(g) + 2\Delta H_f°\ H_2O(l) - \Delta G_f°\ C_2H_4(g) - 3\Delta G_f°\ O_2(g)$

$= 2(-394.4) + 2(-237.13) - 68.11 - 3(0) = -1331.2$ kJ

The system can accomplish at most 1331.2 kJ of work per mole of C_2H_4 on the surroundings.

Free Energy and Equilibrium

19.61 *Analyze/Plan.* We are given a chemical reaction and asked to predict the effect of the partial pressure of $O_2(g)$ on the value of ΔG for the system. Consider the relationship $\Delta G = \Delta G° + RT \ln Q$ where Q is the reaction quotient. *Solve:*

(a) $O_2(g)$ appears in the denominator of Q for this reaction. An increase in pressure of O_2 decreases Q and ΔG becomes smaller or more negative. Increasing the concentration of a reactant increases the tendency for a reaction to occur.

(b) $O_2(g)$ appears in the numerator of Q for this reaction. Increasing the pressure of O_2 increases Q and ΔG becomes more positive. Increasing the concentration of a product decreases the tendency for the reaction to occur.

(c) $O_2(g)$ appears in the numerator of Q for this reaction. An increase in pressure of O_2 increases Q and ΔG becomes more positive. Since pressure of O_2 is raised to the third power in Q, an increase in pressure of O_2 will have the largest effect on ΔG for this reaction.

19.62 Consider the relationship $\Delta G = \Delta G° + RT \ln Q$ where Q is the reaction quotient.

(a) $H_2(g)$ appears in the denominator of Q for this reaction. An increase in pressure of H_2 decreases Q and ΔG becomes smaller or more negative. Increasing the concentration of a reactant increases the tendency for a reaction to occur.

(b) $H_2(g)$ appears in the numerator of Q for this reaction. Increasing the pressure of H_2 increases Q and ΔG becomes more positive. Increasing the concentration of a product decreases the tendency for the reaction to occur.

(c) $H_2(g)$ appears in the denominator of Q for this reaction. An increase in pressure of H_2 decreases Q and ΔG becomes smaller or more negative.

19.63 *Analyze/Plan.* Given a chemical reaction, we are asked to calculate $\Delta G°$ from Appendix C data, and ΔG for a given set of initial conditions. Use Equation 19.13 to calculate $\Delta G°$, and Equation 19.21 to calculate ΔG. Follow the logic in Sample Exercise 19.12 when calculating ΔG. *Solve:*

(a) $\Delta G° = \Delta G°\ N_2O_4(g) - 2\Delta G°\ NO_2(g) = 98.28 - 2(51.84) = -5.40\ kJ$

(b) $\Delta G = \Delta G° + RT \ln P_{N_2O_4} / P^2_{NO_2}$

$$= -5.40\ kJ + \frac{8.314 \times 10^{-3}\ kJ}{K \bullet mol} \times 298\ K \times \ln[1.60/(0.40)^2] = 0.3048 = 0.30\ kJ$$

19.64 (a) $\Delta G° = 2\Delta G°\ HF(g) - [\Delta G°\ H_2(g) + \Delta G°\ F_2(g)] = 2(-270.70) - [0 + 0] = -541.40\ kJ$

(b) $\Delta G = \Delta G° + RT \ln P^2_{HF} / P_{H_2} \times P_{F_2}$

$$= -541.40 + \frac{8.314 \times 10^{-3}\ kJ}{K \bullet mol} \times 298\ K \ln[(0.36)^2/8.0 \times 4.5] = -555.3\ kJ$$

19.65 *Analyze/Plan.* Given a chemical reaction, we are asked to calculate K_{eq} using ΔG_f° data from Appendix C. Follow the logic in Sample Exercise 19.13. $\Delta G° = -RT \ln K_{eq}$, Equation 19.22; $\ln K_{eq} = -\Delta G°/RT$ *Solve:*

(a) $\Delta G° = 2\Delta G°\ HI(g) - \Delta G°\ H_2(g) - \Delta G°\ I_2(g)$

$= 2(1.30) - 0 - 19.37 = -16.77\ kJ$

$$\ln K_{eq} = \frac{-(-16.77\ kJ) \times 10^3\ J/kJ}{8.314\ J/K \times 298\ K} = 6.76876 = 6.769;\ K_{eq} = 870$$

(b) $\Delta G° = \Delta G°\ C_2H_4(g) + \Delta G°\ H_2O(g) - \Delta G°\ C_2H_5OH(g)$

$= 68.11 - 228.57 - (-168.5) = 8.04 = 8.0\ kJ$

$$\ln K_{eq} = \frac{-8.04\ kJ \times 10^3\ J/kJ}{8.314\ J/K \times 298\ K} = -3.24511 = -3.2;\ K_{eq} = 0.04$$

(c) $\Delta G° = \Delta G°\ C_6H_6(g) - 3\Delta G°\ C_2H_2(g) = 129.7 - 3(209.2) = -497.9\ kJ$

$$\ln K_{eq} = \frac{-\Delta G°}{RT} = \frac{-(-497.9\ kJ) \times 10^3\ J/kJ}{8.314\ J/K \times 298\ K} = 200.963 = 201.0;\ K_{eq} = 2 \times 10^{87}$$

19.66 $\Delta G° = -RT \ln K_{eq}$; $\ln K_{eq} = -\Delta G° / RT$; at 298 K, $RT = 2.4776 = 2.478\ kJ$

(a) $\Delta G° = \Delta G°\ NaOH(s) + \Delta G°\ CO_2(g) - \Delta G°\ NaHCO_3(s)$

$= -379.5 + (-394.4) - (-851.8) = +77.9\ kJ$

$$\ln K_{eq} = \frac{-\Delta G°}{RT} = \frac{-77.9\ kJ}{2.478\ kJ} = -31.442 = -31.4;\ K_{eq} = 2 \times 10^{-14}$$

$K_{eq} = P_{CO_2} = 2 \times 10^{-14}$

(b) $\Delta G° = 2\Delta G°\ HCl(g) + \Delta G°\ Br_2(g) - 2\Delta G°\ HBr(g) - \Delta G°\ Cl_2(g)$
$= 2(-95.27) + 3.14 - 2(-53.22) - 0 = -80.96$ kJ

$$\ln K_{eq} = \frac{-(-80.96)}{2.4776} = +32.68;\ K_{eq} = 1.6 \times 10^{14}$$

$$K_{eq} = \frac{P_{HCl}^2 \times P_{Br_2}}{P_{HBr}^2 \times P_{Cl_2}} = 1.6 \times 10^{14}$$

(c) From Exercise 19.45(a), $\Delta G°$ at 298 K = -140.0 kJ.

$$\ln K_{eq} = \frac{-\Delta G°}{RT} = \frac{-(-140.0)}{2.4776} = 56.51;\ K_{eq} = 3.5 \times 10^{24}$$

$$K_{eq} = \frac{P_{SO_3}^2}{P_{SO_2}^2 \times P_{O_2}} = 3.5 \times 10^{24}$$

19.67 *Analyze/Plan.* Given a chemical reaction and thermodynamic data in Appendix C, calculate the equilibrium pressure of $CO_2(g)$ at two temperatures. $K_{eq} = P_{CO_2}$. Calculate $\Delta G°$ at the two temperatures using $\Delta G° = \Delta H° - T\Delta S°$ and then calculate K_{eq} and P_{CO_2}. *Solve:*

$\Delta H° = \Delta H°\ BaO(s) + \Delta H°\ CO_2(g) - \Delta H°\ BaCO_3(s)$
$= -553.5 + -393.5 - (-1216.3) = +269.3$ kJ

$\Delta S° = S°\ BaO(s) + S°\ CO_2(g) - S°\ BaCO_3(s)$
$= 70.42 + 213.6 - 112.1 = 171.92$ J/K $= 0.1719$ kJ/K

(a) ΔG at 298 K = 269.3 kJ - 298 K (0.17192 kJ/K) = 218.07 = 218.1 kJ

$$\ln K_{eq} = \frac{-\Delta G°}{RT} = \frac{-218.07 \times 10^3\ J}{8.314\ J/K \times 298\ K} = -88.017 = -88.02$$

$K_{eq} = 6.0 \times 10^{-39}$; $P_{CO_2} = 6.0 \times 10^{-39}$ atm

(b) ΔG at 1100 K = 269.3 kJ - 1100 K (0.17192 kJ) = 80.19 = +80.2 kJ

$$\ln K_{eq} = \frac{-\Delta G°}{RT} = \frac{-80.19 \times 10^3\ J}{8.314\ J/K \times 1100\ K} = -8.768 = -8.77$$

$K_{eq} = 1.6 \times 10^{-4}$; $P_{CO_2} = 1.6 \times 10^{-4}$ atm

19.68 $K_{eq} = P_{CO_2}$. Calculate $\Delta G°$ at the two temperatures using $\Delta G° = \Delta H° - T\Delta S°$ and then calculate K_{eq} and P_{CO_2}.

$\Delta H° = \Delta H°\ PbO(s) + \Delta H°\ CO_2(g) - \Delta H°\ PbCO_3(s)$
$= -217.3 - 393.5 + 699.1 = 88.3$ kJ

$\Delta S° = S°\ PbO(s) + S°\ CO_2(g) - S°\ PbCO_3(s)$
$= 68.70 + 213.6 - 131.0 = 151.3$ J/K or 0.1513 kJ/K

(a) $\Delta G° = \Delta H° - T\Delta S°$. At 393 K, $\Delta G°$ = 88.3 kJ - 393 K(0.1513 kJ/K) = 28.84 = 28.8 kJ

$$\ln K_{eq} = \frac{-\Delta G°}{RT} = \frac{-28.84 \times 10^3 \text{ J}}{8.314 \text{ J/K} \times 418 \text{ K}} = -8.29842 = -8.83$$

$K_{eq} = P_{CO_2} = 2.5 \times 10^{-4}$ atm

(b) $\Delta G° = \Delta H° - T\Delta S°$. At 753 K, $\Delta G°$ = 88.3 kJ - 753 K (0.1513 kJ) = -25.629 = -25.6 kJ

$$\ln K_{eq} = \frac{-(-25.629 \times 10^3 \text{ J})}{8.314 \text{ J/K} \times 753 \text{ K}} = 4.0938 = 4.09;\ K_{eq} = P_{CO_2} = 60 \text{ atm}$$

19.69 *Analyze/Plan.* Given an acid dissociation equilibrium and the corresponding K_a value, calculate $\Delta G°$ and ΔG for a given set of concentrations. Use Equation 19.22 to calculate $\Delta G°$ and Equation 19.21 to calculate ΔG. *Solve:*

(a) $HNO_2(aq) \rightleftharpoons H^+(aq) + NO_2^-(aq)$

(b) $\Delta G° = -RT \ln K_a = -(8.314 \times 10^{-3})(298) \ln (4.5 \times 10^{-4}) = 19.0928 = 19.1$ kJ

(c) $\Delta G = 0$ at equilibrium

(d) $\Delta G = \Delta G° + RT \ln Q$

$$= 19.09 \text{ kJ} + (8.314 \times 10^{-3})(298) \ln \frac{(5.0 \times 10^{-2})(6.0 \times 10^{-4})}{0.20} = -2.72 \text{ kJ}$$

19.70 (a) $CH_3NH_2(aq) + H_2O(l) \rightleftharpoons CH_3NH_3^+(aq) + OH^-(aq)$

(b) $\Delta G° = -RT \ln K_b = -(8.314 \times 10^{-3})(298) \ln (4.4 \times 10^{-4}) = 19.148 = 19.1$ kJ

(c) $\Delta G = 0$ at equilibrium

(d) $\Delta G = \Delta G° + RT \ln Q$; $[OH^-] = 1 \times 10^{-14}/1.5 \times 10^{-8} = 6.7 \times 10^{-7}$

$$= 19.148 + (8.314 \times 10^{-3})(298) \ln \frac{(5.5 \times 10^{-4})(6.67 \times 10^{-7})}{0.120} = -29.4 \text{ kJ}$$

Additional Exercises

19.71 (a) False. The essential question is whether the reaction proceeds far to the right before arriving at equilibrium. The position of equilibrium, which is the essential aspect, is not only dependent on ΔH but on the entropy change as well.

(b) True.

(c) True.

(d) False. **Non**spontaneous processes in general require that work be done to force them to proceed. Spontaneous processes occur without application of work.

(e) False. Such a process **might** be spontaneous, but would not necessarily be so. Spontaneous processes are those that are exothermic and/or that lead to increased disorder in the system.

19.72 The situation described is the isothermal compression of a gas.

(a) Since $nR\Delta T$ is zero, ΔE is zero. (ΔE is zero for all isothermal expansions or compressions.)

(b) $\Delta S = nR\ln(V_2/V_1)$. Since $V_2 < V_1$, $\ln(V_2/V_1)$ is negative and ΔS is negative. Qualitatively, a smaller volume means less motional freedom for the gas particles and a lower entropy.

19.73

Process	ΔH	ΔS
(a)	+	+
(b)	-	-
(c)	+	+
(d)	+	+
(e)	-	+

19.74 There is no inconsistency. The second law states that in any spontaneous process there is an increase in the entropy of the universe. While there may be a decrease in entropy of the system, as in the present case, this decrease is more than offset by an increase in entropy of the surroundings.

19.75 (a) An *isolated* system is a system that does not exchange matter or energy with its surroundings.

(b) $\Delta E = 0$, $q = 0$, $w = 0$

(c) The second law is: $\Delta S_{sys} + \Delta S_{surr} \geq 0$. But for an isolated system $\Delta S_{surr} = 0$, so $\Delta S_{sys} = 0$ for a reversible process; $\Delta S_{sys} > 0$ for an irreversible process.

19.76 Melting = -126.5°C; boiling = 97.4°C.

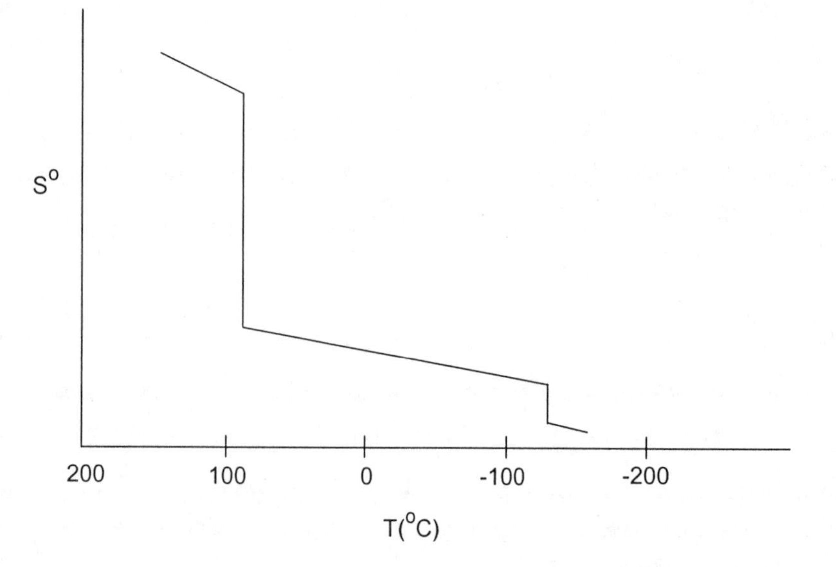

19.77 Propylene will have a higher S° at 25°C. At this temperature, both are gases, so there are no lattice effects (see Solution 19.78). Since they have the same molecular formula, only the details of their structures are different. In propylene, there is free rotation around the C–C single bond, while in cyclopropane the 3-membered ring severely limits rotation. The greater motional freedom of the propylene molecule leads to a higher absolute entropy.

19.78 (a) C(diamond), S° = 2.43 J/mol•K; C(graphite), S° = 5.69 J/mol•K. Diamond is a network covalent solid with each C atom tetrahedrally bound to four other C atoms. Graphite consists of sheets of fused planar 6-membered rings with each C atom bound in a trigonal planar arrangement to three other C atoms. The internal entropy in graphite is greater because there is translational freedom among the planar sheets of C atoms while there is very little translational or vibrational freedom within the covalent-network diamond lattice.

(b) S° for buckminsterfullerene will be ≥ 10 J/mol•K. S° for graphite is twice S° for diamond, and S° for the fullerene should be higher than that of graphite. The 60-atom "bucky" balls have more flexibility than graphite sheets. Also, the balls have translational freedom in three dimensions, while graphite sheets have it in only two directions. Because of the ball structure, there is more empty space in the fullerene lattice than in graphite or diamond; essentially, 60 C-atoms in fullerene occupy a larger volume than 60 C-atoms in graphite or diamond. Thus, the fullerene has additional "molecular" complexity, more degrees of translational freedom, and occupies a larger volume, all features that point to a higher absolute entropy.

19.79 (a) Formation reactions are the synthesis of 1 mole of compound from elements in their standard states.

$1/2\ N_2(g) + 3/2\ H_2(g) \rightarrow NH_3(g)$

$C(s) + 2Cl_2(g) \rightarrow CCl_4(l)$

$K(s) + 1/2\ N_2(g) + 3/2\ O_2(g) \rightarrow KNO_3(s)$

In each of these formation reactions, there are fewer moles of gas in the products than the reactants, so we expect ΔS° to be negative. If $\Delta G_f^\circ = \Delta H_f^\circ - T\Delta S_f^\circ$ and ΔS_f° is negative, $-T\Delta S_f^\circ$ is positive and ΔG_f° is more positive than ΔH_f°.

(b) $C(s) + 1/2\ O_2(g) \rightarrow CO(g)$

In this reaction, there are more moles of gas in products, ΔS_f° is positive, $-T\Delta S_f^\circ$ is negative and ΔG_f° is more negative than ΔH_f°.

19.80 (a) (i) $2RbCl(s) + 3O_2(g) \rightarrow 2RbClO_3(s)$

$\Delta H^\circ = 2\Delta H^\circ\ RbClO_3(s) - 3\Delta H^\circ\ O_2(g) - 2\Delta H^\circ\ RbCl(s)$

$= 2(-392.4) - 3(0) - 2(-430.5) = +76.2$ kJ

$\Delta S^\circ = 2(152) - 3(205.0) - 2(92) = -495$ J/K $= -0.495$ kJ/K

$\Delta G^\circ = 2(-292.0) - 3(0) - 2(-412.0) = +240.0$ kJ

(ii) $C_2H_2(g) + 4Cl_2(g) \rightarrow 2CCl_4(l) + H_2(g)$

$\Delta H° = 2\Delta H°\ CCl_4(l) + \Delta H°\ H_2(g) - \Delta H°\ C_2H_2(g) - 4\Delta H°\ Cl_2(g)$

$= 2(-139.3) + 0 - (226.7) - 4(0) = -505.3$ kJ

$\Delta S° = 2(214.4) + 130.58 - (200.8) - 4(222.96) = -533.3$ J/K $= -0.5333$ kJ/K

$\Delta G° = 2(-68.6) + 0 - (209.2) - 4(0) = -346.4$ kJ

(iii) $TiCl_4(l) + 2H_2O(l) \rightarrow TiO_2(s) + 4HCl(aq)$

$\Delta H° = \Delta H°\ TiO_2(s) + 4\Delta H°\ HCl(aq) - \Delta H°\ TiCl_4(l) - 2\Delta H°\ H_2O(l)$

$= -944.7 + 4(-167.2) - (-804.2) - 2(-285.83) = -237.6$ kJ

$\Delta S° = 50.29 + 4(56.5) - 221.9 - 2(69.91) = -85.43$ J/K $= -0.0854$ kJ/K

$\Delta G° = -889.4 + 4(-131.2) - (-728.1) - 2(-237.13) = -211.8$ kJ

(b) (i) $\Delta G°$ is (+), nonspontaneous

(ii) $\Delta G°$ is (-), spontaneous

(iii) $\Delta G°$ is (-), spontaneous

(c) In each case the manner in which free energy change varies with temperature depends mainly on ΔS: $\Delta G = \Delta H - T\Delta S$. When ΔS is substantially positive, ΔG becomes more negative as temperature increases. When ΔS is substantially negative, ΔG becomes more positive as temperature increases.

(i) $\Delta S°$ is negative, $\Delta G°$ becomes more positive with increasing temperature.

(ii) $\Delta S°$ is negative, $\Delta G°$ becomes more positive with increasing temperature. (The reaction will become nonspontaneous at some temperature.)

(iii) $\Delta S°$ is negative, $\Delta G°$ becomes more positive with increasing temperature. (The reaction will become nonspontaneous at some temperature.)

19.81 $\Delta G = \Delta G° + RT \ln Q$

(a) $$Q = \frac{P^2_{NH_3}}{P_{N_2} \times P^3_{H_2}} = \frac{(1.2)^2}{(2.6)(5.9)^3} = 2.697 \times 10^{-3} = 2.7 \times 10^{-3}$$

$\Delta G° = 2\Delta G°\ NH_3(g) - \Delta G°\ N_2(g) - 3\Delta G°\ H_2(g)$

$= 2(-16.66) - 0 - 3(0) = -33.32$ kJ

$$\Delta G = -33.32 \text{ kJ} + \frac{8.314 \times 10^{-3} \text{ kJ}}{\text{K} \bullet \text{mol}} \times 298 \text{ K} \times \ln(2.69 \times 10^{-3})$$

$\Delta G = -33.32 - 14.66 = -47.98$ kJ

(b) $Q = \frac{P_{N_2}^3 \times P_{H_2O}^4}{P_{N_2H_4}^2 \times P_{NO_2}^2} = \frac{(0.5)^3(0.3)^4}{(5.0 \times 10^{-2})^2(5.0 \times 10^{-2})^2} = 162 = 2 \times 10^2$

$\Delta G° = 3\Delta G°\ N_2(g) + 4\Delta G°\ H_2O(g) - 2\Delta G°\ N_2H_4(g) - 2\Delta G°\ NO_2(g)$

$= 3(0) + 4(-228.57) - 2(159.4) - 2(51.84) = -1336.8$ kJ

$\Delta G = -1336.8$ kJ $+ 2.478 \ln 162 = -1324.2$ kJ

(c) $Q = \frac{P_{N_2} \times P_{H_2}^2}{P_{N_2H_4}} = \frac{(1.5)(2.5)^2}{0.5} = 18.75 = 2 \times 10^1$

$\Delta G° = \Delta G°\ N_2(g) + 2\Delta G°\ H_2(g) - \Delta G°\ N_2H_4(g)$

$= 0 + 2(0) - 159.4 = -159.4$ kJ

$\Delta G = -159.4$ kJ $+ 2.478 \ln 18.75 = -152.1$ kJ

19.82

Reaction	(a) Sign of $\Delta H°$	(a) Sign of $\Delta S°$	(b) $K_{eq} > 1$?	(c) Variation in K_{eq} as Temp. Increases
(i)	-	-	yes	decrease
(ii)	+	+	no	increase
(iii)	+	+	no	increase
(iv)	+	+	no	increase

(a) Note that at a particular temperature, positive $\Delta H°$ leads to a smaller value of K, while positive $\Delta S°$ increases the value of K.

19.83 (a) $K_{eq} = \frac{\chi_{CH_3COOH}}{\chi_{CH_3OH}\, P_{CO}}$

$\Delta G° = -RT \ln K_{eq}$; $\ln K_{eq} = -\Delta G/RT$

$\Delta G° = \Delta G°\ CH_3COOH(l) - \Delta G°\ CH_3OH(l) - \Delta G°\ CO(g)$

$= -392.4 - (-166.23) - (-137.2) = -89.0$ kJ

$\ln K_{eq} = \frac{-(-89.0 \text{ kJ})}{(8.314 \times 10^{-3} \text{ kJ/K})(298 \text{ K})} = 35.922 = 35.9$; $K_{eq} = 4 \times 10^{15}$

(b) $\Delta H° = \Delta H°\ CH_3COOH(l) - \Delta H°\ CH_3OH(l) - \Delta H°\ CO(g)$

$= -487.0 - (-238.6) - (-110.5) = -137.9$ kJ

The reaction is exothermic, so the value of K_{eq} will decrease with increasing temperature, and the mole fraction of CH_3COOH will also decrease. Elevated temperatures must be used to increase the speed of the reaction. Thermodynamics cannot predict the rate at which a reaction reaches equilibrium.

(c) $\Delta G° = -RT \ln K_{eq}$; $K_{eq} = 1$, $\ln K_{eq} = 0$, $\Delta G° = 0$

$\Delta G° = \Delta H° - T\Delta S°$; when $\Delta G° = 0$, $\Delta H° = T\Delta S°$

$\Delta S° = S° \; CH_3COOH(l) - S° \; CH_3OH(l) - S° \; CO(g)$

$= 159.8 - 126.8 - 197.9 = -164.9 \; J/K = -0.1649 \; kJ/K$

$-137.9 \; kJ = T(-0.1649 \; kJ/K), \; T = 836.3 \; K$

The equilibrium favors products up to 836 K or 563 °C, so the elevated temperatures to increase the rate of reaction can be safely employed.

19.84 (a) First calculate $\Delta G°$ for each reaction:

For $C_6H_{12}O_6(s) + 6O_2(g) \rightleftharpoons 6CO_2(g) + 6H_2O(l)$ (A)

$\Delta G° = 6(-237.13) + 6(-394.4) - (-910.4) + 6(0) = -2878.8 \; kJ$

For $C_6H_{12}O_6(s) \rightleftharpoons 2C_2H_5OH(l) + 2CO_2(g)$ (B)

$\Delta G° = 2(-394.4) + 2(-174.8) - (-910.4) = -228.0 \; kJ$

For (A), $\ln K_{eq} = 2879 \times 10^3/(8.314)(298) = 1162; \quad K_{eq} = 5 \times 10^{504}$

For (B), $\ln K_{eq} = 228 \times 10^3/(8.314)(298) = 92.026 = 92.0; \quad K_{eq} = 9 \times 10^{39}$

(b) Both these values for K_{eq} are unimaginably large. However, K_{eq} for reaction (A) is larger, because $\Delta G°$ is more negative. The magnitude of the work that can be accomplished by coupling a reaction to its surroundings is measured by ΔG. According to the calculations above, considerably more work can in principle be obtained from reaction (A), because $\Delta G°$ is more negative.

19.85 (a) $\Delta G° = -RT \ln K_{eq}$ (Equation 19.22); $\ln K_{eq} = -\Delta G°/RT$

Use $\Delta G° = \Delta H° - T\Delta S°$ to get $\Delta G°$ at the two temperatures. Calculate $\Delta H°$ and $\Delta S°$ using data in Appendix C.

$2CH_4(g) \rightarrow C_2H_6(g) + H_2(g)$

$\Delta H° = \Delta H° \; C_2H_6(g) + \Delta H° \; H_2(g) - 2\Delta H° \; CH_4(g) = -84.68 + 0 - 2(-74.8) = 64.92$
$= 64.9 \; kJ$

$\Delta S° = S° \; C_2H_6(g) + S° \; H_2(g) - 2S° \; CH_4(g) = 229.5 + 130.58 - 2(186.3) = -12.52$
$= -12.5 \; J/K$

at 298 K, $\Delta G = 64.92 \; kJ - 298 \; K(-12.52 \times 10^{-3} \; kJ/K) = 68.65 = 68.7 \; kJ$

$$\ln K_{eq} = \frac{-68.65 \; kJ}{(8.314 \times 10^{-3} \; kJ/K)(298 \; K)} = -27.709 = -27.7, \; K_{eq} = 9.25 \times 10^{-13} = 9 \times 10^{-13}$$

at 773 K, $\Delta G = 64.9 \; kJ - 773 \; K(-12.52 \times 10^{-3} \; J/K) = 74.598 = 74.6 \; kJ$

$$\ln K_{eq} = \frac{-74.598 \; kJ}{(8.314 \times 10^{-3} \; kJ/K)(773 \; K)} = -11.607 = -11.6, \; K_{eq} = 9.1 \times 10^{-6} = 9 \times 10^{-6}$$

Because the reaction is endothermic, the value of K_{eq} increases with an increase in temperature.

$2CH_4(g) + 1/2 \; O_2(g) \rightarrow C_2H_6(g) + H_2O(g)$

$\Delta H° = \Delta H°\ C_2H_6(g) + \Delta H°\ H_2O(g) - 2\Delta H°\ CH_4(g) - 1/2\ \Delta H°\ O_2(g)$

$= -84.68 + (-241.82) - 2(-74.8) - 1/2\ (0) = -176.9$ kJ

$\Delta S° = S°\ C_2H_6(g) + S°\ H_2O(g) - 2S°\ CH_4(g) - 1/2\ S°\ O_2(g)$

$= 229.5 + 188.83 - 2(186.3) - 1/2\ (205.0) = -56.77 = -56.8$ J/K

at 298 K, $\Delta G = -176.9\text{ kJ} - 298\text{ K}(-56.77 \times 10^{-3}\text{ kJ/K}) = -159.98 = -160.0$ kJ

$$\ln K_{eq} = \frac{-(-159.98\text{ kJ})}{(8.314 \times 10^{-3}\text{ kJ/K})(298\text{ K})} = 64.571 = 64.6;\ K_{eq} = 1 \times 10^{28}$$

at 773 K, $\Delta G = -176.9\text{ kJ} - 773\text{ K}(-56.77 \times 10^{-3}\text{ kJ/K}) = -133.02 = -133.0$ kJ

$$\ln K_{eq} = \frac{-(-133.02\text{ kJ})}{(8.314 \times 10^{-3}\text{ kJ/K})(773\text{ K})} = 20.698 = 20.7;\ K_{eq} = 1 \times 10^{9}$$

Because this reaction is exothermic, the value of K_{eq} decreases with increasing temperature.

(b) The difference in $\Delta G°$ for the two reactions is primarily enthalpic; the first reaction is endothermic and the second exothermic. Both reactions have $-\Delta S°$, which inhibits spontaneity.

(c) This is an example of coupling a useful but nonspontaneous reaction with a spontaneous one to spontaneously produce a desired product.

$2CH_4(g) \rightarrow C_2H_6(g) + H_2(g)$ $\quad \Delta G°_{298} = +68.7$ kJ, nonspontaneous

$H_2(g) + 1/2\ O_2(g) \rightarrow H_2O(g)$ $\quad \Delta G°_{298} = -228.57$ kJ, spontaneous

$2CH_4(g) + 1/2\ O_2(g) \rightarrow C_2H_6(g) + H_2O(g)$ $\quad \Delta G°_{298} = -159.9$ kJ, spontaneous

(d) $CH_4(g) + 2O_2(g) \rightarrow CO_2(g) + 2H_2O(g)$

19.86 $\Delta G°$ for the metabolism of glucose is:

$6\Delta G°\ CO_2(g) + 6\Delta G°\ H_2O(l) - \Delta G°\ C_6H_{12}O_6(s) - 6\Delta G°\ O_2(g)$

$\Delta G° = 6(-394.4) + 6(-237.13) - (-910.4) + 6(0) = -2878.8$ kJ

moles ATP = $-2878.8\text{ kJ} \times 1\text{ mol ATP} / (-30.5\text{ kJ}) = 94.4$ mol ATP / mol glucose

19.87 (a) The equilibrium of interest here can be written as:

$K^+\text{ (plasma)} \rightleftharpoons K^+\text{ (muscle)}$

Since an aqueous solution is involved in both cases, assume that the equilibrium constant for the above process is exactly 1, that is, $\Delta G° = 0$. However, ΔG is not zero because the concentrations are not the same on both sides of the membrane. Use Equation [19.21] to calculate ΔG:

$$\Delta G = \Delta G° + RT \ln \frac{[K^+(\text{muscle})]}{[K^+(\text{plasma})]}$$

$$= 0 + (8.314)(310) \ln \frac{(0.15)}{(5.0 \times 10^{-3})} = 8.77 \text{ kJ}$$

(b) Note that ΔG is positive. This means that work must be done on the system (blood plasma plus muscle cells) to move the K^+ ions "uphill," as it were. The minimum amount of work possible is given by the value for ΔG. This value represents the minimum amount of work required to transfer one mole of K^+ ions from the blood plasma at 5×10^{-3} *M* to muscle cell fluids at 0.15 *M*, assuming constancy of concentrations. In practice, a larger than minimum amount of work is required.

19.88 (a) To obtain $\Delta H°$ from the equilibrium constant data, graph $\ln K_{eq}$ at various temperatures vs $1/T$, being sure to employ absolute temperature. The slope of the linear relationship that should result is $-\Delta H°/R$; thus, $\Delta H°$ is easily calculated.

(b) Use $\Delta G° = \Delta H° - T\Delta S°$ and $\Delta G° = -RT \ln K_{eq}$. Substituting the second expression into the first, we obtain

$$-RT \ln K = \Delta H° - T\Delta S°; \quad \ln K = \frac{-\Delta H°}{RT} - \frac{-\Delta S°}{R} = \frac{-\Delta H°}{RT} + \frac{\Delta S°}{R}$$

Thus, the constant in the equation given in the exercise is $\Delta S°/R$.

19.89 (a) Both equations describe the entropy change of the system when a gas expands at constant temperature.

$\Delta S = nR \ln(V_2/V_1)$; $\Delta S = q_{rev}/T$ (Equation 19.1)

$q_{rev}/T = nR \ln(V_2/V_1)$; $q_{rev} = nRT \ln(V_2/V_1)$

(b) $n = 0.50$ mol, $V_1 = 10.0$ L, $V_2 = 75.0$ L

$\Delta S = 0.50 \text{ mol } (8.314 \text{ J/mol•K}) \ln (75.0 \text{ L}/10.0 \text{ L}) = 8.376 = 8.4 \text{ J/K}$

(c) When a gas expands, there are more possible arrangements for the particles, and entropy increases. The positive sign for ΔS in part (b) is consistent with this prediction.

(d) $n = 8.5$; $V_2 = 1/8\ V_1$; $V_2/V_1 = 1/8$

$\Delta S = 8.5 \text{ mol } (8.314 \text{ J/mol•K}) \ln (1/8) = -146.95 = -1.5 \times 10^2 \text{ J/K}$

19.90 $S = k \ln W$ (Equation 19.8), $k = R/N$, $W \propto V^m$

$\Delta S = S_2 - S_1$; $S_1 = k \ln W_1$, $S_2 = k \ln W_2$

$\Delta S = k \ln W_2 - k \ln W_1$; $W_2 = cV_2^{\,m}$; $W_1 = cV_1^{\,m}$

(The number of particles, m, is the same in both states.)

$\Delta S = k \ln cV_2^m - k \ln cV_1^m$; $\ln a^b = b \ln a$

$\Delta S = k\, m \ln cV_2 - k\, m \ln cV_1$; $\ln a - \ln b = \ln (a/b)$

$$\Delta S = k\, m \ln \left(\frac{cV_2}{cV_1}\right) = k\, m \ln \left(\frac{V_2}{V_1}\right) = \frac{R}{N} m \ln \left(\frac{V_2}{V_1}\right)$$

$$\frac{m}{N} = \frac{\text{particles}}{6.022 \times 10^{23}} = n(\text{mol}); \quad \Delta S = nR \ln \left(\frac{V_2}{V_1}\right)$$

19.91 Absolute entropy is a fundamental property of matter at a specified set of conditions, that is, a state. In order to lower the entropy of the fuel, either the structure of the molecules or the conditions (temperature, pressure, amount) must be changed. Any of these changes would require energy, which would reduce the amount of energy available to drive the car, not increase it.

Integrative Exercises

19.92 (a) At the boiling point, vaporization is a reversible process, so $\Delta S^\circ_{vap} = \Delta H^\circ_{vap}/T$.

acetone: $\Delta S^\circ_{vap} = \Delta H^\circ_{vap}/T$ = (29.1 kJ/mol) / 329.25 K = 88.4 J/mol•K

dimethyl ether: ΔS°_{vap} =(21.5 kJ/mol) / 248.35 K = 86.6 J/mol•K

ethanol: ΔS°_{vap} = (38.6 kJ/mol) / 351.6 K = 110 J/mol•K

octane: ΔS°_{vap} = (34.4 kJ/mol) / 398.75 K = 86.3 J/mol•K

pyridine: ΔS°_{vap} = (35.1kJ/mol) / 388.45 K = 90.4 J/mol•K

(b) Ethanol is the only liquid listed that doesn't follow *Trouton's rule* and it is also the only substance that exhibits hydrogen bonding in the pure liquid. Hydrogen bonding leads to more ordering in the liquid state and a greater than usual increase in entropy upon vaporization. The rule appears to hold for liquids with London dispersion forces (octane) and ordinary dipole-dipole forces (acetone, dimethyl ether, pyridine), but not for those with hydrogen bonding.

(c) Owing to strong hydrogen bonding interactions, water probably does not obey Trouton's rule.

From Appendix B, ΔH°_{vap} at 100°C = 40.67 kJ/mol.

ΔS°_{vap} = (40.67 kJ/mol) / 373.15 K = 109.0 J/mol•K

(d) Use ΔS°_{vap} = 88 J/mol•K, the middle of the range for Trouton's rule, to estimate ΔH°_{vap} for chlorobenzene.

$\Delta H^\circ_{vap} = \Delta S^\circ_{vap} \times T$ = 88 J/mol•K × 404.95 K = 36 kJ/mol

19.93 (a) Polymerization is the process of joining many small molecules (monomers) into a few very large molecules (polymers). Polyethylene in particular can have extremely high molecular weights. In general, reducing the number of particles in a system reduces entropy, so ΔS_{poly} is expected to be negative.

(b) $\Delta G_{poly} = \Delta H_{poly} - T\Delta S_{poly}$. If the polymerization of ethylene is spontaneous, ΔG_{poly} is negative. If ΔS_{poly} is negative, $-T\Delta S_{poly}$ is positive, so ΔH_{poly} must be negative for ΔG_{poly} to be negative. The enthalpy of polymerization must be exothermic.

(c) According to Equation 12.1, polymerization of ethylene requires breaking one C=C and forming 2C–C per monomer (1C–C between the C-atoms of the monomer and 2 × 1/2 C–C to two other monomers).

$\Delta H = D(C{=}C) - 2D(C{-}C) = 614 - 2(348) = -82$ kJ/mol C_2H_4

$$\frac{-82\text{ kJ}}{\text{mol } C_2H_4} \times \frac{1\text{ mol}}{6.022 \times 10^{23}\text{ molecules}} \times \frac{1000\text{ J}}{1\text{ kJ}} = 1.36 \times 10^{-19}\text{ J}/C_2H_4\text{ monomer}$$

(d) The products of a condensation polymerization are the polymer and a small molecule, typically H_2O; there is usually one small molecule formed per monomer unit. Unlike addition polymerization, the total number of particles is not reduced. A condensation polymer does impose more order on the monomer or monomers than an addition polymer. If there is a single monomer, it has different functional groups at the two ends and only one end can react to join the polymer, so orientation is required. If there are two different monomers, as in nylon, the monomers alternate in the polymer, so only the correct monomer can react to join the polymer. In terms of structure, the condensation polymer imposes more order on the monomer(s) than an addition polymer. But, condensation polymerization does not lead to a reduction in the number of particles in the system, so ΔS_{poly} will be less negative than for addition polymerization.

19.94 The activated complex in Figure 14.13 is a single "particle" or entity that contains 4 atoms. It is formed from an atom A and a triatomic molecule, ABC, that must collide with exactly the correct energy and orientation to form the single entity. There are many fewer degrees of freedom for the activated complex than the separate reactant particules, so the *entropy of activation* is negative.

19.95 (a) $O_2(g) \xrightarrow{h\nu} 2O(g)$; ΔS increases because there are more moles of gas in the products.

(b) $O_2(g) + O(g) \rightarrow O_3(g)$, ΔS decreases because there are fewer moles of gas in the products.

(c) ΔS increases as the gas molecules diffuse into the larger volume of the stratosphere; there are more possible positions and therefore more motional freedom.

(d) $NaCl(aq) \rightarrow NaCl(s) + H_2O(l)$; ΔS decreases as the mixture (seawater, greater disorder) is separated into pure substances (fewer possible arrangements, more order).

19.96 (a) 16 e^-, 8 e^- pairs. The C-S bond order is approximately 2.

$$:\ddot{S}=C=\ddot{S}:$$

(b) 2 e^- domains around C, linear e^- domain geometry, linear molecular structure

(c) $CS_2(l) + 3O_2(g) \rightarrow CO_2(g) + 2SO_2(g)$

(d) $\Delta H° = \Delta H°\ CO_2(g) + 2\Delta H°\ SO_2(g) - \Delta H°\ CS_2(l) - 3\ \Delta H\ \Delta H°\ O_2(g)$

$= -393.5 + 2(-296.9) - (89.7) - 3(0) = -1077.0$ kJ

$\Delta G° = \Delta G°\ CO_2(g) + 2\Delta G°\ SO_2(g) - \Delta G°\ CS_2(l) - 3\ \Delta G°\ O_2(g)$

$= -394.4 + 2(-300.4) - (65.3) - 3(0) = -1060.5$ kJ

The reaction is exothermic ($-\Delta H°$) and spontaneous ($-\Delta G°$) at 298 K.

(e) vaporization: $CS_2(l) \rightarrow CS_2(g)$

$\Delta G°_{vap} = \Delta H°_{vap} - T\Delta S°_{vap}$; $\Delta S°_{vap} = (\Delta H°_{vap} - \Delta G°_{vap})/T$

$\Delta G°_{vap} = \Delta G°\ CS_2(g) - \Delta G°\ CS_2(l) = 67.2 - 65.3 = 1.9$ kJ

$\Delta H°_{vap} = \Delta H°\ CS_2(g) - \Delta H°\ CS_2(l) = 117.4 - 89.7 = 27.7$ kJ

$\Delta S°_{vap} = (27.7 - 1.9)$ kJ/298 K = 0.086577 = 0.0866 kJ/K = 86.6 J/K

ΔS_{vap} is always positive, because the gas phase occupies a greater volume, has more motional freedom and a larger absolute entropy than the liquid.

(f) At the boiling point, $\Delta G = 0$ and $\Delta H_{vap} = T_b\Delta S_{vap}$.

$T_b = \Delta H_{vap}/\Delta S_{vap} = 27.7$ kJ/0.086577 kJ/K = 319.9 = 320 K

$T_b = 320$ K = 47°C. CS_2 is a liquid at 298 K, 1 atm

19.97 (a) $Ag(s) + 1/2\ N_2(g) + 3/2\ O_2(g) \rightarrow AgNO_3(s)$; ΔS decreases because there are fewer moles of gas in the product.

(b) $\Delta G°_f = \Delta H°_f - T\Delta S°_f$; $\Delta S°_f = (\Delta G°_f - \Delta H°_f) / (-T) = (\Delta H°_f - \Delta G°_f) / T$

$\Delta S°_f = -124.4$ kJ - (-33.4 kJ) / 298 K = -0.305 kJ/K = -305 J/K

$\Delta S°_f$ is relatively large and negative, as anticipated from part (a).

(c) Dissolving of $AgNO_3$ can be expressed as

$AgNO_3(s) \rightarrow AgNO_3$ (aq, 1 m)

$\Delta H° = \Delta H°\ AgNO_3(aq) - \Delta H°\ AgNO_3(s) = -101.7 - (-124.4) = +22.7$ kJ

$\Delta H° = \Delta H°\ MgSO_4(aq) - \Delta H°\ MgSO_4(s) = -1374.8 - (-1283.7) = -91.1$ kJ

Dissolving $AgNO_3(s)$ is endothermic ($+\Delta H°$), but dissolving $MgSO_4(s)$ is exothermic ($-\Delta H°$).

(d) $AgNO_3$: $\Delta G° = \Delta G°_f\ AgNO_3(aq) - \Delta G°_f\ AgNO_3(s) = -34.2 - (-33.4) = -0.8$ kJ

$\Delta S° = (\Delta H° - \Delta G°) / T = [22.7$ kJ - (-0.8 kJ)] / 298 K = 0.0789 kJ/K = 78.9 J/K

$MgSO_4$: $\Delta G° = \Delta G°_f\ MgSO_4(aq) - \Delta G°_f\ MgSO_4(s) = -1198.4 - (-1169.6) = -28.8$ kJ

$\Delta S° = (\Delta H° - \Delta G°) / T = [-91.1$ kJ - (-28.8 kJ)] / 298 K = -0.209 kJ/K = -209 J/K

(e) In general, we expect dissolving a crystalline solid to be accompanied by an increase in positional disorder and an increase in entropy; this is the case for $AgNO_3$ ($\Delta S°$= + 78.9 J/K). However, for dissolving $MgSO_4$(s), there is a substantial decrease in entropy (ΔS = -209 J/K). According to Section 13.5, ion-pairing is a significant phenomenon in electrolyte solutions, particularly in concentrated solutions where the charges of the ions are greater than 1. According to Table 13.5, a 0.1 m $MgSO_4$ solution has a van't Hoff factor of 1.21. That is, for each mole of $MgSO_4$ that dissolves, there are only 1.21 moles of "particles" in solution instead of 2 moles of particles. For a 1 m solution, the factor is even smaller. Also, the exothermic enthalpy of mixing indicates substantial interactions between solute and solvent. Substantial ion-pairing coupled with ion-dipole interactions with H_2O molecules lead to a decrease in entropy for $MgSO_4$(aq) relative to $MgSO_4$(s).

19.98 (a) $K_{eq} = P^2_{NO_2} / P_{N_2O_4}$

Assume equal amounts means equal number of moles. For gases, P = n(RT/V). In an equilibrium mixture, RT/V is a constant, so moles of gas is directly proportional to partial pressure. Gases with equal partial pressures will have equal moles of gas present. The condition $P_{NO_2} = P_{N_2O_4}$ leads to the expression $K_{eq} = P_{NO_2}$. The value of K_{eq} then depends on P_T for the mixture. For any particular value of P_T, the condition of equal moles of the two gases can be achieved at some temperature. For example, $P_{NO_2} = P_{N_2O_4} = 1.0$ atm, $P_T = 2.0$ atm.

$$K_{eq} = \frac{(1.0)^2}{1.0} = 1.0;\ \ln K_{eq} = 0;\ \Delta G° = 0 = \Delta H° - T\Delta S°;\ T = \Delta H°/\Delta S°$$

$$\Delta H° = 2\Delta H°\ NO_2(g) - \Delta H°\ N_2O_4(g) = 2(33.84) - 9.66 = +58.02 \text{ kJ}$$

$$\Delta S° = 2S°\ NO_2(g) - S°\ N_2O_4(g) = 2(240.45) - 304.3 = 0.1766 \text{ kJ/K}$$

$$T = \frac{58.02 \text{ kJ}}{0.1766 \text{ kJ/K}} = 328.5 \text{ K or } 55.5°\text{C}$$

(b) $P_T = 1.00$ atm; $P_{N_2O_4} = x$, $P_{NO_2} = 2x$; $x + 2x = 1.00$ atm

$x = P_{N_2O_4} = 0.3333 = 0.333$ atm; $P_{NO_2} = 0.6667 = 0.667$ atm

$$K_{eq} = \frac{(0.6667)^2}{0.3333} = 1.334 = 1.33;\ \Delta G° = -RT \ln K_{eq} = \Delta H° - T\Delta S°$$

$$-(8.314 \times 10^{-3} \text{ kJ/K})(\ln 1.334)\ T = 58.02 \text{ kJ} - (0.1766 \text{ kJ/K})\ T$$

$$(-0.00239 \text{ kJ/K})\ T + (0.1766 \text{ kJ/K})\ T = 58.02 \text{ kJ}$$

$$(0.1742 \text{ kJ/K})\ T = 58.02 \text{ kJ};\ T = 333.0 \text{ K}$$

(c) $P_T = 10.00$ atm; $x + 2x = 10.00$ atm

$x = P_{N_2O_4} = 3.3333 = 3.333$ atm; $P_{NO_2} = 6.6667 = 6.667$ atm

$$K_{eq} = \frac{(6.6667)^2}{3.3333} = 13.334 = 13.33;\ -RT \ln K_{eq} = \Delta H° - T\Delta S°$$

$-(8.314 \times 10^{-3}$ kJ/K)(ln 13.334) T = 58.02 kJ - (0.1766 kJ/K) T

(-0.02154 kJ/K) T + (0.1766 kJ/K) T = 58.02 kJ

(0.15506 kJ/K) T = 58.02 kJ; T = 374.2 K

(d) The reaction is endothermic, so an increase in the value of K_{eq} as calculated in parts (b) and (c) should be accompanied by an increase in T.

19.99 (a) $\Delta G° = 3\Delta G_f^\circ\ S(s) + 2\Delta G_f^\circ\ H_2O(g) - \Delta G_f^\circ\ SO_2(g) - 2\Delta G_f^\circ\ H_2S(g)$

= 3(0) + 2(-228.57) - (-300.4) - 2(-33.01) = -90.72 = -90.7 kJ

$$\ln K_{eq} = \frac{-\Delta G°}{RT} = \frac{-(-90.72\ \text{kJ})}{(8.314 \times 10^{-3}\ \text{kJ/K})(298\ \text{K})} = 36.6165 = 36.6;\ K_{eq} = 7.99 \times 10^{15} = 8 \times 10^{15}$$

(b) The reaction is highly spontaneous at 298 K and feasible in principle. However, use of $H_2S(g)$ produces a severe safety hazard for workers and the surrounding community.

(c) $$P_{H_2O} = \frac{25\ \text{torr}}{760\ \text{torr/atm}} = 0.033\ \text{atm}$$

$$K_{eq} = \frac{P_{H_2O}^2}{P_{SO_2} \times P_{H_2S}^2};\ P_{SO_2} = P_{H_2S} = x\ \text{atm}$$

$$K_{eq} = 7.99 \times 10^{15} = \frac{(0.033)^2}{x(x)^2};\ x^3 = \frac{(0.033)^2}{7.99 \times 10^{15}}$$

$x = 5 \times 10^{-7}$ atm

(d) $\Delta H° = 3\Delta H_f^\circ\ S(s) + 2\Delta H_f^\circ\ H_2O(g) - \Delta H_f^\circ\ SO_2(g) - 2\Delta H_f^\circ\ H_2S(g)$

= 3(0) + 2(-241.82) - (-296.9) - 2(-20.17) = -146.4 kJ

$\Delta S° = 3S°\ S(s) + 2S°\ H_2O(g) - S°\ SO_2(g) - 2S°\ H_2S(g)$

= 3(31.88) + 2(188.83) - 248.5 - 2(205.6) = -186.4 J/K

The reaction is exothermic ($-\Delta H$), so the value of K_{eq} will decrease with increasing temperature. The negative $\Delta S°$ value means that the reaction will become nonspontaneous at some higher temperature. The process will be less effective at elevated temperatures.

19.100 (a) When the rubber band is stretched, the molecules become more ordered, so the entropy of the system decreases, ΔS_{sys} is negative.

(b) $\Delta S_{sys} = q_{rev}/T$. Since ΔS_{sys} is negative, q_{rev} is negative and heat is evolved by the system.

20 Electrochemistry

Oxidation-Reduction Reactions

20.1 (a) *Oxidation* is the loss of electrons.

(b) The electrons appear on the products side (right side) of an oxidation half-reaction.

(c) The *oxidant* is the reactant that is reduced; it gains the electrons that are lost by the substance being oxidized.

20.2 (a) *Reduction* is the gain of electrons.

(b) The electrons appear on the reactants side (left side) of a reduction half-reaction.

(c) The *reductant* is the reactant that is oxidized; it provides the electrons that are gained by the substance being reduced.

20.3 *Analyze/Plan.* Given a chemical equation, we are asked to indicate which elements undergo a change in oxidation number and the magnitude of the change. Assign oxidation numbers according to the rules given in Section 4.4. Note the changes and report the magnitudes. *Solve*:

(a) I is reduced from +5 to 0; C is oxidized from +2 to +4.

(b) Hg is reduced from +2 to 0; N is oxidized from -2 to 0.

(c) N is reduced from +5 to +2; S is oxidized from -2 to 0.

(d) Cl is reduced from +4 to +3; O is oxidized from -1 to 0.

20.4 (a) No oxidation-reduction

(b) I is oxidized from -1 to +5; Cl is reduced from +1 to -1.

(c) S is oxidized from +4 to +6; N is reduced from +5 to +2.

(d) S is reduced from +6 to +4;. Br is oxidized from -1 to 0.

20.5 *Analyze/Plan.* Write the balanced chemical equation and assign oxidation numbers. The substance oxidized is the reductant and the substance reduced is the oxidant. *Solve*:

(a) $TiCl_4(g) + 2Mg(l) \rightarrow Ti(s) + 2MgCl_2(l)$

(b) Mg(l) is the reductant; $TiCl_4(g)$ is the oxidant.

20.6 (a) $2N_2H_4(g) + N_2O_4(g) \rightarrow 3N_2(g) + 4H_2O(g)$

(b) N_2O_4 serves as the oxidizing agent; it is itself reduced. N_2H_4 serves as the reducing agent; it is itself oxidized.

20.7 *Analyze/Plan.* Follow the logic in Sample Exercises 20.2 and 20.3. If the half-reaction occurs in basic solution, balance as in acid, then add OH^- to each side. *Solve:*

(a) $Sn^{2+}(aq) \rightarrow Sn^{4+}(aq) + 2e^-$, oxidation

(b) $TiO_2(s) + 4H^+(aq) + 2e^- \rightarrow Ti^{2+}(aq) + 2H_2O(l)$, reduction

(c) $ClO_3^-(aq) + 6H^+(aq) + 6e^- \rightarrow Cl^-(aq) + 3H_2O(l)$, reduction

(d) $4OH^-(aq) \rightarrow O_2(g) + 2H_2O(l) + 4e^-$, oxidation

(e) $SO_3^{2-}(aq) + 2OH^-(aq) \rightarrow SO_4^{2-}(aq) + H_2O(l) + 2e^-$, oxidation

20.8 (a) $Mo^{3+}(aq) + 3e^- \rightarrow Mo(s)$, reduction

(b) $H_2SO_3(aq) + H_2O(l) \rightarrow SO_4^{2-}(aq) + 4H^+(aq) + 2e^-$, oxidation

(c) $NO_3^-(aq) + 4H^+(aq) + 3e^- \rightarrow NO(g) + 2H_2O(l)$, reduction

(d) $Mn^{2+}(aq) + 4OH^-(aq) \rightarrow MnO_2(s) + 2H_2O(l) + 2e^-$, oxidation

(e) $Cr(OH)_3(s) + 5OH^-(aq) \rightarrow CrO_4^{2-}(aq) + 4H_2O(l) + 3e^-$, oxidation

20.9 *Analyze/Plan.* Follow the logic in Sample Exercises 20.2 and 20.3 to balance the given equations. Use the method in Sample Exercise 20.1 to identify oxidizing and reducing agents.

Solve:

(a) $Cr_2O_7^{2-}(aq) + I^-(aq) + 8H^+ \rightarrow 2Cr^{3+}(aq) + IO_3^-(aq) + 4H_2O(l)$

oxidizing agent, $Cr_2O_7^{2-}$; reducing agent, I^-

(b) The half-reactions are:

$4[MnO_4^-(aq) + 8H^+(aq) + 5e^- \rightarrow Mn^{2+}(aq) + 4H_2O(l)]$

$5[CH_3OH(aq) + H_2O(l) \rightarrow HCO_2H(aq) + 4H^+(aq) + 4e^-]$

$4MnO_4^-(aq) + 5CH_3OH(aq) + 12H^+(aq) \rightarrow 4Mn^{2+}(aq) + 5HCO_2H(aq) + 11H_2O(l)$

oxidizing agent, MnO_4^-; reducing agent, CH_3OH

(c) $I_2(s) + 6H_2O(l) \rightarrow 2IO_3^-(aq) + 12H^+(aq) + 10e^-$

$5[OCl^-(aq) + 2H^+(aq) + 2e^- \rightarrow Cl^-(aq) + H_2O(l)]$

$I_2(s) + 5OCl^-(aq) + H_2O(l) \rightarrow 2IO_3^-(aq) + 5Cl^- + 2H^+(aq)]$

oxidizing agent, OCl^-; reducing agent, I_2

(d) $As_2O_3(s) + 5H_2O(l) \rightarrow 2H_3AsO_4(aq) + 4H^+(aq) + 4e^-$

$2NO_3^-(aq) + 6H^+(aq) + 4e^- \rightarrow N_2O_3(aq) + 3H_2O(l)$

$As_2O_3(s) + 2NO_3^-(aq) + 2H_2O(l) + 2H^+(aq) \rightarrow 2H_3AsO_4(aq) + N_2O_3(aq)$

oxidizing agent, NO_3^-; reducing agent, As_2O_3

(e) $2[MnO_4^-(aq) + 2H_2O(l) + 3e^- \rightarrow MnO_2(s) + 4OH^-]$

$Br^-(aq) + 6OH^-(aq) \rightarrow BrO_3^-(aq) + 3H_2O(l) + 6e^-$

$2MnO_4^-(aq) + Br^-(aq) + H_2O(l) \rightarrow 2MnO_2(s) + BrO_3^-(aq) + 2OH^-(aq)$

oxidizing agent, MnO_4^-; reducing agent, Br^-

(f) $Pb(OH)_4^{2-}(aq) + ClO^-(aq) \rightarrow PbO_2(s) + Cl^-(aq) + 2OH^-(aq) + H_2O(l)$

oxidizing agent, ClO^-; reducting agent, $Pb(OH)_4^{2-}$

20.10 (a) $3[NO_2^-(aq) + H_2O(l) \rightarrow NO_3^-(aq) + 2H^+(aq) + 2e^-]$

$Cr_2O_7^{2-}(aq) + 14H^+(aq) + 6e^- \rightarrow 2Cr^{3+}(aq) + 7H_2O(l)$

Net: $3NO_2^-(aq) + Cr_2O_7^{2-}(aq) + 8H^+(aq) \rightarrow 3NO_3^-(aq) + 2Cr^{3+}(aq) + 4H_2O(l)$

oxidizing agent, $Cr_2O_7^{2-}$; reducing agent, NO_2^-

(b) $4[As(s) + 3H_2O(l) \rightarrow H_3AsO_3(aq) + 3H^+(aq) + 3e^-]$

$3[ClO_3^-(aq) + 5H^+(aq) + 4e^- \rightarrow HClO(aq) + 2H_2O(l)]$

$4As(s) + 3ClO_3^-(aq) + 6H_2O(l) + 3H^+(aq) \rightarrow 4H_3AsO_3(aq) + 3HClO(aq)$

oxidizing agent, ClO_3^-; reducing agent, As

(c) $2[Cr_2O_7^{2-}(aq) + 14H^+(aq) + 6e^- \rightarrow 2Cr^{3+}(aq) + 7H_2O(l)]$

$3[CH_3OH(aq) + H_2O(l) \rightarrow HCO_2H(aq) + 4H^+(aq) + 4e^-]$

Net: $2Cr_2O_7^{2-}(aq) + 3CH_3OH(aq) + 16H^+(aq) \rightarrow 4Cr^{3+}(aq) + 3HCO_2H(aq) + 11H_2O(l)$

oxidizing agent, $Cr_2O_7^{2-}$; reducing agent, CH_3OH

(d) $2[MnO_4^-(aq) + 8H^+(aq) + 5e^- \rightarrow Mn^{2+}(aq) + 4H_2O(l)]$

$5[2Cl^-(aq) \rightarrow Cl_2(aq) + 2e^-]$

Net: $2MnO_4^-(aq) + 10Cl^-(aq) + 16H^+(aq) \rightarrow 2Mn^{2+}(aq) + 5Cl_2(g) + 8H_2O(l)$

oxidizing agent, MnO_4^-; reducing agent, Cl^-

(e) $H_2O_2(aq) + 2e^- \rightarrow O_2(g) + 2H^+(aq)$

Since the reaction is in base, the H^+ can be "neutralized" by adding $2OH^-$ to each side of the equation to give $H_2O_2(aq) + 2OH^-(aq) \rightarrow O_2(g) + 2H_2O(l) + 2e^-$
The other half reaction is $2[ClO_2(aq) + e^- \rightarrow ClO_2^-(aq)]$.

Net: $H_2O_2(aq) + 2ClO_2(aq) + 2OH^-(aq) \rightarrow O_2(g) + 2ClO_2^-(aq) + 2H_2O(l)$

oxidizing agent, ClO_2; reducing agent, H_2O_2

(f) $4[H_2O_2(aq) + 2OH^-(aq) \rightarrow O_2(g) + 2H_2O(l) + 2e^-]$

$Cl_2O_7(aq) + 3H_2O(l) + 8e^- \rightarrow 2ClO_2^-(aq) + 6OH^-(aq)$

Net: $Cl_2O_7(aq) + 4H_2O_2(aq) + 2OH^-(aq) \rightarrow 2ClO_2^-(aq) + 4O_2(g) + 5H_2O(l)$

oxidizing agent, Cl_2O_7; reducing agent, H_2O_2

Voltaic Cells; Cell Potential

20.11 (a) The reaction $Cu^{2+}(aq) + Zn(s) \rightarrow Cu(s) + Zn^{2+}(aq)$ is occurring in both Figures. In Figure 20.3, the reactants are in contact, and the concentrations of the ions in solution aren't specified. In Figure 20.4, the oxidation half-reaction and reduction half-reaction are occurring in separate compartments, joined by a porous connector. The concentrations of the two solutions are initially 1.0 *M*. In Figure 20.4, electrical current is isolated and flows through the voltmeter. In Figure 20.3, the flow of electrons cannot be isolated or utilized.

(b) In the cathode compartment of the voltaic cell in Figure 20.5, Cu^{2+} cations are reduced to Cu atoms, decreasing the number of positively charged particles in the compartment. Na^+ cations are drawn into the compartment to maintain charge balance as Cu^{2+} ions are removed.

20.12 (a) The porous glass dish in Figure 20.4 provides a mechanism by which ions not directly involved in the redox reaction can migrate into the anode and cathode compartments to maintain charge neutrality of the solutions. Ionic conduction within the cell, through the glass disk, completes the cell circuit.

(b) In the anode compartment of Figure 20.5, Zn atoms are oxidized to Zn^{2+} cations, increasing the number of positively charged particles in the compartment. NO_3^- anions migrate into the compartment to maintain charge balance as Zn^{2+} ions are produced.

20.13 *Analyze/Plan*. Follow the logic in Sample Exercise 20.4. *Solve*:

(a) $Ag^+(aq) + 1e^- \rightarrow Ag(s)$; $Fe(s) \rightarrow Fe^{2+}(aq) + 2e^-$

(b) Fe(s) is the anode, Ag(s) is the cathode.

(c) Fe(s) is negative; Ag(s) is positive.

(d) Electrons flow from the Fe(-) electrode toward the Ag(+) electrode.

(e) Cations migrate toward the Ag(s) cathode; anions migrate toward the Fe(s) anode.

20.14 (a) $Al(s) \rightarrow Al^{3+}(aq) + 3e^-$; $Ni^{2+}(aq) + 2e^- \rightarrow Ni(s)$

(b) Al(s) is the anode; Ni(s) is the cathode.

(c) Al(s) is negative (-); Ni(s) is positive (+).

(d) Electrons flow from the Al(-) electrode toward the Ni(+) electrode.

(e) Cations migrate toward the Ni(s) cathode; anions migrate toward the Al(s) anode.

20.15 (a) *Electromotive force*, emf, is the driving force that causes electrons to flow through the external circuit of a voltaic cell. It is the potential energy difference between an electron at the anode and an electron at the cathode.

(b) One *volt* is the potential energy difference required to impart 1 J of energy to a charge of 1 coulomb. 1 V = 1 J/C.

(c) *Cell potential*, E_{cell}, is the emf of an electrochemical cell.

20.16 (a) The Zn electrode, the anode, on the right of figure 20.4 has the higher potential energy for electrons.

(b) The units of electrical potential are volts. A potential of one volt imparts one joule of energy to one coulomb of charge.

(c) A *standard* cell potential describes the potential of an electrochemical cell where all components are present at standard conditions: elements in their standard states, gases at 1 atm pressure and 1 *M* aqueous solutions.

20.17 (a) $2H^+(aq) + 2e^- \rightarrow H_2(g)$

(b) A *standard* hydrogen electrode is a hydrogen electrode where the components are at standard conditions, 1 *M* $H^+(aq)$ and $H_2(g)$ at 1 atm.

(c) The platinum foil in an SHE serves as an inert electron carrier and a solid reaction surface.

20.18 (a) $H_2(g) \rightarrow 2H^+(aq) + 2e^-$

(b) The platinum electrode serves as a reaction surface; the greater the surface area, the more H_2 or H^+ that can be adsorbed onto the surface to facilitate the flow of electrons.

(c)

20.19 (a) A *standard reduction potential* is the relative potential of a reduction half-reaction measured at standard conditions, 1 *M* aqueous solutions and 1 atm gas pressure.

(b) E°_{red} = 0 V for a standard hydrogen electrode.

(c) The reduction of $Ag^+(aq)$ to Ag(s) is much more energetically favorable, because it has a substantially more positive E°_{red} (0.799 V) than the reduction of $Sn^{2+}(aq)$ to Sn(s) (-0.136 V).

20.20 (a) It is not possible to measure the standard reduction potential of a single half-reaction because each voltaic cell consists of two half-reactions and only the potential of a complete cell can be measured.

(b) The standard reduction potential of a half-reaction is determined by combining it with a reference half-reaction of known potential and measuring the cell potential. Assuming the half-reaction of interest is the reduction half-reaction:

$E^\circ_{cell} = E^\circ_{red}(\text{cathode}) - E^\circ_{red}(\text{anode}) = E^\circ_{red}(\text{unknown}) - E^\circ_{red}(\text{reference})$;

$E^\circ_{red}(\text{unknown}) = E^\circ_{cell} + E^\circ_{red}$ (reference).

(c) $Cd^{2+}(aq) + 2e^- \rightarrow Cd(s)$ $E° = -0.403$ V

$Ca^{2+}(aq) + 2e^- \rightarrow Ca(s)$ $E° = -2.87$ V

The reduction of $Ca^{2+}(aq)$ to Ca(s) is the more energetically unfavorable reduction because it has a more negative E° value.

20.21 *Analyze/Plan.* Follow the logic in Sample Exercise 20.5. *Solve*:

(a) The two half-reactions are:

$Tl^{3+}(aq) + 2e^- \rightarrow Tl^+(aq)$ cathode $E^\circ_{red} = ?$

$2[Cr^{2+}(aq) \rightarrow Cr^{3+}(aq) + e^-]$ anode $E^\circ_{red} = -0.41$

(b) $E^\circ_{cell} = E^\circ_{red}$ (cathode) $- E^\circ_{red}$ (anode); $1.19\text{ V} = E^\circ_{red} - (-0.41\text{ V})$;

$E^\circ_{red} = 1.19\text{ V} - 0.41\text{ V} = 0.78\text{ V}$

(c)

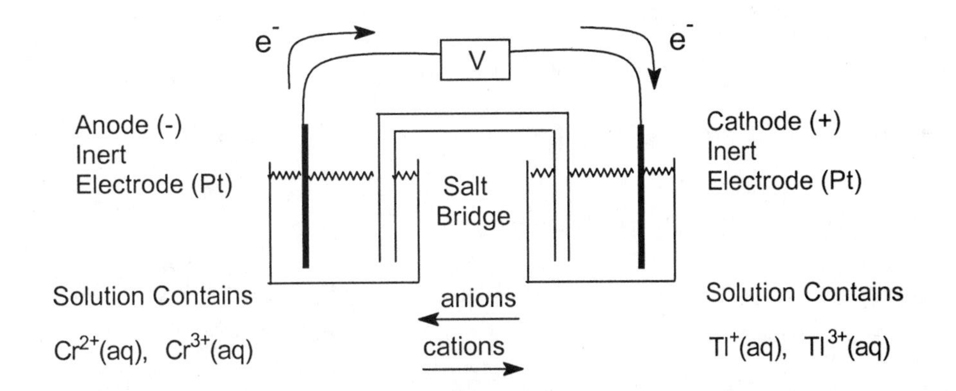

Note that because $Cr^{2+}(aq)$ is readily oxidized, it would be necessary to keep oxygen out of the left-hand cell compartment.

20.22 (a) $PdCl_4^{2-}(aq) + 2e^- \rightarrow Pd(s) + 4Cl^-$ cathode $E^\circ_{red} = ?$

$Cd(s) \rightarrow Cd^{2+}(aq) + 2e^-$ anode $E^\circ_{red} = -0.403$ V

(b) $E^\circ_{cell} = E^\circ_{red}$ (cathode) $- E^\circ_{red}$ (anode); $1.03\text{ V} = E^\circ_{red} - (-0.403\text{ V})$;

$E^\circ_{red} = 1.03\text{ V} - 0.403 = 0.63\text{ V}$

(c)

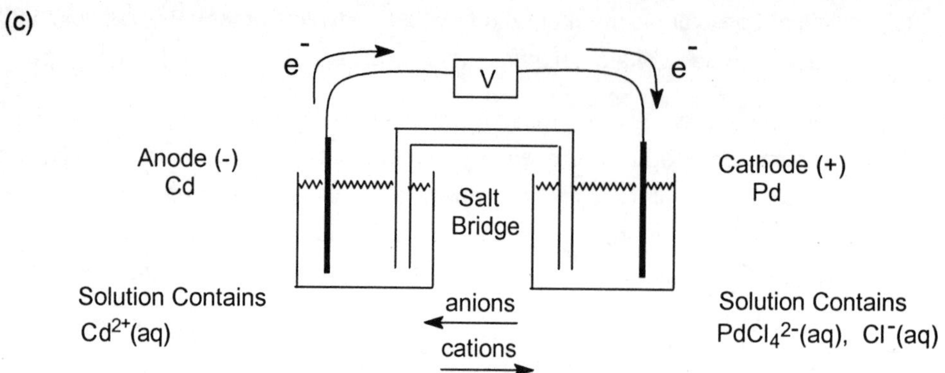

20.23 *Analyze/Plan.* Follow the logic in Sample Exercise 20.6. *Solve:*

a) $Cl_2(g) \rightarrow 2Cl^-(aq) + 2e^-$ $E^\circ_{red} = 1.359$ V

$I_2(s) + 2e^- \rightarrow 2I^-(aq)$ $E^\circ_{red} = 0.536$ V

$E^\circ = 1.359$ V - 0.536 V = 0.823 V

(b) $Ni(s) \rightarrow Ni^{2+}(aq) + 2e^-$ $E^\circ_{red} = -0.28$ V

$2[Ce^{4+}(aq) + 1e^- \rightarrow Ce^{3+}(aq)]$ $E^\circ_{red} = 1.61$ V

$E^\circ = 1.61$ V - (-0.28 V) = 1.89 V

(c) $Fe(s) \rightarrow Fe^{2+}(aq) + 2e^-$ $E^\circ_{red} = -0.440$ V

$2[Fe^{3+}(aq) + 1e^- \rightarrow Fe^{2+}(aq)]$ $E^\circ_{red} = +0.771$ V

$E^\circ = 0.771$ V - (-0.440 V) = 1.211 V

(d) $3[Ca(s) \rightarrow Ca^{2+}(aq) + 2e^-]$ $E^\circ_{red} = -2.87$ V

$2[Al^{3+}(aq) + 3e^- \rightarrow Al(s)]$ $E^\circ_{red} = -1.66$ V

$E^\circ = -1.66$ V - (-2.87 V)] = 1.21 V

20.24 (a) $F_2(g) + 2e^- \rightarrow 2F^-(aq)$ $E^\circ_{red} = 2.87$ V

$H_2(g) \rightarrow 2H^+(aq) + 2e^-$ $E^\circ_{red} = 0.00$ V

$E^\circ = 2.87$ V - 0.00 V = 2.87 V

(b) $Cu(s) \rightarrow Cu^{2+}(aq) + 2e^-$ $E^\circ_{red} = 0.337$ V

$Ba^{2+}(aq) + 2e^- \rightarrow Ba(s)$ $E^\circ_{red} = -2.90$ V

$E^\circ = -2.90$ V - (0.337 V) = -3.24 V

(c) $Fe^{2+}(aq) + 2e^- \rightarrow Fe(s)$ $E^\circ_{red} = -0.440$ V

$2[Fe^{2+}(aq) \rightarrow Fe^{3+}(aq) + 1e^-]$ $E^\circ_{red} = 0.771$ V

$E^\circ = -0.440$ V - 0.771 V = -1.211 V

(d) $Hg_2^{2+}(aq) + 2e^- \rightarrow 2Hg(l)$ $E^\circ_{red} = 0.789$ V

$2[Cu^+(aq) \rightarrow Cu^{2+}(aq) + 1e^-]$ $E^\circ_{red} = 0.153$ V

$E^\circ = 0.789$ V - 0.153 V = 0.636 V

20.25 *Analyze/Plan.* Given four half-reactions, find E°_{red} from Appendix E and combine them to obtain a desired E_{cell}. (a) The largest E_{cell} will combine the half-reaction with the most positive E°_{red} as the cathode reaction and the one with the most negative E°_{red} as the anode reaction. (b) The smallest positive E°_{cell} will combine two half-reactions whose E°_{red} values are closest in magnitude **and** sign. *Solve:*

(a) $3[Ag^+(aq) + 1e^- \rightarrow Ag(s)]$ $E^\circ_{red} = 0.799$

$Cr(s) \rightarrow Cr^{3+}(aq) + 3e^-$ $E^\circ_{red} = -0.74$

$3Ag^+(aq) + Cr(s) \rightarrow 3Ag(s) + Cr^{3+}(aq)$ $E^\circ = 0.799 - (-0.74) = 1.54$ V

(b) Two of the combinations have essentially equal E° values.

$2[Ag^+(aq) + 1e^- \rightarrow Ag(s)]$ $E^\circ_{red} = 0.799$ V

$Cu(s) \rightarrow Cu^{2+}(aq) + 2e^-$ $E^\circ_{red} = 0.337$ V

$2Ag^+(aq) + Cu(s) \rightarrow 2Ag(s) + Cu^{2+}(aq)$ $E^\circ = 0.799\text{ V} - 0.337\text{ V} = 0.462$ V

$3[Ni^{2+}(aq) + 2e^- \rightarrow Ni(s)]$ $E^\circ_{red} = -0.28$ V

$2[Cr(s) \rightarrow Cr^{3+}(aq) + 3e^-]$ $E^\circ_{red} = -0.74$ V

$3Ni^{2+}(aq) + 2Cr(s) \rightarrow 3Ni(s) + 2Cr^{3+}(aq)$ $E^\circ = -0.28\text{ V} - (-0.74\text{ V}) = 0.46$ V

20.26 (a) $2[Au(s) + 4Br^-(aq) \rightarrow AuBr_4^-(aq) + 3e^-]$ $E^\circ_{red} = -0.858$ V

$3[2e^- + IO^-(aq) + H_2O(l) \rightarrow I^-(aq) + 2OH^-(aq)]$ $E^\circ_{red} = 0.49$ V

$2Au(s) + 8Br^-(aq) + 3IO^-(aq) + 3H_2O(l) \rightarrow 2AuBr_4^-(aq) + 3I^-(aq) + 6OH^-(aq)$

$E^\circ = 0.49 - (-0.858) = 1.35$ V

(b) $2[Eu^{2+}(aq) \rightarrow Eu^{3+}(aq) + 1e^-]$ $E^\circ_{red} = -0.43$ V

$Sn^{2+}(aq) + 2e^- \rightarrow Sn(s)$ $E^\circ_{red} = -0.14$ V

$2Eu^{2+}(aq) + Sn^{2+}(aq) \rightarrow 2Eu^{3+}(aq) + Sn(s)$ $E^\circ = -0.14 -(-0.43) = 0.29$ V

20.27 *Analyze/Plan.* follow the logic in Sample Exercise 20.7. *Solve:*

(a) $MnO_4^-(aq) + 8H^+(aq) + 5e^- \rightarrow Mn^{2+}(aq) + 4H_2O(l)$ $E^\circ_{red} = 1.51$ V

(b) Because the half-reaction in part (a) is the more favorable reduction, it is the cathode reaction.

(c) $Sn^{2+}(aq) \rightarrow Sn^{4+}(aq) + 2e^-$ $E^\circ_{red} = 0.154$ V

(d) Balance electrons by multiplying the cathode reaction by 2 and the anode reaction by 5. $5Sn^{2+}(aq) + 2MnO_4^-(aq) + 16H^+(aq) \rightarrow 5Sn^{4+}(aq) + 2Mn^{2+}(aq) + 8H_2O(l)$

(e) $E° = 1.51\ V - 0.154\ V = 1.356 = 1.36\ V$

20.28 (a) The half-reactions are:

$2H^+(aq) + 2e^- \rightarrow H_2(g)$ $E^°_{red} = 0.00\ V$

$Al(s) \rightarrow Al^{3+}(aq) + 3e^-$ $E^°_{red} = -1.66\ V$

Because it has the larger $E^°_{red}$ (0.00 V vs -1.66 V), the first half-reaction is the reduction half-reaction in a voltaic cell. The standard hydrogen electrode (SHE) is the cathode and the Al strip is the anode.

(b) The Al strip will lose mass as the reaction proceeds, because Al(s) is transformed to $Al^{3+}(aq)$.

(c) Balance electrons by multiplying the cathode reaction by 3 and the anode reaction by 2. $6H^+(aq) + 2Al(s) \rightarrow 3H_2(g) + 2Al^{3+}(aq)$

(d) $E° = 0.00\ V - (-1.66\ V) = 1.66\ V$

20.29 *Analyze/Plan.* Given the description of a voltaic cell, answer questions about this cell. Combine ideas in Sample Exercises 20.4 and 20.7. The reduction half-reactions are:

$Cu^{2+}(aq) + 2e^- \rightarrow Cu(s)$ $E° = 0.337\ V$

$Sn^{2+}(aq) + 2e^- \rightarrow Sn(s)$ $E° = -0.136\ V$

Solve:

(a) It is evident that Cu^{2+} is more readily reduced. Therefore, Cu serves as the cathode, Sn as the anode.

(b) The copper electrode gains mass as Cu is plated out, the Sn electrode loses mass as Sn is oxidized.

(c) The overall cell reaction is $Cu^{2+}(aq) + Sn(s) \rightarrow Cu(s) + Sn^{2+}(aq)$

(d) $E° = 0.337\ V - (-0.136\ V) = 0.473\ V$

20.30 (a) The two half-reactions are:

$Pb^{2+}(aq) + 2e^- \rightarrow Pb(s)$ $E° = -0.126\ V$

$Cl_2(g) + 2e^- \rightarrow 2Cl^-(aq)$ $E° = 1.359\ V$

Because E° for the reduction of Cl_2 is higher, the reduction of Cl_2 occurs at the Pt cathode. The Pb electrode is the anode.

(b) The Pb anode loses mass as $Pb^{2+}(aq)$ is produced.

(c) $Cl_2(g) + Pb(s) \rightarrow Pb^{2+}(aq) + 2Cl^-(aq)$

(d) $E° = 1.359\ V - (-0.126\ V) = 1.485\ V$

Oxidizing and Reducing Agents; Spontaneity

20.31 *Analyze/Plan.* Use the definitions of oxidizing agent, reducing agent and the convention for writing reduction half-reactions to answer the stated questions. *Solve:*

(a) Negative. A strong reductant is likely to be oxidized, thus having a negative reduction potential.

(b) Right. Reducing agents are likely to be oxidized, and thus to be in a low oxidation state; the products of reduction half-reactions are in lower oxidation states than reactants.

20.32 (a) Top. The reduction half-reactions near the top of Table 20.1 are most likely to occur; a strong oxidant is most likely to be reduced.

(b) Left. An oxidant is reduced, so it is a reactant in a reduction half-reaction.

20.33 *Analyze/Plan.* Follow the logic in Sample Exercise 20.8. In each case, choose the half-reaction with the more positive reduction potential and with the given substance on the left. *Solve:*

(a) $Cl_2(g)$ (1.359 V vs. 1.065 V) (b) $Ni^{2+}(aq)$ (-0.28V vs. -0.403 V)

(c) $BrO_3^-(aq)$ (1.52 V vs. 1.195 V) (d) $O_3(g)$ (2.07 V vs. 1.776 V)

20.34 The more readily a substance is oxidized, the stronger it is as a reducing agent. In each case choose the half-reaction with the more negative reduction potential and the given substance on the right.

(a) Mg(s) (-2.37 V vs. -0.44 V) (b) Ca(s) (-2.87 V vs. -1.66 V)

(c) H_2(g, acidic) (0.000 V vs. 0.141 V) (d) $H_2C_2O_4(aq)$ (-0.49 V vs. 0.17 V)

20.35 *Analyze/Plan.* If the substance is on the left of a reduction half-reaction, it will be an oxidant; if it is on the right, it will be a reductant. The sign and magnitude of the E°_{red} determines whether it is strong or weak. *Solve:*

(a) $Cl_2(aq)$: strong oxidant (on the left, E°_{red} = 1.359 V)

(b) MnO_4^-(aq, acidic): strong oxidant (on the left, E°_{red} = 1.51 V)

(c) Ba(s): strong reductant (on the right, E°_{red} = -2.90 V)

(d) Zn(s): reductant (on the right, E°_{red} = -0.763 V)

20.36 If the substance is on the left of a reduction half-reaction, it will be an oxidant; if it is on the right, it will be a reductant. The sign and magnitude of the E°_{red} determines whether it is strong or weak.

(a) Na(s): strong reductant (on the right, E°_{red} = -2.71 V)

(b) $O_3(g)$: strong oxidant (on the left, E°_{red} = 2.07 V)

(c) $Ce^{3+}(aq)$: very weak reductant (on the right, E°_{red} = 1.61 V)

(d) $Sn^{2+}(aq)$: reductant (on the right, E°_{red} = 0.154 V) **or** weak oxidant (on the left, -0.136 V)

20.37 *Analyze/Plan.* Follow the logic in Sample Exercise 20.8. *Solve:*

(a) Arranged in order of increasing strength as oxidizing agents (and increasing reduction potential):

$Cu^{2+}(aq) < O_2(g) < Cr_2O_7^{2-}(aq) < Cl_2(g) < H_2O_2(aq)$

(b) Arranged in order of increasing strength as reducing agents (and decreasing reduction potential):

$H_2O_2(aq) < I^-(aq) < Sn^{2+}(aq) < Zn(s) < Al(s)$

20.38 (a) The strongest oxidizing agent is the species most readily reduced, as evidenced by a large, positive reduction potential. That species is H_2O_2. The weakest oxidizing agent is the species that least readily accepts an electron. We expect that it will be very difficult to reduce Zn(s); indeed, Zn(s) acts as a comparatively strong **reducing** agent. No potential is listed for reduction of Zn(s), but we can safely assume that it is less readily reduced than any of the other species present.

(b) The strongest reducing agent is the species most easily oxidized (the largest negative reduction potential). Zn, E°_{red} = -0.76 V, is the strongest reducing agent and F^-, E°_{red} = 2.87 V, is the weakest.

20.39 *Analyze/Plan.* In order to reduce Eu^{3+} to Eu^{2+}, we need an oxidizing agent, one of the reduced species from Table 20.1 or Appendix E. It must have a greater tendency to be oxidized than Eu^{3+} has to be reduced. That is, E°_{red} must be more negative than -0.43 V. *Solve:*

Any of the **reduced** species in Table 20.1 or Appendix E from a half-reaction with a reduction potential more negative than -0.43 V will reduce Eu^{3+} to Eu^{2+}. From the list of possible reductants in the Exercise, Al and $H_2C_2O_4$ will reduce Eu^{3+} to Eu^{2+}.

20.40 Any oxidized species from Table 20.1 or Appendix E with a reduction potential greater than 0.59 V will oxidize RuO_4^{2-} to RuO_4^-. From the list of possible oxidants in the Exercise, $Cr_2O_7^{2-}(aq)$ and $ClO^-(aq)$ will oxidize RuO_4^{2-} to RuO_4^-.

20.41 *Analyze/Plan.* Follow the logic in Sample Exercises 20.9 and 20.10. *Solve:*

(a) The more positive the emf of a reaction the more spontaneous the reaction.

(b) Reactions (a), (b), (c) and (d) in Exercise 20.23 have positive E° values and are spontaneous.

(c) $\Delta G^\circ = -nFE^\circ$; F = 96,500 J/V•mol e^- = 96.5 kJ/V•mol e^-

20.23 (a) $\Delta G^\circ = -2 \text{ mol } e^- \times \dfrac{96.5 \text{ kJ}}{\text{V}\bullet\text{mol } e^-} \times 0.823 \text{ V} = -158.839 = -159 \text{ kJ}$

20.23 (b) $\Delta G^\circ = -2(96.5)(1.89) = -364.77 = -365$ kJ

20.23 (c) $\Delta G^\circ = -2(96.5)(1.211) = -233.72 = -234$ kJ

20.23 (d) $\Delta G^\circ = -6(96.5)(1.21) = -700.59 = -701$ kJ

20.42 (a) $\Delta G = -nFE$. The more positive the emf of a reaction, the more negative the value of ΔG.

(b) Reactions (a) and (d) in Exercise 20.24 have positive E° values and are spontaneous.

(c) $\Delta G^\circ = -nFE^\circ$; F = 96,500 J/V•mol e^- = 96.5 kJ/V•mol e^-

20.24 (a) $\Delta G^\circ = -2 \text{ mol } e^- \times \dfrac{96.5 \text{ kJ}}{\text{V}\bullet\text{mol } e^-} \times 2.87 \text{ V} = -553.91 = -554 \text{ kJ}$

20.24 (b) $\Delta G^\circ = -2(96.5)(-3.24) = 625.32 = 625$ kJ

20.24 (c) $\Delta G^\circ = -2(96.5)(-1.21) = 233.53 = 234$ kJ

20.24 (d) $\Delta G^\circ = -2(96.5)(0.636) = -122.75 = -123$ kJ

20.43 *Analyze/Plan.* In each reaction, $Fe^{2+} \rightarrow Fe^{3+}$ will be the oxidation half-reaction and one of the other given half-reactions will be the reduction half-reaction. Follow the logic in Sample Exercise 20.10 to calculate E° and ΔG° for each reaction. *Solve:*

(a) $2Fe^{2+}(aq) + S_2O_6^{2-}(aq) + 4H^+(aq) \rightarrow 2Fe^{3+}(aq) + 2H_2SO_3(aq)$

$E^\circ = 0.60 \text{ V} - 0.77 \text{ V} = -0.17 \text{ V}$

$2Fe^{2+}(aq) + N_2O(aq) + 2H^+(aq) \rightarrow 2Fe^{3+}(aq) + N_2(g) + H_2O(l)$

$E^\circ = -1.77 \text{ V} - 0.77 \text{ V} = -2.54 \text{ V}$

$Fe^{2+}(aq) + VO_2^+(aq) + 2H^+(aq) \rightarrow Fe^{3+}(aq) + VO^{2+}(aq) + H_2O(l)$

$E^\circ = 1.00 \text{ V} - 0.77 \text{ V} = +0.23 \text{ V}$

(b) $\Delta G^\circ = -nFE^\circ$ For the first reaction,

$\Delta G^\circ = -2 \text{ mol} \times \dfrac{96{,}500 \text{ J}}{1 \text{ V}\bullet\text{mol}} \times (-0.17 \text{ V}) = 3.3 \times 10^5 \text{ J or } 33 \text{ kJ}$

For the second reaction, $\Delta G^\circ = -2(96{,}500)(-2.54) = 4.90 \times 10^2$ kJ

For the third reaction, $\Delta G^\circ = -1(96{,}500)(0.23) = -22$ kJ

20.44 (a)

$2I^-(aq) \rightarrow I_2(s) + 2e^-$ $\quad E^\circ_{red} = 0.536$ V

$Hg_2^{2+}(aq) + 2e^- \rightarrow 2Hg(l)$ $\quad E^\circ_{red} = 0.789$ V

$2I^-(aq) + Hg_2^{2+}(aq) \rightarrow I_2(s) + 2Hg(l)$ $\quad E^\circ = 0.789 - 0.536 = 0.253$ V

$\Delta G^\circ = -nFE^\circ = -2 \text{ mol } e^- \times \dfrac{96.5 \text{ kJ}}{\text{V}\bullet\text{mol } e^-} \times 0.253 \text{ V} = -48.8 \text{ kJ}$

(b)

$$3[Cu^+(aq) \rightarrow Cu^{2+}(aq) + 1e^-] \qquad E^\circ_{red} = 0.153 \text{ V}$$
$$NO_3^-(aq) + 4H^+(aq) + 3e^- \rightarrow NO(g) + H_2O(l) \qquad E^\circ_{red} = 0.96 \text{ V}$$

$$3Cu^+(aq) + NO_3^-(aq) + 4H^+(aq) \rightarrow 3Cu^{2+}(aq) + NO(g) + 2H_2O(l)$$

$E° = 0.96 - 0.153 = 0.81$ V; $\Delta G° = -3(96.5)(0.81) = -2.3 \times 10^2$ kJ

(c)

$$2[Cr(OH)_3(s) + 5OH^-(aq) \rightarrow CrO_4^{2-}(aq) + 4H_2O(l) + 3e^-] \qquad E^\circ_{red} = -0.13 \text{ V}$$
$$[ClO^-(aq) + H_2O(l) + 2e^- \rightarrow Cl^-(aq) + 2OH^-(aq)] \qquad E^\circ_{red} = 0.89 \text{ V}$$

$$2Cr(OH)_3(s) + 3ClO^-(aq) + 4OH^-(aq) \rightarrow 2CrO_4^{2-}(aq) + 3Cl^-(aq) + 5H_2O(l)$$

$E° = 0.89 - (-0.13) = 1.02$ V; $\Delta G° = -6(96.5)(1.02) = -591$ kJ

EMF and Concentration

20.45 (a) The *Nernst equation* is applicable when the components of an electrochemical cell are at nonstandard conditions.

(b) Q = 1 if all reactants and products are at standard conditions.

(c) If concentration of reactants increases, Q decreases, and E increases.

20.46 (a) No. As the spontaneous chemical reaction of the voltaic cell proceeds, the concentrations of products increase and the concentrations of reactants decrease, so standard conditions are not maintained.

(b) $$0.0592 = 2.303 \frac{RT}{F} = 2.303 \times \frac{8.314 \text{ J}}{\text{mol} \bullet \text{K}} \times 298 \text{ K} \times \frac{\text{V} \bullet \text{mol}}{96{,}500 \text{ J}}$$

The "2.303" is a factor to change natural logs into base-10 logs. The "0.0592 V" assumes the cell operates at 298 K and that base-10 logs are used.

(c) If concentration of products increases, Q increases, and E decreases.

20.47 *Analyze/Plan.* Given a circumstance, determine its effect on cell emf. Each circumstance changes the value of Q. An increase in Q reduces emf; a decrease in Q increases emf. *Solve*:

$$Zn(s) + 2H^+(aq) \rightarrow Zn^{2+}(aq) + H_2(g); \; E = E° - \frac{0.0592}{n} \log Q; \; Q = \frac{[Zn^{2+}] P_{H_2}}{[H^+]^2}$$

(a) P_{H_2} increases, Q increases, E decreases

(b) $[Zn^{2+}]$ increases, Q increases, E decreases

(c) $[H^+]$ decreases, Q increases, E decreases

(d) No effect; does not appear in the Nernst equation

20.48 $$Al(s) + 3Ag^+(aq) \rightarrow Al^{3+}(aq) + 3Ag(s); \; E = E° - \frac{0.0592}{n} \log Q; \; Q = \frac{[Al^{3+}]}{[Ag^+]^3}$$

Any change that causes the reaction to be less spontaneous (that causes Q to increase and ultimately shifts the equilibrium to the left) will result in a less positive value for E°.

(a) Increases E by decreasing $[Al^{3+}]$ on the right side of the equation.

(b) No effect; the "concentrations" of pure solids and liquids do not influence the value of K_{eq} for a heterogeneous equilibrium.

(c) No effect; the concentration of Ag^+ and the value of Q are unchanged.

(d) Decreases E; forming AgCl(s) decreases the concentration of Ag^+, which increases Q.

20.49 *Analyze/Plan.* Follow the logic in Sample Exercise 20.11. *Solve:*

(a)

$$Ni^{2+}(aq) + 2e^- \rightarrow Ni(s) \qquad E^\circ_{red} = -0.28\ V$$

$$Zn(s) \rightarrow Zn^{2+}(aq) + 2e^- \qquad E^\circ_{red} = -0.763\ V$$

$$Ni^{2+}(aq) + Zn(s) \rightarrow Ni(s) + Zn^{2+}(aq) \qquad E^\circ = -0.28 - (-0.763) = 0.483 = 0.48\ V$$

(b)

$$E = E^\circ - \frac{0.0592}{n} \log \frac{[Zn^{2+}]}{[Ni^{2+}]};\ n = 2$$

$$E = 0.483 - \frac{0.0592}{2} \log \frac{(0.100)}{(3.00)} = 0.483 - \frac{0.0592}{2} \log (0.0333)$$

$$E = 0.483 - \frac{0.0592\,(-1.477)}{2} = 0.483 + 0.0437 = 0.527 = 0.53\ V$$

(c)

$$E = 0.483 - \frac{0.0592}{2} \log \frac{(0.900)}{(0.200)} = 0.483 - 0.0193 = 0.464 = 0.46\ V$$

20.50 (a)

$$3[Ce^{4+}(aq) + 1e^- \rightarrow Ce^{3+}(aq)] \qquad E^\circ_{red} = 1.61\ V$$

$$Cr(s) \rightarrow Cr^{3+}(aq) + 3e^- \qquad E^\circ_{red} = -0.74\ V$$

$$3Ce^{4+}(aq) + Cr(s) \rightarrow 3Ce^{3+}(aq) + Cr^{3+}(aq) \qquad E^\circ = 1.61 - (-0.74) = 2.35\ V$$

(b)

$$E = E^\circ - \frac{0.0592}{n} \log \frac{[Ce^{3+}]^3 [Cr^{3+}]}{[Ce^{4+}]^3};\ n = 3$$

$$E = 2.35 - \frac{0.0592}{3} \log \frac{(0.010)^3 (0.010)}{(2.0)^3} = 2.35 - \frac{0.0592}{3} \log (1.250 \times 10^{-9})$$

$$E = 2.35 - \frac{0.0592\,(-8.903)}{3} = 2.35 + 0.176 = 2.53\ V$$

(c)

$$E = 2.35 - \frac{0.0592}{3} \log \frac{(0.85)^3 (1.2)}{(0.35)^3} = 2.35 - 0.0244 = 2.33\ V$$

20.51 *Analyze/Plan.* Follow the logic in Sample Exercise 20.11. *Solve:*

(a)

$$4[Fe^{2+}(aq) \rightarrow Fe^{3+}(aq) + 1e^-] \qquad E^\circ_{red} = 0.771\ V$$

$$O_2(g) + 4H^+(aq) + 4e^- \rightarrow 2H_2O(l) \qquad E^\circ_{red} = 1.23\ V$$

$$4Fe^{2+}(aq) + O_2(g) + 4H^+(aq) \rightarrow 4Fe^{3+}(aq) + 2H_2O(l) \qquad E^\circ = 1.23 - 0.771 = 0.459 = 0.46\ V$$

(b)

$$E = E^\circ - \frac{0.0592}{n}\log\frac{[Fe^{3+}]^4}{[Fe^{2+}]^4[H^+]^4 P_{O_2}};\ n = 4,\ [H^+] = 1.00 \times 10^{-3}\ M$$

$$E = 0.459\ V - \frac{0.0592}{4}\log\frac{(0.010)^4}{(3.0)^4(1.0 \times 10^{-3})^4(0.50)} = 0.459 - \frac{0.0592}{4}\log(246.9)$$

$$E = 0.459 - \frac{0.0592}{4}(2.393) = 0.459 - 0.0354 = 0.4236 = 0.42\ V$$

20.52 (a)

$$2[Fe^{3+}(aq) + 1\ e^- \rightarrow Fe^{2+}(aq)] \qquad E^\circ_{red} = 0.771\ V$$

$$H_2(g) \rightarrow 2H^+(aq) + 2e^- \qquad E^\circ_{red} = 0.000\ V$$

$$2Fe^{3+}(aq) + H_2(g) \rightarrow 2Fe^{2+}(aq) + 2H^+(aq) \qquad E^\circ = 0.771 - 0.000 = 0.771\ V$$

(b)

$$E = E^\circ - \frac{0.0592}{n}\log\frac{[Fe^{2+}]^2[H^+]^2}{[Fe^{3+}]^2 P_{H_2}};\ [H^+] = 10^{-pH} = 1.0 \times 10^{-5},\ n = 2$$

$$E = 0.771 - \frac{0.0592}{2}\log\frac{(0.0010)^2(1.0 \times 10^{-5})^2}{(1.50)^2(0.50)} = 0.771 - \frac{0.0592}{2}\log(8.9 \times 10^{-17})$$

$$E = 0.771 - \frac{0.0592(-16.05)}{2} = 0.771 + 0.475 = 1.246\ V$$

20.53 *Analyze/Plan.* We are given a concentration cell with Zn electrodes. Use the definition of a concentration cell in Section 20.6 to answer the stated questions. Use Equation 20.16 to calculate the cell emf. For a concentration cell, Q = [dilute]/[concentrated]. *Solve:*

(a) The compartment with the more dilute solution will be the anode. That is, the compartment with $[Zn^{2+}] = 1.00 \times 10^{-2}\ M$ is the anode.

(b) Since the oxidation half-reaction is the opposite of the reduction half-reaction, E° is zero.

(c)

$$E = E^\circ - \frac{0.0592}{n}\log Q;\quad Q = [Zn^{2+}, \text{dilute}] / [Zn^{2+}, \text{conc.}]$$

$$E = 0 - \frac{0.0592}{2}\log\frac{(1.00 \times 10^{-2})}{(5.00)} = 0.0799\ V$$

(d) In the anode compartment, $Zn(s) \rightarrow Zn^{2+}(aq)$, so $[Zn^{2+}]$ increases from $1.00 \times 10^{-2}\ M$. In the cathode compartment, $Zn^{2+}(aq) \rightarrow Zn(s)$, so $[Zn^{2+}]$ decreases from 5.00 *M*.

20.54 (a) The compartment with 0.0150 *M* $Cl^-(aq)$ is the cathode.

(b) E° = 0 V

(c) $E = E° - \frac{0.0592}{n} \log Q$; $Q = [Cl^-, \text{dilute}] / [Cl^-, \text{conc.}]$

$$E = 0 - \frac{0.0592}{1} \log \frac{(0.0150)}{(2.55)} = 0.132 \text{ V}$$

(d) In the anode compartment, $[Cl^-]$ will decrease from 2.55 *M*. In the cathode, $[Cl^-]$ will increase from 0.0150 *M*.

20.55 *Analyze/Plan*. Follow the logic in Sample Exercise 20.12. *Solve*:

$$E = E° - \frac{0.0592}{2} \log \frac{[P_{H_2}][Zn^{2+}]}{[H^+]^2}; \; E° = 0.0 \text{ V} - (-0.763 \text{ V}) = 0.763 \text{ V}$$

$$0.684 = 0.763 - \frac{0.0592}{2} \times (\log [P_{H_2}][Zn^{2+}] - 2 \log [H^+]) = 0.763 - \frac{0.0592}{2} \times (-0.5686 - 2 \log [H^+])$$

$$0.684 = 0.763 + 0.0168 + 0.0592 \log [H^+]; \; \log [H^+] = \frac{0.684 - 0.0168 - 0.763}{0.0592}$$

$\log [H^+] = -1.6188 = -1.6$; $[H^+] = 0.0241 = 0.02$ *M*; pH = 1.6

20.56 (a) $E° = -0.136 \text{ V} - (-0.126 \text{ V}) = -0.010 \text{ V}$; $n = 2$

$$0.22 = -0.010 - \frac{0.0592}{2} \log \frac{[Pb^{2+}]}{[Sn^{2+}]} = -0.010 - \frac{0.0592}{2} \log \frac{[Pb^{2+}]}{1.00}$$

$$\log [Pb^{2+}] = \frac{-0.23\,(2)}{0.0592} = -7.770 = -7.8; \; [Pb^{2+}] = 1.7 \times 10^{-8} = 2 \times 10^{-8} \; M$$

(b) For $PbSO_4(s)$, $K_{sp} = [Pb^{2+}][SO_4^{2-}] = (1.0)(1.7 \times 10^{-8}) = 1.7 \times 10^{-8}$

20.57 *Analyze/Plan*. Follow the logic in Sample Exercise 20.14. $E° = \frac{0.0592 \text{ V}}{n} \log K_{eq}$; $\log K_{eq} = \frac{nE°}{0.0592 \text{ V}}$. *Solve*:

(a) $E° = -0.28 - (-0.440) = 0.16$ V, $n = 2$ ($Ni^{2+} + 2e^- \rightarrow Ni$)

$$\log K_{eq} = \frac{2(0.16)}{0.0592} = 5.4054 = 5.4; \; K_{eq} = 2.54 \times 10^5 = 3 \times 10^5$$

(b) $E° = 0 - (-0.277) = 0.277$ V; $n = 2$ ($2H^+ + 2e^- \rightarrow H_2$)

$$\log K_{eq} = \frac{2(0.277)}{0.0592} = 9.358 = 9.36; \; K_{eq} = 2.3 \times 10^9$$

(c) $E° = 1.51 - 1.065 = 0.445 = 0.45$ V; $n = 10$ ($2MnO_4^- + 10e^- \rightarrow 2Mn^{+2}$)

$$\log K_{eq} = \frac{10(0.445)}{0.0592} = 75.169 \approx 75; \; K_{eq} = 1.5 \times 10^{75} = 10^{75}$$

20.58 $E° = \frac{0.0592 \text{ V}}{n} \log K_{eq}$; $\log K_{eq} = \frac{nE°}{0.0592 \text{ V}}$

(a) $E° = 1.00\ V - (-0.799\ V) = 0.201 = 0.20\ V$; $n = 2$ ($Ni^{+2} + 2e^- \rightarrow Ni$)

$$\log K_{eq} = \frac{2(0.201\ V)}{0.0592\ V} = 6.7905 = 6.8;\ K_{eq} = 6.17 \times 10^6 = 6 \times 10^6$$

(b) $E° = 1.61\ V - 0.32\ V = 1.29\ V$; $n = 3$ ($3Ce^{4+} + 3e^- \rightarrow 3Ce^{3+}$)

$$\log K_{eq} = \frac{3(1.29)}{0.0592} = 65.372 = 65.4;\ K_{eq} = 2.35 \times 10^{65} = 2 \times 10^{65}$$

(c) $E° = 0.36\ V - (-0.23\ V) = 0.59\ V$; $n = 4$ ($4Fe(CN)_6^{3-} + 4e^- \rightarrow 4Fe(CN)_6^{4-}$)

$$\log K_{eq} = \frac{4(0.59)}{0.0592} = 39.865 = 40;\ K_{eq} = 7 \times 10^{39} = 10^{40}$$

20.59 *Analyze/Plan.* Follow the logic in Sample Exercise 20.14. $E° = \frac{0.0592\ V}{n} \log K_{eq}$; $\log K_{eq} = \frac{nE°}{0.0592\ V}$. *Solve*:

(a) $$\log K_{eq} = \frac{1(0.177\ V)}{0.0592\ V} = 2.9899 = 2.99;\ K_{eq} = 9.8 \times 10^2$$

(b) $$\log K_{eq} = \frac{2(0.177\ V)}{0.0592\ V} = 5.9797 = 5.98;\ K_{eq} = 9.5 \times 10^5$$

(c) $$\log K_{eq} = \frac{3(0.177\ V)}{0.0592\ V} = 8.9696 = 8.97;\ K_{eq} = 9.32 \times 10^8 = 9.3 \times 10^8$$

20.60 $E° = \frac{0.0592\ V}{n} \log K_{eq}$; $n = \frac{0.0592\ V}{E°} \log K_{eq}$

$$n = \frac{0.0592\ V}{0.17\ V} \log 5.5 \times 10^5;\ n = 2$$

Batteries; Corrosion

20.61 (a) A *battery* is a portable, self-contained electrochemical power source composed of one or more voltaic cells.

(b) A *primary* battery is not rechargeable, while a *secondary* battery can be recharged.

(c) No. No single voltaic cell is capable of producing 7.5 V. If a single voltaic cell could be designed to produce 2.5 V, three of these cells connected in series would produce the desired voltage.

20.62 (a) The emf of a battery decreases as it is used. This happens because the concentrations of products increase and the concentrations of reactants decrease. According to the Nernst equation, these changes increase Q and decrease E_{cell}.

(b) The major difference between AA- and D-size batteries is the amount of reactants present. The additional reactants in a D-size battery enable it to provide power for a longer time.

20.63 *Analyze/Plan.* Given mass of a reactant (Pb), calculate mass of product (PbO_2). This is a stoichiometry problem; we need the balanced equation for the chemical reaction that occurs in the lead-acid battery. Then, g Pb → mol Pb → mol PbO_2 → g PbO_2. *Solve:*

The overall cell reaction (Equation [20.19]) is:

$$Pb(s) + PbO_2(s) + 2H^+(aq) + 2HSO_4^-(aq) \rightarrow 2PbSO_4(s) + 2H_2O(l)$$

$$382 \text{ g Pb} \times \frac{1 \text{ mol Pb}}{207.2 \text{ g Pb}} \times \frac{1 \text{ mol } PbO_2}{1 \text{ mol Pb}} \times \frac{239.2 \text{ g } PbO_2}{1 \text{ mol } PbO_2} = 441 \text{ g } PbO_2$$

20.64 The overall cell reaction is:

$$2MnO_2(s) + Zn(s) + 2H_2O(l) \rightarrow 2MnO(OH)(s) + Zn(OH)_2(s)$$

$$12.9 \text{ g Zn} \times \frac{1 \text{ mol Zn}}{65.39 \text{ g Zn}} \times \frac{2 \text{ mol } MnO_2}{1 \text{ mol Zn}} \times \frac{86.94 \text{ g } MnO_2}{1 \text{ mol } MnO_2} = 34.3 \text{ g } MnO_2 \text{ reduced}$$

20.65 *Analyze/Plan.* We are given a redox reaction and asked to write half-reactions, calculate E°, and indicate whether Li(s) is the anode or cathode. Determine which reactant is oxidized and which is reduced. Separate into half-reactions, find E°_{red} for the half-reactions from Appendix E and calculate E°. *Solve:*

(a) Li(s) is oxidized at the anode.

(b)

$$Ag_2CrO_4(s) + 2e^- \rightarrow 2Ag(s) + CrO_4^{2-}(aq) \qquad E^\circ_{red} = 0.446 \text{ V}$$

$$2[Li(s) \rightarrow Li^+(aq) + 1e^-] \qquad E^\circ_{red} = -3.05 \text{ V}$$

$$Ag_2CrO_4(s) + 2Li(s) \rightarrow 2Ag(s) + CrO_4^-(aq) + 2Li^+(aq)$$

$$E^\circ = 0.446 \text{ V} - (-3.05 \text{ V}) = 3.496 = 3.50 \text{ V}$$

(c) The emf of the battery, 3.5 V, is exactly the cell potential calculated in part (b).

20.66 (a) $HgO(s) + Zn(s) \rightarrow Hg(l) + ZnO(s)$

(b) $E^\circ_{cell} = E^\circ_{red}$ (cathode) - E°_{red} (anode); E°_{red} (anode) = $E^\circ_{red} - E^\circ_{cell} = 0.098 - 1.35$ $= -1.25$ V

(c) E°_{red} is different from $Zn^{2+}(aq) + 2e^- \rightarrow Zn(s)$ (-0.76 V) because in the battery the process happens in the presence of base and Zn^{2+} is stabilized as ZnO(s). Stabilization of the product of a reaction increases the driving force for the reaction, so E° is more positive.

20.67 *Analyze/Plan.* (a) Consider the function of Zn in an alkaline battery. What effect would it have on the redox reaction and cell emf if Cd replaces Zn? (b) Both batteries contain Ni. What is the difference in environmental impact between Cd and the metal hydride? *Solve:*

(a) E°_{red} for Cd (-0.40 V) is less negative than E°_{red} for Zn (-0.76 V), so E_{cell} will have a smaller (less positive) value.

(b) NiMH batteries use an alloy such as $ZrNi_2$ as the anode material. This eliminates the use and concomitant disposal problems associated with Cd, a toxic heavy metal.

20.68 (a) The alkali metal Li has much greater metallic character than Zn, Cd, Pb or Ni. The reduction potential for Li is thus more negative, leading to greater overall cell emf for the battery. Also, Li is less dense than the other metals, so greater total energy for a battery can be achieved for a given total mass of material. One disadvantage is that Li is very reactive and the cell reactions are difficult to control.

(b) Li has a much smaller molar mass (6.94 g/mol) than Ni (58.69 g/mol). A Li-ion battery can have many more charge-carrying particles than a Ni-based battery with the same mass. That is, Li-ion batteries have a greater *energy density* than Ni-based batteries.

20.69 *Analyze/Plan.* (a) Decide which reactant is oxidized and which is reduced. Write the balanced half-reactions and assign the appropriate one as anode and cathode. (b) Write the balanced half-reaction for $Fe^{2+}(aq) \rightarrow Fe_2O_3 \cdot 3H_2O$. Use the reduction half-reaction from part (a) to obtain the overall reaction. *Solve:*

(a) anode: $Fe(s) \rightarrow Fe^{2+}(aq) + 2e^-$
cathode: $O_2(g) + 4H^+(aq) + 4e^- \rightarrow 2H_2O(l)$

(b) $2Fe^{2+}(aq) + 6H_2O(l) \rightarrow Fe_2O_3 \cdot 3H_2O(s) + 6H^+(aq) + 2e^-$

$O_2(g) + 4H^+(aq) + 4e^- \rightarrow 2H_2O(l)$

(Multiply the oxidation half-reaction by two to balance electrons and obtain the overall balanced reaction.)

20.70 (a) Calculate E°_{cell} for the given reactants at standard conditions.

$O_2(g) + 4H^+(aq) + 4\,e^- \rightarrow 2H_2O(l)$ $\quad E^{\circ}_{red} = 1.23$ V

$2[Cu(s) \rightarrow Cu^{2+}(aq) + 2e^-]$ $\quad E^{\circ}_{red} = 0.337$ V

$2Cu(s) + O_2(g) + 4H^+(aq) \rightarrow 2Cu^{2+}(aq) + 2H_2O(l)$ $\quad E° = 1.23 - 0.337 = 0.89$ V

At standard conditions with $O_2(g)$ and $H^+(aq)$ present, the oxidation of Cu(s) has a positive E° value and is spontaneous. Cu(s) will oxidize (corrode) in air in the presence of acid.

(b) Fe^{2+} has a more negative reduction potential (-0.440 V) than Cu^{2+} (+0.337 V), so Fe(s) is more readily oxidized than Cu(s). If the two metals are in contact, Fe(s) would act as a sacrificial anode and oxidize (corrode) in preference to Cu(s); this would weaken the iron support skeleton of the statue. The teflon spacers prevent contact between the two metals and insure that the iron skeleton doesn't corrode when the Cu(s) skin comes in contact with atmospheric $O_2(g)$ and $H^+(aq)$.

20.71 *Analyze/Plan*. Follow the logic in Sample exercise 20.15. *Solve*:

(a) Zn^{2+} has a more negative reduction potential than Fe^{2+}, so Zn(s) is more readily oxidized. If Zn and Fe are both available for oxidation by O_2 (corrosion), Zn will be oxidized and Fe will not; Zn acts as a sacrificial anode.

(b) During the corrosion of galvanized iron, Zn acts as the anode and Fe acts as the inert cathode at which O_2 is reduced. Zn protects Fe by making it the cathode in the electrochemical process; this is called *cathodic protection*.

20.72 (a) A *sacrificial anode* is a metal that is oxidized in preference to another when the two metals are coupled in an electrochemical cell; the sacrificial anode has a more negative E°_{red} than the other metal. In this case, E°_{red} for Mg^{2+} is -2.37 V, more negative than most metals present in pipes, including Fe (E°_{red} = -0.44 V) and Zn (E°_{red} = -0.763 V).

(b) No. To afford cathodic protection, a metal must be more difficult to reduce (have a more negative reduction potential) than Fe^{2+}. E°_{red} Co^{2+} = -0.28 V, E°_{red} Fe^{2+} = -0.44 V.

Electrolysis; Electrical Work

20.73 (a) *Electrolysis* is an electrochemical process driven by an outside energy source.

(b) Electrolysis reactions are, by definition, nonspontaneous.

(c) $2Cl^-(l) \rightarrow Cl_2(g) + 2e^-$

20.74 (a) An electrolytic cell is the vessel in which electrolysis occurs. It consists of a power source and two electrodes in a molten salt or aqueous solution.

(b) It is the cathode. In an electrolysis cell, as in a voltaic cell, electrons are consumed (via reduction) at the cathode. Electrons flow from the negative terminal of the voltage source and then to the cathode.

(c) A small amount of $H_2SO_4(aq)$ present during the electrolysis of water acts as a change carrier, or supporting electrolyte. This facilitates transfer of electrons through the solution and at the electrodes, speeding up the reaction. (Considering $H^+(aq)$ as the substance reduced at the cathode changes the details of the half-reactions, but not the overall E° for the electrolysis. $SO_4^{2-}(aq)$ cannot be oxidized.)

20.75 *Analyze/Plan.* (a) If the products in the two environments are different, one or both of the half-reactions must be different. Consider available reactants other than $MgCl_2$. (b) Write balanced equations for the two redox reactions. (c) Follow the logic in Sample Exercise 20.16. Solve:

(a) The products are different because in aqueous electrolysis water is reduced in preference to Mg^{2+}.

(b) $MgCl_2(l) \rightarrow Mg(l) + Cl_2(g)$

$2Cl^-(aq) + 2H_2O(l) \rightarrow Cl_2(g) + H_2(g) + 2OH^-(aq)$

The aqueous solution electrolysis is entirely analogous to that for NaCl(aq), Section 20.9.

(c) $Mg^{2+}(aq) + 2e^- \rightarrow Mg(s)$ $\quad E^\circ_{red} = -2.37$ V

$2Cl^-(aq) \rightarrow Cl_2(g) + 2e^-$ $\quad E^\circ_{red} = 1.359$ V

$MgCl_2(aq) \rightarrow Mg(s) + Cl_2(g)$ $\quad E^\circ = -2.37 - 1.359 = -3.73$ V

$H_2O(l) + 2e^- \rightarrow H_2(g) + 2OH^-(aq)$ $\quad E^\circ_{red} = -0.83$ V

$2Cl^-(aq) \rightarrow Cl_2(g) + 2e^-$ $\quad E^\circ_{red} = 1.359$ V

$2Cl^-(aq) + 2H_2O(l) \rightarrow Cl_2(g) + H_2(g) + 2OH^-(aq)$ $\quad E^\circ = -0.83 - 1.359 = -2.19$ V

The minus signs mean that voltage must be applied in order for the reaction to occur.

20.76 (a) anode: $2Br^-(l) \rightarrow Br_2(l) + 2e^-$

cathode: $Al^{3+}(l) \rightarrow Al(l) + 3e^-$ (Al(s) melts at a lower temperature than AlBr(s).)

(b) anode: $2Br^-(l) \rightarrow Br_2(l) + 2e^-$

cathode: $2H_2O(l) + 2e^- \rightarrow H_2(g) + 2OH^-(aq)$

The reduction potential for water is less negative than that for $Al^{3+}(aq)$, so water is reduced in preference to Al^{3+}.

(c) The minimum emf values for the two possibilities in aqueous solution are:

Al^{3+} is reduced: -1.66 V - 1.065 V = -2.73 V

H_2O is reduced: -0.83 V - 1.065 V = -1.90 V

The minus signs indicate that voltage must be applied.

20.77 *Analyze/Plan.* Write the balanced half-reactions for the electrolysis of $CuCl_2(aq)$. Assign the oxidation process to the anode and reduction process to the cathode. Indicate the direction of electron flow and ion flow. *Solve:*

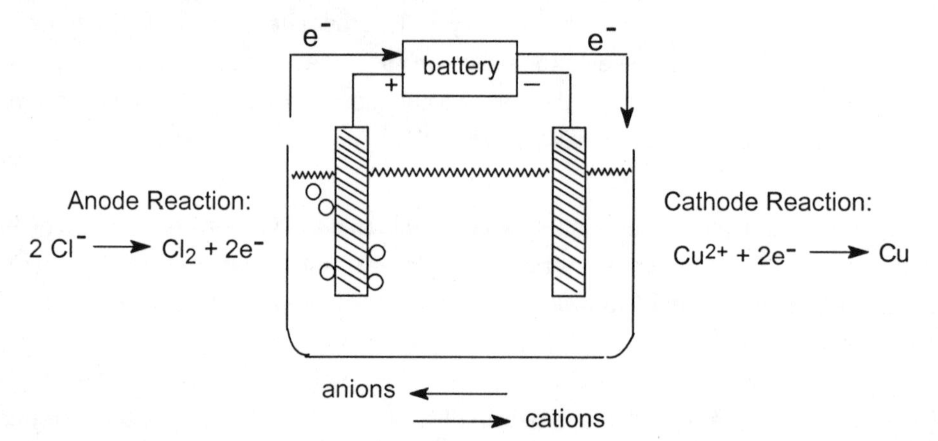

Cl^- is oxidized in preference to water because production of Cl_2 is kinetically favored.

20.78

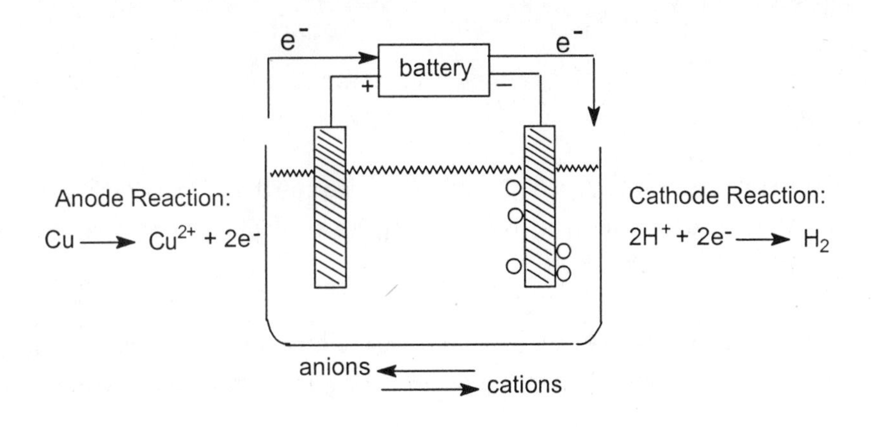

Overall cell reaction: $Cu(s) + 2H^+(aq) \rightarrow Cu^{2+}(aq) + H_2(g)$ $E° = -0.34$ V
Cu(s) is oxidized in preference to Br^- because of the smaller $E°_{red}$ for Cu(s).
E_{min}, the **minimum** voltage required to produce the cell reaction under standard conditions, is 0.34 V. Note, however, that this value is at standard conditions $[H^+] = 1$ *M*, and $[Cu^{2+}] = 1$ *M*. This value does not include any additional voltage required to overcome kinetic effects.

20.79 *Analyze/Plan*. Follow the logic in Sample Exercise 20.17, paying close attention to units. Coulombs = amps•s; since this is a $3e^-$ reduction, each mole of Cr(s) requires 3 Faradays. *Solve*:

(a) $$7.75 \text{ A} \times 1.50 \text{ d} \times \frac{24 \text{ hr}}{1 \text{ d}} \times \frac{60 \text{ min}}{1 \text{ hr}} \times \frac{60 \text{ s}}{1 \text{ min}} \times \frac{1 \text{ C}}{1 \text{ amp} \cdot \text{s}} \times \frac{1 \text{ F}}{96{,}500 \text{ C}} \times \frac{1 \text{ mol Cr}}{3 \text{ F}} \times \frac{52.00 \text{ g Cr}}{1 \text{ mol Cr}} = 180 \text{ g Cr(s)}$$

(b) $$0.250 \text{ mol Cr} \times \frac{3 \text{ F}}{1 \text{ mol Cr}} \times \frac{96{,}500 \text{ C}}{\text{F}} \times \frac{1 \text{ amp} \cdot \text{s}}{1 \text{ C}} \times \frac{1}{8.00 \text{ hr}} \times \frac{1 \text{ hr}}{60 \text{ min}} \times \frac{1 \text{ min}}{60 \text{ s}} = 2.51 \text{ A}$$

20.80 Coulombs = amps•s; since this is a $2e^-$ reduction, each mole of Mg(s) requires 2 Faradays.

(a) $$5.25 \text{ A} \times 2.50 \text{ d} \times \frac{24 \text{ hr}}{1 \text{ d}} \times \frac{60 \text{ min}}{1 \text{ hr}} \times \frac{60 \text{ s}}{1 \text{ min}} \times \frac{1 \text{ C}}{1 \text{ amp} \cdot \text{s}} \times \frac{1 \text{ F}}{96{,}500 \text{ C}} \times \frac{1 \text{ mol Mg}}{2 \text{ F}} \times \frac{24.31 \text{ g Mg}}{1 \text{ mol Mg}} = 143 \text{ g Mg}$$

(b) $$10.00 \text{ g Mg} \times \frac{1 \text{ mol Mg}}{24.31 \text{ g Mg}} \times \frac{2 \text{ F}}{1 \text{ mol Mg}} \times \frac{96{,}500 \text{ C}}{\text{F}} \times \frac{1 \text{ amp} \cdot \text{s}}{\text{C}} \times \frac{1 \text{ min}}{60 \text{ s}} \times \frac{1}{3.50 \text{ A}} = 378 \text{ min}$$

20.81 *Analyze/Plan*. Follow the logic in Sample Exercise 20.17, paying close attention to units. Coulombs = amps • s; since $2Cl^- \rightarrow Cl_2$ is a $2e^-$ oxidation, each mole of Cl_2 requires 2 Faradays. *Solve*:

(a) $$16.8\text{ A} \times 90.0\text{ min} \times \frac{60\text{ s}}{1\text{ min}} \times \frac{1\text{ C}}{1\text{ amp}\bullet\text{s}} \times \frac{1\text{ F}}{96{,}500\text{ C}} \times \frac{1\text{ mol Cl}_2}{2\text{ F}}$$

$$\times \frac{22.400\text{ L Cl}_2}{1\text{ mol Cl}_2} = 10.5\text{ L Cl}_2$$

(b) From the balanced equation (Section 20.9), we see that 2 mol NaOH are formed per mol Cl_2. Proceeding as in (a), but replacing the last factor by (2 mol NaOH/1 mol Cl_2), we obtain 0.940 mol NaOH.

20.82 (a) *Plan.* Use the ideal gas equation to calculate moles $H_2(g)$. Follow the progression in Figure 20.31, starting at moles of substance and moving to the left. *Solve:*

$$n = \frac{PV}{RT} = 725\text{ torr} \times \frac{1\text{ atm}}{760\text{ torr}} \times \frac{5.0\text{ L}}{296\text{ K}} \times \frac{\text{K}\bullet\text{mol}}{0.08206\text{ L}\bullet\text{atm}} = 0.19637 = 0.20\text{ mol H}_2$$

$$0.19637\text{ mol H}_2 \times \frac{2\text{ F}}{1\text{ mol H}_2} \times \frac{96{,}500\text{ C}}{\text{F}} \times \frac{1\text{ amp}\bullet\text{s}}{\text{C}} \times \frac{1}{1.5\text{ A}} = 2.527 \times 10^4$$

$$= 2.5 \times 10^4\text{ s (7.0 hr)}$$

(b) $2H_2O(l) \rightarrow 2H_2(g) + O_2(g)$

$$0.19637\text{ mol H}_2 \times \frac{1\text{ mol O}_2}{2\text{ mol H}_2} \times \frac{32.00\text{ g O}_2}{\text{mol O}_2} = 3.142 = 3.1\text{ g O}_2$$

20.83 *Analyze/Plan.* Given a spontaneous chemical reaction, calculate the maximum possible work for a given amount of reactant at standard conditions. Separate the equation into half-reactions and calculate cell emf. Use Equation 20.21, $w_{max} = -nFE$, to calculate maximum work. At standard conditions, $E = E°$. Solve:

$I_2(s) + 2e^- \rightarrow 2I^-(aq)$ $E°_{red} = 0.536$ V

$Sn(s) \rightarrow Sn^{2+}(aq) + 2e^-$ $E°_{red} = -0.136$ V

$I_2(s) + Sn(s) \rightarrow 2I^-(aq) + Sn^{2+}(aq)$ $E° = 0.536 - (-0.136) = 0.672$ V

$w_{max} = -2(96.5)(0.672) = -129.7 = -130$ kJ/mol Sn

$$\frac{-129.7\text{ kJ}}{\text{mol Sn(s)}} \times 0.850\text{ mol Sn} = -110\text{ kJ}$$

20.84 For this cell at standard conditions, $E° = 1.10$ V.

$w_{max} = \Delta G° = -nFE° = -2(96.5)(1.10) = -212.3 = -212$ kJ/mol Cu

$$50.0\text{ g Cu} \times \frac{1\text{ mol Cu}}{63.55\text{ g Cu}} \times \frac{-212.3\text{ kJ}}{\text{mol Cu}} = -167\text{ kJ}$$

20.85 *Analyze/Plan*. Follow the logic in Sample Exercise 20.18, paying close attention to units.

Solve:

(a) $7.5 \times 10^4 \text{ A} \times 24 \text{ hr} \times \frac{3600 \text{ s}}{1 \text{ hr}} \times \frac{1 \text{ C}}{1 \text{ amp} \cdot \text{s}} \times \frac{1 \text{ F}}{96{,}500 \text{ C}} \times \frac{1 \text{ mol Li}}{1 \text{ F}}$

$\times \frac{6.94 \text{ g Li}}{1 \text{ mol Li}} \times 0.85 = 3.961 \times 10^5 = 4.0 \times 10^5 \text{ g Li}$

(b) If the cell is 85% efficient, $\frac{96{,}500 \text{ C}}{\text{F}} \times \frac{1 \text{ F}}{0.85 \text{ mol}} = 1.135 \times 10^5$

$= 1.1 \times 10^5 \text{ C/mol Li required}$

$\text{Energy} = 7.5 \text{ V} \times \frac{1.135 \times 10^5 \text{ C}}{\text{mol Li}} \times \frac{1 \text{ J}}{1 \text{ C} \cdot \text{V}} \times \frac{1 \text{ kWh}}{3.6 \times 10^6 \text{ J}} = 0.24 \text{ kWh/mol Li}$

20.86 (a) $6.5 \times 10^3 \text{ A} \times 48 \text{ hr} \times \frac{3600 \text{ s}}{1 \text{ hr}} \times \frac{1 \text{ C}}{1 \text{ amp} \cdot \text{s}} \times \frac{1 \text{ F}}{96{,}500 \text{ C}} \times \frac{1 \text{ mol Ca}}{2 \text{ F}}$

$\times \frac{40.08 \text{ g Ca}}{1 \text{ mol Ca}} \times 0.68 = 1.586 \times 10^5 = 1.6 \times 10^5 \text{ g Ca}$

(b) If the cell is 68% efficient, $\frac{96{,}500 \text{ C}}{\text{F}} \times \frac{2 \text{ F}}{0.68 \text{ mol Ca}} = 2.838 \times 10^5$

$= 2.8 \times 10^5 \text{ C/mol Ca required}$

$\text{Energy} = 5.00 \text{ V} \times \frac{2.838 \times 10^5 \text{ C}}{\text{mol Ca}} \times \frac{1 \text{ J}}{\text{C} \cdot \text{V}} \times \frac{1 \text{ kWh}}{3.6 \times 10^6 \text{ J}} = 0.3942 = 0.39 \text{ kWh}$

Additional Exercises

20.87 (a)

$MnO_4^{2-}(aq) + 4H^+(aq) + 2e^- \rightarrow MnO_2(s) + 2H_2O(l)$

$2[MnO_4^{2-}(aq) \rightarrow MnO_4^-(aq) + 1e^-]$

$3MnO_4^{2-}(aq) + 4H^+(aq) \rightarrow 2MnO_4^-(aq) + MnO_2(s) + 2H_2O(l)$

(b)

$H_2SO_3(aq) + 4H^+(aq) + 4e^- \rightarrow S(s) + 3H_2O(l)$

$2[H_2SO_3(aq) + H_2O(l) \rightarrow HSO_4^-(aq) + 3H^+(aq) + 2e^-]$

$3H_2SO_3(aq) \rightarrow S(s) + 2HSO_4^-(aq) + 2H^+(aq) + H_2O(l)$

(c)

$Cl_2(aq) + 2H_2O(l) \rightarrow 2ClO^-(aq) + 4H^+(aq) + 2e$

$4OH^-(aq) \qquad + 4OH^-(aq)$

$Cl_2(aq) + 4OH^-(aq) \rightarrow 2ClO^-(aq) + 2H_2O(l) + 2e^-$

$Cl_2(aq) + 2e^- \rightarrow 2Cl^-(aq)$

$1/2[2Cl_2(aq) + 4OH^-(aq) \rightarrow 2Cl^-(aq) + 2ClO^-(aq) + 2H_2O(l)]$

$Cl_2(aq) + 2OH^-(aq) \rightarrow Cl^-(aq) + ClO^-(aq) + H_2O(l)$

20.88 (a)

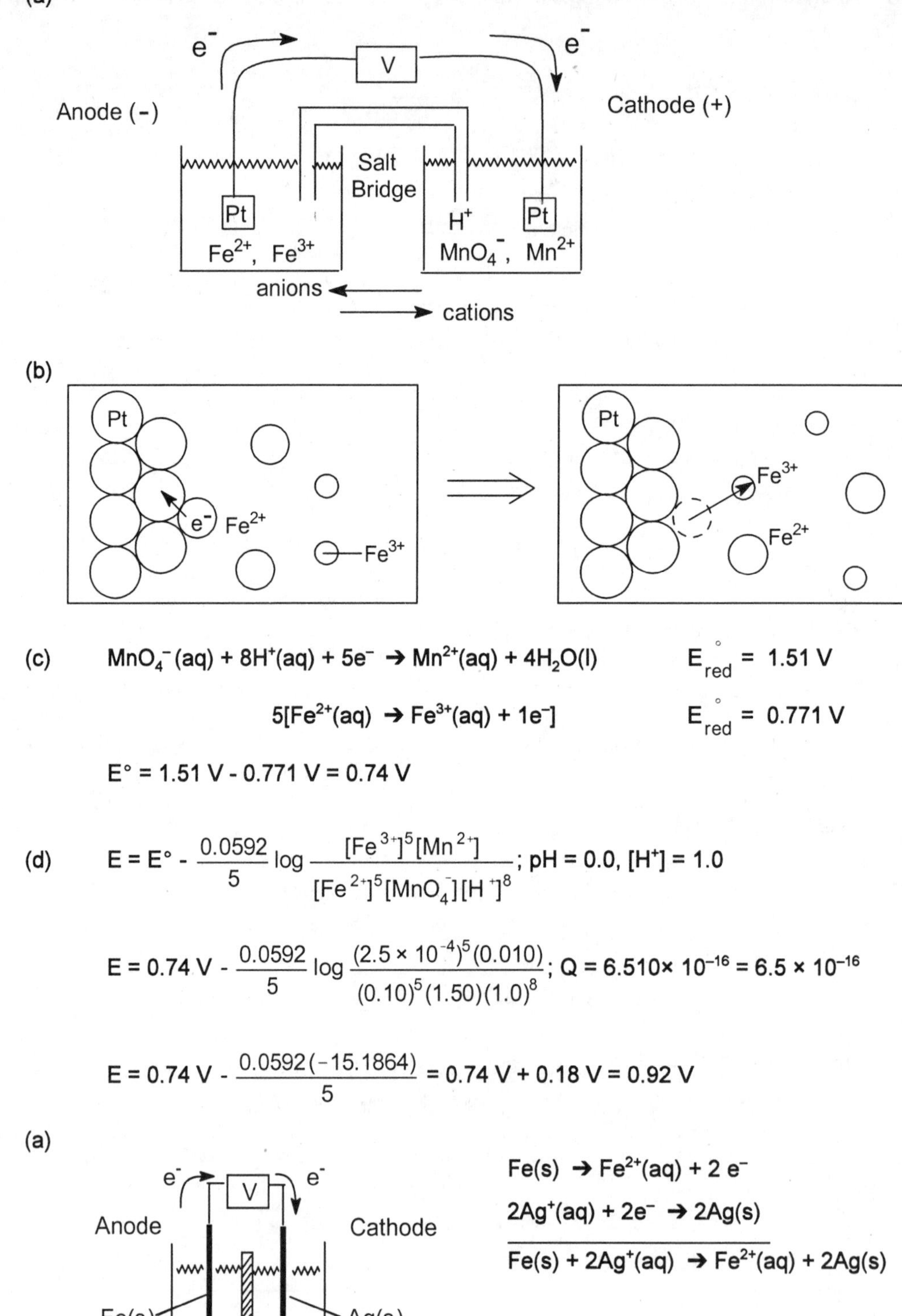

(b)

(c) $MnO_4^-(aq) + 8H^+(aq) + 5e^- \rightarrow Mn^{2+}(aq) + 4H_2O(l)$ $\qquad E^\circ_{red} = 1.51\ V$

$5[Fe^{2+}(aq) \rightarrow Fe^{3+}(aq) + 1e^-]$ $\qquad E^\circ_{red} = 0.771\ V$

$E^\circ = 1.51\ V - 0.771\ V = 0.74\ V$

(d) $E = E^\circ - \dfrac{0.0592}{5} \log \dfrac{[Fe^{3+}]^5[Mn^{2+}]}{[Fe^{2+}]^5[MnO_4^-][H^+]^8}$; pH = 0.0, $[H^+] = 1.0$

$E = 0.74\ V - \dfrac{0.0592}{5} \log \dfrac{(2.5 \times 10^{-4})^5(0.010)}{(0.10)^5(1.50)(1.0)^8}$; $Q = 6.510 \times 10^{-16} = 6.5 \times 10^{-16}$

$E = 0.74\ V - \dfrac{0.0592(-15.1864)}{5} = 0.74\ V + 0.18\ V = 0.92\ V$

20.89 (a)

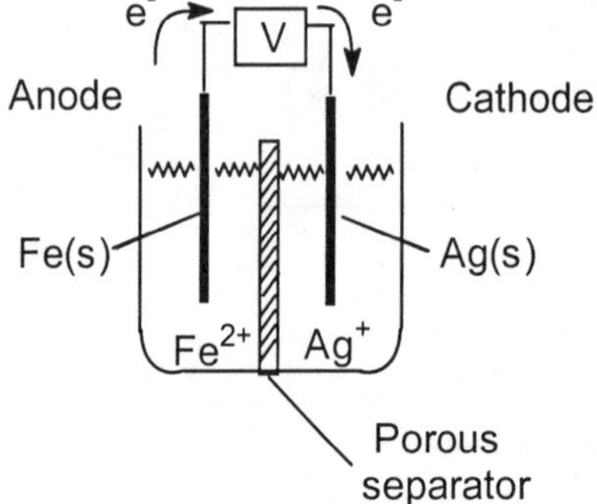

$Fe(s) \rightarrow Fe^{2+}(aq) + 2\,e^-$

$2Ag^+(aq) + 2e^- \rightarrow 2Ag(s)$

$Fe(s) + 2Ag^+(aq) \rightarrow Fe^{2+}(aq) + 2Ag(s)$

(b) $Zn(s) \rightarrow Zn^{2+}(s) + 2e^-$

$2H^+(aq) + 2e^- \rightarrow H_2(g)$

$Zn(s) + 2H^+(aq) \rightarrow Zn^{2+}(aq) + H_2(g)$

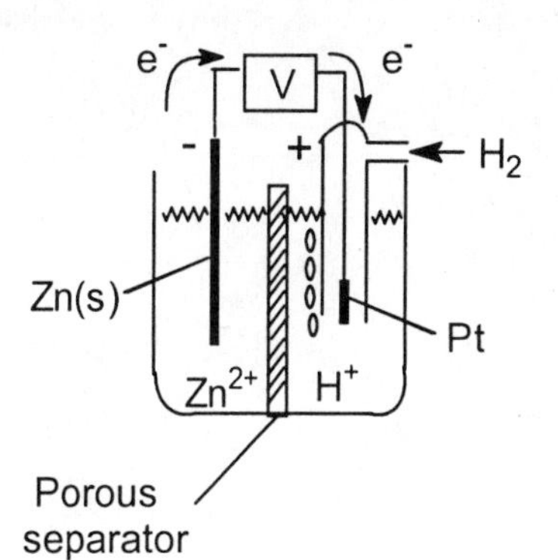

(c) Cu|Cu^{2+}||ClO$_3^-$, Cl$^-$| Pt Here, both the oxidized and reduced forms of the cathode solution are in the same phase, so we separate them by a comma, and then indicate an inert electrode.

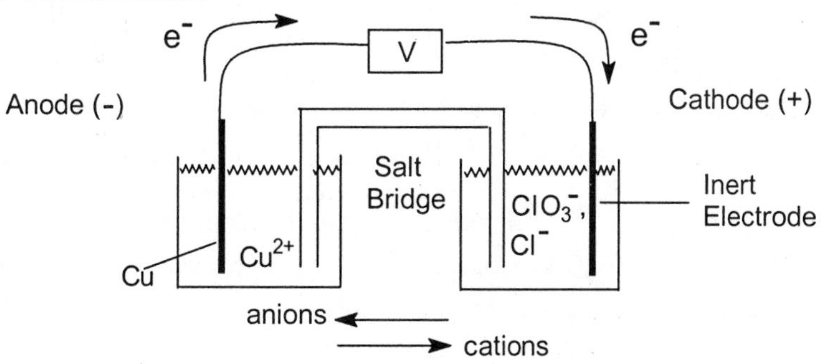

20.90 $2[Rh^{3+}(aq) + 3e^- \rightarrow Rh(s)]$ $E^\circ_{red} = ?$

$3[Cd(s) \rightarrow Cd^{2+}(aq) + 2e^-]$ $E^\circ_{red} = -0.403$ V

$2Rh^{3+}(aq) + 3Cd(s) \rightarrow 2Rh(s) + 3Cd^{2+}(aq)$ $E^\circ = 1.20$ V

(b) Cd(s) is the anode, and Rh(s) is the cathode.

(c) The cell is at standard conditions. $E^\circ_{cell} = E^\circ_{red}$ (cathode) - E°_{red} (anode)

$E^\circ_{red} = E^\circ_{cell} + E^\circ_{red}$ (anode) = 1.20 V - 0.403 V = 0.80 V

(d) $\Delta G^\circ = -nFE^\circ = -6(96.5)(1.20) = -695$ kJ

20.91 We need in each case to determine whether E° is positive (spontaneous) or negative (nonspontaneous).

(a) $I_2(s) + 2e^- \rightarrow 2I^-(aq)$ $E^\circ_{red} = 0.536$ V

$Sn(s) \rightarrow Sn^{2+}(aq) + 2e^-$ $E^\circ_{red} = -0.136$

$Sn(s) + I_2(s) \rightarrow Sn^{2+}(aq) + 2I^-(aq)$ $E^\circ = 0.536 - (-0.136) = 0.672$ V, spontaneous

(b) $Ni^{2+}(aq) + 2e^- \rightarrow Ni(s)$ $E^\circ_{red} = -0.28$ V

$2I^-(aq) \rightarrow I_2(s) + 2e^-$ $E^\circ_{red} = 0.536$ V

$Ni^{2+}(aq) + 2I^-(aq) \rightarrow Ni(s) + I_2(s)$ $E^\circ = -0.28 - 0.536 = -0.82$ V, nonspontaneous

(c) $2[Ce^{4+}(aq) + 1e^- \rightarrow Ce^{3+}(aq)]$ $E^\circ_{red} = 1.61$ V

$H_2O_2(aq) \rightarrow O_2(g) + 2H^+(aq) + 2e^-$ $E^\circ_{red} = 0.68$ V

$2Ce^{4+}(aq) + H_2O_2(aq) \rightarrow 2Ce^{3+}(aq) + O_2(g) + 2H^+(aq)$ $E^\circ = 1.61 - 0.68 = 0.93$ V, spontaneous

(d) $Cu^{2+}(aq) + 2e^- \rightarrow Cu(s)$ $E^\circ_{red} = 0.337$ V

$Sn^{2+}(aq) \rightarrow Sn^{4+}(aq) + 2e^-$ $E^\circ_{red} = 0.154$ V

$Cu^{2+}(aq) + Sn^{2+}(aq) \rightarrow Cu(s) + Sn^{4+}(aq)$ $E^\circ = 0.337 - 1.54 = 0.183$ V, spontaneous

20.92 (a) The reduction potential for $O_2(g)$ in the presence of acid is 1.23 V. $O_2(g)$ cannot oxidize Au(s) to $Au^+(aq)$ or $Au^{3+}(aq)$, even in the presence of acid.

(b) The substances need a reduction potential greater than 1.50 V. These include $Co^{3+}(aq)$, $F_2(g)$, $H_2O(aq)$ and $O_3(g)$. Marginal oxidizing agents (those with reduction potential near 1.50 V) from Appendix E are $BrO_3^-(aq)$, $Ce^{4+}(aq)$, HClO(aq), $MnO_4^-(aq)$ and $PbO_2(s)$.

(c) $3[Au^+(aq) + 1e^- \rightarrow Au(s)]$ $E^\circ_{red} = 1.69$ V

$Au(s) \rightarrow Au^{3+}(aq) + 3e^-$ $E^\circ_{red} = 1.50$ V

$3Au^+(aq) \rightarrow 2Au(s) + Au^{3+}(aq)$ $E^\circ = 1.69 - 1.50 = +0.19$ V

Since E° is positive, ΔG° is negative and the disproportionation of $Au^+(aq)$ is spontaneous.

(d) Since the Au^+ spontaneously disproportionates, the product of the reaction is AuF_3.

$3[F_2(g) + 2e^- \rightarrow 2F^-(aq)]$ $E^\circ_{red} = 2.87$ V

$2[Au(s) \rightarrow Au^{3+}(aq) + 3e^-]$ $E^\circ_{red} = 1.50$ V

$2Au(s) + 3F_2(g) \rightarrow 2AuF_3(aq)$ $E^\circ = 2.87 - 1.50 = 1.37$ V

20.93 (a) $2[Ag^+(aq) + 1e^- \rightarrow Ag(s)]$ $E^\circ_{red} = 0.80$ V

$Ni(s) \rightarrow Ni^{2+}(aq) + 2e^-$ $E^\circ_{red} = -0.28$ V

$2Ag^+(aq) + Ni(s) \rightarrow 2Ag(s) + Ni^{2+}(aq)$ $E^\circ = 0.80 - (-0.28) = 1.08$ V

(b) As the reaction proceeds, $Ni^{2+}(aq)$ is produced, so $[Ni^{2+}]$ increases as the cell operates.

(c) $E = E° - \frac{0.0592}{n} \log K_{eq}$; $1.12 = 1.08 - \frac{0.0592}{2} \log \frac{[Ni^{2+}]}{[Ag^+]^2}$

$-\frac{0.04(2)}{0.0592} = \log(0.0100) - \log[Ag^+]^2$; $\log[Ag^+]^2 = \log(0.0100) + \frac{0.04(2)}{0.0592}$

$\log[Ag^+]^2 = -2.000 + 1.351 = -0.649$; $[Ag^+]^2 = 0.255\ M$; $[Ag^+] = 0.474 = 0.5\ M$

(Strictly speaking, [E - E°] having only one sig fig leads (after several steps) to the answer having only one sig fig. This is not a very precise or useful result.)

20.94 (a)

$I_2(s) + 2e^- \rightarrow 2I^-(aq)$ $\quad E°_{red} = 0.536$ V

$2[Cu(s) \rightarrow Cu^+(aq) + 1\ e^-]$ $\quad E°_{red} = 0.521$ V

$I_2(s) + 2Cu(s) \rightarrow 2Cu^+(aq) + 2\ I^-(aq)$ $\quad E° = 0.536 - 0.521 = 0.015$ V

$$E = E° - \frac{0.0592}{n} \log Q = 0.015 - \frac{0.0592}{2} \log [Cu^+]^2 [I^-]^2$$

$$E = +0.015 - \frac{0.0592}{2} \log (2.5)^2(3.5)^2 = +0.015 - 0.056 = -0.041\text{V}$$

(b) Since the cell potential is negative at these concentration conditions, the cell would be spontaneous in the opposite direction and the inert electrode in the I_2/I^- compartment would be the anode; Cu(s) would be the cathode.

(c) No. At standard conditions the cell reaction is as written in part (a) and Cu(s) is the anode.

(d) $E = 0$, $+0.015 = \frac{0.0592}{2} \log (1.4)^2 [I^-]^2$; $\frac{2(0.015)}{0.0592} = \log (1.4)^2 + 2 \log [I^-]$;

$\log[I^-] = 0.107 = 0.11$; $[I^-] = 10^{0.107} = 1.28 = 1.3\ M\ I^-$

20.95

$Cu^+(aq) \rightarrow Cu^{2+}(aq) + 1e^-$ $\quad E°_{red} = +0.153$ V

$1e^- + Cu^+(aq) \rightarrow Cu°(s)$ $\quad E°_{red} = +0.521$ V

$2Cu^+(aq) \rightarrow Cu°(s) + Cu^{2+}(aq)$ $\quad E° = +0.521 - 0.153 = 0.368$ V

$$E° = \frac{0.0592}{n} \log K_{eq};\ \log K_{eq} = \frac{nE°}{0.0592} = \frac{1(0.368)}{0.0592} = 6.216 = 6.22$$

$K_{eq} = 10^{6.216} = 1.6 \times 10^6$

20.96 In order for the same electrode reactions that "normally" produce 1.5 V to produce 9 V, six of the 1.5-V cells must be connected in series; inside the 9-V battery there are six individual Zn, OH^- / MnO_2 cells connected in series. Since the 9-V battery is rectangular and the D-size battery is a cylinder, the shape of the cells and arrangement of the electrodes in the two batteries is probably different.

20.97 (a) In discharge: $Cd(s) + 2NiO(OH)(s) + 2H_2O(l) \rightarrow Cd(OH)_2(s) + 2Ni(OH)_2(s)$
In charging, the reverse reaction occurs.

(b) $E° = 0.49\ V - (-0.76\ V) = 1.25\ V$

(c) The 1.25 V calculated in part (b) is the standard cell potential, E°. The concentrations of reactants and products inside the battery are adjusted so that the cell output is greater than E°. Note that most of the reactants and products are pure solids or liquids, which do not appear in the Q expression. It must be $[OH^-]$ that is other than 1.0 *M*, producing an emf of 1.30 rather than 1.25.

20.98 (a) To act as a sacrificial anode, a metal must have a more negative reduction potential (be easier to oxidize) than Fe^{2+}. E°_{red} Sn^{2+} = -0.14 V, E°_{red} Fe^{2+} = -0.44 V. Tin does not give cathodic protection to iron.

(b) Tin gives protection to iron by providing a protective coating to keep out air and water. If this coating is broken, not only does corrosion occur, it occurs more readily because tin acts as the cathode in the electrochemical cell.

20.99 The ship's hull should be made negative. By keeping an excess of electrons in the metal of the ship, the tendency for iron to undergo oxidation, with release of electrons, is diminished. The ship, as a negatively charged "electrode," becomes the site of reduction, rather than oxidation, in an electrolytic process.

20.100 It is well established that corrosion occurs most readily when the metal surface is in contact with water. Thus, moisture is a requirement for corrosion. Corrosion also occurs more readily in acid solution, because O_2 has a more positive reduction potential in the presence of $H^+(aq)$. SO_2 and its oxidation products dissolve in water to produce acidic solutions, which encourage corrosion. The anodic and cathodic reactions for the corrosion of Ni are:

$$Ni(s) \rightarrow Ni^{2+}(aq) + 2e^- \quad E^\circ_{red} = -0.28\ V$$

$$O_2(g) + 4H^+(aq) + 4e^- \rightarrow 2H_2O(l) \quad E^\circ_{red} = 1.23\ V$$

Nickel(II) oxide, NiO(s), can form by the dry air oxidation of Ni. This NiO coating serves to protect against further corrosion. However, NiO dissolves in acidic solutions such as those produced by SO_2 or SO_3, according to the reaction: $NiO(s) + 2H^+(aq) \rightarrow Ni^{2+}(aq) + H_2O(l)$
This exposes Ni(s) to further wet corrosion.

20.101 A battery is a voltaic cell, so the cathode compartment contains the positive terminal of the battery. In the alkaline, Ni–Cd and NiMH batteries, $OH^-(aq)$ is produced at the cathode. The wire that turns the indicator pink is in contact with $OH^-(aq)$, so the rightmost wire is connected to the positive terminal of the battery. [The battery could not be a lead-acid battery because $OH^-(aq)$ is not present in either compartment, so neither wire would turn the indicator pink.]

20.102 (a) Total volume of Cr = 2.5×10^{-4} m × 0.32 m^2 = 8.0×10^{-5} m^3

$$\text{mol Cr} = 8.0 \times 10^{-5}\ m^3\ \text{Cr} \times \frac{100^3\ cm^3}{1\ m^3} \times \frac{7.20\ g\ Cr}{1\ cm^3} \times \frac{1\ mol\ Cr}{52.0\ g\ Cr} = 11.077 = 11\ \text{mol Cr}$$

The electrode reaction is:

$CrO_4^{2-}(aq) + 4H_2O(l) + 6e^- \rightarrow Cr(s) + 8OH^-(aq)$

$$\text{Coulombs required} = 11.077\ \text{mol Cr} \times \frac{6\ F}{1\ mol\ Cr} \times \frac{96{,}500\ C}{1\ F} = 6.41 \times 10^6 = 6.4 \times 10^6\ C$$

(b) $$6.41 \times 10^6\ C \times \frac{1\ amp \cdot s}{1\ C} \times \frac{1}{10.0\ s} = 6.4 \times 10^5\ amp$$

(c) If the cell is 60% efficient, (6.41×10^6 / 0.65) = 9.867×10^6 = 9.9×10^6 C are required to plate the bumper.

$$6.0\ V \times 9.867 \times 10^6\ C \times \frac{1\ J}{1\ C \cdot V} \times \frac{1\ kWh}{3.6 \times 10^6\ J} = 16.445 = 16\ kWh$$

20.103 $$3.20\ amp \times 40\ min \times \frac{60\ s}{1\ min} \times \frac{1\ C}{1\ amp \cdot s} \times \frac{1\ F}{96{,}500\ C} \times \frac{1\ mol\ e^-}{1\ F} = 0.0796 = 0.080\ mol\ e^-$$

$$\frac{4.57 g\ In}{0.0796\ mol\ e^-} \times \frac{1\ mol\ In}{114.8\ g\ In} = 0.50\ mol\ In/1\ mol\ e^-$$

This result tells us that In must be in the +2 oxidation state in the molten halide.

20.104 (a) The work obtainable is given by the product of the voltage, which has units of J/C, times the number of Coulombs of electricity produced:

$$w_{max} = 300\ amp \cdot hr \times \frac{3600\ s}{1\ hr} \times \frac{1\ C}{1\ amp \cdot s} \times \frac{6\ J}{1\ C} \times \frac{1\ kWh}{3.6 \times 10^6\ J} = 1.8\ kWh \approx 2\ kWh$$

(b) This maximum amount of work is never realized because some of the electrical energy is dissipated in overcoming the internal resistance of the battery; because the cell voltage does not remain constant as the reaction proceeds; because the systems to which the electrical energy is delivered are not capable of completely converting electrical energy into work.

20.105 (a) $$7 \times 10^8\ mol\ H_2 \times \frac{2\ F}{1\ mol\ H_2} \times \frac{96{,}500\ C}{1\ F} = 1.35 \times 10^{14} = 1 \times 10^{14}\ C$$

(b)

$2H_2O(l) \rightarrow O_2(g) + 4H^+(aq) + 4e^-$ $\quad E^\circ_{red} = 1.23\ V$

$2[H^+(aq) + 2e^- \rightarrow H_2(g)]$ $\quad E^\circ_{red} = 0\ V$

$2H_2O(l) \rightarrow O_2(g) + 2H_2(g)$ $\quad E^\circ = 0.00 - 1.23 = -1.23\ V$

$$E = E^\circ - \frac{0.0592}{4} \log (P_{O_2} \times P_{H_2}^2) = -1.23\ V - \frac{0.0592}{4} \log (300)^3$$

$E = -1.23\ V - 0.11\ V = -1.34\ V;\ E_{min} = 1.34\ V$

(c) Energy = nFE = $2(7 \times 10^8 \text{ mol})(1.34 \text{ V}) \dfrac{96{,}500 \text{ J}}{\text{V} \bullet \text{mol}} = 1.81 \times 10^{14} = 2 \times 10^{14}$ J

(d) $1.81 \times 10^{14} \text{ J} \times \dfrac{1 \text{ kWh}}{3.6 \times 10^6 \text{ J}} \times \dfrac{\$0.23}{\text{kWh}} = \$1.16 \times 10^7 = \$1 \times 10^7$

It will cost more than ten million dollars for the electricity alone.

Integrative Exercises

20.106 $2[NO_3^-(aq) + 4H^+(aq) + 3e^- \rightarrow NO(g) + 2H_2O(l)]$ $E^\circ_{red} = 0.96$ V

$3[Cu(s) \rightarrow Cu^{2+}(aq) + 2e^-]$ $E^\circ_{red} = 0.34$ V

$3Cu(s) + 2NO_3^-(aq) + 8H^+(aq) \rightarrow 3Cu^{2+}(aq) + 2NO(g) + 4H_2O(l)$

$E^\circ = 0.96 - 0.34 = 0.62$ V

$2H^+(aq) + 2e^- \rightarrow H_2(g)$ $E^\circ_{red} = 0$ V

$Cu(s) \rightarrow Cu^{2+}(aq) + 2e^-$ $E^\circ_{red} = 0.34$ V

$Cu(s) + 2H^+(aq) \rightarrow Cu^{2+}(aq) + H_2(g)$ $E^\circ = 0 - 0.34 = -0.34$ V

The overall cell potential for the oxidation of Cu(s) by HNO_3 is positive and the reaction is spontaneous. The cell potential for the oxidation of Cu(s) by HCl is negative and the reaction is nonspontaneous. Note that in the reaction with HNO_3, it is NO_3^- that is reduced, not H^+.

20.107 *Analyze/Plan.* We are given a reaction that is spontaneous at standard conditions and asked if it will be spontaneous at a different set of conditions. A reaction is spontaneous if the sign the cell emf is positive. $E = E^\circ - \dfrac{0.0592}{n} \log Q$ [Equation 20.16]. Calculate E° from data in Appendix E. Write the Q expression, which will contain a term for $[H^+]$. Calculate $[H^+]$ for the buffer listed. Finally, calculate the value of ΔG and evaluate the sign. *Solve:*

The half-reactions are:

$O_2(g) + 4H^+(aq) + 4e^- \rightarrow 2H_2O(l)$ $E^\circ_{red} = 1.23$ V

$2[2Br^-(aq) \rightarrow Br_2(l) + 2e^-]$ $E^\circ_{red} = 1.065$ V

$E^\circ = 1.23 - 1.065 = 0.165 = 0.17$ V

Calculate $[H^+]$. K_a for benzoic acid is 6.3×10^{-5}. $pK_a = 4.20$

$pH = pK_a + \log[\text{base}]/[\text{acid}] = 4.20 + \log(0.12/0.10) = 4.28$. $[H^+] = 10^{-4.28} = 5.25 \times 10^{-5}$ *M*

Calculate E. $Q = 1/P_{O_2} \times [H^+]^4 \times [Br^-]^4$. At standard conditions, $P_{O_2} = 1$ atm and $[Br^-] = 1$ *M*; only $[H^+]$ has been adjusted.

$E = E^\circ - \dfrac{0.0592}{n} \log Q = 0.165 - \dfrac{0.0592}{4} \log (1/1 \times [H^+]^4 \times 1)$

$E = 0.165 - \dfrac{0.0592}{4} \log [1/(5.25 \times 10^{-5})^4] = 0.165 - \dfrac{0.0592}{4}(17.120)$

$E = 0.165 - 0.2534 = -0.0884 = -0.088$ V

The sign of E is negative, so the reaction is not spontaneous in this buffer.

Comment: The same result can be obtained by calculating $\Delta G°$ from the data in Appendix C and using the $\Delta G = \Delta G° - RT \ln Q$ [Equation 19.21]. The sign of ΔG must be positive for a reaction to be spontaneous. The two methods are related by Equation 20.12, $\Delta G° = -nFE°$.

20.108 (a) The oxidation potential of A is equal in magnitude but opposite in sign to the reduction potential of A^+.

(b) Li(s) has the highest oxidation potential, Au(s) the lowest.

(c) The relationship is reasonable because both oxidation potential and ionization energy describe removing electrons from a substance. Ionization energy is a property of gas phase atoms or ions, while oxidation potential is a property of the bulk material.

20.109 (a)

$$NO_3^-(aq) + 4H^+(aq) + 3e^- \rightarrow NO(g) + 2H_2O(l) \qquad E^\circ_{red} = 0.96\ V$$
$$Au(s) \rightarrow Au^{3+}(aq) + 3e^- \qquad E^\circ_{red} = 1.498\ V$$

$$Au(s) + NO_3^-(aq) + 4H^+(aq) \rightarrow Au^{3+}(aq) + NO(g) + 2H_2O(l)$$

$E° = 0.96 - 1.498 = -0.54$ V; E° is negative, the reaction is not spontaneous.

(b)

$$3[2H^+(aq) + 2e^- \rightarrow H_2(g)] \qquad E^\circ_{red} = 0.000\ V$$
$$2[Au(s) + 4Cl^-(aq) \rightarrow AuCl_4^-(aq) + 3e^-] \qquad E^\circ_{red} = 1.002\ V$$

$$2Au(s) + 6H^+(aq) + 8Cl^-(aq) \rightarrow 2AuCl_4^-(aq) + 3H_2(g)$$

$E° = 0.000 - 1.002 = -1.002$ V; E° is negative, the reaction is not spontaneous.

(c)

$$NO_3^-(aq) + 4H^+(aq) + 3e^- \rightarrow NO(g) + 2H_2O(l) \qquad E^\circ_{red} = 0.96\ V$$
$$Au(s) + 4Cl^-(aq) \rightarrow AuCl_4^-(aq) + 3e^- \qquad E^\circ_{red} = 1.002\ V$$

$$Au(s) + NO_3^-(aq) + 4Cl^-(aq) + 4H^+(aq) \rightarrow AuCl_4^-(aq) + NO(g) + 2H_2O(l)$$

$E° = 0.96 - 1.002 = -0.04$; E° is small but negative, the process is not spontaneous.

(d)

$$E = E° - \frac{0.0592}{3} \log \frac{[AuCl_4^-]\, P_{NO}}{[NO_3^-][Cl^-]^4[H^+]^4}$$

If $[H^+]$, $[Cl^-]$ and $[NO_3^-]$ are much greater than 1.0 *M*, the log term is negative and the correction to E° is positive. If the correction term is greater than 0.042 V, the value of E is positive and the reaction at nonstandard conditions is spontaneous.

20.110 (a)

$$Ag^+(aq) + e^- \rightarrow Ag(s) \qquad E^\circ_{red} = 0.799\ V$$
$$Fe^{2+}(aq) \rightarrow Fe^{3+}(aq) + 1e^- \qquad E^\circ_{red} = 0.771\ V$$

$$Ag^+(aq) + Fe^{2+}(aq) \rightarrow Ag(s) + Fe^{3+}(aq) \qquad E° = 0.799\ V - 0.771\ V = 0.028\ V$$

(b) $Ag^+(aq)$ is reduced at the cathode and $Fe^{2+}(aq)$ is oxidized at the anode.

(c) $\Delta G° = -nFE° = -(1)(96.5)(0.028) = -2.7$ kJ

$\Delta S° = S°\ Ag(s) + S°\ Fe^{3+}(aq) - S° Ag^{+}(aq) - S°\ Fe^{2+}(aq)$

$= 42.55\ J + 293.3\ J - 73.93\ J - 113.4\ J = 148.5\ J$

$\Delta G° = \Delta H° - T\Delta S°$ Since $\Delta S°$ is positive, $\Delta G°$ will become more negative and E° will become more positive as temperature is increased.

20.111 (a) $\Delta H° = 2\Delta H°\ H_2O(l) - 2\Delta H°\ H_2(g) - \Delta H°\ O_2(g) = 2(-285.83) - 2(0) - 0 = -571.66$ kJ

$\Delta S° = 2S°\ H_2O(l) - 2S°\ H_2(g) - \Delta S°\ O_2(g)$

$= 2(69.91) - 2(130.58) - (205.0) = -326.34$ J

(b) Since $\Delta S°$ is negative, $-T\Delta S$ is positive and the value of ΔG will become more positive as T increases. The reaction will become nonspontaneous at a fairly low temperature, because the magnitude of $\Delta S°$ is large.

(c) $\Delta G = w_{max}$. The larger the negative value of ΔG, the more work the system is capable of doing on the surroundings. As the magnitude of ΔG decreases with increasing temperature, the usefulness of H_2 as a fuel decreases.

(d) The combustion method increases the temperature of the system, which quickly decreases the magnitude of the work that can be done by the system. Even if the effect of temperature on this reaction could be controlled, only about 40% of the energy from any combustion can be converted to electrical energy, so combustion is intrinsically less efficient than direct production of electrical energy via a fuel cell.

20.112 First balance the equation:

$4CyFe^{2+}(aq) + O_2(g) + 4H^{+}(aq) \rightarrow 4CyFe^{3+}(aq) + 2H_2O(l)$; $E = +0.60$ V; $n = 4$

(a) From Equation [20.11] we can calculate ΔG for the process under the conditions specified for the measured potential E:

$$\Delta G = -nFE = -(4\ \text{mol}\ e^-) \times \frac{96.5\ \text{kJ}}{1\ \text{V} \bullet \text{mol}\ e^-}(0.60\ \text{V}) = -231.6 = -232\ \text{kJ}$$

(b) The moles of ATP synthesized per mole of O_2 is given by:

$$\frac{231.6\ \text{kJ}}{O_2\ \text{molecule}} \times \frac{1\ \text{mol ATP formed}}{37.7\ \text{kJ}} = \text{approximately 6 mol ATP/mol } O_2$$

20.113 $AgSCN(s) + e^- \rightarrow Ag(s) + SCN^-(aq)$ $\quad E°_{red} = 0.0895$ V

$Ag(s) \rightarrow Ag^{+}(aq) + e^-$ $\quad E°_{red} = 0.799$ V

$AgSCN(s) \rightarrow Ag^{+}(aq) + SCN^-(aq)$ $\quad E° = 0.0895 - 0.799 = -0.710$ V

$$E° = \frac{0.0592}{n}\log K_{sp};\ \log K_{sp} = \frac{(-0.710)(1)}{0.0592} = -11.993 = -12.0$$

$K_{sp} = 10^{-11.993} = 1.02 \times 10^{-12} = 1 \times 10^{-12}$

20.114 The reaction can be written as a sum of the steps:

$$Pb^{2+}(aq) + 2e^- \rightarrow Pb(s) \qquad E^\circ_{red} = -0.126 \text{ V}$$

$$PbS(s) \rightarrow Pb^{2+}(aq) + S^{2-}(aq) \qquad \text{"}E^\circ\text{"} = ?$$

$$PbS(s) + 2e^- \rightarrow Pb(s) + S^{2-}(aq) \qquad E^\circ_{red} = ?$$

"E°" for the second step can be calculated from K_{sp}.

$$E^\circ = \frac{0.0592}{n} \log K_{sp} = \frac{0.0592}{2} \log (8.0 \times 10^{-28}) = \frac{0.0592}{2} (-27.10) = -0.802 \text{ V}$$

E° for the half-reaction = -0.126 V + (-0.802 V) = -0.928 V

Calculating an imaginary E° for a nonredox process like step 2 may be a disturbing idea. Alternatively, one could calculate K_{eq} for step 1 (5.4×10^{-5}), K_{eq} for the reaction in question ($K_{eq} = K_1 \times K_{sp} = 4.4 \times 10^{-32}$) and then E° for the half-reaction. The result is the same.

20.115 (a)

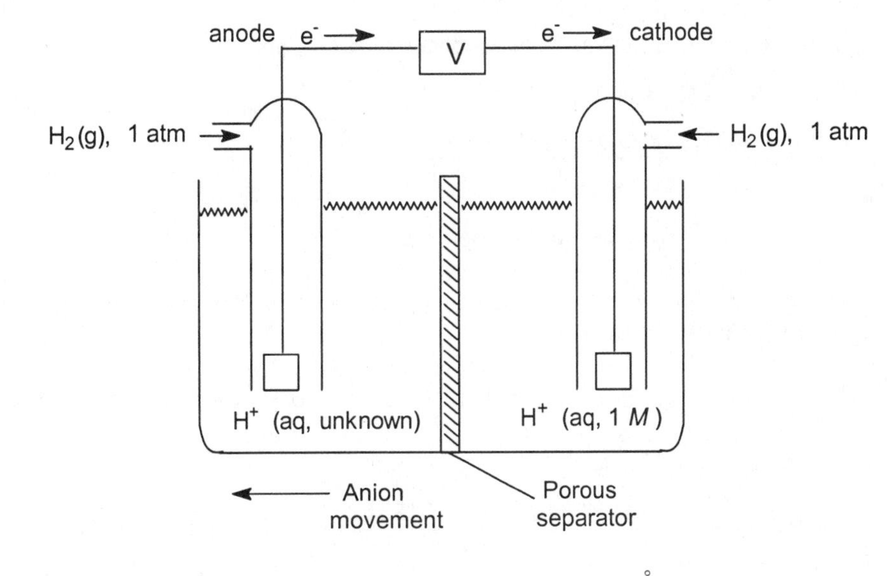

(b)

$$2H^+(aq, 1\,M) + 2e^- \rightarrow H_2(g) \qquad E^\circ_{red} = 0$$

$$H_2(g) \rightarrow 2H^+(aq, 1\,M) + 2e^- \qquad E^\circ_{red} = 0$$

$$2H^+(aq, 1\,M) + H_2(g) \rightarrow 2H^+(aq, 1\,M) + H_2(g) \qquad E^\circ = 0$$

(c) At standard conditions, $[H^+] = 1\,M$, pH = 0

(d)

$$E = E^\circ - \frac{0.0592}{2} \log \frac{[H^+(\text{unknown})]^2 P_{H_2}}{[H^+(1\,M)]^2 P_{H_2}}$$

$$E = 0 - \frac{0.0592}{2} \log [H^+]^2 = \frac{0.0592}{2} \times 2(-\log [H^+]) = 0.0592(\text{pH}) = 0.0592(5.0) = 0.30 \text{ V}$$

(e) E_{cell} changes 0.0592(0.01) = 0.000592 = 0.0006 V for each 0.01 pH unit. The voltmeter would have to be precise to at least 0.001 V to detect a charge of 0.01 pH units.

20.116 The cell reaction is presumed to be:

$Sn^{2+}(aq) + H_2O(l) \rightarrow Sn(s) + 1/2\ O_2(g) + 2H^+(aq)$

Calculate mol Sn^{2+} reduced:

$$4.50 \text{ amp} \times 25.00 \text{ min} \times \frac{60 \text{ s}}{1 \text{ min}} \times \frac{1 \text{ C}}{1 \text{ amp}\bullet\text{s}} \times \frac{1 \text{ F}}{96{,}500 \text{ C}} \times \frac{1 \text{ mol Sn}^{2+}}{2 \text{ F}} = 0.03497$$
$$= 0.0350 \text{ mol Sn}^{2+}$$

Initially there were $\frac{0.600 \text{ mol}}{\text{L}} \times (0.500 \text{ L}) = 0.300$ mol. Thus, following electrolysis, there are $0.300 - 0.0350 = 0.265$ mol Sn^{2+}.

$$[Sn^{2+}] = \frac{0.265 \text{ mol}}{0.500 \text{ L}} = 0.530\ M$$

Electrolysis also produces 2(0.0350) = 0.0700 mol $H^+(aq)$.

Thus $[H^+] = \frac{0.0700 \text{ mol}}{0.500 \text{ L}} = 0.140\ M$. The concentration of SO_4^{2-} remains unchanged.

20.117 The two half-reactions in the electrolysis of $H_2O(l)$ are:

$2[2H_2O(l) + 2e^- \rightarrow H_2(g) + 2OH^-]$

$2H_2O(l) \rightarrow O_2(g) + 4H^+ + 4e^-$

$2H_2O(l) \rightarrow 2H_2(g) + O_2(g)$

4 mol e^-/2 mol $H_2(g)$ or 2 mol e^-/mol $H_2(g)$

Using partial pressures and the ideal-gas law, calculate the mol $H_2(g)$ produced, and the current required to do so.

$P_T = P_{H_2} + P_{H_2O}$. From Appendix B, P_{H_2O} at 25.5°C is approximately 24.5 torr.

P_{H_2} = 768 torr - 24.5 torr = 743.5 = 744 torr

$$n = PV/RT = \frac{(743.5/760) \text{ atm} \times 0.0123 \text{ L}}{298.5 \text{ K} \times 0.08206 \text{ L}\bullet\text{atm/mol}\bullet\text{K}} = 4.912 \times 10^{-4} = 4.91 \times 10^{-4} \text{ mol H}_2$$

$$4.912 \times 10^{-4} \text{ mol H}_2 \times \frac{2 \text{ mol e}^-}{\text{mol H}_2} \times \frac{96{,}500 \text{ C}}{1 \text{ mol e}^-} \times \frac{1 \text{ amp}\bullet\text{s}}{1 \text{ C}} \times \frac{1 \text{ min}}{60 \text{ s}} \times \frac{1}{2.00 \text{ min}}$$
$$= 0.790 \text{ amp}$$

21 Nuclear Chemistry

Radioactivity

21.1 *Analyze/Plan.* Given various nuclide descriptions, determine the number of protons and neutrons in each nuclide. The left superscript is the mass number, protons plus neutrons. If there is a left subscript, it is the atomic number, the number of protons. Protons can always be determined from chemical symbol; all isotopes of the same element have the same number of protons. A number following the element name, as in part (c) is the mass number. *Solve:*

p = protons, n = neutrons, e = electrons; number of protons = atomic number; number of neutrons = mass number - atomic number

(a) ${}^{55}_{25}Mn$: 25p, 30n (b) ${}^{201}Hg$: 80p, 121n (c) ${}^{39}K$: 19p, 20n

21.2 p = protons, n = neutrons, e = electrons; number of protons = atomic number; number of neutrons = mass number - atomic number

(a) ${}^{126}_{55}Cs$: 55p, 71n (b) ${}^{119}Sn$: 50p, 69n (c) ${}^{141}Ba$: 56p, 85n

21.3 *Analyze/Plan.* See definitions in Section 21.1. In each case, the left superscript is mass number, the left subscript is related to atomic number. *Solve:*

(a) ${}^{1}_{1}p$ or ${}^{1}_{1}H$ (b) ${}^{0}_{1}e$ (c) ${}^{0}_{-1}\beta$ or ${}^{0}_{-1}e$

21.4 (a) ${}^{1}_{0}n$ (b) ${}^{0}_{-1}e$ or ${}^{0}_{-1}\beta$ (c) ${}^{4}_{2}He$ or ${}^{4}_{2}\alpha$

21.5 *Analyze/Plan.* Follow the logic in Sample Exercises 21.1 and 21.2. Pay attention to definitions of decay particles and conservation of mass and charge. *Solve:*

(a) ${}^{214}_{83}Bi \rightarrow {}^{214}_{84}Po + {}^{0}_{-1}e$ (b) ${}^{195}_{79}Au + {}^{0}_{-1}e$ (orbital electron) $\rightarrow {}^{195}_{78}Pt$

(c) ${}^{38}_{19}K \rightarrow {}^{38}_{18}Ar + {}^{0}_{1}e$ (d) ${}^{242}_{94}Pu \rightarrow {}^{238}_{92}U + {}^{4}_{2}He$

21.6 (a) ${}^{141}_{60}Nd + {}^{0}_{-1}e$ (orbital electron) $\rightarrow {}^{141}_{59}Pr$ (b) ${}^{201}_{79}Au \rightarrow {}^{201}_{80}Hg + {}^{0}_{-1}\beta$

(c) ${}^{81}_{34}Se \rightarrow {}^{81}_{35}Br + {}^{0}_{-1}\beta$ (d) ${}^{83}_{38}Sr \rightarrow {}^{83}_{37}Rb + {}^{0}_{1}e$

21.7 *Analyze/Plan*. Using definitions of the decay processes and conservation of mass number and atomic number, work backwards to the reactants in the nuclear reactions. *Solve*:

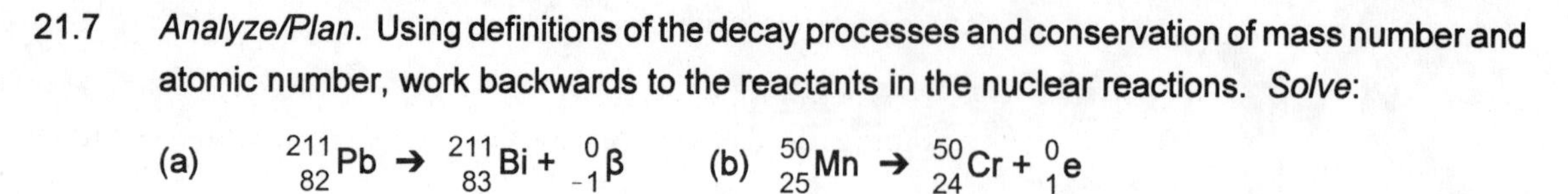

(a) $^{211}_{82}Pb \rightarrow ^{211}_{83}Bi + ^{0}_{-1}\beta$ (b) $^{50}_{25}Mn \rightarrow ^{50}_{24}Cr + ^{0}_{1}e$

(c) $^{179}_{74}W + ^{0}_{-1}e \rightarrow ^{179}_{73}Ta$ (d) $^{230}_{90}Th \rightarrow ^{226}_{88}Ra + ^{4}_{2}He$

21.8 (a) $^{24}_{11}Na \rightarrow ^{24}_{12}Mg + ^{0}_{-1}e$; a β particle is produced

(b) $^{188}_{80}Hg \rightarrow ^{188}_{79}Au + ^{0}_{1}e$; a positron is produced

(c) $^{122}_{53}I \rightarrow ^{122}_{54}Xe + ^{0}_{-1}e$; a β particle is produced

(d) $^{242}_{94}Pu \rightarrow ^{238}_{92}U + ^{4}_{2}He$; an α particle is produced

21.9 *Analyze/Plan*. Given the starting and ending nuclides in a nuclear decay sequence, we are asked to determine the number of alpha and beta emissions. Use the total change in A and Z, along with definitions of alpha and beta decay, to answer the question. *Solve*:

The total mass number change is (235-207) = 28. Since each α particle emission decreases the mass number by four, whereas emission of a β particle does not correspond to a mass change, there are 7 α particle emissions. The change in atomic number in the series is 10. Each α particle results in an atomic number lower by two. The 7 α particle emissions alone would cause a decrease of 14 in atomic number. Each β particle emission raises the atomic number by one. To obtain the observed lowering of 10 in the series, there must be 4 β emissions.

21.10 This decay series represents a change of (232-208 =) 24 mass units. Since only alpha emissions change the nuclear mass, and each changes the mass by four, there must be a total of 6 α emissions. Each alpha emission causes a decrease of two in atomic number. Therefore, the 6 alpha emissions, by themselves, would cause a decrease in atomic number of 12. The series as a whole involves a decrease of 8 in atomic number. Thus, there must be a total of 4 β emissions, each of which increases atomic number by one. Overall, there are 6 α emissions and 4 β emissions.

Nuclear Stability

21.11 *Analyze/Plan*. Follow the logic in sample Exercise 21.3, paying attention to the guidelines for neutron-to-proton ratio. *Solve*:

(a) $^{8}_{5}B$ - low neutron/proton ratio, positron emission (for low atomic numbers, positron emission is more common than orbital electron capture)

(b) $^{68}_{29}Cu$ - high neutron/proton ratio, beta emission

(c) $^{241}_{93}Np$ - high neutron/proton ratio, beta emission

(Even though ^{241}Np has an atomic number $\geq$ 84, the most common decay pathway for nuclides with neutron/proton ratios higher than the isotope listed on the periodic chart is beta decay.)

(d) $^{39}_{17}Cl$ - high neutron/proton ratio, beta emission

21.12 (a) $^{66}_{32}Ge$ - low neutron/proton ratio, positron emission

(b) $^{105}_{45}Rh$ - high neutron/proton ratio, beta emission

(c) $^{137}_{53}I$ - high neutron/proton ratio, beta emission

(d) $^{133}_{58}Ce$ - low neutron/proton ratio, positron emission

21.13 *Analyze/Plan*. For each nuclide, determine the number of protons and neutrons and find the location on Figure 21.2. If the nuclide does not lie in the belt of stability, refer to the guidelines for neutron-to-proton ratio to determine likely decay modes. *Solve*:

(a) No - high neutron/proton ratio; should be a beta emitter.

(b),(c) No - low neutron/proton ratio; should be a positron emitter, or possibly undergo orbital electron capture.

(d) No - high atomic number; it should be an alpha emitter.

21.14 (a) Yes - within the belt of stability

(b) Close - With 51 neutrons and 43 protons, this nuclide is close to the lower edge of the stability belt. The neutron/proton ratio is somewhat low; if decay occurs, it would be positron emission or orbital electron capture.

(c) Close - slightly low neutron/proton ratio; if decay occurs, it would be positron emission or orbital electron capture.

(d) Yes - within the belt of stability

21.15 *Analyze/Plan*. Use the criteria listed in Table 21.3. *Solve*:

(a) Stable: $^{39}_{19}K$ odd proton, even neutron more abundant than odd proton, odd neutron; 20 neutrons is a magic number.

(b) Stable: $^{209}_{83}Bi$ odd proton, even neutron more abundant than odd proton, odd neutron; 126 neutrons is a magic number.

(c) Stable: $^{25}_{12}Mg$ even though $^{24}_{10}Ne$ is an even proton, even neutron nuclide, it has a very high neutron/proton ratio and lies outside the band of stability.

21.16 Use criteria listed in Table 21.3.

(a) $^{112}_{48}Cd$ even, even more abundant

(b) $^{27}_{13}Al$ odd proton, even neutron more abundant

(c) $^{106}_{46}Pd$ even, even more abundant

(d) $^{128}_{54}Xe$ even proton, even neutron much more abundant than odd proton; odd neutron

21.17 *Analyze/Plan.* For each nuclide, determine the number of protons and neutrons and decide if they are magic numbers. *Solve*:

(a) $^{4}_{2}He$ (c) $^{40}_{20}Ca$ (e) $^{208}_{82}Pb$

(d) $^{58}_{28}Ni$ has a magic number of protons, but not neutrons.

21.18 $^{112}_{50}Sn$ has a magic number of protons and even numbers of protons and neutrons, good indications of nuclear stability. $^{112}_{49}In$ has no magic numbers and odd numbers of protons and neutrons, indicators of nuclear instability or radioactivity.

21.19 *Analyze/Plan.* For each nuclide, determine the number of protons and neutrons and find the location on Figure 21.2. Rationalize the location based on magic numbers, neutron-to-proton ratio and Z value. Predict radioactivity (nonstability of nucleus). *Solve*:

Radioactive: $^{14}_{8}O$, $^{115}_{52}Te$ – low neutron/proton ratio; $^{208}_{84}Po$ – atomic number $\geq$ 84

Stable: $^{32}_{16}S$, $^{78}_{34}Se$ – even proton, even neutron, stable neutron/proton ratio

21.20 The criterion employed in judging whether the nucleus is likely to be radioactive is the position of the nucleus on the plot shown in Figure 21.2. If the neutron/proton ratio is too high or low, or if the atomic number exceeds 83, the nucleus will be radioactive. Radioactive: (b) low neutron/proton ratio; (e) high atomic number; stable: (a), (c) and (d).

Nuclear Transmutations

21.21 Protons and alpha particles are positively charged and must be moving very fast to overcome electrostatic forces which would repel them from the target nucleus. Neutrons are electrically neutral and not repelled by the nucleus.

21.22 A major difference is that the charge on the nitrogen nucleus, +7, is much smaller than on the gold nucleus, +79. Thus, the alpha particle could more easily penetrate the coulomb barrier (that is, the repulsive energy barrier due to like charges) to make contact with the nitrogen nucleus than the gold nucleus. Rutherford used alpha particles that were being emitted from some radioactive source. He did not have access to machines that can accelerate particles to very high energy. It would be necessary to do just that to observe reaction of an alpha particle with a gold nucleus.

21.23 *Analyze/Plan*. Determine A and Z for the missing particle by conservation principles. Find the appropriate symbol for the particle. *Solve*:

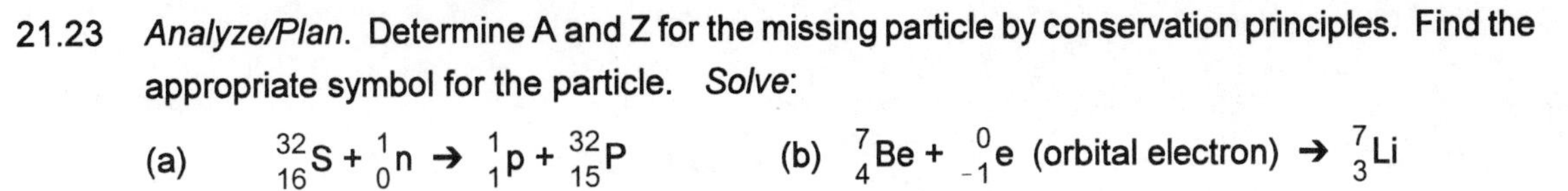

(a) $^{32}_{16}S + ^{1}_{0}n \rightarrow ^{1}_{1}p + ^{32}_{15}P$ (b) $^{7}_{4}Be + ^{0}_{-1}e$ (orbital electron) $\rightarrow ^{7}_{3}Li$

(c) $^{187}_{75}Re \rightarrow ^{187}_{76}Os + ^{0}_{-1}e$ (d) $^{98}_{42}Mo + ^{2}_{1}H \rightarrow ^{1}_{0}n + ^{99}_{43}Tc$

(e) $^{235}_{92}U + ^{1}_{0}n \rightarrow ^{135}_{54}Xe + ^{99}_{38}Sr + 2\, ^{1}_{0}n$

21.24 (a) $^{252}_{98}Cf + ^{10}_{5}B \rightarrow 3\, ^{1}_{0}n + ^{259}_{103}Lr$ (b) $^{2}_{1}H + ^{3}_{2}He \rightarrow ^{4}_{2}He + ^{1}_{1}H$

(c) $^{1}_{1}H + ^{11}_{5}B \rightarrow 3\, ^{4}_{2}He$ (d) $^{122}_{53}I \rightarrow ^{122}_{54}Xe + ^{0}_{-1}e$

(e) $^{59}_{26}Fe \rightarrow ^{0}_{-1}e + ^{59}_{27}Co$

21.25 *Analyze/Plan*. Follow the logic in Sample Exercise 21.5, paying attention to conservation of A and Z. *Solve*:

(a) $^{238}_{92}U + ^{1}_{0}n \rightarrow ^{239}_{92}U + ^{0}_{0}\gamma$ (b) $^{14}_{7}N + ^{1}_{1}H \rightarrow ^{11}_{6}C + ^{4}_{2}He$

(c) $^{18}_{8}O + ^{1}_{0}n \rightarrow ^{19}_{9}F + ^{0}_{-1}e$

21.26 (a) $^{238}_{92}U + ^{4}_{2}H \rightarrow ^{241}_{94}Pu + ^{1}_{0}n$ (b) $^{14}_{7}N + ^{4}_{2}He \rightarrow ^{17}_{8}O + ^{1}_{1}H$

(c) $^{59}_{26}Fe + ^{4}_{2}He \rightarrow ^{63}_{29}Cu + ^{0}_{-1}e$

Rates of Radioactive Decay

21.27 Chemical reactions do not affect the character of atomic nuclei. The energy changes involved in chemical reactions are much too small to allow us to alter nuclear properties via chemical processes. Therefore, the nuclei that are formed in a nuclear reaction will continue to emit radioactivity regardless of any chemical changes we bring to bear. However, we can hope to use chemical means to separate radioactive substances, or remove them from foods or a portion of the environment.

21.28 The suggestion is not reasonable. The energies of nuclear states are very large relative to ordinary temperatures. Thus, merely changing the temperature by less than 100 K would not be expected to significantly affect the behavior of nuclei with regard to nuclear decay rates.

21.29 *Analyze/Plan*. Follow the logic in Sample Exercise 21.6. *Solve*:

After 12.3 yr, one half-life, there are (1/2)48.0 = 24.0 mg. 49.2 yr is exactly four half-lives. There are then $(48.0)(1/2)^4$ = 3.0 mg tritium remaining.

21.30 Calculate the decay constant, k, and then $t_{1/2}$.

$$k = \frac{-1}{t} \ln \frac{N_t}{N_o} = \frac{-1}{5.2 \text{ min}} \ln \frac{0.250 \text{ g}}{1.000 \text{ g}} = 0.2666 = 0.27 \text{ min}^{-1}$$

Using Equation 21.20, $t_{1/2} = 0.693/k = 0.693/0.0.2666 \text{ min}^{-1} = 2.599 = 2.6 \text{ min}$

21.31 *Analyze/Plan.* Given decay time (t), N_o and N_t, calculate the half-life of curium-243. Use the given information with Equation 21.19 to calculate the rate constant, k. Then use Equation 21.20 to calculate half-life. *Solve:*

Using Equation 21.19, $k = \frac{-1}{t} \ln \frac{N_t}{N_o} = \frac{-1}{1.00 \text{ yr}} \times \ln \frac{2921}{3012} = 0.03068 = 0.0307 \text{ yr}^{-1}$

Using Equation 21.20, $t_{1/2} = 0.693/k = 0.693/(0.03068 \text{ yr}^{-1}) = 22.6 \text{ yr}$

21.32 Using Equation 21.19, $k = \frac{-1}{t} \ln \frac{N_t}{N_o} = \frac{-1}{120 \text{ hr}} \times \ln \frac{457}{2310} = 0.01350 = 0.0135 \text{ hr}^{-1}$

Using Equation 21.20, $t_{1/2} = 0.693/k = 0.693/(0.01350 \text{ hr}^{-1}) = 51.3 \text{ hr}$

21.33 *Analyze/Plan.* Follow the logic in Sample Exercise 21.7. In this case, we are given initial sample mass as well as mass at time t, so we can proceed directly to calculate k [Equation 21.20] and then t [Equation 21.19]. *Solve:*

$k = 0.693 / t_{1/2} = 0.693/27.8 \text{ d} = 0.02493 = 0.0249 \text{ d}^{-1}$

$$t = \frac{-1}{k} \ln \frac{N_t}{N_o} = \frac{-1}{0.02493 \text{ d}^{-1}} \ln \frac{1.50}{5.75} = 53.9 \text{ d}$$

21.34 $k = 0.693 / t_{1/2} = 0.693/5.26 \text{ yr} = 0.1317 = 0.132 \text{ yr}^{-1}$; $N_t / N_o = 0.75$

$$t = \frac{-1}{k} \ln \frac{N_t}{N_o} = -(1/0.1317 \text{ yr}^{-1}) \ln (0.75) = 2.18 \text{ yr}$$

2.18 yr = 26.2 mo = 797 d. The source will have to be replaced sometime in the fall of 2002, probably in October.

21.35 (a) *Analyze/Plan.* ${}^{226}_{88}\text{Ra} \rightarrow {}^{222}_{86}\text{Rn} + {}^{4}_{2}\text{He}$

1 α particle is produced for each ^{226}Ra that decays. Calculate the mass of ^{226}Ra remaining after 1.0 min, calculate by subtraction the mass that has decayed, and use Avogadro's number to get the number of ${}^{4}_{2}$He particles. *Solve:*

Calculate k in min^{-1}. $1600 \text{ yr} \times \frac{365 \text{ d}}{1 \text{ yr}} \times \frac{24 \text{ hr}}{1 \text{ d}} \times \frac{60 \text{ min}}{1 \text{ hr}} = 8.410 \times 10^8 \text{ min}$

$$k = \frac{0.693}{t_{1/2}} = \frac{0.693}{8.410 \times 10^8 \text{ min}} = 8.241 \times 10^{-10} \text{ min}^{-1}$$

$$\ln \frac{N_t}{N_o} = -kt = (-8.241 \times 10^{-10} \text{ min}^{-1})(1.0 \text{ min}) = -8.241 \times 10^{-10}$$

$\frac{N_t}{N_o} = e^{-8.241 \times 10^{-10}} = (1.000 - 8.241 \times 10^{-10})$; (don't round here!)

$N_t = 5.0 \times 10^{-3}$ g $(1.00 - 8.241 \times 10^{-10})$ The amount that decays is $N_o - N_t$:

5.0×10^{-3} g $- [5.0 \times 10^{-3} (1.00 - 8.241 \times 10^{-10})] = 5.0 \times 10^{-3}$ g (8.241×10^{-10})

$= 4.120 \times 10^{-12} = 4.1 \times 10^{-12}$ g Ra

$$[N_o - N_t] = 4.120 \times 10^{-12} \text{ g Ra} \times \frac{1 \text{ mol Ra}}{226.0 \text{ g Ra}} \times \frac{6.022 \times 10^{23} \text{ Ra atoms}}{1 \text{ mol Ra}} \times \frac{1\ {}_2^4\text{He}}{1 \text{ Ra atom}}$$

$= 1.098 \times 10^{10} = 1.1 \times 10^{10}$ α particles emitted in 1 min

(b) Plan. The result from (a) is disintegrations/min. Change this to dis/s and apply the definition 1 Ci = 2.7×10^{10} dis/s.

$$1.098 \times 10^{10} \frac{\text{dis}}{\text{min}} \times \frac{1 \text{ min}}{60 \text{ s}} \times \frac{1 \text{ Ci}}{3.7 \times 10^{10} \text{ dis/s}} \times \frac{1000 \text{ mCi}}{\text{Ci}} = 4.945 = 4.9 \text{ mCi}$$

21.36 (a) Proceeding as in Solution 21.35, calculate k in s^{-1}.

$$5.26 \text{ yr} \times \frac{365 \text{ d}}{1 \text{ yr}} \times \frac{24 \text{ hr}}{1 \text{ d}} \times \frac{3600 \text{ sec}}{1 \text{ hr}} = 1.659 \times 10^8 = 1.66 \times 10^8 \text{ s}$$

$$k = \frac{0.693}{t_{1/2}} = \frac{0.693}{1.659 \times 10^8} = 4.178 \times 10^{-9} = 4.18 \times 10^{-9} \text{ s}^{-1}$$

$$\ln \frac{N_t}{N_o} = -kt = -(4.178 \times 10^{-9} \text{ s}^{-1})(45.5 \text{ s}) = -1.901 \times 10^{-7} = -1.90 \times 10^{-7}$$

$\frac{N_t}{N_o} = e^{-1.90 \times 10^{-7}} = (1.000 - 1.90 \times 10^{-7})$; $N_t = 2.44 \times 10^{-3}$ g $(1.000 - 1.90 \times 10^{-7})$

The amount that decays is $N_o - N_t$:

2.44×10^{-3} g $- [2.44 \times 10^{-3}$ g $(1.000 - 1.90 \times 10^{-7})] = 2.44 \times 10^{-3}$ g (1.90×10^{-7})

$= 4.638 \times 10^{-10} = 4.64 \times 10^{-10}$ g Co

$$N_o - N_t = 4.638 \times 10^{-10} \text{ g Co} \times \frac{1 \text{ mol Co}}{60 \text{ g Co}} \times \frac{6.022 \times 10^{23} \text{ Co atoms}}{1 \text{ mol Co}} \times \frac{1\ \beta}{1 \text{ Co atom}}$$

$= 4.655 \times 10^{12} = 4.66 \times 10^{12}$ β particles emitted in 45.5 seconds

(b) $$\frac{4.655 \times 10^{12} \text{ dis}}{45.5 \text{ s}} \times \frac{1 \text{ Bq}}{1 \text{ dis/s}} = 1.02 \times 10^{11} \text{ Bq}$$

21.37 *Analyze/Plan.* Follow the logic in Sample Exercise 21.7. *Solve:*

$$t = \frac{-1}{k} \ln \frac{N_t}{N_o};\ k = 0.693/5715 \text{ yr} = 1.213 \times 10^{-4} = 1.21 \times 10^{-4} \text{ yr}^{-1}$$

$$t = \frac{-1}{1.213 \times 10^{-4} \text{ yr}^{-1}} \ln \frac{24.9}{32.5} = 2.20 \times 10^3 \text{ yr}$$

21.38 Calculate k in yr^{-1} and solve Equation 21.19 for t. N_o = 15.2/min/g, N_t = 8.9/min/g

$k = 0.693/t_{1/2} = 0.693/5715 \text{ yr} = 1.213 \times 10^{-4} = 1.21 \times 10^{-4} \text{ yr}^{-1}$

$$t = \frac{-1}{k} \ln \frac{N_t}{N_o} = \frac{-1}{1.213 \times 10^{-4} \text{ yr}^{-1}} \ln \frac{8.9}{15.2} = 4.414 \times 10^3 = 4.4 \times 10^3 \text{ yr}$$

21.39 *Analyze/Plan.* Follow the procedure outlined in Sample Exercise 21.7. The original quantity of ^{238}U is 50.0 mg plus the amount that gave rise to 14.0 mg of ^{206}Pb. This amount is 14.0(238/206) = 16.2 mg. *Solve:*

$k = 0.693/4.5 \times 10^9 \text{ yr} = 1.54 \times 10^{-10} = 1.5 \times 10^{-10} \text{ yr}^{-1}$

$$t = \frac{-1}{k} \ln \frac{N_t}{N_o} = \frac{-1}{1.54 \times 10^{-10} \text{ yr}^{-1}} \ln \frac{50.0}{66.2} = 1.8 \times 10^9 \text{ yr}$$

21.40 $k = 0.693/1.27 \times 10^9 \text{ yr} = 5.457 \times 10^{-10} = 5.46 \times 10^{-10} \text{ yr}^{-1}$

If the mass of ^{40}Ar is 3.6 times that of ^{40}K, then the original mass of ^{40}K must have been 3.6 + 1 = 4.6 times that now present.

$$t = \frac{-1}{5.457 \times 10^{-10} \text{ yr}^{-1}} \times \ln \frac{1}{(4.6)} = 2.8 \times 10^9 \text{ yr}$$

Energy Changes

21.41 *Analyze/Plan.* Given an energy change, find the corresponding change in mass. Use Equation 21.22, $E = mc^2$. *Solve:*

$\Delta E = c^2 \Delta m$; $\Delta m = \Delta E/c^2$; $1 \text{ J} = \text{kg} \cdot \text{m}^2/\text{s}^2$

$$\Delta m = \frac{393.5 \times 10^3 \text{ kg} \cdot \text{m}^2/\text{s}^2}{(2.9979 \times 10^8 \text{ m/s})^2} \times \frac{1000 \text{ g}}{1 \text{ kg}} = 4.378 \times 10^{-9} \text{ g}$$

21.42 $\Delta E = c^2 \Delta m = (3.0 \times 10^8 \text{ m/s})^2 \times 0.1 \text{ mg} \times \frac{1 \text{ g}}{1000 \text{ mg}} \times \frac{1 \text{ kg}}{1000 \text{ g}} \times \frac{1 \text{ kJ}}{1000 \text{ J}} = 9 \times 10^6 \text{ kJ}$

21.43 *Analyze/Plan.* Given the mass of a ^{23}Na nucleus, find the energy required to separate the nucleus into protons and neutrons. This corresponds to the binding energy of the nucleus. Calculate the total mass of the separate particles and subtract the mass of the nucleus. Convert the difference to energy using Equation 21.22. Use Avogadro's number to calculate energy per mole of nuclei. *Solve:*

Δm = mass of individual protons and neutrons - mass of nucleus

$\Delta m = 11(1.0072765 \text{ amu}) + 12(1.0086649 \text{ amu}) - 22.983733 \text{ amu} = 0.2002873$
$= 0.200287 \text{ amu}$

$$\Delta E = (2.9979246 \times 10^8 \text{ m/s})^2 \times 0.2002873 \text{ amu} \times \frac{1 \text{ g}}{6.0221421 \times 10^{23} \text{ amu}} \times \frac{1 \text{ kg}}{1 \times 10^3 \text{ g}}$$

$= 2.989123 \times 10^{-11} = 2.98912 \times 10^{-11}$ J / ^{23}Na nucleus required

$$2.989123 \times 10^{-11} \frac{\text{J}}{\text{nucleus}} \times \frac{6.0221421 \times 10^{23} \text{ atoms}}{\text{mol}} = 1.80009 \times 10^{13} \text{ J/mol } ^{23}\text{Na}$$

21.44 Δm = mass of individual protons and neutrons - mass of nucleus

$\Delta m = 10(1.0072765 \text{ amu}) + 11(1.0086649 \text{ amu}) - 20.98846 \text{ amu} = 0.1796189$
$= 0.17962 \text{ amu}$

$$\Delta E = (2.9979246 \times 10^8 \text{ m/s})^2 \times 0.1796189 \text{ amu} \times \frac{1 \text{ g}}{6.0221421 \times 10^{23} \text{ amu}} \times \frac{1 \text{ kg}}{1000 \text{ g}}$$

$= 2.680664 \times 10^{-11} = 2.6807 \times 10^{-11}$ J / ^{31}P nucleus required

$$2.680664 \times 10^{-11} \frac{\text{J}}{\text{nucleus}} \times \frac{6.0221421 \times 10^{23} \text{ nuclei}}{\text{mol}}$$

$= 1.6143 \times 10^{13}$ J/mol ^{31}P binding energy

21.45 *Analyze/Plan*. In each case, calculate the mass defect (Δm), total nuclear binding energy and then binding energy per nucleon. *Solve*:

(a) $\Delta m = 6(1.0072765) + 6(1.0086649) - 11.996708 = 0.0989404 = 0.098940 \text{ amu}$

$$\Delta E = 0.0989404 \text{ amu} \times \frac{1 \text{ g}}{6.0221421 \times 10^{23} \text{ amu}} \times \frac{1 \text{ kg}}{1000 \text{ g}} \times \frac{8.987551 \times 10^{16} \text{ m}^2}{\text{s}^2}$$

$= 1.476604 \times 10^{-11} = 1.4766 \times 10^{-11}$ J

binding energy/nucleon $= 1.476604 \times 10^{-11}$ J /12 $= 1.2305 \times 10^{-12}$ J/nucleon

(b) $\Delta m = 17(1.0072765) + 20(1.0086649) - 36.956576 = 0.3404225 = 0.340423 \text{ amu}$

$$\Delta E = 0.3404225 \text{ amu} \times \frac{1 \text{ g}}{6.0221421 \times 10^{23} \text{ amu}} \times \frac{1 \text{ kg}}{1000 \text{ g}} \times \frac{8.987551 \times 10^{16} \text{ m}^2}{\text{s}^2}$$

$= 5.080525 \times 10^{-11} = 5.08053 \times 10^{-11}$ J

binding energy/ nucleon $= 5.080525 \times 10^{-11}$ J / 37 $= 1.37312 \times 10^{-12}$ J/nucleon

(c) Calculate the nuclear mass by subtracting the electron mass from the atomic mass.

$136.905812 \text{ amu} - 56(5.485799 \times 10^{-4} \text{ amu}) = 136.875092 \text{ amu}$

$\Delta m = 56(1.0072765) + 81(1.0086649) - 136.875092 = 1.2342489 = 1.234249 \text{ amu}$

$$\Delta E = 1.2342489 \text{ amu} \times \frac{1 \text{ g}}{6.0221421 \times 10^{23} \text{ amu}} \times \frac{1 \text{ kg}}{1000 \text{ g}} \times \frac{8.987551 \times 10^{16} \text{ m}^2}{\text{s}^2}$$

$= 1.842014 \times 10^{-10}$ J

binding energy/nucleon = 1.842014×10^{-10} J / 137 = 1.344536×10^{-12} J/nucleon

21.46 In each case, calculate the mass defect, total nuclear binding energy and then binding energy per nucleon.

(a) $\Delta m = 7(1.0072765) + 7(1.0086649) - 13.999234 = 0.1123558 = 0.112356$ amu

$$\Delta E = 0.1123558 \text{ amu} \times \frac{1 \text{ g}}{6.0221421 \times 10^{23} \text{ amu}} \times \frac{1 \text{ kg}}{1000 \text{ g}} \times \frac{8.987551 \times 10^{16} \text{ m}^2}{\text{s}^2}$$

$= 1.676817 \times 10^{-11} = 1.67682 \times 10^{-11}$ J

binding energy/nucleon = 1.676817×10^{-11} J / 14 = 1.19773×10^{-12} J/nucleon

(b) $\Delta m = 22(1.0072765) + 26(1.0086649) - 47.935878 = 0.4494924 = 0.449492$ amu

$$\Delta E = 0.4494924 \text{ amu} \times \frac{1 \text{ g}}{6.0221421 \times 10^{23} \text{ amu}} \times \frac{1 \text{ kg}}{1000 \text{ g}} \times \frac{8.987551 \times 10^{16} \text{ m}^2}{\text{s}^2}$$

$= 6.708304 \times 10^{-11} = 6.70830 \times 10^{-11}$ J

binding energy/nucleon = 6.708304×10^{-11} J / 48 = 1.39756×10^{-12} J/nucleon

(c) Calculate the nuclear mass by subtracting the electron mass from the atomic mass.

$200.970277 - 80(5.485799 \times 10^{-4} \text{ amu}) = 200.926391$ amu

$\Delta m = 80(1.0072765) + 121(1.0086649) - 200.926391 = 1.7041819 = 1.704182$ amu

$$\Delta E = 1.7041819 \text{ amu} \times \frac{1 \text{ g}}{6.0221421 \times 10^{23} \text{ amu}} \times \frac{1 \text{ kg}}{1000 \text{ g}} \times \frac{8.987551 \times 10^{16} \text{ m}^2}{\text{s}^2}$$

$= 2.543351 \times 10^{-10}$ J

binding energy/nucleon = 2.543351×10^{-10} J / 201 = 1.265348×10^{-12} J/nucleon

21.47 *Analyze/Plan.* Use Equation 21.22 to calculate the mass equivalence of the solar radiation. *Solve:*

(a) $$\frac{1.07 \times 10^{16} \text{ kJ}}{1 \text{ min}} \times \frac{60 \text{ min}}{1 \text{ hr}} \times \frac{24 \text{ hr}}{1 \text{ day}} = 1.541 \times 10^{19} \frac{\text{kJ}}{\text{day}} = 1.54 \times 10^{22} \text{ J/day}$$

$$\Delta m = \frac{1.541 \times 10^{22} \text{ kg} \cdot \text{m}^2/\text{s}^2/\text{d}}{(2.998 \times 10^8 \text{ m/s})^2} = 1.714 \times 10^5 = 1.71 \times 10^5 \text{ kg/d}$$

(b) *Analyze/Plan.* Calculate the mass change in the given nuclear reaction, then a conversion factor for g ^{235}U to mass equivalent. *Solve:*

$\Delta m = 140.8833 + 91.9021 + 2(1.0086649) - 234.9935 = -0.19077 = -0.1908$ amu

Converting from atoms to moles and amu to grams, it requires 1.000 mol or 235.0 g ^{235}U to produce energy equivalent to a change in mass of 0.1908 g.

0.10% of 1.714×10^5 kg is 1.714×10^2 kg = 1.714×10^5 g

$$1.714 \times 10^5 \text{ g} \times \frac{235.0 \text{ g } ^{235}\text{U}}{0.1908 \text{ g}} = 2.111 \times 10^8 = 2.1 \times 10^8 \text{ g } ^{235}\text{U}$$

(This is about 230 tons of ^{235}U **per day**.)

21.48 (a) $\Delta m = 4.00260 + 1.0086649 - 3.01605 - 2.01410 = -0.0188851 = -0.01889$ amu

$$\Delta E = 0.0188851 \text{ amu} \times \frac{1 \text{ g}}{1 \text{ amu}} \times \frac{1 \text{ kg}}{10^3 \text{ g}} \times (2.99792458 \times 10^8 \text{ m/sec})^2$$

$$= 1.697 \times 10^{12} \text{ J/mol}$$

(b) $\Delta m = 3.01603 + 1.0086649 - 2(2.01410) = -3.5051 \times 10^{-3} = -3.51 \times 10^{-3}$ amu

$\Delta E = 3.15 \times 10^{11}$ J/mol

(c) $\Delta m = 4.00260 + 1.00782 - 3.01603 - 2.01410 = -1.971 \times 10^{-2}$ amu

$\Delta E = 1.771 \times 10^{12}$ J/mol

21.49 We can use Figure 21.13 to see that the binding energy per nucleon (which gives rise to the mass defect) is greatest for nuclei of mass numbers around 50. Thus (a) $^{59}_{27}Co$ should possess the greatest mass defect per nucleon.

21.50 In a fission reactor absorption of neutrons causes a single heavy nucleus to undergo fission, producing two medium mass nuclei. These are seen in Figure 21.13 to have a larger total mass defect than the starting single heavier nucleus, so energy is released.

Effects and Uses of Radioisotopes

21.51 The ^{59}Fe would be incorporated into the diet component, which in turn is fed to the rabbits. After a time blood samples could be removed from the animals, the red blood cells separated, and the radioactivity of the sample measured. If the iron in the dietary compound has been incorporated into blood hemoglobin, the blood cell sample should show beta emission. Samples could be taken at various times to determine the rate of iron uptake, rate of loss of the iron from the blood, and so forth.

21.52 (a) Add ^{36}Cl to water as a chloride salt. Then dissolve ordinary CCl_3COOH. After a time, distill the volatile materials away from the salt; CCl_3COOH is volatile, and will distill with water. Count radioactivity in the volatile material. If chlorine exchange has occurred, there will be radioactivity.

(b) Prepare a saturated solution of $BaCl_2$ containing a small amount of solid $BaCl_2$. Add to this solution solid $BaCl_2$ containing ^{36}Cl. If the solid-solution equilibrium is dynamic, some of the ^{36}Cl in the solid will find itself in solution as chloride ion. After allowing some time for equilibrium to become established, filter the solution, measure radioactivity in the solution that is separated from the solid. If there were no dynamic equilibrium, the $^{36}Cl^-$ would remain in the added solid, since the solution is already saturated before the addition of more solid.

(c) Utilize ^{36}Cl in soils of various pH values; grow plants for a given period of time. Remove plants, and directly measure radioactivity in samples from stems, leaves and so forth, or reduce the volume of plant sample by some form of digestion and evaporation of solution to give a dry residue that can be counted.

21.53 (a) *Control rods* control neutron flux so that there are enough neutrons to sustain the chain reaction but not so many that the core overheats.

(b) A *moderator* slows neutrons so that they are more easily captured by fissioning nuclei.

21.54 (a) In a *chain reaction*, one neutron initiates a nuclear transformation that produces more than one neutron. The product neutrons initiate more transformations, so that the reaction is self-sustaining.

(b) *Critical mass* is the mass of fissionable material required to sustain a chain reaction so that only one product neutron is effective at initiating a new transformation.

21.55 *Analyze/Plan.* Use conservation of A and Z to complete the equations, keeping in mind the symbols and definitions of various decay products. *Solve:*

(a) $^{235}_{92}U + ^{1}_{0}n \rightarrow ^{160}_{62}Sm + ^{72}_{30}Zn + 4\,^{1}_{0}n$ (b) $^{239}_{94}Pu + ^{1}_{0}n \rightarrow ^{144}_{58}Ce + ^{94}_{36}Kr + 2\,^{1}_{0}n$

21.56 (a) $^{2}_{1}H + ^{2}_{1}H \rightarrow ^{3}_{2}He + ^{1}_{0}n$ (b) $^{233}_{92}U + ^{1}_{0}n \rightarrow ^{133}_{51}Sb + ^{98}_{41}Nb + 3\,^{1}_{0}n$

21.57 The extremely high temperature is required to overcome the electrostatic charge repulsions between the nuclei so that they come together to react.

21.58 (a) If the spent fuel rods are more radioactive than the original rods, the products of fission must lie outside the belt of stability and be radioactive themselves.

(b) The heavy ($Z > 83$) nucleus has a high neutron/proton ratio. The lighter radioactive fission products, (e.g., barium-142 and krypton-91) also have high neutron/proton ratios, since only 2 or 3 free neutrons are produced during fission. The preferred decay mode to reduce the neutron/proton ratio is β decay, which has the effect of converting a neutron into a proton. Both barium-142 (86 n, 56 p) and krypton-91 (55 n, 36 p) undergo β decay.

21.59 •OH is a free radical; it contains an unpaired (free) electron, which makes it an extremely reactive species. (As an odd electron molecule, it violates the octet rule.) It can react with almost any particle (atom, molecule, ion) to acquire an electron and become OH^-. This often starts a disruptive chain of reactions, each producing a different free radical.

Hydroxide ion, OH^-, on the other hand, will be attracted to cations or the positive end of a polar molecule. Its most common reaction is ubiquitous and innocuous: $H^+ + OH^- \rightarrow H_2O$. The acid-base reactions of OH^- are usually much less disruptive to the organism than the chain of redox reactions initiated by •OH radical.

21.60 $[H_2\dot{O}\cdot]^+ + H_2\ddot{O}: \longrightarrow [H_3O]^+ + \cdot\ddot{O}-H$

H_2O^+ is a free radical because it contains seven valence electrons. Around the central O atom there are two bonding and one nonbonding electron pairs and a single unpaired electron (3(2) + 1 = 7 valence electrons).

21.61 *Analyze/Plan.* Use definitions of the various radiation units and conversion factors to calculate the specified quantities. Pay particular attention to units. *Solve:*

(a) 1 Ci = 3.7×10^{10} disintegrations(dis)/s; 1 Bq = 1 dis/s

$$8.7 \text{ mCi} \times \frac{1 \text{ Ci}}{1000 \text{ mCi}} \times \frac{3.7 \times 10^{10} \text{ dis/s}}{\text{Ci}} = 3.22 \times 10^8 = 3.2 \times 10^8 \text{ dis/s} = 3.2 \times 10^8 \text{ Bq}$$

(b) 1 rad = 1×10^{-2} J/kg; 1 Gy = 1 J/kg = 100 rad. From part (a), the activity of the source is 3.2×10^8 dis/s.

$$3.22 \times 10^8 \text{ dis/s} \times 2.0 \text{ s} \times 0.65 \times \frac{9.12 \times 10^{-13} \text{ J}}{\text{dis}} \times \frac{1}{0.250 \text{ kg}} = 1.53 \times 10^{-3}$$

$$= 1.5 \times 10^{-3} \text{ J/kg}$$

$$1.5 \times 10^{-3} \text{ J/kg} \times \frac{1 \text{ rad}}{1 \times 10^{-2} \text{ J/kg}} \times \frac{1000 \text{ mrad}}{\text{rad}} = 1.5 \times 10^2 \text{ mrad}$$

$$1.5 \times 10^{-3} \text{ J/kg} \times \frac{1 \text{ Gy}}{1 \text{ J/kg}} = 1.5 \times 10^{-3} \text{ Gy}$$

(c) rem = rad (RBE); Sv = Gy (RBE) = 100 rem

mrem = 1.53×10^2 mrad (9.5) = 1.45×10^3 = 1.5×10^3 mrem (or 1.5 rem)

Sv = 1.53×10^{-3} Gy (9.5) = 1.45×10^{-2} = 1.5×10^{-2} Sv

21.62 (a) 1 Ci = 3.7×10^{10} dis/s; 1 Bq = 1 dis/s

$$21 \text{ mCi} \times \frac{1 \text{ Ci}}{1000 \text{ mCi}} \times 3.7 \times 10^{10} \text{ dis/s} = 7.77 \times 10^8 = 7.8 \times 10^8 \text{ dis/s} = 7.8 \times 10^8 \text{ Bq}$$

(b) 1 Gy = 1 J/kg; 1 Gy = 100 rad

$$7.77 \times 10^8 \text{ dis/s} \times 116 \text{ s} \times 0.065 \times \frac{8.75 \times 10^{-14} \text{ J}}{\text{dis}} \times \frac{1}{65 \text{ kg}} = 7.887 \times 10^{-6}$$

$$= 7.9 \times 10^{-6} \text{ J/kg}$$

$$7.9 \times 10^{-6} \text{ J/kg} \times \frac{1 \text{ Gy}}{1 \text{ J/kg}} = 7.9 \times 10^{-6} \text{ Gy};\ 7.9 \times 10^{-6} \text{ Gy} \times \frac{100 \text{ rad}}{1 \text{ Gy}} = 7.9 \times 10^{-4} \text{ rad}$$

(c) rem = rad (RBE); Sv = Gy (RBE)

7.9×10^{-4} rad (1.0) = 7.9×10^{-4} rem $\times \frac{1000 \text{ mrem}}{1 \text{ rem}}$ = 0.79 mrem

7.9×10^{-6} Gy (1.0) = 7.9×10^{-6} Sv

(d) From Figure 21.23, the average annual background radiation is 360 mrem, or about 1 mrem/day. This 0.79 mrem exposure is less than the average background radiation for a day.

Additional Exercises

21.63 $^{222}_{86}Rn \rightarrow X + 3\,^{4}_{2}He + 2\,^{0}_{-1}\beta$

This corresponds to a reduction in mass number of (3 × 4 =) 12 and a reduction in atomic number of (3 × 2 - 2) = 4. The stable nucleus is $^{210}_{82}Pb$. [This is part of the sequence in Figure 21.4.]

21.64 (a) $^{1}_{0}n \rightarrow {}^{1}_{1}p + {}^{0}_{-1}e$ (or $^{0}_{-1}\beta$)

The other product of neutron decay is a β particle (with the mass and charge of an electron).

(b) Neutrons in atomic nuclei do not decay at this rate because they are stabilized by the strong forces among subatomic particles in a nucleus. Evidence for strong forces in the nucleus includes nuclear binding energies and the coexistence of like-charged protons in the very small volume of the nucleus.

21.65 The most massive radionuclides will have the highest neutron/proton ratios. Thus, they are most likely to decay by a process that lowers this ratio, beta emission. The least massive nuclides, on the other hand, will decay by a process that increases the neutron/proton ratio, positron emission or orbital electron capture.

21.66 (a) $^{36}_{17}Cl \rightarrow {}^{36}_{18}Ar + {}^{0}_{-1}e$

(b) According to Table 21.3, nuclei with even numbers of both protons and neutrons, or an even number of one kind of nucleon, are more stable. ^{35}Cl and ^{37}Cl both have an odd number of protons **but** an even number of neutrons. ^{36}Cl has an odd number of protons and neutrons (17 p, 19 n), so it is less stable than the other two isotopes. Also, ^{37}Cl has 20 neutrons, a nuclear closed shell.

21.67 (a) $^{6}_{3}Li + {}^{56}_{28}Ni \rightarrow {}^{62}_{31}Ga$

(b) $^{40}_{20}Ca + {}^{248}_{96}Cm \rightarrow {}^{288}_{116}X$

(c) $^{88}_{38}Sr + {}^{84}_{36}Kr \rightarrow {}^{116}_{46}Pd + {}^{56}_{28}Ni$

(d) $\ ^{40}_{20}Ca + \ ^{238}_{92}U \rightarrow \ ^{70}_{30}Zn + 4\ ^{1}_{0}n + 2\ ^{102}_{41}Nb$

21.68 This is similar to Solutions 21.35 and 21.36.

$$^{212}_{86}Rn \rightarrow \ ^{208}_{84}Po + \ ^{4}_{2}He$$

Each ^{212}Rn nucleus that decays is 1 disintegration. Calculate the mass of ^{212}Rn remaining after 1.0 s, calculate by subtraction the mass that has decayed, and use Avogadro's number to get the number of nuclei that have decayed.

Calculate k in s^{-1}. $25\ \text{min} \times \dfrac{60\ \text{s}}{1\ \text{min}} = 1.5 \times 10^3\ \text{s}$

$k = 0.693 / t_{1/2} = 0.693/1.5 \times 10^3\ \text{s} = 4.62 \times 10^{-4} = 4.6 \times 10^{-4}\ \text{s}^{-1}$

$\ln(N_t / N_o) - kt = -(4.62 \times 10^{-4}\ \text{s}^{-1})(1.0\ \text{s}) = -4.62 \times 10^{-4} = -4.6 \times 10^{-4}$

$N_t / N_o = e^{-4.62 \times 10^{-4}} = (1.00 - 4.62 \times 10^{-4})$; $N_t = 1.0 \times 10^{-12}\ \text{g}\ (1.000 - 4.62 \times 10^{-4})$

The amount that decays is $N_o - N_t$:

$1.0 \times 10^{-12}\ \text{g} - [1.0 \times 10^{-12}\ \text{g}\ (1.000 - 4.62 \times 10^{-4})] = 1.0 \times 10^{-12}\ \text{g}\ (4.62 \times 10^{-4})$

$= 4.62 \times 10^{-16} = 4.6 \times 10^{-16}\ \text{g}\ ^{212}Rn$

$$N_o - N_t = 4.62 \times 10^{-16}\ \text{g Rn} \times \frac{1\ \text{mol Rn}}{212\ \text{g Rn}} \times \frac{6.022 \times 10^{23}\ \text{Rn atoms}}{1\ \text{mol Rn}} = 1.31 \times 10^6 = 1.3 \times 10^6\ \text{dis}$$

This is 1.3×10^6 disintegrations in 1.0 s, or approximately 1.3×10^6 α particles/s

$$1.31 \times 10^6\ \text{dis/s} \times \frac{1\ \text{Ci}}{3.7 \times 10^{10}\ \text{dis/s}} = 3.547 \times 10^{-5} = 3.5 \times 10^{-5}\ \text{Ci}$$

21.69

Time (hr)	N_t (dis/min)	ln N_t
0	180	5.193
2.5	130	4.868
5.0	104	4.644
7.5	77	4.34
10.0	59	4.08
12.5	46	3.83
17.5	24	3.18

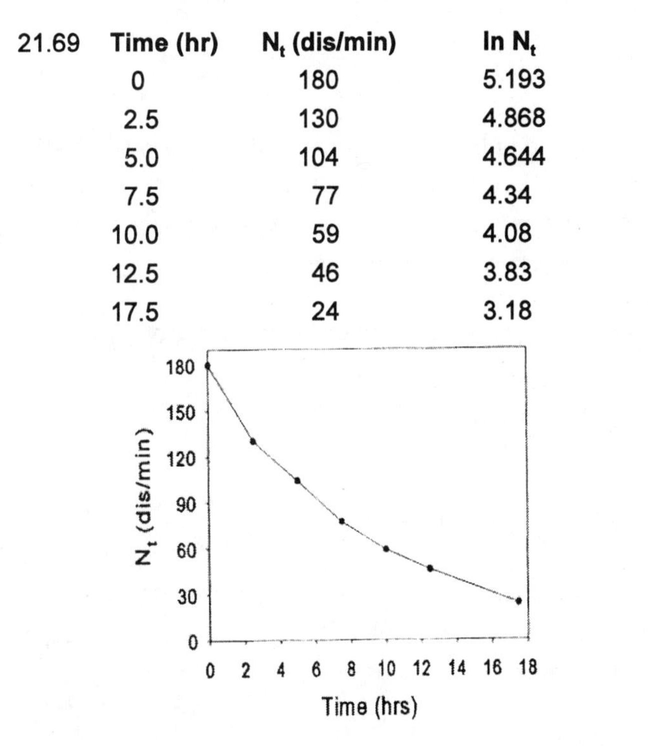

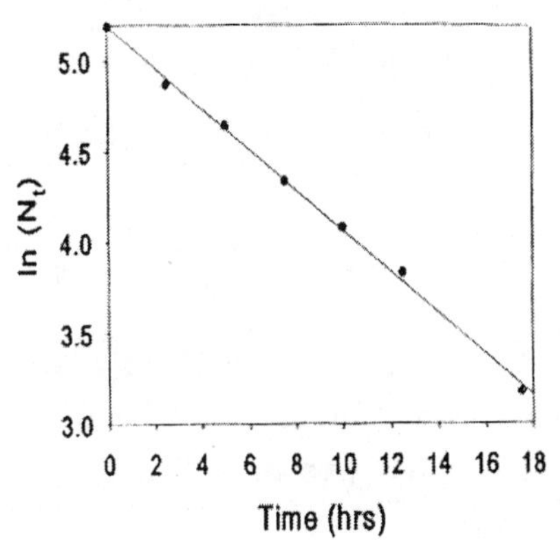

The plot on the left is a graph of activity (disintegrations per minute) vs. time. Choose $t_{1/2}$ at the time where $N_t = 1/2\ N_o = 90$ dis/min. $t_{1/2} \approx 6.0$ hr.

Rearrange Equation 21.19 to obtain the linear relationship shown on the right.

$\ln(N_t / N_o) = -kt$; $\ln N_t - \ln N_o = -kt$; $\ln N_t = -kt + \ln N_o$

The slope of this line = $-k = -0.11$; $t_{1/2} = 0.693/0.11 = 6.3$ hr.

21.70 $1 \times 10^{-6}\ \text{curie} \times \dfrac{3.7 \times 10^{10}\ \text{dis/s}}{\text{curie}} = 3.7 \times 10^4$ dis/s

rate = 3.7×10^4 nuclei/s = kN

$$k = \frac{0.693}{t_{1/2}} = \frac{0.693}{28.8\ \text{yr}} \times \frac{1\ \text{yr}}{365 \times 24 \times 3600\ \text{sec}} = 7.630 \times 10^{-10} = 7.63 \times 10^{-10}\ \text{s}^{-1}$$

3.7×10^4 nuclei/s = $(7.63 \times 10^{-10}/\text{s})$ N; N = $4.849 \times 10^{13} = 4.8 \times 10^{13}$ nuclei

$$\text{mass}\ ^{90}\text{Sr} = 4.849 \times 10^{13}\ \text{nuclei} \times \frac{90\ \text{g Sr}}{6.022 \times 10^{23}\ \text{nuclei}} = 7.2 \times 10^{-9}\ \text{g Sr}$$

21.71 First calculate k in s^{-1}

$$k = \frac{0.693}{2.4 \times 10^4\ \text{yr}} \times \frac{1\ \text{yr}}{365 \times 24 \times 3600\ \text{s}} = 9.16 \times 10^{-13} = 9.2 \times 10^{-13}\ \text{s}^{-1}$$

Now calculate N:

$$N = 0.500\ \text{g Pu} \times \frac{1\ \text{mol Pu}}{239\ \text{g Pu}} \times \frac{6.022 \times 10^{23}\ \text{Pu atoms}}{1\ \text{mol Pu}} = 1.26 \times 10^{21} = 1.3 \times 10^{21}\ \text{Pu atoms}$$

rate = $(9.16 \times 10^{-13}\ \text{s}^{-1})(1.26 \times 10^{21}\ \text{Pu atoms}) = 1.15 \times 10^9 = 1.2 \times 10^9$ dis/s

21.72 The C–OH bond of the acid and the O–H bond of the alcohol break in this reaction. Initially, ^{18}O is present in the C–^{18}OH group of the alcohol. In order for ^{18}O to end up in the ester, the ^{18}O–H bond of the alcohol must break. This requires that the C–OH bond in the acid also breaks. The unlabeled O from the acid ends up in the H_2O product.

21.73 Assume that no depletion of iodide from the water due to plant uptake has occurred. Then the activity after 32 days would be:

$k = 0.693/t_{1/2} = 0.693/8.04\ \text{d} = 0.0862 = 0.086\ \text{d}^{-1}$

$$\ln \frac{N_t}{N_o} = -(0.0862\ \text{d}^{-1})(32\ \text{d}) = -2.758 = -2.8; \quad \frac{N_t}{N_o} = 0.0634 = 0.06$$

We thus expect $N_t = 0.0634(175) = 11.1$ counts/min. We can assume that the plants did not absorb iodide, because absorption would have resulted in an observed level of remaining activity that was **lower** than the theoretical value of 11.1 counts/min.

21.74 First, calculate k in s^{-1}

$$k = \frac{0.693}{12.3 \text{ yr}} \times \frac{1 \text{ yr}}{365 \text{ d}} \times \frac{1 \text{ d}}{24 \text{ hr}} \times \frac{1 \text{ hr}}{3600 \text{ sec}} = 1.7866 \times 10^{-9} = 1.787 \times 10^{-9} \text{ s}^{-1}$$

From Equation 21.18, $1.50 \times 10^3 \text{ s}^{-1} = (1.7866 \times 10^{-9} \text{ s}^{-1})(N)$;

$N = 8.396 \times 10^{11} = 8.40 \times 10^{11}$ In 26.0 g of water, there are

$$26.0 \text{ g } H_2O \times \frac{1 \text{ mol } H_2O}{18.02 \text{ g } H_2O} \times \frac{6.022 \times 10^{23} H_2O}{1 \text{ mol } H_2O} \times \frac{2 \text{ H}}{1 H_2O} = 1.738 \times 10^{24}$$

$= 1.74 \times 10^{24}$ H atoms

The mole fraction of ${}^3_1H$ atoms in the sample is thus

$8.396 \times 10^{11}/1.738 \times 10^{24} = 4.831 \times 10^{-13} = 4.83 \times 10^{-13}$

21.75 Because of the relationship $\Delta E = \Delta mc^2$, the mass defect (Δm) is directly related to the binding energy (ΔE) of the nucleus.

${}^7Be$: 4p, 3n; 4(1.0072765) + 3(1.0086649) = 7.05510 amu

Total mass defect = 7.0551 - 7.0147 = 0.0404 amu

0.0404 amu/7 nucleons = 5.77×10^{-3} amu/nucleon

$$\Delta E = \Delta m \times c^2 = \frac{5.77 \times 10^{-3} \text{ amu}}{\text{nucleon}} \times \frac{1 \text{ g}}{6.022 \times 10^{23} \text{ amu}} \times \frac{1 \text{ kg}}{1 \times 10^3 \text{ g}} \times \frac{8.988 \times 10^{16} \text{ m}^2}{\text{sec}^2}$$

$$= \frac{5.77 \times 10^{-3} \text{ amu}}{\text{nucleon}} \times \frac{1.4925 \times 10^{-10} \text{ J}}{1 \text{ amu}} = 8.612 \times 10^{-13} = 8.61 \times 10^{-13} \text{ J/nucleon}$$

${}^9Be$: 4p, 5n; 4(1.0072765) + 5(1.0086649) = 9.07243 amu

Total mass defect = 9.0724 - 9.0100 = 0.06243 = 0.0624 amu

0.0624 amu/9 nucleons = $6.937 \times 10^{-3} = 6.94 \times 10^{-3}$ amu/nucleon

6.937×10^{-3} amu/nucleon $\times 1.4925 \times 10^{-10}$ J/amu $= 1.035 \times 10^{-12} = 1.04 \times 10^{-12}$ J/nucleon

${}^{10}Be$: 4p, 6n; 4(1.0072765) + 6(1.0086649) = 10.0811 amu

Total mass defect = 10.0811 - 10.0113 = 0.0698 amu

0.0698 amu/10 nucleons = 6.98×10^{-3} amu/nucleon

6.98×10^{-3} amu/nucleon $\times 1.4925 \times 10^{-10}$ J/amu $= 1.042 \times 10^{-12} = 1.04 \times 10^{-12}$ J/nucleon

The binding energies/nucleon for ${}^9Be$ and ${}^{10}Be$ are very similar; that for ${}^{10}Be$ is slightly higher.

21.76 (a) $\Delta m = \Delta E/c^2$; $\Delta m = \dfrac{3.9 \times 10^{26} \text{ J/s}}{(3.00 \times 10^8 \text{ m/s})^2} \times \dfrac{1 \text{ kg} \cdot \text{m}^2/\text{s}^2}{1 \text{ J}} = 4.3 \times 10^9 \text{ kg /s}$

The rate of mass loss is 4.3×10^9 kg/s.

(b) The mass loss arises from fusion reactions that produce more stable nuclei from less stable ones, e.g. Equations 21.26-21.29.

21.77 $1000 \text{ Mwatts} \times \frac{1 \times 10^6 \text{ watts}}{1 \text{ Mwatt}} \times \frac{1 \text{ J}}{1 \text{ watt} \cdot \text{s}} \times \frac{1\ ^{235}\text{U atom}}{3 \times 10^{-11} \text{ J}} \times \frac{1 \text{ mol U}}{6.02 \times 10^{23} \text{ atoms}}$

$\times \frac{235 \text{ g U}}{1 \text{ mol}} \times \frac{3600 \text{ s}}{1 \text{ hr}} \times \frac{24 \text{ hr}}{1 \text{ d}} \times \frac{365 \text{ d}}{1 \text{ yr}} \times \frac{40}{100} (\text{efficiency}) = 1.64 \times 10^5$

$= 2 \times 10^5 \text{ g U/ yr} = 200 \text{ kg U/yr}$

21.78 $2 \times 10^{-12} \text{ curies} \times \frac{3.7 \times 10^{10} \text{ dis/s}}{1 \text{ curie}} = 7.4 \times 10^{-2} = 7 \times 10^{-2} \text{ dis/s}$

$\frac{7.4 \times 10^{-2} \text{ dis/s}}{75 \text{ kg}} \times \frac{8 \times 10^{-13} \text{ J}}{\text{dis}} \times \frac{1 \text{ rad}}{1 \times 10^{-2} \text{ J/g}} \times \frac{3600 \text{ s}}{\text{hr}} \times \frac{24 \text{ hr}}{1 \text{ d}}$

$\times \frac{365 \text{ d}}{1 \text{ yr}} = 2.49 \times 10^{-6} = 2 \times 10^{-6} \text{ rad/yr}$

Recall that there are 10 rem/rad for alpha particles.

$\frac{2.49 \times 10^{-6} \text{ rad}}{1 \text{ yr}} \times \frac{10 \text{ rem}}{1 \text{ rad}} = 2.49 \times 10^{-5} = 2 \times 10^{-5} \text{ rem/yr}$

Integrative Exercises

21.79 Calculate the molar mass of $NaClO_4$ that contains 31% ^{36}Cl. Atomic mass of the enhanced Cl is 0.31(36) + 0.69(35.453) = 35.62. The molar mass of $NaClO_4$ is then (22.99 + 35.62 + 64.00) = 122.61. Calculate N, the number of ^{36}Cl nuclei, the value of k in s^{-1}, and the activity in dis/s.

$49.5 \text{ mg NaClO}_4 \times \frac{1 \text{ g}}{1000 \text{ mg}} \times \frac{1 \text{ mol NaClO}_4}{122.61 \text{ g NaClO}_4} \times \frac{1 \text{ mol Cl}}{1 \text{ mol NaClO}_4} \times \frac{6.022 \times 10^{23} \text{ Cl atoms}}{\text{mol Cl}}$

$\times \frac{31\ ^{36}\text{Cl atoms}}{100 \text{ Cl atoms}} = 7.537 \times 10^{19} = 7.54 \times 10^{19}\ ^{36}\text{Cl atoms}$

$k = 0.693/t_{1/2} = \frac{0.693}{3.0 \times 10^5 \text{ yr}} \times \frac{1 \text{ yr}}{365 \times 24 \times 3600 \text{ s}} = 7.32 \times 10^{-14} = 7.3 \times 10^{-14} \text{ s}^{-1}$

rate = kN = $(7.32 \times 10^{-14} \text{ s}^{-1})(7.547 \times 10^{19} \text{ nuclei}) = 5.52 \times 10^6 = 5.5 \times 10^6$ dis/s

21.80 Calculate the amount of energy produced by the nuclear fusion reaction, the enthalpy of combustion, $\Delta H°$, of C_3H_8, and then the mass of C_3H_8 required.

Δm for the reaction $4\ ^1_1\text{H} \rightarrow\ ^4_2\text{He} + 2\ ^0_1\text{e}$ is $4(1.00782) - 4.00260 \text{ amu} - 2(5.4858 \times 10^{-4} \text{ amu})$

$= 0.027583 = 0.02758 \text{ amu}$

$\Delta E = \Delta mc^2 = 0.027583 \text{ amu} \times \frac{1 \text{ g}}{6.02214 \times 10^{23} \text{ amu}} \times \frac{1 \text{ kg}}{1000 \text{ g}} \times (2.9979246 \times 10^8 \text{ m/s})^2$

$= 4.11654 \times 10^{-12} = 4.117 \times 10^{-12} \text{ J / 4 }^1\text{H nuclei}$

$$1.0 \text{ g } ^1\text{H} \times \frac{1\ ^1\text{H nucleus}}{1.00782 \text{ amu}} \times \frac{6.02214 \times 10^{23} \text{ amu}}{\text{g}} \times \frac{4.11654 \times 10^{-12} \text{ J}}{4\ ^1\text{H nuclei}}$$

$= 6.1495 \times 10^{11}$ J $= 6.1 \times 10^{8}$ kJ produced by the fusion of 1.0 g ^1H.

$C_3H_8(g) + 5O_2(g) \rightarrow 3CO_2(g) + 4H_2O(g)$

$\Delta H° = 3(-393.5 \text{ kJ}) + 4(-241.82 \text{ kJ}) - (-103.85) - (0) = -2043.9 \text{ kJ}$

$$6.1495 \times 10^{8} \text{ kJ} \times \frac{1 \text{ mol } C_3H_8(g)}{2043.9 \text{ kJ}} \times \frac{44.094 \text{ g } C_3H_8}{\text{mol } C_3H_8} = 1.327 \times 10^{7} \text{ g} = 1.3 \times 10^{4} \text{ kg } C_3H_8$$

13,000 kg $C_3H_8(g)$ would have to be burned to produce the same amount of energy as fusion of 1.0 g ^1H.

21.81 (a) $$0.18 \text{ Ci} \times \frac{3.7 \times 10^{10} \text{ dis/s}}{\text{Ci}} \times \frac{3600 \text{ s}}{\text{hr}} \times \frac{24 \text{ hr}}{\text{d}} \times 235 \text{ d} = 1.35 \times 10^{17}$$

$= 1.4 \times 10^{17}$ α particles

(b) $$P = nRT/V = 1.35 \times 10^{17} \text{ He atoms} \times \frac{1 \text{ mol He}}{6.022 \times 10^{23} \text{ atoms}} \times \frac{295 \text{ K}}{0.0150 \text{ L}} \times \frac{0.08206 \text{ L} \cdot \text{atm}}{\text{K} \cdot \text{mol}}$$

$= 3.62 \times 10^{-4} = 3.6 \times 10^{-4}$ atm $= 0.28$ torr

21.82 Calculate N_t in dis/min/g C from 1.5×10^{-2} dis/0.788 g $CaCO_3$. $N_o = 15.3$ dis/min/g C. Calculate k from $t_{1/2}$, calculate t from $\ln(N_t / N_o) = -kt$.

$C(s) + O_2(g) \rightarrow CO_2(g) + Ca(OH_2)(aq) \rightarrow CaCO_3(s) + H_2O(l)$

1 C atom $\rightarrow$ 1 $CaCO_3$ molecule

$$\frac{1.5 \times 10^{-2} \text{ Bq}}{0.788 \text{ g } CaCO_3} \times \frac{1 \text{ dis/s}}{1 \text{ Bq}} \times \frac{60 \text{ s}}{1 \text{ min}} \times \frac{100.1 \text{ g } CaCO_3}{12.01 \text{ g C}} = 9.52 = 9.5 \text{ dis/min/g C}$$

$k = 0.693/t_{1/2} = 0.693/5.715 \times 10^{3} \text{ yr} = 1.213 \times 10^{-4} = 1.21 \times 10^{-4} \text{ yr}^{-1}$

$$t = -\frac{1}{k} \ln \frac{N_t}{N_o} = \frac{-1}{1.213 \times 10^{-4} \text{ yr}^{-1}} \ln \frac{9.52 \text{ dis/min/g C}}{15.3 \text{ dis/min/g C}} = 3.91 \times 10^{3} \text{ yr}$$

21.83 Calculate the energy required to ionize one H_2O molecule, and then the wavelength associated with this energy.

$$\frac{1216 \text{ kJ}}{\text{mol } H_2O} \times \frac{1000 \text{ J}}{1 \text{ kJ}} \times \frac{1 \text{ mol } H_2O}{6.02214 \times 10^{23} \text{ molecules}} = 2.0192 \times 10^{-18}$$

$= 2.019 \times 10^{-18}$ J/molecule

$$\lambda = hc/E = \frac{6.626 \times 10^{-34} \text{ J} \cdot \text{s} \times 2.998 \times 10^{8} \text{ m/s}}{2.0192 \times 10^{-18} \text{ J}} = 9.838 \times 10^{-8} \text{ m} = 98.38 \text{ nm}$$

21.84 Determine the wavelengths of the photons by first calculating the energy equivalent of the mass of an electron or positron. (Since **two** photons are formed by annihilation of **two** particles of equal mass, we need to calculate the energy equivalent of just one particle.) The mass of an electron is 9.109×10^{-31} kg.

$\Delta E = (9.109 \times 10^{-31} \text{ kg}) \times (2.998 \times 10^{8} \text{ m/s})^2 = 8.187 \times 10^{-14} \text{ J}$

Also, $\Delta E = h\nu$; $\Delta E = hc/\lambda$; $\lambda = hc/\Delta E$

$$\lambda = \frac{(6.626 \times 10^{-34} \text{ J}\bullet\text{s})(2.998 \times 10^{8} \text{ m/s})}{8.187 \times 10^{-14} \text{ J}} = 2.426 \times 10^{-12} \text{ m} = 2.426 \times 10^{-3} \text{ nm}$$

This is a very short wavelength indeed; it lies at the short wavelength end of the range of observed gamma ray wavelengths (see Figure 6.4).

21.85 (a) $Ba(NO_3)_2(aq) + Na_2SO_4(aq) \rightarrow BaSO_4(s) + 2NaNO_3(aq)$

(b) 1.25 mmol Ba^{2+} + 1.25 mmol SO_4^{2-} $\rightarrow$ 1.25 mmol $BaSO_4$

Neither reactant is in excess, so the activity of the filtrate is due entirely to $[SO_4^{2-}]$ from dissociation of $BaSO_4(s)$. Calculate $[SO_4^{2-}]$ in the filtrate by comparing the activity of the filtrate to the activity of the reactant.

$$\frac{0.050\ M\ SO_4^{2-}}{1.22 \times 10^{6}\ \text{Bq/mL}} = \frac{x\ M \text{ filtrate}}{250\ \text{Bq/mL}}$$

$[SO_4^{2-}]$ in the filtrate = $1.0246 \times 10^{-5} = 1.0 \times 10^{-5}\ M$

$K_{sp} = [Ba^{2+}][SO_4^{2-}]$; $[SO_4^{2-}] = [Ba^{2+}]$

$K_{sp} = (1.0246 \times 10^{-5})^2 = 1.0498 \times 10^{-10} = 1.0 \times 10^{-10}$

22 Chemistry of the Nonmetals

Periodic Trends and Chemical Reactions

22.1 *Analyze/Plan.* Use the color coded periodic chart on the front-inside cover of the text to classify the given elements. *Solve:*

Metals: (b) Sr, (c) Ce, (e) Rh; nonmetals: (d) Se, (f) Kr; metalloid: (a) Sb

22.2 Metals: (a) Re, (d) Zr, (f) Ga nonmetals: (c) Ar metalloid: (b) As, (e) Te

22.3 *Analyze/Plan.* Follow the logic in Sample Exercise 22.1. *Solve:*

(a) Cl (b) K

(c) K in the gas phase (lowest ionization energy), Li in aqueous solution (most positive E° value)

(d) Ne; Ne and Ar are difficult to compare because they do not form compounds and their radii are not measured in the same way as other elements. However, Ne is several rows to the right of C and surely has a smaller atomic radius. The next smallest is C.

(e) C

22.4 (a) O (b) Br (c) Ba (d) O (e) Co

22.5 *Analyze/Plan.* Use the position of the specified elements on the periodic chart, periodic trends and the arguments in Sample Exercise 22.1 to explain the observations. *Solve:*

(a) Nitrogen is too small to accommodate five fluorine atoms about it. The P and As atoms are larger. Furthermore, P and As have available 3d and 4d orbitals, respectively, to form hybrid orbitals that can accommodate more than an octet of electrons about the central atom.

(b) Si does not readily form π bonds, which would be necessary to satisfy the octet rule for both atoms in SiO.

(c) A reducing agent is a substance that readily loses electrons. As has a lower electronegativity than N; that is, it more readily gives up electrons to an acceptor and is more easily oxidized.

22.6 (a) Nitrogen is a highly electronegative element. In HNO_3 it is in its highest oxidation state, +5, and thus is more readily reduced than phosphorus, which forms stable P-O bonds.

(b) The difference between the third row element and the second lies in the smaller size of C as compared with Si, and the fact that Si has 3d orbitals available to form an sp^3d^2 hybrid set that can accommodate more than an octet of electrons.

(c) Two of the carbon compounds, C_2H_4 and C_2H_2, contain C-C π bonds. Si does not readily form π bonds (to itself or other atoms), so Si_2H_4 and Si_2H_2 are not known as stable compounds.

22.7 *Analyze/Plan.* Follow the logic in Sample Exercise 22.2. *Solve:*

(a) $LiN_3(s) + H_2O(l) \rightleftharpoons HN_3(aq) + LiOH(aq)$

(b) $2C_3H_7OH(l) + 9O_2(g) \rightarrow 6CO_2(g) + 8H_2O(l)$

(c) $NiO(s) + C(s) \rightarrow CO(g) + Ni(s)$ or $2NiO(s) + C(s) \rightarrow CO_2(g) + 2Ni(s)$

(d) $AlP(s) + 3H_2O(l) \rightarrow PH_3(g) + Al(OH)_3(s)$

(e) $Na_2S(s) + 2HCl(aq) \rightarrow H_2S(g) + 2NaCl(aq)$

22.8 (a) $NaOCH_3(s) + H_2O(l) \rightarrow NaOH(aq) + CH_3OH(aq)$

(b) $CuO(s) + 2HNO_3(aq) \rightarrow Cu(NO_3)_2(aq) + H_2O(l)$

(c) $WO_3(s) + 3H_2(g) \rightarrow W(s) + 3H_2O(g)$

(d) $4NH_2OH(l) + O_2(g) \rightarrow 6H_2O(l) + 2N_2(g)$

(e) $Al_4C_3(s) + 12H_2O(l) \rightarrow 4Al(OH)_3(s) + 3CH_4(g)$

Hydrogen, the Noble Gases, and the Halogens

22.9 *Analyze/Plan.* Use information on the isotopes of hydrogen in Section 22.2 to list their symbols, names and relative abundances. *Solve:*

a) ${}^1_1H$ - protium; ${}^2_1H$ - deuterium; ${}^3_1H$ - tritium

b) The order of abundance is proteum > deuterium > tritium.

22.10 Tritium is radioactive. ${}^3_1H \rightarrow {}^3_2He + {}^{0}_{-1}e$

22.11 *Analyze/Plan.* Consider the electron configuration and electronegativity of hydrogen and the halogens. *Solve:*

Like other elements in group 1A, hydrogen has only one valence electron. Like other elements in group 7A, hydrogen needs only one electron to complete its valence shell. The most common oxidation number of H is +1, like the group 1A elements; H can also exist in the -1 oxidation state, a state common to the group 7A elements.

22.12 Hydrogen does not have a strictly comparable electronic arrangement. There are no closed shells of electrons underlying the valence shell. When hydrogen completes its valence shell it has only two electrons therein, a characteristic it shares only with He.

22.13 *Analyze/Plan.* Use information on the descriptive chemistry of hydrogen in Section 22.2 to formulate the required equations. Steam is $H_2O(g)$. *Solve:*

(a) $Mg(s) + 2H^+(aq) \rightarrow Mg^{2+}(aq) + H_2(g)$

(b) $C(s) + H_2O(g) \xrightarrow{1000\,^{\circ}C} CO(g) + H_2(g)$

(c) $CH_4(g) + H_2O(g) \xrightarrow{1100\,^{\circ}C} CO(g) + 3H_2(g)$

22.14 (a) Electrolysis of brine; reaction of carbon with steam; reaction of methane with steam; byproduct in petroleum refining

(b) Synthesis of ammonia; synthesis of methanol; reducing agent; hydrogenation of unsaturated vegetable oils

22.15 *Analyze/Plan.* Use information on the descriptive chemistry of hydrogen given in Section 22.2 to complete and balance the equations. *Solve:*

(a) $NaH(s) + H_2O(l) \rightarrow NaOH(aq) + H_2(g)$

(b) $Fe(s) + H_2SO_4(aq) \rightarrow Fe^{2+}(aq) + H_2(g) + SO_4^{2-}(aq)$

(c) $H_2(g) + Br_2(g) \rightarrow 2HBr(g)$

(d) $2Na(l) + H_2(g) \rightarrow 2NaH(s)$

(e) $PbO(s) + H_2(g) \xrightarrow{\Delta} Pb(s) + H_2O(g)$

22.16 (a) $2Al(s) + 6H^+(aq) \rightarrow 2Al^{3+}(aq) + 3H_2(g)$

(b) $Mg(s) + H_2O(g) \rightarrow MgO(s) + H_2(g)$

(c) $MnO_2(s) + H_2(g) \rightarrow MnO(s) + H_2O(g)$

(d) $CaH_2(s) + 2H_2O(l) \rightarrow Ca(OH)_2(aq) + 2H_2(g)$

22.17 *Analyze/Plan.* If the element bound to H is a nonmetal, the hydride is molecular. If H is bound to a metal with integer stoichiometry, the hydride is ionic; with noninteger stoichiometry, the hydride is metallic. *Solve:*

(a) Molecular (b) ionic (c) metallic

22.18 (a) Ionic (metal hydride) (b) molecular (nonmetal hydride)

(c) metallic (nonstoichiometric transition metal hydride)

22.19 *Analyze/Plan.* Consider the periodic properties of Xe and Ar. *Solve:*

Xenon is larger, and can more readily accommodate an expanded octet. More important is the lower ionization energy of xenon; because the valence electrons are a greater average distance from the nucleus, they are more readily promoted to a state in which the Xe atom can form bonds with fluorine.

22.20 The noble gases are colorless, odorless, diamagnetic, inert gases. They make up a very small fraction of the atmosphere and do not exist in naturally occurring compounds. Because they are rare and unreactive, they were difficult to detect.

22.21 *Analyze/Plan.* Follow the rules for assigning oxidation numbers in Section 4.4 and the logic in Sample Exercise 4.8. *Solve:*

(a) $\mathbf{Br}O_3^-$, +5 (b) H**I**, -1 (c) $\mathbf{BrF_3}$; Br, +3; F, -1

(d) NaO**Cl**, +1 (e) H**Cl**O_4, +7 (f) $\mathbf{XeF_4}$, +4; F, -1

22.22 (a) CaO**Br**, +1 (b) H**Br**O_3, + 5 (c) $\mathbf{Xe}O_3$; Xe, +6

(d) **Cl**O_4^-, +7 (e) H**I**O_2, +3 (f) $\mathbf{IF_5}$; I, +5; F, -1

22.23 *Analyze/Plan.* Review the nomenclature rules and ion names in Section 2.8. *Solve:*

(a) potassium chlorate (b) calcium iodate (c) aluminum chloride

(d) bromic acid (e) paraperiodic acid (f) xenon tetrafluoride

22.24 (a) iron(III) chlorate (b) chlorous acid (c) xenon hexafluoride

(d) bromine pentafluoride (e) xenon oxide tetrafluoride (f) iodic acid

22.25 *Analyze/Plan.* Consider intermolecular forces and periodic properties, including oxidizing power, of the listed substances. *Solve:*

(a) Van der Waals intermolecular attractive forces increase with increasing numbers of electrons in the atoms.

(b) F_2 reacts with water: $F_2(g) + H_2O(l) \rightarrow 2HF(aq) + 1/2\ O_2(g)$. That is, fluorine is too strong an oxidizing agent to exist in water.

(c) HF has extensive hydrogen bonding.

(d) Oxidizing power is related to electronegativity. Electronegativity decreases in the order given.

22.26 (a) The more electronegative the central atom, the greater the extent to which it withdraws charge from oxygen, in turn making the O-H bond more polar, and enhancing ionization of H^+.

(b) HF reacts with the silica which is a major component of glass:
$6HF(aq) + SiO_2(s) \rightarrow SiF_6^{2-}(aq) + 2H_2O(l) + 2H^+(aq)$

(c) Iodide is oxidized by sulfuric acid, as shown in Figure 22.12.

(d) The major factor is size; there is not room about Br for the three chlorides plus the two unshared electron pairs that would occupy the bromine valence shell orbitals.

22.27 *Analyze/Plan.* Use information on the descriptive chemistry of the halogens given in Section 22.4 to complete and balance the equations. *Solve:*

(a) $Br_2(l) + 2OH^-(aq) \rightarrow BrO^-(aq) + Br^-(aq) + H_2O(l)$

(b) $Cl_2(g) + 2I^-(aq) \rightarrow I_2(l) + 2Cl^-(aq)$

22.28 (a) $3CaBr_2(s) + 2H_3PO_4(l) \rightarrow Ca_3(PO_4)_2(s) + 6HBr(g)$

(b) $2HF(aq) + CaCO_3(s) \rightarrow CaF_2(s) + H_2O(l) + CO_2(g)$

22.29 *Analyze/Plan.* For each substance, count valence electrons, draw the correct Lewis structure, and apply the rules of VSEPR to decide electron domain geometry and geometric structure. *Solve:*

(a) square-planar (b) trigonal pyramidal (c) octahedral about the central iodine

(d) linear

22.30 $[:\ddot{F}-\ddot{Br}-\ddot{F}:]^+$ $[SbF_6]^-$

(For clarity, the three unshared electron pairs on each F in the anion are omitted.) The VSEPR model predicts a bent structure for the cation, and an octahedral geometry about Sb for the anion.

Oxygen and the Group 6A Elements

22.31 *Analyze/Plan.* Consider the industrial uses of oxygen and ozone given in Section 22.5. *Solve:*

(a) As an oxidizing agent in steel-making; to bleach pulp and paper; in oxyacetylene torches; in medicine to assist in breathing

(b) Synthesis of pharmaceuticals, lubricants and other organic compounds where C=C bonds are cleaved; in water treatment

22.32 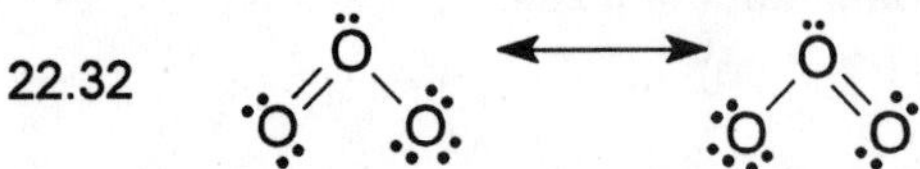

Ozone has two resonance forms (Section 8.7); the molecular structure is bent, with an O–O–O bond angle of approximately 120°. The π bond in ozone is delocalized over the entire molecule; neither individual O–O bond is a full double bond, so the observed O–O distance of 1.28 Å is greater than the 1.21 Å distance in O_2, which has a full O–O double bond.

22.33 *Analyze/Plan.* Use information on the descriptive chemistry of oxygen given in Section 22.5 to complete and balance the equations. *Solve*:

(a) $CaO(s) + H_2O(l) \rightarrow Ca^{2+}(aq) + 2OH^-(aq)$

(b) $Al_2O_3(s) + 6H^+(aq) \rightarrow 2Al^{3+}(aq) + 3H_2O(l)$

(c) $Na_2O_2(s) + 2H_2O(l) \rightarrow 2Na^+(aq) + 2OH^-(aq) + H_2O_2(aq)$

(d) $N_2O_3(g) + H_2O(l) \rightarrow 2HNO_2(aq)$

(e) $2KO_2(s) + 2H_2O(l) \rightarrow 2K^+(aq) + 2OH^-(aq) + O_2(g) + H_2O_2(aq)$

(f) $NO(g) + O_3(g) \rightarrow NO_2(g) + O_2(g)$

22.34 (a) $2HgO(s) \xrightarrow{\Delta} 2Hg(l) + O_2(g)$

(b) $2Cu(NO_3)_2(s) \xrightarrow{\Delta} 2CuO(s) + 4NO_2(g) + O_2(g)$

(c) $PbS(s) + 4O_3(g) \rightarrow PbSO_4(s) + 4O_2(g)$

(d) $2ZnS(s) + 3O_2(g) \rightarrow 2ZnO(s) + 2SO_2(g)$

(e) $2K_2O_2(s) + 2CO_2(g) \rightarrow 2K_2CO_3(s) + O_2(g)$

22.35 *Analyze/Plan.* Oxides of metals are bases, oxides of nonmetals are acids, oxides that act as both acids and bases are amphoteric and oxides that act as neither acids nor bases are neutral. *Solve*:

(a) Neutral (b) acidic (oxide of a nonmetal)

(c) basic (oxide of a metal) (d) amphoteric

22.36 (a) Mn_2O_7 (higher oxidation state of Mn)

(b) SnO_2 (higher oxidation state of Sn)

(c) SO_3 (higher oxidation state of S)

(d) SO_2 (more nonmetallic character of S)

(e) Ga_2O_3 (more nonmetallic character of Ga)

(f) SO_2 (more nonmetallic character of S)

22.37 *Analyze/Plan.* Follow the rules for assigning oxidation numbers in Section 4.4 and the logic in Sample Exercise 4.8. *Solve*:

(a) SeO_3, +6 (b) $Na_2S_2O_3$, +2 (c) SF_4, +4 (d) H_2S, -2 (e) H_2SO_3, +4

Oxygen (a group 6A element) is in the -2 oxidation state in compounds (a), (b) and (e).

22.38 (a) H_2SeO_3, +4 (b) $KHSO_3$, +4 (c) H_2Te, -2 (d) CS_2, -2 (e) $CaSO_4$, +6

22.39 *Analyze/Plan.* The half-reaction for oxidation in all these cases is:

$H_2S(aq) \rightarrow S(s) + 2H^+ + 2e^-$ (The product could be written as $S_8(s)$, but this is not necessary. In fact it is not necessarily the case that S_8 would be formed, rather than some other allotropic form of the element.) Combine this half-reaction with the given reductions to write complete equations. The reduction in (c) happens only in acid solution. The reactants in (d) are acids, so the medium is acidic. *Solve*:

(a) $2Fe^{3+}(aq) + H_2S(aq) \rightarrow 2Fe^{2+}(aq) + S(s) + 2H^+(aq)$

(b) $Br_2(l) + H_2S(aq) \rightarrow 2Br^-(aq) + S(s) + 2H^+(aq)$

(c) $2MnO_4^-(aq) + 6H^+(aq) + 5H_2S(aq) \rightarrow 2Mn^{2+}(aq) + 5S(s) + 8H_2O(l)$

(d) $2NO_3^-(aq) + H_2S(aq) + 2H^+(aq) \rightarrow 2NO_2(aq) + S(s) + 2H_2O(l)$

22.40 (a)

$$2[MnO_4^-(aq) + 8H^+(aq) + 5e^- \rightarrow Mn^{2+}(aq) + 4H_2O(l)]$$
$$5[H_2SO_3(aq) + H_2O(l) \rightarrow SO_4^{2-}(aq) + 4H^+(aq) + 2e^-]$$

$$2MnO_4^-(aq) + 5H_2SO_3(aq) \rightarrow 2MnSO_4(s) + 3SO_4^{2-}(aq) + 3H_2O(l) + 4H^+(aq)$$

(b)

$$Cr_2O_7^{2-}(aq) + 14H^+(aq) + 6e^- \rightarrow 2Cr^{3+}(aq) + 7H_2O(l)$$
$$[H_2SO_3(aq) + H_2O(l) \rightarrow SO_4^{2-}(aq) + 4H^+(aq) + 2e^-]$$

$$Cr_2O_7^{2-}(aq) + 3H_2SO_3(aq) + 2H^+(aq) \rightarrow 2Cr^{3+}(aq) + 3SO_4^{2-}(aq) + 4H_2O(l)$$

(c)

$$Hg_2^{2+}(aq) + 2e^- \rightarrow 2Hg(l)$$
$$H_2SO_3(aq) + H_2O(l) \rightarrow SO_4^{2-}(aq) + 4H^+(aq) + 2e^-$$

$$Hg_2^{2+}(aq) + H_2SO_3(aq) + H_2O(l) \rightarrow 2Hg(l) + SO_4^{2-}(aq) + 4H^+(aq)$$

22.41 *Analyze/Plan.* For each substance, count valence electrons, draw the correct Lewis structure, and apply the rules of VSEPR to decide electron domain geometry and geometric structure. *Solve*:

(a) $[:\ddot{O}-\ddot{Se}-\ddot{O}:]^{2-}$ with a third :Ö: bonded to Se

trigonal pyramidal

(b) :Cl—S—S—Cl:

bent (free rotation around S-S bond)

(c) :O: bonded to S; :Ö—S—Cl:; S bonded to :Ö—H

tetrahedral

22.42 SF_4, 34 e^- SF_5^-, 42 e^-

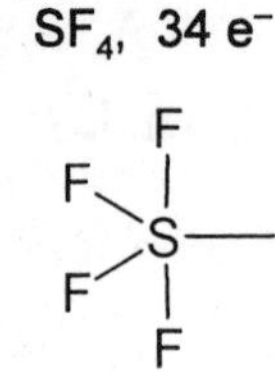

(lone pairs on F atoms omitted for clarity)

trigonal bipyramidal electron pair geometry

see saw molecular geometry

octahedral electron pair geometry

square pyramidal molecular geometry

22.43 *Analyze/Plan.* Use information on the descriptive chemistry of sulfur given in Section 22.6 to complete and balance the equations. *Solve:*

(a) $SO_2(s) + H_2O(l) \rightarrow H_2SO_3(aq) \rightleftharpoons H^+(aq) + HSO_3^-(aq)$

(b) $ZnS(s) + 2HCl(aq) \rightarrow ZnCl_2(aq) + H_2S(g)$

(c) $8SO_3^{2-}(aq) + S_8(s) \rightarrow 8S_2O_3^{2-}(aq)$

(d) $SO_3(aq) + H_2SO_4(l) \rightarrow H_2S_2O_7(l)$

22.44 (a) $Al_2Se_3(s) + 6H^+(aq) \rightarrow 2Al^{3+}(aq) + 3H_2Se(g)$

(b) $Cl_2(aq) + S_2O_3^{2-}(aq) + H_2O(l) \rightarrow 2Cl^-(aq) + S(s) + SO_4^{2-}(aq) + 2H^+(aq)$

Nitrogen and the Group 5A Elements

22.45 *Analyze/Plan.* Follow the rules for assigning oxidation numbers in Section 4.4 and the logic in Sample Exercise 4.8. *Solve:*

(a) $NaNO_2$, +3 (b) NH_3, -3 (c) N_2O, +1 (d) NaCN, -3
(e) HNO_3, +5 (f) NO_2, +4

22.46 (a) HNO_2, +3 (b) N_2H_4, -2 (c) KCN, -3 (d) $NaNO_3$, +5

(e) NH_4Cl, -3 (f) Li_3N, -3

22.47 *Analyze/Plan.* For each substance, count valence electrons, draw the correct Lewis structure, and apply the rules of VSEPR to decide electron domain geometry and geometric structure. *Solve:*

(a) tetrahedral

(b) 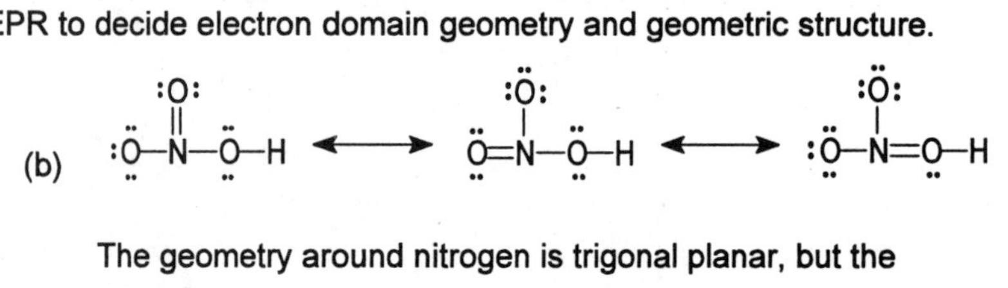

The geometry around nitrogen is trigonal planar, but the hydrogen atom is not required to lie in this plane. The third resonance form makes a much smaller contribution to the structure than the first two.

(c) $:\ddot{N}=N=\ddot{O}: \longleftrightarrow :N\equiv N-\ddot{O}: \longleftrightarrow :\ddot{N}-N\equiv O:$

The molecule is linear. Again, the third resonance form makes less contribution to the structure because of the high formal charges involved.

(d) $\ddot{O}=\dot{N}-\ddot{O}: \longleftrightarrow :\ddot{O}-\dot{N}=\ddot{O}:$

The molecule is bent (nonlinear).

22.48 (a) $:\ddot{O}=\ddot{N}-\ddot{O}-H$

The molecule is bent around the central oxygen and nitrogen atoms; the four atoms need not lie in a plane.

(b) $[:\ddot{N}=N=\ddot{N}:]^- \longleftrightarrow [:N\equiv N-\ddot{N}:]^- \longleftrightarrow [:\ddot{N}-N\equiv N:]^-$

The molecule is linear.

(c) $[H-N(H)(H)-\ddot{N}(H)(H)]^+$

The geometry is tetrahedral around the left nitrogen, trigonal pyramidal around the right.

(d) 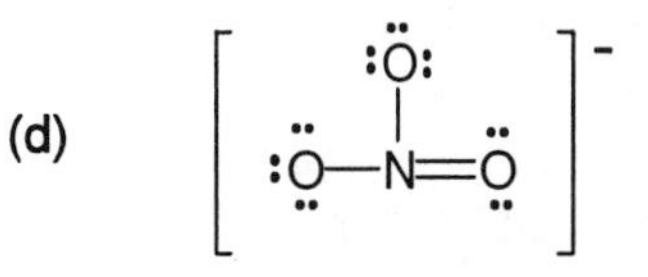

(three equivalent resonance forms)
The ion is trigonal planar.

22.49 *Analyze/Plan.* Use information on the descriptive chemistry of nitrogen given in Section 22.7 to complete and balance the equations. *Solve:*

(a) $Mg_3N_2(s) + 6H_2O(l) \rightarrow 3Mg(OH)_2(s) + 2NH_3(aq)$

(b) $2NO(g) + O_2(g) \rightarrow 2NO_2(g)$

(c) $N_2O_5(g) + H_2O(l) \rightarrow 2H^+(aq) + 2NO_3^-(aq)$

(d) $NH_3(aq) + H^+(aq) \rightarrow NH_4^+(aq)$

(e) $N_2H_4(l) + O_2(g) \rightarrow N_2(g) + 2H_2O(g)$

22.50 (a) $4Zn(s) + 2NO_3^-(aq) + 10H^+(aq) \rightarrow 4Zn^{2+}(aq) + N_2O(g) + 5H_2O(l)$

(b) $4NO_3^-(aq) + S(s) + 4H^+(aq) \rightarrow 4NO_2(g) + SO_2(g) + 2H_2O(l)$

(or $6NO_3^-(aq) + S(s) + 4H^+(aq) \rightarrow 6NO_2(g) + SO_4^{2-}(aq) + 2H_2O(l)$)

(c) $2NO_3^-(aq) + 3SO_2(g) + 2H_2O(l) \rightarrow 2NO(g) + 3SO_4^{2-}(aq) + 4H^+(aq)$

(d) $N_2H_4(g) + 5F_2(g) \rightarrow 2NF_3(g) + 4HF(g)$

(e) $4CrO_4^{2-}(aq) + 3N_2H_4(aq) + 4H_2O(l) \rightarrow 4Cr(OH)_4^-(aq) + 4OH^-(aq) + 3N_2(g)$

22.51 *Analyze/Plan.* Follow the method for writing balanced half-reactions given in Section 20.1 and Sample Exercises 20.2 and 20.3. Find standard reduction potentials in figure 22.30. *Solve:*

(a) $2NO_3^-(aq) + 12H^+(aq) + 10e^- \rightarrow N_2(g) + 6H_2O(l)$ $E^\circ_{red} = +1.25$ V

(b) $2NH_4^+(aq) \rightarrow N_2(g) + 8H^+(aq) + 6e^-$ $E^\circ_{red} = 0.27$ V

22.52 (a) $NO_3^-(aq) + 4H^+(aq) + 3e^- \rightarrow NO(g) + 2H_2O(l)$ $E^\circ_{red} = +0.96$ V

(b) $HNO_2(aq) \rightarrow NO_2(g) + H^+(aq) + 1e^-$ $E^\circ_{red} = 1.12$ V

22.53 *Analyze/Plan.* Follow the rules for assigning oxidation numbers in Section 4.4 and the logic in Sample Exercise 4.8. *Solve:*

(a) H_3PO_4, +5 (b) $H_3\mathbf{As}O_3$, +3 (c) $\mathbf{Sb}_2S_3$, +3 (d) $Ca(H_2PO_4)_2$, +5 (e) $K_3\mathbf{P}$, -3

22.54 (a) H_3PO_3, +3 (b) $H_4P_2O_7$, + 5 (c) $\mathbf{Sb}Cl_3$, +3 (d) $Mg_3\mathbf{As}_2$, +5 (e) $\mathbf{P}_2O_5$, +5

22.55 *Analyze/Plan.* Consider the structures of the componds of interest when explaining the observations. *Solve:*

(a) Phosphorus is a larger atom and can more easily accommodate five surrounding atoms and an expanded octet of electrons than nitrogen can. Also, P has energetically "available" 3d orbitals which participate in the bonding, but nitrogen does not.

(b) Only one of the three hydrogens in H_3PO_2 is bonded to oxygen. The other two are bonded directly to phosphorus and are not easily ionized because the P–H bond is not very polar.

(c) PH_3 is a weaker base than H_2O (PH_4^+ is a stronger acid than H_3O^+). Any attempt to add H^+ to PH_3 in the presence of H_2O merely causes protonation of H_2O.

(d) White phosphorus consists of P_4 molecules, with P–P–P bond angles of 60°. Each P atom has four VSEPR pairs of electrons, so the predicted electron pair geometry is tetrahedral and the preferred bond angle is 109°. Because of the severely strained bond angles in P_4 molecules, white phosphorus is highly reactive.

22.56 (a) Only two of the hydrogens in H_3PO_3 are bound to oxygen. The third is attached directly to phosphorus, and not readily ionized, because the H–P bond is not very polar.

(b) The smaller, more electronegative nitrogen withdraws more electron density from the O–H bond, making it more polar and more likely to ionize.

(c) Phosphate rock consists of $Ca_3(PO_4)_2$, which is only slightly soluble in water. The phosphorus is unavailable for plant use.

(d) N_2 can form stable π bonds to complete the octet of both N atoms. Because phosphorus atoms are larger than nitrogen atoms, they do not form stable π bonds with themselves and must form σ bonds with several other phosphorus atoms (producing P_4 tetrahedral or sheet structures) to complete their octets.

(e) In solution Na_3PO_4 is completely dissociated into Na^+ and PO_4^{3-}. PO_4^{3-}, the conjugate base of the very weak acid HPO_4^{2-}, has a K_b of 2.4×10^{-2} and produces a considerable amount of OH^- by hydrolysis of H_2O.

22.57 *Analyze/Plan.* Use information on the descriptive chemistry of phosphorus given in Section 22.8 to complete and balance the equations. *Solve:*

(a) $2Ca_3(PO_4)_2(s) + 6SiO_2(s) + 10C(s) \xrightarrow{\Delta} P_4(g) + 6CaSiO_3(l) + 10CO(g)$

(b) $3H_2O(l) + PCl_3(l) \rightarrow H_3PO_3(aq) + 3H^+(aq) + 3Cl^-(aq)$

(c) $6Cl_2(g) + P_4(s) \rightarrow 4PCl_3(l)$

22.58 (a) $PCl_5(l) + 4H_2O(l) \rightarrow H_3PO_4(aq) + 5HCl(aq)$

(b) $2H_3PO_4(aq) \xrightarrow{\Delta} H_4PO_7(aq) + H_2O(l)$

(c) $P_4O_{10}(s) + 6H_2O(l) \rightarrow 4H_3PO_4(aq)$

Carbon, the Other Group 4A Elements, and Boron

22.59 *Analyze/Plan.* Review the nomenclature rules and ion names in Section 2.8. *Solve:*

(a) HCN (b) SiC (c) $CaCO_3$ (d) CaC_2

22.60 (a) H_2CO_3 (b) $NaCN$ (c) $KHCO_3$ (d) C_2H_2

22.61 *Analyze/Plan.* Use the correct number of valence electrons and satisfy the octet rule for all atoms. *Solve:*

(a) $[:C\equiv N:]^-$ (b) $:C\equiv O:$ (c) $[:C\equiv C:]^{2-}$

(d) $S=C=S$ (e) $O=C=O$ (f) $[O=C(-O)-O]^{2-}$

one of three equivalent resonance structures

22.62 (a) The CH_3 carbon is tetrahedral; it employs an sp^3 hybrid orbital set. The other two carbons have a linear geometry about them; they employ an sp hybrid orbital set.

(b) The carbon in CN^- uses an sp hybrid orbital set. The Lewis structure of CN^- is [:C ≡ N:]⁻. The lone pair and the C-N σ bond electrons occupy the sp hybrid orbitals. The other two p orbitals are employed in π bonding to N.

(c) The carbon in CS_2 has an sp hybrid orbital set, consistent with the linear geometry.

(d) In C_2H_6 each carbon is in a tetrahedral environment of one C–C and three C–H bonds. The C cations employ sp^3 hybrid orbitals.

22.63 *Analyze/Plan.* Use information on the descriptive chemistry of carbon given in Section 22.9 to complete and balance the equations. *Solve:*

(a) $ZnCO_3(s) \xrightarrow{\Delta} ZnO(s) + CO_2(g)$

(b) $BaC_2(s) + 2H_2O(l) \rightarrow Ba^{2+}(aq) + 2OH^-(aq) + C_2H_2(g)$

(c) $C_2H_4(g) + 3O_2(g) \rightarrow 2CO_2(g) + 2H_2O(g)$

(d) $2CH_3OH(l) + 3O_2(g) \rightarrow 2CO_2(g) + 4H_2O(g)$

(e) $NaCN(s) + H^+(aq) \rightarrow Na^+(aq) + HCN(g)$

22.64 (a) $CO_2(g) + OH^-(aq) \rightarrow HCO_3^-(aq)$

(b) $NaHCO_3(s) + H^+(aq) \rightarrow Na^+(aq) + H_2O(l) + CO_2(g)$

(c) $2CaO(s) + 5C(s) \xrightarrow{\Delta} 2CaC_2(s) + CO_2(g)$

(d) $C(s) + H_2O(g) \xrightarrow{\Delta} H_2(g) + CO(g)$

(e) $CuO(s) + CO(g) \rightarrow Cu(s) + CO_2(g)$

22.65 *Analyze/Plan.* Use information on the descriptive chemistry of carbon given in Section 22.9 to complete and balance the equations. *Solve:*

(a) $2CH_4(g) + 2NH_3(g) + 3O_2(g) \xrightarrow[\text{cat}]{800^\circ C} 2HCN(g) + 6H_2O(g)$

(b) $NaHCO_3(s) + H^+(aq) \rightarrow CO_2(g) + H_2O(l) + Na^+(aq)$

(c) $2BaCO_3(s) + O_2(g) + 2SO_2(g) \rightarrow 2BaSO_4(s) + 2CO_2(g)$

22.66 (a) $2Mg(s) + CO_2(g) \rightarrow 2MgO(s) + C(s)$

(b) $6CO_2(g) + 6H_2O(l) \xrightarrow{h\nu} C_6H_{12}O_6(aq) + 6O_2(g)$

(c) $CO_3^{2-}(aq) + H_2O(l) \rightarrow HCO_3^-(aq) + OH^-(aq)$

22.67 *Analyze/Plan.* Follow the rules for assigning oxidation numbers in Section 4.4 and the logic in Sample Exercise 4.8. *Solve:*

(a) H_3BO_3, +3 (b) $SiBr_4$, +4 (c) $PbCl_2$, +2 or $PbCl_4$, + 4

(d) $Na_2B_4O_7 \cdot 10H_2O$, +3 (e) B_2O_3, +3

22.68 (a) SiO_2, +4 (b) $GeCl_4$, +4 (c) $NaBH_4$, +3 (d) $SnCl_2$, +2 (e) B_2H_6, +3

22.69 *Analyze/Plan.* Consider periodic trends within a family, particularly metallic character. *Solve*:

(a) Carbon (b) lead (c) silicon

22.70 (a) Carbon (b) lead (c) germanium

22.71 *Analyze/Plan.* Consider the structural chemistry of silicates discussed in Section 22.10 and shown in Figures 22.51-22.53. *Solve*:

(a) SiO_4^{4-} (b) SiO_3^{2-} (c) SiO_3^{2-}

22.72 (a)

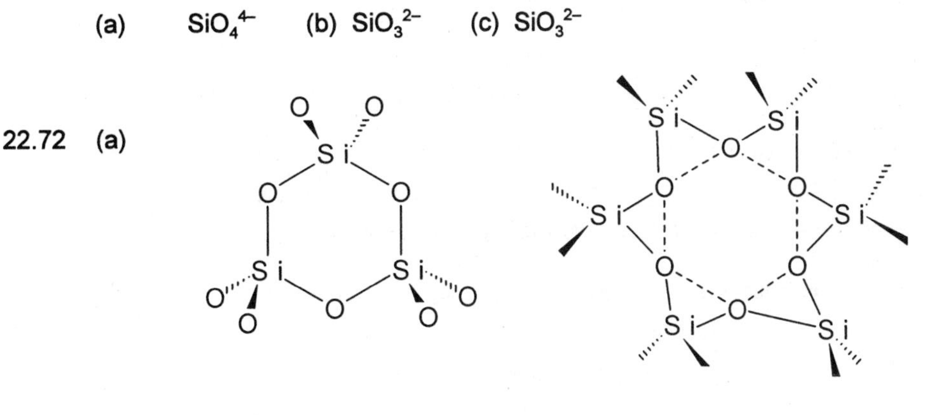

(b) $Si_3O_9^{6-}$ $Si_6O_{18}^{12-}$

22.73 (a) Diborane (Figure 22.55 and below) has bridging H atoms linking the two B atoms. The structure of ethane shown below has the C atoms bound directly, with no bridging atoms.

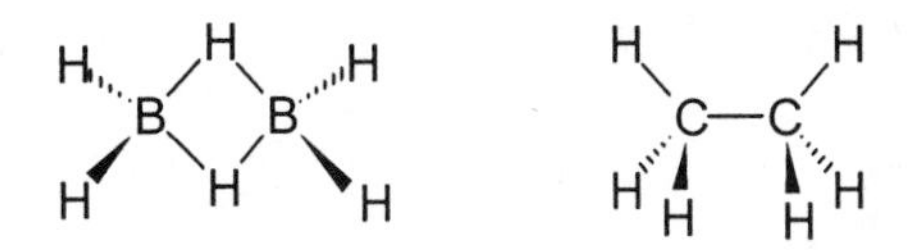

(b) B_2H_6 is an electron deficient molecule. It has 12 valence electrons, while C_2H_6 has 14 valence electrons. The 6 valence electron pairs in B_2H_6 are all involved in B–H sigma bonding, so the only way to satisfy the octet rule at B is to have the bridging H atoms shown in Figure 22.55.

(c) A hydride ion, H^-, has two electrons while an H atom has one. The term *hydridic* indicates that the H atoms in B_2H_6 have more than the usual amount of electron density for a covalently bound H atom.

22.74 (a) $B_2H_6(g) + 6H_2O(l) \rightarrow 2H_3BO_3(aq) + 6H_2(g)$

(b) $4H_3BO_3(s) \xrightarrow{\Delta} H_2B_4O_7(s) + 5H_2O(g)$

(c) $B_2O_3(s) + 3H_2O(l) \rightarrow 2H_3BO_3(aq)$

Additional Exercises

22.75 (a) *Isotopes* are atoms of the same element with different masses; their nuclei have the same number of protons but different numbers of neutrons.

(b) *Allotropes* are different structural forms of the same element. They are composed of atoms of a single element bound into different structures. For example, graphite, diamond and buckey balls are all allotropes of carbon.

(c) *Disproportionation* is an oxidation-reduction process where the same element is both oxidized and reduced.

(d) *Interhalogen* is a compound formed from atoms of two or more halogens.

(e) *Frasch* process is a mining technique for extracting deposits of elemental sulfur. Superheated water is forced into the deposit and melts the sulfur. Compressed air forces the molten sulfur to the surface, where it solidifies.

(f) *Ostwald* process is a commercial process for converting ammonia to nitric acid. It begins with the catalytic oxidation of NH_3 to NO, followed by oxidation of NO to NO_2 and hydration of NO_2 to form HNO_3.

(g) A *condensation reaction* is the combination of two molecules to form a large molecule and a small one such as H_2O or HCl.

22.76 (a) $$10.0 \text{ lb FeTi} \times \frac{453.6 \text{ g}}{1 \text{ lb}} \times \frac{1 \text{ mol FeTi}}{103.7 \text{ g FeTi}} \times \frac{1 \text{ mol } H_2}{1 \text{ mol FeTi}} \times \frac{2.016 \text{ g H}}{1 \text{ mol } H_2} = 88.18$$

$$= 88.2 \text{ g H}$$

(b) $$V = \frac{88.18 \text{ g } H_2}{2.016 \text{ g/mol } H_2} \times \frac{0.08206 \text{ L} \bullet \text{atm}}{\text{mol} \bullet \text{K}} \times \frac{273 \text{ K}}{1 \text{ atm}} = 979.9 = 980 \text{ L}$$

22.77 (a) React an ionic nitride with D_2O, e.g.,
$Mg_3N_2(s) + 6D_2O(l) \rightarrow 2ND_3(aq) + 3Mg(OD)_2(s)$

(b) React SO_3 with D_2O: $SO_3(g) + D_2O(l) \rightleftharpoons D_2SO_4(aq)$

(c) React Na_2O with D_2O: $Na_2O(s) + D_2O(l) \rightarrow 2NaOD(aq)$

(d) Dissolve $N_2O_5(g)$ in D_2O: $N_2O_5(g) + D_2O(l) \rightarrow 2DNO_3(aq)$

(e) React CaC_2 with D_2O: $CaC_2(s) + 2D_2O(l) \rightarrow Ca^{2+}(aq) + 2OD^-(aq) + C_2D_2(g)$

(f) Add NaCN to the D_2SO_4 solution prepared in (b):

$NaCN(s) + D^+(aq) \xrightarrow{\Delta} DCN(aq) + Na^+(aq)$

The DCN can be removed as gas from the reaction.

22.78 $BrO_3^-(aq) + XeF_2(aq) + H_2O(l) \rightarrow Xe(g) + 2HF(aq) + BrO_4^-(aq)$

22.79 Substances that will burn in O_2: SiH_4, CO, Mg.

The others, SiO_2, CO_2 and CaO, have Si, C and Ca in maximum oxidation states, so O_2 cannot act as an oxidizing agent.

22.80 (a) $SO_2(g) + H_2O(l) \rightleftharpoons H_2SO_3(aq)$

(b) $Cl_2O(g) + H_2O(l) \rightleftharpoons 2HClO(aq)$

(c) $Na_2O(s) + H_2O(l) \rightarrow 2Na^{2+}(aq) + 2OH^-(aq)$

(d) $BaC_2(s) + 2H_2O(l) \rightarrow Ba^{2+}(aq) + 2OH^-(aq) + C_2H_2(g)$

(e) $2RbO_2(s) + 2H_2O(l) \rightarrow 2Rb^+(aq) + 2OH^-(aq) + O_2(g) + H_2O_2(aq)$

(f) $Mg_3N_2(s) + 6H_2O(l) \rightarrow 3Mg(OH)_2(s) + 2NH_3(g)$

(g) $Na_2O_2(s) + 2H_2O \rightarrow H_2O_2(aq) + 2NaOH(aq)$

(h) $NaH(s) + H_2O \rightarrow NaOH(aq) + H_2(g)$

22.81 (a) $H_2SO_4 - H_2O \rightarrow SO_3$ (b) $2HClO_3 - H_2O \rightarrow Cl_2O_5$

(c) $2HNO_2 - H_2O \rightarrow N_2O_3$ (d) $H_2CO_3 - H_2O \rightarrow CO_2$

(e) $2H_3PO_4 - 3H_2O \rightarrow P_2O_5$

22.82 $8Fe(s) + S_8(s) \rightarrow 8FeS(s)$

$S_8(s) + 16F_2(g) \rightarrow 8SF_4(g)$ or $S_8(s) + 24F_2(g) \rightarrow 8SF_6(g)$

$S_8(s) + 8O_2(g) \rightarrow 8SO_2(g)$

$S_8(s) + 8H_2(g) \rightarrow 8H_2S(g)$

Sulfur acts as an oxidizing agent in reactions with Fe or H_2 and as a reducing agent in reactions with O_2 or F_2. Incidentally, these reactions are often written using the symbol S rather than S_8 for sulfur.

22.83

$S(g) + O_2(g) \rightarrow SO_2(g)$	$\Delta H = -296.9$ kJ	(1)
$SO_2(g) + 1/2\ O_2(g) \rightarrow SO_3(g)$	$\Delta H = -98.3$ kJ	(2)
$SO_3(g) + H_2O(l) \rightarrow H_2SO_4(aq)$	$\Delta H = -130$ kJ	(3)
$S(g) + 3/2\ O_2(g) + H_2O(l) \rightarrow H_2SO_4(aq)$	$\Delta H = -525$ kJ	

$$1 \text{ ton } H_2SO_4 \times \frac{2000 \text{ lb}}{\text{ton}} \times \frac{453.6 \text{ g}}{1 \text{ lb}} \times \frac{1 \text{ mol } H_2SO_4}{98.09 \text{ g}} \times \frac{-525 \text{ kJ}}{\text{mol } H_2SO_4}$$

$$= -4.86 \times 10^6 \text{ kJ of heat/ton } H_2SO_4$$

22.84 (a) PO_4^{3-}, + 5; NO_3^-, + 5

(b) The Lewis structure for NO_4^{3-} would be:

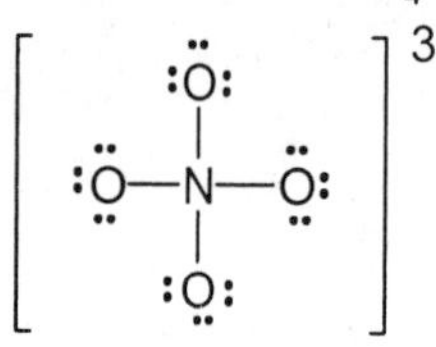

The formal charge on N is +1 and on each O atom is -1. The four electronegative oxygen atoms withdraw electron density, leaving the nitrogen deficient. Since N can form a maximum of four bonds, it cannot form a π bond with one or more of the O atoms to regain electron density, as the P atom in PO_4^{3-} does. Also, the short N–O distance would lead to a tight tetrahedron of O atoms subject to steric repulsion.

22.85 (a) P_4, P_4O_6 and P_4O_{10} all contain a tetrahedron of phosphorus atoms with P–P–P angles of approximately 60°. In the acids containing phosphorus in the +5 oxidation state, H_3PO_4, $H_4P_2O_7$, and $(HPO_3)_n$, P is bound to four O atoms with tetrahedral geometry and approximate 109° bond angles around phosphorus.

(b)

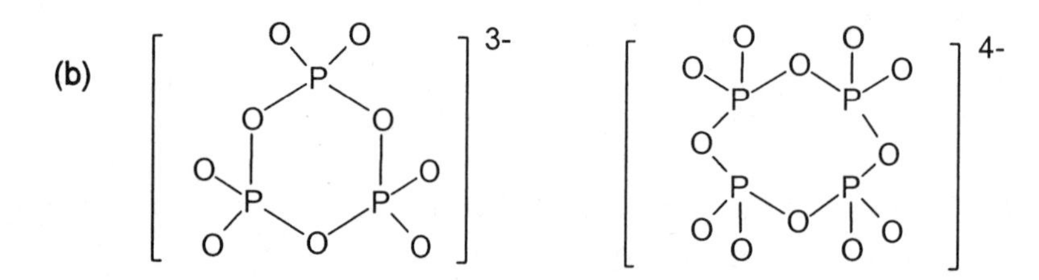

In both structures there are unshared pairs on all oxygens to give octets and the geometry around each P is approximately tetrahedral.

22.86 (a) Use the formula that relates the length of a side of a triangle to the lengths of the other two sides and the opposite angle: $a^2 = b^2 + c^2 - 2bc \cos A$. In this case, a = b = the P–O distance, and A is the P–O–P angle:

For P_4O_6, $a^2 = 2b^2(1 - \cos A) = 2(1.65)^2(1 - \cos 127.5°)$ a = 2.96 Å

For P_4O_{10}, $a^2 = 2b^2(1 - \cos A) = 2(1.60)^2(1 - \cos 124.5°)$ a = 2.83 Å

(b) The shorter P–P distance in P_4O_{10} is due both to the sharper angle and to the shorter P–O distance. The shorter P–O distance may be due to the higher oxidation state and concomitant higher effective charge of P in P_4O_{10}.

22.87 $GeO_2(s) + C(s) \xrightarrow{\Delta} Ge(l) + CO_2(g)$

$Ge(l) + 2Cl_2(g) \rightarrow GeCl_4(l)$

$GeCl_4(l) + 2H_2O(l) \rightarrow GeO_2(s) + 4HCl(g)$

$GeO_2(s) + 2H_2(g) \rightarrow Ge(s) + 2H_2O(l)$

22.88 (a)

$$2[5e^- + MnO_4^-(aq) + 8H^+(aq) \rightarrow Mn^{2+}(aq) + 4H_2O(l)]$$
$$5[H_2O_2(aq) \rightarrow O_2(g) + 2H^+(aq) + 2e^-]$$

$$2MnO_4^-(aq) + 5H_2O_2(aq) + 6H^+(aq) \rightarrow 2Mn^{2+}(aq) + 5O_2(g) + 8H_2O(l)$$

(b)

$$2[Fe^{2+}(aq) \rightarrow Fe^{3+}(aq) + e^-]$$
$$H_2O_2(aq) + 2H^+(aq) + 2e^- \rightarrow 2H_2O(l)$$

$$2Fe^{2+}(aq) + H_2O_2(aq) + 2H^+(aq) \rightarrow 2Fe^{3+}(aq) + 2H_2O(l)$$

(c)

$$2I^-(aq) \rightarrow I_2(s) + 2e^-$$
$$H_2O_2(aq) + 2H^+(aq) + 2e^- \rightarrow 2H_2O(l)$$

$$2I^-(aq) + H_2O_2(aq) + 2H^+(aq) \rightarrow I_2(s) + 2H_2O(l)$$

(d)

$$MnO_2(s) + 4H^+(aq) + 2e^- \rightarrow Mn^{2+}(aq) + 2H_2O(l)$$
$$H_2O_2(aq) \rightarrow O_2(g) + 2H^+(aq) + 2e^-$$

$$MnO_2(s) + 2H^+(aq) + H_2O_2(aq) \rightarrow Mn^{2+}(aq) + 2H_2O(l) + O_2(g)$$

(e)

$$2I^-(aq) \rightarrow I_2(s) + 2e^-$$
$$O_3(g) + H_2O(l) + 2e^- \rightarrow O_2(g) + 2OH^-(aq)$$

$$2I^-(aq) + O_3(g) + H_2O(l) \rightarrow O_2(g) + I_2(s) + 2OH^-(aq)$$

22.89 Assume that the reactions occur in basic solution. The half-reaction for reduction of H_2O_2 is in all cases $H_2O_2(aq) + 2e^- \rightarrow 2OH^-(aq)$.

(a) $H_2O_2(aq) + S^{2-}(aq) \rightarrow 2OH^-(aq) + S(s)$

(b) $SO_2(g) + 2OH^-(aq) + H_2O_2(aq) \rightarrow SO_4^{2-}(aq) + 2H_2O(l)$

(c) $NO_2^-(aq) + H_2O_2(aq) \rightarrow NO_3^-(aq) + H_2O(l)$

(d) $As_2O_3(s) + 2H_2O_2(aq) + 6OH^-(aq) \rightarrow 2AsO_4^{3-}(aq) + 5H_2O(l)$

(e) This reaction must be occurring in acidic solution, since $Fe(OH)_3$ would form if the solution were basic. The half-reactions are:

$$2H^+(aq) + H_2O_2(aq) + 2e^- \rightarrow 2H_2O(l)$$
$$2[Fe^{2+}(aq) \rightarrow Fe^{3+}(aq) + e^-$$

$$2Fe^{2+}(aq) + H_2O_2(aq) + 2H^+(aq) \rightarrow 2Fe^{3+}(aq) + 2H_2O(l)$$

22.90 (a) $Li_3N(s) + 3H_2O(l) \rightarrow 3Li^+(aq) + 3OH^-(aq) + NH_3(aq)$

(b) $NH_3(aq) + H_2O(l) \rightleftharpoons NH_4^+(aq) + OH^-(aq)$

(c) $3NO_2(g) + H_2O(l) \rightarrow NO(g) + 2H^+(aq) + 2NO_3^-(aq)$

(d) $2NO_2(g) \rightleftharpoons N_2O_4(g)$

(e) $4NH_3(g) + 5O_2(g) \xrightarrow{\text{catalyst}} 4NO(g) + 6H_2O(g)$

(f) $2CO(g) + O_2(g) \rightarrow 2CO_2(g)$

(g) $H_2CO_3(aq) \xrightarrow{\Delta} H_2O(g) + CO_2(g)$

(h) $Ni(s) + CO(g) \rightarrow NiO(s) + C(s)$

(i) $CS_2(g) + O_2(g) \rightarrow CO_2(g) + S_2(g)$

(j) $CaO(s) + SO_2(g) \rightarrow CaSO_3(s)$

(k) $2Na(s) + 2H_2O(l) \rightarrow 2NaOH(aq) + H_2(g)$

(l) $CH_4(g) + H_2O(g) \xrightarrow{\Delta} CO(g) + 3H_2(g)$

(m) $LiH(s) + H_2O(l) \rightarrow LiOH(aq) + H_2(g)$

(n) $Fe_2O_3(s) + 3H_2(g) \rightarrow 2Fe(s) + 3H_2O(g)$

Integrative Exercises

22.91 $2XeO_3(s) \rightarrow 2Xe(g) + 3O_2(g)$

$$0.500 \text{ g } XeO_3 \times \frac{1 \text{ mol } XeO_3}{179.1 \text{ g } XeO_3} \times \frac{5 \text{ mol gas}}{2 \text{ mol } XeO_3} = 6.979 \times 10^{-3} = 6.98 \times 10^{-3} \text{ mol gas}$$

$$P = \frac{(6.979 \times 10^{-3} \text{ mol})(0.08206 \text{ L} \bullet \text{atm/mol} \bullet \text{K})(303 \text{ K})}{1.00 \text{ L}} = 0.17354 = 0.174 \text{ atm}$$

22.92 From Appendix C, we need only ΔH_f° for F(g), so that we can estimate ΔH for the process:

$F_2(g) \rightarrow F(g) + F(g)$; $\Delta H^\circ = +160$ kJ.

$XeF_2(g) \rightarrow Xe(g) + F_2(g)$ $-\Delta H_f^\circ = +109$ kJ

$XeF_2(g) \rightarrow Xe(g) + 2F(g)$ $\Delta H^\circ = +269$ kJ

The average Xe-F bond enthalpy is thus 269/2 = 134 kJ. Similarly,

$XeF_4(g) \rightarrow Xe(g) + 2F_2(g)$ $-\Delta H_f^\circ = +218$ kJ

$2F_2(g) \rightarrow 4F(g)$ $\Delta H^\circ = 320$ kJ

$XeF_4(g) \rightarrow Xe(g) + 4F(g)$ $\Delta H^\circ = 538$ kJ

Average Xe-F bond energy = 538/4 = 134 kJ

$XeF_6(g) \rightarrow Xe(g) + 3F_2(g)$ $-\Delta H_f^\circ = 298$ kJ

$3F_2(g) \rightarrow 6F(g)$ $\Delta H^\circ = 480$ kJ

$XeF_6(g) \rightarrow Xe(g) + 6F(g)$ $\Delta H^\circ = 778$ kJ

Average Xe-F bond energy = 778/6 = 130 kJ

The average bond enthalpies are: XeF_2: 134 kJ, XeF_4: 134 kJ, XeF_6: 130 kJ. They are remarkably constant in the series.

22.93 (a) $H_2(g) + 1/2\ O_2(g) \rightarrow H_2O(l)$; $\Delta H = -285.83$ kJ

$CH_4(g) + 2O_2(g) \rightarrow CO_2(g) + 2H_2O(l)$

$\Delta H = 2(-285.83) - 393.5 - (-74.8) = -890.4$ kJ

(b) for H_2: $\dfrac{-285.83\text{ kJ}}{1\text{ mol }H_2} \times \dfrac{1\text{ mol }H_2}{2.0159\text{ g }H_2} = -141.79\text{ kJ/g }H_2$

for CH_4: $\dfrac{-890.4\text{ kJ}}{1\text{ mol }CH_4} \times \dfrac{1\text{ mol }CH_4}{16.043\text{ g }CH_4} = -55.50\text{ kJ/g }CH_4$

(c) Find the number of moles of gas that occupy 1 m^3 at STP:

$$n = \frac{1\text{ atm} \times 1\text{ m}^3}{273\text{ K}} \times \frac{1\text{ K}\bullet\text{mol}}{0.08206\text{ L}\bullet\text{atm}} \times \left[\frac{100\text{ cm}}{1\text{ m}}\right]^3 \times \frac{1\text{ L}}{10^3\text{ cm}^3} = 44.64\text{ mol}$$

for H_2: $\dfrac{-285.83\text{ kJ}}{1\text{ mol }H_2} \times \dfrac{44.64\text{ mol }H_2}{1\text{ m}^3\ H_2} = 1.276 \times 10^4\text{ kJ/m}^3\ H_2$

for CH_4: $\dfrac{-890.4\text{ kJ}}{1\text{ mol }CH_4} \times \dfrac{44.64\text{ mol }CH_4}{1\text{ m}^3\ CH_4} = 3.975 \times 10^4\text{ kJ/m}^3\ CH_4$

22.94 First calculate the molar solubility of Cl_2 in water.

$$n = \frac{1\text{ atm }(0.310\text{ L})}{\frac{0.08206\text{ L}\bullet\text{atm}}{1\text{ mol}\bullet\text{K}} \times 273\text{ K}} = 0.01384 = 0.0138\text{ mol }Cl_2;\ M = \frac{0.01384\text{ mol}}{0.100\text{ L}} = 0.1384$$

$= 0.138\ M$

$$K_{eq} = \frac{[Cl^-][HOCl][H^+]}{[Cl_2]} = 4.7 \times 10^{-4}$$

$[Cl^-] = [HOCl] = [H^+]$ Let this quantity = x. Then, $\dfrac{x^3}{(0.1384 - x)} = 4.7 \times 10^{-4}$

Assuming that x is small compared with 0.1384:

$x^3 = (0.1384)(4.7 \times 10^{-4}) = 6.504 \times 10^{-5}$; $x = 0.0402 = 0.040\ M$

We can correct the denominator using this value, to get a better estimate of x:

$\dfrac{x^3}{0.1384 - 0.0402} = 4.7 \times 10^{-4}$; $x = 0.0359 = 0.036\ M$

One more round of approximation gives $x = 0.0364 = 0.036\ M$. This is the equilibrium concentration of HClO.

22.95 (a) $N_2H_4(g) + O_2(g) \rightarrow N_2(g) + 2H_2O(l)$

(b) $\Delta H^\circ = \Delta H_f^\circ\ N_2(g) + 2\Delta H_f^\circ\ H_2O(l) - \Delta H_f^\circ\ N_2H_4(aq) - \Delta H_f^\circ\ O_2(g)$

$= 0 + 2(-285.83) - 95.40 - 0 = -667.06$ kJ

(c) $\frac{9.1 \text{ g } O_2}{1 \times 10^6 \text{ g } H_2O} \times \frac{1.0 \text{ g } H_2O}{1 \text{ mL } H_2O} \times \frac{1000 \text{ mL}}{1 \text{ L}} \times 3.0 \times 10^4 \text{ L} = 273 = 2.7 \times 10^2 \text{ g } O_2$

$2.73 \times 10^2 \text{ g } O_2 \times \frac{1 \text{ mol } O_2}{32.00 \text{ g } O_2} \times \frac{1 \text{ mol } N_2H_4}{1 \text{ mol } O_2} \times \frac{32.05 \text{ g } N_2H_4}{1 \text{ mol } N_2H_4} = 2.7 \times 10^2 \text{ g } N_2H_4$

22.96 (a) $SO_2(g) + 2H_2S(s) \rightarrow 3S(s) + 2H_2O(g)$ or, if we assume S_8 is the product, $8SO_2(g) + 16H_2S(g) \rightarrow 3S_8(s) + 16H_2O(g)$.

(b) $2000 \text{ lb coal} \times \frac{0.035 \text{ lb S}}{1 \text{ lb coal}} \times \frac{453.6 \text{ g S}}{1 \text{ lb S}} \times \frac{1 \text{ mol S}}{32.07 \text{ g S}} \times \frac{1 \text{ mol } SO_2}{1 \text{ mol S}} \times \frac{2 \text{ mol } H_2S}{1 \text{ mol } SO_2}$

$= 1.98 \times 10^3 = 2.0 \times 10^3 \text{ mol } H_2S$

$V = \frac{1.98 \times 10^3 \text{ mol } (0.08206 \text{ L} \bullet \text{atm/mol} \bullet \text{K})(300 \text{ K})}{(740/760) \text{ atm}} = 5.01 \times 10^4 = 5.0 \times 10^4 \text{ L}$

(c) $1.98 \times 10^3 \text{ mol } H_2S \times \frac{3 \text{ mol S}}{2 \text{ mol } H_2S} \times \frac{32.07 \text{ g S}}{1 \text{ mol S}} = 9.5 \times 10^4 \text{ g S}$

This is about 210 lb S per ton of coal combusted. (However, two-thirds of this comes from the H_2S, which was presumably also obtained from coal.)

22.97 *Plan.* vol air → kg air → g H_2S → g FeS. Use the ideal-gas equation to change volume of air to mass of air, (assuming 1.00 atm, 298 K and an average molar mass ($\mathcal{M}$) for air of 29.0 g/mol. Use 20 mg H_2S/kg air to find the mass of H_2S in the given mass of air.

$V_{air} = 2.7 \text{ m} \times 4.3 \text{ m} \times 4.3 \text{ m} \times \frac{(100)^3 \text{ cm}^3}{1 \text{ m}^3} \times \frac{1 \text{ L}}{1000 \text{ cm}^3} = 4.9923 \times 10^4 = 5.0 \times 10^4 \text{ L}$

$g_{air} = \frac{PV\mathcal{M}}{RT}$; assume P = 1.00 atm, T = 298 K, $\mathcal{M}_{air}$ = 29.0 g/mol

$g_{air} = \frac{1.00 \text{ atm} \times 4.9923 \times 10^4 \text{ L} \times 29.0 \text{ g/mol}}{298 \text{ K}} \times \frac{\text{K} \bullet \text{mol}}{0.08206 \text{ L} \bullet \text{atm}} = 59{,}204 = 5.9 \times 10^4 \text{ g air}$

$5.9203 \times 10^4 \text{ g air} \times \frac{1 \text{ kg}}{1000 \text{ g}} \times \frac{20 \text{ mg } H_2S}{1 \text{ kg air}} \times \frac{1 \text{ g}}{1000 \text{ mg}} = 1.184 = 1.2 \text{ g } H_2S$

$FeS(s) + 2HCl(aq) \rightarrow FeCl_2(aq) + H_2S(g)$

$1.184 \text{ g } H_2 \times \frac{1 \text{ mol } H_2}{34.08 \text{ g } H_2S} \times \frac{1 \text{ mol FeS}}{1 \text{ mol } H_2S} \times \frac{87.91 \text{ g FeS}}{1 \text{ mol FeS}} = 3.054 = 3.1 \text{ g FeS}$

22.98 The reactions can be written as follows:

$H_2(g) + X(\text{std state}) \rightarrow H_2X(g)$	ΔH°_f
$2H(g) \rightarrow H_2(g)$	ΔH°_f (H–H)
$X(g) \rightarrow X(\text{std state})$	ΔH_3
Add: $2H(g) + X(g) \rightarrow H_2X(g)$	$\Delta H = \Delta H^\circ_f + \Delta H^\circ_f(\text{H–H}) + \Delta H_3$

These are all the necessary ΔH values. Thus,

Compound	ΔH	D H–X
H_2O	ΔH = -242 kJ - 436 kJ - 248 kJ = -926 kJ	463 kJ
H_2S	ΔH = -20 kJ - 436 kJ - 277 kJ = -733 kJ	367 kJ
H_2Se	ΔH = +30 kJ - 436 kJ - 227 kJ = -633 kJ	316 kJ
H_2Te	ΔH = +100 kJ - 436 kJ - 197 kJ = -533 kJ	266 kJ

The average H–X bond energy in each case is just half of ΔH. The H–X bond energy decreases steadily in the series. The origin of this effect is probably the increasing size of the orbital from X with which the hydrogen 1s orbital must overlap.

22.99 *Plan.* Calculate the required $[H^+]$ from the given pH. Then determine the concentration and amount of H_2SO_4 needed to achieve this $[H^+]$. The first dissociation step of H_2SO_4 is complete, but the second is not, so we must solve an equilibrium problem. Let x equal the initial concentration of H_2SO_4. *Solve:*

$[H^+] = 10^{-pH} = 10^{-3.5} = 3.16 \times 10^{-4} = 3 \times 10^{-4}\ M\ H^+$

$H_2SO_4(aq) \rightarrow H^+(aq) + HSO_4^-(aq)$

$HSO_4^-(aq) \rightleftharpoons H^+(aq) + SO_4^{2-}(aq) \qquad K_a = 1.2 \times 10^{-2}$

	HSO_4^-	H^+	SO_4^{2-}
initial	x	x	0
change	- y	+ y	+ y
equil.	x - y	x + y	y

$$K_a = \frac{[H^+][SO_4^{2-}]}{[HSO_4^-]} = \frac{(x+y)(y)}{(x-y)};\ [H^+] = x + y = 3.16 \times 10^{-4}\ M;\ y = 3.16 \times 10^{-4} - x$$

$$1.2 \times 10^{-2} = \frac{(3.16 \times 10^{-4})(3.16 \times 10^{-4} - x)}{[x - (3.16 \times 10^{-4} - x)]} = \frac{1.00 \times 10^{-7} - 3.16 \times 10^{-4}\,x}{2x - 3.16 \times 10^{-4}}$$

$1.2 \times 10^{-2}\,(2x - 3.16 \times 10^{-4}) = 1.00 \times 10^{-7} - 3.16 \times 10^{-4}\,x;$

$2.4 \times 10^{-2}\,x - 3.79 \times 10^{-6} = 1.00 \times 10^{-7} - 3.16 \times 10^{-4}\,x$

$2.43 \times 10^{-2}\,x = 3.89 \times 10^{-6};\ x = 1.60 \times 10^{-4}\ M = 2 \times 10^{-4}\ M\ [H^+]$

$$\frac{1.60 \times 10^{-4}\ \text{mol}\ H_2SO_4}{1\ \text{L}\ H_2SO_4} \times 1.00 \times 10^3\ \text{L seawater} \times \frac{98.09\ \text{g}\ H_2SO_4}{\text{mol}} = 15.7\ \text{g} = 2 \times 10^1\ \text{g}\ H_2SO_4$$

Check. The calculated concentration should be slightly greater than one-half the desired $[H^+]$, since the second dissociation of H_2SO_4 is not quite complete.

1.60×10^{-4} *M* > $(3.16 \times 10^{-4})/2 = 1.58 \times 10^{-2}$ *M*. Substituting into the K_a expression:

$$\frac{(3.16 \times 10^{-4})(1.56 \times 10^{-4})}{(1.60 \times 10^{-4} - 1.56 \times 10^{-4})} = 1.2 \times 10^{-2}$$

Also note that the pH indicates 1 sig fig for the result, but this level of precisionis not very useful during calculations.

Calculate the mass of Br^- present. L seawater $\xrightarrow{\text{density}}$ g seawater $\xrightarrow[Br^-]{\text{ppm}}$ mass Br^-.

Write the balanced equation for the reaction of Cl_2 with Br^- to form Br_2 and Cl^-. Solve the stoichiometry problem to get mass Cl_2 needed. Add 15%.

$$1.00 \times 10^3 \text{ L seawater} \times \frac{1.03 \text{ g}}{\text{mL}} \times \frac{1000 \text{ mL}}{1 \text{ L}} = 1.03 \times 10^6 \text{ g seawater}$$

$$67 \text{ ppm } Br^- = \frac{67 \text{ g } Br^-}{1 \times 10^6 \text{ g seawater}}; \quad \frac{67 \text{ g } Br^-}{1 \times 10^6 \text{ g seawater}} \times 1.03 \times 10^6 \text{ g seawater}$$

$$= 69.01 = 69 \text{ g } Br^- \text{ total}$$

$$2\,Br^-(aq) + Cl_2(aq) \rightarrow Br_2(aq) + 2Cl^-(aq)$$

$$69.01 \text{ g } Br^- \times \frac{1 \text{ mol } Br^-}{79.904 \text{ g } Br^-} \times \frac{1 \text{ mol } Cl_2}{2 \text{ mol } Br^-} \times \frac{70.906 \text{ g } Cl_2}{\text{mol } Cl_2} = 30.619 = 31 \text{ g } Cl_2$$

Accounting for the 15% excess: $0.15(30.619 \text{ g } Cl_2) = 4.593 \text{ g } Cl_2$ excess;

total g $Cl_2 = 30.619 + 4.593 = 35.212 = 35$ g Cl_2

22.100 $N_2H_5^+(aq) \rightarrow N_2(g) + 5H^+(aq) + 4e^-$ $\quad E^\circ_{red} = -0.23$ V

Reduction of the metal should occur when E°_{red} of the metal ion is more positive than about -0.15 V. This is the case for (b) Sn^{2+} (marginal), (c) Cu^{2+} and (d) Ag^+.

22.101 $(CH_3)_2N_2H_2(g) + 2N_2O_4(g) \rightarrow 2CO_2(g) + 3N_2(g) + 4H_2O(g)$

$$4.0 \text{ tons } (CH_3)_2N_2H_2 \times \frac{2000 \text{ lb}}{1 \text{ ton}} \times \frac{453.6 \text{ g}}{1 \text{ lb}} \times \frac{1 \text{ mol } (CH_3)_2N_2H_2}{60.10 \text{ g } (CH_3)_2N_2H_2}$$

$$\times \frac{2 \text{ mol } N_2O_4}{1 \text{ mol } (CH_3)_2N_2H_2} \times \frac{92.02 \text{ g } N_2O_4}{1 \text{ mol } N_2O_4} \times \frac{1 \text{ lb}}{453.6 \text{ g}} \times \frac{1 \text{ ton}}{2000 \text{ lb}} = 12 \text{ tons } N_2O_4$$

22.102 First write the balanced equation to give the number of moles of gaseous products per mole of hydrazine.

(A) $(CH_3)_2NNH_2 + 2N_2O_4 \rightarrow 3N_2(g) + 4H_2O(g) + 2CO_2(g)$

(B) $(CH_3)HNNH_2 + 5/4\ N_2O_4 \rightarrow 9/4\ N_2(g) + 3H_2O(g) + CO_2(g)$

In case (A) there are nine moles gas per one mole $(CH_3)_2NNH_2$ plus two moles N_2O_4. The total mass of reactants is 60 + 2(92) = 244 g. Thus, there are

$$\frac{9 \text{ mol gas}}{244 \text{ g reactants}} = \frac{0.0369 \text{ mol gas}}{1 \text{ g reactants}}$$

In case (B) there are 6.25 moles of gaseous product per one mole $(CH_3)HNNH_2$ plus 1.25 moles N_2O_4. The total mass of this amount of reactants is 46.0 + 1.25(92.0) = 161 g.

$$\frac{6.25 \text{ mol gas}}{161 \text{ g reactants}} = \frac{0.0388 \text{ mol gas}}{1 \text{ g reactants}}$$

Thus the methylhydrazine (B) has marginally greater thrust.

22.103 (a) $HOOC{-}CH_2{-}COOH \xrightarrow{P_2O_5} C_3O_2 + 2H_2O$

(b) 24 valence e^-, 12 e^- pair $\quad$ $\ddot{O}{=}C{=}C{=}C{=}\ddot{O}$

(c) C=O, about 1.23 Å; C=C, 1.34 Å or less. Since consecutive C=C bonds require sp hybrid orbitals on C (as in allene, C_3H_4), we might expect the orbital overlap requirements of this bonding arrangement to require smaller than usual C=C distances.

(d) The product has the formula $C_3H_4O_2$.

28 valence e^-, 14 e^- pr

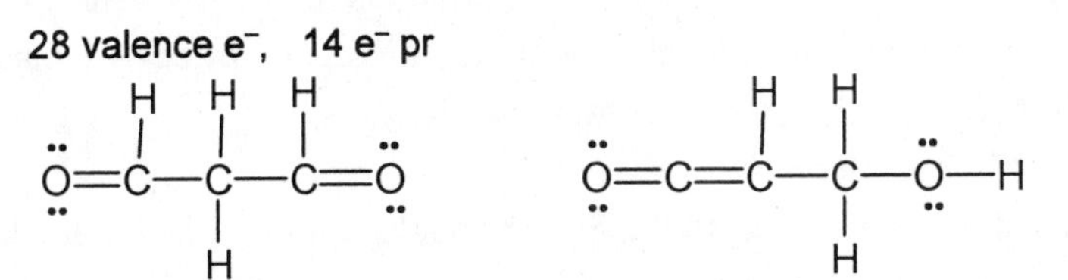

Two possibilities are shown above. The O=C=C group in the structure on the right is uncommon and less likely than the symmetrical structure on the left.

22.104 BN has the same number of valence electrons per formula unit as carbon. (Three from B, five from N, for an average of four per atom.) To the extent that we can neglect the difference in nuclear charges between B and N, we can think of BN as carbon-like. Indeed, BN takes on the same structural forms as carbon. However, because the B–N bonds are somewhat polar, BN is in fact even harder than diamond.

23 Metals and Metallurgy

Metallurgy

23.1 *Analyze/Plan*. Use Table 23.1 and other information in Section 23.1 to find important natural sources of Al and Fe. Use the rules for assigning oxidation numbers in Section 4.4 to determine the oxidation state of the metal in each natural source. *Solve*:

The important sources of iron are **hematite** (Fe_2O_3) and **magnetite** (Fe_3O_4). The major source of aluminum is **bauxite** ($Al_2O_3 \cdot xH_2O$). In ores, iron is present as the +3 ion, or in both the +2 and +3 states, as in magnetite. Aluminum is always present in the +3 oxidation state.

23.2 (a) +4

(b) $MnO_2(s) + 4H^+(aq) + 2e^- \rightarrow Mn^{2+}(aq) + 2H_2O(l) \quad E^\circ_{red} = 1.23\ V$

$Mn^{2+}(aq) + 2e^- \rightarrow Mn(s) \quad E^\circ_{red} = -1.18\ V$

Standard reduction potentials indicate that a very strong reducing agent is required to reduce the ore to Mn(s), at least if $Mn^{2+}(aq)$ is an intermediate product. According to Appendix E, only Group I and Group II metals (Li, Na, Mg, etc.) are strong enough to reduce Mn^{2+} to Mn. In practice, $MnO_2(s)$ is reduced by coke in blast furnaces, into which MnO_2 is added to incorporate Mn into steel.

23.3 An ore consists of a little bit of the stuff we want, (chalcopyrite, $CuFeS_2$) and lots of other junk (gangue).

23.4 (a) *Calcination* is heating an ore to decompose the mineral of interest into a simple solid and volatile compound. Calcination usually produces a metal oxide and a gas that is a nonmetal oxide.

(b) *Leaching* is dissolving the mineral of interest to remove it from an ore. The solvent is usually water or an aqueous solution of acid, base or salt.

(c) *Smelting* is heating an ore, often in a reducing atmosphere, to a very high temperature so that two immiscible liquid layers form. The layers are usually the molten metal or metals of interest and slag.

(d) *Slag* is the unwanted layer of the smelting process. It contains molten silicate, aluminate, phosphate or fluoride compounds.

23.5 *Analyze/Plan*. Use principles of writing and balancing chemical equations from Chapter 3 to complete and balance the given reactions. The Δ above each arrow indicates that the reactions take place at elevated temperature. Information in Section 23.2 on *pyrometallurgy* will probably be useful. *Solve*:

(a) $2PbS(s) + 3O_2(s) \xrightarrow{\Delta} 2PbO(s) + 2SO_2(g)$

(b) $PbCO_3(s) \xrightarrow{\Delta} PbO(s) + CO_2(g)$

(c) $WO_3(s) + 3H_2(g) \xrightarrow{\Delta} W(s) + 3H_2O(g)$

(d) $ZnO(s) + CO(g) \xrightarrow{\Delta} Zn(l) + CO_2(g)$

23.6 (a) $2CdS(s) + 3O_2(g) \xrightarrow{\Delta} 2CdO(s) + 2SO_2(g)$

(b) $CoCO_3(s) \xrightarrow{\Delta} CoO(s) + CO_2(g)$

(c) $Cr_2O_3(s) + 6Na(l) \rightarrow 2Cr(s) + 3Na_2O(s)$

(d) $VCl_3(g) + 3K(l) \rightarrow V(s) + 3KCl(s)$

(e) $3BaO(s) + P_2O_5(l) \rightarrow Ba_3(PO_4)_2(l)$

23.7 *Analyze/Plan*. Use information on *pyrometallurgy* in Section 23.2, along with principles of writing and balancing equations to provide the requested information. *Solve*:

(a) $SO_3(g)$

(b) $CO(g)$ provides a reducing environment for the transformation of Pb^{2+} to Pb.

(c) $PbSO_4(s) \rightarrow PbO(s) + SO_3(g)$

$PbO(s) + CO(g) \rightarrow Pb(s) + CO_2(g)$

23.8 (a) $CoO(s) + CO_2(g)$ (b) $CO(g)$

(c) $CoCO_3(s) \xrightarrow{\Delta} CoO(s) + CO_2(g)$; $CoO(s) + CO(g) \rightarrow Co(s) + CO_2(g)$

23.9 *Analyze/Plan*. Use information on *pyrometallurgy* in Section 23.2, along with principles of writing and balancing equations to provide the requested information. *Solve*:

$FeO(s) + H_2(g) \rightarrow Fe(s) + H_2O(g)$

$FeO(s) + CO(g) \rightarrow Fe(s) + CO_2(g)$

$Fe_2O_3(s) + 3H_2(g) \rightarrow 2Fe(s) + 3H_2O(g)$

$Fe_2O_3(s) + 3CO(g) \rightarrow 2Fe(s) + 3CO_2(g)$

23.10 The major reducing agent is CO, formed by partial oxidation of the coke (C) with which the furnace is charged.

$Fe_2O_3(s) + 3CO(g) \rightarrow 2Fe(l) + 3CO_2(g)$

$Fe_3O_4(s) + 4CO(g) \rightarrow 3Fe(l) + 4CO_2(g)$

23.11 *Analyze/Plan.* Use information on *pyrometallurgy* in Section 23.2, along with principles of writing and balancing equations to provide the requested information. *Solve:*

(a) Air serves primarily to oxidize coke (C) to CO, the main reducing agent in the blast furnace. This exothermic reaction also provides heat for the furnace.

$2C(s) + O_2(g) \rightarrow 2CO(g) \quad \Delta H = -221 \text{ kJ}$

(b) Limestone, $CaCO_3$, is the source of basic oxide for slag formation.

$CaCO_3(s) \xrightarrow{\Delta} CaO(s) + CO_2(g)$; $CaO(l) + SiO_2(l) \rightarrow CaSiO_3(l)$

(c) Coke is the fuel for the blast furnace, and the source of CO, the major reducing agent in the furnace.

$2C(s) + O_2(g) \rightarrow 2CO(g)$; $4CO(g) + Fe_3O_4(s) \rightarrow 4CO_2(g) + 3Fe(l)$

(d) Water acts as a source of hydrogen, and as a means of controlling temperature. (see Equation [23.8]). $C(s) + H_2O(g) \rightarrow CO(g) + H_2(g) \quad \Delta H = +131 \text{ kJ}$

23.12 (a) In the *converter*, oxidation of C, Si and metals by O_2 are exothermic reactions that raise the temperature.

(b) $2C(s) + O_2(g) \rightarrow 2CO(g)$; $S(s) + O_2(g) \rightarrow SO_2(g)$; $Si(s) + O_2(g) \rightarrow SiO_2(l)$

23.13 *Analyze/Plan.* Consider the information on hydrometallurgy in Section 23.3 to provide the requested information on the Bayer process. *Solve:*

(a) The Bayer process is necessary to separate the unwanted iron-containing solids from bauxite before electroreduction.

(b) The Bayer process takes advantage of the fact that Al^{3+} is amphoteric, but Fe^{3+} is not. Because it is amphoteric, Al^{3+} reacts with excess OH^- to form the soluble complex ion $Al(OH)_4^-$ while the Fe^{3+} solids cannot. This allows separation of the iron-containing solids by filtration.

23.14 Gold appears in elemental form in ores. The $O_2(g)$ in air oxidizes Au(s) to $Au^+(aq)$ and the $CN^-(aq)$ stabilizes the Au^+ in the form of the soluble complex ion $Au(CN)_2^-(aq)$. This enables the unwanted solids (gangue) to be filtered away from the gold-containing solution.

23.15 *Analyze/Plan.* Use information on the *electrometallurgy* of Cu as a model for describing how electrometallurgy can be employed to purify pure Co. Compare the ease of oxidation and reduction of cobalt with that of water. *Solve:*

Cobalt could be purified by constructing an electrolysis cell in which the crude metal was the anode and a thin sheet of pure cobalt was the cathode. The electrolysis solution is aqueous with a soluble cobalt salt such as $CoSO_4 \cdot 7H_2O$ serving as the electrolyte. (Other soluble salts with anions that do not participate in the cell reactions could be used.) Anode reaction: $Co(s) \rightarrow Co^{2+}(aq) + 2e^-$; cathode reaction: $Co^{2+}(aq) + 2e^- \rightarrow Co(s)$. Although E° for reduction of $Co^{2+}(aq)$ is slightly negative (-0.277 V), it is less than the standard reduction potential for $H_2O(l)$, -0.83 V.

23.16 $SnO_2(s) + C(s) \xrightarrow{\Delta} Sn(l) + CO_2(g)$

$Sn(s) \rightarrow Sn^{2+}(aq) + 2e^-$ (anode)

$Sn^{2+}(aq) + 2e^- \rightarrow Sn(s)$ (cathode)

Metals and Alloys

23.17 *Analyze/Plan.* Compare the bonding characteristics of metallic sodium and ionic sodium chloride and use them to explain the difference in malleability. *Solve:*

Sodium is metallic; each atom is bonded to many nearest neighbor atoms by metallic bonding involving just one electron per atom, and delocalized over the entire three-dimensional structure. When sodium metal is distorted, each atom continues to have bonding interactions with many nearest neighbors. In NaCl the ionic forces are strong, and the arrangement of ions in the solid is very regular. When subjected to physical stress, the three-dimensional lattice tends to cleave along the very regular lattice planes, rather than undergo the large distortions characteristic of metals.

23.18 Since Si has the same crystal structure as diamond, it is a covalent-network solid with its bonding electrons localized between Si atoms. Since there is no significant delocalization in Si, it is not likely to have the metallic properties of malleability, ductility and high electrical and thermal conductivity. It is likely to be hard and high-melting like other covalent-network solids.

23.19 *Analyze/Plan.* Apply the description of the electron-sea model of metallic bonding given in Section 23.5 to the conductivity of silver. *Solve:*

In the electron-sea model for metallic bonding, the valence electrons of the silver atoms move about the three-dimensional metallic lattice, while the silver atoms maintain regular lattice positions. Under the influence of an applied potential the electrons can move throughout the structure, giving rise to high electrical conductivity. The mobility of the electrons facilitates the transfer of kinetic energy and leads to high thermal conductivity.

23.20 (a) Cr: $[Ar]4s^13d^5$, Z = 24; Se: $[Ar]4s^23d^{10}4p^4$, Z = 34

Both elements have the [Ar] core configuration and both have six valence electrons. The orbital locations of the six valence electrons is different in the two elements, because Se has more electrons.

(b) Different Z and Z_{eff} for the two elements, and the different orbital locations of the valence electrons are the main factors that lead to the differences in properties. In Cr, the 4s and 3d electrons are the valence electrons. Its Z and Z_{eff} are smaller than those of Se and it is not likely to gain enough electrons to achieve a noble-gas configuration. Thus, Cr loses electrons when it forms ions, acting like a metal. Se is in the same row of the periodic table as Cr, but its 3d subshell is filled, so its valence electrons are in 4s and 4p. Because Se has a larger Z and Z_{eff}, it is more likely to hold its own valence electrons and gain other electrons when it forms ions. That Se needs only two additional electrons to achieve the noble-gas configuration of Kr is also a driving force for it gaining electrons when it forms ions, acting like a nonmetal.

23.21 *Analyze/Plan.* Consider trends in atomic mass and volume of the elements listed to explain the variation in density. *Solve:*

The variation in densities reflects shorter metal-metal bond distances. These shorter distances suggest that the extent of metal-metal bonding increases in the series. Thus, it would appear that all the valence electrons in these elements (1, 2, 3 and 4, respectively) are involved in metallic bonding.

23.22 The ΔH_{atom} is an indication of the strength of metallic bonding in the metals. According to the molecular orbital or band theory of metallic bonding, the strength of metallic bonding increases with the number of electrons in bonding molecular orbitals, up to a maximum of six valence electrons. As the number of valence electrons in the series increases from two for Ca to five for V, the ΔH_{atom} and the strength of metallic bonding increases.

23.23 *Analyze/Plan.* Consider the definitions of insulators, conductors and semiconductors given in 'A Closer Look' in Section 23.5. *Solve:*

According to band theory, an *insulator* has a completely filled valence band and a large energy gap between the valence band and the nearest empty band; electrons are localized within the lattice. A *conductor* must have a partially filled energy band; a small excitation will promote electrons to previously empty levels within the band and allow them to move freely throughout the lattice, giving rise to the property of conduction. A *semiconductor* has a filled valence band, but the gap between the filled and empty bands is small enough to jump to the empty conduction band. The presence of an impurity may also place an electron in an otherwise empty band (producing an n-type semiconductor), or create a vacancy in an otherwise full band (producing a p-type semiconductor), providing a mechanism for conduction.

23.24 Germanium doped with arsenic (Ge/As) is a better conductor than germanium. Ge, a semiconductor, has a filled valence band but an energetically accessible conduction band. In Ge/As, the additional valence electrons of the doped As atoms cannot fit into the filled valence band and occupy the conduction band. These electrons have access to the vacant orbitals within the conduction band and serve as carriers of electrical current. Ge/As is an *n-type* semiconductor.

23.25 *Analyze/Plan.* Recall the diamond and closest-packed structures described in Sections 11.7 and 11.8. Use these structures to draw conclusions about Sn–Sn distance and electrical conductivity in the two allotropes. *Solve*:

White tin, with a characteristic metallic structure, is expected to be more metallic in character. The electrical conductivity of the white allotropic form is higher because the valence electrons are shared with 12 nearest neighbors rather than being localized in four bonds to nearest neighbors as in gray tin. The Sn–Sn distance should be longer in white tin; there are only four valence electrons from each atom, and 12 nearest neighbors. The **average** tin–tin bond order can, therefore, be only about 1/3, whereas in gray tin the bond order is one. (In gray tin the Sn–Sn distance is 2.81 Å in white tin it is 3.02 Å.)

23.26 According to the MO model of metallic bonding, electrical and thermal conductivity are high when electrons can be delocalized throughout the substance. In graphite, the layers are sheets of interconnected benzene rings, with extensive delocalization of π electrons. This facilitates conductivity within the layers, but not in the direction perpendicular to them.

23.27 *Analyze/Plan.* Use information in Section 23.6 to define *alloy*, and compare the various types of alloys. Solve:

An *alloy* contains atoms of more than one element and has the properties of a metal. *Solution alloys* are homogeneous mixtures with different kinds of atoms dispersed randomly and uniformly. In *heterogeneous alloys* the components (elements or compounds) are not evenly dispersed and their properties depend not only on composition but methods of preparation. In an *intermetallic compound* the component elements have interacted to form a compound substance, for example, Cu_3As. As with more familiar compounds, these are homogeneous and have definite composition and properties.

23.28 Substitutional and interstitial alloys are both solution alloys. In a *substitutional* alloy, the atoms of the "solute" take positions normally occupied by the "solvent." Substitutional alloys tend to form when solute and solvent atoms are of comparable size and have similar bonding characteristics. In an *interstitial* alloy, the atoms of the "solute" occupy the holes or interstitial positions between "solvent" atoms. Solute atoms are necessarily much smaller than solvent atoms.

Transition Metals

23.29 *Analyze/Plan.* Consider the definitions of the properties listed (Chapter 7 and Chapter 23) and whether they refer to single, isolated atoms or bulk material. *Solve*:

Of the properties listed, (b) the first ionization energy, (c) atomic radius and (f) electron affinity are characteristic of isolated atoms. Electrical conductivity (a), melting point (d) and heat of vaporization (e) are properties of the bulk metal.

23.30 (b) NiCo alloy and (c) W will have metallic properties. The lattices of these substances are composed of neutral metal atoms. Delocalization of valence electrons via the MO model of metallic bonding produces metallic properties.

(d) Ge is a metalloid, not a metal. (a) $TiCl_4$ is an ionic compound and (e) Hg_2^{2+} is a metal ion. In ions and ionic compounds, electrons are localized on the individual ions, precluding metallic properties.

23.31 *Analyze/Plan.* Define lanthanide contraction (Section 23.7). Based on the definition, list properties related to atomic radius. *Solve:*

The *lanthanide contraction* is the name given to the decrease in atomic size due to the build-up in effective nuclear charge as we move through the lanthanides (elements 58-71) and beyond them. This effect offsets the expected increase in atomic size going from the second to the third transition series. The lanthanide contraction affects size-related properties such as ionization energy, electron affinity and density.

23.32 Zr: $[Kr]5s^24d^2$, Z = 40; Hf: $[Kr]6s^24f^{14}5d^2$, Z = 72

Moving down a family of the periodic chart, atomic size increases because the valence electrons are in a higher principle quantum level (and thus further from the nucleus) and are more effectively shielded from the nuclear charge by a larger core electron cloud. However, the build-up in Z that accompanies the filling of the 4f orbitals causes the valence electrons in Hf to experience a much greater relative nuclear charge than those in La, its neighbor to the left. This increase in Z offsets the usual effect of the increase in *n* value of the valence electrons and the radii of Zr and Hf atoms are similar.

23.33 *Analyze/Plan.* Use Figure 23.24 to determine the highest oxidation state of each metal. Write formulas of the metal fluorides, given that fluoride ion is F^-. *Solve:*

(a) ScF_3 (b) CoF_3 (c) ZnF_2

23.34 (a) CdO (b) WO_3 (c) Nb_2O_5

23.35 *Analyze/Plan.* Consider the electron configurations of Cr and Al to rationize observed oxidation states. *Solve:*

Chromium, $[Ar]4s^13d^5$, has six valence-shell electrons, some or all of which can be involved in bonding, leading to multiple stable oxidation states. By contrast, aluminum, $[Ne]3s^23p^1$, has only three valence electrons which are all lost or shared during bonding, producing the +3 state exclusively.

23.36 V: $[Ar]4s^23d^3$, (V^{2+}: $[Ar]3d^3$); Sc: $[Ar]4s^23d^1$; (Sc^{2+}: $[Ar]3d^1$)

V has a slightly larger Z (23) and Z_{eff} than Sc (Z > 21), so the 3d electrons in V are more tightly held than those in Sc. Also, losing the lone 3d electron in Sc^{2+} leads to the stable noble-gas configuration of [Ar] for Sc^{3+}.

23.37 *Analyze/Plan.* Write electron configurations for the neutral elements and their positive ions recalling that valence electrons are last in order of descending *n*-value. *Solve:*

(a) Cr^{3+}: $[Ar]3d^3$ (b) Au^{3+}: $[Xe]4f^{14}5d^8$ (c) Ru^{2+}: $[Kr]4d^6$

(d) Cu^+: $[Ar]3d^{10}$ (e) Mn^{4+}: $[Ar]3d^3$ (f) Ir^{3+}: $[Xe]4f^{14}5d^6$

23.38 (a) Ti^{3+}: $[Ar]3d^2$ (b) Co^{3+}: $[Ar]3d^6$ (c) Pd^{2+}: $[Kr]4d^8$

(d) Mo^{3+}: $[Kr]4d^3$ (e) Ru^{3+}: $[Kr]4d^5$ (f) Ni^{4+}: $[Ar]3d^6$

23.39 *Analyze/Plan.* Oxidation is loss of electrons. Which periodic trend determines how tightly a valence electron is held in a particular atom or ion? *Solve:*

Ease of oxidation decreases from left to right across a period (owing to increasing effective nuclear charge); Ti^{2+} should be more easily oxidized than Ni^{2+}.

23.40 The stronger reducing agent is more easily oxidized; Cr^{2+} is more easily oxidized (see Solution 23.39), so it is the stronger reducing agent.

23.41 *Analyze/Plan.* Consider Equation 23.26 regarding the oxidation states of iron. *Solve:*

Fe^{2+} is a reducing agent that is readily oxidized to Fe^{3+} in the presence of O_2 from air.

23.42 Chromate ion, CrO_4^{2-}, is bright yellow. Dichromate, $Cr_2O_7^{2-}$, is orange and more stable in acid solution than CrO_4^{2-} because of the equilibrium $2CrO_4^{2-}(aq) + 2H^+(aq) \rightleftharpoons Cr_2O_7^{2-}(aq) + H_2O(l)$.

23.43 *Analyze/Plan.* Consider information on the descriptive chemistry of iron in Section 23.8. *Solve:*

(a) $Fe(s) + 2HCl(aq) \rightarrow FeCl_2(aq) + H_2(g)$

(b) $Fe(s) + 4HNO_3(aq) \rightarrow Fe(NO_3)_3(aq) + NO(g) + 2H_2O(l)$

(See net ionic equation, Equation 23.28) In concentrated nitric acid, the reaction can produce $NO_2(g)$ according to the reaction:

$Fe(s) + 6HNO_3(aq) \rightarrow Fe(NO_3)_3(aq) + 3NO_2(g) + 3H_2O(l)$

23.44 (a) $MnO_2(s) + 4HCl(aq) \rightarrow MnCl_2(aq) + Cl_2(g) + 2H_2O(l)$

(b) Yes. MnO_2 is the oxidizing agent; HCl is the reducing agent.

23.45 *Analyze/Plan.* Consider the definitions of *paramagnetic* and *diamagnetic.* Solve:

The unpaired electrons in a *paramagnetic* material cause it to be weakly attracted into a magnetic field. A *diamagnetic* material, where all electrons are paired, is very weakly repelled by a magnetic field.

23.46 (a) *Ferromagnetic* materials can form "permanent" magnets, whereas *paramagnetic* materials cannot.

(b) In order for a substance to be ferromagnetic, the magnetic moments on sites throughout the lattice must interact with one another. That is, the sites must be physically close and the overlap of orbitals must enable the individual magnetic sites to couple forming a much larger magnetic moment throughout the solid. Because of these interactions, the sustained existence of the magnetic moment does not require application of an external magnetic field.

(c) No. Other metals (Co and Ni, for example), alloys and some oxides (CrO_2) are also ferromagnetic.

Additional Exercises

23.47 $PbS(s) + O_2(g) \rightarrow Pb(l) + SO_2(g)$

Regardless of the metal of interest, $SO_2(g)$ is a product of roasting sulfide ores. In an oxygen rich environment, $SO_2(g)$ is oxidized to $SO_3(g)$, which dissolves in $H_2O(l)$ to form sulfuric acid, $H_2SO_4(aq)$. Because of its corrosive nature, $SO_2(g)$ is a dangerous environmental pollutant (Section 18.4) and cannot be freely released into the atmosphere. A sulfuric acid plant near a roasting plant would provide a means for disposing of $SO_2(g)$ that would also generate a profit.

23.48 Al^{3+}, Mg^{2+} and Na^+ all have large negative reduction potentials (Al, Mg and Na are very active metals). A substance with a more negative reduction potential would have to be used to chemically reduce them. All such substances are more expensive and difficult to obtain than Al, Mg and Na. Electrolysis is thus the most cost efficient way to reduce Al^{3+}, Mg^{2+} and Na^+ to their metallic states.

23.49 $CO(g)$: $Pb(s)$; $H_2(g)$: $Fe(s)$; $Zn(s)$: $Au(s)$

23.50 (a) $2VCl_3(s) + O_2(g) \rightarrow 2VOCl_3(s)$

(b) $Nb_2O_5(s) + 5H_2(g) \rightarrow 2Nb(s) + 5H_2O(l)$

(c) $2Fe^{3+}(aq) + Zn(s) \rightarrow 2Fe^{2+}(aq) + Zn^{2+}(aq)$

(d) $NbCl_5(s) + 3H_2O(l) \rightarrow HNbO_3(s) + 5HCl(aq)$

23.51 (a) $NiO(s) + 2H^+(aq) \rightarrow Ni^{2+}(aq) + H_2O(l)$

(b) The simple answer is that the solid is subjected to acid hydrolysis:
$CuCo_2S_4(s) + 8H^+(aq) \rightarrow Cu^{2+}(aq) + 2Co^{3+}(aq) + 4H_2S(g)$
However, in the absence of a strong complexing ligand, Co^{3+} is not stable in water. It oxidizes water according to the following reaction:
$4Co^{3+}(aq) + 2H_2O(l) \rightarrow 4Co^{2+}(aq) + O_2(g) + 4H^+(aq)$

(c) $TiO_2(s) + C(s) + 2Cl_2(g) \rightarrow TiCl_4(g) + CO_2(g)$

(d) In this reaction O_2 is reduced and sulfide is oxidized. Writing the sulfur product as S_8, the balanced equation is:

$$8ZnS(s) + 4O_2(g) + 16H^+(aq) \rightarrow 8Zn^{2+}(aq) + S_8(s) + 8H_2O(l)$$

23.52 Because selenium and tellurium are both nonmetals, we expect them to be difficult to oxidize. Thus, both Se and Te are likely to accumulate as the free elements in the so-called anode slime, along with noble metals that are not oxidized.

23.53 All transition metals have the generic electron configuration $ns^2(n\text{-}1)d^x$. Regardless of the number of d electrons, each transition metal has 2 ns valence electrons that are the first electrons lost when metal ions are formed. Thus, almost every transition metal has a stable +2 oxidation state.

After the 2 ns electrons are lost, a varying number of $(n\text{-}1)$d electrons can be lost, depending on the identity of the transition metal. The availability of different numbers of d electrons leads to a wide variety of accessible oxidation states for the transition metals.

23.54 Assuming that SO_2 and N_2 are the nonmetallic products, two half-reactions can be written:

$$5[MoS_2(s) + 7H_2O(l) \rightarrow MoO_3(s) + 2SO_2(g) + 14H^+(aq) + 14e^-]$$

$$7[12H^+(aq) + 2NO_3^-(aq) + 10e^- \rightarrow N_2(g) + 6H_2O(l)]$$

$$5MoS_2(s) + 14H^+(aq) + 14NO_3^-(aq) \rightarrow 5MoO_3(s) + 10SO_2(g) + 7N_2(g) + 7H_2O(l)$$

$$MoO_3(s) + 2NH_3(aq) + H_2O(l) \rightarrow (NH_4)_2MoO_4(s)$$

$$(NH_4)_2MoO_4(s) \xrightarrow{\Delta} 2NH_3(g) + H_2O(g) + MoO_3(s)$$

$$MoO_3(s) + 3H_2(g) \xrightarrow{\Delta} Mo(s) + 3H_2O(g)$$

23.55 (a) Substitutional alloys and intermetallic compounds are both homogeneous solution alloys. Intermetallic compounds have a definite stoichiometry and properties, while substitutional alloys have a range of compositions.

(b) A paramagnetic substance has unpaired electrons and is attracted into a magnetic field. A diamagnetic substance has only paired electrons and is weakly repelled by a magnetic field.

(c) Insulators have a filled valence band with a large energy gap between the valence and the conduction band, making delocalization difficult. Semiconductors have a filled valence band but a smaller band gap, so that some electrons can move to the conduction band.

(d) In metallic conduction, metal atoms are stationary while a few valence electrons are mobile and available to carry charge throughout the substance. In electrolytic conduction, mobile ions carry charge throughout the liquid.

23.56 Silicon has the diamond structure. As with carbon, the four valence electrons of silicon are completely involved in the four localized bonds to its neighbors. There are thus no electrons free to migrate throughout the solid. Titanium exists as a close-packed lattice; each Ti atom has twelve equivalent nearest neighbors. The valence shell electrons cannot be localized between pairs of atoms; rather they are delocalized and mobile throughout the structure. In terms of the band model described in Figure 23.15, Ti has an incompletely occupied allowed energy band. The origin of the different behaviors with regard to structure has to do with the extent of the orbitals in space. Electrons in Si feel a greater attraction to the nucleus than electrons in Ti, so they are more localized.

23.57 The equilibrium of interest is $[ZnL_4] \rightleftharpoons Zn^{2+}(aq) + 4L \quad K = 1/K_f$
Since $Zn(H_2O)_4^{2+}$ is $Zn^{2+}(aq)$, its reduction potential is -0.763 V. As the stability (K_f) of the complexes increases, K decreases. Since E° is directly proportional to log K_{eq} (Equation 20.18), E° values for the complexes will become more negative as K_f increases.

23.58 In a paramagnetic substance, the unpaired electrons on one atom are not affected by (coupled to) the unpaired electrons on adjacent atoms. The unpaired electrons are in random orientations. The external magnetic field causes rough alignment and therefore weak attraction to the magnetic field. In a ferromagnetic substance, unpaired electrons on adjacent atoms are coupled and aligned in the same direction. This permanent magnetic moment causes the ferromagnetic substance to be strongly attracted into the magnetic field.

23.59 (a) Nb^{5+}: [Ar]; diamagnetic, no unpaired electrons

(b) Cr^{2+}: $[Ar]3d^4$ paramagnetic, unpaired electrons

(c) Cu^{+}: $[Ar]3d^{10}$; diamagnetic, no unpaired electrons

(d) Ru^{8+}: [Ar]; diamagnetic, no unpaired electrons

(e) Ni^{2+}: $[Ar]3d^8$; paramagnetic, unpaired electrons

23.60 In a ferromagnetic solid, the magnetic centers are coupled such that the spins of all unpaired electrons are parallel. As the temperature of the solid increases, the average kinetic energy of the atoms increases until the energy of motion overcomes the force aligning the electron spins. The substance becomes paramagnetic; it still has unpaired electrons, but their spins are no longer aligned.

23.61 (a) $Mn(s) + 2HNO_3(aq) \rightarrow Mn(NO_3)_2(aq) + H_2(g)$

(b) $Mn(NO_3)_2(s) \xrightarrow{\Delta} MnO_2(s) + 2NO_2(g)$

(c) $3MnO_2(s) \xrightarrow{\Delta} Mn_3O_4(s) + O_2(g)$

(d) $2MnCl_2(s) + 9F_2(g) \rightarrow 2MnF_3(s) + 4ClF_3(g)$

23.62 (a) Nothing. As noted in Section 20.8, a basic environment (OH^-) inhibits oxidation of Fe^{2+} to Fe^{3+}, even in the presence of $O_2(g)$.

(b) $Cu(NO_3)_2(aq) + 2KOH(aq) \rightarrow Cu(OH)_2(s) + 2KNO_3(aq)$

$Cu(OH)_2(s)$ precipitates. Cu^{2+} forms a soluble complex ion with $NH_3(aq)$, but not $OH^-(aq)$.

(c) The color of the solution changes from orange ($Cr_2O_7^{2-}$) to yellow (CrO_4^{2-}). The equilibrium is

$Cr_2O_7^{2-}(aq) + H_2O(l) \rightleftharpoons 2CrO_4^{2-}(aq) + 2H^+(aq)$

As $OH^-(aq)$ is added, it reacts with and removes $H^+(aq)$ from solution, shifting the equilibrium to the right in favor of the yellow CrO_4^{2-}.

23.63 (a) $2NiS(s) + 3O_2(g) \rightarrow 2NiO(s) + 2SO_2(g)$

(b) $2C(s) + O_2(g) \rightarrow 2CO(g)$; $C(s) + H_2O(g) \rightarrow CO(g) + H_2(g)$

$NiO(s) + CO(g) \rightarrow Ni(s) + CO_2(g)$; $NiO(s) + H_2(g) \rightarrow Ni(s) + H_2O(g)$

(c) $Ni(s) + 2HCl(aq) \rightarrow NiCl_2(aq) + H_2(g)$

(d) $NiCl_2(aq) + 2NaOH(aq) \rightarrow Ni(OH)_2(s) + 2NaCl(aq)$

(e) $Ni(OH)_2(s) \xrightarrow{\Delta} NiO(s) + H_2O(g)$

23.64 (a) insulator (b) p-type semiconductor (c) metallic conductor

(d) metallic conductor (e) insulator (f) metallic conductor

Integrative Exercises

23.65 *Analyze/Plan.* Given the mass of Fe produced, calculate the mass of C required. Write the balanced equations for the reaction of Fe_2O_3 with CO, and for the formation of CO. Solve the stoichiometry problem, using mole ratios from the balanced equations and paying attention to units. *Solve*:

$Fe_2O_3(s) + 3CO(g) \rightarrow 2Fe(s) + 3CO_2(g)$

$2C(s) + O_2(g) \rightarrow 2CO(g)$

$$9.00 \times 10^3 \text{ tan Fe} \times \frac{2000 \text{ lb}}{1 \text{ ton}} \times \frac{453.6 \text{ g}}{1 \text{ lb}} \times \frac{1 \text{ mol Fe}}{55.845 \text{ g Fe}} \times \frac{3 \text{ mol CO}}{3 \text{ mol Fe}} \times \frac{2 \text{ mol C}}{2 \text{ mol CO}} \times \frac{12.011 \text{ g C}}{1 \text{ mol C}}$$

$$= 2.634 \times 10^9 = 2.63 \times 10^9 \text{ g}$$

This amount can also be expressed as 2.63×10^6 kg or 2.90×10^3 ton.

23.66 (a) Calculate mass Cu_2S and FeS, then mass SO_2 from each.

3.3×10^6 kg sample $\times 0.27 = 8.91 \times 10^5 = 8.9 \times 10^5$ kg $= 8.9 \times 10^8$ g Cu_2S

3.3×10^6 kg sample $\times 0.13 = 4.29 \times 10^5 = 4.3 \times 10^5$ kg $= 4.3 \times 10^8$ g FeS

$$8.91 \times 10^8 \text{ g } Cu_2S \times \frac{1 \text{ mol } Cu_2S}{159.1 \text{ g } Cu_2S} \times \frac{1 \text{ mol } SO_2}{1 \text{ mol } Cu_2S} \times \frac{64.07 \text{ g } SO_2}{1 \text{ mol } SO_2} = 3.588 \times 10^8$$

$$= 3.6 \times 10^8 \text{ g } SO_2$$

$$4.29 \times 10^8 \text{ g FeS} \times \frac{1 \text{ mol FeS}}{87.9 \text{ g FeS}} \times \frac{1 \text{ mol } SO_2}{1 \text{ mol FeS}} \times \frac{64.07 \text{ g } SO_2}{1 \text{ mol } SO_2} = 3.127 \times 10^8 \text{ g} = 3.1 \times 10^8 \text{ g } SO_2$$

$\text{g } SO_2 = 3.588 \times 10^8 + 3.127 \times 10^8 = 6.715 \times 10^8 = 6.7 \times 10^8 \text{ g } SO_2$

(b) Calculate mol Cu, mol Fe and mole ratio Cu: Fe.

$$8.91 \times 10^8 \text{ g } Cu_2S \times \frac{1 \text{ mol } Cu_2S}{159.1 \text{ g } Cu_2S} \times \frac{2 \text{ mol Cu}}{1 \text{ mol } Cu_2S} = 1.12 \times 10^7 = 1.1 \times 10^7 \text{ mol Cu}$$

$$4.29 \times 10^8 \text{ g FeS} \times \times \frac{1 \text{ mol FeS}}{87.9 \text{ g FeS}} \times \frac{1 \text{ mol Fe}}{1 \text{ mol FeS}} = 4.88 \times 10^6 = 4.9 \times 10^6 \text{ mol Fe}$$

1.12×10^7 mol Cu/4.88×10^6 mol Fe = 2.3 mol Cu/mol Fe

(c) The oxidizing environment of the converter is likely to produce CuO and Fe_2O_3.

(d) $Cu_2S(s) + 2O_2(g) \rightarrow 2CuO(s) + SO_2(g)$

$4FeS(s) + 7O_2(g) \rightarrow 2Fe_2O_3(s) + 4SO_2(g)$

23.67 Recall from the discussion in Chapter 13 that like substances tend to be soluble in one another, whereas unlike substances do not. Molten metal consists of atoms that continue to be bound to one another by metallic bonding, even though the substance is liquid. In a slag, on the other hand, the attractive forces are those between ions. The slag phase is a highly polar, ionic medium, whereas the metallic phase is nonpolar, and the attractive interactions are due to metallic bond formation. There is little driving force for materials with such different characteristics to dissolve in one another.

23.68 The first equation indicates that one mole Ni^{2+} is formed from passage of two moles of electrons, and the second equation indicates the same thing. Thus, the simple ratio (1 mol Ni^{2+}/2F).

$$67 \text{ A} \times 11.0 \text{ hr} \times \frac{3600 \text{ s}}{1 \text{ hr}} \times \frac{1 \text{ C}}{1 \text{ A} \bullet \text{s}} \times \frac{1 \text{ F}}{96{,}500 \text{ C}} \times \frac{1 \text{ mol } Ni^{2+}}{2 \text{ F}} \times \frac{58.7 \text{ g } Ni^{2+}}{1 \text{ mol } Ni^{2+}} \times \frac{0.90 \text{ g Ni actual}}{1.00 \text{ g Ni theoretical}} = 7.3 \times 10^2 \text{ g } Ni^{2+}(aq)$$

23.69 (a) $\Delta G° = \Delta H° - T\Delta S°$ (assume $\Delta H°$ and $S°$ are constant with changes in temperature)

$Si(s) + 2MnO(s) \rightarrow SiO_2(s) + 2Mn(s)$

$\Delta H° = \Delta H_f° \; SiO_2(s) + 2\Delta H_f° \; Mn(s) - 2\Delta H_f° \; MnO(s) - \Delta H_f° \; Si(s)$

$\Delta H° = -910.9 + 2(0) - 2(-385.2) + 0 = -140.5 \text{ kJ}$

$\Delta S° = S° \; SiO_2(s) + 2S° \; Mn(s) - 2S° \; MnO(s) - S° \; Si(s)$

$= 41.84 + 2(32.0) - 2(59.7) - 18.7 = -32.26 = -32.3 \text{ J/K}$

$\Delta G° = -140.5 \text{ kJ} - 1473 \text{ K}(-0.03226 \text{ kJ/K}) = -93.0 \text{ kJ}$

(b) At 1473 K, the reactants and products are all solids, so they are in their standard states. Since $\Delta G°$ is negative at this temperature, the reaction should be spontaneous and thus feasible.

23.70 (a) According to Section 20.8, the reduction of O_2 during oxidation of Fe(s) to Fe_2O_3 requires H^+. Above pH 9, iron does not corrode. At the high temperature of the converter, it is unlikely to find H_2O or H^+ in contact with the molten Fe. Also, the basic slag (CaO(l)) that is present to remove phosphorus will keep the environment basic rather than acidic. Thus, the H^+ necessary for oxidation of Fe in air is not present in the converter.

(b) $C + O_2(g) \rightarrow CO_2(g)$

$S + O_2(g) \rightarrow SO_2(g)$

$P + O_2(g) \rightarrow P_2O_5(l)$; $P_2O_5(l) + 3CaO(l) \rightarrow Ca_3(PO_4)(l)$

$Si + O_2(g) \rightarrow SiO_2$

$M + O_2(g) \rightarrow M_xO_y(l)$; $M_xO_y + SiO_2 \rightarrow$ silicates

CO_2 and SO_2 escape as gases. P_2O_5 reacts with CaO(l) to form $Ca_3(PO_4)_2(l)$, which is removed with the basic slag layer. SiO_2 and metal oxides can combine to form other silicates; SiO_2, M_xO_y and complex silicates are all removed with the basic slag layer.

23.71
$$2[Cu^+(aq) + 1e^- \rightarrow Cu(s)] \qquad E^\circ_{red} = 0.521\ V$$
$$Cu(s) \rightarrow Cu^{2+}(aq) + 2e^- \qquad E^\circ_{red} = 0.337\ V$$
$$2Cu^+(aq) \rightarrow Cu^{2+}(aq) + Cu(s) \qquad E^\circ = (0.521\ V - 0.337\ V) = 0.184\ V$$

According to Equation 20.18, $\log K_{eq} = \dfrac{nE^\circ}{0.0592}$; $n = 2$

$$\log K_{eq} = \frac{2(0.184)}{0.0592} = 6.2162 = 6.22;\ K_{eq} = 1.6 \times 10^6$$

23.72 $\Delta G^\circ = -RT \ln K$; $\Delta G^\circ = \Delta H^\circ - T\Delta S^\circ$

Calculate ΔH° and ΔS° using data from Appendix C, assuming ΔH° and ΔS° remain constant with changing temperature. Then calculate ΔG° and K at the two temperatures.

$$\Delta H^\circ = 2\Delta H_f\ CO(g) - \Delta H^\circ_f\ C(s) - \Delta H^\circ_f\ CO_2(g)$$
$$\Delta H^\circ = 2(-110.5) - 0 - (-393.5) = +172.5\ kJ$$

$$\Delta S^\circ = 2S^\circ\ CO(g) - S^\circ\ C(s) - S^\circ\ CO_2(g)$$
$$= 2(197.9) - 5.69 - 213.6 = +176.5\ J/K = 0.1765\ kJ/K$$

$$\Delta G^\circ_{298} = 172.5\ kJ - 298\ K(0.1765\ kJ/K) = +119.9\ kJ$$

$$\ln K_{eq} = \frac{\Delta G^\circ}{-RT} = \frac{119.9\ kJ}{-(8.314 \times 10^{-3}\ kJ/K)(298\ K)} = -48.3942 = -48.39;\ K_{eq} = 9.6 \times 10^{-22}$$

$$\Delta G^\circ_{2000} = 172.5\ kJ - 2000\ K\ (0.1765\ kJ/K) = -180.5\ kJ$$

$$\ln K_{eq} = \frac{-180.5}{-(8.314 \times 10^{-3}\ kJ/K)(2000\ K)} = 10.8552 = 10.86;\ K_{eq} = 5.2 \times 10^4$$

23.73 (a) The standard reduction potential for $H_2O(l)$ is much greater than that of $Mg^{2+}(aq)$(-0.83 V vs. -2.37 V). In aqueous solution, $H_2O(l)$ would be preferentially reduced and no Mg(s) would be obtained.

(b) $$97{,}000\text{ A} \times 24\text{ hr} \times \frac{3600\text{ s}}{1\text{ hr}} \times \frac{1\text{ C}}{1\text{ A}\bullet\text{s}} \times \frac{1\text{ F}}{96{,}500\text{ C}} \times \frac{1\text{ mol Mg}}{2\text{ F}} \times \frac{24.31\text{ g Mg}}{1\text{ mol Mg}} \times 0.96$$

$$= 1.0 \times 10^6\text{ g Mg} = 1.0 \times 10^3\text{ kg Mg}$$

23.74 (a) The very low melting and boiling points for VF_5 indicate that it is molecular rather than ionic, and that the intermolecular forces are probably weak London-dispersion forces. In order for the molecule to experience only London-dispersion forces, it must be nonpolar covalent, which requires the symmetrical trigonal bipyramidal structure shown below. PF_5 also has this structure.

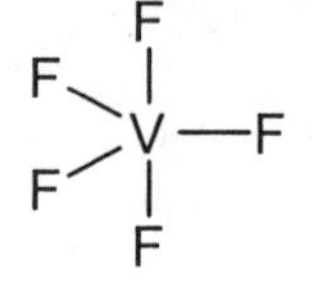

(b) $VCl_3(s) + 3HF(g) \xrightarrow{\Delta} VF_3(s) + 3HCl(g)$

(c) V(V) has a relatively small covalent radius. F is the smallest and most electronegative halogen. The steric repulsions associated with placing five larger halogens around the small V(V) central atom would be substantial. Also, the extreme electron attracting nature of F might be required to coax V into the +5 oxidation state.

23.75 Calculate the mass of Zn(s) that will be deposited.

$$2.0\text{ m} \times 80\text{ m} \times \frac{(100)^2\text{ cm}^2}{1\text{ m}^2} \times 0.49\text{ mm} \times \frac{1\text{ cm}}{10\text{ mm}} \times \frac{7.1\text{ g}}{\text{cm}^3} \times 2\text{ sides}$$

$$= 1.113 \times 10^6 = 1.1 \times 10^6\text{ g Zn}$$

$$1.113 \times 10^6\text{ g Zn} \times \frac{1\text{ mol Zn}}{65.39\text{ g Zn}} \times \frac{2\text{ F}}{0.90\text{ mol Zn}} \times \frac{96{,}500\text{ C}}{\text{F}} = 3.651 \times 10^9 = 3.7 \times 10^9\text{ C}$$

(2 F/0.90 mol Zn takes the 90% efficiency into account.)

$$3.651 \times 10^9\text{ C} \times 3.5\text{ V} \times \frac{1\text{ J}}{\text{C}\bullet\text{V}} \times \frac{1\text{ kWh}}{3.6 \times 10^6\text{ J}} = 3{,}550 = 3.6 \times 10^3\text{ kWh}$$

$$3.550 \times 10^3\text{ kWh} \times \frac{\$0.082}{1\text{ kWh}} = \$291.06 \rightarrow \$291$$

23.76 (a) (See Solution 17.49)

$$Ag_2S(s) \rightleftharpoons 2Ag^+(aq) + S^{2-}(aq) \qquad K_{sp}$$

$$2[Ag^+(aq) + 2CN^-(aq) \rightleftharpoons Ag(CN)_2^-] \qquad K_f^2$$

$$Ag_2S(s) + 4CN^-(aq) \rightleftharpoons 2Ag(CN)_2^-(aq) + S^{2-}(aq)$$

$$K = K_{sp} \times K_f^2 = [Ag^+]^2[S^{2-}] \times \frac{[Ag(CN)_2^-]^2}{[Ag^+]^2[CN^-]^4} = (6 \times 10^{-51})(1 \times 10^{21})^2 = 6 \times 10^{-9}$$

(b) The equilibrium constant for the cyanidation of Ag_2S, 6×10^{-9}, is much less than one and favors the presence of reactants rather than products. The process is not practical.

(c)

$$AgCl(s) \rightleftharpoons Ag^+(aq) + Cl^-(aq) \qquad K_{sp}$$

$$Ag^+(aq) + 2CN^-(aq) \rightleftharpoons Ag(CN)_2^-(aq) \qquad K_f$$

$$AgCl(s) + 2CN^-(aq) \rightleftharpoons Ag(CN)_2^-(aq) + Cl^-(aq)$$

$$K = K_{sp} \times K_f = [Ag^+][Cl^-] \times \frac{[Ag(CN)_2^-]}{[Ag^+][CN^-]^2} = (1.8 \times 10^{-10})(1 \times 10^{21}) = 2 \times 10^{11}$$

Since K >> 1 for this process, it is potentially useful for recovering silver from horn silver. However the magnitude of K says nothing about the rate of reaction. The reaction could be slow and require heat, a catalyst or both to be practical.

23.77 (a) $M(s) \rightarrow M(g)$. The process of atomization is essentially breaking the "metallic bonds" in the solid metal and separating the particles into isolated gas-phase atoms. This requires relocalizing electrons from the solid lattice onto the individual metal atoms.

(b) ΔH_{atom} is the difference between the energy of a mole of gaseous metal atoms, isolated from one another, and a mole of the metal, with all its metal-metal bonding. The difference will be smaller if: 1) the gaseous atoms have some special stability relative to other metallic elements or 2) the metal-metal bonding in the solid is weaker.

The data indicate that Cr and Mn, in the middle of the first transition series, and Cu at the end, have smaller ΔH_{atom} than their neighbors. The electron configurations for these elements are: Cr, $[Ar]4s^13d^5$ (exception); Mn, $[Ar]4s^23d^5$; Cu, $[Ar]4s^13d^{10}$. The gaseous atoms of each of these elements have special stability due to either full or half-full subshells. Assuming relatively constant metal-metal bond strength, the special stability of the gaseous atoms reduces ΔH_{atom} for these elements, relative to their neighbors.

The lower values of ΔH_{atom} for Fe, Co and Ni relative to the elements around V (after taking account of the variations in stability of the gaseous atoms) is likely due to decreasing metal-metal bond strength. Moving to the right across the transition series from the middle onward, effective nuclear charge increases, the radial extension of the d-orbitals decreases and the strength of metallic bonding decreases. This is somewhat of a trend.

24 Chemistry of Coordination Compounds

Introduction to Metal Complexes

24.1 (a) A *metal complex* consists of a central metal ion bonded to a number of surrounding molecules or ions. The number of bonds formed by the central metal ion is the *coordination number.* The surrounding molecules or ions are the *ligands.*

(b) A Lewis acid is an electron pair acceptor and a Lewis base is an electron pair donor. All ligands have at least one unshared pair of valence electrons. Metal ions have empty valence orbitals (d, s or p) that can accommodate donated electron pairs. Ligands act as electron pair donors, or Lewis bases, and metal ions act as electron pair acceptors, or Lewis acids, via their empty valence orbitals.

24.2 (a) In Werner's theory, *primary valence* is the charge of the metal cation at the center of the complex. *Secondary valence* is the number of atoms bound or coordinated to the central metal ion. The modern terms for these concepts are oxidation state and coordination number, respectively. (Note that 'oxidation state' is a broader term than ionic charge, but Werner's complexes contain metal ions where cation charge and oxidation state are equal.)

(b) Ligands are the Lewis base in metal-ligand interactions (see Solution 24.1(b)). As such, they must possess at least one unshared electron pair. NH_3 has an unshared electron pair but BH_3, with less than 8 electrons about B, has no unshared electron pair and cannot act as a ligand. In fact, BH_3 acts as a Lewis acid, an electron pair acceptor, because it is electron-deficient.

24.3 *Analyze/Plan.* Follow the logic in Sample Exercises 24.1 and 24.2. *Solve*:

(a) This compound is electrically neutral, and the NH_3 ligands carry no charge, so the charge on Ni must balance the -2 charge of the 2 Br^- ions. The charge and oxidation state of Ni is +2.

(b) Since there are 6 NH_3 molecules in the complex, the likely coordination number is 6. In some cases Br^- acts as a ligand, so the coordination number could be other than 6.

(c) Assuming that the 6 NH_3 molecules are the ligands, 2 Br^- ions are not coordinated to the Ni^{2+}, so 2 mol AgBr(s) will precipitate. (If one or both of the Br^- act as a ligand, the mol AgBr(s) would be different.)

24.4 (a) Yes. There are 6 possible ligands, 3 H_2O molecules and 3 Cl^- ions. Any Cl^- ions that are not coordinated to the metal will form AgCl(s) precipitate when the complex is treated with $AgNO_3$(aq). Absence of AgCl(s) would mean all Cl^- ions were ligands and a coordination number of 6. One mole of AgCl(s) per mole of complex would mean one uncoordinated ligand, and so on. This assumes that all 3 H_2O molecules act as ligands. In fact, they could serve as water of hydration (Section 13.1, A Closer Look: Hydrates). Reaction with $AgNO_3$ gives no information about the nature of H_2O molecules.

(b) Yes. Conductivity is directly related to the number of ions in a solution. The lower the conductivity, the more Cl^- ions that act as ligands. Conductivity measurements on a set of standard solutions with various moles of ions per mole of complex would provide a comparative method for quantitative determination of the number of free and bound Cl^- ions.

24.5 (a) Coordination number = 4, oxidation number = +2

(b) 5, +4 (c) 6, +3 (d) 5, +2 (e) 6, +3 (f) 4, +2

24.6 (a) Coordination number = 6, oxidation number = +3

(b) 6, +2 (c) 6, +4 octahedral (d) 6, +2 (e) 6, +3 (f) 6, +3

24.7 *Analyze/Plan*. Given the formula of a coordination compound, determine the number and kinds of donor atoms. The ligands are enclosed in the square brackets. Decide which atom in the ligand has an unshared electron pair it is likely to donate. *Solve*:

(a) 4 Cl^- (b) 4 Cl^-, 1 O^{2-} (c) 4 N, 2 Cl^-

(d) 5 C. In CN^-, both C and N have an unshared electron pair. C is less electronegative and more likely to donate its unshared pair.

(e) 6 O. $C_2O_4^{2-}$ is a bidentate ligand; each ion is bound through 2 O atoms for a total of 6 O donor atoms.

(f) 4 N. en is a bidentate ligand bound through 2 N atoms.

24.8 (a) 6 C (see Solution 24.7(d)) (b) 5 O, 1 Br^- (c) 3 N, 3 Br^- (d) 2 O, 4 N

(e) 2 O, 2 S (f) 4 N, 2 F

Polydendate ligands; Nomenclature

24.9 (a) A monodendate ligand binds to a metal in through one atom, a bidendate ligand binds through two atoms.

(b) If a bidentate ligand occupies two coordination sites, three bidentate ligands fill the coordination sphere of a six-coordinate complex.

(c) A tridentate ligand has at least three atoms with unshared electron pairs in the correct orientation to simultanously bind one or more metal ions.

24.10 (a) 2 coordination sites, 2 N donor atoms

(b) 2 coordination sites, 2 N donor atoms

(c) 2 coordination sites, 2 O donor atoms (Although there are four potential O donor atoms in $C_2O_4^{2-}$, it is geometrically impossible for more than two of these to be bound to a single metal ion.

(d) 4 coordination sites, 4 N donor atoms

(e) 6 coordination sites, 2 N and 4 O donor atoms

24.11 *Analyze/Plan.* Given the formula of a coordination compound, determine the number of coordination sites occupied by the polydentate ligand. The coordination number of the complexes is either 4 or 6. Note the number of monodentate ligands and determine the number of coordination sites occupied by the polydentate ligands. *Solve*:

(a) *ortho*-phenanthroline, *o*-phen, is bidentate

(b) oxalate, $C_2O_4^{2-}$, is bidentate

(c) ethylenediaminetetraacetate, EDTA, is pentadentate

(d) ethylenediamine, en, is bidentate

24.12 (a) 4 (b) 4 (c) 6 (d) 6

24.13 (a) The term *chelate effect* means there is a special stability associated with formation of a metal complex containing a polydentate (chelate) ligand relative to a complex containing only monodentate ligands.

(b) When a single chelating ligand replaces two or more monodendate ligands, the number of free molecules in the system increases and the entropy of the system increases. Chemical reactions with $+\Delta S$ tend to be spontaneous, have negative ΔG, and large positive values of K_{eq}.

(c) Polydentate ligands can be used to bind metal ions and prevent them from undergoing unwanted chemical reactions without removing them from solution. The polydentate ligand thus hides or *sequesters* the metal ion.

24.14 (a) Monodentate; py has only one N donor atom.

(b) K_{eq} for this reaction will be less than one. Two free pyridine molecules are replaced by one free bipy molecule. There are more moles of particles in the reactants than products, so ΔS is predicted to be negative. Processes with a net decrease in entropy are usually nonspontaneous, have positive ΔG, and values of K_{eq} less than one. This equilibrium is likely to be spontaneous in the reverse direction.

24.15 *Analyze/Plan.* Given the name of a coordination compound, write the chemical formula. Refer to Table 24.2 to find ligand formulas. Place the metal complex (metal ion + ligands) inside square brackets and the counter ion (if there is one) outside the brackets. *Solve*:

(a) $[Cr(NH_3)_6](NO_3)_3$ (b) $[Co(NH_3)_4CO_3]_2SO_4$ (c) $[Pt(en)_2Cl_2]Br_2$

(d) $K[V(H_2O)_2Br_4]$ (e) $[Zn(en)_2][HgI_4]$

24.16 (a) $[Mn(H_2O)_5Br]SO_4$ (b) $[Ru(bipy)_3](NO_3)_2$ (c) $[Fe(o\text{-phen})_2Cl_2]ClO_4$

(d) $Na[Co(en)Br_4]$ (e) $[Ni(NH_3)_6]_3[Cr(ox)_3]_2$

24.17 Analyze/Plan. Follow the logic in Sample Exercise 24.4, paying attention to naming rules in Section 24.3. *Solve*:

(a) tetraamminedichlororhodium(III) chloride

(b) potassium hexachlorotitanate(IV)

(c) tetrachlorooxomolybdenum(VI)

(d) tetraaqua(oxalato)platinum(IV) bromide

24.18 (a) trichloroethylenediamineniobium(V) sulfate

(b) tricarbonyltripyridinemolybdenum(0)

(c) ammonium tetrachloroaurate(III)

(d) tetraamminediaquairidium(III) nitrate

Isomerism

24.19 *Analyze/Plan.* Consider the definitions of the various types of isomerism, and which of the complexes could exhibit isomerism of the specified type. *Solve*:

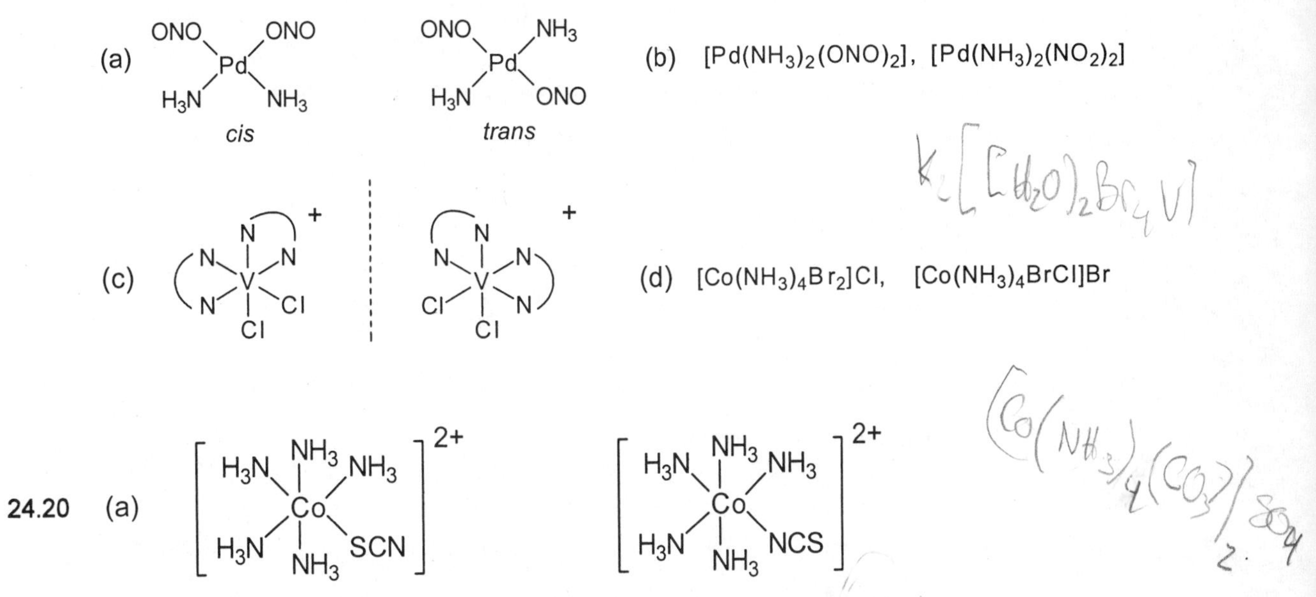

coordination sphere isomerism

24.20 (continued)

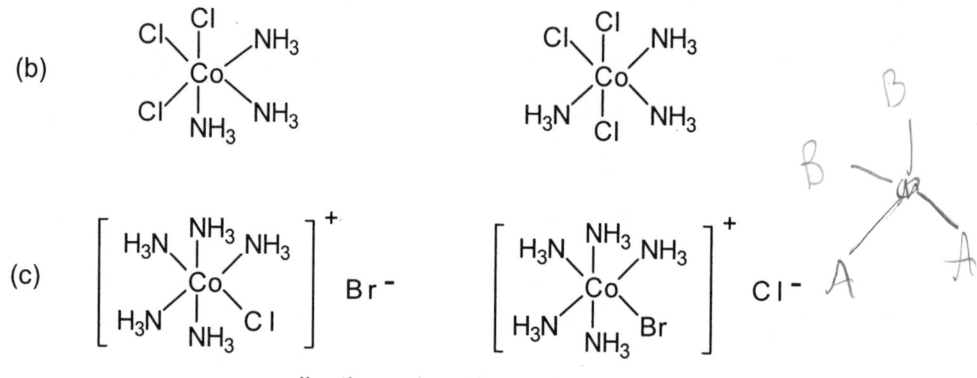

coordination sphere isomerism

24.21 Yes. A tetrahedral complex of the form MA_2B_2 would have neither structural nor stereoisomers. For a tetrahedral complex, no differences in connectivity are possible for a single central atom, so the terms *cis* and *trans* do not apply. No optical isomers with tetrahedral geometry are possible because M is not bound to four different groups. The complex must be square planar with *cis* and *trans* geometric isomers.

24.22 Two geometric isomers are possible for an octahedral MA_3B_3 complex (see below). All other arrangements, including mirror images, can be rotated into these two structures. Neither isomer is optically active.

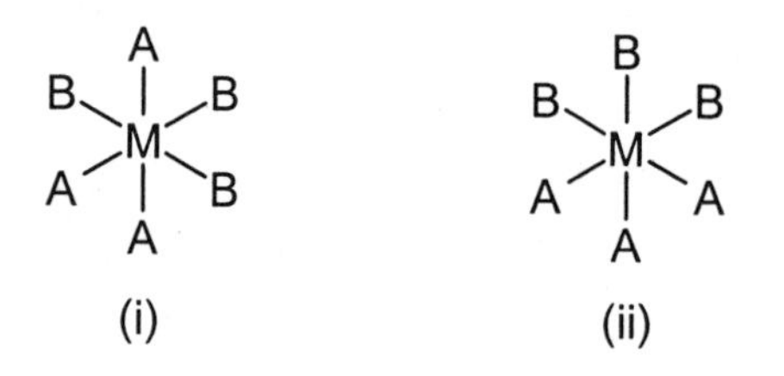

24.23 *Analyze/Plan.* Follow the logic in Sample Exercises 24.5 and 24.6. *Solve*:

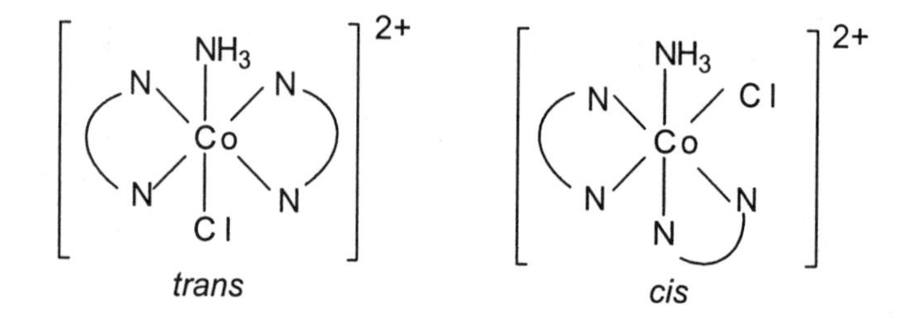

The *cis* isomer is chiral.

24.23 (continued)

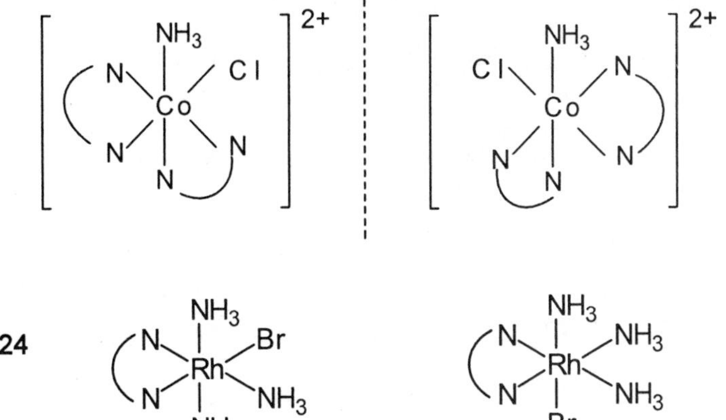

24.24 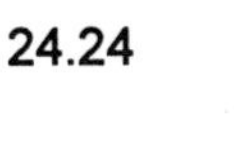 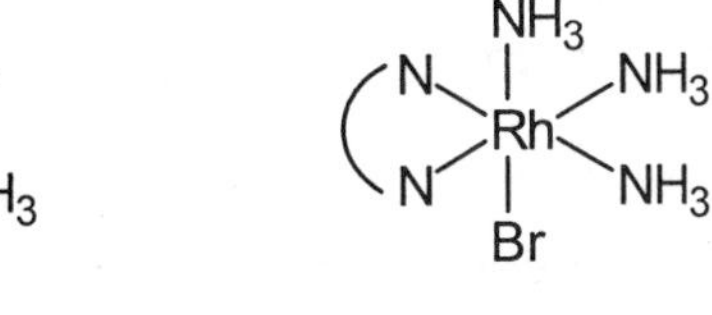

The symbol N N represents the bidentate ligand (bipy).

There are no optical isomers. The mirror image of each structural isomer can be superimposed on the structure above by a 180° rotation.

24.25 *Analyze/Plan*. Follow the logic in Sample Exercise 24.5 and 24.6. *Solve*:

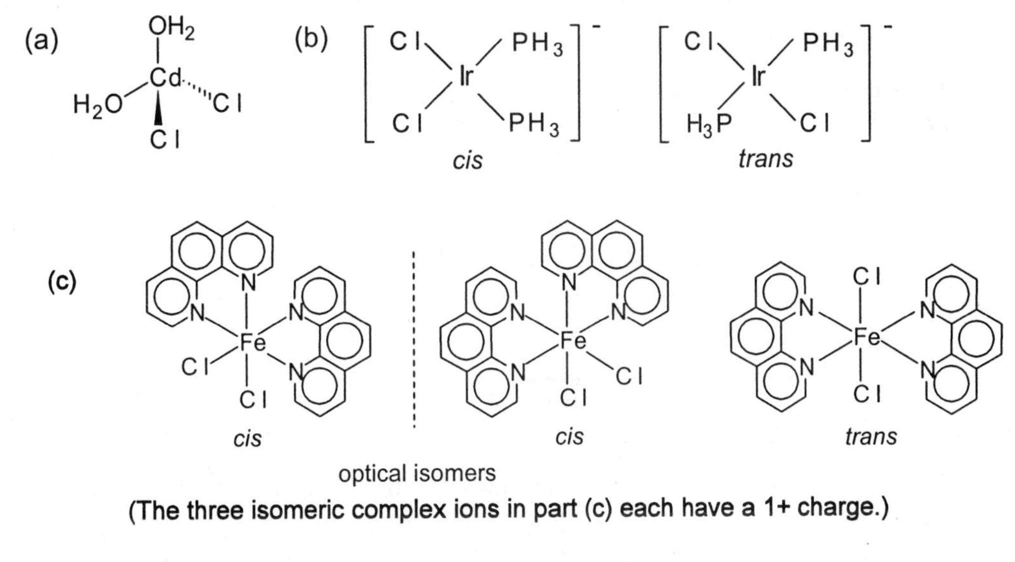

(The three isomeric complex ions in part (c) each have a 1+ charge.)

24.26

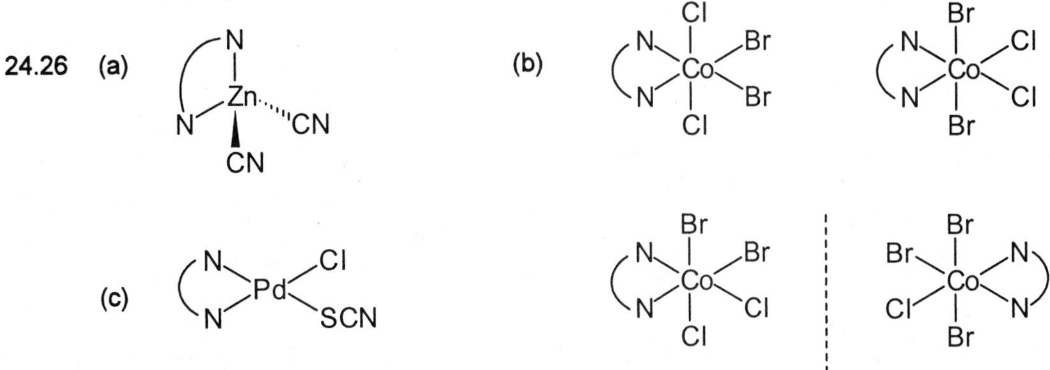

Color, Magnetism; Crystal-Field Theory

24.27 (a) Visible light has wavelengths between 400 and 700 nm.

(b) *Complementary* colors are opposite each other on a color wheel such as Figure 24.25.

(c) A colored metal complex absorbs visible light of its complementary color. For example, a red complex absorbs green light.

24.28 (a) Yes. A complex that absorbs visible light of one wavelength or color will appear as the complementary color. This complex absorbs green light and will appear red.

(b) No. A solution can appear green by transmitting or reflecting only green light (the situation stated in the exercise) **or** by absorbing red light, the complementary color of green.

(c) A visible absorption spectrum shows the amount of light absorbed at a given wavelength. It is a plot of absorbance (dependent variable, y-axis) vs. wavelength (independent variable, x-axis).

24.29 *Analyze/Plan.* A compound that absorbs visible light of one color appears as the complementary color. *Solve*:

Blue to blue-violet (Figure 24.25)

24.30

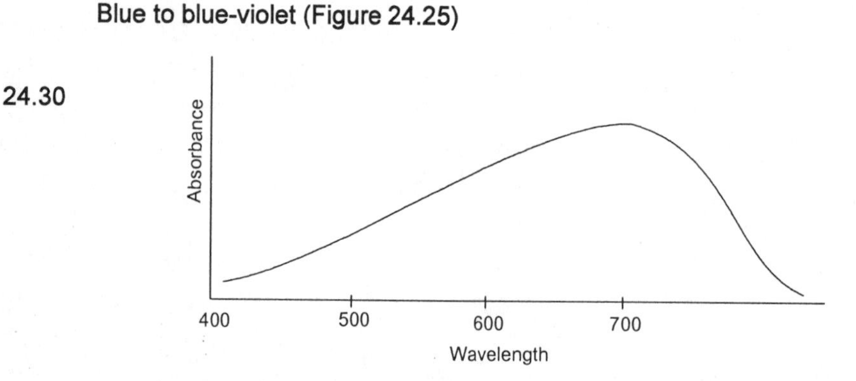

24.31 Most of the electrostatic interaction between a metal ion and a ligand is the attractive interaction between a positively charged metal cation and the full negative charge of an anionic ligand or the partial negative charge of a polar covalent ligand. Whether the interaction is ion-ion or ion-dipole, the ligand is strongly attracted to the metal center and can be modeled as a point negative charge.

24.32 Six ligands in an octahedral arrangement are oriented along the x, y and z axes of the metal. These negatively charged ligands (or the negative end of ligand dipoles) have greater electrostatic repulsion with valence electrons in metal orbitals that also lie along these axes, the d_{z^2}, and $d_{x^2-y^2}$. The d_{xy}, d_{xz} and d_{yz} metal orbitals point between the x, y, and z axes, and electrons in these orbitals experience less repulsion with ligand electrons. Thus, in the presence of an octahedral ligand field, the d_{xy}, d_{xz} and d_{xy} metal orbitals are lower in energy than the $d_{x^2-y^2}$ and d_{z^2}.

24.33 (a)

$$\begin{array}{c} \text{— —} \quad d_{x^2-y^2},\ d_{z^2} \\ \uparrow \\ \Delta \\ \downarrow \\ \text{— — —} \quad d_{xy},\ d_{xz},\ d_{yz} \end{array}$$

(b) The magnitude of Δ and the energy of the d-d transition for a d^1 complex are equal.

(c) The spectrochemical series is an ordering of ligands according to their ability to increase the energy gap Δ.

24.34 (a) $$\Delta E = hc/\lambda = \frac{6.626 \times 10^{-34}\ \text{J}\bullet\text{s} \times 2.998 \times 10^{8}\ \text{m/s}}{500 \times 10^{-9}\ \text{m}} = 3.973 \times 10^{-19}$$

$$= 3.97 \times 10^{-19}\ \text{J/photon}$$

$$\Delta = 3.973 \times 10^{-19}\ \text{J/photon} \times \frac{6.022 \times 10^{23}\ \text{photons}}{1\ \text{mol}} \times \frac{1\ \text{kJ}}{1000\ \text{J}} = 239.25 = 239\ \text{kJ/mol}$$

(b) If H_2O is replaced by NH_3, the magnitude of Δ would increase because NH_3 is higher in the spectrochemical series and creates a stronger ligand field.

24.35 *Analyze/Plan.* Consider the relationship between the color of a complex, the wavelength of absorbed light and the position of a ligand in the spectrochemical series. *Solve*:

Cyanide is a strong field ligand. The d-d electronic transitions occur at relatively high energy, because Δ is large. A yellow color corresponds to absorption of a photon in the violet region of the visible spectrum, between 430 and 400 nm. H_2O is a weaker field ligand than CN^-. The blue or green colors of aqua complexes correspond to absorptions in the region of 620 nm. Clearly, this is a region of lower energy photons than those with characteristic wavelengths in the 430 to 400 nm region. These are very general and imprecise comparisons. Other factors are involved, including whether the complex is high spin or low spin.

24.36 The ions absorb the complement of the color they appear. Green $[Ni(H_2O)_6]^{2+}$ absorbs red light, 650-800 nm. Purple $[Ni(NH_3)_6]^{2+}$ absorbs yellow light, 560-580 nm. Thus, $[Ni(NH_3)_6]^{2+}$ absorbs light with the shorter wavelength. This agrees with the spectrochemical series, which indicates that H_2O will produce a smaller d orbital splitting (Δ) than NH_3. Thus, $[Ni(H_2O)_6]^{2+}$ should absorb light with a smaller energy and longer wavelength.

24.37 *Analyze/Plan.* Determine the charge on the metal ion, subtract it from the row number (3-12) of the transition metal, and the remainder is the number of d-electrons. *Solve*:

(a) Ru^{3+}, d^5 (b) Cu^{2+}, d^9 (c) Co^{3+}, d^6 (d) Mo^{5+}, d^1 (e) Re^{3+}, d^4

24.38 (a) Fe^{3+}, d^5 (b) Mn^{2+}, d^5 (c) Co^{2+}, d^7 (d) Ag^+, d^{10} (e) Sr^{2+}, d^0

24.39 *Analyze/Plan.* Follow the logic in Sample Exercise 24.9. *Solve:*

(a) Mn: $[Ar]4s^23d^5$
Mn^{3+}: $[Ar]3d^4$

(b) Ru: $[Kr]5s^14d^7$
Ru^{3+}: $[Kr]4d^5$

(c) Rh: $[Kr]5s^14d^8$
Rh^{3+}: $[Kr]4d^6$

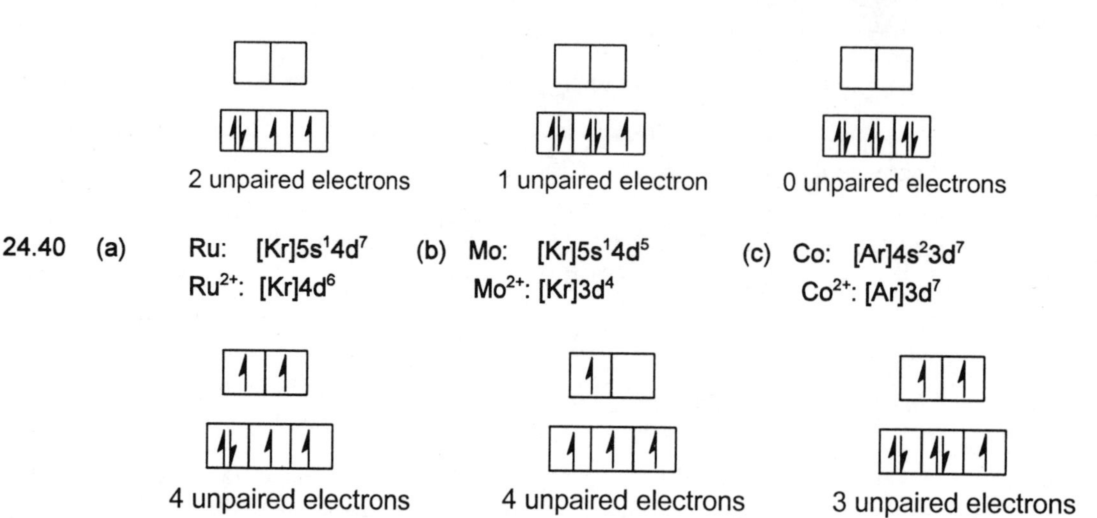

24.40 (a) Ru: $[Kr]5s^14d^7$
Ru^{2+}: $[Kr]4d^6$

(b) Mo: $[Kr]5s^14d^5$
Mo^{2+}: $[Kr]3d^4$

(c) Co: $[Ar]4s^23d^7$
Co^{2+}: $[Ar]3d^7$

24.41 *Analyze/Plan.* All complexes in this exercise are six-coordinate octahedral. Use the definitions of high-spin and low-spin along with the orbital diagram from Sample Exercise 24.9 to place electrons for the various complexes. *Solve:*

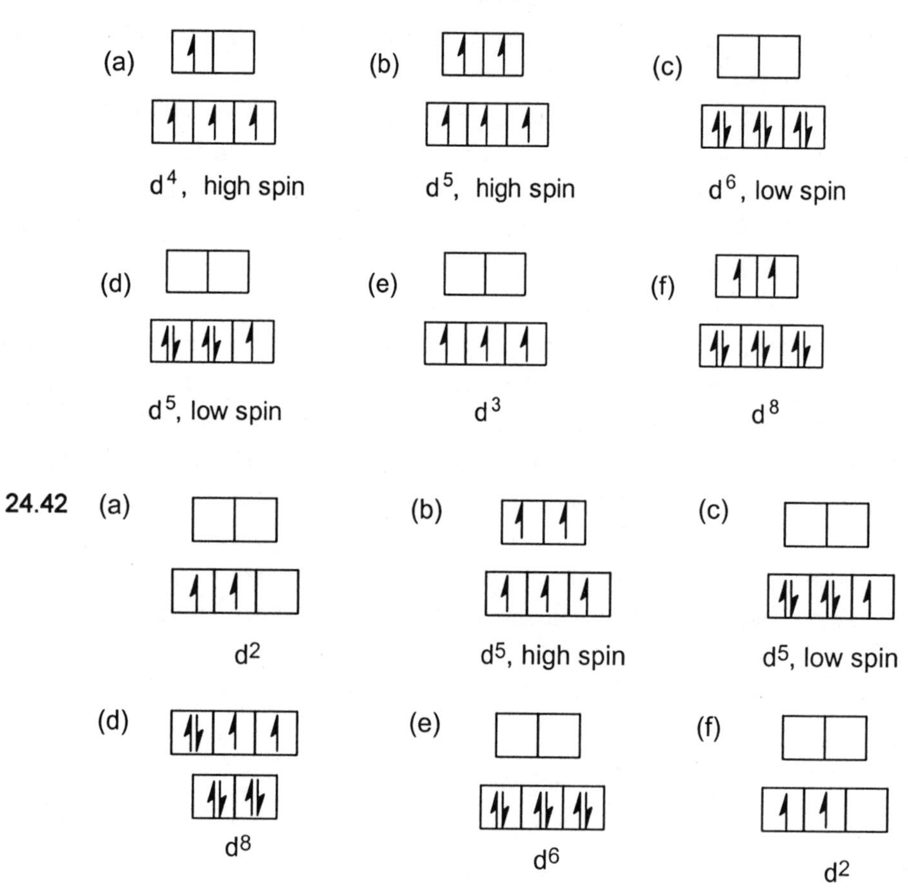

24.43 *Analyze/Plan.* Follow the ideas but reverse the logic in Sample Exercise 24.9. *Solve:*

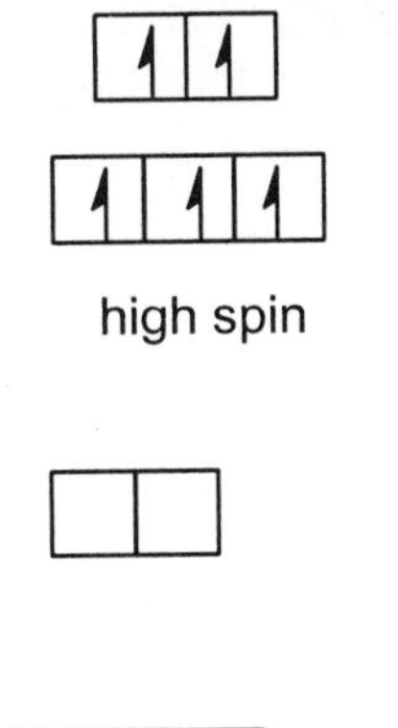

high spin

24.44

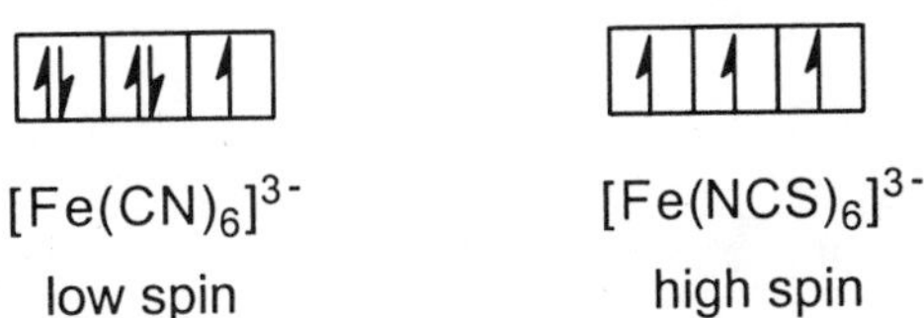

$[Fe(CN)_6]^{3-}$ low spin $[Fe(NCS)_6]^{3-}$ high spin

Both complexes contain Fe^{3+}, a d^5 ion. CN^-, a strong field ligand, produces such a large Δ that the splitting energy is greater than the pairing energy, and the complex is low spin. NCS^- produces a smaller Δ, so it is energetically favorable for d electrons to be unpaired in the higher energy d orbitals. NCS^- is a much weaker-field ligand than CN^-. It is probably weaker than NH_3 and near H_2O in the spectrochemical series.

Additional Exercises

24.45 (a) $[Ni(en)_2Cl_2]$; $[Ni(en)_2(H_2O)_2]Cl_2$

(b) $K_2[Ni(CN)_4]$; $[Zn(H_2O)_4](NO_3)_2$; $[Cu(NH_3)_4]SO_4$

(c) $[CoF_6]^{3-}$, high spin; $[Co(NH_3)_6]^{3+}$ or $[Co(CN)_6]^{3-}$, low spin

(d) thiocyanate, SCN^- or NCS^-; nitrite, NO_2^- or ONO^-

(e) $[Co(en)_2Cl_2]Cl$; see Exercise 24.25(c) for another example.

(f) $[Co(en)_3]Cl_3$, $K_3[Fe(ox)_3]$

24.46 $[Pt(NH_3)_6]Cl_4$; $[Pt(NH_3)_4Cl_2]Cl_2$; $[Pt(NH_3)_3Cl_3]Cl$; $[Pt(NH_3)_2Cl_4]$; $K[Pt(NH_3)Cl_5]$

24.47 (a)

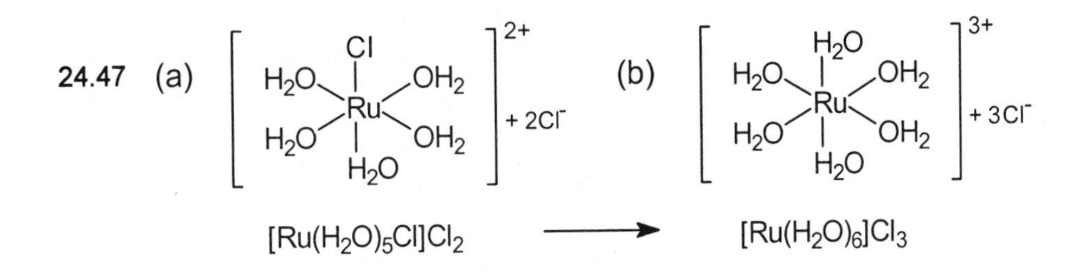

24.48 (a) (b) (c) (d)

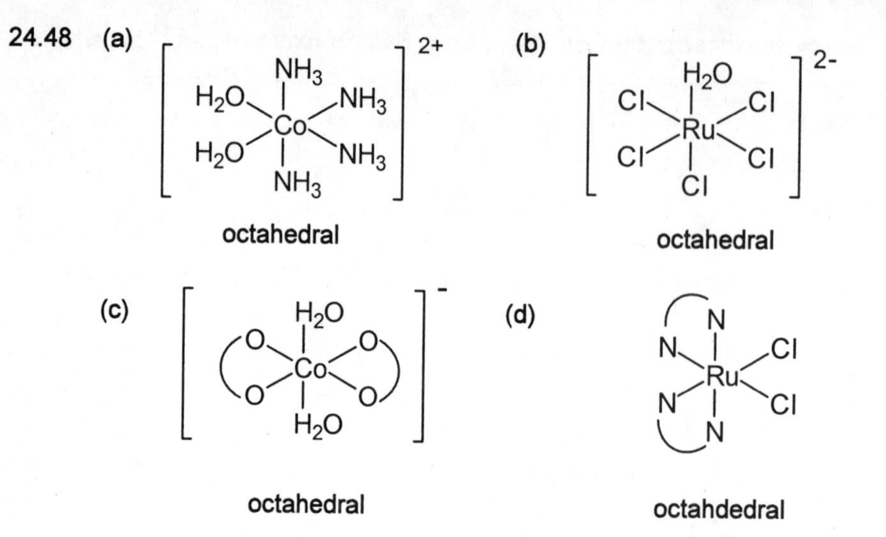

octahedral   octahdedral

24.49 (a) [24.48(a)] *cis*-tetraamminediaquacobalt(II) nitrate

[24.48(b)] sodium aquapentachlororuthenate(III)

[24.48(c)] ammonium *trans*-diaquabisoxalatocobaltate(III)

[24.48(d)] *cis*-dichlorobisethylenediamineruthenium(II)

(b) Only the complex in 24.48(d) is optically active. The mirror images of (a)-(c) can be superimposed on the original structure. The chelating ligands in (d) prevent its mirror images (enantiomers) from being superimposable.

24.50 (a) Valence electrons: 2P + 6C + 16H = 10 + 24 + 16 = 50 e^-, 25 e^- pr

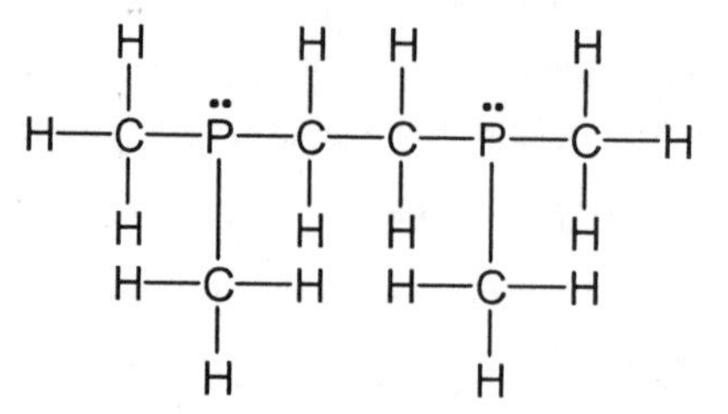

(b) Both CO and dmpe are neutral molecules, so the oxidation state of Mo must be zero.

(c)

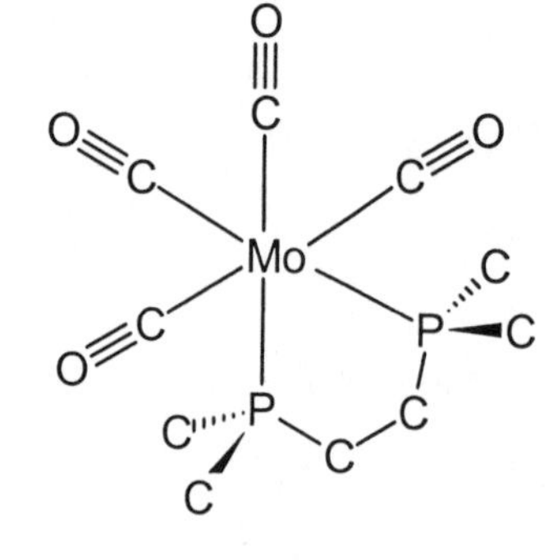

(H atoms omitted for clarity)

C≡O coordinates through C, because it is less electronegative than O, and a better electron pair donor. The molecule has only a single isomer. The dmpe ligand cannot span *trans* positions, so there are no geometric isomers. The mirror image of the structure above is easily superimposable, so there are no optical isomers.

24.51 (a) In a square planar complex such as $[Pt(en)Cl_2]$, if one pair of ligands is *trans*, the remaining two coordination sites are also *trans* to each other. Ethylenediamine is a relatively short bidentate ligand that cannot occupy *trans* coordination sites, so the *trans* isomer is unknown.

(b) A polydentate ligand such as EDTA necessarily occupies *trans* positions in an octahedral complex. The minimum steric requirement for a bidentate ligand is a medium-length chain between the two coordinating atoms that will occupy the *trans* positions. In terms of reaction rate theory, it is unlikely that a flexible bidentate ligand will be in exactly the right orientation to coordinate *trans*. The polydentate ligand has a much better chance of occupying *trans* positions, because it locks the metal ion in place with multiple coordination sites (and shields the metal ion from competing ligands present in the solution).

24.52 We will represent the end of the bidentate ligand containing the CF_3 group by a shaded oval, the other end by an open oval:

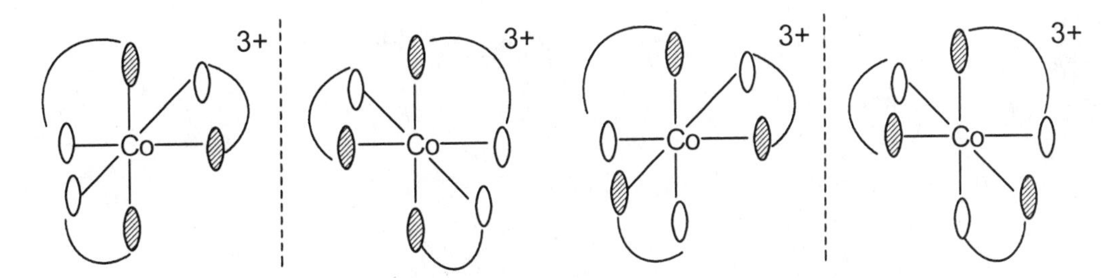

24.53 (a) *Hemoglobin* is the iron-containing protein that transports O_2 in human blood.

(b) *Chlorophylls* are magnesium-containing porphyrins in plants. They are the key components in the conversion of solar energy into chemical energy that can be used by living organisms.

(c) *Siderophores* are iron-binding compounds or ligands produced by a microorganism. They compete on a molecular level for iron in the medium outside the organism and carry needed iron into the cells of the organism.

24.54 (a) $AgCl(s) + 2NH_3(aq) \rightarrow [Ag(NH_3)_2]^+(aq) + Cl^-(aq)$

(b) $[Cr(en)_2Cl_2]Cl(aq) + 2H_2O(l) \rightarrow [Cr(en_2)(H_2O)_2]^{3+}(aq) + 3Cl^-(aq)$

green brown-orange

$3Ag^+(aq) + 3Cl^-(aq) \rightarrow 3AgCl(s)$

$[Cr(en)_2(H_2O)_2]^{3+}$ and $3NO_3^-$ are spectator ions in the second reaction.

(c) $Zn(NO_3)_2(aq) + 2NaOH(aq) \rightarrow Zn(OH)_2(s) + 2NaNO_3(aq)$

$Zn(OH)_2(s) + 2NaOH(aq) \rightarrow [Zn(OH_3)_4]^{2-}(aq) + 2Na^+(aq)$

(d) $Co^{2+}(aq) + 4Cl^-(aq) \rightarrow [CoCl_4]^{2-}(aq)$

24.55 (a) pentacarbonyliron(0)

(b) Since CO is a neutral molecule, the oxidation state of iron must be zero.

(c) $[Fe(CO)_4CN]^-$ has two geometric isomers. In a trigonal bipyramid, the axial and equatorial positions are not equivalent and not superimposable. One isomer has CN in an axial position and the other has it in an equatorial position.

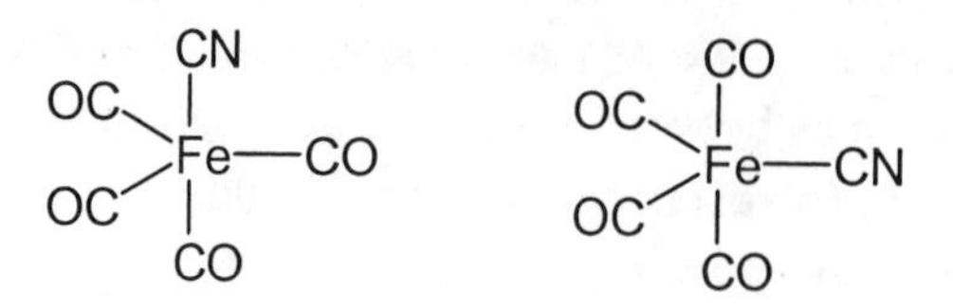

24.56 The ligands that possess the greatest ability to interact with the central metal atom cause the largest splitting. The properties of ligands that are important are **charge** (usually the more negatively charged ligands produce larger splittings) and **polarizability**, which measures the ability of the ligand to distort its charge distribution as it interacts with the positively charged metal ion.

24.57 (a) left shoe (c) wood screw (e) a typical golf club

24.58 (a)

Δ

d^2

(b) These complexes are colored because the crystal-field splitting energy, Δ, is in the visible portion of the electromagnetic spectrum. Visible light with $\lambda = hc/\Delta$ is absorbed, promoting one of the d electrons into a higher energy d orbital. The remaining wavelengths of visible light are reflected or transmitted; the combination of these wavelengths is the color we see.

(c) $[VF_6]^{3-}$ will absorb light with a longer wavelength and lower energy. As a weak-field ligand in the spectrochemical series, F^- causes a smaller Δ than H_2O. Since Δ and λ are inversely related, smaller Δ corresponds to longer λ.

24.59 (a) Formally, the two Ru centers have different oxidation states; one is +2 and the other is +3.

(b) Ru^{2+}, d^6 Ru^{3+}, d^5

(c) There is extensive bonding-electron delocalization in the isolated pyrazine molecule. When pyrazine acts as a bridging ligand, its delocalized molecular orbitals provide a pathway for delocalization of the "odd" d electron in the Creutz-Taube ion. The two metal ions appear equivalent because the odd d electron is delocalized across the pyrazine bridge.

24.60 According to the spectrochemical series, the order of increasing Δ for the ligands is Cl^- < H_2O < NH_3. (The tetrahedral Cl^- complex will have an even smaller Δ than an octahedral one.) The smaller the value of Δ, the longer the wavelength of visible light absorbed. The color of light absorbed is the complement of the observed color. A blue complex absorbs orange light (580-650 nm), a pink complex absorbs green light (490-560 nm) and a yellow complex absorbs violet light (400-430 nm). Since $[CoCl_4]^{2-}$ absorbs the longest wavelength, it appears blue. $[Co(H_2O)_6]^{2+}$ absorbs green and appears pink, and $[Co(NH_3)_6]^{3+}$ absorbs violet and appears yellow.

24.61

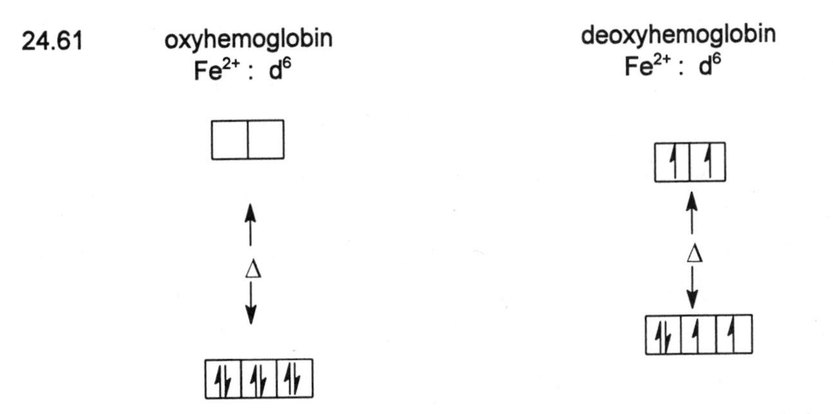

In general, the crystal field splitting, Δ, is greater in low spin than high spin complexes. The energy, Δ, corresponds to the wavelength of light absorbed by the complex and determines its color. Since oxyhemoglobin absorbs higher energy, shorter wavelength light, longer wavelengths remain and the sample appears red. Deoxyhemoglobin absorbs lower energy (orange-red) light, and the sample appears blue.

24.62

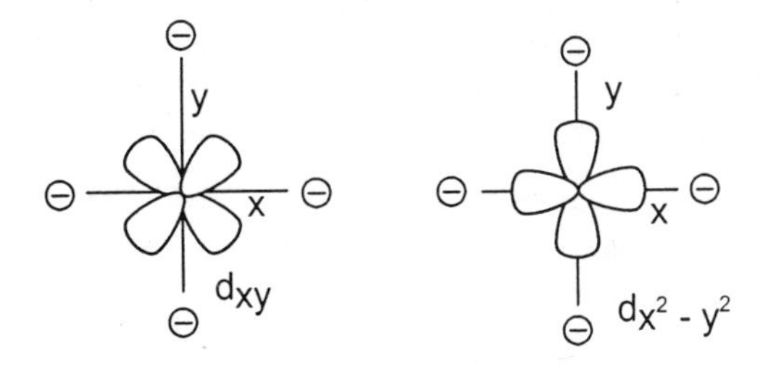

The lobes of the d_{xy} orbital point between the *x* and *y* axes, while the lobes of the $d_{x^2-y^2}$ orbital point along the axes. In a square-planar crystal field with ligands along the axes, an electron in a $d_{x^2-y^2}$ orbital is closer to the electron cloud of the ligands. Electron-electron repulsion with ligand electrons is greater for an electron in the $d_{x^2-y^2}$ orbital of the metal, so the $d_{x^2-y^2}$ orbital has a higher energy than the d_{xy} orbital.

24.63 (a) $[FeF_6]^{4-}$. Both complexes contain the same metal ion, Fe^{2+}; F^- is a weak-field ligand that imposes a smaller Δ and longer λ for the complex ion.

(b) $[V(H_2O)_6]^{2+}$. Both complexes contain the same ligand, H_2O. V^{2+} has a lower charge, so the interaction with the ligand will produce a weaker field, a smaller Δ and a longer absorbed wavelength.

(c) $[CoCl_4]^{2-}$. Both complexes contain the same metal ion, Co^{2+}; Cl^- is a weak-field ligand that imposes a smaller Δ and a longer λ for the complex ion.

24.64 (a) The term *isoelectronic* means that the three ions have the same number of valence electrons and the same electron configuration.

(b) In each ion, the metal is in its maximum oxidation state and has a d^0 electron configuration. That is, the metal ions have no *d*-electrons, so there should be no *d-d* transitions.

(c) A ligand-metal charge transfer transition occurs when an electron in a filled ligand orbital is excited to an empty d-orbital of the metal.

(d) Absorption of 565 nm yellow light by MnO_4^- causes the compound to appear violet, the complementary color. CrO_4^{2-} appears yellow, so it is absorbing violet light of approximately 420 nm. The wavelength of the LMCT transition for chromate, 420 nm, is shorter than the wavelength of LCMT transition in permanganate, 565 nm. This means that there is a larger energy difference between filled ligand and empty metal orbitals in chromate than in permanganate.

(e) Yes. A white compound indicates that no visible light is absorbed. Going left on the periodic chart from Mn to Cr, the absorbed wavelength got shorter and the energy difference between ligand and metal orbitals increased. The 420 nm absorption by CrO_4^- is at the short wavelength edge of the visible spectrum. It is not surprising that the ion containing V, further left on the chart, absorbs at a still shorter wavelength in the ultraviolet region and that VO_4^{3-} appears white.

24.65 Application of pressure would result in shorter metal ion-oxide distances. This would have the effect of increasing the ligand-electron repulsions, and would result in a larger splitting in the d orbital energies. Thus, application of pressure should result in a shift in the absorption to a higher energy and shorter wavelength.

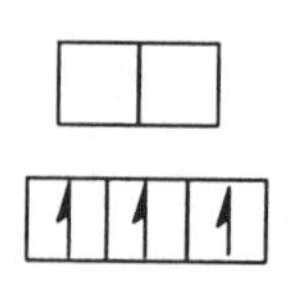

24.66 (a)

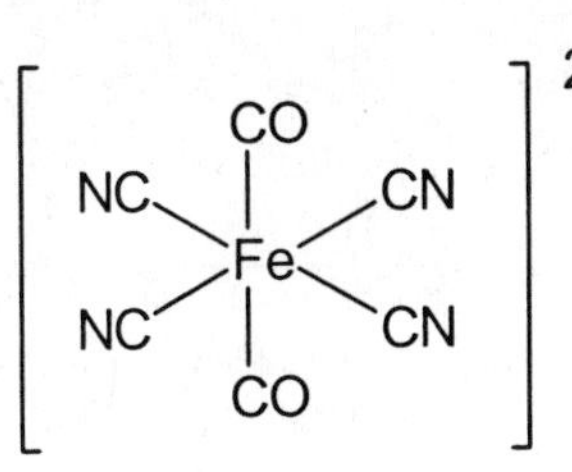

(b) sodium dicarbonyltetracyanoferrate(II)

(c) +2, 6 d electrons

(d) We expect the complex to be low spin. Cyanide (and carbonyl) are high on the spectrochemical series, which means the complex will have a large Δ splitting characteristic of low spin complexes.

24.67

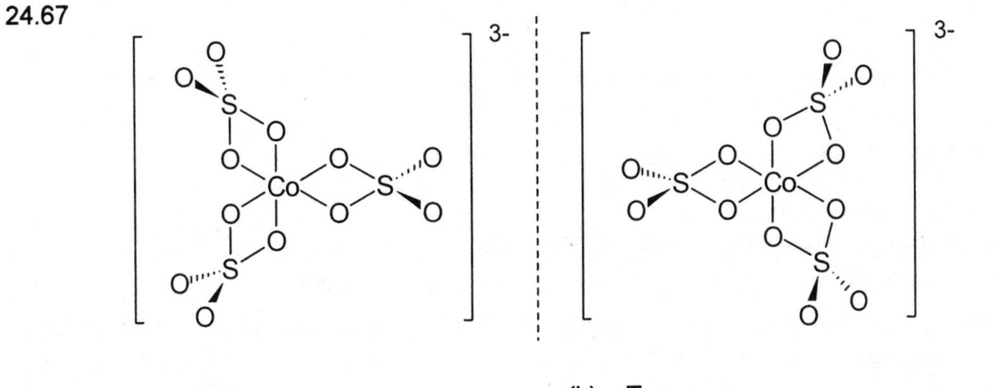

24.68 (a) Only one

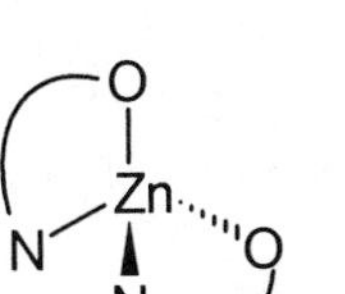

(b) Two

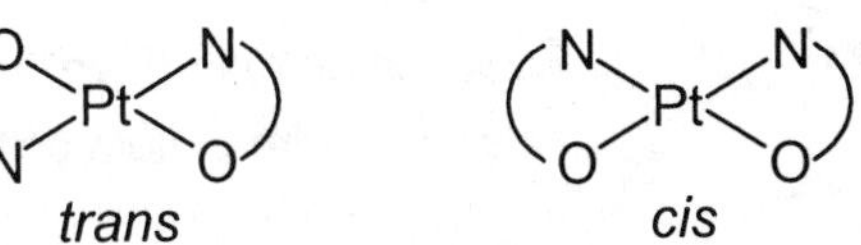

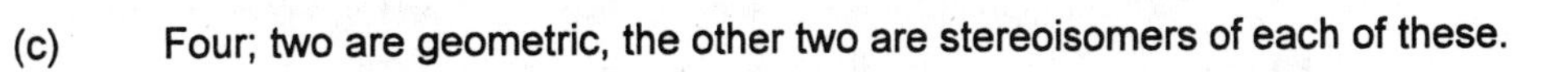

(c) Four; two are geometric, the other two are stereoisomers of each of these.

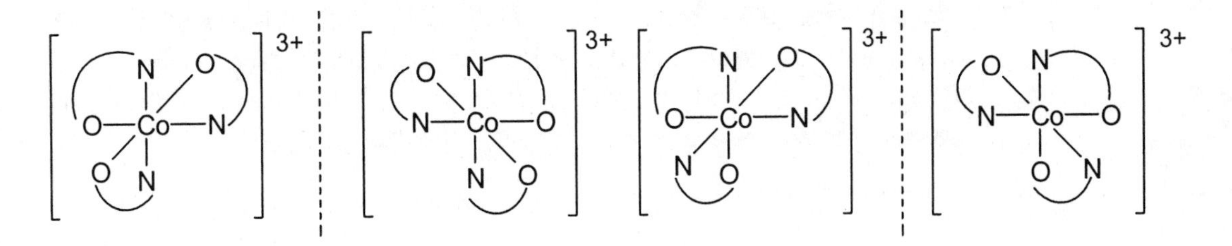

24.69

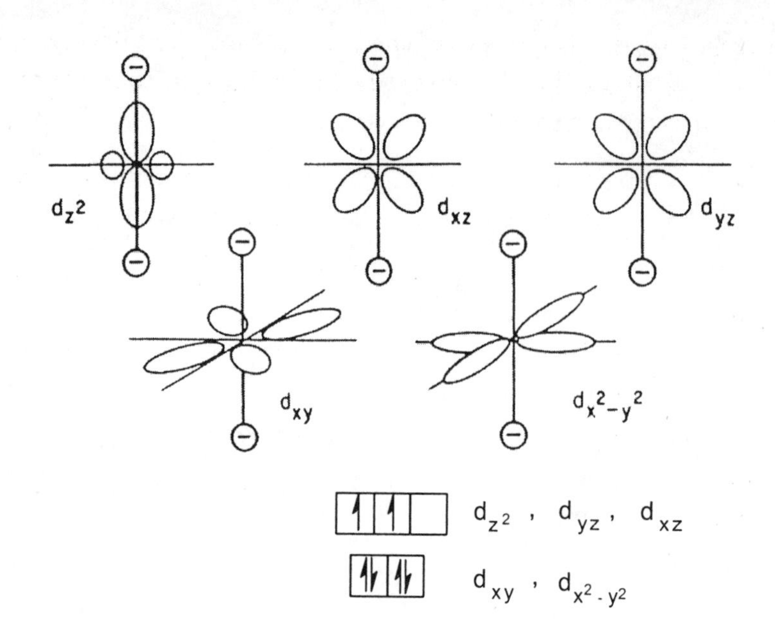

[↑][↑][ ] d_{z^2}, d_{yz}, d_{xz}

[↑↓][↑↓] d_{xy}, $d_{x^2-y^2}$

For a d^6 metal ion in a strong ligand field, there would be two unpaired electrons.

Integrative Exercises

24.70 In a complex ion, the transition metal is an electron pair acceptor, a Lewis acid; the ligand is an electron pair donor, a Lewis base. In carbonic anhydrase, the Zn^{2+} ion withdraws electron density from the O atom of water. The electronegative oxygen atom compensates by withdrawing electron-density from the O–H bond. The O–H bond is polarized and H becomes more ionizable, more acidic than in the bulk solvent. This is similar to the effect of an electronegative central atom in an oxyacid such as H_2SO_4.

24.71 (a) Both compounds have the same general formulation, so Co is in the same (+3) oxidation state in both complexes.

(b) Cobalt(III) complexes are generally inert; that is, they do not rapidly exchange ligands inside the coordination sphere. Therefore, the ions that form precipitates in these two cases are probably outside the coordination sphere. The dark violet compound A forms a precipitate with $BaCl_2(aq)$ but not $AgNO_3(aq)$, so it has SO_4^{2-} outside the coordination sphere and coordinated Br^-, $[Co(NH_3)_5Br]SO_4$. The red-violet compound B forms a precipitate with $AgNO_3(aq)$ but not $BaCl_2(aq)$ so it has Br^- outside the coordination sphere and coordinated SO_4^{2-}, $[Co(NH_3)_5SO_4]Br$.

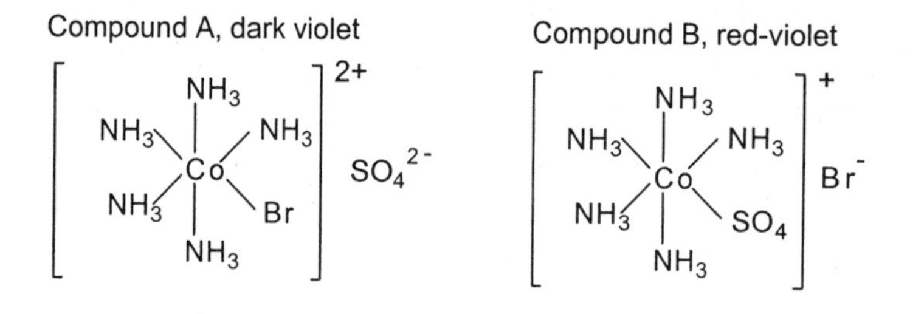

(c) Compounds A and B have the same formula but different properties (color, chemical reactivity), so they are isomers. They vary by which ion is inside the coordination sphere, so they are *coordination sphere isomers.*

(d) Compound A is an ionic sulfate and compound B is an ionic bromide, so both are strong electrolytes. According to the solubility rules in Table 4.1, both should be water-soluble.

24.72 (a)

$$[Cd(CH_3NH_2)_4]^{2+} \rightleftharpoons Cd^{2+}(aq) + 4CH_3NH_2(aq) \qquad \Delta G° = 37.2 \text{ kJ}$$

$$Cd^{2+}(aq) + 2en(aq) \rightleftharpoons [Cd(en)_2]^{2+}(aq) \qquad \Delta G° = -60.7 \text{ kJ}$$

$$Cd(CH_3NH_2)_4]^{2+} + 2en(aq) \rightleftharpoons [Cd(en)_2]^{2+}(aq) + 4CH_3NH_2(aq) \qquad \Delta G° = -23.5 \text{ kJ}$$

$\Delta G° = -RT\ln K_{eq}$; $-23.5 \text{ kJ} = -2.35 \times 10^4 \text{ J}$

$$-2.35 \times 10^4 \text{ J} = \frac{-8.314 \text{ J}}{\text{K} \cdot \text{mol}} \times 298 \text{ K} \times \ln K_{eq}; \ \ln K_{eq} = 9.485, \ K_{eq} = 1.32 \times 10^4$$

(b) The magnitude of K_{eq} is large, so the reaction favors products. The bidentate chelating ligand en will spontaneously replace the monodentate ligand CH_3NH_2. This is an illustration of the chelate effect.

(c) Using the stepwise construction from part (a),

$\Delta H° = 57.3 \text{ kJ} - 56.5 \text{ kJ} = 0.8 \text{ kJ}$

$\Delta S° = 67.3 \text{ J/K} + 14.1 \text{ J/K} = 81.4 \text{ J/K}$

$-T\Delta S = -298 \text{ K} \times 81.4 \text{ J/K} = -2.43 \times 10^4 \text{ J} = -24.3 \text{ kJ}$

The chelate effect is mainly the result of entropy. The reaction is spontaneous due to the increase in the number of free particles and corresponding increase in entropy going from reactants to products. The enthalpic contribution is essentially zero because the bonding interactions of the two ligands are very similar and the reaction is not 'downhill' in enthalpy.

(d) $\Delta H°$ will be very small and negative. When NH_3 replaces H_2O in a complex (Closer Look Box), the tighter bonding of the NH_3 ligand causes a substantial negative $\Delta H°$ for the substitution reaction. When a bidentate amine ligand replaces a monodentate amine ligand of similar bond strength, $\Delta H°$ is very small and either positive (part (c)) or negative (Closer Look Box). In the case of NH_3 replacing CH_3NH_2, the bonding characteristics are very similar. The presence of CH_3 groups in CH_3NH_2 produces some steric hinderance in $[Cd(CH_3NH_2)_4]^{2+}$. This complex is at a slightly higher energy than $[Cd(NH_3)_4]^{2+}$, which experiences no steric hinderance, so $\Delta H°$ will have a negative sign but a very small magnitude. Relief of steric hindrance leads to a very small negative $\Delta H°$ for the substitution reaction.

24.73 First determine the empirical formula, assuming that the remaining mass of complex is Pd.

$$37.6 \text{ g Br} \times \frac{1 \text{ mol Br}}{79.904 \text{ g Br}} = 0.4706 \text{ mol Br; } 0.4706/0.2361 = 2$$

$$28.3 \text{ g C} \times \frac{1 \text{ mol C}}{12.01 \text{ g C}} = 2.356 \text{ mol C; } 2.356/0.2361 = 10$$

$$6.60 \text{ g N} \times \frac{1 \text{ mol N}}{14.01 \text{ g N}} = 0.4711 \text{ mol N; } 0.4711/0.2361 = 2$$

$$2.37 \text{ g H} \times \frac{1 \text{ mol H}}{1.008 \text{ g H}} = 2.351 \text{ mol H; } 2.351/0.2361 = 10$$

$$25.13 \text{ g Pd} \times \frac{1 \text{ mol Pd}}{106.42 \text{ g Pd}} = 0.2361 \text{ mol Pd; } 0.2361/0.2361 = 1$$

The chemical formula is $[Pd(NC_5H_5)_2Br_2]$. This should be a neutral square-planar complex of Pd(II), a nonelectrolyte. Because the dipole moment is zero, we can infer that it must be the *trans*-isomer.

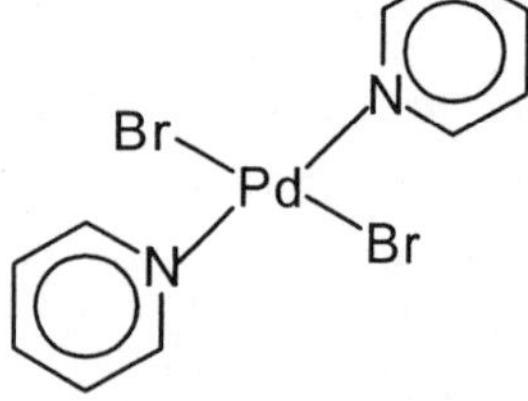

24.74 Determine the empirical formula of the complex, assuming the remaining mass is due to oxygen, and a 100 g sample.

$$10.0 \text{ g Mn} \times \frac{1 \text{ mol Mn}}{54.94 \text{ g Mn}} = 0.1820 \text{ mol Mn; } 0.182 / 0.182 = 1$$

$$28.6 \text{ g K} \times \frac{1 \text{ mol K}}{39.10 \text{ g K}} = 0.7315 \text{ mol K; } 0.732 / 0.182 = 4$$

$$8.8 \text{ g C} \times \frac{1 \text{ mol C}}{12.0 \text{ g C}} = 0.7327 \text{ mol C; } 0.733 / 0.182 = 4$$

$$29.2 \text{ g Br} \times \frac{1 \text{ mol Br}}{79.904 \text{ g Br}} = 0.3654 \text{ mol Br; } 0.365 / 0.182 = 2$$

$$23.4 \text{ g O} \times \frac{1 \text{ mol O}}{16.00 \text{ g O}} = 1.463 \text{ mol O; } 1.46 / 0.182 = 8$$

There are 2 C and 4 O per oxalate ion, for a total of two oxalate ligands in the complex. To match the conductivity of $K_4[Fe(CN)_6]$, the oxalate and bromide ions must be in the coordination sphere of the complex anion. Thus, the compound is $K_4[Mn(ox)_2Br_2]$.

24.75 (a) The reaction that occurs increases the conductivity of the solution by producing a greater number of charged particles, particles with higher charges, or both. It is likely that H_2O from the bulk solvent exchanges with a coordinated Br^- according to the reaction below. This reaction would convert the 1:1 electrolyte, $[Co(NH_3)_4Br_2]Br$, to a 1:2 electrolyte, $[Co(NH_3)_3(H_2O)Br]Br_2$.

(b) $[Co(NH_3)_4Br_2]^+(aq) + H_2O(l) \rightarrow [Co(NH_3)_4(H_2O)Br]^{2+}(aq) + Br^-(aq)$

(c) Before the exchange reaction, there is one mole of free Br^- per mole of complex. mol Br^- = mol Ag^+

M = mol/L; L $AgNO_3$ = mol $AgNO_3$/M $AgNO_3$

$$\frac{3.87 \text{ g complex}}{0.500 \text{ L soln}} \times \frac{1 \text{ mol complex}}{366.77 \text{ g complex}} \times 0.02500 \text{ L soln used} = 5.276 \times 10^{-4} = 5.28 \times 10^{-4} \text{ mol complex}$$

$$5.276 \times 10^{-4} \text{ mol complex} \times \frac{1 \text{ mol Br}^-}{1 \text{ mol complex}} \times \frac{1 \text{ mol Ag}^+}{1 \text{ mol Br}^-} \times \frac{1 \text{ L Ag}^+(aq)}{0.0100 \text{ mol Ag}^+(aq)}$$

$$= 0.05276 \text{ L} = 52.8 \text{ mL AgNO}_3(aq)$$

(d) After the exchange reaction, there are 2 mol free Br^- per mol of complex. Since M $AgNO_3(aq)$ and volume of complex solution are the same for the second experiment, the titration after conductivity changes will require twice the volume calculated in part (c), 105.52 = 106 mL of 0.0100 M $AgNO_3(aq)$.

24.76 Calculate the concentration of Mg^{2+} alone, and then the concentration of Ca^{2+} by difference. M × L = mol

$$\frac{0.0104 \text{ mol EDTA}}{1 \text{ L}} \times 0.0187 \text{L} \times \frac{1 \text{ mol Mg}^{2+}}{1 \text{ mol EDTA}} \times \frac{24.31 \text{ g Mg}^{2+}}{1 \text{ mol Mg}^{2+}} \times \frac{1000 \text{ mg}}{\text{g}} \times \frac{1}{0.100 \text{ L H}_2\text{O}} = 47.28 = 47.3 \text{ mg Mg}^{2+}/\text{L}$$

0.0104 M EDTA × 0.0315 L = mol (Ca^{2+} + Mg^{2+})

0.0104 M EDTA × 0.0187 L = mol Mg^{2+}

0.0104 M EDTA × 0.0128 L = mol Ca^{2+}

$$0.0104 \text{ } M \text{ EDTA} \times 0.0128 \text{ L} \times \frac{1 \text{ mol Ca}^{2+}}{1 \text{ mol EDTA}} \times \frac{40.08 \text{ g Ca}^{2+}}{1 \text{ mol Ca}^{2+}} \times \frac{1000 \text{ mg}}{\text{g}} \times \frac{1}{0.100 \text{ L H}_2\text{O}}$$

$$= 53.35 = 53.4 \text{ mg Ca}^{2+}/\text{L}$$

24.77 $$\frac{182 \times 10^3 \text{ J}}{1 \text{ mol}} \times \frac{1 \text{ mol}}{6.022 \times 10^{23} \text{ molecules}} = 3.022 \times 10^{-19} = 3.02 \times 10^{-19} \text{ J/photon}$$

$\Delta E = h\nu = 3.02 \times 10^{-19}$ J; $\nu = \Delta E/h$

$\nu = 3.022 \times 10^{-19} \text{ J}/6.626 \times 10^{-34} \text{ J} \cdot \text{s} = 4.561 \times 10^{14} = 4.56 \times 10^{14} \text{ s}^{-1}$

$$\lambda = \frac{2.998 \times 10^8 \text{ m/s}}{4.561 \times 10^{14} \text{ s}^{-1}} = 6.57 \times 10^{-7} \text{ m} = 657 \text{ nm}$$

We expect that this complex will absorb in the visible, at around 660 nm. It will thus exhibit a blue-green color (Figure 24.25).

24.78 The process can be written:

$$H_2(g) + 2e \rightarrow 2H^+(aq) \qquad E^\circ_{red} = 0.0\text{ V}$$

$$Cu(s) \rightarrow Cu^{2+} + 2e^- \qquad E^\circ_{red} = 0.337\text{ V}$$

$$Cu^{2+}(aq) + 4NH_3(aq) \rightarrow [Cu(NH_3)_4]^{2+}(aq) \qquad \text{“}E^\circ_f\text{”} = ?$$

$$H_2(g) + Cu(s) + 4NH_3(aq) \rightarrow 2H^+(aq) + [Cu(NH_3)_4]^{2+}(aq) \qquad E = 0.08\text{ V}$$

$$E = E^\circ - RT \ln K_{eq};\; K_{eq} = \frac{[H^+]^2[Cu(NH_3)_4^{2+}]}{P_{H_2}[NH_3]^4}$$

P_{H_2} = 1 atm, $[H^+]$ = 1 *M*, $[NH_3]$ = 1 *M*, $[Cu(NH_3)_4]^{2+}$ = 1 *M*, K = 1

$E = E^\circ - RT \ln(1);\; E = E^\circ - RT(0);\; E = E^\circ = 0.08\text{ V}$

Since we know E° values for two steps and the overall reaction, we can calculate "E°" for the formation reaction and then K_f, using $E^\circ = \frac{0.0592}{n} \log K_f$ for the step.

$E_{cell} = 0.08\text{ V} = 0.0\text{ V} - 0.337\text{ V} + \text{“}E^\circ_f\text{”} \qquad \text{“}E^\circ_f\text{”} = -0.08\text{ V} + 0.337\text{ V} = 0.417\text{ V} = 0.42\text{ V}$

$$\text{“}E^\circ_f\text{”} = \frac{0.0592}{n}\log K_f;\; \log K_f = \frac{n(E^\circ_f)}{0.0592} = \frac{2(0.417)}{0.0592} = 14.0878 = 14.09$$

$K_f = 10^{14.0878} = 1.2 \times 10^{14}$

24.79 (a) The units of the rate constant and the rate dependence on the identity of the second ligand show that the reaction is second order. Therefore, the rate-determining step cannot be a dissociation of water, since that would be independent of the concentration and identity of the incoming ligand. The alternative mechanism, a bimolecular association of the incoming ligand with the complex, is indicated.

(b) The relative values of rate constant are a reflection of the kinetic basicities of the three ligands: Pyridine > SCN^- > CH_3CN.

(c) Ru(III) is a d^5 ion. In a low-spin d^5 octahedral complex, there is one unpaired electron.

25 The Chemistry of Life: Organic and Biological Chemistry

Introduction to Organic Compounds; Hydrocarbons

25.1 *Analyze/Plan.* Given a condensed structural formula, determine the bond angles and hybridization about each carbon atom in the molecule. Visualize the number of electron domains about each carbon. State the bond angle and hybridization based on electron domain geometry.

Solve:

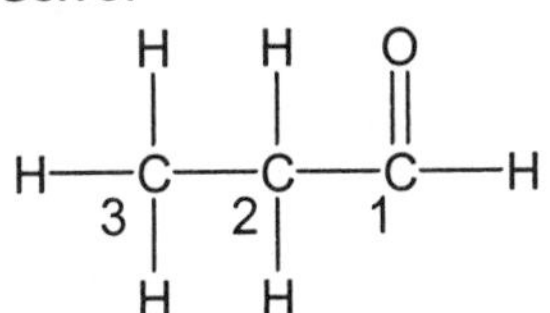

C2 and C3 both have tetrahedral electron domain geometry, 109° bond angles and sp^3 hybridization. C1 has trigonal planar electron domain geometry, 120° bond angles and sp^2 hybridization.

25.2

```
          H   H   H   H   OH  O
          |   |   |   |   |   ||
N≡C — C — C — C = C — C — C — H
  7   6|  5|  4   3  2|   1
       H   H          H
```

(a) C2, C5 and C6 have sp^3 hybridization (4 e^- domains around C)

(b) C7 has sp hybridization (2 e^- domains around C)

(c) C1, C3 and C4 have sp^2 hybridization (3 e^- domains around C)

25.3 Carbon (of course), hydrogen, oxygen, nitrogen, sulfur, phosphorus, chlorine (and other halogens). According to periodic trends and Figure 8.6, oxygen, nitrogen and chlorine are more electronegative than carbon. Sulfur has the same electronegativity as carbon.

25.4 In an organic molecule, bonds with unequal charge distribution are reactive, because they attract reactants that are either electron deficient (electrophilic) or electron rich (nucleophilic). These bonds are either polar bonds with permanent dipole moments, or polarizable bonds with electron clouds that are easily disturbed, or polarized. Of the bonds listed, C–O and C–Cl are reactive. C–O is polar; C–Cl is slightly polar, and polarizable due to the diffuse electron cloud about Cl.

25.5 (a) A *straight-chain hydrocarbon* has all carbon atoms connected in a continuous chain; no carbon atom is bound to more than two other carbon atoms. A *branched-chain hydrocarbon* has a branch; at least one carbon atom is bound to three or more carbon atoms.

(b) An *alkane* is a complete molecule composed of carbon and hydrogen in which all bonds are single (sigma) bonds. An *alkyl group* is a substituent formed by removing a hydrogen atom from an alkane.

(c) Alkanes are said to be *saturated* because they contain only single bonds. Multiple bonds that enable addition of H_2 or other substances are absent. The bonding capacity of each carbon atom is fulfilled with single bonds to C or H.

25.6 All the classifications listed are hydrocarbons; they contain only the elements hydrogen and carbon.

(a) *Alkanes* are hydrocarbons that contain only single bonds.

(b) *Cycloalkanes* contain at least one ring of three or more carbon atoms joined by single bonds. Because it is a type of alkane, all bonds in a cycloalkane are single bonds.

(c) *Alkenes* contain at least one C=C double bond.

(d) *Alkynes* contain at least one C≡C triple bond.

(e) A *saturated hydrocarbon* contains only single bonds. Alkanes and cycloalkanes fit this definition.

(f) An *aromatic hydrocarbon* contains one or more planar, six-membered rings of carbon atoms with delocalized π-bonding throughout the ring.

25.7 *Analyze/Plan.* Consider the definition of the stated classification and apply it to a compound containing 5 C atoms. *Solve:*

(a) $CH_3CH_2CH_2CH_2CH_3$, C_5H_{12}

(b) cyclic structure: CH_2 at top, bonded to H_2C and CH_2, which are bonded to $H_2C—CH_2$ at bottom; C_5H_{10}

(c) $CH_2=CHCH_2CH_2CH_3$, C_5H_{10}

(d) $HC\equiv CCH_2CH_2CH_3$, C_5H_8

saturated: (a), (b); unsaturated: (c), (d)

25.8

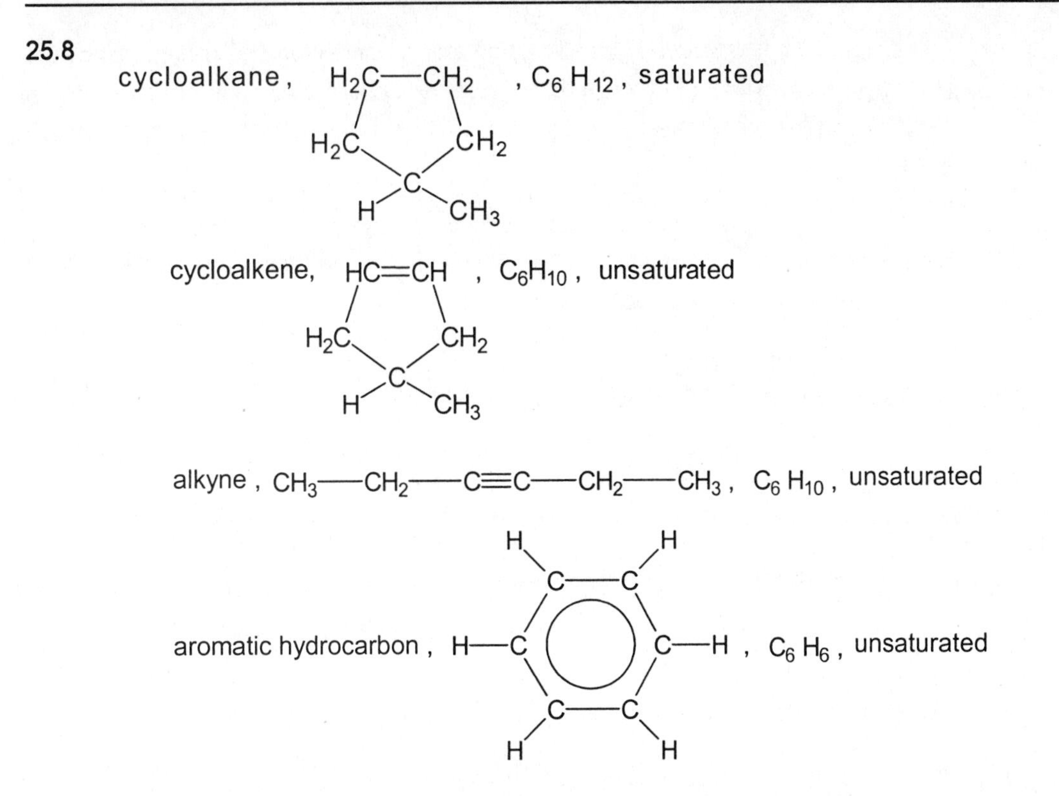

25.9 *Analyze/Plan*. The general formula of an alkane is C_nH_{2n+2}. For an alkene, with 2 fewer H atoms, the general formula is C_nH_{2n}. *Solve*:

A dialkene has one more C=C and thus two fewer H atoms than an alkene. The general formula is C_nH_{2n-2} .

25.10 C_nH_{2n-2}

25.11 *Analyze/Plan*. Follow the logic in Sample Exercise 25.3. *Solve*:

$CH_3-CH_2-CH_2-CH{=}CH_2$

pentene

$CH_3-CH_2-CH{=}CH-CH_3$

2-pentene

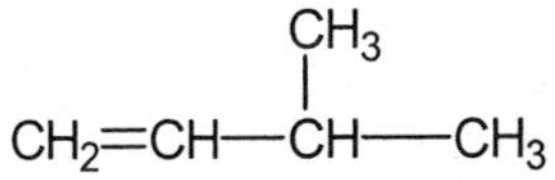

$CH_2{=}CH-CH(CH_3)-CH_3$

3-methyl-1-butene

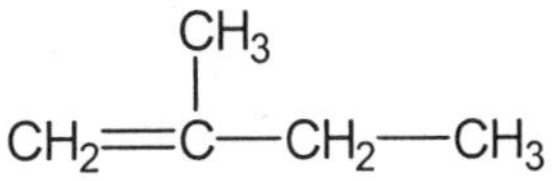

$CH_2{=}C(CH_3)-CH_2-CH_3$

2-methyl-1-butene

$CH_3-C(CH_3){=}CH-CH_3$

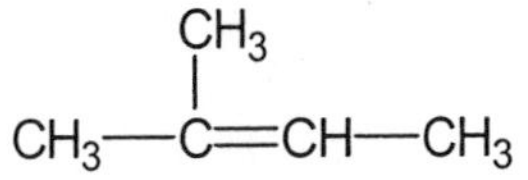

2-methyl-2-butene

25.12

$CH_3CH_2CH_2CH_2C{\equiv}CH$

$CH_3CH_2CH_2C{\equiv}CCH_3$

$CH_3CH_2C{\equiv}CCH_2CH_3$

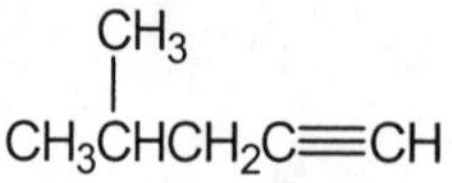

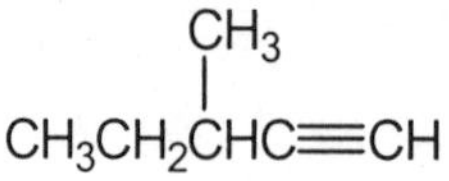

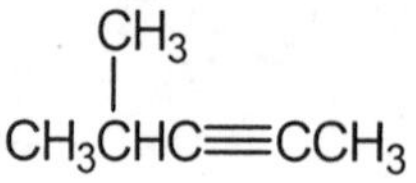

CH_3
$CH_3CHCH_2C{\equiv}CH$

CH_3
$CH_3CH_2CHC{\equiv}CH$

CH_3
$CH_3CHC{\equiv}CCH_3$

$CH_3CH_2CH{=}CHCH{=}CH_2$

$CH_3CH{=}CHCH_2CH{=}CH_2$

$CH_2{=}CHCH_2CH_2CH{=}CH_2$

$CH_3CH{=}CHCH{=}CHCH_3$

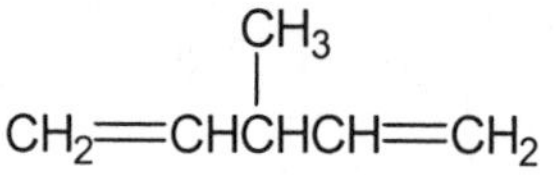

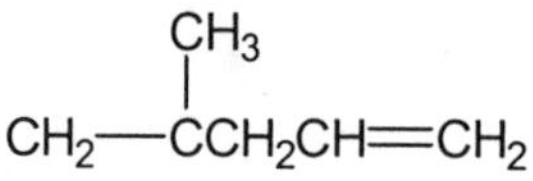

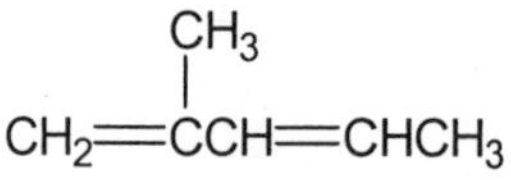

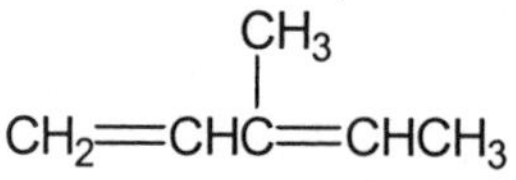

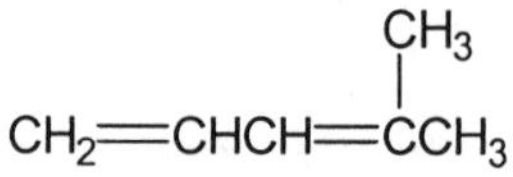

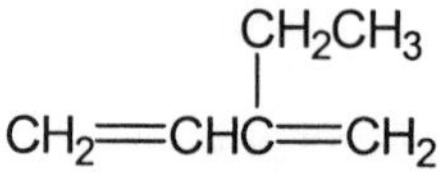

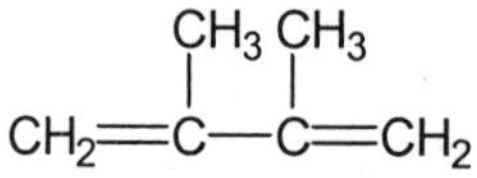

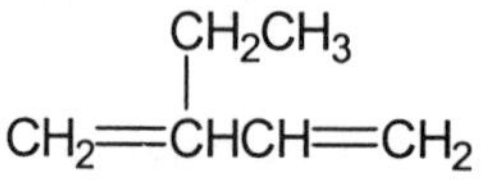

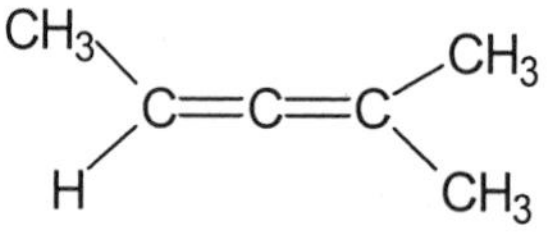

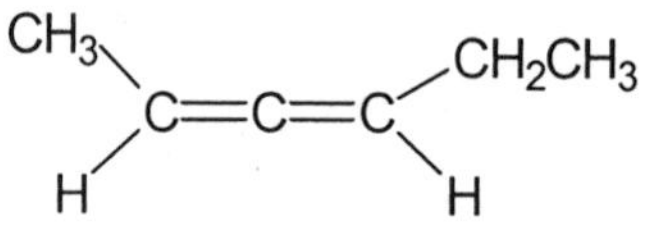

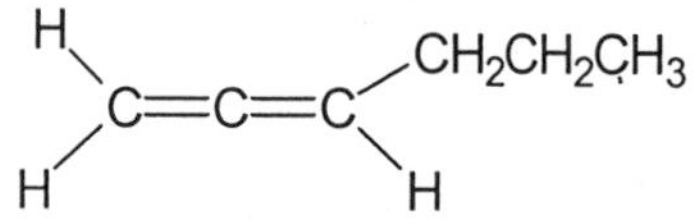

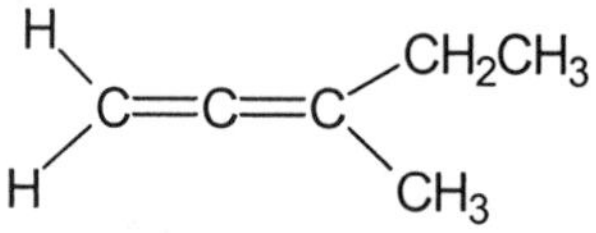

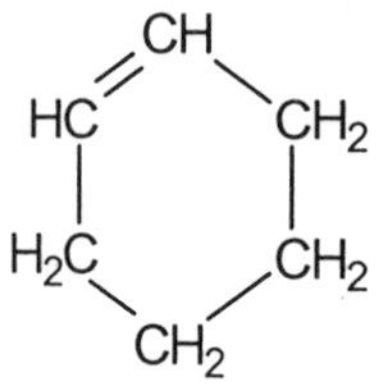

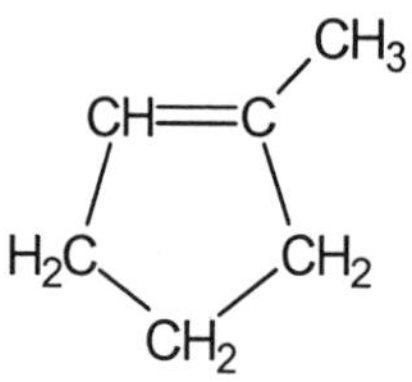

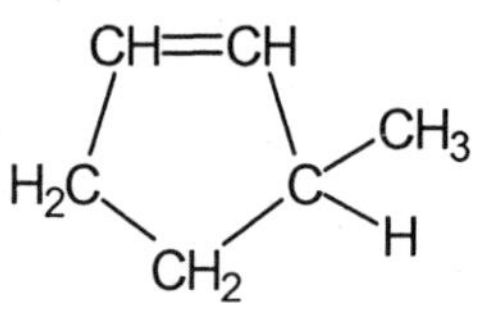

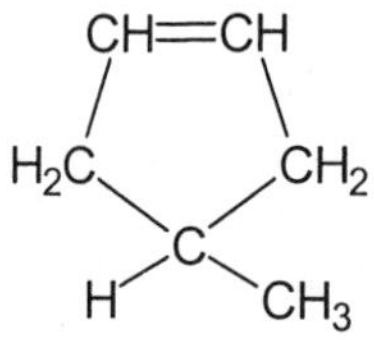

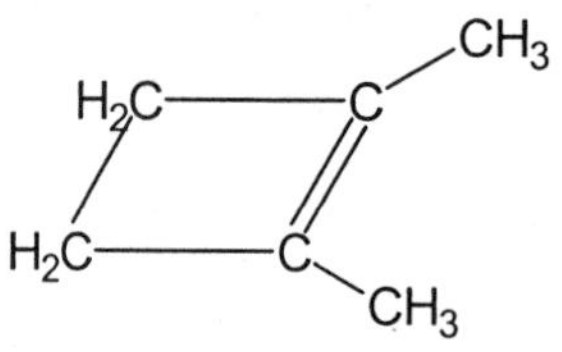

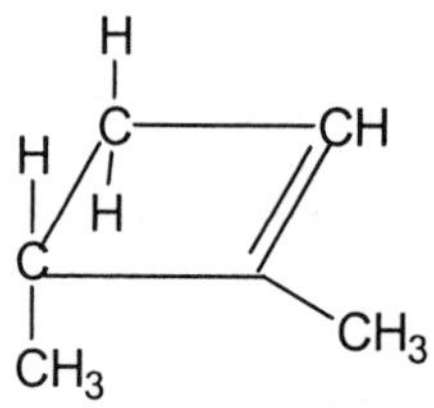

CH_3
$CH_2{=}CHCHCH{=}CH_2$

CH_3
$CH_2{-}CCH_2CH{=}CH_2$

CH_3
$CH_2{=}CCH{=}CHCH_3$

CH_3
$CH_2{=}CHC{=}CHCH_3$

CH_3
$CH_2{=}CHCH{=}CCH_3$

CH_2CH_3
$CH_2{=}CHC{=}CH_2$

CH_3 CH_3
$CH_2{=}C{-}C{=}CH_2$

CH_2CH_3
$CH_2{=}CHCH{=}CH_2$

$(CH_3)(H)C{=}C{=}C(CH_3)(CH_3)$

$(CH_3)(H)C{=}C{=}C(CH_2CH_3)(H)$

$(H)(H)C{=}C{=}C(CH_2CH_2CH_3)(H)$

$(H)(H)C{=}C{=}C(CH_2CH_3)(CH_3)$

CH, HC, CH_2, H_2C, CH_2, CH_2

CH_3, $CH{=}C$, H_2C, CH_2, CH_2

$CH{=}CH$, CH_3, H_2C, C, H, CH_2

$CH{=}CH$, H_2C, CH_2, C, H, CH_3

CH_3, H_2C, C, H_2C, C, CH_3

H, C, CH, H, H, C, CH_3, CH_3

25.12 (continued)

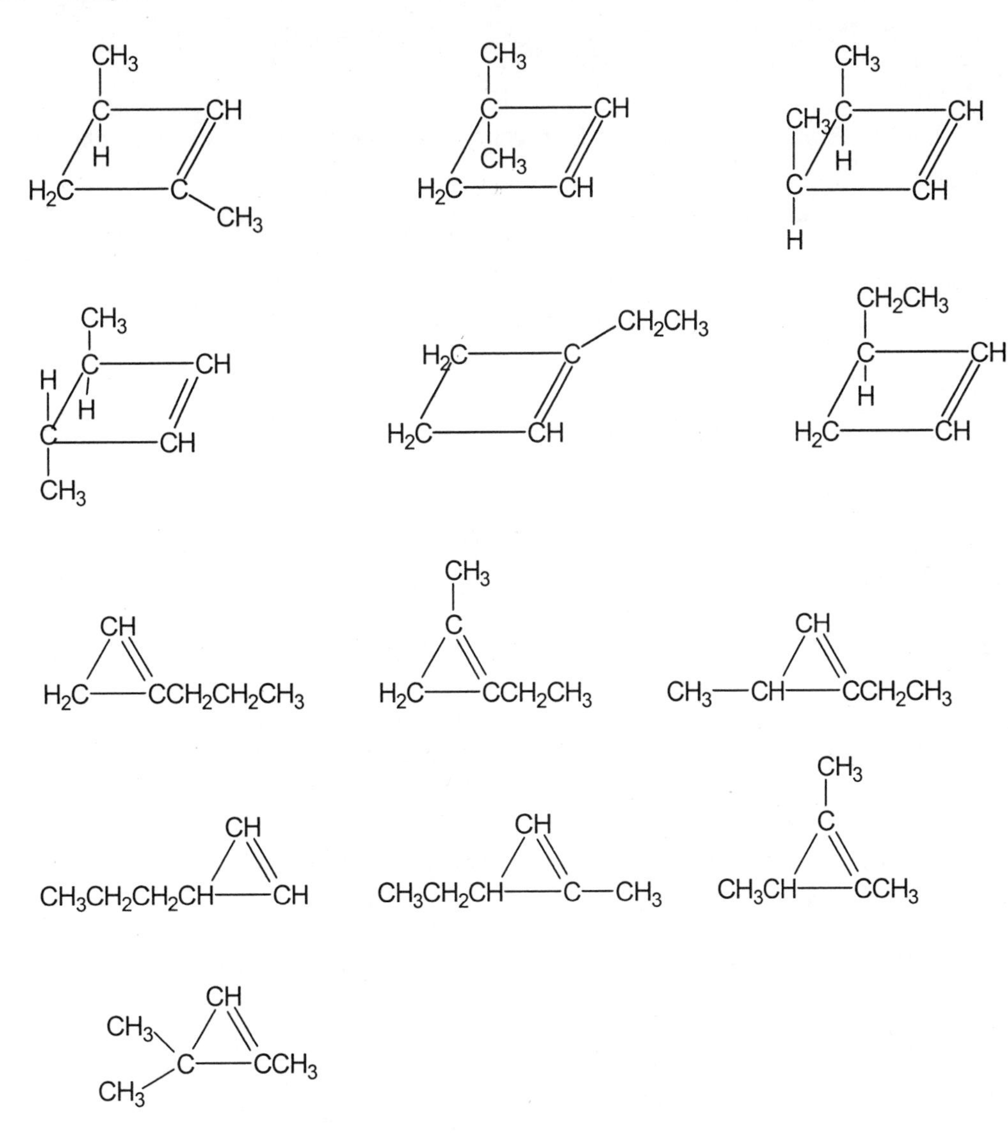

25.13 *Analyze/Plan*. Follow the logic in Sample Exercise 25.1 to name each compound. Decide which structures are the same compound. *Solve*:

(a) 2,2,4-trimethylpentane (b) 3-ethyl-2-methylpentane

(c) 2,3,4-trimethylpentane (d) 2,3,4-trimethylpentane

(c) and (d) are the same molecule

25.14 (a) 3-ethylpentane (b) 2,3-dimethylpentane

(c) 3,3-dimethylpentane (d) 2,3-dimethylpentane

(b) and (d) are the same substance

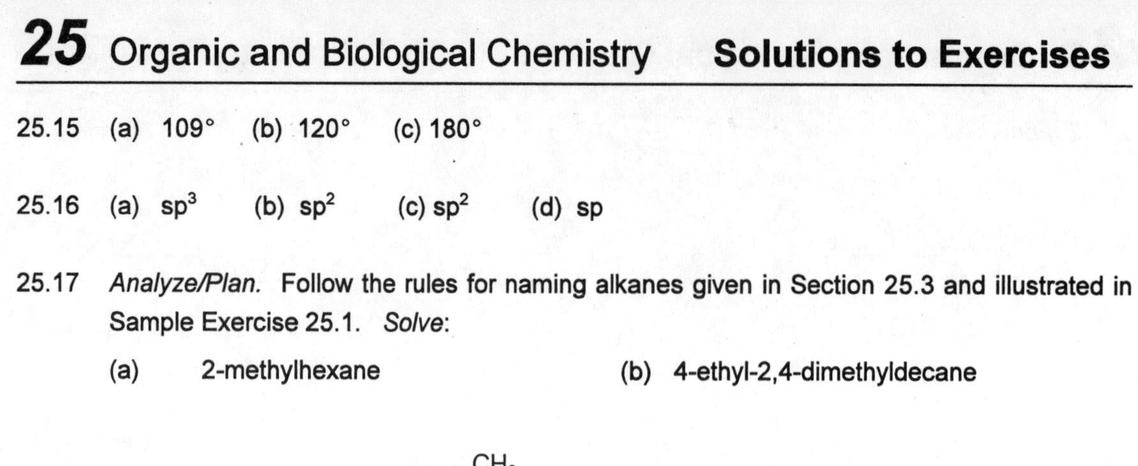

25.15 (a) 109° (b) 120° (c) 180°

25.16 (a) sp^3 (b) sp^2 (c) sp^2 (d) sp

25.17 *Analyze/Plan.* Follow the rules for naming alkanes given in Section 25.3 and illustrated in Sample Exercise 25.1. *Solve*:

(a) 2-methylhexane (b) 4-ethyl-2,4-dimethyldecane

(c) $CH_3—CH_2—CH_2—CH(CH_3)—CH_2—CH_3$

(d)

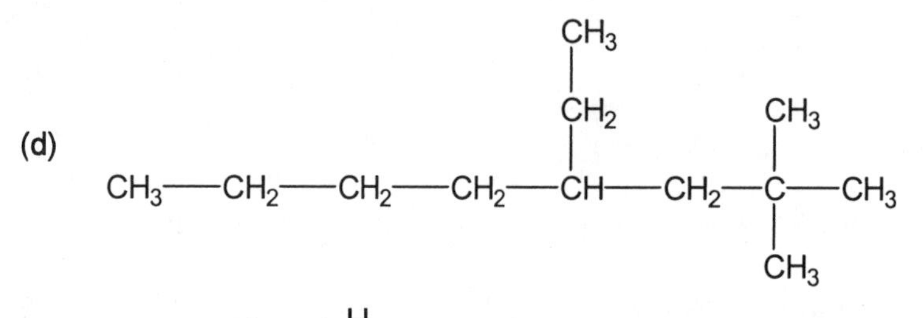

(e)

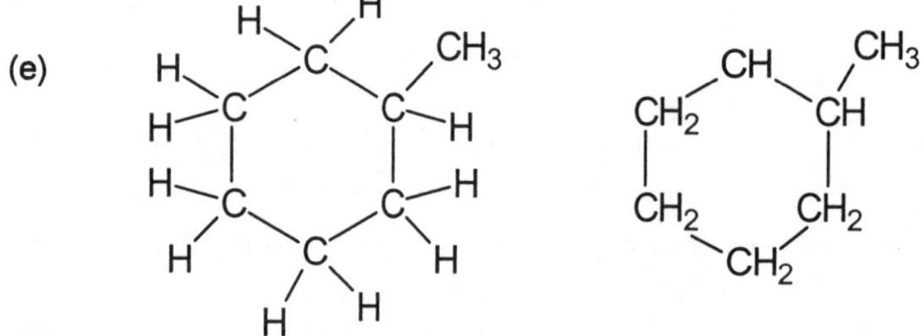

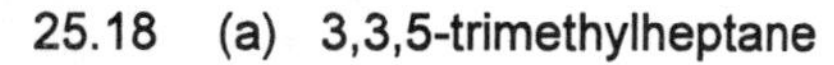

25.18 (a) 3,3,5-trimethylheptane (b) 3,4,4-trimethylheptane

(c)

(d)

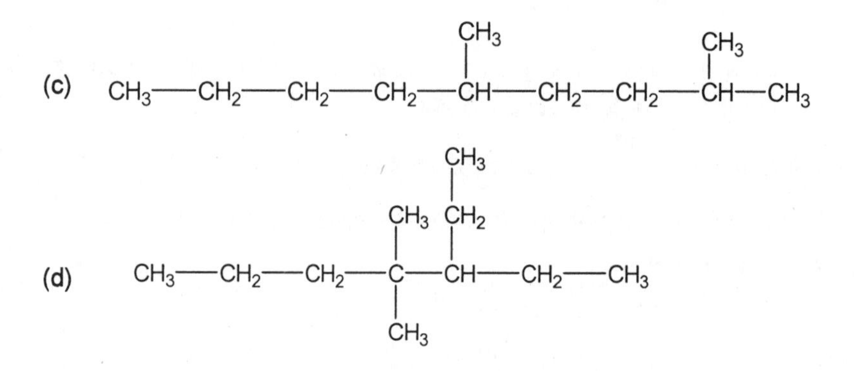

(e)

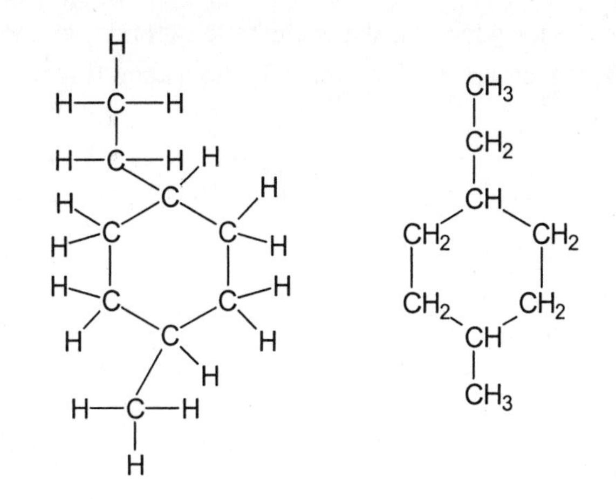

25.19 *Analyze/Plan*. Follow the logic in Sample Exercises 25.1 and 25.4. *Solve*:

(a) 2,3-dimethylheptane (b) *cis*-6-methyl-3-octene (c) *para*-dibromobenzene

(d) 4,4-dimethyl-1-hexyne (e) methylcyclobutane

25.20 (a) 1,4-dichlorocyclohexane (b) 3-chloro-1-propyne (c) *trans*-2-hexene

(d) 1-chloro-2-methyl-2-phenyl-butane or (1-chloro-2-methyl)-2-butylbenzene

(e) *cis*-5-chloro-1,3-pentadiene

25.21 Each doubly bound carbon atom in an alkene has two unique sites for substitution. These sites cannot be interconverted because rotation about the double bond is restricted; geometric isomerism results. In an alkane, carbon forms only single bonds, so the three remaining sites are interchangeable by rotation about the single bond. Although there is also restricted rotation around the triple bond of an alkyne, there is only one additional bonding site on a triply bound carbon, so no isomerism results.

25.22 Butene is an alkene, C_4H_8. There are two possible placements for the double bond:

CH_2=$CHCH_2CH_3$ or CH_3CH=$CHCH_3$
1-butene 2-butene

These two compounds are structural isomers. For 2-butene, there are two different, noninterchangeable ways to construct the carbon skeleton (owing to the absence of free rotation around the double bond). These two compounds are geometric isomers.

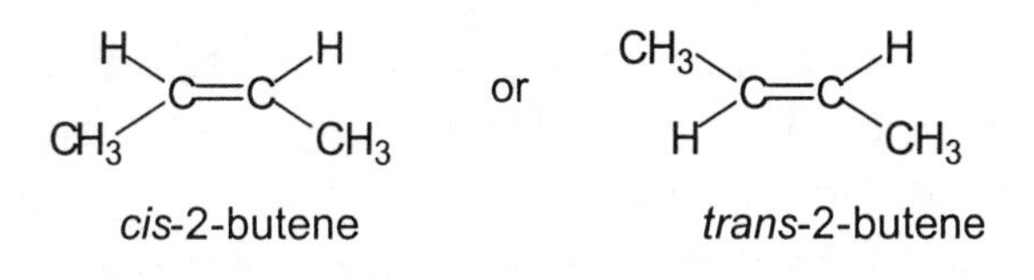

cis-2-butene *trans*-2-butene

25.23 *Analyze/Plan.* In order for geometrical isomerism to be possible, the molecule must be an alkene with two different groups bound to each of the alkene C atoms.

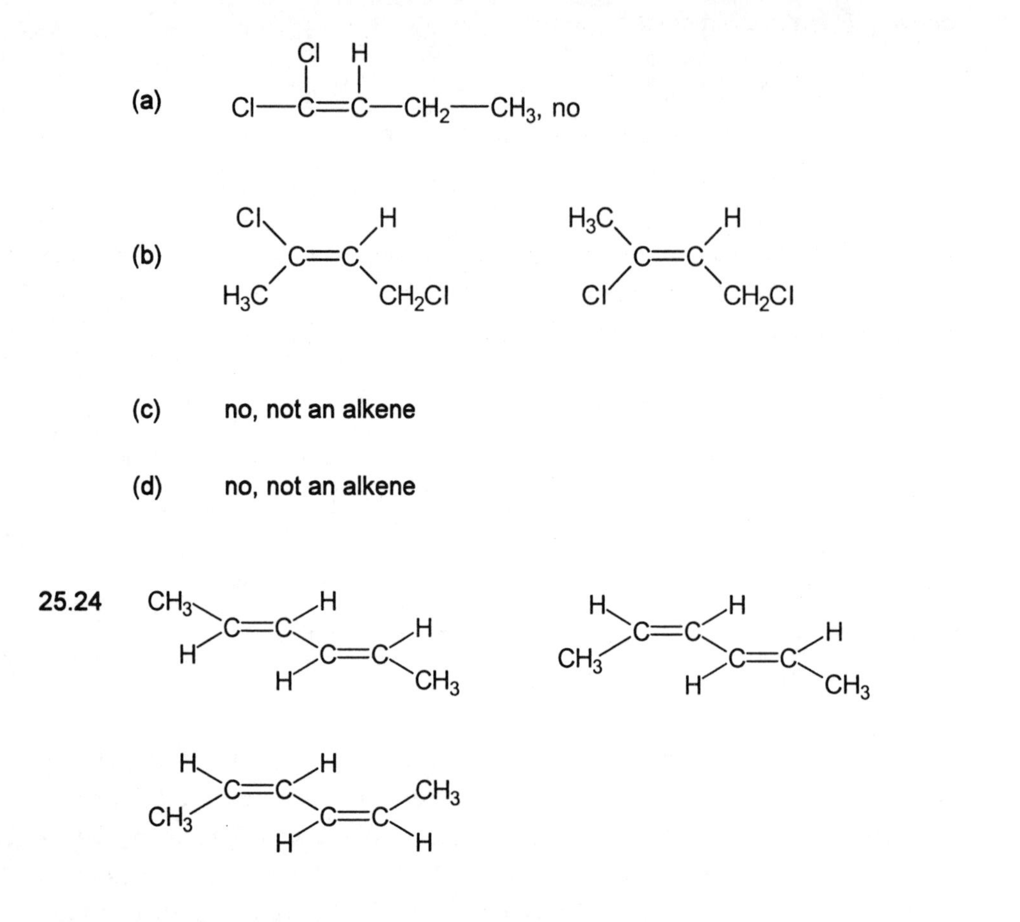

25.25 Assuming that each component retains its effective octane number in the mixture (and this isn't always the case), we obtain: octane number = 0.35(0) + 0.65(100) = 65.

25.26 Octane number can be increased by increasing the fraction of branched-chain alkanes or aromatics, since these have high octane numbers. This can be done by cracking. The octane number also can be increased by adding an anti-knock agent such as tetraethyl lead, $Pb(C_2H_5)_4$ (no longer legal), methyl t-butyl ether, MTBE, or an alcohol, methanol or ethanol.

Reactions of Hydrocarbons

25.27 (a) An addition reaction is the addition of some reagent to the two atoms that form a multiple bond. In a substitution reaction, one atom or group of atoms replaces (substitutes for) another atom or group of atoms. In an addition reaction, two atoms and a multiple bond on the target molecule are altered; in a substitution reaction, the environment of one atom in the target molecule changes. Alkenes typically undergo addition, while aromatic hydrocarbons usually undergo substitution.

(b) *Plan.* consider the general form of addition across a double bond. The π bond is broken and one new substituent (in this case two Br atoms) adds to each of the C atoms involved in the π bond. *Solve*:

$$CH_3{-}\underset{}{\overset{CH_3}{\overset{|}{C}}}{=}\overset{CH_3}{\overset{|}{C}}{-}CH_3 + Br_2 \longrightarrow CH_3{-}\underset{Br}{\underset{|}{\overset{CH_3}{\overset{|}{C}}}}{-}\underset{Br}{\underset{|}{\overset{CH_3}{\overset{|}{C}}}}{-}CH_3$$

(c) *Plan.* Consider the general form of a substitution reaction. A Cl atom will replace one of the H atoms on the benzene ring. In the target molecule, all H atoms are equivalent, so no choice of position is required. *Solve:*

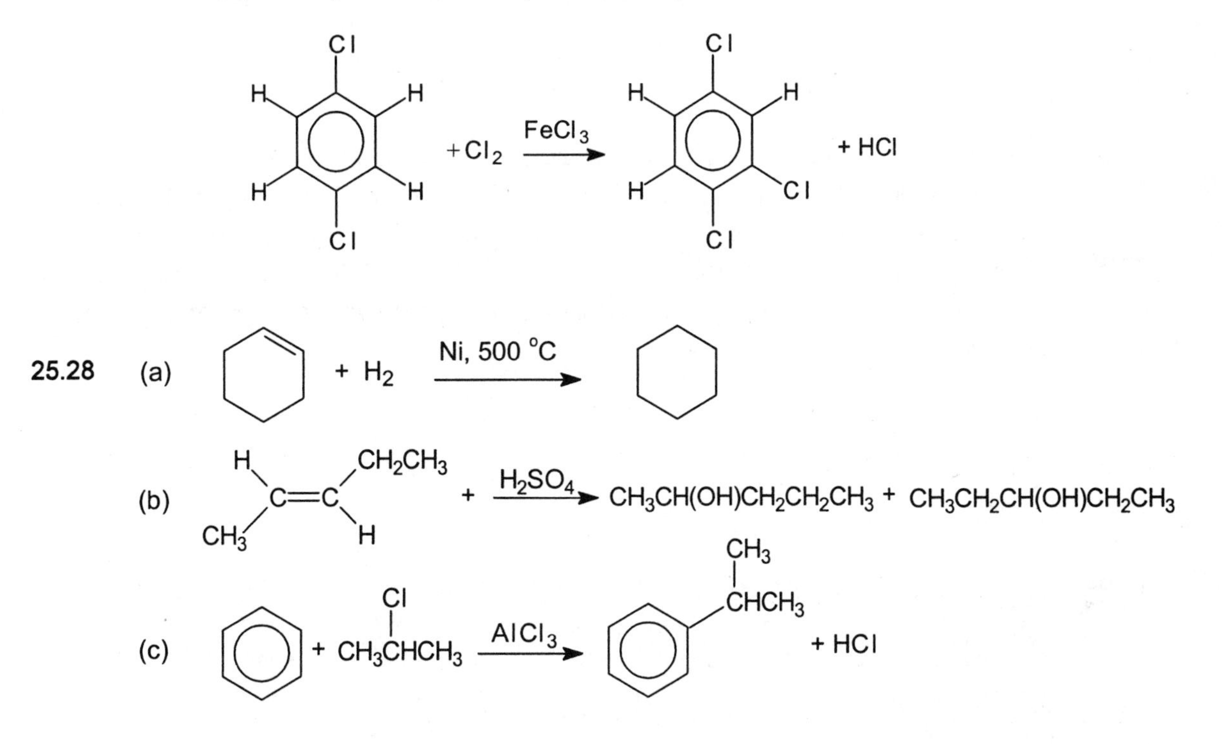

25.29 (a) *Plan.* Consider the structures of cyclopropane, cyclopentane and cyclohexane. *Solve*:

The small 60° C-C-C angles in the cyclopropane ring cause strain that provides a driving force for reactions that result in ring-opening. There is no comparable strain in the five- or six-membered rings.

(b) *Plan.* First form an alkyl halide: $C_2H_4(g) + HBr(g) \rightarrow CH_3CH_2Br(l)$; then carry out a Friedel-Crafts reaction. *Solve*:

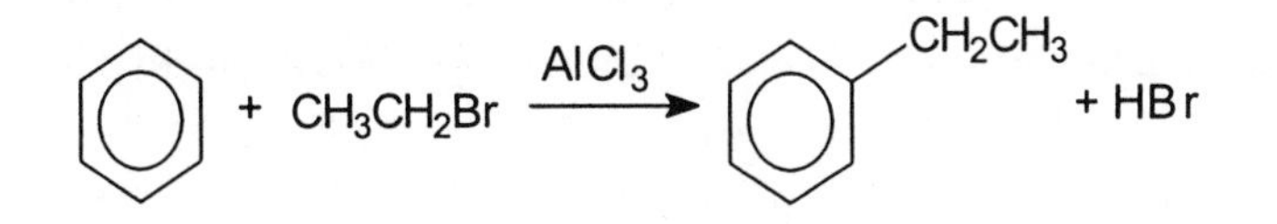

25.30 (a) The reaction of Br_2 with an alkene to form a colorless halogenated alkane is an addition reaction. Aromatic hydrocarbons do not readily undergo addition reactions, because their π-electrons are stabilized by delocalization.

(b) *Plan.* Both the *ortho* and *para* isomers are formed; they must be separated by distillation or some other technique. *Solve:*

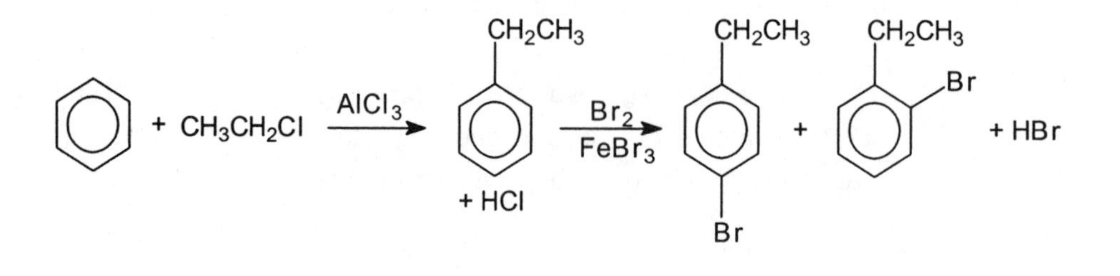

25.31 The partially positive end of the hydrogen halide, $\overset{\delta^+}{H}—\overset{\delta^-}{X}$, is attached to the π electron cloud of the alkene, cyclohexene. The electrons that formed the π bond in cyclohexene form a sigma bond to the H atom of HX, leaving a halide ion, X^-. The intermediate is a carbocation; one of the C atoms formerly involved in the π bond is now bound to a second H atom. The other C atom formerly involved in the π bond carries a full positive charge and forms only three sigma bonds, two to adjacent C atoms and one to H.

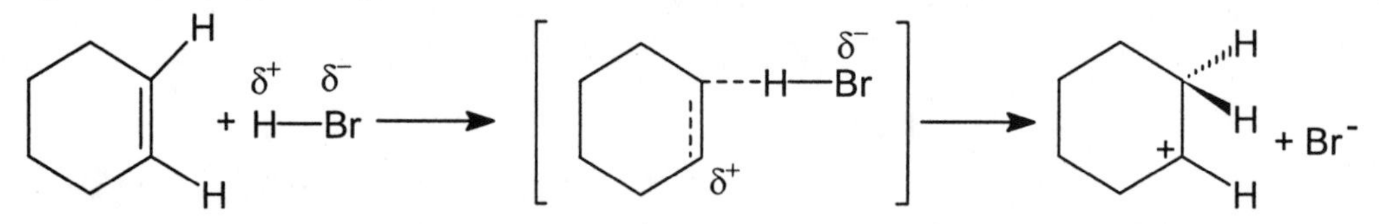

25.32 Not necessarily. That the rate laws are both first order in both reactants and second order overall indicates that the activated complex in the rate-determining step in each mechanism is bimolecular and contains one molecule of each reactant. This is usually an indication that the mechanisms are the same, but it does not rule out the possibility of different fast steps, or a different order of elementary steps.

25.33 *Analyze/Plan.* Both combustion reactions produce CO_2 and H_2O:

$C_3H_6(g) + 9/2\ O_2(g) \rightarrow 3CO_2(g) + 3H_2O(l)$

$C_5H_{10}(g) + 15/2\ O_2(g) \rightarrow 5CO_2(g) + 5H_2O(l)$

Thus, we can calculate the ΔH_{comb} / CH_2 group for each compound. *Solve:*

$$\frac{\Delta H_{comb}}{CH_2\ \text{group}} = \frac{2089\ \text{kJ/mol}\ C_3H_6}{3\ CH_2\ \text{groups}} = \frac{696.3\ \text{kJ}}{\text{mol}\ CH_2};\quad \frac{3317\ \text{kJ/mol}\ C_5H_{10}}{5\ CH_2\ \text{groups}} = 663.4\ \text{kJ/mol}\ CH_2$$

$\Delta H_{comb}/CH_2$ group for cyclopropane is greater because C_3H_6 contains a strained ring. When combustion occurs, the strain is relieved and the stored energy is released during the reaction.

25.34

	ΔH
$C_{10}H_8(l) + 12O_2(g) \rightarrow 10CO_2(g) + 4H_2O(l)$	-5157 kJ
$-[C_{10}H_{18}(l) + 29/2\ O_2(g) \rightarrow 10CO_2(g) + 9H_2O(l)]$	-(-6286) kJ
$C_{10}H_8(l) + 5H_2O(l) \rightarrow C_{10}H_{18}(l) + 5/2\ O_2(g)$	+1129 kJ
$5/2\ O_2(g) + 5H_2(g) \rightarrow 5H_2O(l)$	5(-285.8) kJ
$C_{10}H_8(l) + 5H_2(g) \rightarrow C_{10}H_{18}(l)$	-300 kJ

Compare this with the heat of hydrogenation of ethylene:

$C_2H_4(g) + H_2(g) \rightarrow C_2H_6(g)$; ΔH = -84.7 - (52.3) = -137 kJ. This value applies to just one double bond. For five double bonds, we would expect about -685 kJ. The fact that hydrogenation of napthalene yields only -300 kJ indicates that the overall energy of the napthalene molecule is lower than expected for five isolated double bonds and that there must be some special stability associated with the aromatic system in this molecule.

Functional Groups and Chirality

25.35 (a) ketone (b) carboxylic acid (c) alcohol (d) ester (e) amide (f) amine

25.36

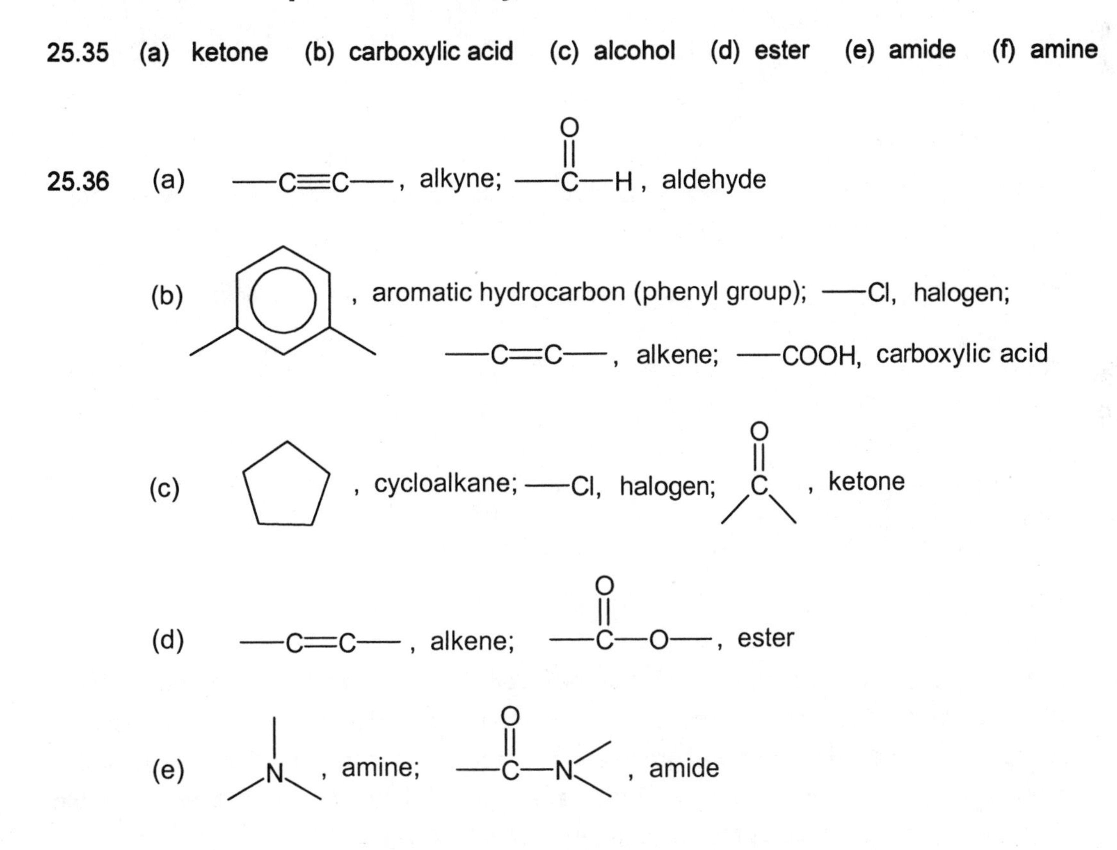

25.37 *Analyze/Plan.* Given the name of a molecule, write the structural formula of an isomer that contains a specified functional group. Consider the definition of isomer, write the molecular formula of the given molecule, draw the structural formula of a molecule with the same formula that contains the specified functional group. *Solve*:

(a) The formula of acetone is C_3H_6O. An aldehyde contains the group $-C(=O)H$

An aldehyde that is an isomer of acetone is propionaldehyde (or propanal),

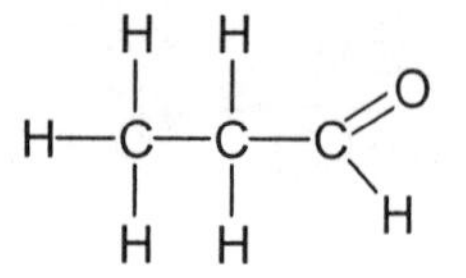

(b) The formula of 1-propanol is C_3H_8O. An ether contains the group –O–. An ether that is an isomer of 1-propanol is:

ethylmethyl ether,

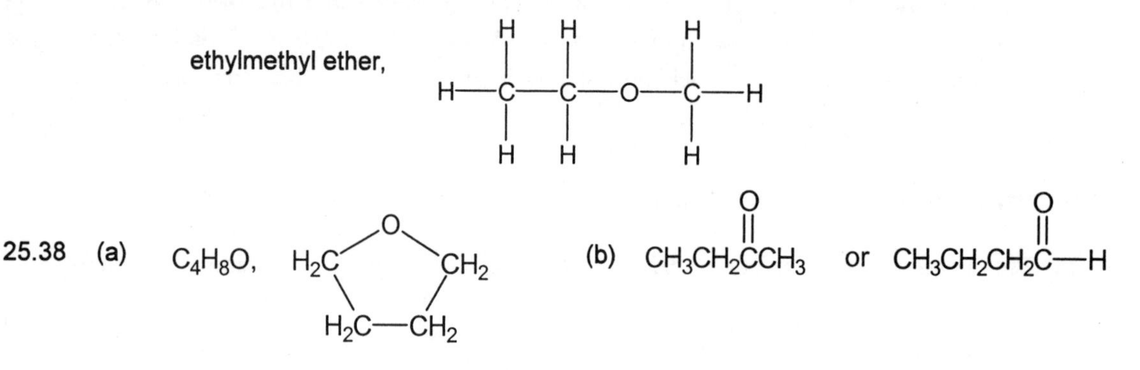

25.38 (a) C_4H_8O, (cyclic ether, H_2C–O–CH_2 ring with H_2C—CH_2) (b) $CH_3CH_2C(O)CH_3$ or $CH_3CH_2CH_2C(O)$—H

25.39 *Analyze/Plan.* Count the number of C atoms in each chain, including the carboxyl C atom. Name the chain and the acid. *Solve*:

(a) methanoic acid (b) butanoic acid (c) 3-methylpentanoic acid

25.40 (a) $CH_3CH_2C(O)$—H (b) $CH_3CH_2CH_2C(O)CH_3$

(c) $CH_3CH(CH_3)C(O)CH_3$ (d) $CH_3CH_2CH(CH_3)C(O)$—H

25.41 *Analyze/Plan.* In a condensation reaction between an alcohol and a carboxylic acid, the alcohol loses its –OH hydrogen atom and the acid loses its –OH group. The alkyl group from the acid is attached to the carbonyl group and the alkyl group from alcohol is attached to the ether oxygen of the ester. The name of the ester is the alkyl group from the alcohol plus the alkyl group from the acid plus the suffix *-oate*. *Solve*:

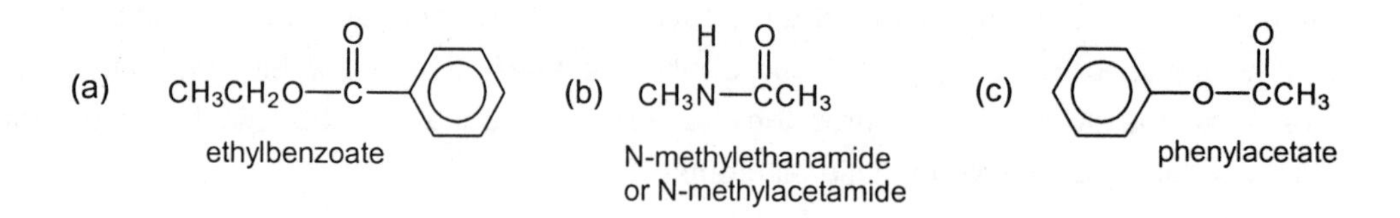

(a) ethylbenzoate (b) N-methylethanamide or N-methylacetamide (c) phenylacetate

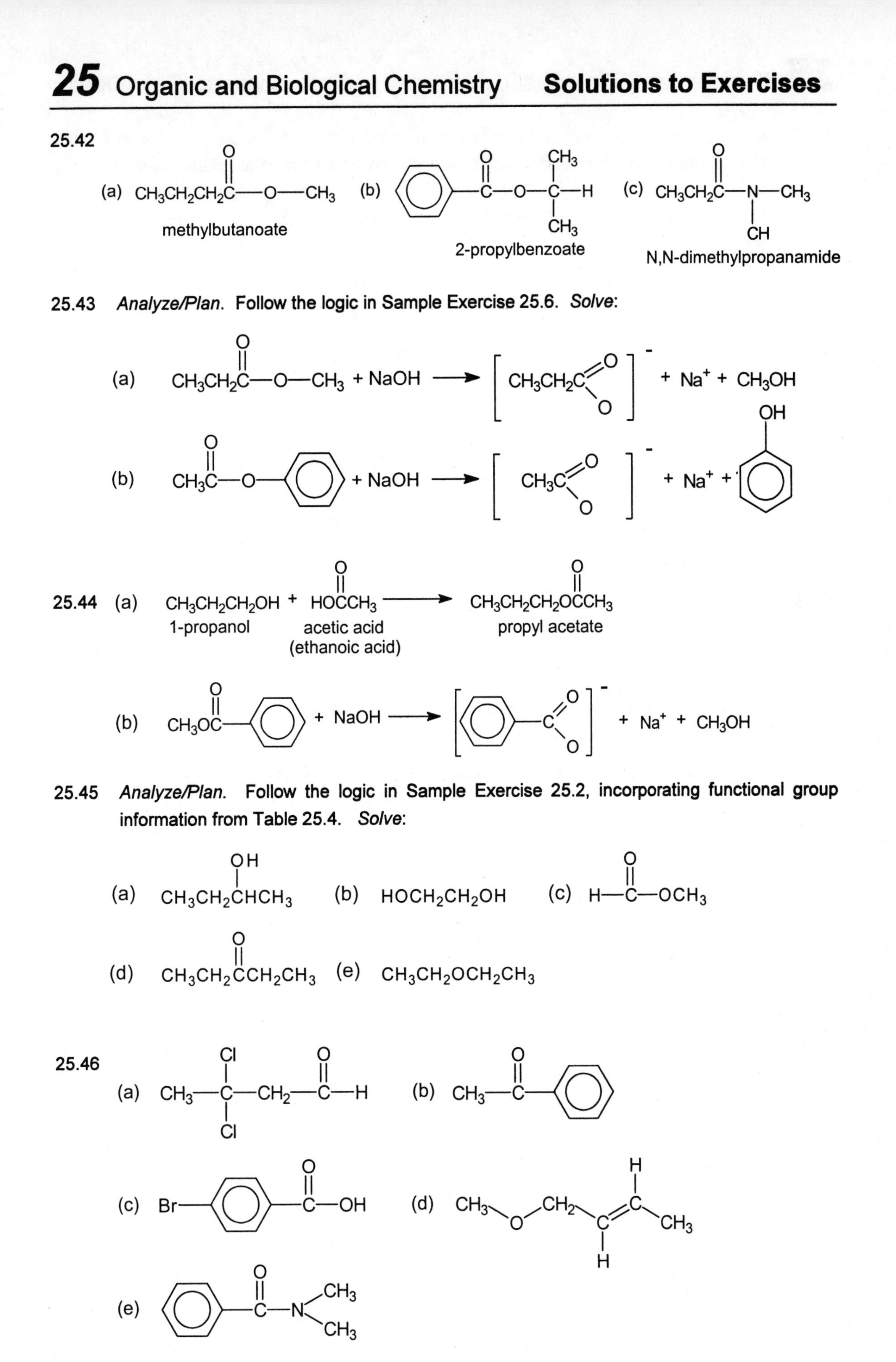
25 Organic and Biological Chemistry
Solutions to Exercises
25.42
(a) CH3CH2CH2C—O—CH3
methylbutanoate
(b)
2-propylbenzoate
(c) CH3CH2C—N—CH3
N,N-dimethylpropanamide
25.43 Analyze/Plan. Follow the logic in Sample Exercise 25.6. Solve:
(a) CH3CH2C—O—CH3 + NaOH
+ Na+ + CH3OH
(b) + NaOH
+ Na+ +
25.44 (a) CH3CH2CH2OH + HOCCH3
1-propanol
acetic acid
(ethanoic acid)
CH3CH2CH2OCCH3
propyl acetate
(b) CH3OC + NaOH
+ Na+ + CH3OH
25.45 Analyze/Plan. Follow the logic in Sample Exercise 25.2, incorporating functional group information from Table 25.4. Solve:
(a) CH3CH2CHCH3
(b) HOCH2CH2OH
(c) H—C—OCH3
(d) CH3CH2CCH2CH3
(e) CH3CH2OCH2CH3
25.46
(a) CH3—C—CH2—C—H
(b) CH3—C
(c) Br—C—OH
(d) CH3 O CH2 C C CH3
(e)
641

25.47 *Analyze/Plan*. Reverse the rules for naming alkanes and haloalkanes; draw the structures. That is, draw the carbon chain indicated by the root name, place substituents, fill remaining positions with H atoms. Each C atom attached to four different groups is chiral. *Solve*:

(a)

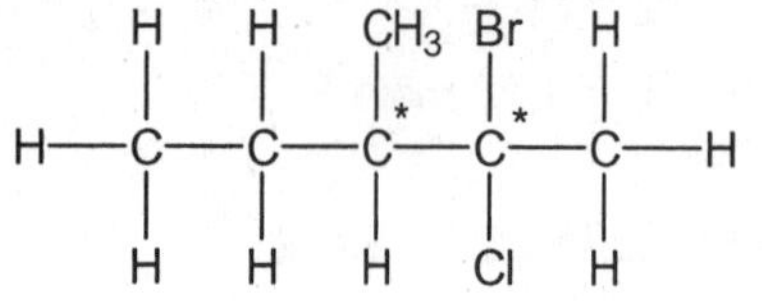

```
      H   H   CH3 Br  H
      |   |   |*  |*  |
   H—C—C—C—C—C—H
      |   |   |   |   |
      H   H   H   Cl  H
```

* chiral C atoms

C2 is obviously attached to four different groups. C3 is chiral because the substituents on C2 render the C1-C2 group different than the C4-C5 group.

(b)

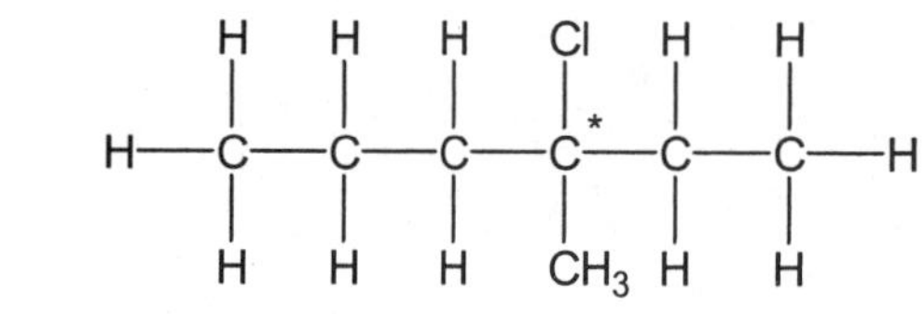

Yes, the molecule has optical isomers. The chiral carbon atom is attached to chloro, methyl, ethyl and propyl groups. [If the root was a 5-carbon chain, the molecule would not have optical isomers because two of the groups would be ethyl groups.]

25.48 (a) Compounds in 25.18:

25.18(a) Yes, C5 is attached to H, methyl, ethyl and a complex group.

25.18(b) Yes, C3 is attached to H, methyl, ethyl and a complex group.

25.18(c) Yes, C5 is attached to H, methyl and two different complex groups.

25.18(d) No, C3 is attached to two ethyl groups, C4 to two methyl groups.

25.18(e) No, the ring C atoms with substituents are not chiral because the ring is symmetrical in both directions from these C atoms.

(b) Compounds in 25.30:

No. These compounds are all aromatic hydrocarbons, which do not have optical isomers. None of the substituent groups have chiral C atoms.

Proteins

25.49 (a) An α-amino acid contains an NH_2 group attached to the carbon that is bound to the carbon of the carboxylic acid function.

(b) In forming a protein, amino acids undergo a condensation reaction between the amino group and carboxylic acid:

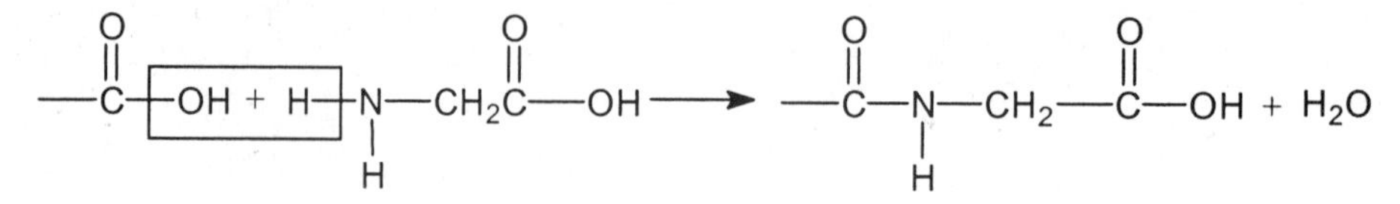

25.50 The side chains possess three characteristics that may be of importance. They may be bulky (e.g., the phenyl group in phenylalanine) and thus impose restraints on where and how the amino acid can undergo reaction. Secondly, the side chain may possess a polar group (e.g., the -OH group in serine), that will in part determine solubility. Finally, the side chain may contain an acidic (e.g., the -COOH group in glutamic acid) or basic (e.g., the $-NH_2$ group in lysine) functional group that will partially determine solubility in acidic or basic media, and that may become involved in hydrogen-bonding with other amino acids.

25.51 *Analyze/Plan.* Either peptide can have the terminal carboxyl group or the terminal amino group. *Solve:*

Two dipeptides are possible:

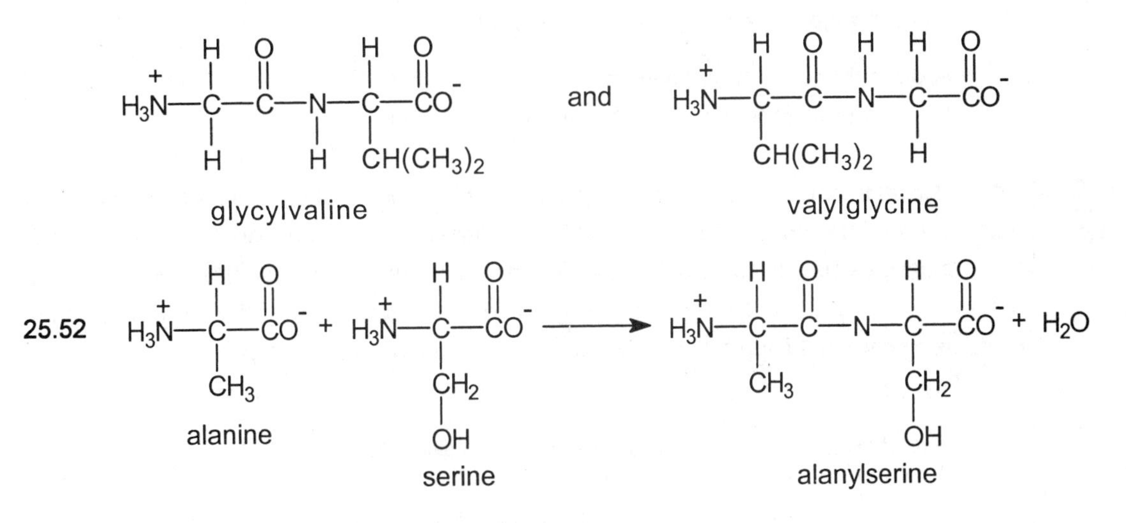

25.53 *Analyze/Plan.* Follow the logic in Sample Exercise 25.7. *Solve:*

(a)

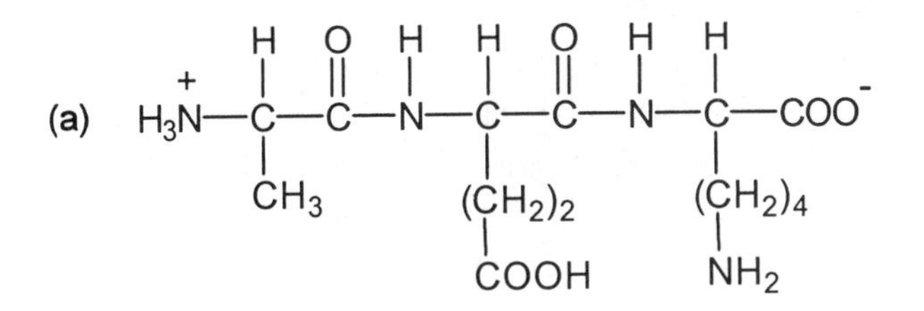

(b) Eight: Ser-Ser-Ser; Ser-Ser-Phe; Ser-Phe-Ser; Phe-Ser-Ser; Ser-Phe-Phe; Phe-Ser-Phe; Phe-Phe-Ser; Phe-Phe-Phe

25.54 (a) Valine, serine, glutamic acid

(b) Six: Gly-Ser-Glu; Gly-Glu-Ser; Ser-Gly-Glu; Ser-Glu-Gly; Glu-Ser-Gly; Glu-Gly-Ser

25.55 The *primary structure* of a protein refers to the sequence of amino acids in the chain. Along any particular section of the protein chain the configuration may be helical, or it may be an open chain, or arranged in some other way. This is called the *secondary structure*. The overall shape of the protein molecule is determined by the way the segments of the protein chain fold together, or pack. The interactions which determine the overall shape are referred to as the *tertiary structure*.

25.56 It is quite evident from Figure 25.24 that the hydrogen bonds between an NH group along the chain and the unshared electron pairs of a carbonyl group further along are responsible for maintaining the helix. Indeed, the pitch and general shape of the helix are determined by what specific interactions produce a good hydrogen-bonding arrangement.

Carbohydrates

25.57 (a) Carbohydrates, or sugars, are composed of carbon, hydrogen and oxygen. From a chemical viewpoint, they are polyhydroxyaldehydes or ketones. Carbohydrates are primarily derived from plants and are a major food source for animals.

(b) A monosaccharide is a simple sugar molecule that cannot be decomposed into smaller sugar molecules by (acid) hydrolysis.

(c) A disaccharide is a carbohydrate composed of two simple sugar units. Hydrolysis breaks the disaccharides into two monosaccharides.

25.58 Glucose exists in solution as a cyclic structure in which the aldehyde function on carbon 1 reacts with the OH group of carbon 5 to form what is called a hemiacetal, Figure 25.27. Carbon atom 1 carries an OH group in the hemiacetal form; in α-glucose this OH group is on the opposite side of the ring as the CH_2OH group on carbon atom 5. In the β (beta) form the OH group on carbon 1 is on the same side of the ring as the CH_2OH group on carbon 5.

The condensation product of two glucose units looks like this:

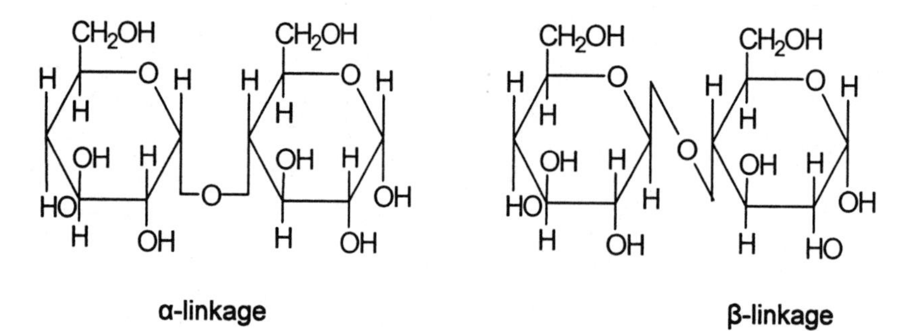

25.59 (a) In the linear form of galactose, the aldehydic carbon is C1. Carbon atoms 2, 3, 4 and 5 are chiral because they each carry four different groups. Carbon 6 is not chiral because it contains two H atoms.

(b) The structure is best deduced by comparing galactose with glucose, and inverting the configurations at the appropriate carbon atoms. Recall from Solution 25.58 that both the β-form (shown here) and the α-form (OH on carbon 1 on the opposite side of ring as the CH_2OH on carbon 5) are possible.

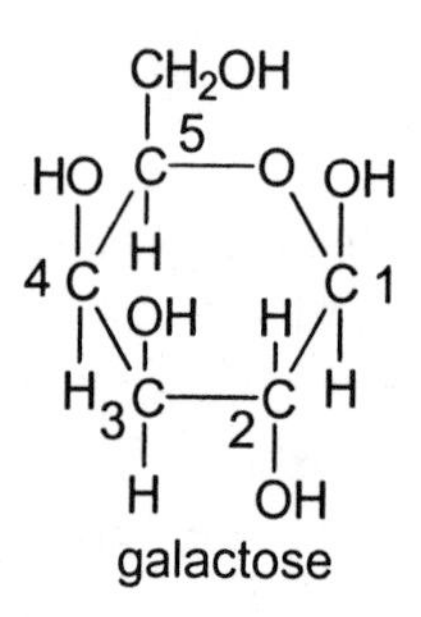

25.60 (a) In the linear form of mannose, the aldehydic carbon is C1. Carbon atoms 2, 3, 4 and 5 are chiral because they each carry four different groups. Carbon 6 is not chiral because it contains two H atoms.

(b) Both the α (left) and β (right) forms are possible.

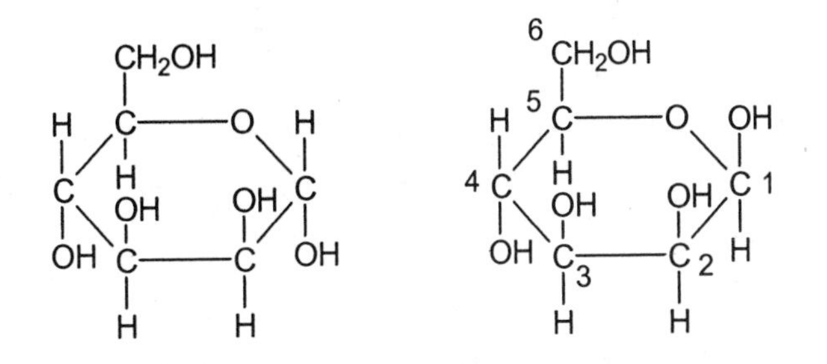

25.61 The empirical formula of glycogen is $C_6H_{10}O_5$. The six-membered ring form of glucose is the unit that forms the basis of glycogen. The monomeric glucose units are joined by α linkages.

25.62 The empirical formula of cellulose is $C_6H_{10}O_5$. As in glycogen, the six-membered ring form of glucose forms the monomer unit that is the basis of the polymer cellulose. In cellulose, glucose monomer units are joined by β linkages.

Nucleic Acids

25.63 A *nucleotide* consists of a nitrogen-containing aromatic compound, a sugar in the furanose (5-membered) ring form, and a phosphoric acid group. The structure of deoxycytidine monophosphate is shown at right.

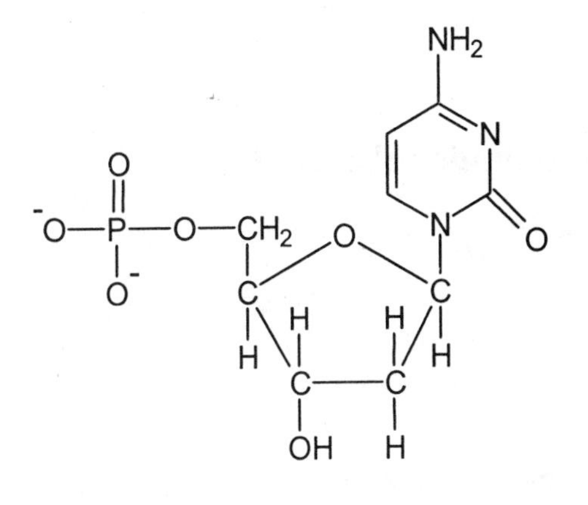

25.64

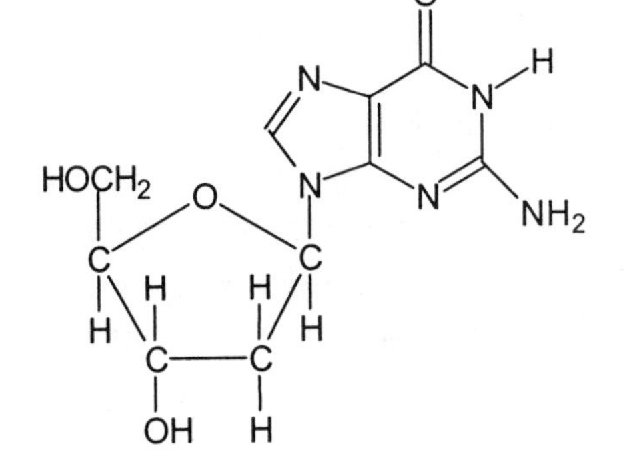

25.65 $C_4H_7O_3CH_2OH + HPO_4^{2-} \rightarrow C_4H_7O_3CH_2\text{-O-}PO_3^{2-} + H_2O$

25.66

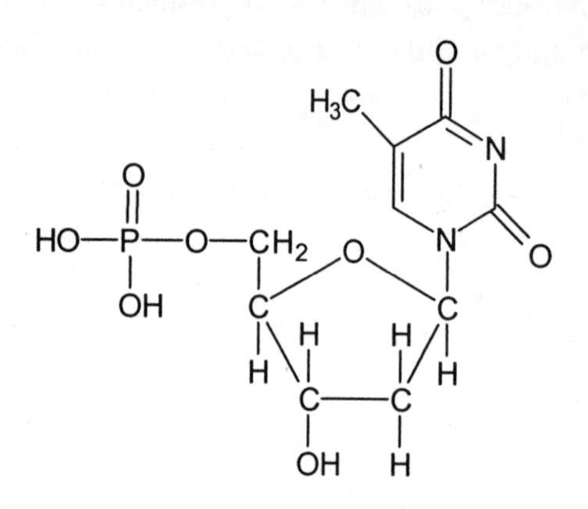

25.67 In the helical structure for DNA, the strands of the polynucleotides are held together by hydrogen-bonding interactions between particular pairs of bases. It happens that adenine and thymine form an especially effective base pair, and that guanine and cytosine are similarly related. Thus, each adenine has a thymine as its opposite number in the other strand, and each guanine has a cytosine as its opposite number. In the overall analysis of the double strand, total adenine must then equal total thymine, and total guanine equals total cytosine.

25.68

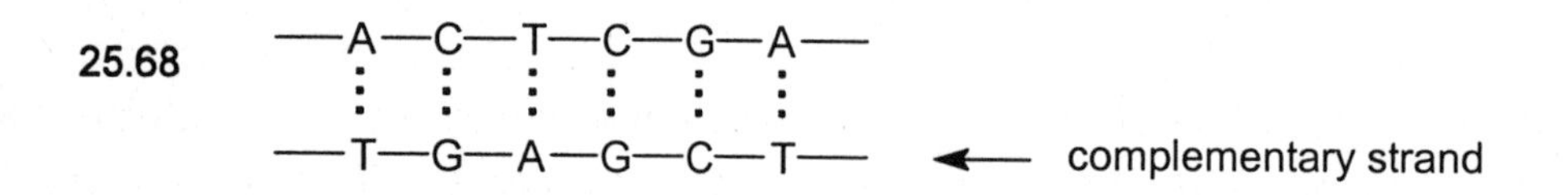

Additional Exercises

25.69

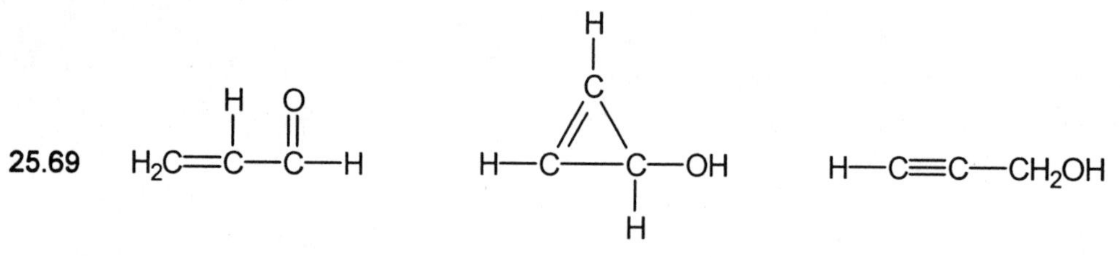

25.70 *Analyze/Plan.* We are asked the number of structural isomers for two specified carbon chain lengths and a certain number of double bonds. Structural isomers have different connectivity. Since the chain length is specified, we can ignore structural isomers created by branching. We are not asked about geometrical isomers, so we ignore those as well. The resulting question is: How many ways are there to place the specified number of double bonds along the specified C chain? *Solve:*

5 C chain with one double bond: 2 structural isomers

C=C–C–C–C   C–C=C–C–C

6 C chain with two double bonds: 6 structural isomers

C=C–C=C–C–C   C=C–C–C=C–C   C=C–C–C–C=C

C–C=C–C=C–C   C=C=C–C–C–C   C–C=C=C–C–C

25.71 Because of the strain in bond angles about the ring, cyclic alkynes with less than eight carbons are not stable. Alkyne carbon atoms preferentially have 180° bond angles; this requires a linear four-carbon group in the ring. Three additional carbons in the ring does not provide enough flexibility to make this possible without gross bond length or angle distortions. It is possible that a ring with eight or more carbons could accommodate an alkyne linkage. To test this with models, construct a linear four-carbon group and then add tetrahedral C-atoms until you can complete the ring and it stays together without intervention. This is a good indication of the minimum number of carbon atoms in a stable ring that contains an alkyne linkage.

25.72

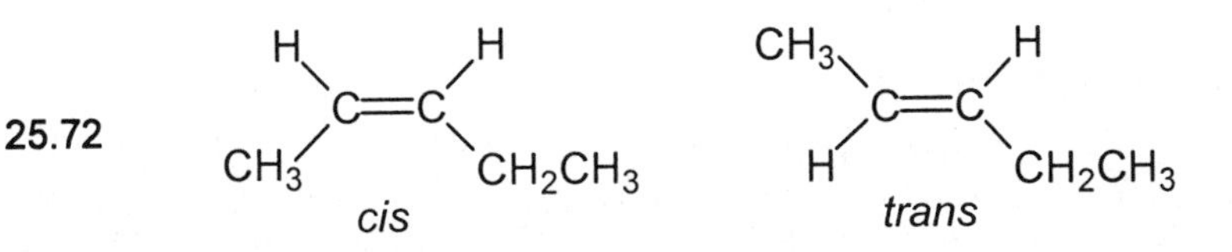

Cyclopentene does not show *cis-trans* isomerism because the existence of the ring demands that the C-C bonds be *cis* to one another.

25.73 The structural features that produce *cis-trans* isomerism are restricted rotation about a bond and two different groups attached to the two atoms that form the rigid bond. Alkanes have only C–C single bonds, and there is free rotation of attached groups about the single bond. Alkanes fail the criterion of restricted rotation. Alkynes have restricted rotation about the triple bond, but only one group is attached to the C atoms that form the bond. There is only one relative orientation (linear) for the two attached groups, so *cis* and *trans* isomers are not possible.

25.74 The C-Cl bonds in the *trans* compound are pointing in exactly opposite directions. Thus, the C-Cl bond dipoles cancel (Section 9.3). This is not the case in the *cis* compound, as can be seen by drawing the structure:

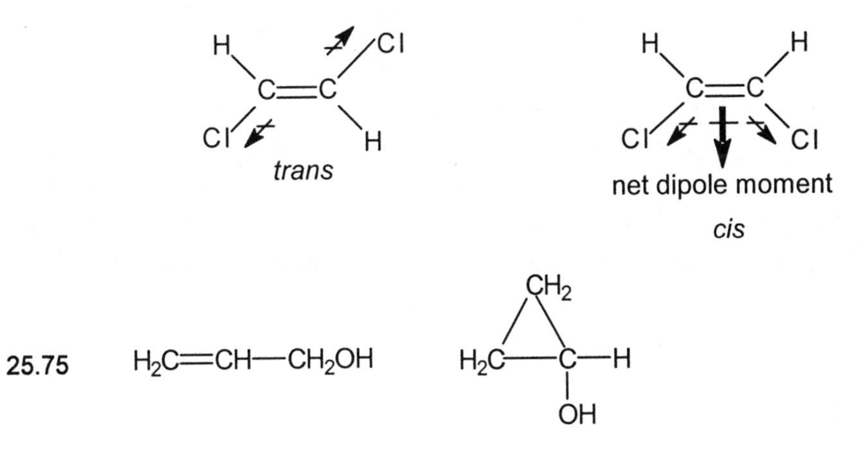

(The -OH group cannot be attached to an alkene carbon atom; these molecules are called "vinyl alcohols" and are unstable.)

25.76 One. One molecule of HBr would add to the C=C according to the reaction

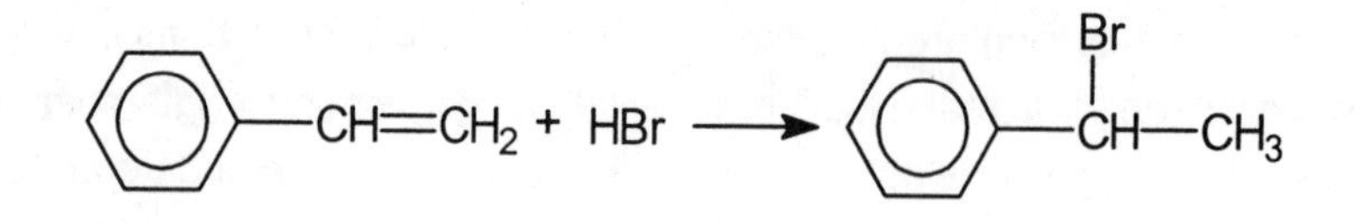

Because the π electrons in the phenyl ring are delocalized, the group is particularly stable and resistant to addition reactions. The phenyl group could undergo substitution with HBr, but only with a catalyst or special conditions.

25.77 Two plausible decomposition reactions are:

(i) $CH_2Cl_2(l) \rightarrow C(s) + H_2(g) + Cl_2(g)$

(ii) $CH_2(NO_2)_2(l) \rightarrow N_2(g) + CO_2(g) + H_2O(g) + 1/2O_2(g)$

Use bond dissociation energies (Table 8.4) to evaluate approximate ΔH values for each reaction.

(i) $\Delta H = 2D(C\text{-}H) + 2D(C\text{-}Cl) - D(H\text{-}H) - D(Cl\text{-}Cl)$

$= 2(413) + 2(328) - 436 - 242 = +804$ kJ

(ii) $\Delta H = 2D(C\text{-}H) + 2D(C\text{-}N) + 2D(N{=}O) + 2D(N\text{-}O) - D(N{\equiv}N) - 2D(C{=}O) - 2D(O\text{-}H) - 1/2D(O{=}O)$

$= 2(413) + 2(293) + 2(607) + 2(201) - 941 - 2(799) - 2(463) - 1/2(495)$

$\Delta H = -685$ kJ

Clearly, the decomposition of $CH_2(NO_2)_2$ is thermodynamically favorable, while the decomposition of CH_2Cl_2 is not. In particular, this is because of the stability of N_2 and CO_2 relative to $CH_2(NO_2)_2$. For CH_2Cl_2, no oxygen atoms are available to form stable products such as CO_2 and H_2O.

25.78 (a) Ether, C—O—C ; alkene, —CH=CH$_2$

(b) carboxylic acid, —C(=O)—OH ; ester, CH$_3$C(=O)—O— ; aromatic, (benzene ring)

(c) ketone, —C(=O)— ; alkene, —CH=CH— ; alcohol —C—OH

25.79

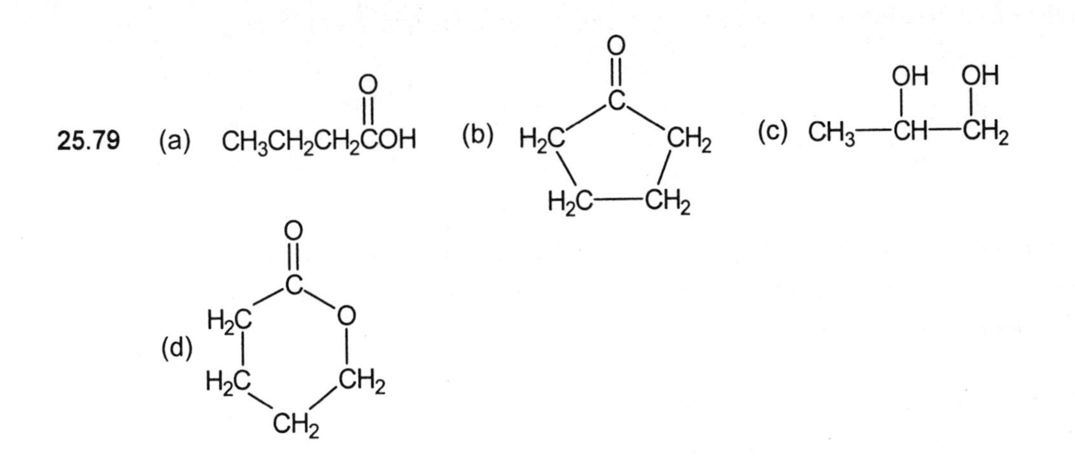

25.80 The difference between an alcoholic hydrogen and a carboxylic acid hydrogen lies in the carbon to which the –OH is attached. In a carboxylic acid, the electronegative carbonyl oxygen withdraws electron density from the O–H bond, rendering the bond more polar and the H more ionizable. In an alcohol no electronegative atoms are bound to the carbon that holds the –OH group, and the H is tightly bound to the O.

25.81 (a) $CH_3C(=O)$—OH, C_6H_5OH (b) $C_6H_5C(=O)$—OH, CH_3OH

25.82 In order for indole to be planar, the N atom must be sp^2 hybridized. The nonbonded electron pair on N is in a pure p orbital perpendicular to the plane of the molecule. The electrons that form the π bonds in the molecule are also in pure p orbitals perpendicular to the plane of the molecule. Thus, each of these p orbitals is in the correct orientation for π overlap; the delocalized π system extends over the entire molecule and includes the "nonbonded" electron pair on N. The reason that indole is such a weak base (H^+ acceptor), is that the nonbonded electron pair is delocalized and a H^+ ion does not feel the attraction of a full localized electron pair.

25.83 (a) None

(b) The carbon bearing the secondary –OH has four different groups attached, and is thus chiral.

(c) The carbon bearing the $-NH_2$ group and the carbon bearing the CH_3 group are both chiral.

25.84 (a)

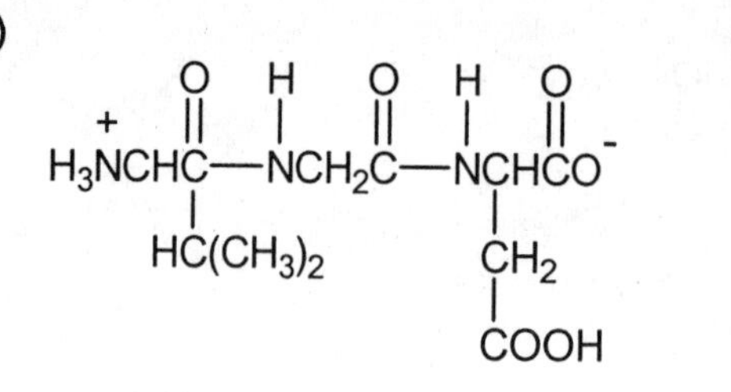

(b)

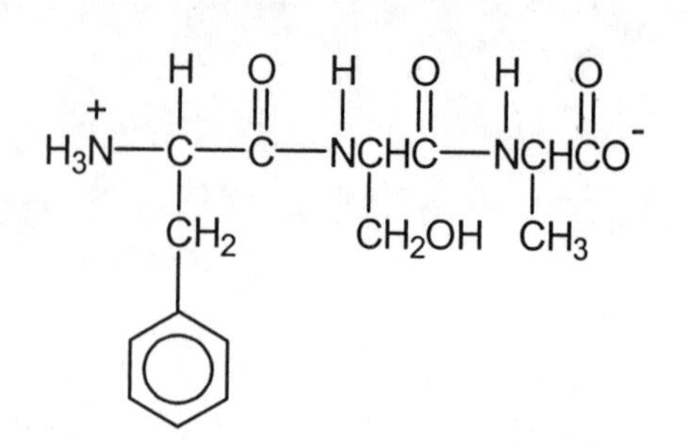

25.85 Glu-Cys-Gly is the only possible structure.

25.86 Starch, glycogen and cellulose are all biopolymers built by linking glucose monomers. Starch and glycogen have alpha (α) glucose linkages, where the bridging O atom is on the same side of the ring as the CH_2OH group. The smallest repeating unit in starch and glycogen is a single glucose unit.

Cellulose has beta (β) glucose linkages, where the bridging O atom is on the same side of one of the rings as the CH_2OH group and on the opposite side of the CH_2OH group on the second ring. The geometry of the β linkage requires that the two linked glucose units have different orientations and that the smallest repeating unit in cellulose is two glucose units with a β linkage.

The molecular weight of a polymer is an indication of the number of monomer units present. Starch, glycogen and cellulose all have a range of molecular weights. Glycogen has the widest range of molecular weights, 5,000-5,000,000 amu, and is potentially the largest polymer. Cellulose is intermediate in size with an average molar mass of 500,000 amu. Starch and cellulose are produced in plants, while glycogen is produced in animals and serves as an energy storage mechanism.

25.87 Both glucose and fructose contain six C atoms, so both are hexoses. Glucose contains an aldehyde group at C1, so it is an aldohexose. Fructose has a ketone at C2, so it is a ketohexose.

25.88

```
—G—G—T—A—C—T—
 :  :  :  :  :  :
—C—C—A—T—G—A— ←— complementary strand
```

Integrative Exercises

25.89 CH_3CH_2OH — ethanol; CH_3-O-CH_3 — dimethyl ether

Ethanol contains –O–H bonds which form strong intermolecular hydrogen bonds, while dimethyl ether experiences only weak dipole-dipole and dispersion forces.

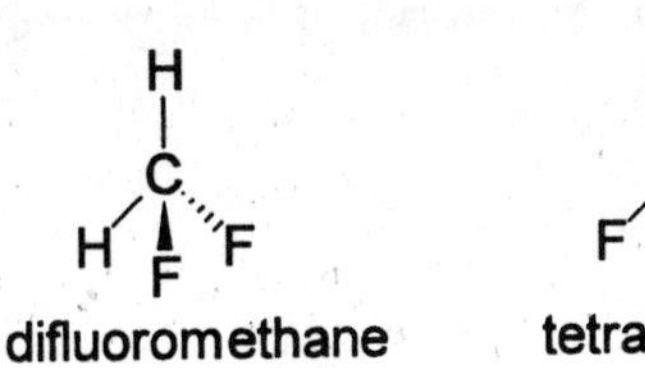

CH_2F_2 is a polar molecule, while CF_4 is nonpolar. CH_2F_2 experiences dipole-dipole and dispersion forces, while CF_4 experiences only dispersion forces.

In both cases, stronger intermolecular forces lead to the higher boiling point.

25.90 The compound is clearly an alcohol. Its slight solubility in water is consistent with the properties expected of a secondary alcohol with a five-carbon chain. The fact that oxidation results in a ketone rather than an aldehyde and subsequently a carboxylic acid tells us that it is a secondary alcohol:

$CH_3CH(OH)CH_2CH_2CH_3$ $\quad$ $CH_3CH_2CH(OH)CH_2CH_3$ $\quad$ $CH_3CH(OH)CH(CH_3)_2$

25.91 Determine the empirical formula, molar mass and thus molecular formula of the compound. Confirm with physical data.

$$66.7 \text{ g C} \times \frac{1 \text{ mol C}}{12.01 \text{ g C}} = 5.554 \text{ mol C}; \; 5.554 / 1.388 = 4$$

$$11.2 \text{ g H} \times \frac{1 \text{ mol H}}{1.008 \text{ g H}} = 11.11 \text{ mol H}; \; 11.11 / 1.388 = 8$$

$$22.2 \text{ g O} \times \frac{1 \text{ mol O}}{16.00 \text{ g O}} = 1.388 \text{ mol O}; \; 1.388 / 1.388 = 1$$

The empirical formula is C_4H_8O. Using Equation 10.11 ($\mathcal{M}$ = molar mass):

$$\mathcal{M} = \frac{(2.28 \text{ g/L})(0.08206 \text{ L} \cdot \text{atm/mol} \cdot \text{K})(373 \text{ K})}{0.970 \text{ atm}} = 71.9$$

The formula weight of C_4H_8O is 72, so the molecular formula is also C_4H_8O. Since the compound has a carbonyl group and cannot be oxidized to an acid, the only possibility is 2-butanone.

$CH_3C(=O)CH_2CH_3$

The boiling point of 2-butanone is 79.6°C, confirming the identification.

25.92 Determine the empirical formula, molar mass and thus molecular formula of the compound. Confirm with physical data.

$$85.7 \text{ g C} \times \frac{1 \text{ mol C}}{12.01 \text{ g C}} = 7.136 \text{ mol C}; \; 7.136 / 7.136 = 1$$

$$14.3 \text{ g H} \times \frac{1 \text{ mol H}}{1.008 \text{ g H}} = 14.19 \text{ mol H}; \; 14.19 / 7.136 \approx 2$$

Empirical formula is CH_2. Using Equation 10.11 ($\mathcal{M}$ = molar mass):

$$\mathcal{M} = \frac{(2.21 \text{ g/L})(0.08206 \text{ L}\bullet\text{atm/mol}\bullet\text{K})(373 \text{ K})}{(735/760) \text{ atm}} = 69.9 \text{ g/mol}$$

The molecular formula is thus C_5H_{10}. The absence of reaction with aqueous Br_2 indicates that the compound is not an alkene, so the compound is probably the cycloalkane, cyclopentane. According to the *Handbook of Chemistry and Physics*, the boiling point of cyclopentane is 49°C at 760 torr. This confirms the identity of the unknown.

25.93 The reaction is: $2NH_2CH_2COOH(aq) \rightarrow NH_2CH_2CONHCH_2COOH(aq) + H_2O(l)$

$\Delta G° = (-488) + (-285.83) - 2(-369) = -35.8 = -36$ kJ

25.94 (a) A = adenosine = $C_{10}H_{10}O_3N_5$

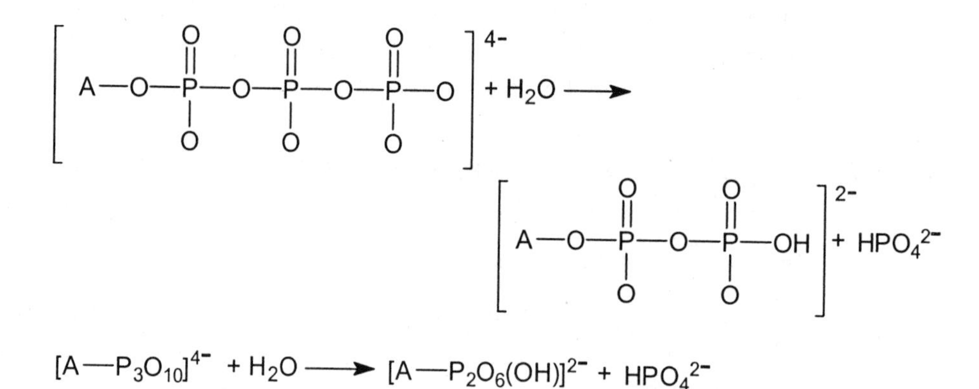

$$[A—P_3O_{10}]^{4-} + H_2O \longrightarrow [A—P_2O_6(OH)]^{2-} + HPO_4^{2-}$$

(b) If the hydrolysis reaction is spontaneous, the sign of ΔG must be negative.

(c) Adenosine monophosphate (AMP) + inorganic phosphate

$$[A—O—P(=O)(O)—OH]^{-} + HPO_4^{2-}$$

(The placement of the H^+ in these reactions is somewhat arbitrary; H^+ is attracted to the strongest base, but the equilibria are complex.)

25.95 (a) At low pH, the amine and carboxyl groups are protonated.

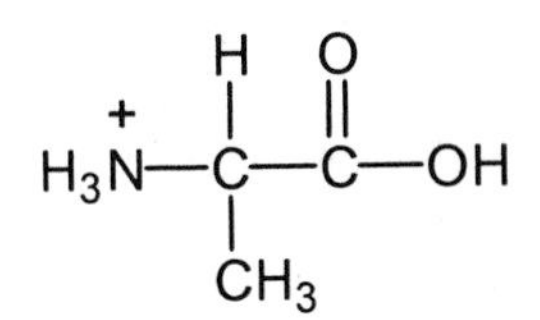

At high pH, the amine and carboxyl groups are deprotonated.

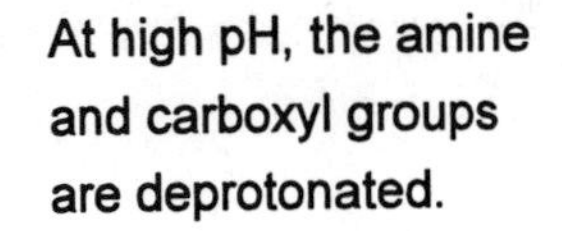

(b) $CH_3C(=O){-}OH(aq) \longrightarrow CH_3C(=O){-}O^-(aq) + H^+(aq)$

$K_a = 1.8 \times 10^{-5}$, $pK_a = -\log(1.8 \times 10^{-5}) = 4.74$

The conjugate acid of NH_3 is NH_4^+.

$NH_4^+(aq) \rightarrow NH_3(aq) + H^+(aq)$

$K_a = K_w / K_b = 1 \times 10^{-14} / 1.8 \times 10^{-5} = 5.55 \times 10^{-10} = 5.6 \times 10^{-10}$

$pK_a = -\log(5.55 \times 10^{-10}) = 9.26$

In general, a –COOH group is a stronger acid than a $-NH_3]^+$ group. The lower pK_a value for amino acids is for the ionization (deprotonation) of the –COOH group and the higher pK_a is for the deprotonation of the $[NH_3]^+$ group.

25.96 (a) The native form has a lower, more negative free energy than the denatured form.

(b) ΔS is negative in going from the denatured form to the folded (native) form; the native protein is more ordered.

(c) The four S-S linkages are strong covalent links holding the chain in place in the folded structure. A folded structure without these links would be less stable (more positive ΔG) and have more motional freedom (more positive entropy).

(d) At first glance, the answer seems the same as part (c). However, the eight S–H linkages are capable of forming hydrogen-bond-like interactions, which would decrease ΔG and S. Although these interactions are not as strong as S–S covalent bonds, there are more of them, so this reduction might not have much effect on ΔG and S.

25.97 $AMPOH^-(aq) \rightleftharpoons AMPO^{2-}(aq) + H^+(aq)$

$pK_a = 7.21$; $K_a = 10^{-pK_a} = 6.17 \times 10^{-8} = 6.2 \times 10^{-8}$

$K_a = \dfrac{[AMPO^{2-}][H^+]}{[AMPOH^-]} = 6.2 \times 10^{-8}$. When pH = 7.40, $[H^+] = 3.98 \times 10^{-8} = 4.0 \times 10^{-8}$.

Then $\dfrac{[AMPOH^-]}{[AMPO^{2-}]} = 3.98 \times 10^{-8} / 6.17 \times 10^{-8} = 0.65$